KINETICS OF
NONHOMOGENEOUS
PROCESSES

KINETICS OF NONHOMOGENEOUS PROCESSES

A PRACTICAL INTRODUCTION FOR CHEMISTS, BIOLOGISTS, PHYSICISTS, AND MATERIALS SCIENTISTS

Edited by
Gordon R. Freeman
University of Alberta

A WILEY-INTERSCIENCE PUBLICATION
JOHN WILEY & SONS
New York • Chichester • Brisbane • Toronto • Singapore

7344.4091

CHEMISTRY

Library of Congress Cataloging in Publication Data:

Kinetics of nonhomogeneous processes.

   "A Wiley-Interscience publication."

   Includes bibliographies and indexes.

   1. Materials—Effect of radiation on. 2. Polymers and
polymerization. 3. Radiobiology. 4. Chemical reaction,
Conditions and laws of. I. Freeman, Gordon R.
II. Title: Nonhomogeneous processes.

TA418.6.K55 1986   620.1   86-15869
ISBN 0-471-81324-9

Printed in the United States of America

10 9 8 7 6 5 4 3 2 1

CONTRIBUTORS

ALOKE CHATTERJEE, Lawrence Berkeley Laboratory, University of California, Berkeley, California

MOHAMED DAOUD, Laboratoire Léon Brillouin, Centre d'Études Nucléaires, Saclay, France

GORDON R. FREEMAN, Department of Chemistry, University of Alberta, Edmonton, Canada

ANDRIES HUMMEL, Interuniversitair Reactor Instituut, Delft, The Netherlands

ALBRECHT M. KELLERER, Institut für Medizinische Strahlenkunde der Universität Würzburg, Würzburg, Federal Republic of Germany

JOHN L. MAGEE, Lawrence Berkeley Laboratory, University of California, Berkeley, California

LOUIS K. MANSUR, Metals and Ceramics Division, Oak Ridge National Laboratory, Oak Ridge, Tennessee

WILLIAM W. MERRILL, Department of Chemical Engineering and Materials Science, University of Minnesota, Minneapolis, Minnesota

DIETMAR MÖBIUS, Max-Planck-Institut für Biophysicalische Chemie, Göttingen, Federal Republic of Germany

JAAN NOOLANDI, Xerox Research Center of Canada, Mississauga, Ontario, Canada

HERWIG G. PARETZKE, GSF-Institut für Strahlenschutz, Neuherberg, Federal Republic of Germany

KENNETH SHOWALTER, Department of Chemistry, West Virginia University, Morgantown, West Virginia

MASANORI TACHIYA, National Chemical Laboratory for Industry, Yatabe, Japan

MATTHEW TIRRELL, Department of Chemical Engineering and Materials Science, University of Minnesota, Minneapolis, Minnesota

à nos femmes

PREFACE

It is exciting to find new relationships between things that previously seemed to have little in common. During the past two decades I have gradually become aware of patches of information about unconventional kinetics that seemed to be somehow related, growing in relatively isolated areas. I have waited for the time when they would touch each other and gel into a network. This book is the gel point. The network is a new branch of kinetics, the kinetics of nonhomogeneous processes. It spans physics, chemistry, biology, and materials science. Half of the book describes processes induced by radiation. The other half describes thermal processes in structured systems that model parts of nature.

The physical arrangement (structure) of a system has an enormous effect on its behavior. That truth is self-evident for an object as small as a single molecule and for a very large object such as an animal. For objects of intermediate size, such as agglomerations of a few tens to a few millions of molecules, the effects of arrangement can also be very large and are only now receiving great attention.

Historically, the subject as defined by the contents of this book can be recognized in an 80-year-old study of α particles, which ionize gases in a subtly peculiar way (Chapter 2). A larger book on nonhomogeneous kinetics would include unimolecular reaction theory; inklings of the needed concepts appeared a century ago in a paper by Lord Rayleigh, "On the Resultant of a large Number of Vibrations of the same Pitch and of arbitrary Phase" [*Phil. Mag.*, 5th Ser., **10**, 73 (1880)]. Indeed, one can read into the last page of that paper a seed of the great contribution that Heisenberg made to physical theory nearly half a century later, the Uncertainty principle, which still fascinates us.

The present volume describes dynamic processes that occur in network-forming polymerization systems, in percolation, in the spread of a disease, and in the growth of a cancer. It describes the kinetics of physical, chemical, and biological changes caused by high-energy radiation. It describes dynamic processes in photoconductors, in membrane-mimetic systems of micelles and vesicles, in monolayers and membranes, in polymer welding, and in chemical reactions that oscillate in space or time. The book is intended for all radiation research people (physicists, chemists, biologists, and oncologists); people concerned with radiation-induced changes in metals, engineers designing nuclear fission and fusion reactors; photoconductor and semiconductor people who work in the electronics and photoimaging industries; biochemical kineticists; microbiologists and cell biologists; people who work with microemulsions, catalysis chemists; and people who design VLSI microelectronics systems for spacecraft.

We have tried to write a "How To" book at the level of new graduate students or specialized upper-year undergraduates. We introduce the concepts of nonhomogeneous systems and processes, and practical methods of treating them. Each chapter begins at a descriptive level and develops to include material useful to workers at the front of their fields. We have tried to use language that is understandable in all disciplines. Some specialized jargon was unavoidable, but all nondictionary terms have been defined in their context. The level of mathematics has been restrained to increase general understanding, but it meets the level required for practical application. The authors are all leaders in their fields.

Concepts of a nonhomogeneous system are introduced quantitatively by using a familiar but seemingly unlikely entity for such a purpose, the hydrogen atom. Much is known about the process of excitation and ionization of an isolated hydrogen atom; the process is greatly modified in the liquid phase. The concepts are then applied to actual experimental systems, and the book develops from there, ending with chemical waves.

Deterministic and stochastic models are used. Scientific literacy in the future will imply a grasp of both determinacy and stochasticity. It will also imply knowledge of their different realms of applicability and the regions where they blend into each other.

High-energy radiation passing through matter deposits energy along tracks. The reactive intermediates produced by the energy are therefore not distributed randomly in space but in strings. Conventional kinetics involves randomly distributed reactants, so the equations that describe the kinetics of radiation-induced processes are different from those found in conventional textbooks. Three major approaches to radiation chemical kinetics are presented in three consecutive chapters. Radiation effects in metals and in biological systems are treated separately, although the mechanisms are formally similar to some of those used in the chemistry.

Is it an omen for the future that microelectronic circuits are becoming so small and intricate that a single high-energy particle can create a false signal or even damage the device? A model of radiation effects on living cells has been applied to explain and reduce cosmic-ray-induced malfunctions of VLSI electronic devices in spacecraft.

Nonhomogeneous processes are also generated by low-energy photons and by thermal energy. Photosynthesis and photocopying are nonhomogeneous processes. Some of the thermal systems are initially highly structured. In others, structure grows from randomness. Membranes, micelles, vesicles, and biological cells are highly structured and have properties in common. Reactions that involve them are nonhomogeneous. The formation of these structures from random distributions of monomeric molecules, and the production of chemical waves in certain homogeneous chemical systems, bring us close to wondering about the processes of life. The study of the kinetics of nonhomogeneous processes in thermal systems, the realm of the second half of this book, probably will grow rapidly during the next decade.

The concept of fractals was conceived in mathematics; it now permeates parts of physics and will rapidly spread through areas of chemistry and biology. Keep an eye out for growth at tips!

Nonhomogeneous processes occur throughout nature. We focus attention on special properties that nonhomogeneity can create. The most magnificent special property, which remains far beyond understanding, is life itself.

GORDON R. FREEMAN

Edmonton, Alberta, Canada
November 1986

ACKNOWLEDGMENTS

I happily acknowledge global advice obtained during the organization of this global project. It is gratifying to have such highly accomplished coauthors.

The nuclear misfortune at Chernobyl, the Ukraine, in April 1986, touched the lives of some of us.

I would like to thank the members of my group and especially Norman Gee of the Radiation Research Center, University of Alberta, for many comments on my chapters, and Mary Waters for typing the manuscripts and other material.

The indexes were prepared with the kind assistance of Gardner Indexing Service, Edmonton, and Phyllis Freeman. The comprehensive subject index, in addition to its usual function, interrelates diverse topics in the book. It also indicates that a greater uniformity of terminology will assist communication and the growth of the subject.

I express special gratitude to my wife Phyllis for her fascination, and to Kedey's Island in Fitzroy Harbour for its beauty.

G.R.F.

CONTENTS

1 INTRODUCTION

Gordon R. Freeman

Department of Chemistry
University of Alberta
Edmonton, Canada

A structured system is one in which at least one component is not randomly distributed in space. Such systems are common in nature. Examples are a partly cloudy sky, a bacterium, and a piece of agate. Processes that occur in structured systems are strongly affected by details of the structure. The kinetics of such processes have equations different from the types usually given in textbooks.

The effects of structure have usually been ignored or removed from kinetics treatments, by making gross simplifying assumptions. Techniques that involve less severe simplifying assumptions can take effects of structure into account. They have been available for many years, but during the past decade they have evolved and spread in application. The amount of attention paid to effects of microscopic structure in chemical and physical systems is increasing rapidly as techniques to investigate them become better known.

Figure 1.1 shows the discharge pattern of electrical breakdown in an insulator. The pattern is beautifully structured. It was produced by uniformly irradiating one face of the insulator with energetic electrons that would penetrate half way through

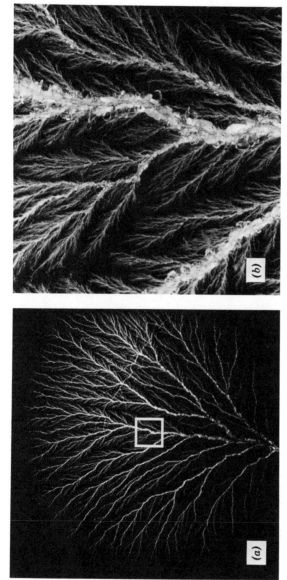

Figure 1.1. (*a*) Dielectric breakdown pattern in Lucite (polymethylmethacrylate) formed by injecting electrons broadside into the insulator plate until spontaneous discharge occurred. This is an example of formation of a structured system. The pattern is 0.1 cm thick and ~ 80 cm² area, an essentially two-dimensional layer in the middle of the plate. The $10 \times 10 \times 0.6$-cm plate was marked (stressed) with a center punch at a point in the middle of an edge, then suspended by a waxed string in front of a 2-MeV van de Graaff accelerator. The 2-MeV electrons in the broad beam penetrated halfway through the plate, where they accumulated until dielectric breakdown occurred and the electrons discharged into the air. The electric field strength in the Lucite caused by the accumulated electrons was ~ 10^9 V/m (~0.1 C/m²) at the time of breakdown. The breakdown discharge began at the stressed point on the edge and grew inside the plate in a branched pattern as the electrons bled out. The violent process was accompanied by a flash of light and an audible snap. When the discharge began at the stressed point, polymethyl-methacrylate molecules in that site were bombarded by the exiting electrons. Some of the molecules decomposed to hydrogen and other products. Enough heat and gas were formed to soften the solid and produce fine bubbles near the discharge point. The gas bubbles then provided an easier conduction path than did the solid, so they became the main conductor. Electrons from adjacent regions discharged into the bubbles, extending the string of bubbles in a branched pattern and enlarging the initial string into a broader channel. The channel pattern grew in this fractal fashion until the plate was sufficiently discharged that the remaining internal field could not continue the process. The visible gas channels provide a permanent record of the paths along which the more violent transport took place. The fractal dimensionality is about 1.7 (see Section 1.5). (*b*) Tenfold magnification of the square region marked in (*a*).

the plate and become trapped there. The electrons accumulated in a layer, making a sort of electron sandwich. The concentration of trapped electrons increased as the irradiation continued. As the mean distance between the trapped electrons became smaller, the force of electrostatic repulsion between them grew, until finally the insulator could no longer keep them immobile against the repulsive force. Then a mini lightning bolt occurred inside the plate. The electrical breakdown of the insulator started at a previously stressed point on the edge of the plate and grew inward from there.

Patterns somewhat similar to that in Figure 1.1 occur widely in nature, in sizes that differ by many orders of magnitude, generated by processes that might seem to have nothing in common. On a ten million times larger scale the pattern is similar to that of a stream and river system formed by terrestrial water drainage, as shown by a geographic map (Fig. 1.2). On a 10-fold larger scale, but in three dimensions, the pattern is similar to that of a shrub. It is fascinating that there are broad similarities in the dynamics of processes that form such diverse things. When

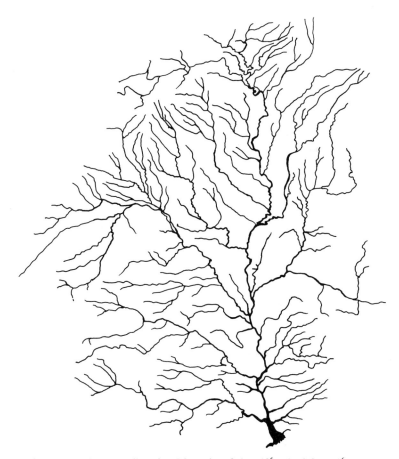

Figure 1.2. Amazon river and tributaries. Dimensions 3.4×10^6 m by 3.0×10^6 m. Age 10^7 years.

one considers that the time scales of formation of the dielectric breakdown pattern, the shrub, and the river system differ by factors of up to 10^{21}, the fascination increases. The times are respectively on the order of 10^{-6}, 10^{8}, and 10^{15} s.

There are many other types of structured systems, in all imaginable sizes. They include the cosmos, as illustrated by the Shane–Wirtanen map of a million galaxies (1); the distributions of people and of other living things on the earth's surface; and the geographic distributions of information and technology. On a microscopic scale structured systems include colloidal dispersions. Another example is the spatial distribution of reactive intermediates produced in a substance subjected to high-energy radiation. In progressively smaller dimensions one finds structure on the intramolecular, intraatomic, and intranuclear scales. The ends are not in sight, literally or figuratively, going inward or outward.

Mechanical and electronic devices are designed and built for specific uses. It is self-evident that the dynamics of a device are governed by its structure. On the scale of single molecules, spectroscopic properties indicate that intramolecular processes are strongly dependent on the structure of the molecule. However, on the intermediate scale of a few dozen molecules the effects of intermolecular configurations are less obvious. Until recently, few techniques have been available to examine and describe the intermolecular behavior. Details of the intermolecular structure are still mainly inferred. Early beginnings occurred in radiation physics and chemistry, where techniques for treating the dynamics of processes in nonrandom systems were proposed 80 years ago (2, 3). Recently progress has been rapid.

The purpose of this book is to bring together several facets of the subject and introduce it to a broad audience. The range of topics included is limited for practical reasons. Descriptions are begun at an elementary level, so that the reader can grasp the spirit of the subject. The limits of the field continue to expand. In naming the subject it was necessary to leave room for major developments in the future. Types of systems that will eventually be included extend beyond those now commonly conceived to be ''structured.'' For example, a solvent is usually approximated as a continuum. Methods are being developed to account for the great influence of the few dozen solvent molecules in the immediate vicinity of a species whose behavior is being examined (Marcus, Chandler, Wolynes, and others; Refs. 4–6 are a sample). The usual interpretation of *structured* makes it too narrow a term for our purpose. The adjective *nonrandom* is also too restrictive because the processes to be discussed include random steps. Furthermore, a nonrandom spatial distribution of species is a starting point for some processes and an endpoint for others. The name *kinetics of nonhomogeneous processes* has the desired flavor.

1.1. NEW CLASSIFICATION OF KINETICS SYSTEMS

The types of reaction systems currently described in kinetics textbooks may be divided into two categories: homogeneous and heterogeneous. With the addition of the intermediary nonhomogeneous category, the following distinctions are made between the type of systems:

Homogeneous—single phase; components are distributed randomly.

Heterogeneous—two or more phases; macroscopic heterogeneity; reaction occurs at interface between phases.

Nonhomogeneous—single phase or microheterogeneous; at least one component is not randomly distributed; a *nonhomogeneous process* is defined as one that does not occur randomly in space.

The prefix *non-* in nonhomogeneous was picked in preference to *un-*, *in-*, or *a-*. The simple negation *non-* has the appropriate flavor for the broad variety of systems and processes to which the term will be applied.

The kinetics of nonhomogeneous processes do not conform to conventional kinetics treatments. Concepts have been developed to describe the rates and relative probabilities of competing processes in nonhomogeneous systems. For example, when a dilute solution of nitrous oxide in liquid ethane is subjected to γ radiation (7), the reactions may be represented by the following simplified scheme (see Chapter 6 for a more detailed discussion):

$$C_2H_6 \rightsquigarrow C_2H_6^+ + e^- \tag{1.1}$$

$$C_2H_6^+ + e^- \longrightarrow \text{products} \tag{1.2}$$

$$e^- + N_2O \longrightarrow N_2 + O^- \tag{1.3}$$

$$O^- + C_2H_6^+ \longrightarrow \text{products}' \tag{1.4}$$

The wiggly arrow means upon absorption of radiation. Through reactions (1.2) and (1.3) the cations and nitrous oxide compete for the electrons. If all three species were homogeneously distributed in the system, the yield of nitrogen would have been expected to vary with nitrous oxide concentration in the manner of the dashed line in Figure 1.3. But the ions and electrons are generated in correlated pairs along the radiation particle tracks. The system is nonhomogeneous. An electron has a greater than expected probability of reacting with its correlated cation than with one of the nitrous oxide molecules that are distributed randomly in the system. The concentration of nitrous oxide required to capture the electrons was nearly five orders of magnitude greater than would be expected in a homogeneous system in which the same number of electrons and ions had been produced per unit time.

The calculation of the curve for the nonhomogeneous system in Figure 1.3 is described in Chapter 6.

1.2. HISTORICAL SKETCH

1.2.1. Early

Development of the kinetics of nonhomogeneous processes appears to have originated in radiation physics and chemistry around 1900. This field has generated a

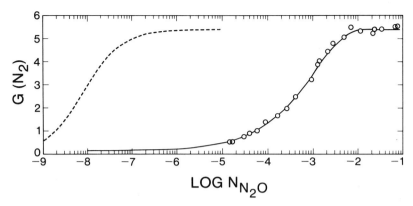

Figure 1.3. Yield of nitrogen as a function of nitrous oxide concentration in γ-irradiated liquid ethane at 183 K. Here $G(N_2)$ is the number of nitrogen molecules formed per 100 eV of energy absorbed by the ethane and N_{N_2O} is the mole fraction of nitrous oxide. Experimental points from Robinson and Freeman (7). ——, values calculated for a nonhomogeneous system, discussed in Chapter 6. ----, expectation values for a homogeneous system: steady rate of energy absorption $= 3 \times 10^{21}$ eV/m³·s, $G(N_2)_{max} = 5.4$,

$$k_{1.2} \approx 4\pi r_c(D_e + D_+) \approx \xi\mu_e/\varepsilon_0\epsilon \approx 6 \times 10^{-13} \text{ m}^3/\text{ptl·s}$$

$$k_{1.3} \approx 4\pi(1 \times 10^{-10} \text{ m})D_e \approx 1 \times 10^{-15} \text{ m}^3/\text{ptl·s}$$

$$k_{1.4} \approx 4\pi r_c(D_- + D_+) \approx 3\xi\mu_+/\varepsilon_0\epsilon \approx 4 \times 10^{-15} \text{ m}^3/\text{ptl·s}$$

$$N_{N_2O} = 1.0 \times 10^{-3} \text{ is equivalent to } 1.1 \times 10^{25} \text{ N}_2\text{O molecule/m}^3$$

$$\frac{1}{G(N_2)} = \frac{1}{5.4}\left(1 + \frac{k_{1.2}[C_2H_6^+]}{k_{1.3}[N_2O]}\right) \approx \frac{1}{5.4} + \frac{1.5 \times 10^{-9}}{N_{N_2O}}$$

number of ideas that affected or formed other areas of inquiry. For example, the concept of chain reactions arose in 1905 from early work in radiation chemistry (8–11) but grew most vigorously between about 1930 and 1960 under the attention of free radical chemists. The rapid development of nonhomogeneous kinetics is occurring only now in fields outside radiation research.

In the early years after the discovery of radioactivity there was a fascination with the ionizing power of the emanations. Alpha particles caused the greatest effect and were easiest to experiment with in gases. The amount of ionization was measured by applying an electric field across the gas and collecting the ions. The current collected at the electrodes as a function of field strength did not completely agree with theory (2, 3). It was expected that above a certain field strength all of the ions would be collected and the current would be constant, equal to the initial ionization rate. At low fields strengths the current increased with field as expected, but at high fields it did not become constant. The current continued to increase with field strength, as though ions were being pulled from a more reluctant source. It was concluded that the action of α particles does not completely separate all of the ions from their parent molecules and that some of the ions "fall back on to their parent molecules" (2). The effect was less extensive when β or γ rays were used instead of α rays.

The nonhomogeneous distributions of ionization in gases irradiated by α, β, and X rays were photographed by Wilson in 1912 (12) using the cloud chamber for which he won the Nobel Prize in 1927. Example tracks are shown in Figure 1.4. The ionization tracks were made visible by condensing water molecules onto the ions, making microdroplets, and scattering light off the "clouds" so formed. Separate droplets can be seen in the magnified photograph, Figure 1.4c.

A schematic representation of a portion of the column of ionization along the α track in Figure 1.4a is shown in Figure 1.5 (13). An electron knocked from a molecule by the α particle suffers collisions with other molecules and loses its excess energy. After a number of collisions the average kinetic energy of the electron becomes the same as that of the molecules, and the electron is said to be thermalized. The radius of the outer shaded portion in Figure 1.5 is approximately the distance traveled by the electrons away from the ions during thermalization, and the times given in the legend are the "thermalization times." If the thermalization distance of an electron is not too large, the coulombic attraction between the ion and electron might draw them back to recombine. At the same time random motions caused by collisions with the molecules tend to drive the electron and ion apart. Recombination and escape are in competition.

An α particle interacts so strongly with molecules that it ionizes every one it comes near. It produces a column of ionization consisting of a core of positive ions surrounded by a sheath of electrons (Figure 1.5). The combined attraction of

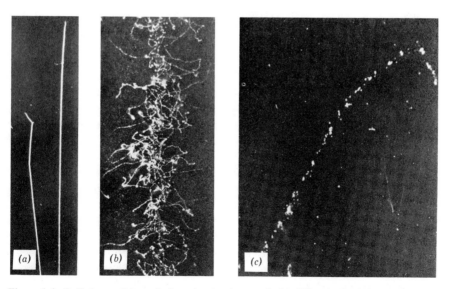

Figure 1.4. Radiation particle tracks in moist air, photographed in Wilson's cloud chamber (12). (a) α particles, magnification 3.5×. (b) 3-mm-diameter X-ray beam, magnification 3.1×. (c) 31× magnification of part of an X-ray beam track. The tracks in (b) and (c) are produced by the energetic electrons formed when molecules are ionized by X rays. Ionizations in tracks of high-energy β particles are more widely separated than those in (c). (Reproduced, with permission, from the Proceedings of the Royal Society, London.)

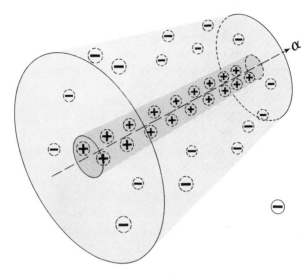

Figure 1.5. Schematic of a portion of an α particle track shortly after passage of the α particle (13). In air the radius of the outer shaded portion is ~ 3 μm, and the evolution time is $\sim 10^{-10}$ s after passage of the particle. In a thousand times denser substance such as water the evolution time and distance are about a thousand times smaller, namely $\sim 10^{-13}$ s and ~ 3 nm. (Reproduced, with permission, from the *Annual Review of Physical Chemistry*, vol. 34. © 1983 by Annual Reviews Inc.)

an electron toward the several positive ions in an adjacent segment of the core is usually so great that only a small fraction of the electrons wanders away. However, more electrons can be attracted away from the core by application of an external electric field.

The electrons and ions that escape initial recombination become free ions diffusing at random in the system. They can be collected at electrodes and counted, so the effect of applied field strength on escape from the column can be measured.

In 1913 Jaffé (14, 15) devised an approximate equation that describes the fraction of electrons that escapes from the ionized columns produced by α rays, and the manner in which the fraction increases with applied field. The equation is given in Chapter 2, Section 2.4.1.

There are sociological analogs to the preceding model. A correlated pair of ions in a liquid tends to become separated by the random action of collisions with molecules. Random action tends to separate any correlated pair, including a husband and wife. An analog of the external electric field effect on charge separation is used in the business of espionage. Application of external fields of money and sex weakens the bond of loyalty to country and family.

The concept of ion atmospheres in electrolyte solutions seems to have been inspired by the early treatments of the behavior of ionized gases. In 1912 Milner (16, 17) developed the ion atmosphere model of electrolyte solutions to overcome shortcomings of the Arrhenius model of incomplete dissociation (18). The latter did not give the correct concentration dependence of osmotic pressure, freezing point lowering, or electrical conductivity for strong electrolytes. Milner assumed that strong electrolytes are completely dissociated but that there is a slightly non-random spatial distribution of the ions. The electrical forces between the ions increase the chance of finding oppositely charged (attracting) ions near to each other and decrease the chance of finding like charged (repelling) ions near to each other. Milner used the Boltzmann principle and a stationary stochastic (probabilistic, see Section 1.3) treatment to determine the potential energy due to the nonrandom distribution of ions, the so-called ionic atmosphere. The potential energy is a measure of the deviation of the behavior from ideality. It is equal to the work the electrical forces are capable of doing as the ions are moved to infinite distances apart (infinite dilution). Milner's model was successful, but his mathematical technique was beyond the scope of most chemists. The work was ignored for a decade.

Eleven years later Debye and Hückel (19, 20) streamlined the mathematics in Milner's model, with the aid of Poisson's equation to relate the electrostatic potential to the charge density at any point. They obtained simple equations in the limit of extreme dilution. The simple equations are easy to use, so the model and equations usually bear the name Debye–Hückel. The more correct designation, Milner–Debye–Hückel, might increase in popularity as stochastic techniques become more commonplace.

Milner's work is one of the historic steps in the conceptual development of the dynamics of processes in nonhomogeneous systems. He used a stochastic model, whereas the models mentioned earlier in this history were deterministic. The conceptual difference between stochastic and deterministic models is discussed in Section 1.3.

The next major development occurred in 1924 in a different field, radiation biology, more than 10 years after the works of Milner and Jaffé. It was the target theory for the lethal effect of radiation on living cells (21–26). The sensitive zone in a cell is a small fraction of the total cell volume. To inactivate a cell, a certain minimum of energy must be deposited in the sensitive zone. Radiation particles deposit energy discontinuously along their tracks, as shown in Figure 1.4c. The probability that enough energy is deposited in the sensitive zone depends on the probability that one or more particles pass through the zone and the probability that they deposit enough cumulative energy at that place. It is a stochastic model.

The target theory has been applied to several biological effects, including gene mutations, inactivation of viruses, and the killing of different kinds of cells. The amount of energy required in the sensitive zone depends on the nature of the cell, the type of effect, and the ionizing properties of the radiation particles (26). Biological effects of radiation are discussed in Chapter 7.

A decade later, in 1938, Onsager treated the initial recombination of ions and electrons formed in a dense gas by X or β rays (27). These relatively weakly ionizing radiations tend to produce ions and electrons in isolated pairs (Figure 1.4c). An electron produced in a dense gas or liquid has a greater probability of escaping from its isolated parent ion in a β particle track than from the dense positive core of an α particle track. Onsager's method was to consider the Brownian motion of the positive and negative particles in their mutual coulombic field, and the effect of an externally applied field. He had earlier solved a similar problem in the theory of weak electrolytes (28), so inspiration flowed in the opposite direction to that in Milner's work.

Onsager's theory is described in Chapter 2 and is applied to the electrical conductivity induced in insulators by the action of high-energy radiation.

In 1951 Magee added a new direction to the development of kinetics of non-homogeneous processes (29–31). He made a model of the chemical reactions that occur in a particle track. The ions and electrons are chemically reactive. They may react with each other, with solutes added to the system (32), or perhaps with the solvent itself. The reactive species produced by the radiation are initially distributed in little bunches along the particle track. A little bunch of reactive intermediates is called a spur, a name that had been given by Wilson (12) to the tiny projection from the side of each α particle track in Figure 1.4a. Magee applies the name to separate little clusters such as those in the electron track in Figure 1.4c. The spatial distribution of radiation-produced species is distinctly nonrandom, whereas that of solute molecules is random. For a solute to be able to appreciably intercept the recombination of the species in the track, the bulk concentration of solute should be similar to the initial microscopic concentration of radiation-produced species in the spurs. Random diffusion of the species in a spur tends to move them away from each other. Therefore, ions and electrons that escape recombination for a time t after formation have a progressively decreasing probability of recombination as t increases, and the probability that they will ultimately react with solutes increases. This problem was formulated in terms of continuum dynamics, using Fick's second law of diffusion and reaction rate concepts for homogeneous systems. The model and further developments of it are described in Chapter 4.

A practical step that inspired further development of theory of nonhomogeneous processes was the estimation of electron thermalization distances in irradiated liquids. It was done by electrically measuring the fractions of ions that escape initial recombination in different liquids, and the field dependences of the fractions (33–38). An extension of Onsager's theory (34, 38, 39) was used to extract the thermalization distances from the conductance data. The theory and data are presented in Chapter 2.

This brief history must omit dozens of interesting works.

The next major stage of growth was a stochastic treatment of reaction probabilities of the electrons, ions, and solutes in γ-irradiated liquids (40). The volume of a spur and the number of reactive species in it are small. Continuum dynamics is not appropriate for such systems, so molecular dynamics was introduced. The model was inspired by the observations that only a few percent of the ions and electrons escape initial recombination in liquid hexane (33, 34), but much larger fractions of them can react with moderately low concentrations of solutes (41, 42). The stochastic model employed a more intimate examination of the processes at a molecular level than had been used previously. It is discussed in Chapter 6.

A semiempirical model of reaction between electrons, ions, and scavengers, to be described in Chapter 5, employs continuum dynamics but has the advantage of simpler equations (43, 44). It has been extensively used.

Radiation-induced reactions occur in solids as well as in the other phases. Microscopic structure changes that occur in metals and alloys are of crucial importance to the nuclear power industry. Models that interpret the changes in metals under irradiation are presented in Chapter 8.

The nature and extent of change in any system under irradiation is strongly affected by the details of the spatial distribution of energy deposition. Recent advances in knowledge of radiation track structure are the subject of Chapter 3.

1.2.2. Growth in Other Fields

Many types of systems have at least one component that is not randomly distributed in space. One large category is microheterogeneous systems, including micelles, vesicles, and microemulsions. A micelle is a particle in a colloidal dispersion, often a solid or liquid microcrystal. Nowadays the word refers mainly to clusters of surface-active molecules that form when the concentration of surfactant in solution is greater than a critical value of a few weight percent. A micelle can accommodate a small number of compatible molecules on its surface or in its interior. The reactivities of some molecules depend greatly on whether or not they are attached to a micelle and on how many molecules are attached to a given micelle (45, 46). A vesicle is similar to a micelle but has a bit of solvent inside it. The kinetics of processes in micelles and vesicles are suited to stochastic treatments (45–49). Kinetics in microemulsions remains to be treated. Behavior in micelles and vesicles is described in Chapter 11.

Dispersive charge transport is an important characteristic of amorphous semiconductors. It causes tailing of a pulsed conductance signal and is interpreted by a stochastic hopping model (50–52). A stochastic model can also be used to describe fast ion transport in solid electrolytes (53), materials that will appear in many future devices.

Nonhomogeneous kinetics models have been discussed in positronium (54, 55) and muonium (56, 57) chemistry. These pseudoatoms live only a few millionths of a second or less, but they are used in studies ranging from hydrogen isotope effects in chemical reactions to lattice defects in solids.

The burning of the fine droplets in a fuel spray in a furnace is also a nonhomogeneous process (58). Hydrocarbon molecules evaporate from the droplets and react with oxygen as they mix with the surrounding hot air.

Several types of models are required to encompass the broad range of nonhomogeneous processes. Different types of models have sometimes been constructed to emphasize different aspects of the same process.

1.3. MODELS

Models may be classified into two general categories, stochastic and deterministic, although many models contain components of both types. Classifications are useful because they convey the nature of the main approximations included in the models. Better classification terms may arise as the subject matures.

1.3.1. Stochastic and Deterministic: Molecular and Continuum Dynamics

The distinction between stochastic and deterministic models in kinetics is not sharp, so the terms require discussion. For example, Brownian motion is a stochastic model of diffusion. Fick's laws, which consider matter sliding down a chemical potential gradient, are a deterministic model of diffusion.

Stochastic is a word that emerges from probability theory (59–61). It was used during the sixteenth to eighteenth centuries, but then fell out of usage until about 55 years ago (62). A stochastic model considers probability as a function of time. It usually involves a step-by-step process. The roots of the word are Greek: *stochastikos*, which means skillful at aiming, guessing; *stochazesthai*, to aim at, guess at; *stochos*, an aim (shot), guess. In a reaction system, when the number of reactive species in a tiny zone is small, the probable fate of the species is strongly affected by the actual number initially in the zone and by their separation distances. If two of the species react together, the probable fate of the others is also affected. For example, in reactions (1.2)–(1.4), if the electron reacts with nitrous oxide, the ultimate products of the cation are also affected because reaction (1.2) is replaced by (1.4).

Deterministic implies that events are the inevitable result of antecedent conditions: Latin *de-* means from and *terminus* means boundary. In the context of kinetics a deterministic model is one based on macroscopic concepts. It deals with net changes in averages. Although diffusion equations and conventional rate equations for homogeneous systems can be derived from the stochastic point of view, they have a deterministic form: diffusion rate $= -D(dc/dx) \, dt$, where D is the diffusion coefficient, c is the concentration, x is the position coordinate, t is the time; first-order reaction rate $= k_r c_A$, where k_r is the rate constant and c_A is the concentration of reactant A.

Both stochastic and deterministic models are useful in treating the kinetics of nonhomogeneous processes. The former are best adapted to systems that contain only small numbers of reactive particles per microzone. They are usually based on molecular dynamics. The latter apply best when there are many reactive particles per microzone. They are based on continuum dynamics. Many models are hybrids (40, 63, 64).

1.3.2. Deterministic

These models are characterized by a rate equation of the form

$$\frac{\partial c_i}{\partial t} = D_i \, \nabla^2 c_i - \sum_{j,m} k_{jm} c_j c_m \tag{1.5}$$

where c_i is the concentration of reactant i, D_i is the diffusion coefficient of i, ∇^2 is the Laplacian operator, k_{jm} is the reaction rate constant between species j and m, and the summation is taken over all pairs that destroy or produce i in the system, including $j = m = i$ if appropriate. The sign of k_{jm} is positive if either j or m is i and negative if reaction between j and m produces i. The term $D_i \, \nabla^2 c_i$ corresponds to Fick's second law of diffusion. For example, if the cations in the central core in Figure 1.5 were actually neutral free radicals, the radicals would tend to diffuse outward. The total volume occupied by the core would therefore increase, thereby decreasing the average concentration of radicals even if none reacted.

A solution to Eq. (1.5) can be attained after making approximations. For example, random diffusion of the free radicals without reaction in the above-mentioned core would produce a Gaussian radial distribution of c_i. Although the initial distribution was probably not Gaussian and combination reactions would prevent the distribution from becoming Gaussian, to obtain a solution to Eq. (1.5) it is usually assumed that the distribution is Gaussian at all times. This is the so-called prescribed diffusion approximation (14, 31).

Equations similar to (1.5) are used extensively in later chapters in conjunction with rate equations of the type

$$\frac{dc_p}{dt} = k_{im} \, c_i c_m \tag{1.6}$$

where c_p is the concentration of product p.

1.3.3. Stochastic

The emphasis in stochastic models is often the calculation of relative probabilities of competing reactions. The absolute reaction rates are included implicitly, and may be estimated explicitly by using appropriate dynamics equations.

There are several types of models. No single one can serve to illustrate procedures in general. A model usually focuses on only one or two aspects of a system. Simplifying assumptions tend to separate the portion being examined from the system as a whole, so the model usually magnifies the relative importance of the portion. Each type of model has a different appearance, much as the form of a human face is approximated differently in portraits by Amedeo Modigliani and Joan Miró. As an example, consider the behavior of transient ions produced in liquids or solids exposed to ionizing radiation. Recombination of some of the ions and electrons can be intercepted by sweeping them out of the system with an externally applied electric field (Chapter 2) or by allowing one or both of them to react with a solute (Chapter 6). Models for these processes are conceptually similar but have different appearances.

Spatial correlation between species in a system is sometimes indicated by putting square brackets around them (40). A generalization of reactions (1.1)–(1.4) involving correlated ion–electron pairs and randomly distributed electron scavenger molecules S is

$$M \rightsquigarrow [M^+ + e^-] \tag{1.7}$$

$$[M^+ + e^-] \longrightarrow P_1 \tag{1.8}$$

$$[M^+, e^-, S] \begin{cases} \nearrow P_1 + S \\ \searrow P_2 \end{cases} \begin{matrix} (1.9) \\ (1.10) \end{matrix}$$

Reactions (1.9) and (1.10) indicate that occasionally the ion and electron are generated in the vicinity of an S; there is a finite probability that the electron will react with S to form P_2 rather than with M^+ to form P_1.

A stochastic equation for the fraction ϕ of electrons that reacts with S is

$$\phi = 1 - (1 - fN_s)^{\nu\beta} \tag{1.11}$$

where f is the encounter efficiency of the $e^- + S$ reaction, N_s is the mole fraction of S in the solution, the parameter ν relates to the inital separation distance between the thermalized electron and ion, and β relates to the diffusion jump distances of the species, among other things.

The stochastic equation (1.11) is quite different from the deterministic equation (1.6). The former attempts a more intimate view of the system than does the latter.

Stochastic treatments have been developed for electron–hole transport in amorphous materials (Chapter 9), reactions in micelles (Chapter 11), and the dynamics of polymerization and of polymeric structures (Chapters 12 and 13). Proteins are a living example of dynamic polymeric structures (65).

Microelectronic systems using very large scale integration (VLSI) of circuits and detectors on semiconductor chips sometimes rival in complexity a living cell. Cosmic rays can produce noise and damage in VLSI devices in spacecraft. A stochastic model developed to interpret radiation damage to cells (Chapter 7) has been applied to improve the devices (66).

1.4. ORGANIZED SYSTEMS

A biological membrane is an example of an organized system. It is a double layer of long-chain molecules and has some of the properties of a two-dimensional liquid crystal. Embedded in the double layer are other large molecules; some modify the structural properties of the membrane and others are catalytic agents (enzymes).

Membranes are not yet well understood. Cooperative effects are of major importance in the physics and chemistry of their behavior. The study of membranes can be approached by using monolayers of long-chain fatty acids as models, with test molecules embedded in or adsorbed on them. Such systems are called monolayer organizates. They are discussed in Chapter 10.

A different kind of molecular organization develops spontaneously during certain chemical reactions. An initially homogeneous system can become nonhomogeneous through a complex reaction mechanism that generates periodic variations in the concentrations of species. The phenomenon is called chemical waves and is discussed in Chapter 14. The subject has a large future that relates to the generation of biological systems.

A topological approach to general reaction dynamics, involving network analysis, shows great promise (67). It includes chemical waves, explosions, and other stable and unstable reaction systems.

The transformation of less organized into more organized material is a common concern of thermodynamics, but the kinetics of the processes have been difficult to study. Dramatic examples are the formation of intricately shaped crystals by cooling a melted material or by precipitation from a supersaturated solution. The growth of order, with the associated decrease of entropy, must be nucleated; ordered growth continues easily once a template has been formed. Nucleation theory remains a challenge in the kinetics of nonhomogeneous processes.

The requirement of a nucleation event for the growth of a highly ordered system is carried to its most remarkable level in the sperm–egg nucleation of animal reproduction. To speculate about the driving force to nucleate is beyond the scope of this book.

1.5. FRACTALS

The formation of an object that has fractional dimensionality is a nonhomogeneous process. A system that has fractional dimensionality is a fractal (68). A line, a square, and a cube are structures with Euclidean dimensionalities equal to 1, 2, and 3, respectively. Euclidean geometry has only integer dimensionalities. The surface area of a square or circle varies as the 2.0 power of the length of a side or the diameter. Many objects in nature have surface areas that vary as a fractional power of their diameter. A snowflake has a ramified structure that does not fill all the space within its circumference; the actual surface area varies approximately as the 1.5 power of the diameter, so the fractal dimensionality of snowflakes is about 1.5 (68, 69).

The recent flurry of publications about fractals (70) is part of the wider rapid growth of interest in nonhomogeneous systems. The initial appeal of fractals is so strong that one tends to become as a pup that has just learned to lift its hind leg, running around squirting a few drops on everything that he thinks is waiting to receive them. The boundaries of the territory claimed in this way might require adjustment, but they are wide. Popular interest in the nonanalytic behavior of systems in nature grows with the easy availability of mathematical techniques to treat them.

The dielectric breakdown pattern in Figure 1.1 is a fractal. The total length of all the branches inside a circle of radius r varies approximately as $r^{1.7}$, so the fractal dimensionality is $D \approx 1.7$ (71, 72). The Euclidean dimensionality of the pattern is $d = 2$, and the topological dimensionality is $D_t = 1$ (lines). Details of the pattern are probably governed by such factors as (a) fluctuations in the microscopic density and stresses within the insulating material; (b) once a gas channel has been formed by a local discharge, it serves as a conduction path to drain charges from solid regions around it; (c) the greatest electrical potential gradient occurs at the tip of a channel, so branching occurs at or near the tip; and (d) drainage of charge into a channel from the surrounding region prevents a discharge from joining two existing channels, so cross-linking does not occur. Breakdown begins at a stressed *point* on the edge. A microscopic zone of low density in the solid is an insipient bubble with lower than average dielectric strength, which facilitates formation of a real bubble by breakdown at that place. This gives a random component to the pattern.

Theories of "homogeneous turbulence" fall short of describing natural processes, and the concept of fractals has been applied (68). What has been called intermittent turbulence (for example, wind comes in gusts) is actually a nonhomogeneous process that need only be called turbulence.

Fractal objects have growth rates that vary as fractional powers of time (68, 69). Fractional powers of time are familiar in many fields, including diffusion, percolation, and radiation effects.

1.6. FUTURE TOPICS

Topics that fall naturally into the area of kinetics of nonhomogeneous processes but are not included in this book include detonations and shock waves, instabilities induced by high electric fields (switching and lasing in semiconductors, dielectric breakdown in insulators), turbulence (including weather forecasting and solar coronal arches), and possibly quantum mechanics (73). The following chapters convey the flavor of nonhomogeneous processes and offer a working knowledge of several approaches to them. It is hoped that the book provides a unifying concept for a number of previously uncorrelated areas of the scientific literature.

REFERENCES

1. E. J. Groth, P. J. E. Peebles, M. Seldner, and R. M. Soneira, *Sci. Am.* **237**(5), 76 (1977), 5th and 11th figures.

2. W. H. Bragg and R. D. Kleemann, *Philos. Mag.* **11**, 466 (1906).

3. R. D. Kleemann, *Philos. Mag.* **12**, 273 (1906).

4. (a) R. A. Marcus and many others, *Faraday Disc. Chem. Soc. (London)* **74**, 7–405 (1982), *Electron and Proton Transfer.* (b) D. G. Truhlar, W. L. Hase, and J. T. Hynes, *J. Phys. Chem.* **87**, 2664 (1983), and refs. therein.

5. (a) J. A. Montgomery, S. L. Holmgren, and D. Chandler, *J. Chem. Phys.* **73**, 3688 (1980). (b) M. J. Thompson, K. S. Schweizer, and D. Chandler, *J. Chem. Phys.* **76**, 1128 (1982). (c) D. Chandler, *J. Phys. Chem.* **88**, 3400 (1984). (d) A. L. Nichols and D. Chandler, *J. Chem. Phys.* **84**, 398 (1986). (e) D. Chandler, "Roles of Classical Dynamics and Quantum Dynamics on Activated Processes Occurring in Liquids," *J. Stat. Phys.* **42**, 49 (1986).

6. (a) D. F. Calef and P. G. Wolynes, *J. Chem. Phys.* **78**, 470, 4145 (1983) and earlier work of Wolynes. (b) R. Kapral, *J. Phys. Chem.* **86**, 1259 (1982).

7. M. G. Robinson and G. R. Freeman, *J. Chem. Phys.* **55**, 5644 (1971).

8. W. P. Jorissen and W. E. Ringer, *Ber. Dtsch. Chem. Ges.* **38**, 899 (1905); **39**, 2093 (1906).

9. S. C. Lind, *J. Phys. Chem.* **16**, 564 (1912).

10. M. Bodenstein-Hannover, *Z. Elektrochem.* **19**, 836 (1913).

11. K. M. Bansal and G. R. Freeman, *Radiat. Res. Rev.* **3**, 209 (1971).

12. C. T. R. Wilson, *Proc. Roy. Soc. (London)* **A87**, 277 (1912).

13. G. R. Freeman, *Ann. Rev. Phys. Chem.* **34**, 463 (1983).

14. G. Jaffé, *Ann. Physik IV* **42**, 303 (1913).

15. H. A. Kramers, *Physica* **18**, 655 (1952).

16. S. R. Milner, *Philos. Mag.* **23**, 551 (1912).

17. S. R. Milner, *Philos. Mag.* **25**, 742 (1913).

18. S. Arrhenius, *Z. Phys. Chem.* **1**, 631 (1887).

19. P. Debye and E. Hückel, *Phys. Z.* **24**, 185, 305, 335 (1923).

20. P. Debye, *Phys. Z.* **25**, 97 (1924).

21. J. A. Crowther, *Proc. Roy. Soc. (London)* **B96**, 207 (1924); **B100**, 390 (1926).

22. F. Howleck, *Compt. Rend. (Paris)* **188**, 197 (1929).

23. A. Lacassagne, *Compt. Rend. (Paris)* **188**, 200 (1929).

24. Mme. P. Curie, *Compt. Rend. (Paris)* **188**, 202 (1929).

25. R. Glocker, *Z. Phys.* **77**, 653 (1932).

26. D. E. Lea, *Actions of Radiations on Living Cells,* 2nd ed., University Press, Cambridge, 1955, Chapter 3.

27. L. Onsager, *Phys. Rev.* **54**, 554 (1938).

28. L. Onsager, *J. Chem. Phys.* **2**, 599 (1934).

29. J. L. Magee, *J. Am. Chem. Soc.* **73**, 3270 (1951).

30. A. H. Samuel and J. L. Magee, *J. Chem. Phys.* **21**, 1080 (1953).

31. A. K. Ganguly and J. L. Magee, *J. Chem. Phys.* **25**, 129 (1956).

32. R. L. Platzman, *Physical and Chemical Aspects of Basic Mechanisms in Radiobiology,* National Academy of Sciences, NRC Publication 305, Washington DC, 1953, pp. 22–50.

33. G. R. Freeman, *J. Chem. Phys.* **38**, 1022 (1963); **39**, 988 (1963).

34. G. R. Freeman, *J. Chem. Phys.* **39**, 1580 (1963).

35. A. Hummel and A. O. Allen, *Disc. Faraday Soc.* **36**, 95 (1963).

36. G. R. Freeman and J. M. Fayadh, *J. Chem. Phys.* **43**, 86 (1965).

37. W. F. Schmidt and A. O. Allen, *J. Chem. Phys.* **72**, 3730 (1968).

38. J.-P. Dodelet, K. Shinsaka, U. Kortsch, and G. R. Freeman, *J. Chem. Phys.* **59**, 2376 (1973).

39. J. Terlecki and J. Fiutak, *Int. J. Radiat. Phys. Chem.* **4**, 469 (1972).

40. G. R. Freeman, *J. Chem. Phys.* **43**, 93 (1965); **46**, 2822 (1967).

41. G. Scholes and M. Simic, *Nature* **202**, 895 (1964).

42. F. Williams, *J. Am. Chem. Soc.* **86**, 3954 (1964).

43. A. Hummel, *J. Chem. Phys.* **48**, 3628 (1968).

44. J. M. Warman, K.-D. Asmus, and R. H. Schuler, *J. Phys. Chem.* **73**, 931 (1969).

45. M. Tachiya, *Chem. Phys. Lett.* **33**, 289 (1975).

46. M. Grätzel, A. Henglein, and E. Janata, *Ber. Bunsenges. Phys. Chem.* **79**, 476 (1975).

47. Sz. Vass, *Chem. Phys. Lett.* **70**, 135 (1980).

48. H. Sano and M. Tachiya, *J. Chem. Phys.* **75**, 2870 (1981).

49. M. Tachiya and M. Almgren, *J. Chem. Phys.* **75**, 865 (1981).

50. H. Scher and E. W. Montroll, *Phys. Rev. B* **12**, 3455 (1975).

51. J. Noolandi, *Phys. Rev. B* **16**, 4474 (1977).

52. M. J. Schaffman and M. Silver, *Phys. Rev. B* **19**, 4116 (1979).

53. S. H. Jacobsen, A. Nitzan, and M. A. Ratner, *J. Chem. Phys.* **72**, 3712 (1980).

54. P. Jansen and O. E. Mogensen, *Chem. Phys.* **25**, 75 (1977).

55. V. M. Byakov, V. I. Grafutin, O. V. Koldaeva, E. V. Minaichev, and O. P. Stepanova, *Inst. Theor. Exp. Phys. (Moscow)*, Preprint 35.

56. P. W. Percival, *J. Chem. Phys.* **72**, 2901 (1980).

57. D. C. Walker, *Hyperfine Interactions* **8**, 329 (1981).

58. (a) R. Samson, D. Bedeaux, M. J. Saxton, and J. M. Deutch, *Combust. Flame* **31**, 215 (1978). (b) N. Peters, *Ber. Bunsenges. Phys. Chem.* **87**, 989 (1983).

59. A. Kolmogorov, *Math. Ann.* **104**, 415 (1931).

60. W. Feller, *Probability Theory and its Applications,* 2 vols., Wiley, New York, 1968.

61. W. Mendenhall, R. L. Scheaffer, and D. D. Wackerly, *Mathematical Statistics with Applications,* 2nd ed., Duxburg, Boston, 1981.

62. J. R. McGregor, private communication; see also ref. 59.

63. R. M. Noyes, *J. Chem. Phys.* **22**, 1349 (1954); *Prog. Reac. Kinet.* **1**, 129 (1961).

64. L. Monchick, *J. Chem. Phys.* **24**, 381 (1956).

65. J. A. McCammon and M. Karplus, *Acc. Chem. Res.* **16**, 187 (1983).

66. P. J. McNulty, *Phys. Today* **36**(9), 108 (1983); **39**(1), 50 (1986).

67. (a) R. B. King (Ed.), *Chemical Applications of Topology and Graph Theory,* Elsevier, Amsterdam, 1983, articles by O. Sinanoglu (p. 57), P. G. Mezey (p. 75), H. G. Othmer (p. 285), B. L. Clarke (p. 322), and O. E. Rossler and J. L. Hudson (p. 358). (b) B. L. Clarke and C. J. Jeffries, *J. Chem. Phys.* **82**, 3107 (1985). (c) R. Larter and B. L. Clarke, *J. Chem. Phys.* **83**, 108 (1985).

68. B. B. Mandelbrot, *The Fractal Geometry of Nature,* Freeman, San Francisco, 1983.

69. P. Meakin and J. M. Deutch, *J. Chem. Phys.* **80**, 2115 (1984).

70. 1983–1985 issues of *Phys. Rev. A, Phys. Rev. B, Phys. Rev. Lett.,* and *J. Phys. A: Math. Gen.*

71. H. G. E. Hentschel, *Phys. Rev. Lett.* **52**, 212 (1984).

72. N. Niemeyer, L. Pietronero, and H. J. Wiesmann, *Phys. Rev. Lett.* **52**, 1033 (1984).

73. The seeds of stochastic quantum mechanics are Born's statistical interpretation of quantum theory (1926) and Schrödinger's pointing out the similarity between his wave equation and equations for diffusion and Brownian motion (1931). The subject is still controversial, but interest has grown recently. Six references provide a sample, of which the last is especially interesting and challenging. (a) E. Nelson, *Phys. Rev.* **150**, 1079 (1966); *Quantum Fluctuations,* Princeton University Press, Princeton, N.J., 1985. (b) M. Jammer, *The Philosophy of Quantum Mechanics,* Wiley, New York, 1974, Chapter 9. (c) C. De Witt-Morette and K. D. Elworthy (Eds.), *Phys. Rep.* **77**C(3), 1981, *New Stochastic Methods in Physics,* contains reviews. (d) D. de Falco, S. De Martino, and S. De Siena, *Phys. Rev. Lett.* **49**, 181 (1982). (e) E. Prugovečki, *Stochastic Quantum Mechanics and Quantum Spacetime,* Reidel, Dordrecht, 1984.

2 IONIZATION AND CHARGE SEPARATION IN IRRADIATED MATERIALS

Gordon R. Freeman

Department of Chemistry
University of Alberta
Edmonton, Canada

Ionization is such a common reaction that it is one of the first chemical changes
we learn about in school. Most people therefore take it for granted. However, new
things are now being learned from a close examination of the ionization process.
The inquiry ultimately leads to questions about the nature of the electron, questions that have attracted new attention every decade or two during the past 70
years. The central question, "What is an electron, precisely?" has never been
answered, although we continue to learn more about its behavior.

Ionization in dense media involves nonrandom distributions of the positive and
negative species. The kinetics of ionization and charge separation provides an excellent, quantitative introduction to the kinetics of nonhomogeneous processes.

The term *ion* is taken to mean a separately distinguishable, charged particle.
Most particles in nature are combined into neutral groupings, such as molecules
or crystals. The term *ionization* is defined as a process that increases the number
of ions in a closed system. A closed system is one where there is no transfer of
matter between the system and its surroundings.

Ionization usually requires the absorption of energy, which permits spatially
correlated charged species to separate from each other against their mutual coulombic attraction:

$$A \longrightarrow B^+ + C^- \tag{2.1}$$

The energy can be supplied thermally or by various forms of radiation or by the
action of a high electric field.

In this chapter the C^- of Eq. (2.1) is usually an electron, and the main source
of energy is high-energy radiation:

$$A \rightsquigarrow A^+ + e^- \tag{2.2}$$

A single α particle that initially possesses 6 MeV (1.0 pJ) of kinetic energy can
create about 2×10^5 ionizations in a system. Addition of an α particle, which is
a He^{2+} ion, means that the system is not completely closed. However, the two
charges that enter on the α particle are negligible by comparison with the 200,000
ion–electron pairs produced by it. The α particle, or any other high-energy radiation, can be treated simply as an energy source.

Ionization is a relatively well defined process in low-density gases. It is more
complex in dense phases because the electron must traverse other matter while
moving away from the cation. The matter can interfere with the motion of the

electron. It can also modify the energy states of the cation and electron. The ionization threshold is therefore not sharp in dense phases.

To assist in understanding the ionization process in liquids, it is helpful to first visualize electron behavior in a low-density gas. We try to *visualize* an electron bound in an atom, and how its behavior changes as the atom absorbs energy. We can then consider an atom embedded in a dense medium and repeat the process.

2.1. LOW-DENSITY GAS

In a low-density gas the interaction between molecules is negligible except during collisions. A molecule spends only a small fraction of its time in collision. Photoexcitation therefore occurs, to a first approximation, through interaction of a photon with an isolated molecule.

2.1.1. Electron Behavior in a Hydrogen Atom

For simplicity consider a hydrogen atom and the changes that occur in it upon photoexcitation. The expectation value of the potential energy of the electron in the ground-state atom is $\langle V \rangle_{1s} = -\xi^2/4\pi\varepsilon_0 a_0 = -4.36 \times 10^{-18}$ J, where ξ is the proton charge, ε_0 is the permittivity of a vacuum, and a_0 is the most probable distance of the electron from the proton (1). The familiar name of this energy is the hartree, and it is expressed as -4.36 aJ (attojoule), or -27.2 eV. The expectation value of the kinetic energy of the electron in the ground state atom is $\langle T \rangle_{1s} = \xi^2/8\pi\varepsilon_0 a_0 = 2.18$ aJ, or 13.6 eV (1a). The total energy is $E_{1s} = \langle T \rangle_{1s} + \langle V \rangle_{1s} = 0.5\langle V \rangle_{1s} = -2.18$ aJ, so the electron is bound to the proton with an energy of 2.18 aJ (1 rydberg).

Visualization of the behavior of the bound electron can be approached by comparing three quantum mechanical models.

The probability function for the three-dimensional spatial distribution of the electron in the ground-state hydrogen atom is

$$\psi_{1s}^2 = (\pi a_0^3)^{-1} \exp\left(\frac{-2r}{a_0}\right) \tag{2.3}$$

Equation (2.3) is spherically symmetric, has a maximum density at the center, $r = 0$, and extends tenuously to $r = \infty$. The probability of the electron being at a certain distance r from the center, independent of direction, is given by the radial distribution function

$$P_{1s}(r) = 4\pi r^2 \psi_{1s}^2$$
$$= \frac{4r^2}{a_0^3} \exp\left(\frac{-2r}{a_0}\right) \tag{2.4}$$

The *most probable distance* of the electron from the nucleus, calculated by solving $d[P_{1s}(r)]/dr = 0$, is a_0. The average distance of the electron from the nucleus is often called the *expectation value of r*, to denote quantum mechanical averaging. It is

$$\langle r \rangle_{1s} = \int_0^\infty r\, P_{1s}(r)\, dr$$

$$= 1.5a_0 \tag{2.5}$$

The quantum mechanical model does not provide completely separate descriptions of momentum and position, so the detailed behavior of the electron in the orbital cannot be specified. However, the electron spends most of its time within a distance of about $2a_0$ of the nucleus.

From a slightly different point of view, one may consider the de Broglie wavelength λ of an electron,

$$\lambda = \frac{h}{mv} \tag{2.6}$$

where h is Planck's constant, $\mathbf{m}$ is the electron mass, and v is its speed. The value of λ for an electron with kinetic energy equal to that in the $1s$ state ($\langle T \rangle_{1s} = 2.18$ aJ) is $\lambda_{1s} = 2\pi a_0$. This is the distance around the Bohr orbit. The quantity $a_0 = \lambda_{1s}/2\pi$ is commonly called the Bohr radius. In general, the quantity $\lambdabar = \lambda/2\pi = (\hbar/mv)$ may be called the *de Broglie radianlength* of the particle. Under most circumstances λ does not correspond to the circumference of an orbit, but it does always correspond to the length of a wave or cycle. A complete wave or cycle is equivalent to the argument of a sine or cosine passing through 2π radians. A *radianlength* is equivalent to the argument passing through 1 radian, which in the circular model is equivalent to an arc equal in length to the radius. In this way we can visualize $\lambdabar$ as being related to the radius of the zone occupied by a bound electron; $\lambdabar_{1s} = a_0$. For an unbound particle $\lambdabar$ can be related to the impact radius.

The third approach is to consider the Heisenberg Uncertainty of radial position, Δr_{1s}, of the electron in the ground-state hydrogen atom. The entire momentum of the 2.18-aJ electron is the Δp_{1s} in the Uncertainty relation, because the direction of motion is not known at any specified place or time. One therefore obtains (1b)

$$\Delta r_{1s} \approx \frac{\hbar}{\Delta p_{1s}}$$

$$= \frac{\hbar}{mv_{1s}}$$

$$= \lambdabar_{1s} \tag{2.7}$$

Thus, for the electron in the ground-state hydrogen atom, $\Delta r_{1s} \approx a_0$.

The Heisenberg Uncertainty of position of the electron in the ground-state atom corresponds to the de Broglie radianlength of the electron and to the most probable distance of the electron from the center of the orbital. These agreements arise because the Heisenberg Uncertainty* principle and the de Broglie concept of particle wavelength were ''givens'' in the construction of the Heisenberg and Schrödinger formulations of quantum mechanics, respectively. For this reason, quantum mechanics cannot provide more detailed descriptions than allowed by the Heisenberg Uncertainty principle or the de Broglie radianlength. The problem of more detailed descriptions has titillated scientists for 60 years (2) and is likely to continue to do so for some time.

Now consider the change in behavior of the electron as energy is added to the hydrogen atom. Energy is absorbed in steps. In the Rydberg level n_R (1) the expectation value of the potential energy is

$$\langle V \rangle_{n_R} = \frac{-\xi^2}{n_R^2 \, 4\pi\varepsilon_0 a_0} \tag{2.8}$$

and that of the kinetic energy is

$$\langle T \rangle_{n_R} = \frac{\xi^2}{n_R^2 8\pi\varepsilon_0 a_0} \tag{2.9a}$$

$$= -E_{n_R} \tag{2.9b}$$

where $n_R = 1, 2, 3, \ldots$. Increasing n_R makes the potential energy less negative and the kinetic energy of the bound electron less positive. As the kinetic energy and therefore the momentum of the electron decrease, its de Broglie radianlength and the Heisenberg Uncertainty of its position increase:

$$\Delta r_{n_R} \approx \lambdabar_{n_R} = \frac{n_R \hbar}{\xi} \left(\frac{4\pi\varepsilon_0 a_0}{m} \right)^{1/2}$$

$$= n_R a_0 \tag{2.10}$$

It should be emphasized that the Uncertainty of position is approximately equal to the de Broglie radianlength, not to the wavelength. The Uncertainty relation $\Delta r \, \Delta p_r \approx \hbar$ [Eq. (2.7)] is an approximation of the more correct relation $\Delta r \, \Delta p_r \geq \hbar/2$ (1b). The common practice of using h instead of $\hbar$ in discussions of the Uncertainty principle (1a) is too crude for present purposes. The concept of de Broglie radianlength is not yet well known, but its use will increase as people stop ignoring the important factor 2π and *visualize* the difference between a cycle and a radian. To this end it would be useful to complete the SI units of h and $\hbar$. At present they are both listed as joule seconds. However, h and $\hbar$ refer to a single oscillator passing through one cycle and one radian, respectively. A unit oscillator, whether a single particle or photon or a more complex

---

*The Heisenberg Uncertainty is connected with the not completely understood wave nature of particles. It is not necessarily similar to the uncertainty about whether I will drink apple juice or beer with dinner tonight. To indicate that it is special, I use a capital U in Uncertainty when it is the Heisenberg kind.

construction, is designated an entity. Thus the units of h should be joule times second per entity times cycle (J·s/ent·cy), and those of $\hbar$ should be joule times second per entity times radian (J·s/ent·rad). The SI units already include joule, second, and radian and should be expanded to include cycle and entity.

This inset could be the subject of a chapter of a different book.

While Δr_{n_R} and λ_{n_R} increase as the first power of n_R, the most probable distance r_{mp,n_R} of the electron from the nucleus increases as n_R^2 (1). For the hydrogen orbitals with no radial nodes ($1s$, $2p$, $3d$, . . .), r_{mp,n_R} is simply

$$r_{mp,n_R} = n_R^2 a_0 \qquad (2.11)$$

The expectation value of the distance of the electron from the nucleus is slightly larger. For an electron in an orbital specified by the quantum numbers (n_R, l),

$$\langle r \rangle_{n_R,l} = \left(1 + \frac{1}{2(l+1)}\right) n_R^2 a_0 \qquad (2.12)$$

where the azimuthal quantum number $l = 0, 1, 2, . . . , n_R - 1$ specifies the number of angular nodes in the orbital. The total number of angular plus radial nodes is $n_R - 1$, so the number of radial nodes is $n_R - 1 - l$. The number of radial nodes in a $2s$ ($n_R = 2$, $l = 0$) orbital is 1, in a $3s$ orbital the number is 2, and so on.

The radial and angular nodes specify regions in space where the electron cannot be, so the uncertainty of position Δr_{n_R} does not increase as rapidly with n_R as does $\langle r \rangle_{n_R}$. For all values of n_R and l the expectation values of the electron–nuclear distance are in the range

$$\langle r \rangle_{n_R} = (1.0\text{-}1.5)n_R^2 a_0 \qquad (2.13)$$

From Eqs. (2.10) and (2.13) one obtains

$$\frac{\Delta r_{n_R}}{\langle r \rangle_{n_R}} \approx \frac{\lambda_{n_R}}{\langle r \rangle_{n_R}}$$
$$\approx \frac{1}{n_R} \qquad (2.14)$$

Equation (2.14) indicates that for a quantum system in high-energy states the relative Uncertainty of position becomes smaller, so it becomes more similar to a classical system.

When the atom is excited exactly to the ionization potential, one has $n_R = \infty$, $\langle V \rangle_\infty = 0 = \langle T \rangle_\infty$ and $\Delta r_\infty = \lambda_\infty = \infty = \langle r \rangle_\infty$. The Uncertainty of location of the electron is infinite. The electron has been ejected from the atom, but its position is not known, so one may visualize $\Delta r_\infty = \infty$. Visualization of λ_∞ and λ_∞ has not been achieved.

If a ground-state hydrogen atom absorbs energy E that is greater than the ionization potential 2.18 aJ, the electron is ejected with excess kinetic energy. One obtains a free electron with $V = 0$, $T = E - 2.18$ aJ and $\lambda < \infty$. The Uncertainty of position Δr becomes indefinite and connected to a measurement of p or r. The relationship (2.10) between λ and Δr only applies to a bound particle.

2.1.2. Molecules

The preceding discussion of a hydrogen atom serves as a general model for Rydberg states of a molecule. The radii of orbitals with $n_R \geq 3$ are larger than the van der Waals radii of small polyatomic molecules made of atoms lighter than neon. This includes water, ammonia, and simple organic compounds. In such a molecule an electron in an orbital with $n_R \geq 3$ is mainly "outside" the ionic core. Molecules containing heavier atoms require larger values of n_R for their Rydberg states.

The molecular cation is polarizable, so the electrostatic binding energy of the excited electron to the positive core might be smaller than that of an electron in a hydrogen orbital of the same n_R.

Orbitals near the ionization limit have very large radii. The binding energy $-E_{n_R}$ of an electron in a hydrogen atom is equal to the thermal energy $1.5k_B T = 6 \times 10^{-21}$ J at 300 K for $n_R = 19$, where k_B is Boltzmann's constant and T is the temperature. The average distance of the electron from the nucleus in the $n_R = 19$ state is ~ 19 nm. In a gas other molecules within this radius can interfere with the electron in the excited state, so effects of gas density on high Rydberg states might be expected to begin at a density of $\sim 3 \times 10^{22}$ molecules/m^3, or a pressure of ~ 100 Pa (10^{-3} atm) at 300 K.

Most ionization occurs by direct excitation to above the ionization potential. Large density effects are not expected until the mean free path of the electrons leaving the molecules is of the order of the radius of a high Rydberg state, < 100 nm. Electron scattering cross sections σ of small molecules are typically a few times 10^{-20} m^2, so the mean free path $(\sigma n)^{-1}$ is < 100 nm when the molecular density n is greater than a few times 10^{26} molecules/m^3. This corresponds to gas pressures at 300 K of $\gtrsim 10^6$ Pa ($\gtrsim 10$ atm).

2.2. LIQUIDS

2.2.1. Weakly Scattering Liquids

The molecular density in a liquid is typically $\sim 10^{28}$ molecules/m^3. The gaps between molecules are $\leq 10^{-10}$ m. A Rydberg state with $n_R \geq 3$ would have an average radius that would encompass one or more layers of surrounding molecules. Most of the states have nodes in the central part of the radial distribution of the excited electron, so the electron usually has one or more molecules between it and the positive core of the excited species. The intervening molecules are electronically polarized by the slowly moving electron in the high Rydberg state, and they partially shield the electron from the positive core. The potential energy of the state is thereby reduced.

This model applies only if the electron moves freely in an orbital and is not appreciably scattered by the liquid molecules (3, 4).

2.2.1.1. Electron Behavior in a Hydrogen Atom.

The expectation value of the potential energy of the electron in a Rydberg state of a hydrogen atom in a weakly scattering liquid would be

$$
\langle V \rangle_{nR} \approx \frac{-\xi^2}{n_R^2 4\pi\varepsilon_0\epsilon\mathbf{a}}
$$

$$
= \frac{-\xi^2}{n_R^2 4\pi\varepsilon_0\epsilon^2 a_0} \tag{2.15}
$$

where ϵ is the average dielectric constant of the medium between the electron and the positive core, $n_R^2\mathbf{a}$ is the average distance of the excited electron from the nucleus, and $\mathbf{a} = \epsilon a_0$ is the equivalent of the Bohr radius in the dielectric medium. Equation (2.8) is recovered by putting $\epsilon = 1$ in Eq. (2.15), corresponding to a vacuum between the electron and the proton. The appropriate value of ϵ in Eq. (2.15) for $n_R \geq 3$ is the high-frequency (optical) dielectric constant, which is equal to the square of the refractive index, $\mathbf{n}^2$.

The expectation value of the kinetic energy of the electron in the Rydberg state is

$$
\langle T \rangle_{nR} \approx \frac{\xi^2}{n_R^2 8\pi\varepsilon_0\epsilon^2 a_0} \tag{2.16}
$$

for $n_R \geq 3$.

The model is reasonably accurate only when the mean free path $\langle fp \rangle$ of the electron is equivalent to many circuits in the orbital: $\langle fp \rangle \gg 2\pi n_R^2\epsilon a_0$, say, > 10 nm. The mean free path of an electron in a high Rydberg orbital can be estimated as described below, using the drift mobility μ_v of extra electrons in the liquid. The electrons must have random speeds v equivalent to the speed of the electron in the orbital.

An extra unattached electron in the liquid will drift in the direction of an applied electric field. The velocity of drift, v_d, is related to the applied field strength E and the drift mobility μ_v:

$$
v_d = \mu_v E \tag{2.17}
$$

The random speed v is due to the total kinetic energy, including thermal energy, and is usually much greater than the drift velocity v_d in the direction of the applied field. Drift occurs because the electron is accelerated by the coulombic force $E\xi$ in the direction of the field. When the electron collides with a molecule, the electron tends to be scattered in a random direction and the acceleration has to start over. Acceleration occurs again until the next collision, and so on. The average drift velocity is given by (5)

$$v_d \approx \frac{E\xi\tau_v}{\mathbf{m}} \tag{2.18}$$

where τ_v is the mean time between collisions that randomize the direction of motion of the electron and is also called the relaxation time; τ_v depends on the total speed v. From Eqs. (2.17) and (2.18) one obtains

$$\mu_v \approx \frac{\xi\tau_v}{\mathbf{m}} \tag{2.19}$$

The mean free path is also dependent on v:

$$\langle fp \rangle_v \approx \langle v \rangle \tau_v \tag{2.20}$$

Combination of Eqs. (2.20) and (2.19) gives

$$\langle fp \rangle_v \approx \frac{\langle v \rangle \mu_v \mathbf{m}}{\xi} \tag{2.21}$$

By putting the values of $\mathbf{m}$ and ξ into eq. (2.21), one obtains

$$\langle fp \rangle_v \approx \frac{\langle v \rangle \mu_v}{1.8 \times 10^{11}} \tag{2.22}$$

where $\langle fp \rangle_v$ is in meters, $\langle v \rangle$ is in m/s, and μ_v is in $m^2/V \cdot s$. For a Maxwellian distribution of speeds at temperature T the average speed is

$$\langle v \rangle = \left(\frac{8k_BT}{\pi\mathbf{m}}\right)^{1/2}$$
$$= 6.2 \times 10^3 T^{1/2} \quad (m/s) \tag{2.23}$$

and the electron mean free path is

$$\langle fp \rangle_v \approx \frac{\mu_v T^{1/2}}{2.8 \times 10^7} \quad (m) \tag{2.24}$$

A more exact derivation gives $\langle fp \rangle_v$ larger by a factor of $3\pi/8$,

$$\langle fp \rangle_v = \frac{\mu_v T^{1/2}}{2.4 \times 10^7}$$

Collision of a Rydberg electron with a molecule of the medium is assumed to disturb the Rydberg state. The detailed behavior of an electron bound in an orbital is not known; the scattering characteristics might be different from those of an unbound electron. The best that can be done at the moment is to assume that the scattering characteristics are similar for bound and unbound electrons of similar kinetic energies. Elastic collisions possibly distort the Rydberg state and alter its characteristic energy. Inelastic collisions can disrupt the state by deexcitation, ionization, or transformation. An example transformation is the trapping of the electron at a site somewhere within the spatial distribution of the Rydberg orbital, thereby forming an electron state that has a wave function not centered on the nucleus. The latter becomes a positive ion. The excited atom has then decomposed to an electron–ion pair that has a potential energy

$$V = \frac{-\xi^2}{4\pi\varepsilon_0\epsilon\, y} \tag{2.25}$$

where y is the distance between the centers of the positive and negative charge distributions.

The mean lifetime τ_{nR} of a Rydberg state in a liquid should be the inverse of the probability per unit time of an inelastic collision of the excited electron if that is smaller than the $\sim 10^{-9}$ s needed to deexcite by emission of a photon. The value of τ_v estimated from the mobility of quasifree electrons in the liquid is governed mainly by elastic collisions, so τ_v is a lower limit for the value of τ_{nR}. For liquids in which $\langle fp \rangle_v \gtrsim 10$ nm, one obtains $\tau_{nR} > 10^{-13}$ s.

Equation (2.24) indicates that in liquids for which $\langle fp \rangle_v \gtrsim 10$ nm the drift mobility μ_v is $\gtrsim 2 \times 10^{-2}$ m^2/V·s at temperatures in the vicinity of 100–300 K. There are relatively few dielectric liquids that permit such high mobilities of electrons (Figure 2.1). Such liquids are composed of molecules that have spherelike external shapes and large polarizabilities. Examples are xenon, krypton, argon, and methane; just below them are tetramethylsilane and tetramethylmethane (also called neopentane or 2,2-dimethylpropane) (6, 7). These are the weakly scattering liquids.

The mobilities μ_e displayed in Figure 2.1 are those of electrons in thermal equilibrium with the liquid. Mobilities $\leq 10^{-3}$ m^2/V·s indicate that the liquids scatter electrons strongly and that the extra electrons spend most of their time in localized states (8, 9). An electron that spends all of its time in a localized state is equivalent to an anion and has a drift mobility of 10^{-7}–10^{-8} m^2/V·s in the types of liquid displayed in Figure 2.1. For $\mu_e \leq 10^{-3}$ m^2/V·s the mean path at 300 K is ≤ 0.6 nm, and one should have $\tau_{nR} < 10^{-13}$ s in such liquids. In the region $10^{-7} < \mu_e$

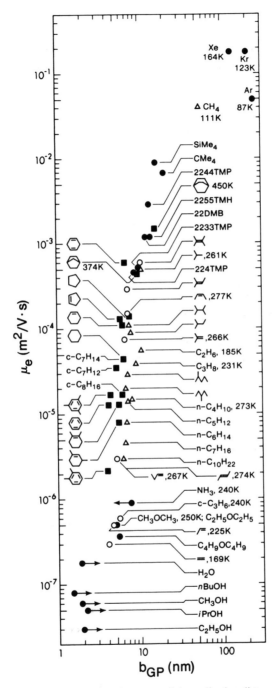

Figure 2.1. Drift mobilities μ_e of thermal electrons and thermalization distances b_{GP} of secondary electrons in dielectric liquids. T = 296 K unless otherwise stated. Temperatures less than 296 K are the normal boiling points of the liquids, except that for Kr, which is 120 K. For additional data see the Appendix. Thermalization distances are discussed in Section 2.4.3.

$< 10^{-2}$ m^2/V·s the electron undergoes repeated transitions between localized and delocalized states, and transport occurs mainly in the delocalized state:

$$e^-_{loc} \rightleftharpoons e^-_{deloc} \tag{2.26}$$

The preceding estimates of Rydberg state lifetimes are crude, but they are adequate to illustrate the points discussed. The main factor that has been neglected is that mean free paths are velocity dependent because the elastic and inelastic scattering cross sections of the medium vary with the electron velocity (5). The ratio of inelastic to elastic cross sections generally increases for a large increase in v. The changes would not alter the order of magnitude of the preceding estimates. The value of $\langle fp \rangle_v$ estimated from the mobility of thermal electrons refers to a mean thermal velocity $\langle v \rangle$ given by Eq. (2.23). The $\langle fp \rangle_v$ relevant to a Rydberg orbital refers to $v = (2 \langle T \rangle_{n_R}/\mathbf{m})^{1/2}$, which is usually larger than the thermal $\langle v \rangle$. The Rydberg orbitals of interest have Bohr circumferences less than $\langle fp \rangle_v$. As an example, in a liquid with $\epsilon = 1.8$ at 300 K, the Rydberg v equals the thermal $\langle v \rangle = 1.1 \times 10^5$ m/s in the $n_R = 11$ orbital, which has a Bohr circumference of ~ 70 nm. Values of $\langle fp \rangle_v$ greater than 70 nm correspond to $\mu_v > 0.11$ m^2/V·s at 300 K. This condition does not exist in any of the liquids displayed in Figure 2.1. Only at lower temperatures in the liquids krypton and xenon are the mean free paths large enough to fulfill the dual conditions Rydberg $v \leq$ thermal $\langle v \rangle$ and $\langle fp \rangle > 2\pi n_R^2 \epsilon a_0$. At densities in the vicinity of 12×10^{27} molecules/m^3, where a maximum occurs in the electron mobility in krypton (10) and xenon (11), the mean free paths are ≥ 180 nm and the conditions are fulfilled. Mobility maxima also occur in argon (12, 13) and methane (14) in this density region, but the mean free paths are too small for orbitals that would fulfill the velocity condition. Thus in the vast majority of systems the Rydberg v is greater than the thermal $\langle v \rangle$, and the estimated values of τ_{n_R} are too large.

2.2.1.2. Energy Level Spacings.

The binding energy of an electron in level n_R of a hydrogen atom in a liquid is

$$E_{n_R} = \langle V \rangle_{n_R} + \langle T \rangle_{n_R}$$

$$= \frac{-\xi^2}{n_R^2 8\pi\varepsilon_0 \epsilon_{n_R}^2 a_0} \tag{2.27}$$

The values of the effective dielectric constants ϵ_{n_R} are not all the same, so the spacings between the levels are not given by a simple Rydberg-like equation. The values increase in the sequence $\epsilon_1 < \epsilon_2 < \epsilon_3 \approx \epsilon_{>3} = \mathbf{n}^2$. Furthermore, the $\epsilon_{n_R}^2$ in the denominator of Eq. (2.27) serves to lower the energy required to reach state n_R, which decreases the state spacings and the ionization potential. A Rydberg-like formula, such as ΔE (eV) $= 13.6 \, [1 - (n_R \epsilon_{n_R})^{-2}]$, would increase the spacings over those in the gas phase and keep the ionization potential the same as in the gas. It is known that the ionization potential is smaller in liquids and solids than in the gas phase (3, 4, 15, 16).

The ϵ_{n_R} in Eq. (2.27) represents electronic polarization of the medium between the electron and nucleus, which occurs *during* the excitation process as the electron moves away from the nucleus toward the outer orbital. As previously discussed, the speed of the electron decreases as it moves to higher excited states. Electronic polarization of the molecules of the medium occurs on the time scale of motion of electrons in the valence shell of the molecules, which is shorter than that of the slower moving excited electron.

Calculation of the energy spacings is not straightforward. Several approximate methods have been devised. We may visualize the effect of solvent polarization on energy level spacings by considering a hydrogen atom in liquid xenon at 200 K. The values of the dielectric constants are $\epsilon_1 = 1.0$, $\epsilon_{\geq 3} = \mathbf{n}^2 = 1.8$, and we guess $\epsilon_2 = 1.1$. The binding energies $E_{n_R}^\epsilon$ are first calculated from Eq. (2.27) for the "naked" levels, with $\epsilon = 1.0$ throughout: $E_1^{1.0} = -13.6$ eV, $E_2^{1.0} = -3.4$ eV, $E_3^{1.0} = -1.5$ eV, $E_4^{1.0} = -0.85$ eV, and so on. The spacings of the naked levels are $\Delta E_{1 \to 2}^{1.0} = 10.2$ eV, $\Delta E_{2 \to 3}^{1.0} = 1.9$ eV, $\Delta E_{3 \to 4}^{1.0} = 0.66$ eV, and the sum of the rest equals $\Delta E_{4 \to \infty}^{1.0} = 0.85$ eV. The sum over all levels equals the ionization potential, 13.6 eV. Next, the states $n_R \geq 2$ are partially "dressed" by increasing $\epsilon_{\geq 2}$ to 1.1: $E_2^{1.1} = -2.8$ eV, so polarization of the xenon decreases the binding energy of the $n_R = 2$ state by $3.4 - 2.8 = 0.6$ eV and $\Delta E_{1 \to 2}^{1.0 \to 1.1} = 10.2 - 0.6 = 9.6$ eV, $E_3^{1.1} = -1.3$ eV and $\Delta E_{2 \to 3}^{1.1} = 2.8 - 1.3 = 1.5$ eV, $E_4^{1.1} = -0.7$ eV and $\Delta E_{3 \to 4}^{1.1} = 0.6$ eV, and $\Delta E_{4 \to \infty}^{1.1} = 0.7$ eV. Then the states $n_R \geq 3$ are dressed completely by increasing $\epsilon_{\geq 3}$ to 1.8: $E_3^{1.8} = -0.5$ eV, so the further polarization of the xenon decreases the binding energy of the $n_R = 3$ level by another $1.3 - 0.5 = 0.8$ eV, and $\Delta E_{2 \to 3}^{1.1 \to 1.8} = 1.5 - 0.8 = 0.7$ eV, $E_4^{1.8} = -0.3$ eV, $\Delta E_{3 \to 4}^{1.8} = 0.2$ eV, and $\Delta E_{4 \to \infty}^{1.8} = 0.3$ eV. The ionization potential in the liquid would therefore be $9.6 + 0.7 + 0.2 + 0.3 = 10.8$ eV, which would be a reduction of 2.8 eV from the gas phase value. The calculation is sensitive to the guessed value of ϵ_2. If $\epsilon_2 = 1.0$ were chosen, one would obtain $\Delta E_{1 \to 2}^{1.0} = 10.2$ eV, $\Delta E_{2 \to 3}^{1.0 \to 1.8} = 0.9$ eV, $\Delta E_{3 \to 4}^{1.8} = 0.2$ eV, and $\Delta E_{4 \to \infty}^{1.8} = 0.3$ eV, with an ionization potential of 11.6 eV, a reduction of 2.0 eV from the gas phase value.

The solvent polarization effect is illustrated in Figure 2.2. The $1 \to 2$ transition is not altered much by the solvent because the average distance of the electron from the nucleus in the second state is 3×10^{-10} m, which does not encompass a shell of xenon atoms. The Wigner–Seitz diameter of the average space occupied by a xenon atom in the normal liquid at 200 K and $n = 1.3 \times 10^{28}$ molecules/m^3 is $2(3/4\pi n)^{1/3} = 5 \times 10^{-10}$ m. In the third state the average distance of the electron from the nucleus is 11×10^{-10} m, which encompasses two layers of xenon atoms. The two layers are assumed to be enough to generate the full dielectric constant, $\epsilon = 1.8$, in the space between the two charges. The spacings between levels above the second are therefore greatly reduced by the solvent.

A more convenient but intuitively less instructive method of estimating the spacings employs the "effective mass" approximation. The levels are counted from the first excited state instead of from the ground state. As a result of the

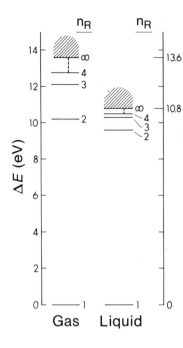

Figure 2.2. Rydberg energy levels of a hydrogen atom in the gas phase and in liquid xenon. The liquid phase values were calculated as described in the text, using $\epsilon_1 = 1.0$, $\epsilon_2 = 1.1$ and $\epsilon_{\geq 3} = 1.8$. Levels between $n_R = 4$ and ∞ are not shown. Above $n_R = \infty$ is a continuum, with no discrete states (hatched). In strongly scattering liquids the excited levels are greatly broadened, which smears them into the continuum.

changed numbering system, the calculated spacings would be about twice too large, so they are reduced by multiplying the spacing parameter by an effective mass ratio, $\mathbf{m}^* = \mathbf{m}_{\text{eff}}/\mathbf{m}$, for the electron translating through the medium. Values of $\mathbf{m}^*$ are usually in the vicinity of 0.5 in weakly scattering media (4, 16), which brings the spacings back near those of Eq. (2.27). This is a grossly oversimplified view of the effective mass model (17), but it makes a useful connection with the Rydberg model. The energy $E_{n_{\text{ex}}}$ required to excite an electron from the ground state to the level n_{ex} is (16)

$$E_{n_{\text{ex}}} = E_G - G/n_{\text{ex}}^2 \qquad n_{\text{ex}} \geq 2 \qquad (2.28)$$

where E_G is the energy gap corresponding to the ionization potential in the liquid and G is the energy level spacing parameter for levels in which the excited electron is sufficiently separated from the positive core. The liquid phase ionization potential is reduced from the gas phase value I_g by the interactions of the ion and electron with the liquid molecules:

$$E_G = I_g + P_+ + V_0 \qquad (2.29)$$

where P_+ is the energy of electronic polarization of the liquid by the positive core and V_0 is the energy of the electron in the bottom of the conduction band in the liquid relative to the potential energy in a vacuum. The spacing parameter G in

Eq. (2.28) is similar to one that could be extracted from Eq. (2.27) except that n_{ex} = $n_R - 1$, and the difference is compensated for by multiplying by $\mathbf{m}^* < 1$:

$$G = \frac{\xi^2 \mathbf{m}^*}{8\pi\varepsilon_0\epsilon^2 a_0}$$

$$= \frac{13.6 \; \mathbf{m}^*}{\epsilon^2} \quad \text{(eV)} \quad\quad\quad (2.30)$$

where $\epsilon = \mathbf{n}^2$ is the optical dielectric constant of the liquid. The electron and the positive core in the ground and $n_{ex} = 1$ states are assumed to be "naked," and those in the $n_{ex} \geq 2$ states are "dressed" in electronically polarized solvent molecules. Equation (2.28) applies only to the dressed states. The $1 \rightarrow 2$ transition energy is assumed to be the same as in the gas phase.

The ionization potential E_G in the liquid phase is smaller than that in the gas phase, I_g; the energy requirement is reduced by the rapid electronic polarization of the medium between the two charges as the electron moves away from the ion. Equation (2.29) divides the polarization energy into two components: P_+ due to polarization of the medium around the localized cation and V_0 due to the more diffuse interaction of the delocalized electron with the medium. The sign of P_+ is always negative (4a, 16–18). The sign of V_0 is negative in weakly scattering liquids such as xenon and krypton and positive in strongly scattering liquids such as helium and hydrogen (16, 19). The latter molecules have such low polarizabilities that the electron–molecule attractive polarization interactions are too weak to overcome the electron–molecule short-range repulsions. In these liquids ionization potentials can be larger than in the gas phase.

Accurate values of P_+ and V_0 cannot yet be determined (16, 20). Crude values of P_+ are usually estimated from

$$P_+ \approx \frac{-\xi^2}{4\pi\varepsilon_0 d}\left(1 - \frac{1}{\epsilon}\right) \quad\quad\quad (2.31)$$

where d is the distance from the center of the ion beyond which the medium is fully polarized. The value of d is usually taken as one or two molecular diameters. The minimum acceptable value of d is probably the ion radius plus the Wigner–Seitz diameter of a solvent molecule. For a hydrogen ion in liquid xenon under its vapor pressure at 200 K, one would have $n = 1.3 \times 10^{28}$ molecules/m^3, $d \geq 5 \times 10^{-10}$ m, and $-P_+ \leq 0.20$ aJ (1.3 eV). The hydrogen ion has a negligible radius. A molecular ion would have a larger d and a smaller P_+.

Equation (2.31) is often written with $8\pi\varepsilon_0$ instead of $4\pi\varepsilon_0$ in the denominator, and d is interpreted as an ion radius (16, 17, 21). The theory is still crude, and only a qualitative understanding of it is required here. Estimated values of P_+ in many of the liquids of interest are in the vicinity of -1 to -2 eV (21).

The energy V_0 of an electron in the bottom of the conduction band is the sum of the potential energy V_p arising from the polarization of the medium by the electron and the zero-point kinetic energy T_0 of the electron caused by scattering (19):

$$V_0 = V_p + T_0 \qquad (2.32)$$

When the molecules of the medium have large polarizabilities and the electron scattering interactions are weak, the negative V_p dominates the positive T_0, so V_0 is negative. In such liquids thermal electrons reside in the conduction band and their mobilities are large ($\gtrsim 10^{-2}$ m^2/V·s, see Figure 2.1); V_0 is the ground-state energy of the excess electron, and the fluid is unperturbed except for electronic polarization. The molecules in all such liquids are spherelike. When the molecules have small polarizabilities or rodlike shapes and the electron scattering interactions are strong, the positive T_0 dominates the negative V_p, and V_0 is positive. In such liquids thermal electrons do not stay in the conduction band but relax into localized states. Their mobilities are low ($\lesssim 10^{-3}$ m^2/V·s, see Figure 2.1). The liquid undergoes orientational and translational adjustments around the localized charge.

Estimated values of V_0 range from -0.6 eV in liquid xenon at 165 K (22) to $+1.1$ eV in liquid helium at 4.2 K (23).

Values of P_+ and V_0 are usually manipulated to fit Eqs. (2.28) and (2.29).

Polarization energies are sometimes referred to as "self-energies" (4a). The latter term is jargon and is less descriptive than the former. When a field of study is mature enough to transfer useful information to another field, jargon terms should be replaced by words that have clearer definitions.

2.2.1.3. Molecules and Atoms Other Than Hydrogen.

Equations of type (2.28) have been applied to exciton and excited impurity states in solid argon, krypton, and xenon (4, 16, 24) and in alkali halide and silver halide crystals (4a). A more rudimentary form of the equation was applied to the optical absorption spectra of alkali metal atoms (25).

Equation (2.27) is preferred to Eqs. (2.28)–(2.30) for the visualization of the effect of a weakly scattering solvent on the process of electronic excitation. Equation (2.29) does not explicitly display the dependence on ϵ^2 needed to match that in Eq. (2.30), and the counting of the levels in Eq. (2.28) seems unnatural.

Ionization potentials of molecules and of atoms other than hydrogen can be roughly accommodated in Eq. (2.27) by adjustment of the effective nuclear charge $z_{eff}\xi$ and of the ground-state radius (25). The ground-state value of n_R is greater than unity for all atoms other than hydrogen and helium. Appropriate values of $z_{eff}\xi$ and ϵ_{n_R} depend on the nature of the orbital, but approximations can be made with the aid of tables of electronic excitation levels (25, 26). For polyatomic molecules ϵ_{n_R} can also be greater than unity for the estimation of the gas phase ionization potential. Details are not needed here because we are ultimately interested in charge separation rather than in excited states.

2.2.1.4. Electron–Ion Separation Distances. An excited species is on the verge of ionization when the binding energy $-E_{n_R}$ of the electron in the excited state is approximately equal to the thermal energy of the system, $1.5k_BT$. For an electron in an excited hydrogen atom in liquid xenon at 200 K, $-E_{n_R}$ nearly equals $1.5k_BT$ $= 4 \times 10^{-21}$ J when $n_R = 13$. The average distance of the electron from the positive center for $n_R = 13$ is ~ 16 nm. Using the Bohr model, the orbital circumference is 100 nm, and the electron velocity is 1×10^5 m/s, so the orbital frequency is 1×10^{-12} s.

The mobility of thermal electrons in liquid xenon at 200 K and $n = 1.3 \times 10^{28}$ molecules/m³ is 0.3 m²/V·s (11). From Eq. (2.24) the mean free path of the electrons is 300 nm, or three orbital circuits at $n_R = 13$. The mean time between elastic collisions is 3 ps. The electron in this state is perturbed relatively little.

Xenon at densities of $1–3 \times 10^{28}$ molecules/m³ is the most weakly scattering liquid known (13). A possible exception is the radioactive liquid radon at a similar number density [near 100 kPa and 211 K (27)]. Perturbation of high Rydberg states is greater in any other liquid. Even in liquid argon at its normal boiling point, which is classed as a weakly scattering liquid (Figure 2.1), the electron in the hydrogen state that has a binding energy of $1.5k_BT$ has a mean free path equal to only one-eighth of the Bohr orbit circumference. (The parameters are T = 87 K, $\epsilon = 1.51$, $n_R = 23$, Bohr radius = 42 nm, $\langle$fp$\rangle$ = 33 nm.) Perturbation of this state is appreciable.

When energy greater than the ionization potential is absorbed by the atom, the electron is ejected and it carries the excess energy into the liquid. The electron undergoes scattering as it moves away from the ion. It transfers energy to the molecules with which it collides and eventually reaches thermal equilibrium with the liquid. The distance that the electron moves away from the ion during thermalization depends on the initial amount of excess energy, the frequency of collision and the amount of energy lost per collision.

The situation is similar for the ionization of a polyatomic molecule, except that some of the excess energy might remain in internal degrees of freedom of the ion.

If there are several ionization events in the same liquid, the electron thermalization distances are probably not the same in each case. The electron scattering process has large random components with respect to where each event occurs, how much energy is lost in a given scattering event, and the direction of deflection. For polyatomic molecules the initial electron energy can be different even for the same energy input because different amounts of energy can be left in the ion. The amounts of energy initially deposited in the molecules are not all the same when high-energy radiation is the energy source (Chapter 3). All these factors contribute to the formation of a relatively broad distribution of thermalization distances.

The form of the distribution function is not precisely known. If it were similar to the radial distribution of a high Rydberg orbital, it would be of the form

$$P_{n_R}(r) \propto r^{2n_R} \exp\left(\frac{-2r}{n_R\epsilon a_0}\right) \tag{2.33}$$

Scattering would shift the distribution toward smaller r, effectively decreasing the value of n_R and distorting the function. Some of the molecules are excited above the ionization limit, and the scattering of the ejected electrons can lead to a Gaussian component of the distribution. The net effect has been approximated in two ways:

1. Reduce the value of the exponent in the preexponential factor and of the dispersion parameter of the exponential,

$$F(y) \propto y^{a_E} \exp\left(\frac{-y}{b_E}\right) \qquad (2.34)$$

where we have switched from the general radial variable r to the thermalization distances y of individual electrons; $F(y)\,dy$ is the fraction of electrons that thermalize at distances between y and $y + dy$ from their parent ions; a_E has been arbitrarily taken as 0, 1, or 2; and b_E is the dispersion parameter of the exponential;

2. Assume that the distribution is dominated by excitations above the ionization limit, that many scattering events are required to remove the excess energy, and that the electron flight direction changes randomly each time it is scattered, thereby generating a three-dimensional Gaussian,

$$F(y) = \frac{4y^2}{\pi^{1/2}b_G^3} \exp\left(\frac{-y^2}{b_G^2}\right) \qquad (2.35)$$

where b_G is the dispersion parameter of the Gaussian and the most probable value of y.

Distributions of these types are discussed later.

2.2.2. Strongly Scattering Liquids

2.2.2.1. Excited States. High Rydberg states cannot be sustained in strongly scattering liquids because the mean free path is too short. Equation (2.24) applies only to quasifree electrons, for which the mean free path is greater than a molecular diameter and greater than the radianlength of the electron. The limiting mobility for quasifree behavior at 300 K would correspond to a minimum mean free path of ~1 nm, which gives $\mu_{min}^{qf} \approx 1 \times 10^{-3}$ m^2/V·s. Mobilities lower than that involve localized states, as depicted by Eq. (2.26).

Strongly scattering liquids are composed of nonspherelike, nonpolar molecules such as ethane or polar molecules such as ammonia (Figure 2.1). The large collision frequency of electrons in highly excited states in these liquids would rapidly reduce them to lower states or cause the excited species to decompose to a localized state of the electron offset from the ion. In either case the time involved would probably be less than 10^{-12} s.

$$M \rightsquigarrow M** \tag{2.36}$$

$$M** + \text{fluid} \longrightarrow M* \tag{2.37}$$

$$M** + \text{fluid} \longrightarrow [M_{ep}^+ + e_{loc,ep}^-] \tag{2.38}$$

where the wiggly arrow signifies absorption of energy from radiation, and the subscript ep designates electronic polarization of the fluid adjacent to the charged species. The square brackets in Eq. (2.38) indicate that the ion and electron are not completely free of each other. They are close enough together that their coulombic attraction is not negligible. They tend to drift back together and recombine. The ion and separated electron are produced as a pair, as twins, so their recombination is called geminate neutralization.

2.2.2.2. Ionization. A molecule excited above its ionization limit behaves similarly to one in a weakly scattering liquid. An electron is ejected with excess energy. However, the electron thermalization distance is shorter in a more strongly scattering liquid. Another difference is that the ground state of an extra electron in a weakly scattering liquid is delocalized or only shallowly localized in a broad potential well, whereas in a strongly scattering liquid it is localized in a deeper, narrower potential well. The ionization process in both kinds of liquid is represented by

$$M \rightsquigarrow [M_{ep}^+ + e_{ep}^-] \tag{2.39}$$

When the electron energy has been sufficiently lowered by collisions, the electron can be captured by a potential fluctuation of appropriate depth and breadth to form a localized state:

$$[e_{ep}^-] \longrightarrow [e_{loc,ep}^-] \tag{2.40}$$

The delocalized electron has to suffer an inelastic collision to drop to a localized state.

The final stages of flight of an energetic electron in a strongly scattering liquid are depicted in Figure 2.3. The interaction energy between an electron and a non-spherical or polar molecule depends on the direction and distance of approach. As the electron moves through the liquid, it encounters fluctuations in molecular orientation and density, so it undergoes fluctuations in potential energy V. The dashed curve in Figure 2.3 qualitatively depicts changes of V relative to the value for an electron in a vacuum. The location of the zero relative to the undulations depends on the properties of the liquid. The magnitudes of the potential fluctuations are also a function of electron energy. The fluctuations are a few tenths of an electronvolt near the end of the electron flight in many liquids. In weakly scattering liquids such as xenon they are a few milli-electronvolts, which is small compared to the

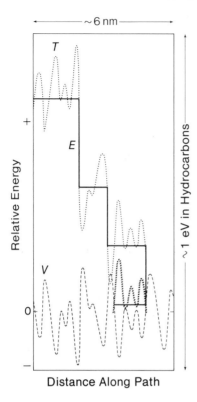

Figure 2.3. Variation of potential energy V, kinetic energy T, and total energy $E = T + V$ of an energetic electron along the last part of its flight in a strongly scattering liquid. The zero is taken as the potential energy of an electron in a vacuum. The distance is depicted along the actual (bent) path, which is much longer than the net penetration distance. The heights and spacings of the potential fluctuations are governed by the shapes, polarities, relative orientations, and spacings of the molecules. The indicated scales are for chainlike hydrocarbons; they would be ~4 eV and ~3 nm in water. The sudden decreases in E are caused by inelastic collisions. The lowest value of E is that of the localized electron; the corresponding T is shown feathered (////) to indicate that a transition has occurred and that the classical significance of T is obscure. For a localized particle T is quantized and similar in magnitude to V. In a weakly scattering liquid the fluctuations in V are smaller than the thermal kinetic energy k_BT of the electron, so a localized state does not form.

thermal agitation energy k_BT, so electron localization does not occur. Xenon liquid must have T $\geq$ 161 K, which gives $k_BT > 14$ meV.

The total energy E of the electron undergoes occasional sudden drops caused by energy transfer to the liquid during inelastic collisions (Figure 2.3). When E falls below the peak values of V, the electron is reflected from the potential barriers and becomes localized in a zone of low potential. Reaction (2.40) represents this step.

The kinetic energy $T = E - V$ of the delocalized electron varies as the mirror image of V when E is constant and undergoes sudden decreases that correspond to the losses of E (Figure 2.3). When the transition to a localized state occurs, the classical visualization of T becomes uncertain. Indeed, there is an ongoing discussion, to which no end is in sight, about the extent to which one may usefully describe electron behavior in classical terms. A related question is the extent to which an electron of energy E in Figure 2.3 could sense the fine structure depicted in V. It is possible that the electron can sense more fine structure than can yet be calculated, but the problem need not be resolved here. For the localized electron $\langle T \rangle$ is quantized and closely related to $\langle V \rangle$, as in the hydrogen atom discussed earlier.

It is sometimes assumed that T of the delocalized electron is reduced to $k_B T$ before the localization event occurs. However, in most strongly scattering liquids this might not be so.

Subsequent to localization the electric field of the electron causes rearrangement of the surrounding molecules. Permanent dipoles and axes of maximum polarizability tend to reorient in the direction of the field. If intermolecular forces such as hydrogen bonds are altered by the charge, the local density tends to change. For example, the local density at an electron site in liquid ammonia is smaller than the average density of the liquid (28). This is attributable to neutralization of the hydrogen bonds and molecular rearrangement. The solvent rearrangements lower the potential energy of the electron (Figure 2.4). The completely relaxed state is called the solvated electron, designated e_s^-. The incompletely relaxed states are collectively designated e_{ir}^-.

Thus, subsequent to step (2.40), if time prior to reaction permits, the solvent rearranges in the field of the electron:

$$[e_{loc, ep}^-] \longrightarrow [e_s^-] \tag{2.41}$$

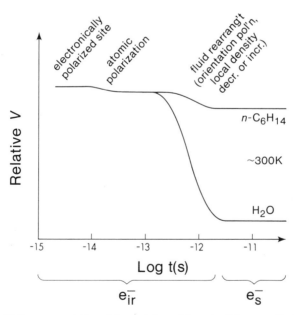

Figure 2.4. Qualitative representation of the variation of the potential energy V of an electron as a function of time after the localization event (2.39). This corresponds to deepening the potential well in which the electron became localized in Figure 2.3. Orientational polarization in a nearly nonpolar liquid such as n-hexane (n-C$_6$H$_{14}$) is only important to the extent that the molecular polarizability is anisotropic. The incompletely relaxed states are e_{ir}^-. The solvated state e_s^- is in thermal equilibrium with the liquid.

Similar rearrangement occurs in the vicinity of the cation. The energy released is dissipated to the fluid. The overall state change after reaction (2.39) is, in this case,

$$[M_{ep}^+ + e_{ep}^-] \longrightarrow [M_s^+ + e_s^-] \tag{2.42}$$

Solvent relaxation analogous to (2.42) can also occur after step (2.38) if time prior to reaction permits.

The dielectric constant of a polar liquid increases as the dipoles orient in the electric field. The time scale is similar to that of Eq. (2.42). While (2.42) is occurring, the dielectric constant of the fluid between the ion and the electron increases from $\epsilon = n^2$ to the static value ϵ_s.

2.2.2.3. Electron–Ion Separation Distances. In strongly scattering liquids the scattered flight of the energetic electron away from its sibling ion is usually terminated by the localization step (2.40). The separation distance between the electron and its sibling ion at this point is usually small enough that the coulombic attraction between them, $-\xi^2/4\pi\varepsilon_0\epsilon y$, is not negligible. The square brackets in Eqs. (2.39) and (2.42) represent the coulombic attraction that correlates the pair. In (2.40) and (2.41) the square brackets indicate that a positive ion is within coulombic interaction distance of the electron.

The scattering mechanism of electron thermalization generates a broad distribution of thermalization distances in a given experiment. The exact form of the distribution function is not known. Forms (2.34) and (2.35) are usually used. However, the scattering mechanism would not likely generate the simple exponential form (2.34). The Gaussian form (2.35), or some modification of it, is favored (29).

2.2.3. The Concept of Ionization in a Dense Phase

Ionization in a single-component low-density gas is well defined. The ionization potential of a gas molecule is the minimum energy required to remove an electron from it and propel the electron beyond the recapturing ability of the coulomb field of the cation that remains behind. The two charges then migrate independently of each other. When an electron has been provided with enough energy to eject it from the molecule, the electron moves away without interference from other particles. The threshold energy to create a spatially uncorrelated electron and ion is therefore quite sharp.

In a liquid the electron suffers collisions while moving away from the ion. Although the minimum energy required to produce an ion–electron pair in a liquid is often lower than in the corresponding low-density gas (Section 2.2.1.2), the electron usually gets stopped by the liquid before escaping the coulombic field of its sibling ion. In practice, it is difficult to distinguish reaction (2.36) followed by (2.38) from (2.39) followed by (2.40').

$$M \rightsquigarrow M^{**} \tag{2.36}$$

$$M^{**} + \text{fluid} \longrightarrow [M_{ep}^+ + e_{loc,ep}^-] \tag{2.38}$$

$$M \rightsquigarrow [M_{ep}^+ + e_{ep}^-] \tag{2.39}$$

$$[M_{ep}^+ + e_{ep}^-] \longrightarrow [M_{ep}^+ + e_{loc,ep}^-] \tag{2.40'}$$

The former set involves a lower excitation energy and probably generates a smaller electron–ion separation distance y than does the latter. However, there is no direct knowledge of y values in either set, so a distinction cannot be made on that basis. In both cases the coulombic attraction tends to pull the charges back together (geminate neutralization). Some of the geminate neutralization can occur on the same time scale as the solvent relaxation step (2.42) (30). Some might even occur before the electron localization step (2.40) takes place. Geminate neutralization can therefore involve the whole spectrum of incompletely relaxed to relaxed (solvated) states. The former are collectively designated e_{ir}^- and M_{ir}^+, while the latter are e_s^- and M_s^+. Both delocalized and localized states are included in e_{ir}^-.

$$M \rightsquigarrow [M_{ir}^+ + e_{ir}^-] \tag{2.39'}$$

$$[M_{ir}^+ + e_{ir}^-] \longrightarrow [M_s^+ + e_s^-] \tag{2.42'}$$

$$[M_{ir}^+ + e_{ir}^-] \longrightarrow M \quad \text{(geminate neutralization)} \tag{2.43a}$$

$$[M_s^+ + e_s^-] \longrightarrow M \quad \text{(geminate neutralization)} \tag{2.43b}$$

The coulombic attraction between the ion and electron tends to draw them back together, which is a nonrandom direction of motion. Thermal agitation tends to drive the ion and electron in random directions, which on average is away from each other. The thermal process is in competition with the coulombic one. If the coulombic energy $-\xi^2/4\pi\varepsilon_0\epsilon y$ is not too great compared to the thermal energy $k_B T$, some of the pairs manage to diffuse apart. Their coulombic attraction becomes negligible. The ion and electron can then diffuse independently of each other, and they are called free ions. The square brackets are removed because the electron and ion are no longer correlated in space:

$$[M_s^+ + e_s^-] \longrightarrow M_s^+ + e_s^- \quad \text{(free ions)} \tag{2.44}$$

As (2.42') occurs, the dielectric constant of the liquid between the charges increases. This decreases the coulombic attraction force and facilitates (2.44). The time scale of (2.44) is sufficiently long that essentially all charges are fully solvated before becoming free ions in strongly scattering liquids. In very weakly scattering liquids the electrons remain in the delocalized state until they react with an ion or molecule. For simplicity we represent free-ion electrons by e_s^- even in weakly scattering liquids, unless otherwise stated.

In the absence of other reactants, the free ions eventually undergo neutralization at random in the bulk fluid:

$$M_s^+ + e_s^- \longrightarrow M \quad \text{(random neutralization)} \quad (2.45)$$

Reaction (2.45) normally occurs on a time scale many orders of magnitude longer than that of (2.43b). The reason is that the steady-state concentration of charged species in the bulk fluid is usually several orders of magnitude lower than the effective concentration in a correlated pair. A given M_s^+ wanders around until it comes within the coulombic correlation range, approximately $\xi^2/4\pi\varepsilon_0\epsilon k_B T$, of an e_s^-. This process is similar to the reverse of (2.44). The newly formed correlated pair can then either redissociate as (2.44) or neutralize as (2.43b).

By applying an electric field E across the system, free ions can be collected at electrodes. Correlated pairs are not collectible, although high electric field strengths increase the fraction of pairs that dissociate. It is therefore possible to measure the extent of (2.44):

$$M_s^+ \xrightarrow{\text{E}} \text{electrode}$$
$$e_s^- \xrightarrow{\text{E}} \text{electrode} \quad (2.46)$$

In low-density gases all ionization events produce free ions. There is no geminate neutralization.

By contrast, in most liquids most electron–ion pairs undergo geminate neutralization. It is therefore not easy to distinguish (2.36) from (2.39). The definition of an ionization event in the liquid phase is not clear-cut. If only spatially uncorrelated charges (free ions) were considered as completed ionization events, the definition would be even more complex.

A practical definition of ionization by electron removal in any phase is: *ionization has occurred when the center of the electron spatial distribution has moved off the molecule or ion to which it was attached.* A minimum practical distance for the center to have moved is one atomic diameter. For example, electron transfer from an atom in one molecule to an atom in an adjacent molecule or group constitutes ionization if charge separation can be designated on the molecular formulas. This definition excludes electron transfers within the metallic state. The metallic state is not discussed in this chapter.

The spatial distribution of the excited electron on M** is highly polarizable. Decomposition of the state to an electron–ion pair can be assisted by application of a strong electric field:

$$M** \xrightarrow{\text{E}} [M_{ir}^+ + e_{ir}^-] \quad (2.38')$$

In the sections that follow some of the geminate pairs designated $[M^+ + e^-]$ might be M^{**}.

Energy levels such as those in Figure 2.2 are broadened by scattering. In strongly scattering liquids the excited levels are greatly broadened. The effect is to smear most of the excited levels and the continuum into each other. One obtains a set of electron–ion pairs with a broad, continuous distribution of intrapair separation distances y (Section 2.4).

2.3. SOLIDS

2.3.1. Amorphous and Crystalline

Solids may be divided into two broad categories: amorphous and crystalline. There are weakly scattering and strongly scattering materials in both categories.

Excitation and ionization processes in amorphous solids behave much like those in liquids. The main difference from liquids is that localized states have much lower mobilities in the rigid phase. Intermolecular rearrangements are also much slower. The time scales of reactions (2.42′)–(2.46) are many orders of magnitude greater in a solid than in a normal liquid. The relaxation process (2.42′) can occur in two ways: (1) molecular reorientation about the sites of the charges and (2) migration of the electron into deeper traps. The migration mechanism occurs to some extent in low-temperature liquid alcohols (30, 31) and possibly in other liquids. It probably dominates in rigid media.

Because solids do not flow in the times under consideration, semiclassical models of electron transport are less useful in solids than in liquids.

In a crystal the long-range periodicity of the structure makes energy band theory convenient to describe the electronic structure (32, 33). Consider a large number of atoms placed singly on sites of an imaginary three-dimensional lattice with spacing s. When s is very large, equal to many atomic mean diameters, the atoms are nearly independent of each other. The electrons are localized in the energy levels of the individual atoms. If the spacing s is gradually decreased, the electrons of one atom come under a progressively stronger influence from neighboring atoms. The electron wave functions overlap more and more. An electron in an outer orbital on a given atom spends more and more time in the vicinity of neighboring nuclei. The electronic levels become progressively modified by the increasing interaction with adjacent nuclei. When s is similar to an atomic mean diameter, the outer electronic levels might be considerably different from those in the isolated atoms. The overlapping orbitals form bands separated by gaps (Figure 2.5). Each band contains a large number of discrete levels that are so closely spaced that they are essentially a continuum within the band; there is one level for each atom in the crystal.

Depending on the electronic structure of the original atoms, the highest band occupied by electrons in the ground state (valence band) might or might not be

Position in Crystal

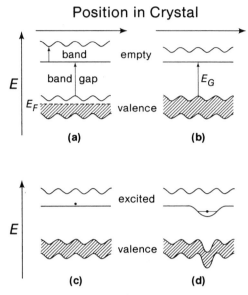

Figure 2.5. Electronic energy levels as a function of position in a simple crystal (not to scale). The wavelength of the undulations in the band surfaces is related to the lattice spacing. Fermi energy E_F is the highest level occupied at 0 K; E_G is the energy gap between bands, similar to that in Eq. (2.28). An unfilled band is a conduction band because electrons in it move easily throughout the crystal. (a) Metal. (b) Insulator, $E_G \sim 1$ aJ (6 eV); semiconductor, $E_G \sim 0.1$ aJ. (c) Delocalized electron formed by excitation from the valence band; the hole is somewhere in the top of the valence band. (d) Electron localized at a crystal imperfection; the hole is somewhere in the top of the valence band.

completely filled. In a completely filled band the electrons are immobile beyond their lattice sites, like a traffic jam at a cross roads. In an unfilled band the electrons are mobile throughout the crystal. When the valence band is completely filled, the material is an insulator. A material with a partly filled valence band is a metal (Figure 2.5).

An unfilled band is a conduction band. Excitation of an electron from a full valence band to a conduction band changes an insulator into a conductor (Figure 2.5c). Removal of an electron from the full band leaves a "positive hole," which might also be mobile. A hole "moves" by an electron falling into it from a neighboring site, leaving a hole on the latter, and so on. A hole moving in one direction is actually a sequence of electron jumps in the opposite direction, analogous to a moving dark spot on a TV screen.

The energy of an electron in a band is quantum mechanically averaged over a distance similar to its de Broglie radianlength λ. The mean radianlength of thermal electrons in a conduction band at T $\leq$ 300 K ($\langle T \rangle \leq 6 \times 10^{-21}$ J $= 40$ meV) is ≥ 1 nm. Lattice spacings in simple crystals are < 1 nm. The energy of an electron in the bottom of a nearly empty band and at the Fermi energy in a partly filled band is essentially independent of position in the crystal. More tightly bound electrons and more energetic delocalized electrons have $\lambda < 1$ nm. Thus the energy of an electron in the bottom of a deeply filled band or at the top of any band is dependent on the position with respect to a lattice site (Figure 2.5). In a perfect crystal a hole in the top of an otherwise full valence band should have a lower mobility than an electron in the bottom of a nearly empty conduction band.

Certain kinds of imperfection in a crystal can capture an electron from the conduction band. The imperfections are dislocations or impurities that provide a potential well of adequate depth and breadth to localize an electron (Figure 2.5d).

If the chemical or physical structure of an insulator is sufficiently irregular, the bottom of the conduction band is everywhere undulated. The potential energy of the bottom of the band is similar to the V curve in Figure 2.3.

2.3.2. Ionization in a Perfect Crystal

In a perfect crystal of a material that does not chemically attach electrons, the regularity of the lattice does not provide an adequate site (potential well) to localize an extra electron. The ionization reaction (2.39) occurs but not the electron localization step (2.40).

An extra electron in a crystal could induce a polarization distortion of the lattice. It was proposed that the lattice distortion could localize the electron (34). Such a localized state is called a Landau polaron. Although the polaron is still invoked occasionally, there seems to be no unambiguous evidence that it exists in any material. Electron localization in a crystal always seems to occur at a preexisting defect.

The positive hole in some crystals is highly mobile, migrating from site to site by electrons hopping sequentially in the opposite direction (32). In other crystals the hole becomes localized, for example, by the ion reacting with a neighbor:

$$[M^+ + M] \longrightarrow [M_2^+] \qquad (2.47)$$

Orientational polarization in a rigid material is very slow, so the positive ion can be represented by the incompletely relaxed species M_{ir}^+ whether it is localized or not.

Irradiation of a perfect crystal generates positive ions, each of which is a defect. The electrons migrate until they encounter a defect and undergo either geminate or random neutralization. The ionization reactions in a perfect crystal are

$$M \rightsquigarrow [M_{ir}^+ + e_{deloc}^-] \qquad (2.48)$$

$$[M_{ir}^+ + e_{deloc}^-] \longrightarrow M \qquad \text{(geminate neutralization)} \qquad (2.49)$$

$$[M_{ir}^+ + e_{deloc}^-] \longrightarrow M_{ir}^+ + e_{deloc}^- \qquad \text{(free ions)} \qquad (2.50)$$

$$M_{ir}^+ + e_{deloc}^- \longrightarrow M \qquad \text{(random neutralization)} \qquad (2.51)$$

The time scales of reactions (2.49)–(2.51) should be similar to those in the normal liquid because the mobility of the delocalized electrons remains high.

2.3.3. Measurement of Free Ion Yields

When the electrons and ions form localized states in solids, the mobilities are too low to allow measurement of free ion yields.

In weakly scattering solids where the electrons remain in the delocalized state, the electrons can be collected at electrodes by applying an electric field across the sample. However, if the positive ions form localized states, they remain in the sample and produce a space charge. The electron free ion yield is measured by generating only a small number of ionizations such that the space charge is small (35). The sample is melted and refrozen between radiation pulses to clear the positive ions.

2.4. HIGH-ENERGY RADIATION

Photons with energies just above the ionization potential of a liquid or solid have high absorption coefficients and are absorbed near the surface. Photoionization of a single-component liquid or solid therefore occurs in a thin surface layer. This process is used in photocopying devices and is discussed in Chapter 9.

Photoionization can be produced throughout a dense phase by having a small concentration of a solute that has an ionization potential below the strong absorption bands of the solvent and using photons with energies below those of the bands.

Ionization can be produced throughout a single-component system by using high-energy radiation, such as γ rays. The disadvantage of high-energy radiation is that it does not deposit energy homogeneously in the system. Energy is deposited along particle tracks (Figures 1.4 and 1.5). An α particle interacts so strongly with materials (36) that even in a normal gas it produces a dense column of ionization, giving a highly nonhomogeneous distribution of ions and electrons in the system.

Reactive species that are distributed nonhomogeneously require appropriate descriptions of their reaction kinetics. Nonhomogeneously distributed species occur in dense phases ionized by photons and particles of all energies above the threshold. They also occur in normal gases ionized by heavy charged particles that produce dense tracks.

2.4.1. Columnar Ionization

A 1-pJ (6-MeV) α particle produces an ionization track similar to that in Figure 1.5. The initial diameter of the positive core in a typical liquid hydrocarbon would be ~1 nm, and the average distance of the electrons from the core would be ~5 nm at the moment of thermalization. Coulombic repulsion between the cations pushes them outward to expand the core. Coulombic attraction between the electrons and ions tends to pull the electrons back to the core. Random diffusion tends to cause the electrons to wander away. Application of an electric field perpendicular to the track assists electrons to escape from the core. The dynamics of the overall process are complex, but an approximate treatment is possible.

The first step toward solving any complex problem is to make simplifying assumptions. The greater the complexity, the greater the simplifications required.

Jaffé (37) estimated the fraction of ions that escape from an α particle track to become free ions. He considered a column of ionization parallel to the z axis with an electric field E applied parallel to the x axis. A number of very crude approximations were necessary: (i) the spatial distributions of the positive (ion) and negative (electron) charges were the same; (ii) positive and negative charges have equal diffusion coefficients D and mobilities μ. The number densities of ions in the track are $n_+ = n_- = n_\pm$. The time dependence of the ion concentration is

$$\frac{\partial n_\pm}{\partial t} = D\left(\frac{\partial^2 n_\pm}{\partial x^2} + \frac{\partial^2 n_\pm}{\partial y^2}\right) \mp \mu E \frac{\partial n_\pm}{\partial x} - \alpha n_\pm^2 \tag{2.52}$$

where α is the neutralization reaction rate constant similar to that used in a homogeneous system. The first term on the right side of the equation corresponds to $D\nabla^2 n_\pm$, which represents Fick's second law of diffusion. Random diffusion of the ions tends to separate them from the column, which decreases their local concentration. The second term corresponds to ion drift in the field and the third term to ion loss by neutralization.

Equation (2.52) is still too complex to solve, so Jaffé further assumed: (iii) the initial distribution of n_+ from the center of the track was Gaussian with a dispersion parameter b_G; (iv) the effect of the neutralization term $\alpha n_\pm^2$ on the shape of the $n_\pm$ distribution was negligible. Approximations (iii) and (iv) comprise what is sometimes called prescribed diffusion (38). The distribution is assumed to be Gaussian at all times. In reality, the neutralization reaction decreases n_+ most where it is largest, so the distribution actually becomes flattened in the center, which in turn reduces the effect of the diffusion term in Eq. (2.52). Jaffé obtained

$$n_\pm = \frac{N}{\pi(4Dt + b_G^2)} \exp\left(-\frac{(x \mp \mu Et)^2 + y^2}{4Dt + b_G^2}\right) \tag{2.53}$$

where N is the number of ion pairs per unit length of column at time t. Equation (2.53) represents a two-dimensional Gaussian skewed by ion drift in the direction of the applied field; the cation distribution is skewed one way and the electron distribution is skewed the opposite way.

The variation of N with t was calculated by putting Eq. (2.53) into (2.52). The value N_∞ at $t = \infty$ is the number of ion pairs that becomes free ions. The fraction $\phi_{\text{fi}}^{\text{E}}$ of ions that becomes free ions under many practical circumstances can be expressed as

$$\phi_{\text{fi}}^{\text{E}} = \frac{N_{\infty}}{N_0} = \phi_{\text{fi}}^0 + A\text{E} \qquad (2.54)$$

where N_0 is the initial number of ion pairs per unit length of column, ϕ_{fi}^0 is the fraction that escapes the track in the absence of an external field, and A is the field coefficient. The values of ϕ_{fi}^0 and A are expected to increase with increasing b_G, T, and dielectric constant ϵ and to decrease with increasing N_0 (37). However, they are complex functions and cannot be calculated accurately. Under most conditions of measurement in liquids, ϕ_{fi}^0 is negligible compared to AE (37).

Kramers (39) modified the treatment of Eq. (2.52) for very low temperatures and intermediate field strengths in liquids. At low temperatures the diffusion term in Eq. (2.52), and hence ϕ_{fi}^0 in (2.54), are negligible. Kramers therefore dropped the diffusion term in (2.52) and kept the neutralization term. Measured values of $\phi_{\text{fi}}^{\text{E}}$ in low-temperature liquids become sublinear in E (second derivative becomes negative) at lower fields than expected from the Jaffé equation (40a, b; Figure 2.6). This behavior is qualitatively accommodated by the Kramers treatment (Figure 2.6 and Table 2.1).

In general, for α particle irradiation of liquids $\phi_{\text{fi}}^0 \leq 1 \times 10^{-3}$.

The main shortcomings of the treatments of columnar ionization are approximations (i) and (ii). In most systems, whether solid, liquid, or gas, the thermalized electrons have much larger values of b_G, D, and μ than do the positive ions. A more exact treatment will have to begin with an equation quite different from (2.52).

2.4.2. Isolated Ionizations

A high-energy electron (say, 100 fJ, 0.6 MeV) passing through a liquid loses energy in small bits (each a few attojoules, tens of electronvolts). The distances between bits are a few hundred nanometers. The few attojoules deposited at a spot cause an electron to be ejected from a molecule. The ejected electron might or might not have enough energy to ionize one or more other molecules in the vicinity. These low-energy electrons, collectively named secondary electrons, lose their energy by collisions with molecules and become thermalized within a few nanometers of their sibling ions. The passage of a high-energy electron through a liquid therefore creates a sequence of activated microzones with diameters of a few nanometers separated by a few hundred nanometers.

In about 75% of the microzones only one ion–electron pair is formed (41–43). In about 15% two pairs are formed, and three or more pairs are formed in about 10%. Hence about 50% of the ionizations are initially in single-pair microzones.

In a microzone that contains more than one pair, the electrons that become thermalized closest to the ion core are quickly drawn back to it by the coulombic attraction (Figure 2.7a). If the number of ions in the core is z_+ and an electron is a distance y from it, the force F_c drawing the electron toward the core is

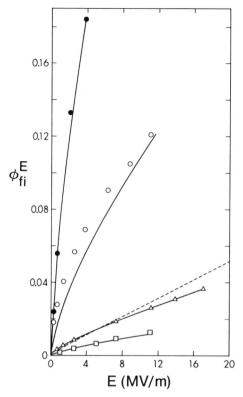

Figure 2.6. Field dependence of the fraction of ions that escapes from a column produced by an α particle in a liquid. Experimental points are ratios of ionization currents measured in the liquid to the saturation current measured in the gas at a density where the α particle range was less than the electrode separation (40a,b): ●, He, 4.0 K; ○, Ar, 88 K; △, H$_2$, 20.4 K; □, N$_2$, 77 K. The full curves are calculated from Kramers's (39) modification of Jaffé's equation using the parameters in Table 2.1. ---, n-hexane, ~295 K, calculated by Jaffé (37) using the parameters in Table 2.1.

TABLE 2.1. Columnar Ionization Parameters for the Curves in Figure 2.6

Liquid	T (K)	N_0 (10^8 pr/m)	b_G (nm)	Reference
He	4.0	6.6	75[a]	40b
Ar	88	40	81[a]	40b
H$_2$	20.4	7.7	1.8[a]	40a
N$_2$	77	26	2.5[a]	40a
n-hexane	~295	43	24[b]	37

[a] Obtained by adjusting Kramer's modification of Jaffé's Eq. (39) to optimally fit the measured data.
[b] Estimated (37). A more realistic value of b_G in n-hexane would be ≤6 nm.

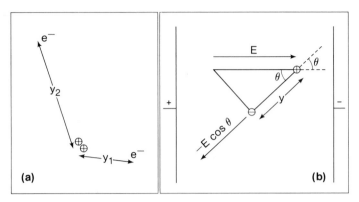

Figure 2.7. (*a*) A liquid phase microzone containing two ion–electron pairs. The electron thermalization distances are y_1 and y_2. (b) A field E applied at an angle θ to the axis of the ion pair exerts a force proportional to $-(+)$ Ecosθ that assists (hinders) separation of the charges. The field vector E indicates magnitude and direction.

$$F_c = \frac{z_+ \xi^2}{4\pi\varepsilon_0\varepsilon\, y^2}$$

(2.55)

Neglecting random diffusion, the time t_{gn} for an electron to undergo geminate neutralization varies as y^3:

$$t_{\text{gn}} \approx \int_y^0 \frac{dr}{\mathrm{E}(\mu_+ + \mu_-)} = \frac{\displaystyle\int_y^0 4\pi\varepsilon_0\varepsilon r^2\, dr}{z_+ \xi(\mu_+ + \mu_-)}$$

$$= \frac{\varepsilon y^3}{4.32 \times 10^{-9} z_+(\mu_+ + \mu_-)} \quad \text{(s)}$$

(2.56)

Thus the electron with the smallest value of y is rapidly captured by an ion. The value of z_+ then decreases by 1. The next closest electron has a larger value of y and interacts with a smaller value of z_+ for most of its lifetime, so the second neutralization takes much longer than the first, and so on. In *n*-hexane, at 300 K, typical times for a two-pair microzone might be 1 ps for the first electron and 10 ps for the second.

In a multipair microzone the number density of charged species rapidly collapses from the center outward. Soon only one cation and the outermost electron remain.

The treatment of electron escape from multipair zones is complex beyond a crude beginning (43). However, the majority of microzones initially contain only one pair, and the rest rapidly evolve into single-pair zones in most liquids. Equations that were developed for isolated ionizations in dense gases (44) are therefore approximately applicable to the ionization of liquids by high-energy electrons. They also apply to ionization in systems irradiated with high-energy photons. A high-energy γ ray, say 100 fJ (0.6 MeV), interacts with matter by transferring about half of its energy to an electron. The electron then transfers the energy in small amounts to molecules here and there along its path, as described above.

To simplify the discussion of geminate neutralization and free ion formation, the electron and solvent relaxation processes represented by (2.42') are neglected in what follows. Thus, the time dependence of the dielectric constant is neglected, although it has been treated (45). Reactions (2.39)–(2.44) reduce to

$$M \rightsquigarrow [M^+ + e^-] \tag{2.39''}$$

$$[M^+ + e^-] \longrightarrow M \quad \text{(geminate neutralization)} \tag{2.43''}$$

$$[M^+ + e^-] \longrightarrow M^+ + e^- \quad \text{(free ions)} \tag{2.44''}$$

The reaction numbers are all double primed for uniformity.

Onsager (44) calculated the relative probabilities of reactions (2.43'') and (2.44'') by treating the Brownian motion of one particle under the combined influence of the collecting field E and the coulombic attraction of the other particle. The combined potential w is initially

$$w = -y\xi E \cos \theta - \frac{\xi^2}{4\pi\varepsilon_0 \epsilon y} \tag{2.57}$$

where θ is the angle between the ion pair axis and the direction of the applied field (Figure 2.7b). The field direction is by convention from the positive to the negative electrode. For the orientation displayed in Figure 2.7b the applied field would assist separation of the ions. If the positions of the positive and negative ions were reversed, the field would help drive them closer together. For a random distribution of orientations the net effect would be zero. However, the most probable of the random orientations is perpendicular to the field. The drift of these ions in the field soon orients the pairs in a direction where E assists separation. The time-averaged effect is therefore field-assisted separation.

The equation for the Brownian motion of the charges is

$$\frac{\partial p}{\partial t} = (D_+ + D_-) \nabla \cdot \left[\nabla p + \frac{p}{k_B T} \nabla w \right]$$

$$= D \nabla^2 p + D \nabla \cdot \left(\frac{p}{k_B T} \nabla w \right) \tag{2.58}$$

where $p(r, t)$ is the probability of finding the electron at position r relative to the ion at time t; D_+ and D_- are the diffusion coefficients of the ion and electron, respectively; and $D = D_+ + D_-$. The first term on the right side of Eq. (2.58) represents Fick's second law of diffusion, as in Eq. (2.52). The second term in (2.58) corresponds to the second and third terms in (2.52), through the two terms in (2.57). However, (2.58) is a two-body problem and is much simpler than (2.52), which is a multibody problem. Equation (2.58) is soluble with difficulty (44, 45). The probability $\phi_{fi}^E(y, \theta)$ that an ion pair of initial separation y and initial angle θ with the field will become free ions is

$$\phi_{fi}^E(y, \theta) = \exp\left[-\frac{r_c}{y} - \beta y(1 + \cos\theta)\right] \sum_{j,m=0}^{\infty} \frac{\beta^{j+m}(1 + \cos\theta)^{j+m} r_c^j y^m}{j!(j + m)!}$$

(2.59)

where $\beta(m^{-1}) = \xi E/2k_B T = 5800E/T$ with E in volts per meter, and $r_c(m) = \xi^2/4\pi\varepsilon_0\epsilon k_B T = 1.67 \times 10^{-5}/\epsilon T$ is the distance at which the coulombic attraction between the ion and electron equals the thermal agitation energy $k_B T$. The charge pairs are initially isotropically oriented, so Eq. (2.59) is averaged over $\cos\theta$ (46–48):

$$\phi_{fi}^E(y) = \int_0^\pi \phi_{fi}^E(y, \theta) \, d\cos\theta \Bigg/ \int_0^\pi d\cos\theta$$

$$= e^{-r_c/y}\left[1 + e^{-2\beta y} \sum_{m=1}^{\infty} \frac{(2\beta y)^m}{(m + 1)!} \sum_{j=0}^{m-1} (m - j) \frac{(r_c/y)^{j+1}}{(j + 1)!}\right] \quad (2.60)$$

Equation (2.60) may be written

$$\phi_{fi}^E(y) = \phi_{fi}^0(y) [1 + f(E, \epsilon, T, y)] \quad (2.60')$$

where

$$\phi_{fi}^0(y) = e^{-r_c/y}$$

$$= \exp\left(\frac{-\xi^2}{4\pi\varepsilon_0 k_B \epsilon T y}\right) \quad (2.61)$$

At relatively low values of ϵ, T, y, and β, $\phi_{fi}^E \ll 1$, and Eq. (2.60') reduces to a form similar to that of the columnar ionization Eq. (2.54). The double summation in (2.60) may then be truncated at $m = 1, j = 0$:

$$\phi_{fi}^{E}(y) \approx \phi_{fi}^{0}(y)[1 + e^{-2\beta y}\beta r_c]$$

$$\approx \phi_{fi}^{0}(y)[1 + (1 - 2\beta y)\beta r_c]$$

$$= \phi_{fi}^{0}(y)\left(1 + \frac{\xi^3 E}{8\pi\varepsilon_0\epsilon(k_B T)^2} - \frac{\xi^4 E^2 y}{8\pi\varepsilon_0(k_B T)^3}\right)$$

$$= \phi_{fi}^{0}(y)\left(1 + 0.097\frac{E}{\epsilon T^2} - 1120\frac{E^2 y}{\epsilon T^3}\right) \tag{2.62}$$

$$\approx \phi_{fi}^{0}(y)\left(1 + \frac{0.097}{\epsilon T^2}E\right) \tag{2.62'}$$

The form of Eq. (2.62') is similar to that of (2.54), but the coefficient A in the latter cannot be accurately derived. The value of the coefficient in Eq. (2.62') has been verified for the ionization of many liquid hydrocarbons by X rays (49, 50).

In liquids where the electron thermalization distance y is relatively large, ϕ_{fi}^{E} becomes sublinear in E at relatively low field strengths (50). The third term in Eq. (2.62) indicates this trend, although the full Eq. (2.60) is needed to adequately describe the behavior.

In any given system there is a distribution of thermalization distances. The overall fraction ϕ_{fi}^{E} of ion pairs that becomes free ions is obtained by averaging (2.60) over the distribution $F(y)\,dy$:

$$\phi_{fi}^{E} = \int_0^\infty \phi_{fi}^{E}(y)F(y)\,dy \tag{2.63}$$

Equation (2.63) is integrated numerically, using distribution function (2.35) or another form.

Free ion yields in radiation chemistry are recorded as G_{fi}^{E}, the number of pairs of free ions formed per 100 eV of energy absorbed by the system:

$$G_{fi}^{E} = \phi_{fi}^{E}G_{tot} \tag{2.64}$$

where G_{tot} is the yield of reaction (2.39''). In the liquid phase G_{tot} is 20–50% greater than in the gas phase (51–54). However, the strengths of the applied fields are usually < 10 MV/m, so they cannot compete effectively with the ≥ 100 MV/m internal fields that cause geminate neutralization in multipair microzones. The central portions of multipair microzones in most liquids undergo geminate neutralization with negligible probability of contributing to G_{fi}^{E}. The appropriate value of G_{tot} to use in Eq. (2.64) is therefore about 30% smaller than the true value. To limit the arbitrariness of the choice, the gas phase value is usually used. In a few cases where $\phi_{fi}^{E} > 0.8$, fitting experimental values of G_{fi}^{E} to Eq. (2.64) provides "experimental" values of both G_{tot} and $F(y)$ (54).

The field dependence of G_{fi}^E in X-irradiated liquid propane at 183 K (50) is shown in Figure 2.8. Taking the gas phase $G_{tot} = 4.1$ as the appropriate value for the single-pair model, ϕ_{fi}^E increases from 0.018 to 0.085 as E increases from zero to 2.6 MV/m.

The calculated curve of G_{fi}^E as a function of E and its T dependence are sensitive to the assumed shape of the $F(y)$ distribution. A two-parameter power function YP is used to illustrate the dependence of the G_{fi}^E curve on the form of $F(y)$:

$$F(y) = 0 \qquad\qquad y \leq y_{min}$$
$$F(y) = xy_{min}^x y^{-(x+1)} \qquad y \geq y_{min} \tag{2.65}$$

If all electrons generated by reaction (2.39″) had the same thermalization distance, $F(y)$ would be a delta function [$x = \infty$ in Eq. (2.65)]. The value $x = \infty$ gives too large a field dependence of G_{fi}^E (Figure 2.8). By contrast, $x = 1.5$ gives too small a dependence. The best fit of YP to the experimental results for propane at 183 K was with $x = 2.5$ and $y_{min} = 6.0$ nm.

The function YP is inadequate for a number of liquids such as neopentane in which $\phi_{fi}^0 \geq 0.2$ (50). Furthermore, a thermalization mechanism that would produce a YP distribution is difficult to imagine. Random scattering would tend to generate the three-dimensional Gaussian (2.35), which we designate YG:

$$F(y) = \frac{4y^2}{\pi^{1/2} b_G^3} \exp\left(-\frac{y^2}{b_G^2}\right) \tag{2.35}$$

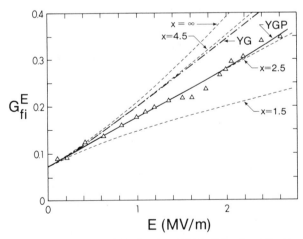

Figure 2.8. Electric field dependence of the free ion yield in X-irradiated liquid propane at 183 K. △, experimental (50). The curves were calculated from eqs. (2.60), (2.63), and (2.64) using initial spatial distributionns (2.35), (2.65), and (2.66), with $G_{tot} = 4.1$. Values of x refer to Eq. (2.65). Best fit: YGP, $b_{GP} = 7.6$ nm; YP, $x = 2.5$, $y_{min} = 6.0$ nm.

This simple Gaussian gives an adequate field dependence of G_{fi}^E in neopentane (50, 55; Figure 2.9) but too large a dependence for propane (Figure 2.8).

A one-parameter hybrid function was found to fit the results from all types of dielectric liquids over wide ranges of T (87–700 K) and E (0–16 MV/m) (54, 56, 57). The distribution, designated YGP, has a Gaussian core that represents most of the pairs and a power tail:

$$F(y) = 0.96\text{YG} \qquad\qquad\qquad y < 2.4b_{GP}$$
$$F(y) = 0.96\text{YG} + 0.48(b_{GP}^2/y^3) \qquad y > 2.4b_{GP}$$

$$(2.66)$$

where 0.96 is a normalization factor. The dispersion parameter in YG is now b_{GP} and is also the most probable value of y. Deviation from YG occurs only for the outermost 7% of the electrons. The power tail refers to the ion–electron separation distance in the last surviving pair in a multipair microzone. The tail is therefore an artifact of the single-pair microzone model.

The difference between estimated thermalization distances obtained with distributions YG and YGP can be illustrated by comparing values for a given ϕ_{fi}^0 [Eq. (2.63) at E = 0]. As $\phi_{fi}^0(y)$ is a simple function of ϵTy [Eq. (2.61)], ϕ_{fi}^0 is a function only of the product ϵTb_G or ϵTb_{GP}. Plots of ϕ_{fi}^0 against these products are given in Figure 2.10. For a given liquid at a given temperature, ϵT is constant. Thus, at $\phi_{fi}^0 < 0.1$, the most probable thermalization distance estimated from YGP

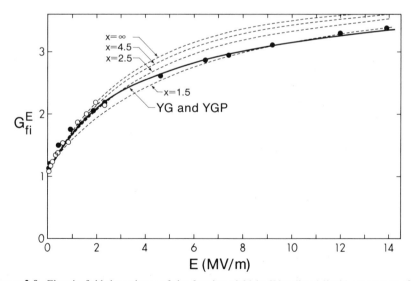

Figure 2.9. Electric field dependence of the free ion yield in X-irradiated liquid neopentane (2,2-dimethylpropane). Experimental ●, 296 K, Ref. 55; ○, 294 K, Ref. 50. The curves were calculated from Eqs. (2.60), (2.63), and (2.64) using initial spatial distributions (2.35), (2.65), and (2.66), with $G_{tot} = 4.3$. Best fit: YGP, $b_{GP} = 21.2$ nm.

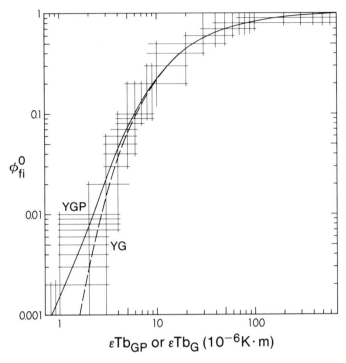

Figure 2.10. Electron escape probability from single-pair microzones at zero applied field strength, averaged over the spatial distribution YG [Eq. (2.35)] or YGP [Eq. (2.66)]. ϕ_{fi}^0 is a function only of the product $\epsilon T b_G$ or $\epsilon T b_{GP}$, respectively.

is somewhat smaller than that estimated from YG. Figure 2.10 can be used to estimate b_G or b_{GP} in any system for which ϕ_{fi}^0, ϵ, and T are known.

The full lines in Figures 2.8 and 2.9 were calculated with YGP.

The exponential distributions represented by Eq. (2.34) do not provide general agreement with measured yields, even when the function contains two adjustable parameters. For example, the distribution (2.67), designated YE, does not provide agreement with the temperature dependence of G_{fi}^E (58, 59).

$$F(y) = 0 \qquad\qquad y \leq y_{min}$$

$$F(y) = \frac{1}{b_E} \exp\left(\frac{y_{min} - y}{b_E}\right) \qquad y \geq y_{min} \qquad (2.67)$$

Nor is it suitable for the photoionization of a solute in a hydrocarbon solvent (60).

Equation (2.67) is of form (2.34) with $a_E = 0$. Forms with $a_E = 1$ (61) and $a_E = 2$ (62) have been used to describe the reduction of photoionization yield caused by the presence of a second solute that is an electron scavenger, but the fits are poor. Solute molecules can affect the electron–ion separation distances in two ways:

(a) they alter the thermalization rate because their scattering properties are different from those of the solvent and (b) they might capture electrons before thermalization is complete, thereby shortening the flight away from the ion. A simple exponential distribution would not be produced by process (a) and would be produced by (b) only if it were energy independent. Both (a) and (b) are energy dependent, so the form (2.34) can be neglected. The form YGP (2.66) is so far preferred.

The three noble liquids xenon, krypton, and argon scatter electrons very weakly and $\phi_{fi}^E > 0.8$ is easily obtained (54; Figure 2.11). Values of both b_{GP} and G_{tot} were estimated by fitting Eqs. (2.60), (2.63), (2.64), and (2.66) to the experimental data. The thermalization distances are much larger than those in liquid hydrocarbons (Table 2.2). The values of G_{tot} in liquid argon, krypton, and xenon are, respectively, 20, 40, and 50% larger than those in the low-density gases (54).

The average amount of energy deposited in argon, krypton, or xenon per ionization event (2.39″) is 1.7 ± 0.1 times the ionization potential in both the liquid and gas phases (54).

For very weakly scattering liquids such as xenon there is some question about the validity of the Brownian motion approximation in the Onsager model (63). If the mean free path of the electron is similar to b_{GP} in magnitude and if its mean kinetic energy is larger than the thermal energy of the liquid molecules, the value of T to use in β and r_c [Eq. (2.59)] is uncertain. A thermal electron in liquid xenon

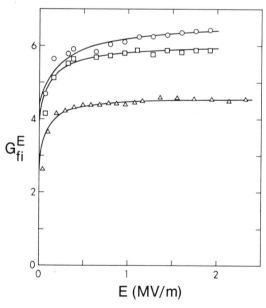

Figure 2.11. Electric field dependence of the free ion yields in X-irradiated liquid xenon at 164 K ($\bigcirc$), krypton at 129 K ($\square$), and argon at 87 K ($\triangle$). Negative applied voltage (54). The curves were calculated from Eqs. (2.60), (2.63), (2.64), and (2.66), using the parameter values in Table 2.2.

at 164 K has a mobility of 0.2 $m^2/V \cdot s$ (11). This indicates a mean free path of 90 nm [Eq. (2.24)], which is a large fraction of the estimated thermalization distance, ~ 134 nm (Table 2.2). However, at the high field strengths used for free ion yield measurements the electrons gain energy from the field. The mobility and mean free path decrease with increasing field and the drift velocity becomes nearly constant. The electron drift velocity in the direction of the field is approximately 3 km/s in xenon at 164 K over the field range displayed in Figure 2.11 (11). This is only a small fraction of the 86 km/s random speed of thermal electrons at 164 K or of the still higher random speed of the epithermal electrons in the field. The mean free path is ~ 1 nm, which is small enough to validate the Brownian motion approximation. Furthermore, the applied field increases G_{fi} by overcoming an internal ion–electron field of similar magnitude. The external field therefore has a relatively small effect on the kinetic energy of the electrons that it attracts away from ions. Electron heating by the field is probably greater in the direction of the field than transverse to it. Counterbalancing the ion–electron field by the external field might actually reduce the electron heating. The Onsager model has not been successfully modified to account for these effects, but it is the best model available at the moment. The error is worst in xenon, much less in argon, and negligible in most liquid hydrocarbons (64).

2.4.3. Secondary Electron Thermalization Distances

The thermalization distances plotted in Figure 2.1 were obtained from measured G_{fi}^E values and Eq. (2.64), except those in the alcohols, water, and ammonia. The latter are limits estimated from electron scavenging studies (Chapter 6).

The correlation between thermalization distance and thermal electron mobility indicates that the major part of the thermalization distance in hydrocarbons and the noble liquids is attained while the electron has a few hundredths of an attojoule of energy (a few tenths of an electronvolt) and is skittering along in the conduction band (6, 9, 65–67a). The electron collisions are gentler and less frequent when inside the band than when above it.

Thermalization distances are a unique source of information about interactions of low-energy (<0.1 aJ, <1 eV) electrons in liquids (9, 56, 58, 68). The distances

TABLE 2.2. Single-Pair Microzone Parameters for the Calculated Curves in Figures 2.8–2.10 Using the Distance Distribution YGP

Liquid	T(K)	ϵ	G_{tot}	G_{fi}^0	b_{GP} (nm)	Reference
Propane, $CH_3CH_2CH_3$	183	1.90	4.1	0.073	7.6	50[a]
Neopentane, $C(CH_3)_4$	294	1.80	4.3	1.1	21.2	50[a]
Xenon, Xe	164	1.93	6.6	4.4	134	54
Krypton, Kr	129	1.63	6.0	4.0	203	54
Argon, Ar	87	1.50	4.5	2.7	247	54

[a]Recalculated using YGP.

are inversely proportional to the liquid density d, so the density-normalized distance $b_{GP}d$ is constant in a series of liquids if the scattering properties are the same. For example, $b_{GP}d$ is essentially constant in the C_5–C_{14} n-alkanes at 296 K (calculated from data in Ref. 69).

2.4.3.1. Molecular Structure Effects in Liquids.

Scattering of low-energy electrons in liquid hydrocarbons is strongly dependent on the molecular structure. The behavior is dominated by two empirical characteristics: (i) scattering is weaker when the molecules are more spherelike (65–67); (ii) alkyl C–H groups are nearly transparent to the electrons, while alkenyl and alkynyl C–H groups scatter progressively more strongly (56). The sphericity effect (i) applies only to dense phases, where the electron interacts with several molecules at a time. In low-density gases, where the electron interacts with only one molecule at a time, scattering at low energies is stronger when the molecule is more spherelike (67b). Thus, the molecular sphericity effect reverses on going from single-body to multibody scattering of the low-energy electron.

The interaction of a low-energy electron (a few times 10^{-3} attojoules or a few times 10^{-2} electronvolts) can penetrate a hydrocarbon molecule to a depth of two C–C bonds in series (65). Groups up to neopentyl (2,2-dimethylpropyl) in size are sensed as a unit. This includes the entire neopentane molecule because the alkyl C–H groups are nearly transparent to the electron.

The interaction of higher energy electrons has a shorter effective range.

The C_1–C_3 hydrocarbon molecules are relatively rigid and are small enough that each molecule can be sensed as a unit by a low-energy electron. The density-normalized thermalization distance nb_{GP} in these liquids (56, 70) decreases with increasing anisotropy of polarizability α^a of the molecules (71–73). The molecular number density n is used in this correlation because α^a is a molecular property. The electron energy transfer efficiency at <0.1 aJ increases with increasing anisotropy of polarizability. On going from methane to ethane, nb_{GP} drops sevenfold, while $\alpha^a/4\pi\varepsilon_0$ increases from zero to 0.7×10^{-30} m^3 (Figure 2.12). A further increase of $\alpha^a/4\pi\varepsilon_0$ to 2.1×10^{-30} m^3 drops nb_{GP} another threefold, but then the thermalization distance changes relatively little on going to the C_3 with the highest anisotropy.

The anisotropy of polarizability of the C–C bonds dominates the electron energy transfer interaction in alkanes. The highly polar $\overset{\delta^-}{\equiv}C\overset{\delta^+}{-}H$ groups in acetylene and methylacetylene (propyne) dominate the electron interaction in those liquids; dimethylacetylene (butyne-2) does not contain a $\equiv C-H$ group and displays a larger value of nb_{GP} in spite of its larger α^a (Figure 2.12). The anisotropy effect saturates at $\alpha^a/4\pi\varepsilon_0 \approx 2 \times 10^{-30}$ m^3 in liquid hydrocarbons.

Molecules that contain more than two C–C bonds in series are not sensed as a unit by the low-energy electrons, so the effective value of n for dimethylacetylene would be larger than the molecular density. The value of $n_{eff}b_{GP}$ for dimethylacetylene would be even higher than the point displayed in Figure 2.12. A preferred method of normalizing for density when the molecules are larger than C_3 is

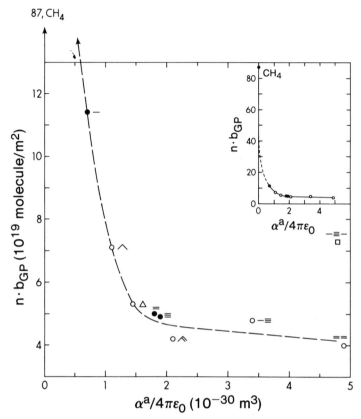

Figure 2.12. Number-density-normalized thermalization distances nb_{GP} of secondary electrons in X-irradiated liquid C_1–C_3 hydrocarbons as a function of the anisotropy of polarizability α^a of the molecules. The anisotropy is plotted as $\alpha^a/4\pi\epsilon_o$ to convert the SI units of α^a ($C^2 \cdot m^2/J$) to the more familiar m^3. The temperature is near the normal boiling point for each liquid. ●: CH_4, H_3C-CH_3, $H_2C=CH_2$, $HC\equiv CH$. ○: $H_3C-CH_2-CH_3$, cyclo$(CH_2)_3$, $H_3C-CH=CH_2$, $H_3C-C\equiv CH$, $H_2C=C=CH_2$. □: $H_3C-C\equiv C-CH_3$ for comparison. The carbon–carbon bonds are designated in the graph. Distances are from Refs. 56 and 70. Anisotropies: △, Ref. 71; CH_3CHCH_2, Ref. 72; all others, Ref. 73.

to multiply b_{GP} by the mass density d. This accounts for all the scattering units in a volume, whether they be separate small molecules or segments of large molecules.

Molecular structure effects are demonstrated by a plot of $b_{GP}d$ against the number of carbon atoms in the molecule (Figure 2.13). The value of $b_{GP}d$ in methane is fourfold larger than that in ethane. This factor is still large, but it is smaller than that for nb_{GP} because methane molecules have less mass than do ethane molecules.

The normalized thermalization distance is constant, 4.5×10^{-6} kg/m², for C_5 and larger n-alkanes (Figure 2.13). This is used as a reference point for comparison of $b_{GP}d$ values. For C_4 and smaller n-alkanes $b_{GP}d$ increases somewhat, reaching 5.7×10^{-6} kg/m² for ethane.

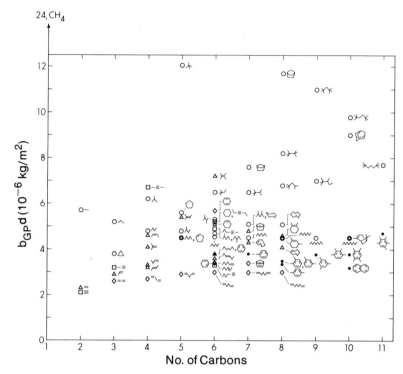

Figure 2.13. Mass-density-normalized thermalization distances $b_{GP}d$ of secondary electrons in liquid hydrocarbons with different molecular structures. The temperatures were mainly near the normal boiling point or near 296 K, whichever is lower. Some of the heavier aromatics and polycyclic compounds were measured at higher temperatures because their melting points are over 296 K (68, 75). O, alkane, cyclo- or polycyclic-alkane. △, alkene, cyclo- or polycyclic-alkene. ◇, diene. □, alkyne. ●, aromatic. More data are given in the Appendix. The bonding in the carbon skeleton identifies each point.

The lowest value of $b_{GP}d$ is that for acetylene, $H-C\equiv C-H$ (Figure 2.13). The hydrogens on acetylene are so electron deficient that they are acidic. The efficiency of electron energy transfer to liquid acetylene is attributed to interaction with the hydrogen (positive) ends of the bond dipoles. Replacement of one H by a CH_3 group increases $b_{GP}d$ from 2.1 to 3.2 (10^{-6} kg/m^2), and replacement of both increases it greatly to 6.7 (10^{-6} kg/m^2). The compounds in this sequence are acetylene, methylacetylene, and dimethylacetylene. The last is the simplest dialkyl-acetylene.

The value of $b_{GP}d$ in a dialkylacetylene $R-C\equiv C-R'$ is similar to that in the analogous alkane $R-R'$; R and R' are methyl or larger alkyl radicals. The value in dimethylacetylene is similar to that in ethane (Figure 2.13). Those in diethyl-acetylene and methyl-n-propylacetylene are similar to that in n-butane. The $-C\equiv C-$ group in $R-C\equiv C-R'$ makes a negligible contribution to energy transfer from the low-energy electrons. The $C\equiv C$ triple bond has a nearly cylin-drical outer shape, due to the four lobes of the two π bonds. An electron approaching the triple bond from the side apparently senses it as a soft, spherelike group similar to an argon atom.

After acetylene, the next shortest thermalization distances are in ethene and the terminal dienes (Figure 2.13). The strength of the electron interaction is attributed to the polarity of the $=C\begin{smallmatrix}H\\\\H\end{smallmatrix}$ group. The latter is less polar than the $\equiv C-H$ group, as demonstrated by the dipole moments of propene (1.2×10^{-24} C·m) and propyne (2.6×10^{-24} C·m) (74).

Replacing all the hydrogens on $H_2C=CH_2$ by CH_3 groups greatly increases $b_{GP}d$, as it did in $HC\equiv CH$ (Figure 2.13). The normalized thermalization distances in tetramethylethene and dimethylacetylene are much greater than those in the n-alkanes with the corresponding numbers of carbon atoms.

Increasing the degree of molecular sphericity by creating one or more cycles in an alkane molecule increases $b_{GP}d$, provided that bond distortion due to strain remains small. The distorted bonds in cyclopropane are strong scatterers (70), so $b_{GP}d$ is relatively small in that liquid. However, the thermalization distance in cyclopentane is 1.2 times that in n-pentane, and that in bicyclooctane is 2.6 times that in n-octane (Figure 2.13).

By contrast, benzene rings and polycyclic aromatics are strong scatterers and give low thermalization distances (75). The normalized distance in naphthalene ($C_{10}H_8$) is only 7% greater than that in a terminal diene (Figure 2.13).

A lot of information about interactions of low-energy electrons with liquids can be derived from Figure 2.13. The reason for mentioning it in this chapter is that the data were obtained from the measurement and kinetics analysis of a non-homogeneous process, radiation-induced ionization in the liquid phase.

Data are compiled in the Appendix.

2.4.3.2. Effect of Liquid Density and the Critical Fluid.

If a liquid is heated under its vapor pressure, the density of the liquid decreases and that of the vapor increases, until at the critical temperature T_c they become the same, d_c. The liquid phase then ceases to exist and the meniscus that characterizes the liquid surface disappears. The fluid at T_c and d_c is called the critical fluid.

It is fascinating to watch the meniscus disappear in the middle of a transparent cell that contains the critical amount of material as T increases past T_c. It is dramatic to watch a gas at d_c be cooled from a temperature above T_c to about half a degree below T_c. If the cell is cooled from the top and the gas supercools a bit and liquid droplets suddenly nucleate, the clear fluid abruptly becomes a thick, opaque cloud with violent-looking turbulences. The exaggerated turbulences are due to the high density and large compressibility of the vapor combined with the complex currents created by the uneven temperature distribution and the revaporization of droplets as they fall to the warmer bottom of the cell. The system is like a micro storm cloud and the processes are nonhomogeneous. The kinetics would be fascinating to study in themselves. The phenomena are easy and safe to observe with xenon ($T_c = 290$ K), carbon dioxide ($T_c = 304$ K), or ethane ($T_c = 306$ K) in a thick-walled glass tube that will withstand an internal pressure of about 10 MPa (100 atm). The critical pressures of these materials are in the range 4.9–7.4 MPa [48–73 atm (76)].

If the electron scattering properties of a liquid did not change as the density was reduced from that at the normal freezing point to that of the critical fluid, the density normalized thermalization distance $b_{GP}d$ would be approximately constant. The actual behavior depends on the degree of sphericity of the molecules. The density effects are illustrated with two differently shaped isomers of pentane: n-pentane (nP), which has a chainlike structure, $CH_3CH_2CH_2CH_2CH_3$, and neo-pentane (NP), a globular tetrahedron, $C(CH_3)_4$, which is spherelike. Isomers were chosen for the illustration so that properties other than molecular shape, such as polarity, mean polarizability, and molecular volume, would be similar.

In the chainlike nP, $b_{GP}d$ is nearly constant in the liquid from the freezing point, 143 K, to about 400 K (59). When discussing the entire liquid range of different substances, it is sometimes convenient to express T as a fraction of T_c [law of corresponding states (77)]. For nP the range 143–400 K corresponds to T/T_c = 0.30–0.85 (78). At T > 400 K, $b_{GP}d$ increases somewhat (Figure 2.14a). At T_c = 470 K, $b_{GP}d$ is 50% above the low-temperature value. The measurements were made in the liquid under its vapor pressure, so an increasing T corresponds to a decreasing density (Figure 2.14b). The energy transfer interaction of low-energy electrons is a bit weaker in the low density liquid where $d/d_c \lesssim 2.1$.

The spherelike NP molecules form a liquid that has a more orderly structure than does liquid nP. The spherelike molecules easily arrange themselves in a crystal lattice, and the liquid freezes at T/T_c = 0.59, compared to 0.30 for n-pentane. The value of $b_{GP}d$ in NP is threefold that in nP (Figure 2.14), and the detailed behavior is more complex. The fractional change of $b_{GP}d$ is relatively small in liquid NP between the freezing point and T/T_c = 0.85 (370 K), then it increases appreciably at higher T, as in nP. However, in NP it passes through a maximum at T/T_c = 0.96, and then decreases. The decrease of $b_{GP}d$ is caused by the breakup of the smooth conduction band in the low-density liquid and the consequent increased scattering of the low-energy electrons. The mobility of thermal electrons goes through a similar maximum in liquid NP (66, 79). Density-normalized mobilities $n\mu$ are plotted in Figure 2.14 for comparison with $b_{GP}d$. The number density n is used by convention to normalize μ, but n is directly proportional to d for the isomers.

In the gas at T > T_c and $d = d_c$ both $b_{GP}d$ and $n\mu$ increase with temperature in both isomers (Figure 2.14a). As T increases above T_c, the compressibility of the gas decreases and the microscopic density fluctuations within the fluid decrease. Scattering is somewhat less intense in the more uniform medium.

The T and d dependences of $n\mu$ are similar to but more marked than those of $b_{GP}d$ in the two isomers. The mobilities involve lower energy electrons than do the thermalization ranges and are more sensitive to the liquid properties.

2.4.3.3. Effect of the Liquid–Solid Phase Transition.

Upon freezing NP at 257 K, the density increases by 10% (35). If $b_{GP}d$ and $n\mu$ were to remain constant,

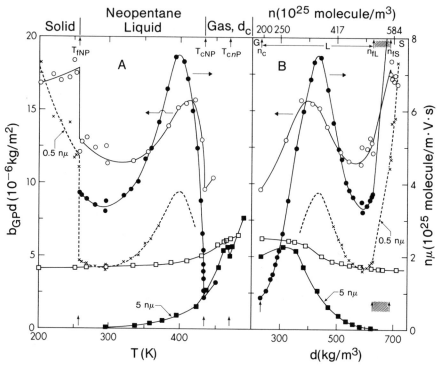

Figure 2.14. Density-normalized thermalization distances $b_{GP}d$ of secondary electrons in isomeric pentanes as functions of temperature, density, and phase: □, chainlike $CH_3CH_2CH_2CH_2CH_3$, nP; ○, spherelike $(CH_3)_4C$, NP. The density-normalized mobilities $n\mu$ of thermal electrons are plotted for comparison: nP, ■ = $5n\mu$; NP, ● = $n\mu$, – – – and × = $0.5n\mu$. Data references: b_{GP} (35, 59, 66); μ (35, 66, 79); liquid and critical d (78); d of solid NP (35). The liquid and solid phases were under their vapor pressure. A: vertical arrows mark the freezing point T_f of NP and the critical temperatures T_c of NP and nP. B: vertical arrows mark densities of NP only; n_c = critical, n_{fL} = liquid at T_f, n_{fS} = solid at T_f; //// indicates the abrupt change of density during liquid–solid phase change. T_c (K), n_c (10^{25} molecules/m^3): nP, 469.6, 198; NP, 433.8, 194.

b_{GP} and μ would have to decrease by 10%. Instead b_{GP} *increases* by about 25% and μ *doubles* (35). Increasing the regularity of the structure of the medium by crystallization decreases the strength of scattering of low-energy electrons. The effect is greater the lower is the energy of the electrons, so μ is more affected than b_{GP}.

2.5. SUMMARY

The energetics and kinetics of ionization and charge separation in dense media are different from those in the low-density gas.

$$M + energy \longrightarrow [M^+ + e^-] \qquad (2.39'')$$

Electronic polarization of the dense medium by the separating charges usually lowers the minimum energy required to ionize M, which is the ionization potential.

The departing electron in reaction (2.39'') is scattered by molecules of the medium and loses energy to them. Some electrons lose energy quickly enough that they become thermalized before escaping the coulombic fields of their sibling ions M^+. They are then usually drawn back to the ions.

$$[M^+ + e^-] \longrightarrow M \quad (geminate\ neutralization) \quad (2.43'')$$

Other electrons are deflected weakly enough or lose energy slowly enough that they travel beyond the effective coulombic fields of the ions. The ions and electrons then diffuse independently of each other.

$$[M^+ + e^-] \longrightarrow M^+ + e^- \quad (free\ ions) \qquad (2.44'')$$

The competition between processes (2.43'') and (2.44'') depends on the separation distance attained by the electron from its sibling ion during thermalization and on the dielectric constant and temperature of the medium. In low-density gases the amount of electron scattering is sufficiently small that all ionizations lead directly to free ions. In dense media the thermalization distance depends on the scattering properties of the medium. Determination of thermalization distances therefore gives information about the interactions of low-energy electrons with dense media. Most of the thermalization distance is attained while the electron is falling through the energy range $\sim 200 \rightarrow 5 \times 10^{-21}$ J, $\sim 1 \rightarrow 0.03$ eV (80–82). Thermalization distances are the main source of information about electron interactions in this important energy region that overlaps the chemically reactive range.

The empirical description of effects of molecular structure on electron thermalization distances is much more advanced than the theoretical interpretation. Data are compiled in the Appendix. A major new theoretical thrust must take into account the molecular shape and the manner of packing of the molecules in the dense phases.

Electron thermalization *times* are related to the thermalization distances. They have been measured in a variety of gases (83) and in the noble liquids (84). In the gases at 1 atm and in the noble liquids, the distances are large and the times are 10^{-10}–10^{-8} s. Theoretical interpretation of the times in terms of energy distributions and scattering cross sections has begun for electrons in mono- and diatomic molecular gases (85–87) and monatomic molecular liquids (88). Thermalization times in most polyatomic liquids are expected to be 10^{-13}–10^{-11} s and are not yet measurable. The times can be estimated by taking them to be roughly proportional to $b_{GP}^2/\langle fp \rangle$. A challenge to theory is to estimate the mean free path as a function of electron energy and liquid properties.

APPENDIX. ELECTRON SCATTERING PROPERTIES OF LIQUIDS: THERMALIZATION DISTANCES OF SECONDARY ELECTRONS AND MOBILITIES OF THERMAL ELECTRONS

The thermalization distance b_{GP} is the dispersion parameter of the distribution function YGP [Eq. (2.66)]. It is the most probable distance that an electron travels away from its sibling ion while transferring its excess energy of ~2 aJ (~10 eV) to the liquid. In most liquids the excess energy is reduced to ~0.2 aJ within ~2 nm of the ion, so b_{GP} mainly reflects the inverse scattering strength of the liquid for electrons that have energies ~0.2–0.02 aJ (80–82). The "scattering strength" involves inelastic and elastic cross sections and the density. An accurate theory of electron scattering at these energies in liquids remains to be developed.

The mobility μ (drift velocity/electric field strength) reflects the interactions of ~0.002–0.02 aJ (~0.01–0.1 eV) electrons with the liquid. The electrons are in thermal equilibrium with the liquid. Mobilities of less than 10^{-2} m^2/V·s involve localized states of the electrons [Eq. (2.26)]. Theories are under development.

Mobilities involve lower energy electrons than do thermalization ranges and are much more sensitive to differences in liquid properties. The combination of b_{GP} and μ provides a probe of the liquid state.

Mobilities of thermal electrons in dielectric liquids lead one to inquire about the nature of the electron itself.

Selection of Data

To display the effects of molecular structure on b_{GP} and μ, values have been selected at the normal boiling point T_b, or near 296 K, or near the freezing point if that is >296 K. Comparisons are made at T_b where possible, to partially normalize for differences in macroscopic properties of the liquids. For example, the viscosity $\eta_b \approx 0.2$ mPa·s (2 mP) and $T_c/T_b = 1.6 \pm 0.1$ for most liquids, where T_c is the critical temperature above which the liquid phase of that material cannot exist.

Electron mobilities in liquid hydrocarbons at T_b can differ by five orders of magnitude, whereas in the critical fluids the mobilities are within one order of magnitude of each other.

Some of the references in the table that follows contain many more data than are displayed here. For example, μ or b_{GP} is sometimes reported for the entire liquid range, from near the freezing point to T_c, and in the gas at densities from d_c downward. References that contain many more numerical data are marked with an asterisk, for example 12*.

When a quantity has been reported in several references, usually only one or two are listed. The intension was to include all pertinent laboratories somewhere in the list. The experimental work required overlap of systems between laboratories. However, many articles such as N. Gee and G. R. Freeman, *Phys. Rev. A*, vol. **23**, p. 1390 (1981), which reports μ in liquid, critical, and gaseous ethene (C_2H_4), have been omitted due to lack of space.

Mono-, Di-, and Triatomic Molecules

Liquid[a]	$\mathfrak{D}^{b}$ (10^{-30} C·m)	T (K)	ϵ^{c}	d (kg/m³)	b_{GP}^{d} (nm)	Reference	μ (10^{-4} m²/V·s)	Reference
He	0	0.274	1.06	145[e]			1350[f]	89
		0.498		145			225[f]	
		0.996		145			3.8[f]	
		2.016	1.06	146			0.65[f]	
		3.02	1.06	141			0.023	
		4.2[g]	1.05	125			0.020	
Ne	0	25	1.23	1444[e,h]			600[h]	90
		25	1.19	1240			0.0016	90
		27[g]	1.19	1208			0.0016	94
Ar	0	82	1.61	1630[e,h]			1000[h]	91
		80					1600[h]	92*
		85	1.51	1400			475	91
		85					625	92*
Kr	0	87[g]	1.51	1390	>247[i]	54	490	12*, 13*[j]
		113	1.79	2830[e,h]			3700[h]	91
		117	1.66	2430			1800	91
		120[g]	1.65	2410			1900	93*[j]
		129	1.63	2340	>203[i]	54	2100	93*
Xe	0	157	2.13	3550[e,h]			~4500[h]	91
		155					2900[h]	92*
		163	1.93	3070			2200	91
		165[g]	1.92	3057			2950	92*
		164	1.93	3060	>134[i]	54	1800	11*[j]
		223	1.76	2620			≥6000[k]	11*
		290[l]	1.28	1100[l]			~4[k]	11*

Liquid[a]	$\mathcal{D}^{b}$ $(10^{-30}\ \mathrm{C \cdot m})$	T (K)	ϵ^{c}	d $(\mathrm{kg/m^3})$	$b_{GP}{}^{d}$ (nm)	Reference	μ $(10^{-4}\ \mathrm{m^2/V \cdot s})$	Reference
H_2	0	20[g]	1.24	71			0.017	94
N_2	0	77[g]	1.44	809			0.0026	95a,b*
		126.2	1.16	313[l]			1.3	95a,b*
O_2	0	87	1.49	1140	11	95c		
CO_2	0	217[m]	1.81	1200	11.6	95d		
OCS	2.4	223[g]	3.7	1200	3.7	95d		
CS_2	0	319[g]	2.57	1230	≥8	95d		
Alkanes								
CH_4	0	77					1500[h]	95e
		87					1100[h]	95e
		98					530	95e
C_2H_6	0	111[g]	1.65	427	>55[i]	67a*[j], 70	400	14*[j]
C_3H_8	0.28	185[g]	1.80	549	10.4	67a*	0.56	96*
$n\text{-}C_4H_{10}$	≤0.17	231[g]	1.78	580	9.0	70	0.39	67b*, 70
		273[g]	1.81	602	8.0	70	0.15	67b*, 70
$i\text{-}C_4H_{10}$	0.44	233	1.87	625	9.9	50	3.2	67b*
		261[g]	1.81	595	10.6	9*	4.9	67b*[k][j]
$n\text{-}C_5H_{12}$	0.24	296	1.84	622	7.1	56, 69[r]	0.12, 0.14	66*, 97*
		309[g]	1.82	609	7.0	66*	0.19	66*
$i\text{-}C_5H_{12}$	0.43	296	1.84	615	7.8	69[r]	0.9	98*
		301[g]	1.83	610			0.9	98*

Compound								
neo-C_5H_{12}	0	202	2.01	715	>24^{h,i}	35*	244^h	35*
		253	1.97	690	>26^{h,i}	35*	165^h	35*
		258	1.85	627	>20^i	35*	71	35*
		283^g	1.80	603	>20^i	35*	68	35*
		296	1.78	588	>20^i	35*, 66*^{j}	68	35*, 66*^{j}
n-C_6H_{14}	0.24	296	1.89	656	6.5	56, 69*^{r}	0.07, 0.08	8, 97*, 99*
		342^g	1.82	614	7.3	56, 100	0.20	8, 99*
i-C_6H_{14}		296	1.88	650	7.1	9*, 69*^{r}	0.29	9*
C_6H_{14}		296	1.90	662	6.7, 7.0	9*, 69*^{r}	0.20	57*
		336^g	1.83	622	6.5	57*	0.6	57*
neo-C_6H_{14}		296	1.87	646	12.0, 10.1	9*, 69*^{r}	12	9*
		323^g	1.83	620	12.6	66*^{j}	13	66*^{j}
C_6H_{14}		296	1.89	659	7.4, 8.0	9*, 69*^{r}	1.1	9*
		331^g	1.83	624	7.0	57*	2.2	57*
n-C_7H_{16}	0.24	296	1.92	681	6.6	69*^{r}	0.046	101
C_7H_{16}		296	1.93	696	6.8	9*	0.10	9*
		367^g	1.84	644	7.5	9*	0.4	9*
C_7H_{16}		296	1.91	668	7.6	69*^{r}		
C_7H_{16}		296	1.93	692	9.3	57*	2.0	57*
		359^g	1.83	634	10.6	57*^{j}	4.4	57*^{j}

Liquid[a]	$\mathfrak{D}^b$ (10^{-30} C·m)	T (K)	ϵ^c	d (kg/m³)	$b_{GP}{}^d$ (nm)	Reference	μ (10^{-4} m²/V·s)	Reference
C_7H_{16}		296	1.92	688	9.4	69[*r]	0.041, 0.040	102, 101
$n\text{-}C_8H_{18}$	0.25	296	1.94	700	6.4	69[*r]		
C_8H_{18}		296	1.96	716	7.3	69[*r]		
"$i\text{-}C_8H_{18}$"		125		881	12[h]	103a	0.11[h]	103a
		155		876	10[h]	103a	0.6[h]	103a
		160		798	10	103a	1.0	103a
		296	1.93	689	8.5, 10.0	57*, 69[r]	4.6, 7	57*, 103
		372[g]	1.81	620	10.6	57*[j]	6.8	57*[j]
C_8H_{18}		379[g]	1.84	654	>12.5[i]	65		
$n\text{-}C_9H_{20}$	0.25	296	1.96	715	6.2	69[*r]	0.047	101
C_9H_{20}		296	2.01	751	7.3	9*	0.80	9*
		419[g]	1.85	656	8.2	9*		
C_9H_{20}		296	2.05	755	9.7	65	5.2	65
		419[g]	1.86	651	10.6	65	11.	65

Compound									
C_9H_{20}	(structure)		296	1.97	717	$>13.1^i$	57*	29	57*
			396^g	1.81	620	$>15.9^i$	57*^{j}	39	57*^{j}
n-$C_{10}H_{22}$	(structure)	0.26	296	1.98	728	6.2	69*^{r}	0.027, 0.038	102, 101
$C_{10}H_{22}$	(structure)		296	1.97	715	$>13.4^i$	65	12	65
			411^g	1.81	632	$>15.8^i$	65	21	65
$C_{11}H_{24}$	(structure)		296	1.97	715	10.8	65		
			428^g	1.78	611	12.7	65		
n-$C_{12}H_{26}$	(structure)	0.27	296					0.024, 0.040	104, 101
$C_{12}H_{26}$	(structure)		316	2.13	795	8.	65		
			423	1.84	640	10.7	65		
n-$C_{14}H_{30}$	(structure)	0.28	296	2.03	761	6.1	69*^{r}	0.038	104
n-$C_{20}H_{42}$			328	2.08	807	5.9	69*^{r}		
$C_{30}H_{62}$ squalane	(structure)		296						
Alkenes									
C_2H_4	(structure)	0	169^g	1.81	570	4.1	70	0.003	70
1-C_3H_6	(structure)	1.22	225^g	2.01	613	4.7	70	0.005	67b*
1-C_4H_8	(structure)	1.13	267^g	1.97	633	5.2, 5.5	56, 105	0.030	106*
tr-2-C_4H_8	(structure)	0	274^g	1.88	630	5.1	56	0.030	106*
cis-2-C_4H_8	(structure)	1.0	277^g	2.04	639	7.2	56	1.5	106*

Liquid[a]	$\mathfrak{D}^b$ (10⁻³⁰ C·m)	T (K)	ϵ^c	d (kg/m³)	$b_{GP}{}^d$ (nm)	Reference	μ (10⁻⁴ m²/V·s)	Reference
i-C₄H₈	1.7	266ᵍ	2.22	626	6.6	56	0.75	106*
(CH₃)₂CCHCH₃		296	1.98	659	7.2	9*	2.9	9*
		312ᵍ	1.94	642	7.4	9*	3.6	9*
1-C₆H₁₂		296	2.07	670	5.2	56		
		337ᵍ	1.97	625	5.5	56		
tr-2-C₆H₁₂		296	1.98	675	5.1	56		
		341ᵍ	1.91	628	5.5	56		
tr-3-C₆H₁₂		296	1.97	674	5.3	56		
		340ᵍ	1.89	629	5.8	56		
cis-3-C₆H₁₂		296	2.07	676	5.6	56		
		340ᵍ	1.97	630	6.2	56		
(CH₃)₃CCHCH₂		296	1.98	650	7.3	9*	1.9	9*
		314ᵍ	1.93	631	7.7	9*	2.2	9*
(CH₃)₂CC(CH₃)₂	0	296	1.98	705	10.2	56	6.0	9*
		346ᵍ	1.91	659	11.1	56	8.0	9*
tr-(CH₃)₃CCHCHCH₃		296	1.97	687	7.2	9*	1.8	9*
		339	1.91	648	7.9	9*	3.6	9*

Compound								
cis-(CH$_3$)$_3$CCHCHCH$_3$		296	2.03	696	7.9	9*	3.5	9*
		342	1.94	657	8.7	9*	5.7	9*
tr-(CH$_3$)$_3$CCHCHC(CH$_3$)$_3$		296	1.96	712	10.4	9*		9*
		359	1.87	658	10.9	9*		9*
(structure)								

Alkynes

Compound								
C$_2$H$_2$	0	198[n]	2.48	608	3.5	56	o	56
CH$_3$CCH	2.6	250[g]	3.20	671	4.8	56	o	
CH$_3$CCCH$_3$	0	300[g]	1.92	680	9.9	56	o	
1-C$_6$H$_{10}$	2.8	253	2.88	755	4.4	56	o	
2-C$_6$H$_{10}$		293	2.00	731	6.8	56	o	
3-C$_6$H$_{10}$		296	1.98	720	7.1	56	o	
		355[g]	1.88	662	8.0	56		
(CH$_3$)$_3$CCCC(CH$_3$)$_3$	0	296	1.97	709	10.9	9*	o	
		359	1.86	646	12.1	9*		

Alkadienes

Compound								
C$_3$H$_4$	0	239[g]	2.14	664	3.9	56		
1,3-C$_4$H$_6$	0	269[g]	2.05	650	3.9	56	o	

Liquid[a]	$\mathcal{D}^{b}$ (10^{-30} C·m)	T (K)	ϵ^{c}	d (kg/m³)	$b_{GP}{}^{d}$ (nm)	Reference	μ (10^{-4} m²/V·s)	Reference
1,4-C₅H₈		299[g]	2.04	655	4.4	56		
1,5-C₆H₁₀		296	2.12	689	4.4	56		
		333[g]	2.01	656	4.6	56		
cis-cis-2,4-C₆H₁₀	0	296	2.16	717	4.2	9*	o	
		353[g]	2.04	660	4.4	9*		
tr-tr-2,4-C₆H₁₀	0	296	2.12	717	3.9	9	o	
		353[g]	2.01	660	4.3	9		
1,6-C₇H₁₂		296	2.15	716	4.2	56		
		358	1.98	658	4.5	56		
1,7-C₈H₁₄		296	2.18	741	4.1	56		
		394	1.93	647	4.4	56		
Alkatriene								
1,3,5-C₆H₈		296	2.28	731	3.3	9*	o	
cis- and tr-mix		351[g]	2.17	700	3.5	9*		
Cycloalk -anes, -enes, and -adienes								
c-C₃H₆	0	240[g]	1.88	675	5.6	70	0.006	70, 107*
c-C₅H₁₀		296	1.96	742	7.5	68	1.4, 1.1	68, 66*
		322[g]	1.92	717	7.8	68	2.2, 1.4	68, 66*

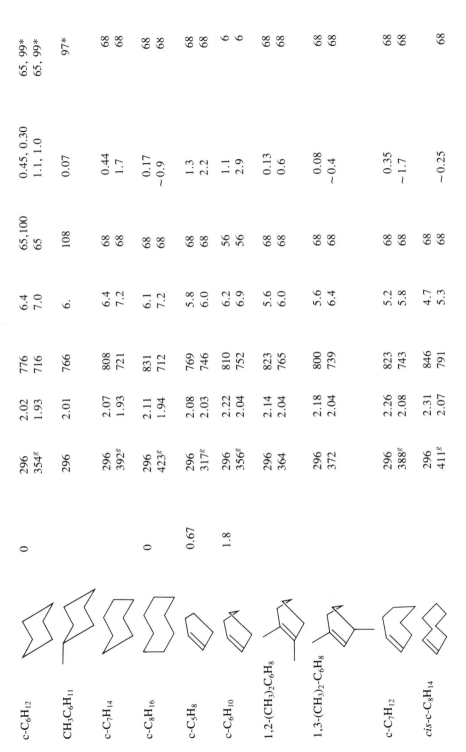

Compound								
c-C₆H₁₂	0	296	2.02	776	6.4	65,100	0.45, 0.30	65, 99*
		354ᵍ	1.93	716	7.0	65	1.1, 1.0	65, 99*
CH₃C₆H₁₁		296	2.01	766	6.	108	0.07	97*
c-C₇H₁₄	0	296	2.07	808	6.4	68	0.44	68
		392ᵍ	1.93	721	7.2	68	1.7	68
c-C₈H₁₆		296	2.11	831	6.1	68	0.17	68
		423ᵍ	1.94	712	7.2	68	~0.9	68
c-C₅H₈	0.67	296	2.08	769	5.8	68	1.3	68
		317ᵍ	2.03	746	6.0	68	2.2	68
c-C₆H₁₀	1.8	296	2.22	810	6.2	56	1.1	6
		356ᵍ	2.04	752	6.9	56	2.9	6
1,2-(CH₃)₂C₆H₈		296	2.14	823	5.6	68	0.13	68
		364	2.04	765	6.0	68	0.6	68
1,3-(CH₃)₂C₆H₈		296	2.18	800	5.6	68	0.08	68
		372	2.04	739	6.4	68	~0.4	68
c-C₇H₁₂		296	2.26	823	5.2	68	0.35	68
		388ᵍ	2.08	743	5.8	68	~1.7	68
cis-c-C₈H₁₄		296	2.31	846	4.7	68		
		411ᵍ	2.07	791	5.3	68	~0.25	68

75

Liquid[a]	$\mathfrak{D}^{b}$ $(10^{-30}$ C·m)	T (K)	ϵ^{c}	d (kg/m³)	$b_{GP}{}^{d}$ (nm)	Reference	μ $(10^{-4}$ m²/V·s)	Reference
1,3-c-C_6H_8	1.5	296 354	2.34 2.24	838 773	4.0 4.0	68 68	[o]	
1,4-c-C_6H_8	0	296 359	2.21 2.08	854 784	6.6 7.1	68 68	6.0 11	68 68
Polycycloalk -anes, -enes, and -adienes								
bicyclo [2,2,1]-heptane		374^{p}	2.00	800	9.5	68	4.4	68
bicyclo[2,2,2]-octane	0	450	2.06	762	$>15.3^{i}$	68	15	68
adamantane		396			9.0^{q}	68		
bicyclo[2,2,1]-heptene		373	2.02	803	6.0	68		
α(+)pinene		296 429^{g}	2.22 1.96	854 742	5.4 6.6	68 68		

Compound			Aromatics					
α(−)pinene		296 429[g]	2.22 1.96	854 742	5.3 6.4	68 68		75, 109* 75, 109*
bicyclo[2,2,1]-hepta-2,5-diene		296 363[g]	2.15 2.04	906 840	3.8 4.1	68 68		75, 109* 109*
C_6H_6	0	296 353[g]	2.28 2.17	876 815	4.2 4.8	76, 69*[r] 75	0.13 1.1	75, 109* 75, 109*
$CH_3C_6H_5$	1.2	296 384[g]	2.38 2.18	864 784	4.3 4.7	75 75	0.075 0.41	75, 109* 109*
$1,2\text{-}(CH_3)_2C_6H_4$	2.0	296 418[g]	2.56 2.24	877 774	4.0 4.4	75 75	0.022 0.18	75 75
$1,3\text{-}(CH_3)_2C_6H_4$	1.3	296 412[g]	2.37 2.14	862 762	4.1 4.6	75 75	0.063 0.54	75 75
$1,4\text{-}(CH_3)_2C_6H_4$	0	296 412[g]	2.26 2.08	859 758	4.1 4.5	75 75	0.066	75
$1,2,3\text{-}(CH_3)_3C_6H_3$	2.0	296 354 449[g]	2.63 2.47 2.22	892 846 770	4.2 4.3 4.8	75 75 75	0.026 0.096	75 75

77

Liquid[a]	$\mathfrak{D}^b$ (10^{-30} C·m)	T (K)	ϵ^c	d (kg/m³)	$b_{GP}{}^d$ (nm)	Reference	μ (10^{-4} m²/V·s)	Reference
1,2,4-(CH$_3$)$_3$C$_6$H$_3$	1.0	296 443^g	2.37 2.09	874 756	4.1 4.8	75 75	0.042 ~0.46	75 75
1,3,5-(CH$_3$)$_3$C$_6$H$_3$	0	296 324 383	2.27 2.19 2.01	862 840 791	4.4 4.6 5.4	75 75 75	0.17 0.26	75 75
1,2,3,4-(CH$_3$)$_4$C$_6$H$_2$	2.0	296 353 413	2.54 2.41 2.29	903 861 816	4.2 4.4 4.7	75 75 75	0.09	75
1,2,4,5-(CH$_3$)$_4$C$_6$H$_2$	0	358^s 402	2.22 2.15	835 800	4.5 4.6	75 75		
(CH$_3$)$_3$CC$_6$H$_5$	2.8	296	2.38	865	4.7	69*^{r}		
(CH$_3$)$_5$C$_6$H	1.3	333^s 414	2.36 2.23	890 822	5.3 5.9	75 75		

Compound								
(CH₃)₆C₆	0	448[t]	2.17	823	7.4	75		111*
Napthalene	0	357[u]	2.54	972	3.3	75		110*
		376	2.51	956	3.3	75		110*
Anthracene	0	500[v]	2.65	964	~2.0	75		
Ethers								
(CH₃)₂O	4.3	165	10.7	814	4.2	110*	0.005	110*
		250[g]	6.2	714	4.9	110*		
(C₂H₅)₂O	3.8	296	4.33	710	5.1, 4.4	110*, 69*[r]	0.0052	110*
		308[g]	4.09	696	5.2	110*	0.0065	
(n-C₃H₇)₂O	4.0	296	3.40	742	5.2	110*	0.0040	110*
		364[g]	2.71	673	5.8	110*	0.027	
(i-C₃H₇)₂O	3.8	296	3.92	734	4.5	69*[r]		110*
(n-C₄H₉)₂O	3.9	296	3.12	767	5.5	110*	0.0037	110*
		415[g]	2.30	656	6.6	110*	0.070	110*
Cyclic Ethers								
Tetrahydrofuran, THF	5.4	205	11.6	983	3.3	110*	0.003	110*
		296	8	882	3.6	110*		
2-methyl-THF		298	7.9	850	1.9?	123		

Liquid[a]	$\mathfrak{D}^b$ (10^{-30} C·m)	T (K)	ϵ^c	d (kg/m³)	$b_{GP}{}^d$ (nm)	Reference	μ (10^{-4} m²/V·s)	Reference
Tetrahydropyran	5.3	296	5.52	878	3.8	110*	0.0031	110*
		361^g					0.016	110*
p-Dioxan	0	296	2.20	1026	4.4, 4.2	110*, 69*^{r}	0.0009	110*
		374^g	2.08	939	4.7	110*	0.005	110*
Halides								
CH$_3$F	6.2	183	27.9	904	>3.7^i	112		
CH$_2$F$_2$	6.6	183	38.9	1300	>3.5^i	112		
CHF$_3$	5.5	183	21.4	1480	>3.6^i	112		
CF$_4$	0	143	1.64	1640	13.2	112		
CCl$_4$	0	296	2.23	1580	5.1	69*^{r}	o	
GeCl$_4$	0	296	2.44	1850	5.2	69*^{r}		
n-C$_5$F$_{12}$	≤0.3	296	1.68		5.1	69*^{r}		
CF$_3$-c-C$_6$F$_{11}$		296	1.85	1788	4.4	69*^{r}	o	
Hydroxy Compounds								
H$_2$O	6.2	296	79	997	>1.9^i	113*	0.0018	114
CH$_3$OH	5.7	298	33	788	>1.9^i	113*	0.0006	115, 113*
C$_2$H$_5$OH	5.6	298	24.6	788	>2.0^i	116*	0.0003	115, 65
(CH$_3$)$_2$CHOH	5.5	298	19.4	781	>2.2^i	116*	0.0005	117
(CH$_3$)$_3$COH	5.	298	12.3	781	>2.2^i	118		
n-C$_4$H$_9$OH	5.5	298	17.1	806	>1.5^i	118, 119	0.0008	119
n-C$_5$H$_{11}$OH		298	14.2	811	>1.6^i	118		
(CH$_3$)$_2$CHCH$_2$CH$_2$OH		293	15	806	>1.4^i	119	0.0010	119
(CH$_3$)$_3$CCH$_2$OH		337	8.3	812	>1.6^i	118		

Amines

Compound	μ[b]	T	ε[c]					
NH_3	4.9	240[g]	22.	682	<7.6[x]	120[w], 121	0.0091	122
$n\text{-}C_4H_9NH_2$	3.3	293	5.	738	3.4	119	0.0027	119
$(C_2H_5)_3N$	2.2	296	2.44	725	5.4	69*[r]		
$(n\text{-}C_4H_9)_3N$		293	5.	777	3.	119		
Pyridine	7.3	298	12.3	978	>1.6[i]	123[w]	0.002	119
$(CH_3)_4X$								
$(CH_3)_4C$	0	296	1.78	588	>20[i]	35*, 66*[j]	68	35*, 66*[j]
$(CH_3)_4Si$	0	133					500[h]	124
		163					340[h]	124
		173					120	124
		223					100	124
$(CH_3)_4Ge$	0	296	1.84	630	>16[i]	69*[r]	100, 90	97*, 69*
$(CH_3)_4Sn$	0	296	2.07	1300	>12[i]	69*[r]	90	125
		296					70	125

[a] Carbon skeletons of hydrocarbons and ethers are shown.

[b] Electric dipole moment, gas phase. 1 debye = 10^{-18} esu·cm = 3.34×10^{-30} C·m. The average values for the n-alkanes are from N. A. Hermiz, J. B. Hasted, and C. Rosenberg, J. Chem. Soc. Faraday Trans. 2, **78**, 147 (1982).

[c] Dielectric constant.

[d] Where values were reported as b_G they have been converted to b_{GP} with the aid of Figure 2.10. Values of ϕ_{fi}^0 can be obtained from those of $\varepsilon T b_{GP}$, using Figure 2.10.

[e] G. A. Cook, Argon, Helium and the Rare Gases, vol. 1, Interscience, New York, 1961.

[f] Superfluid; $T_\lambda = 2.7$ K.

[g] Normal boiling point.

[h] Solid.

[i] The value of ϕ_{fi}^0 is >0.1, for which $\varepsilon T b_{GP}$ is $> 6 \times 10^{-6}$ K·m in the single-pair (SP) microzone model. Use of a multipair (MP) microzone model such as that in Ref. 43 would give a larger value of b_{GP} than the SP value listed. The distance ratio MP/SP would be greatest for water. With the MP model in Ref. 43 and recalculated SP values the ratio was 1.7 for water ($\varepsilon T b_{GP} = 45 \times 10^{-6}$ K·m in SP), 1.3 for $neo\text{-}C_5H_{12}$ ($\varepsilon T b_{GP} = 12 \times 10^{-6}$ K·m in SP), and 1.00 for $1,6\text{-}C_7H_{12}$ ($\varepsilon T b_{GP} = 3 \times 10^{-6}$ K·m in SP).

^{j}A plot of b_{GP} or μ against density contains a maximum similar to that for neopentane in Figure 2.14.

^{k}This thousandfold change from μ_{max} to μ in the critical fluid is the largest change found in any dielectric liquid.

^{l}Critical fluid.

^{m}Melting point.

^{n}Melting point 192 K, vapor pressure of solid 101 kPa at 189 K.

^{o}Liquid reacts rapidly with thermal electrons.

^{p}Melting point 361 K, vapor pressure > 100 kPa.

^{q}Extrapolated from values in a 35-wt. % adamantane-n-heptane solution and pure n-heptane, using a plot of $(b_{GP}d)^{-2}$ against wt. % adamantane. Purified adamantane melts at 541 K, turns yellow, and has a high "intrinsic" conductance attributed to the yellow product.

^{r}Distances given in Refs. 69 and 100 are b_G, but they are nearly the same as the values of b_{GP} from the Edmonton laboratory, probably due to a difference in dosimetry.

^{s}Melting point + 6 K.

^{t}Melting point + 8 K.

^{u}Melting point + 3 K.

^{v}Melting point + 11 K.

^{w}The value of ϕ_{fi}^0 was determined by measuring yields of products from electron scavenging reactions. When the solvent is itself an electron scavenger, such as nitrocompounds (118), the value obtained by this method is probably too low.

^{x}The large free ion yield in liquid ammonia (G_{fi} = 3.2 at 223–258 K, Ref. 120^w) is due to the small encounter efficiency of the e_s^- + NH$_{4,s}^+$ reaction (121), so b_{GP} is smaller than the value estimated by assuming the encounter efficiency to be unity.

82

REFERENCES

1. (a) M. Karplus and R. N. Porter, *Atoms and Molecules*, Benjamin, New York, 1970, Chapters 1–3 and p. 596. (b) R. L. Liboff, *Introductory Quantum Mechanics*, Holden-Day, San Francisco, 1980, pp. 135, 367.

2. (a) A. Einstein, *Rapport et Discussions du 5e Conseil, Institute International de Physique Solvay*, Paris, 1928, p. 253ff, reported by N. Bohr in *Albert Einstein, Philosopher-Scientist*, P. A. Schilpp (Ed.), Library of Living Philosophers, Vol. 7, pt. I, Open Court, LaSalle, IL, 1949, pp. 212–213. (b) A. Einstein, B. Podolsky, and N. Rosen, *Phys. Rev.* **47,** 777 (1935). (c) P. A. M. Dirac, *Proc. Roy. Soc. (London)* **A180,** 1 (1942). (d) D. Bohm, *Phys. Rev.* **85,** 166, 180 (1952). (e) R. P. Feynman and A. R. Hibbs, *Quantum Mechanics and Path Integrals*, McGraw-Hill, New York, 1965, pp. 22–24. (f) J. S. Bell, *Physics* **1,** 195 (1965). (g) J. F. Clauser and A. Shimony, *Rep. Prog. Phys.* **41,** 1881 (1978). (h) A. Aspect, D. Dalibard, and G. Roger, *Phys. Rev. Lett.* **49,** 1804 (1982). (i) A. O. Barut and N. Zanghi, *Phys. Rev. Lett.* **52,** 2009 (1984). (j) L. E. Ballentine, *Am. J. Phys.* **52,** 271 (1984). (k) The question of whether quantum mechanics provides a complete description of reality (2b) is usually discussed at the level of a word game, for example, by using quantum mechanics to define itself (h). The answer seems no closer now than 50 years ago. The stochastic approach to quantum mechanics fits well under the title "Kinetics of Nonhomogeneous Processes" but is not yet well enough developed to include in an introductory book.

3. G. H. Wannier, *Phys. Rev.* **52,** 191 (1937).

4. (a) W. B. Fowler, *Phys. Rev.* **151,** 657 (1966). (b) S. A. Rice and J. Jortner, *J. Chem. Phys.* **44,** 4470 (1966).

5. L. G. H. Huxley and R. W. Crompton, *The Diffusion and Drift of Electrons in Gases*, Wiley, New York, 1974, p. 212.

6. J.-P. Dodelet, K. Shinsaka, and G. R. Freeman, *J. Chem. Phys.* **59,** 1293 (1973).

7. G. R. Freeman, *Ann. Rev. Phys. Chem.* **34,** 463 (1983).

8. R. M. Minday, L. D. Schmidt, and H. T. Davis, (a) *J. Chem. Phys.* **54,** 3112 (1971); (b) *J. Phys. Chem.* **76,** 442 (1972).

9. J.-P. Dodelet, K. Shinsaka, and G. R. Freeman, *Can. J. Chem.* **54,** 744 (1976).

10. H. Schnyders, S. A. Rice, and L. Meyer, *Phys. Rev.* **150,** 127 (1966).

11. S. S.-S. Huang and G. R. Freeman, *J. Chem. Phys.* **68,** 1355 (1978).

12. J. A. Jahnke, L. Meyer, and S. A. Rice, *Phys. Rev. A* **3,** 734 (1971).

13. S. S.-S. Huang and G. R. Freeman, *Phys. Rev. A* **24,** 714 (1981).

14. N. Gee and G. R. Freeman, *Phys. Rev. A* **20,** 1152 (1979).

15. Y. Nakato and H. Tsubomura, *J. Phys. Chem.* **79,** 2135 (1975).

16. J. Jortner and A. Gaathon, *Can. J. Chem.* **55,** 1801 (1977).

17. N. F. Mott and E. A. Davis, *Electronic Processes in Non-crystalline Materials*, 2nd ed., Clarendon Press, Oxford, 1979, Chapters 2–4. In this book m^* represents m_{eff} rather than the ratio. In strongly scattering media $m_{eff} > m$; in weakly scattering media $m_{eff} < m.$

18. N. F. Mott and M. J. Littleton, *Trans. Faraday Soc.* **34,** 485 (1938).

19. B. E. Springett, J. Jortner, and M. H. Cohen, *J. Chem. Phys.* **48,** 2720 (1968).

20. I. Messing and J. Jortner, *Chem. Phys.* **24**, 183 (1977).

21. (a) F.-J. Himpsel, N. Schwentner, and E.-E. Koch, *Phys. Stat. Sol.* (*b*) **71**, 615 (1975). (b) N. Schwentner, E.-E. Koch, and J. Jortner, *Electronic Excitations in Condensed Rare Gases*, Springer Tracts in Modern Physics, Springer-Verlag, Berlin, 1985. (c) R. Reininger, U. Asaf, I. T. Steinberger, V. Saile, and P. Laporte, *Phys. Rev. B* **28**, 3193 (1983).

22. W. K. Tauchert, H. Jungblut, and W. F. Schmidt, *Can. J. Chem.* **55**, 1860 (1977).

23. W. T. Sommer, *Phys. Rev. Lett.* **12**, 271 (1964).

24. G. Baldini, *Phys. Rev.* **128**, 1562 (1962).

25. G. Herzberg, *Atomic Spectra and Atomic Structure*, 2nd ed., Prentice-Hall, New York, 1944, pp. 54–62.

26. C. E. Moore, *Atomic Energy Levels*, Vol. 1, NBS Circ. 467, U.S. Government Printing Office, Washington, DC, 1949.

27. W. Herreman, *Cryogenics* **20**, 133 (1980).

28. W. H. Brendley and E. C. Evers, *Adv. Chem. Ser.* **50**, 111 (1965).

29. M. R. Spiegel, *Probability and Statistics, Shaum's Outline Series in Mathematics*, McGraw-Hill, New York, 1975, Chapter 4.

30. K. Okazaki and G. R. Freeman, *Can. J. Chem.* **56**, 2305 (1978).

31. K. Okazaki and G. R. Freeman, *Can. J. Chem.* **56**, 2313 (1978).

32. C. Kittel, *Introduction to Solid State Physics*, 5th ed., Wiley, New York, 1976.

33. J. Kestin and J. R. Dorfman, *A Course in Statistical Thermodynamics*, Academic Press, New York, 1971, Chapter 9.

34. L. Landau, *Physik. Z. Sowiet Union* **3**, 664 (1933).

35. K. Shinsaka and G. R. Freeman, *Can. J. Chem.* **52**, 3556 (1974).

36. H. A. Bethe and J. Ashkin, in *Experimenntal Nuclear Physics*, Vol. 1, E. Segré (Ed.), Wiley, New York, 1953, pp. 166–357.

37. G. Jaffé, *Ann. Physik* **42**, 303 (1913).

38. A. Mozumder and J. L. Magee, *Radiat. Res.* **28**, 215 (1966).

39. H. A. Kramers, *Physica* **18**, 665 (1952).

40. A. N. Gerritsen, *Physica* **14**, (1948), (a) p. 381, (b) p. 407.

41. A. Ore and A. Larson, *Radiat. Res.* **21**, 331 (1964).

42. H. A. Schwarz, *J. Phys. Chem.* **73**, 1928 (1969).

43. J.-P. Dodelet and G. R. Freeman, *Int. J. Radiat. Phys. Chem.* **7**, 183 (1975).

44. L. Onsager, (a) *Phys. Rev.* **54**, 554 (1938); (b) *J. Chem. Phys.* **2**, 599 (1934).

45. (a) G. R. Freeman, *Adv. Chem. Ser.* **82**, 339 (1968). (b) K. N. Jha and G. R. Freeman, *J. Chem. Phys.* **51**, 2839 (1969).

46. G. R. Freeman, *J. Chem. Phys.* **39**, 1580 (1963).

47. J. Terlecki and J. Fiutak, *Int. J. Radiat. Phys. Chem.* **4**, 469 (1972).

48. G. R. Freeman and J.-P. Dodelet, *Int. J. Radiat. Phys. Chem.* **5**, 371 (1973).

49. A. Hummel and A. O. Allen, *J. Chem. Phys.* **46**, 1602 (1967).

50. J.-P. Dodelet, P. G. Fuochi, and G. R. Freeman, *Can. J. Chem.* **50**, 1617 (1972).

51. G. Buxton, *Proc. Roy. Soc. (London)* **A328**, 9 (1972).

52. G. R. Freeman and T. E. M. Sambrook, *J. Phys. Chem.* **78**, 102 (1974).

53. T. Takahashi, S. Konno, and T. Doke, *J. Phys. C: Solid St.* **7,** 230 (1974).

54. S. S.-S. Huang and G. R. Freeman, *Can. J. Chem.* **55,** 1838 (1977).

55. W. F. Schmidt, *Radiat. Res.* **42,** 73 (1970).

56. J.-P. Dodelet, K. Shinsaka, U. Kortsch, and G. R. Freeman, *J. Chem. Phys.* **59,** 2376 (1973).

57. T. G. Ryan and G. R. Freeman, *J. Chem. Phys.* **68,** 5144 (1978).

58. J.-P. Dodelet and G. R. Freeman, *J. Chem. Phys.* **60,** 4657 (1974).

59. J.-P. Dodelet, *Can. J. Chem.* **55,** 2050 (1977).

60. H. T. Choi, D. S. Sethi, and C. L. Braun, *J. Chem. Phys.* **77,** 6027 (1982).

61. A. Mozumder and M. Tachiya, *J. Chem. Phys.* **62,** 979 (1975).

62. K. Lee and S. Lipsky, *J. Phys. Chem.* **88,** 4251 (1984).

63. J. M. Warman, *Can. J. Chem.* **55,** 1846 (1977).

64. J. M. Warman, *J. Phys. Chem.* **87,** 4353 (1983).

65. J.-P. Dodelet and G. R. Freeman, *Can. J. Chem.* **50,** 2667 (1972).

66. J.-P. Dodelet and G. R. Freeman, *Can. J. Chem.* **55,** 2264 (1977).

67. (a) N. Gee and G. R. Freeman, *Phys. Rev. A* **28,** 3568 (1983). (b) N. Gee and G. R. Freeman, *J. Chem. Phys.* **78,** 1951 (1983).

68. K. Shinsaka, J.-P. Dodelet, and G. R. Freeman, *Can. J. Chem.* **53,** 2714 (1975).

69. W. F. Schmidt and A. O. Allen, *J. Chem. Phys.* **52,** 2345, 4788 (1970).

70. M. G. Robinson and G. R. Freeman, *Can. J. Chem.* **52,** 440 (1974).

71. M. J. Avoey, R. J. W. Le Fevre, W. Luettke, G. L. D. Ritchie, and P. J. Stills, *Aust. J. Chem.* **21,** 2551 (1968).

72. R. J. W. Le Fevre, *Rev. Pure Appl. Chem.* **20,** 67 (1970).

73. M. P. Bogaard, A. D. Buckingham, R. K. Pierens, and A. H. White, *J.C.S. Faraday Trans. I* **74,** 3008 (1978).

74. R. D. Nelson, D. R. Lide, and A. A. Maryott, *NSRDS-NBS 10*, U.S. Government Printing Office, Washington, DC, 1967.

75. K. Shinsaka and G. R. Freeman, *Can. J. Chem.* **52,** 3495 (1974).

76. N. B. Vargaftik, *Tables on the Thermophysical Properties of Liquids and Gases*, 2nd ed., Wiley, New York, 1975.

77. R. C. Reid, J. M. Prausnitz, and T. K. Sherwood, *The Properties of Gases and Liquids*, 3rd ed., McGraw-Hill, New York, 1977, Chapters 1 and 3.

78. T. R. Das, C. O. Reed, and P. T. Eubank, *J. Chem. Eng. Data* **22,** (1977): (a) p. 3, the correct value of ρ_L for nP at T $<$ 323 K is 0.4362; (b) p. 16, NP.

79. S. S.-S. Huang and G. R. Freeman, *J. Chem. Phys.* **69,** 1585 (1978).

80. A. Mozumder and J. L. Magee, *J. Chem. Phys.* **47,** 939 (1967).

81. G. R. Freeman, *Quaderni dell'Area di Ricerca dell'Emilia-Romagna* **2,** 55 (1972).

82. G. C. Abell and K. Funabashi, *J. Chem. Phys.* **58,** 1079 (1973).

83. J. M. Warman and M. C. Sauer, *J. Chem. Phys.* **62,** 1971 (1975).

84. U. Sowada, J. M. Warman and M. P. deHaas, (a) *Phys. Rev. B* **25,** 3434 (1982); (b) *Chem. Phys. Lett.* **90,** 239 (1982).

85. A. Mozumder, *J. Chem. Phys.* **76,** 3277 (1982).

86. K. Koura, *J. Chem. Phys.* **81,** 303 (1984).

87. B. Shizgal and D. R. A. McMahon, *J. Phys. Chem.* **88**, 4854 (1984).

88. A. Mozumder, *J. Electrostat.* **12**, 45 (1982).

89. K. W. Schwarz, (a) *Phys. Rev. A* **6**, 837 (1972); (b) *Adv. Chem. Phys.* **33**, 1 (1975).

90. R. J. Loveland, P. G. Le Comber, and W. E. Spear, *Phys. Lett.* **39A**, 225 (1972).

91. L. S. Miller, S. Howe, and W. E. Spear, *Phys. Rev.* **166**, 871 (1968).

92. E. M. Gushchin, A. A. Kruglov, and I. M. Obodovskii, *Sov. Phys. JETP* **55**, 650 (1982).

93. F. M. Jacobsen, N. Gee, and G. R. Freeman, *Phys. Rev. A* **34**, Sept. (1986).

94. Y. Sakai, H. Böttcher, and W. F. Schmidt, *J. Electrostat.* **12**, 89 (1982).

95. (a) T. Wada and G. R. Freeman, in *1980 Annual Report, Conference on Electrical Insulation and Dielectric Phenomena*, S. A. Boggs (Ed.), National Academy Press, Washington, DC, 1980, p. 386. (b) N. Gee, M. A. Floriano, T. Wada, S. S.-S.-Huang, and G. R. Freeman, *J. Appl. Phys.* **57**, 1097 (1985). (c) P. J. Fuochi and G. R. Freeman, *J. Chem. Phys.* **56**, 2333 (1972). (d) T. E. M. Sambrook and G. R. Freeman, *Can. J. Chem.* **53**, 2822 (1975). (e) Y. Nakamura, K. Shinsaka, and Y. Hatano, *J. Chem. Phys.* **78**, 5820 (1983).

96. N. Gee and G. R. Freeman, *Phys. Rev. A* **22**, 301 (1980).

97. A. O. Allen and R. A. Holroyd, *J. Phys. Chem.* **78**, 796 (1974).

98. I. György and G. R. Freeman, *J. Chem. Phys.* **70**, 4769 (1979).

99. S. S.-S. Huang and G. R. Freeman, *Can. J. Chem.* **56**, 2388 (1978).

100. W. F. Schmidt and A. O. Allen, *J. Phys. Chem.* **72**, 3730 (1968).

101. L. Nyikos, E. Zador and R. Schiller, *4th Tihany Symposium Rad. Chem.*, Akademiai Kiadó, Budapest, 1976, p. 179.

102. J.-P. Dodelet, K. Shinsaka, and G. R. Freeman, *J. Chem. Phys.* **63**, 2765 (1975).

103. (a) T. Tezuka, H. Namba, Y. Nakamura, M. Chiba, K. Shinsaka and Y. Hatano, *Radiat. Phys. Chem.* **21**, 197 (1983). (b) A. Balakin, I. A. Boriev, and B. S. Yakovlev, *High Ener. Chem.* **12**, 171 (1978).

104. I. Kalinowski, J. G. Rabe, and W. F. Schmidt, *4th Tihany Symp. Rad. Chem.*, Akademiai Kiadó, Budapest, 1976, p. 171.

105. M. G. Robinson, P. G. Fuochi, and G. R. Freeman, *Can. J. Chem.* **49**, 3657 (1971).

106. J.-P. Dodelet and G. R. Freeman, *Can. J. Chem.* **55**, 2893 (1977).

107. N. Gee and G. R. Freeman, *Radiat. Phys. Chem.* **15**, 267 (1980).

108. J. H. Baxendale, C. Bell, and P. Wardman, *J. Chem. Soc. Faraday Trans. 1* **69**, 776 (1973). Free ion yields measured by optical absorbance of pyrene anion, normalized to those of Ref. 100, which are about 30% lower than those of Refs. 56, 65, and 66.

109. S. S.-S. Huang and G. R. Freeman, *J. Chem. Phys.* **72**, 2849 (1980).

110. J.-P. Dodelet and G. R. Freeman, *Can. J. Chem.* **53**, 1263 (1975).

111. N. Gee and G. R. Freeman, *Can. J. Chem.* **60**, 1034 (1982).

112. M. G. Robinson and G. R. Freeman, *Can. J. Chem.* **51**, 1010 (1973).

113. J.-P. Dodelet and G. R. Freeman, *Can. J. Chem.* **50**, 2729 (1972).

114. K. H. Schmidt, *Int. J. Radiat. Phys. Chem.* **4**, 439 (1972).

115. P. Fowles, *Trans. Faraday Soc.* **67**, 428 (1971).

116. Calculated from G_{fi} and G_{tot} values reported by K. N. Jha and G. R. Freeman, *J. Chem. Phys.* **51**, 2846 (1969).

117. A. V. Rudnev, A. V. Vannikov, and N. A. Bakh, *High Ener. Chem.* **6**, 416 (1972).

118. C. Capellos and A. O. Allen, *J. Phys. Chem.* **74**, 840 (1970).

119. A. V. Vannikov, E. I. Mal'tzev, V. I. Zolotarevsky, and A. V. Rudnev, *Int. J. Radiat. Phys. Chem.* **4**, 135 (1972).

120. W. A. Seddon, J. W. Fletcher, F. C. Sopchyshyn, and J. Jevcak, *Can. J. Chem.* **52**, 3269 (1974).

121. H. Sano and M. Tachiya, *J. Chem. Phys.* **71**, 1276 (1979).

122. D. S. Burns, *Adv. Chem. Ser.* **50**, 82 (1965). Calculated from the equivalent conductance and transference numbers of solutions of sodium in ammonia.

123. E. Hayon, *J. Chem. Phys.* **53**, 2353 (1970).

124. Y. Nakamura, H. Namba, K. Shinsaka, and Y. Hatano, *Chem. Phys. Lett.* **76**, 311 (1980).

125. W. F. Schmidt, *Can. J. Chem.* **55**, 2197 (1977).

3 RADIATION TRACK STRUCTURE THEORY

Herwig G. Paretzke

GSF—Institut für Strahlenschutz
Neuherberg, Federal Republic of Germany

3.1. INTRODUCTION

3.1.1. Objectives of Track Structure Theory

Ionizing radiation from natural and man-made sources has been part of our environment and our daily life for a long time. Such sources include primordial radionuclides (from the ^{238}U and ^{232}Th series, ^{40}K, ^{87}Rb, and so on), cosmogenic nuclides (^{3}H, ^{7}Be, ^{14}C, and so on), cosmic rays (e^{-}, p, n, mesons, and so on), enhanced levels of radon and decay products in dwellings, medical exposures to diagnostic X rays or radiation for cancer therapy, and so on.

Scientific and technological interest in the production of ionizing radiation, its effects (particularly on man), and its application has steadily increased since the invention of X rays by Röntgen in 1895 and the detection of natural radioactivity by Becquerel in 1896. Interest grew rapidly after the observation of induced nuclear fission by Hahn and Strassmann in 1938. The use of radiation and the protection of humans against its harmful effects require research on (a) methods for the detection and measurement of radiation and (b) the analysis of its physical, chemical, and biological effects.

In both areas knowledge of radiation track structures is needed. For the measurement of radiation, knowledge of track structures is necessary, for example, to identify unknown particles by their tracks left in a detection device. Tracks of charged particles can be made visible and analyzed in gases [in cloud chambers (1, 2) and streamer (3a) and spark (3b) chambers], in liquids [in bubble chambers (4)], and in solids [nuclear emulsions (5) and other solid-state nuclear track detectors (6)]. Many elementary particles (for example, π mesons) and some nuclei (such as superheavy nuclei) were first detected with track registration devices. There are many scientific and applied uses of radiation track detectors.

Ionizing radiation is often used to excite objects from their "ground" state to study the nature and dynamics of the processes that bring the object back to the old or to a new "quasi-equilibrium" state. For the analysis of radiation effects in matter, knowledge of radiation track structures is essential. The spatial distributions of species excited by the radiation affects their reaction probabilities, which in turn govern the final changes in the micro- and macroscopic structures of the material. Examples of radiation track structures are shown in Figure 3.1. The species depicted in a track (7) are ions, electrons, excited molecules, and molecular fragments (free radicals) that have lifetimes longer than, say, 10^{-10} s. These are the species that remain after the decay of plasmonlike excitations and super-

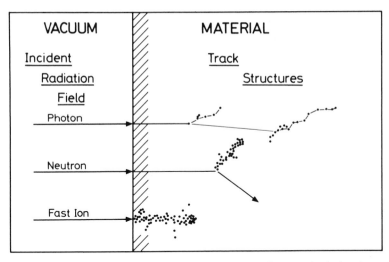

Figure 3.1. Incident radiations are characterized by particle mass, charge, and velocity; the stochastic structures of tracks resulting from interactions of these radiations with matter are described by track matrices $T(S, \mathbf{x}, t)$, where the S is the type of a new chemical species at location $\mathbf{x}$ at time t.

excitations. The new species are the starting points of subsequent physical, chemical, and biological processes. They mark the end of the *physical stage* of radiation action and the beginning of the *chemical stage* (Figure 3.2). During the latter stage diffusion and reaction of primary species with other radiogenic species or with molecules of the irradiated object alter the pattern of the physical track structure. This occurs according to the boundary conditions set in the first stage.

Thus, it is the objective of track structure theory to:

1. Identify the molecular changes of importance for the development of a radiation effect under consideration.
2. Predict and explain the spatial distribution of relevant species with a minimum of assumptions regarding preliminary processes (7b); this spatial distribution should be given in terms of the joint probability to find an event of type S_1 at position $\mathbf{x}_1$ and at the same time an event S_2 at position $\mathbf{x}_2$ and S_3 at $\mathbf{x}_3$, and so on, to account for spatial correlations for further reactions.
3. Identify (for example by correlation studies) the parameters of physical track structure that predominantly determine the nature and magnitude of a final radiation effect and that may also be used to characterize a radiation field with regard to its radiation action when compared to other fields.

The theory that describes the generation of radiation tracks and their structure is still under development. Contributions have been made by Fano (7), Boag (8), Platzman (9), Mozumder and Magee (10), Voltz (11), Ritchie (12), Inokuti (13),

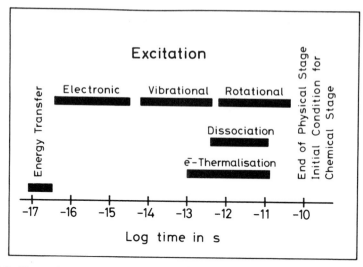

Figure 3.2. Time scale of processes occurring during the physical stage of energy transfer from ionizing radiations to molecules.

Paretzke (14), Lea (15), and Katz (16). Recent progress has been made with new experimental techniques and simulation on fast computers.

The importance of radiation track structure in biological effects was demonstrated already in 1946 (15). Applications of track structure theory include the interpretation of tracks in nuclear emulsions and other solid-state nuclear track detectors (16).

3.1.2. Scope of this Chapter

The three objectives of track structure theory are (i) identification of relevant new species; (ii) prediction of the spatial distribution of such species, that is, of the stochastic track structure; and (iii) evaluation of the structures with respect to their most important parameters. The first step depends on the radiation effect under consideration. Ionizations, dissociations, and the resulting ions, electrons, and radicals play a major role in further chemical or biological reactions, as is discussed in Chapters 2 and 4–8. In this chapter emphasis is given to the second and third steps, namely, to the prediction of track structures and their quantitative evaluation.

We start with a short survey of processes in which photons (Section 3.2) and neutrons (Section 3.3) transfer energy to matter. From this it will become evident that electrons, in particular low-energy electrons, are the porters that distribute most of the energy to the individual molecules in matter and thus create most of the track structures. The interactions of electrons are described in Section 3.4 and those of heavy charged particles in Section 3.5. With consistent sets of interaction cross sections track structures can be simulated on a computer. Methods and results of simulations for electrons, protons, α particles, and heavier ions in water vapor are described in Section 3.6. Section 3.7 outlines concepts for the quantitative evaluation of such stochastic structures for the purpose of characterization and classification.

Because of the wide field of track structure research and of the limited space available for this chapter, a choice had to be made between a more general survey of the whole topic or a more comprehensive discussion of only a few parts of it. Taking account of the scope of the book and its emphasis on the general importance of nonhomogeneous kinetics in science, the form of an introductory survey has been selected to get the reader informed of and interested in this challenging field. Further details are given in the references.

3.2. INTERACTIONS OF PHOTONS

3.2.1. Interaction Processes

Photons play an important role in radiation research and in technical and medical applications of ionizing radiation. A photon imparts a large fraction of its energy to a single electron, which then ionizes many other molecules along its path. The track generated by a photon is therefore an electron track.

A photon of energy $h\nu$ might interact inelastically with a molecule in one of three ways (17, 18):

1. In a photoelectric process where the photon is completely absorbed and an electron is ejected with kinetic energy $E = h\nu - U$, where U is the binding energy of the electron in the molecule. This effect is dominant at low photon energies, below, say, 40 keV in materials with a small atomic number Z. It produces sharp energy spectra of photoelectrons that carry essentially all the photon energy, since the binding potential in low-Z matter is typically only about 500 eV. In addition, often Auger electrons or fluorescent photons are emitted from the atom hit.

2. In an incoherent Compton scattering event where only part of the photon's energy is transferred to an electron. This Compton electron is emitted from its parent atom; the scattered photon continues its flight in a new direction and with reduced energy according to the laws of energy and momentum conservation. In water and other low-Z materials this process is the most important energy loss mechanism for photons in the medium energy range from, say, 0.04 to 10 MeV. It leads to tracks of Compton electrons with a wide energy spectrum, from zero up to a substantial fraction of the primary photon energy $h\nu$ (Section 3.2.2).

3. In a pair production event, when, in the strong field of an atomic nucleus, the photon is transformed into an electron–positron pair. This process can only occur at photon energies above 1.02 MeV, which is the energy equivalent of the masses of an electron and positron with zero kinetic energy. The excess energy $h\nu - 1.02$ MeV is shared as kinetic energy E^+ and E^- between the created positive

and negative particles. Their energy spectrum is rather flat in the energy range from essentially zero up to E^+, $E^- = h\nu - 1.02$ MeV. This interaction process must be taken into account only at high photon energies (above, say, 5 MeV), but it gains importance with increasing energy. For conventional X rays, ^{137}Cs γ rays (0.66 MeV) and ^{60}Co γ rays (1.17 and 1.33 MeV), pair production can be neglected.

A fourth, elastic photon interaction is coherent (Rayleigh) scattering. It can be important at low energies, say, below 0.1 MeV. It leads to a change in direction of the photon and thus influences the location of the next inelastic event. Coherent scattering can be understood as the cooperative action of electrons bound to an atom, which scatter the incident wave into a new direction by interference of their respective elementary waves, but absorbing no energy from the photon. Thus, it does not leave a track in the irradiated matter, although it may contribute up to 10% to the total photon interaction cross section. The characteristic features of Compton scattering, coherent scattering, and the photoelectric effect are depicted in Figure 3.3.

3.2.2. Cross Sections

The probabilities of the four competing interaction processes mentioned above can be expressed as cross sections. For quantitative calculations of track structures

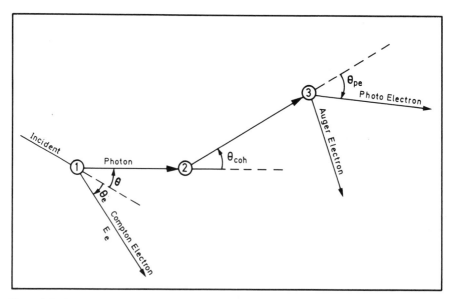

Figure 3.3. Main interaction processes of photons: incoherent scattering (1), coherent scattering (2), and photo effect (3).

absolute values of integral and differential cross sections (17–20) of the important processes are needed. Figure 3.4 shows such cross sections for aluminum as a typical example of low-Z atoms. The strong absorption at low energies (say, below 40 keV) due to the photoeffect is evident, as is the importance of the Compton effect at medium to high energies and of pair production at very high energies.

The photoelectric cross section of an atom, between its resonant absorption edges, varies with the photon energy approximately as $(h\nu)^{-3.5}$ and increases approximately as the fourth power of the atomic number Z. The Compton cross section of an electron, however, is almost independent of the atomic number, and it decreases only slightly with increasing photon energy. Thus, in heterogeneous objects made up by atoms of different Z, a variation in photon energy can lead to a significant change in the relative importance of interaction modes and of the locations of interactions.

Photon track structures are mainly characterized by (a) the kinetic energy spectrum of their secondary electrons and (b) the spacing between two sequential in-

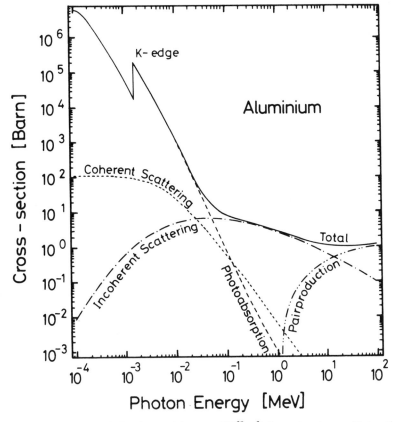

Figure 3.4. Cross sections (in barn/atom; 1 barn $\equiv 10^{-28}$ m^2) for various types of interaction of photons with aluminum, a typical low-Z material (19).

elastic events. Therefore, the differential cross sections with respect to energy and scattering angle are needed. As mentioned above, the kinetic energy E of electrons emitted in photoelectric processes is close to the energy $h\nu$ of the absorbed photon, since the atomic binding potential U is usually small compared to $h\nu$:

$$E = h\nu - U \tag{3.1}$$

Compared to the narrow energy lines of photoelectrons, electrons ejected in incoherent Compton events have a rather wide energy spectrum, $d\sigma/dE$. For the purpose of this chapter it is sufficient to discuss this differential cross section for the case of inelastic photon scattering by a free electron (17), which was first derived by Klein and Nishina (KN):

$$\frac{d\sigma_{KN}}{dE}(h\nu, E) = \frac{\pi r_0^2}{\alpha h\nu}\left\{2 + \left(\frac{E}{h\nu - E}\right)^2\left[\frac{1}{\alpha^2} + \frac{h\nu - E}{h\nu} - \frac{2(h\nu - E)}{\alpha E}\right]\right\} \tag{3.2}$$

where r_0 = electron radius (2.82×10^{-15} m)
       $\alpha = h\nu/m_e c^2$
       $m_e c^2$ = electron rest energy (511 keV)

When the electron binding energy can be ignored, the photon energy $h\nu'$ after the collision is

$$h\nu' = h\nu - E \tag{3.3}$$

This close correspondence between the photon energy after collision and Compton electron kinetic energy is visible in Figure 3.5, which shows the energy spectra of the scattered photons (a) and electrons (b) for photons with incident energy 0.1, 0.5, and 1 MeV. The maximum energy that can be transferred to the ejected electron in Compton collisions depends on the photon energy (17),

$$E_{\max} = \frac{2\alpha}{1 + 2\alpha}\, h\nu \tag{3.4}$$

whereas the minimum energy transfer is $E_{\min} = 0$ (Figure 3.6a).

The Klein–Nishina differential cross section per unit solid angle Ω for scattering of an unpolarized photon into the solid angle bounded by angles θ and $\theta + d\theta$ is

$$\frac{d\sigma_{KN}}{d\Omega}(h\nu, \theta) = \frac{r_0^2}{2}(1 + \cos^2\theta)\left(\frac{1}{1 + \alpha(1 - \cos\theta)}\right)^2$$
$$\times \left(1 + \frac{\alpha^2(1 - \cos\theta)^2}{[1 + \alpha(1 - \cos\theta)](1 + \cos^2\theta)}\right) \tag{3.5}$$

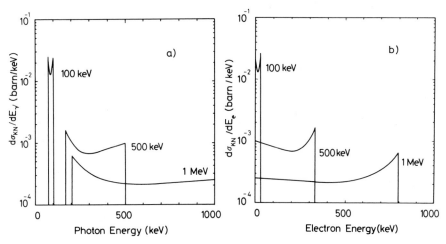

Figure 3.5. Scattered photon (*a*) and corresponding recoil electron (*b*) spectra (in barn/keV per electron) after Compton scattering of 0.1, 0.5, and 1 MeV incident photons.

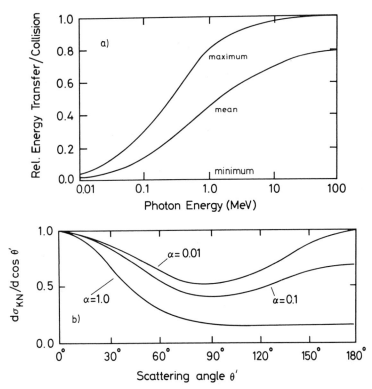

Figure 3.6. (*a*) Fraction of incident photon energy $h\nu$ transferred to recoil electrons in incoherent scattering events; (*b*) normalized (in units of r_0^2) angular distribution of scattered photons; calculated from the Klein–Nishina equation ($\alpha = h\nu/m_ec^2$).

For small values of α, that is, for low photon energies, the angular distribution of photons after incoherent scattering is symmetrical about the angle 90°. Higher energy photons are peaked in the forward direction, leading to a more linear arrangement of subsequent electron tracks in matter.

Electrons set in motion by inelastic photon collisions in turn excite and ionize other molecules. Thus, the original electron can produce an avalanche of higher order generations of electrons with decreasing start energies, which all contribute to the track formation. The number and energy spectra of the higher generation electrons depend on electron collision cross sections, which are discussed in Section 3.4.2. However, for comparison with the Compton electron spectra in Figure

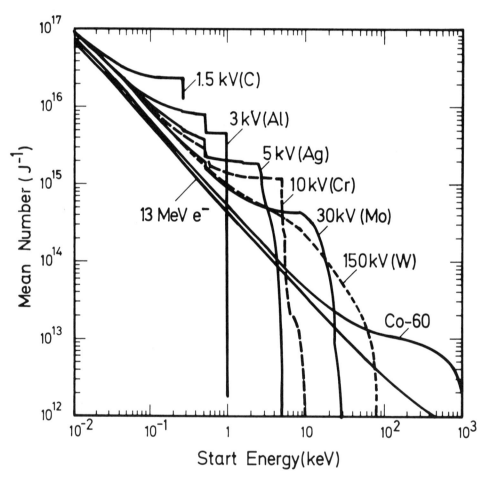

Figure 3.7. Mean number of electrons with start energies larger than E produced by interaction of various radiation fields with water (normalized to an absorbed energy of 1 J); adapted from Ref. 21. Parameters give acceleration voltages and target materials of X-ray tubes.

3.5*b*, Figure 3.7 shows such start spectra for photon fields of widely different energies and for incident fast electrons (21). The subsequent interactions of electrons change the shapes of the spectra from the initial Compton- and photo-effect start spectra.

3.2.3. Energy Deposition

Photon track structures, and thus their energy deposition, are mainly characterized by the start energies of the electrons ejected in inelastic photon collisions and by the distances between subsequent inelastic photon collisions.

The locations and types of energy transfers in inelastic photon events are determined by the respective cross sections of the four processes mentioned above, which are governed by the photon energy and the atomic constitution of the irradiated medium. The lengths of the corresponding electron tracks are generally small (typically nanometers to millimeters in water; Section 3.4.3) compared to the distance between subsequent photon collisions (typically centimeters; Section 3.6.2). Thus the analysis of photon track structures can usually be reduced to the analysis of appropriate electron track structures.

The low density of electron tracks in photon-irradiated matter and its dependence on photon energy can be estimated from Figure 3.8, which shows the frac-

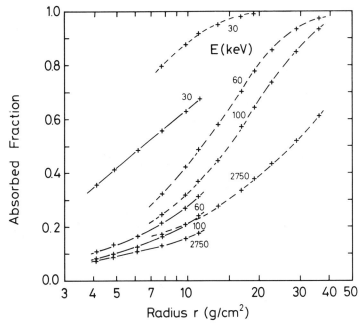

Figure 3.8. Absorbed fractions in spheres for isotropic point sources (---) of photons of various energies and for uniformly distributed sources (——) as a function of the radius (22).

tion of photon energy absorbed within concentric spheres around photon point sources located in the center and for uniformly distributed sources (22). However, this low energy density in single-photon tracks leads to the complicating situation that, with increasing values of energy deposited in finite mass elements, many photons have interacted in that element. Thus the spatial and temporal overlap of track structures of many secondary electrons and their physical and chemical consequences have to be considered. They might cause dose and dose rate effects.

Finally, in extended bodies scattered photons can make a significant contribution to energy deposition. This is usually accounted for by a buildup factor $B(E, \mu r)$, which is used to multiply the energy deposited by unscattered photons. It is illustrated in Figure 3.9 (23) for the case of isotropic point sources of photons in low-Z material. At small distances from a source the contribution of scattered photons can be neglected. At larger depths, and particularly for primary photon energies around 100 keV, the energy deposition of scattered photons is dominant.

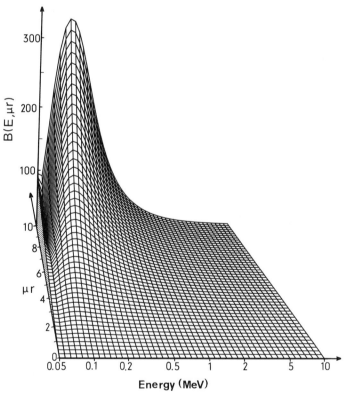

Figure 3.9. Buildup factors for isotropic point sources in air as functions of photon energy and radial distance (in units of mean free paths) (23).

3.3. INTERACTIONS OF NEUTRONS

3.3.1. Interaction Processes

Whereas photons mainly interact with the electrons of molecules, ejecting electrons as charged secondary particles, neutrons collide with atomic nuclei, leading to recoil atomic ions and products of nuclear reactions as charged secondary particles. These reaction processes can be divided into elastic scattering (the most important interaction in this context) producing recoil ions, inelastic scattering (leading to an excitation of the struck nucleus and to a slower neutron), nonelastic scattering (ejecting another particle out of the hit nucleus), neutron capture processes (which are important for slow neutrons and lead to their disappearance), and spallation processes (of importance at neutron energies above, say, 10 MeV) (17b, 24a, 26).

In hydrogen-containing media such as water and biological substances, the most important interaction is elastic scattering with hydrogen nuclei, accounting for typically more than 90% of the energy transferred (see Figure 3.51). Hydrogen nuclei are protons. Neutron collisions produce a wide spectrum of recoil proton energies from zero to the neutron energy.

3.3.2. Cross Sections

The contributions of the various neutron interaction processes to neutron track structures depends on the neutron energy and on the relative abundances of the elements in the irradiated object. Figure 3.10 shows the elastic and inelastic neutron scattering cross sections for hydrogen, carbon, nitrogen, and oxygen. Cross sections for other elements can be found elsewhere (24). The smooth part of the cross section curves represents the contribution of elastic scattering, which is considerably larger for hydrogen than for the other atoms. Therefore, the abundance of hydrogen in an object is of great importance for neutron track structure. The structure in the carbon, nitrogen, and oxygen curves in Figure 3.10 at neutron energies around 1 MeV indicates inelastic collisions with these nuclei, leading usually to the emission of γ rays.

As with photons, the kinetic energy of the charged secondary particles produced by the interactions of neutrons is of main importance in determining the track structure. From energy and momentum conservation in the hard-sphere approximation, it can be shown (17, 18a) that, to a good approximation, the energy spectrum of elastically recoiled nuclei is a rectangular distribution from zero to the maximum energy E_{max} that can be transferred from a neutron of energy E_n to a nucleus of mass A:

$$E_{max} \approx E_n \frac{4A}{(A + 1)^2} \tag{3.6}$$

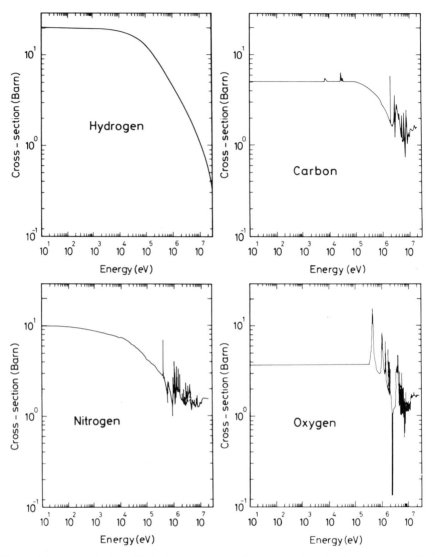

Figure 3.10. Total (elastic and inelastic) neutron scattering cross sections for selected elements (24).

This flat distribution is seen in the initial secondary particle spectra shown in Figure 3.11. The proton contribution around 1.5 MeV for 1-MeV neutrons stems from the (n, p) reaction in nitrogen. The structure in the spectra for 14-MeV neutrons is due to contributions from inelastic and nonelastic processes.

The total and differential cross sections for neutron interactions generally are not as well known as those for photons; this holds particularly for neutron energies above 20 MeV. However, for low to medium-high energy neutrons, which includes most cases of practical importance in radiation research, elastic scattering

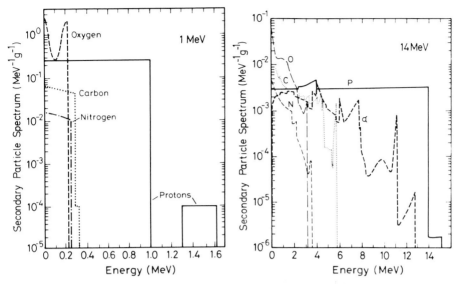

Figure 3.11. Secondary particle start spectrum for 1- and 14-MeV neutrons (per 1 n/cm^2) in ICRU tissue (25).

dominates the energy absorption from neutrons, and these elastic cross sections are sufficiently well known for track structure calculations.

3.3.3. Energy Deposition

Energy deposition by neutrons is determined by the locations of energy transfer and by the type and energy of secondary particles produced. The maximum range of secondary charged particles resulting from elastic or other neutron collisions (Figure 3.12) is small compared to mean distances between subsequent neutron collisions (typically centimeters in low-Z media). Therefore, as with photons, the tracks of secondary charged particles from the same neutron are not likely to overlap. Thus, they can be analyzed separately. This simplification can be used up to doses about 10 times higher than for photons, since the ranges of recoil heavy ions are shorter than those of Compton electrons.

Although most of the energy deposition by fast neutrons in hydrogenic material is via recoil protons, neutron track structures are more complex than those for photons. Proton track structures at low energies are not known because relevant cross sections are lacking. In addition, there are small amounts of low-energy heavy ions, mainly carbon, nitrogen, and oxygen, for which there are essentially no cross sections (Section 3.5). Furthermore, neutrons are often accompanied by energetic photons, either from beam production or from inelastic interactions in the irradiated object. The interaction of these photons in the volume of interest may also have to be taken into account. This mixture of various heavy charged

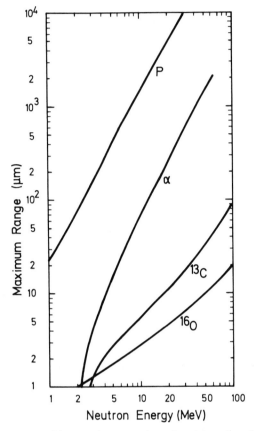

Figure 3.12. Maximum range of the most important charged particles released in tissue by neutrons [p = recoil protons, α and ^{13}C from ^{16}O $(n, \alpha)^{13}$C, ^{16}O from ^{16}O $(n, n')^{16}$O] (26).

particles, electrons, and photons and the corresponding mixture of physical events makes analysis of neutron radiation effects in terms of track structure theory more difficult than for photons or monoenergetic, fast charged particles. This may be the reason why rigorous neutron track structure calculations have not been performed yet. The deficiencies of analytical approximations (25, 27) and of semi-Monte Carlo approaches (28) underline the need for more detailed calculations.

3.4. INTERACTIONS OF ELECTRONS

3.4.1. Interaction Processes

Electrons are of prime importance in the production of track structures. Essentially all energy from photons or primary high-energy electrons is transferred to and transported in irradiated matter by secondary and higher generation electrons. The same is true for a large fraction of the energy lost by fast ions and thus also for neutrons. Because of this importance, much has been written on various aspects of electron interactions with matter (7c, 9a, 29–33), which can be consulted for details. Here a survey of the general characteristics of interactions of low- to medium-energy electrons is given.

The kinetic energy of an electron is mainly transferred to matter through interactions of the electric field of the moving electron with that of electrons bound in the medium. The interactions lead primarily to electronic excitations and ionizations rather than to direct excitations of rotational, translational, or vibrational modes of the molecules affected. It is only at residual energies below, say, 10 eV that the latter modes contribute significantly to electron energy loss. The fate of such subelectronic excitation energy electrons is treated in Chapters 2 and 4–6, so more emphasis is given here to the energy range above 10 eV.

Electron interactions leading to ionization of a molecule (a) cause the ejection of another electron (which in turn might be able to ionize another molecule); (b) usually leave the molecule in an excited state that can decay by dissociation, Auger electron emission, photon emission, and so on; and (c) involve relatively large energy transfers. By contrast with the photoelectric effect, which for high-energy photons is dominated by atomic electrons bound in inner shells, excitation or ionization by electrons occurs preferentially in outer shells. Compared to ionizations, electronic excitations are usually of minor importance, as is the production of bremsstrahlung (X rays) in the strong field of an atomic nucleus (except for high-energy electrons in high-Z media). However, energy transfers of 15–25 eV can produce so-called superexcited states of molecules (9b, 34), the internal energy of which is larger than the first ionization potential and which can decay by autoionization or dissociation into neutral fragments:

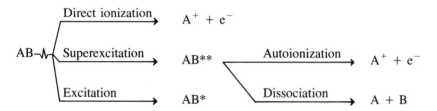

The existence of such states causes the ionization efficiency to be less than unity for energy transfers slightly above the first ionization potential, but it increases gradually to unity at higher energies (say, 25 eV).

Elastic collisions also contribute to electron track structure. Though not leaving a mark in the irradiated matter themselves, they influence the location of the next inelastic event. Elastic collisions gain importance with decreasing electron energy, and below, say, 200 eV, they lead to a very diffusive behavior of electron paths. Because of their large cross sections, electrons in this energy range contribute significantly to the yield of new molecular species.

3.4.2. Cross Sections

For a rigorous calculation of charged particle track structures a set of absolute elastic, excitation, and ionization cross sections is needed for the molecules of the medium.

The ionization cross sections would ideally be at least triply differential, giving the probability of energy transfer, scattering angle, and emission angle. Furthermore, they would contain information about the atomic level affected for estimation of the remaining local binding energy. One also needs information about the rearrangement reactions of the molecular species resulting from the interaction. Complete and accurate data sets are not yet available for any material. However, useful estimates for such cross sections can be derived from singly or doubly differential cross sections, from integral data (such as energy loss or backscattering spectra and the average amount of energy absorbed per ionization produced), from data on low-energy photon absorption, and from theoretical considerations. Estimates for condensed media are usually extrapolated from gas phase data (water vapor, hydrogen, hydrocarbons, rare gases). For electrons a large literature exists on experimental and theoretical cross sections (29–31, 35–37) and on data useful for the derivation of badly known cross sections (7a, 13, 33, 38, 39). General properties of such cross sections and principles for the derivation of badly known cross sections from other pertinent data are indicated below for water vapor.

From photoabsorption (39a) and mass spectroscopy (36b) data the differential oscillator strength spectrum df/dE of a medium can be derived:

$$\frac{df}{dE} = \frac{\sigma}{4\pi\alpha a_0^2 R} \tag{3.7}$$

where σ = photoabsorption cross section
$\quad \alpha$ = fine structure constant
$\quad a_0$ = Bohr radius (5.3×10^{-11} m)
$\quad R$ = Rydberg energy (13.6 eV)

Figure 3.13 shows such a spectrum for water vapor depicting the excitation and ionization part, the contribution of various shells to ionization, and the effect of superexcited states on the ionization efficiency η_i between 12.6 eV (the first ionization potential) and 22 eV. Analysis of photoabsorption and electron scattering data and theoretical considerations about the energy levels of the molecule under consideration (37a) are necessary to derive elastic and inelastic electron scattering cross sections (Figure 3.14) and to check their internal consistency and consistency with theoretical constraints (7a, 38, 41).

A consistency test is indicated in Figure 3.15 (42). According to the Bethe theory (38), total inelastic scattering cross sections for high electron energies T behave asymptotically as (13)

$$\sigma(T) = AT^{-1} \ln T + BT^{-1} + CT^{-2} + \cdots \tag{3.8}$$

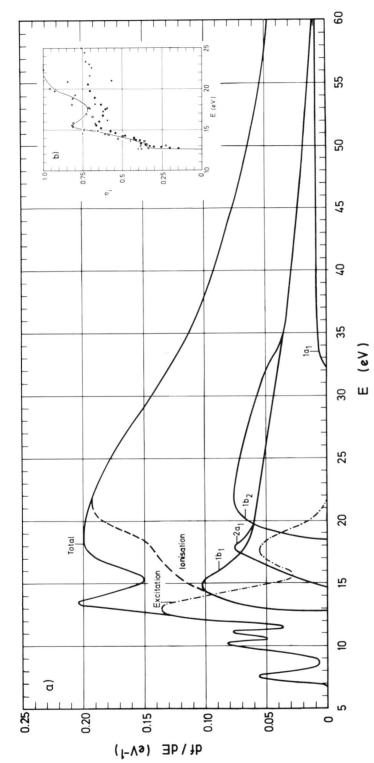

Figure 3.13. Differential optical oscillator strength distribution for water vapor (a) showing the contributions of excitations and of ionizations (for the four outer shells), and the ionization efficiency of photons (b) above the lowest ionization potential.

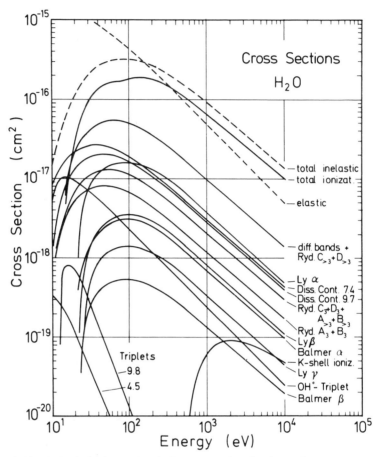

Figure 3.14. Elastic, ionization, and excitation cross sections for electron impact on water vapor.

where A, B, and C are constants determined by properties of the medium. The quantities B and C describe the probability of energy transfers in hard (close) collisions. The quantity A is proportional to the dipole oscillator strength divided by the particular transition energy ($A \propto f_n/E_n$), which expresses the importance of glancing (distant) collisions for inelastic cross sections at high energies T. The electron interaction is similar to that of photons that have a "white" energy spectrum (13). The mathematical relationship between σ and T led Fano (41a) to plot σT versus $\ln T$, which should give a straight line at high energies with the slope A (Figure 3.15). Such a Fano plot represents a means to check the consistency of cross-sectional data. For electron energies above, say, 100 eV, the ionization cross section is significantly larger than the sum of all excitation cross sections (Figure 3.15b).

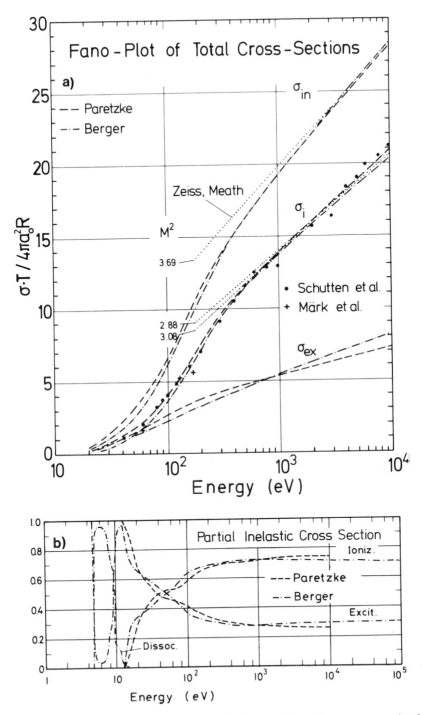

Figure 3.15. (*a*) Fano plot of total inelastic σ_{in}, ionization σ_i, and excitation σ_{ex} cross section for electron impact on water vapor; parmeters are dipole matrix elements squared derived by various authors. (*b*) Contributions of ionizations and excitations to the total inelastic cross section (40).

Electron interaction cross sections have a maximum because of their $T^{-1} \ln T$ shape, and it usually occurs near 100 eV. This leads to a pronounced minimum in the mean free path (the expectation value of the distance between subsequent collisions) for inelastic events near 100 eV (Figure 3.16). Together with the approximate $1/E^2$ spectrum of secondary electrons produced in ionizations, this leads to an approximate $1/E$ start spectrum of all secondary and higher order electrons produced during the complete slowing down of a high-energy electron. This behavior is also responsible for the fact that, even for 100 keV electrons, half of all ionizations are produced by secondaries with momentary kinetic energies below 1 keV. The latter have relatively small distances between two collisions compared to the much larger distances between events of a 100 keV primary electron (Figure 3.16). This is an important feature of the track structure of electrons for chemical yield calculations.

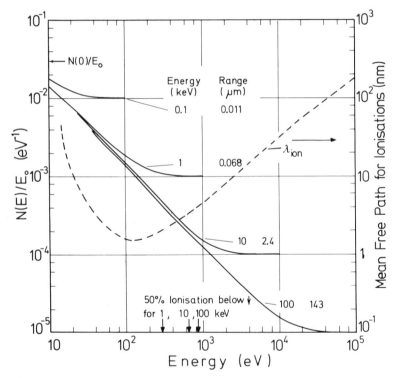

Figure 3.16. Mean number $N(E)$ of electrons with start energies larger than E produced in water divided by the primary electron energy E_0 for $E_0 = 0.1$, 1, 10, and 100 keV (with the respective csda ranges). The dotted line shows the mean free path between primary ionizations as a function of electron energy. Arrows indicate the median electron energies in the slowing-down spectrum, below which 50% of all ionizations due to a primary 1, 10, or 100 keV electron are produced. csda means continuous-slowing-down approximation.

The kinetic energy spectra of secondary electrons is important for radiation track structure theory. Unfortunately, not many absolute experimental data exist for such cross sections (43–46).

Figure 3.17 shows measured doubly differential cross sections for 500 eV electrons in water vapor from Opal et al. (43). They were used with the dipole oscillator strength distribution given in Figure 3.13 to derive by a Platzman–Kim-type approach (39b, c) doubly differential cross sections for other electron energies. In the first Born approximation the singly differential cross section for the production of secondary electrons of energy E is

$$\frac{d\sigma}{dE} = \frac{4\pi a_0^2}{T} \sum_i \left[\frac{df}{dQ_i} \frac{R^2}{Q_i} \ln \left(\frac{4TRC(Q_i)}{Q_i^2} \right) + B(T, Q_i) \right] \qquad (3.9)$$

where $Q_i = E + U_i$ = energy transfer
  U_i = binding energy in ith orbital
  df/dQ_i = dipole oscillator strength density for ionization with energy transfer Q_i

The function $C(Q_i)$ depends only on the target molecule, not on the incident particle. The first term in the square bracket describes the contribution to the cross section of glancing (distant) collisions. It dominates at low secondary electron

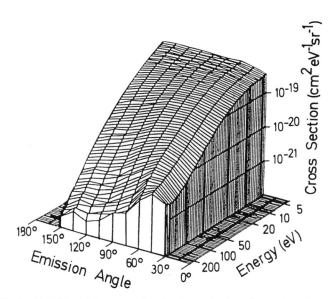

Figure 3.17. Double differential cross section for the production of secondary electrons by 500-eV electron impact on water vapor (43).

energies but falls off at large energy transfers since $Q^{-1}(df/dQ)$ is typically proportional to $Q^{-4.5}$ and thus is of no importance in producing high-energy secondary electrons. Therefore, at low electron energies the shape of the secondary electron spectrum should be similar to the shape of $Q^{-1}(df/dQ)$. This feature is useful for the derivation of electron spectra from photoabsorption data. To display this property, Platzman and Kim plotted the ratio of the actual cross section to the Rutherford cross section $(d\sigma/dE)(Q/R)^2(T/4\pi a_0^2)$ versus R/Q (Figure 3.18). The second term in Eq. (3.9) represents both glancing and hard (close) collisions. It becomes increasingly independent of T with increasing primary energy T, and the total cross section at high secondary electron energies approaches the well-known Mott cross section (39b, c):

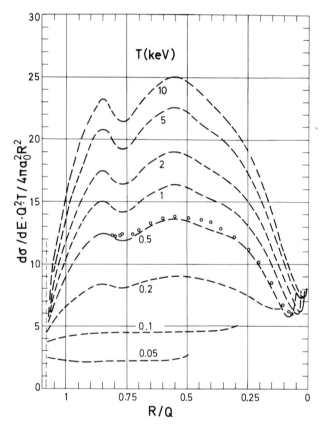

Figure 3.18. Platzman plot of single differential secondary electron cross sections for water vapor as a function of R/E (R = Rydberg energy, E = secondary electron energy) for various primary electron energies; the experimental data points are from Ref. 43. Vertical line indicates ionization potential.

$$\frac{d\sigma}{dE} = \frac{4\pi a_0^2 R^2}{T} \sum_i N_i \left[\frac{1}{Q_i^2} + \frac{1}{(T-E)^2} - \frac{1}{Q_i(T-E)} \right] \tag{3.10}$$

where N_i is the electron occupation number in the ith orbital. In a Platzman plot (Figure 3.18) the knock-on contribution is a rather flat curve at an ordinate value approximately equal to the number of participating molecular electrons. The excess over this almost constant value is proportional to $Q^{-1}(df/dQ)$, and it increases with ln T. The area under a part of the total curve is proportional to the contribution of that part to the total ionization cross section (39). These features can be seen in Figure 3.18, which shows experimental data for 500 eV electrons incident on water vapor (43) and cross sections theoretically derived from the dipole oscillator strength distribution for ionization shown in Figure 3.13. The increasing importance of glancing collisions with increasing primary electron energy T is evident, as well as the asymptotic approach of these cross sections at high secondary electron energies to the Mott cross section.

For further details on electron scattering cross sections the references cited should be consulted.

3.4.3. Energy Deposition, Yields and Ranges

The starting point for a study of the energy deposition of electrons is their stochastic track structure (Section 3.6.3). Input data for track structure calculations are cross sections for all relevant interactions. Averaged quantities such as stopping power (the differential rate of energy loss along a particle path), range (some expectation value or extrapolated value of the length of a particle track), or yields (expectation values of the production of primary, intermediate, or long-lived new species) are also useful in radiation research.

The stopping power dE/dx of a medium can be considered as the sum of all single energy loss processes:

$$\frac{dE}{dx}(T) = N \left[\sum_i Q_i \sigma_i(T) + \sum_j \int_0^\infty Q \frac{d\sigma_j}{dQ} dQ \right] \tag{3.11}$$

where N = number density of molecules in the medium
  Q_i = energy transferred to a molecule in process i
  σ_i = cross section for discrete collision process i
  $d\sigma_j/dQ$ = differential cross section for jth transition into the continuum

Because of this correlation to single processes, a stopping power value known to high accuracy from theory or experiment can be used as a constraint in the estimation of less well known cross sections (in particular of secondary electron spectra). Such a comparison is shown in Figure 3.19, where stopping power values

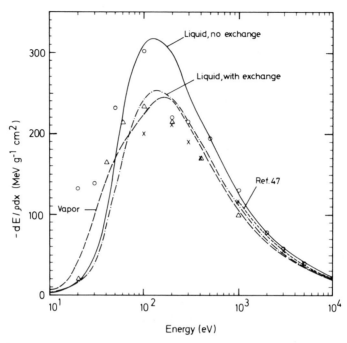

Figure 3.19. Stopping power of electrons in water (‑ ‑, vapor; ———, liquid without exchange term; ‑ · ‑, liquid with exchange term; ○, Ref. 48; △, Ref. 49; ×, Ref. 50; adapted from Ref. 53).

derived by addition of single loss processes for electron impact on water vapor (49, 50, 53) are compared to analytical values derived from the Møller cross section (47, 48, 51). In the latter approach all the combined action of many energy loss channels of a target molecule can be described by its *mean excitation potential* I (47):

$$\frac{dE}{\rho\, dx} = 0.1535 \frac{1}{\beta^2} \frac{Z}{A} \left[\ln\left(\frac{T}{I}\right)^2 + \ln\left(1 + \frac{\tau}{2}\right) + F^-(\tau) - \delta \right] \quad (3.12)$$

$$\ln I = \frac{\displaystyle\int_0^\infty (df/dE)\, \ln E\, dE}{\displaystyle\int_0^\infty (df/dE)\, dE}$$

$$F^-(\tau) = (1 - \beta^2)\left[1 + \tfrac{1}{8}\tau^2 - (2\tau + 1)\ln 2 \right],$$

where $\rho^{-1}(dE/dx)$ = electron collision stopping power per unit mass (MeV cm²/g)

β = electron velocity divided by speed of light c

T = electron kinetic energy
τ = electron kinetic energy divided by its rest energy
δ = Fermi density effect correction (52)
df/dE = dipole oscillator strength

Since T is approximately proportional to β^2, Eq. (3.12) has an approximate T^{-1} · ln T shape at low electron energies (say, below 10 keV). Figure 3.19 displays acceptable agreement between different estimates of stopping power above a few kilo-electronvolts and considerable uncertainty below a few hundred electronvolts.

Above a few hundred electronvolts the kinetic energy transferred to ejected secondary electrons is the largest part of the stopping power (Figure 3.20). The energy going into discrete excitations is only about 10% of that going into ionizations (40). The mean energy of a neutral excited state produced in water vapor (as a typical low-Z material) is about 13 eV independent of incident electron energy (Figure 3.21). The mean potential energy of the molecule after ionization is around

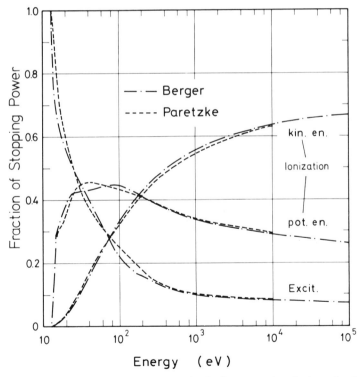

Figure 3.20. Fraction of electron stopping power that leads to neutral excitations, that is used to overcome the ionization potential of molecules, and that appears as kinetic energy of ejected electrons (40).

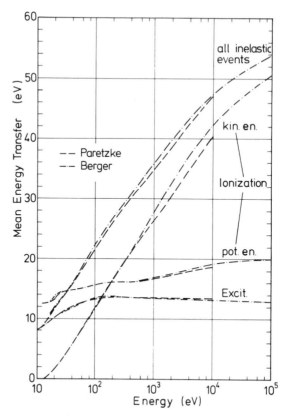

Figure 3.21. Mean energy transfer in collisions of electrons with water molecules leading to excitations or emission of secondary electrons (40).

15 eV for projectile energies below the K shell ionization energy (545 eV in oxygen) and increases to 20 eV for projectile energies well above it. The increase of the maximum transferred energy in an ionizing collision with increasing primary electron energy leads to the observed increase of the mean kinetic energy of secondary electrons (40).

It is useful to introduce here the so-called electron degradation spectrum $Y(E, E_0)$, which gives the sum of the path segment lengths where the degraded primary electron (start energy E_0) and all secondary electrons produced during complete slowing down have an energy between E and $E + dE$ (54a,b):

$$Y(E, E_0) = \frac{N(E, E_0)}{dE/dx} \qquad (3.13)$$

where $N(E, E_0)$ is the absolute number of all secondary and higher order electrons created by complete slowing down of an electron of incident energy E_0, which have kinetic start energy larger than E, plus 1 for the primary electron. The electron degradation spectrum increases steeply with decreasing electron energy (Figure 3.22a), because of the increase in the number of low-energy secondary electrons and the decrease of the stopping power at energies below 100 eV (Figure 3.19).

Figure 3.22b gives degradation spectra for several photon fields and very fast electrons (21). It shows the strong distortion of electron energy spectra during slowing down from their clean Compton- and photo-electron spectra produced by the initial photon interactions (Section 3.2).

Degradation spectra have interesting scaling properties (54c) and are useful for the calculation of the yield G_i of any species i:

$$G_i(E_0) = \int Y(E_0, E)\sigma_i(E) \, dE \qquad (3.14)$$

where $\sigma_i(E)$ is the production cross section of species i by an electron of energy E. Yields (for example of ions, electrons, or light-emitting excitations) produced during the complete slowing down of a primary particle can be measured to a much higher accuracy than total or differential cross sections for single collisions. Therefore, yields represent a means to check predictions of track structure calculations against experiments and thus to improve iteratively on relevant cross sections. Figure 3.23 shows the result of such a calculation for water vapor. The large importance of ionizations at high energies and of excitations below, say, 200 eV is evident. An average of 14 eV are spent per inelastic event ($G_t \approx 7$ events per 100 eV).

Yields, as defined in Eq. (3.14), are expectation values averaged over many primary electron tracks. In reality, however, each track will produce a somewhat different number of new species than another track. In many cases, because of the correlation between energy loss and species production and because of the finite energy available, the frequency distributions of the number of new species is narrower than a Poisson distribution (55). The ratio of the variance of the observed distribution to that of a Poisson distribution is often called the *Fano factor*. Comparison of calculated and experimental yield distributions can also be a sensitive check of consistency of single-collision cross sections (55b). Figure 3.24 gives distributions of ionizations produced in the tracks of electrons in liquid water and in water vapor. From the cumulative distribution in the right-hand panel it appears that such distributions are essentially Gaussian over a wide parameter range. Their mean values (Figure 3.25) show that at high electron energy more ionizations are produced in the liquid and that the outermost electron orbits contribute most to ionization (contrary to irradiation by photons). The calculated total ionization yields for the vapor are in excellent agreement with experimental data (54d). The lower part of Figure 3.25 shows that (a) the frequency distributions for the total ionization yields are smaller in the liquid and (b) the respective distributions for each

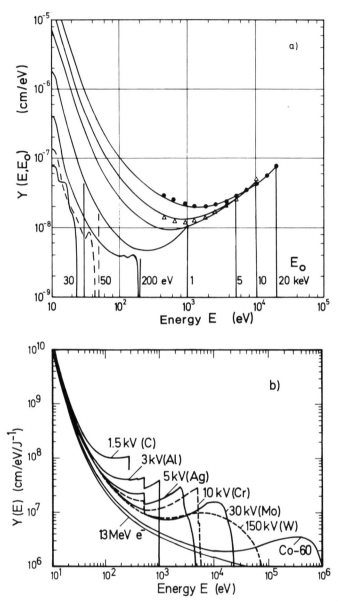

Figure 3.22. Degradation spectrum of electrons (*a*) (own results) and of various low-LET radiation fields (*b*) (adapted from Ref. 21) in water vapor.

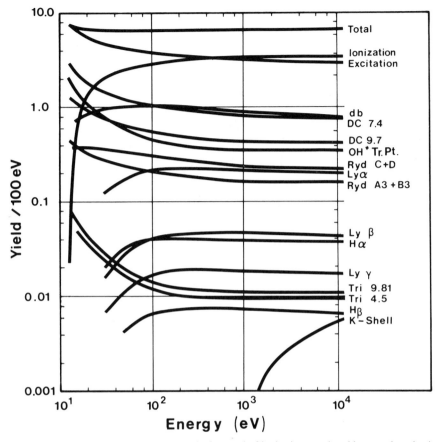

Figure 3.23. Yields of various primary excitations and of ionizations produced by complete slowing down of electrons in water vapor as a function of incident energy.

single shell are much broader than that for all ionizations (i.e., their Fano factor is larger).

Finally some spatial aspects of electron energy deposition shall be considered, since the spatial correlation of all initial events influences the subsequent physical and chemical reactions. The expectation values of electron energy deposition show a number of typical features, some of which are depicted in Figure 3.26 (5 keV electrons vertically incident on a slab of water). The number of transmitted electrons decreases strongly with increasing thickness (compared to similar plots for heavy charged particles, which show almost a step function). However, the kinetic energy per transmitted electron decreases much more slowly. At a thickness equal to their range (obtained in the continuous-slowing-down-approximation [csda, (47)]

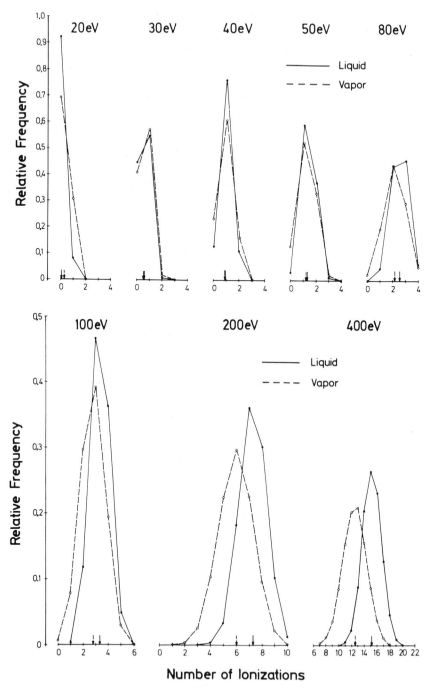

Figure 3.24. Ionization frequency and sum distributions (probability graphs) for complete slowing down of electrons in water vapor (– – –) and in liquid water (——); adapted from Ref. 53. Arrows indicate respective mean values. A straight line on a probability graph indicates the sum of a Gaussian distribution.

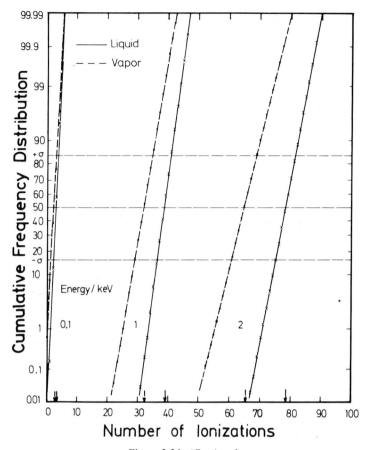

Figure 3.24. (*Continued*)

around 1% of the incident energy and around 5% of the incident number of electrons are still transmitted. With increasing thickness electrons are also reflected (backscattered). Their number fraction at medium energies and for vertical incidence on low-Z matter is around 10%, and the mean energy per reflected electron is around 40–50% of the incident energy. The reflected number is independent of thickness beyond $d/R = 0.4$ because electrons are not appreciably reflected back to the surface from greater depths.

The transmission curves indicate that there are many possibilities for definition of an electron "range" (csda, mean, practical, extrapolated, and so on) (21). The range becomes even less well defined at low energies (≤ 100 eV), where electrons are more or less in complete diffusion, or at subexcitation energies, where electrons can diffuse over long distances because of the reduced number of energy loss

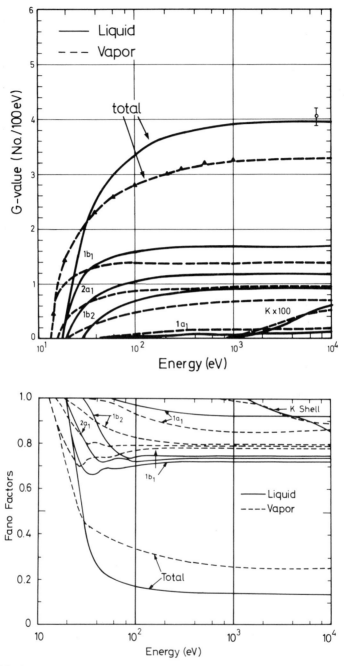

Figure 3.25. *G*-values for ionization from different molecular levels and corresponding Fano factors for electron impact on water vapor (– – –) and liquid water (———) (adapted from Ref. 53).

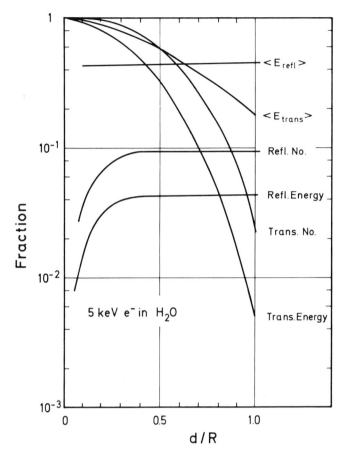

Figure 3.26. Calculated fractions of reflected and transmitted electrons and their energies as a function of slab thickness (in units of the electron range). Vertical incidence, pencil beam, and water vapor. See Figure 3.27 for values of R.

processes (Chapters 2, 5, and 6). Figure 3.27 shows some of these ranges calculated by a Monte Carlo electron track structure program (MOCA-8) (21, 40, 53) in comparison with data from a new ICRU Report (47) and an empirical range–energy relationship for low-Z matter ($\rho = 1$ g/cm^3), which gives reasonably accurate values up to, say, 20 keV:

$$R = 40\,E(1 + 0.5\,E) \quad \text{(nm)} \tag{3.15}$$

where E = electron energy in keV.

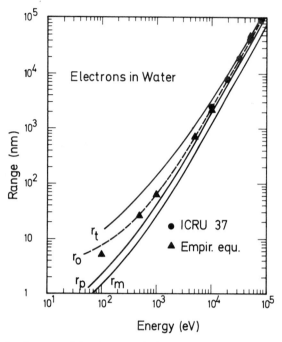

Figure 3.27. Calculated mean track length r_0, practical range r_p, mean range r_m, and total track length r_t for electrons of energy E_0 in water vapor; data from Refs. 21 and 47 and present results.

3.5. INTERACTIONS OF HEAVY CHARGED PARTICLES

3.5.1. Interaction Processes

Energy transfer from fast charged particles to matter occurs primarily through excitations and ionizations. For ions with specific energy E per mass in the range 0.5–100 MeV/u, around 65–75% of the energy lost is transferred to and then transported by secondary electrons, 15–25% is needed to overcome their binding potential, and the residual 5–10% produces neutral excited species. The large fraction transferred to secondary electrons emphasizes the importance of electrons also in this context. For 0.8 MeV/u particles around half of all ionizations are produced by the fast ion itself, and the rest by its secondary electrons. With increasing ion energy two out of three ionizations are ultimately due to secondary electrons and are thus not necessarily located close to the ion path.

There are two new interaction processes for fast ions, compared to electrons. First, heavy charged particles of medium energy (say, 0.1–5 MeV/u) can capture electrons from target molecules into their own continuum states [charge transfer to the continuum (CCT)], which leads to the ejection of additional secondary electrons. These have essentially the same velocity as the ion and travel in essentially the same direction. Their production cross sections are not well known. Second, heavy charged particles are usually incompletely stripped ions that carry electrons in bound states (for example He^+ and Xe^{5+}) and can successively lose electrons and capture target electrons. This complicates the derivation of relevant cross sections for heavy charged particles.

3.5.2. Cross Sections and Effective Charges

The statements in Section 3.4.2 on the derivation and properties of electron scattering cross sections also hold in principle for fast ions. Many total and differential cross sections for protons and a few other fast ions have been measured or derived from theories (56–65). The important singly differential ionization cross sections (except for the CCT effects and attached electron effects mentioned above) show the same general shape as described for primary electrons in Eq. (3.9) (Figure 3.28). This figure gives experimental data for secondary electron impact on various hydrocarbons and on hydrogen. The cross section (a) decreases with increasing ion energy ($\propto T^{-1}$) and (b) falls at medium energies of secondary electrons approximately as E^{-2}. Also, (c) the maximum electron energy increases linearly with the ion energy ($\propto T$). All this is understood from theory and is similar to cross sections for primary electrons. The features (a) and (c) are mainly responsible for the fact that there cannot exist one single parameter characterizing the "quality" of any radiation field in producing a certain radiation action. The absolute value of these cross sections per weakly bound target electron (i.e., for all electrons except for the K orbital electrons) is almost independent on the chemical nature of the target molecule. The same is true for the angular distribution of the ejected secondary electrons (Figure 3.29).

The magnitude of cross sections of charged particles is proportional to the square of their electric charge z. The classical Rutherford cross section is

$$\frac{d\sigma}{dE} = 4\pi a_0^2 \, R^2 \, \frac{z^2}{T} \frac{1}{Q^2} \tag{3.16}$$

where $T = \frac{1}{2} m_e v^2$
  m_e = electron mass
  v = ion velocity
  Q = energy transfer ($= E + U$)

This relationship is approximately correct for bare nuclei or fast ions with tightly bound electrons, but the supposition that this holds also for fast ions with loosely bound electrons is not correct. Figure 3.30 shows an example of this failure; the

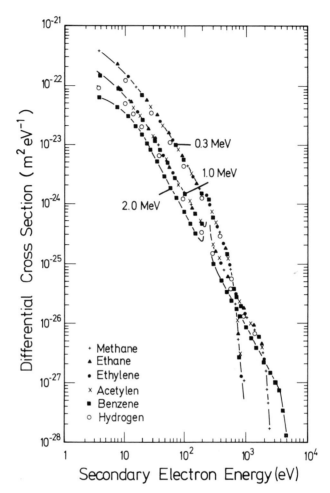

Figure 3.28. Secondary electron ejection cross section for proton impact (0.3, 1.0, and 2.0 MeV) on various target molecules, divided by the number of weakly bound target electrons (56).

ions H_2^+ and H^+ are both singly charged and a constant cross section ratio might be expected, but over a wide secondary electron energy range the contribution of the ejected electron of the H_2^+ molecule shows that such a simple scaling from proton cross sections to heavy-ion cross sections is not obtained. Scaling from proton data to heavier ions with a so-called effective charge z_{eff} (66–69), however, has proven useful in stopping power analysis (see next section). The empirical formula derived by Barkas from nuclear emulsion work (5a) is adequate. His formula is also in agreement with experimental data for the charge state distribution after penetration of heavy ions through gases (Figure 3.31). The higher charge states observed after penetration through solids are most probably due to electron

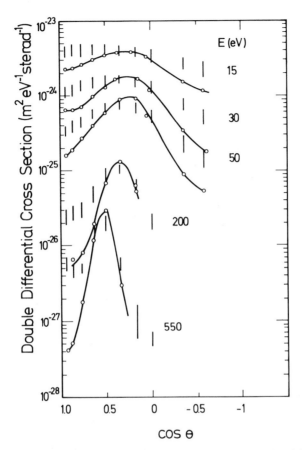

Figure 3.29. Angular distribution of secondary electrons with energies E ejected by 1-MeV proton impact on various molecules divided by the number of weakly bound target electrons. The range of data for hydrocarbons is indicated by vertical slashes (|), data for hydrogen (○) are connected to guide the eye (56).

ejection from the highly excited fast ion after leaving the solid (66a). It might be worthwhile to put more effort into understanding this apparent discrepancy between the meaningfulness of the concept of an effective charge in stopping power and the failure of this concept for those cross sections that essentially determine the stopping power.

3.5.3. Stopping Power, Straggling and Ranges

Tracks of heavy charged particles are often characterized by their mean rate of energy loss and their range. This mean rate, the stopping power, has been the

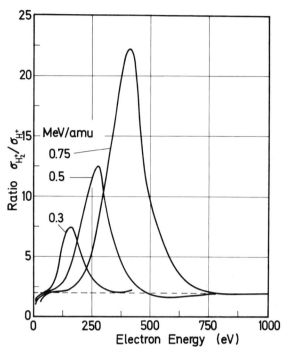

Figure 3.30. Ratio of energy differential cross sections for the production of secondary electrons by H_2^+ and H^+ projectiles of various specific energies impinging on hydrogen gas (57).

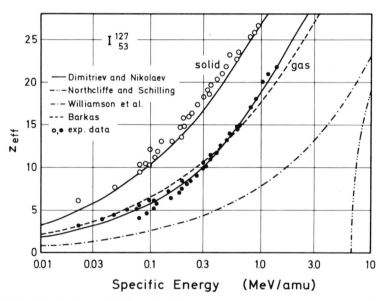

Figure 3.31. Effective charge of iodine projectiles after passing through solid, in gaseous targets, and according to various empirical formulas derived from stopping power analysis [for explanation see text (69)].

object of many theoretical and experimental studies (7a, 38, 70–76), which has led to substantial knowledge about the general underlying principles. However, the accuracy of actual stopping power values for charged particles heavier than α particles and for all particles at low energies is far from satisfactory. In the context of track structure theory it is sufficient to remember the usefulness of this integral quantity for consistency checks as outlined in the section on electrons, the insensitivity of stopping power to the physical phase of the target matter (76), and the principles of the dependency of material and particle parameters in the basic stopping power formula of Bethe (38b, 47):

$$\frac{dE}{dx} = 4\pi r_0^2 \, mc^2 \, \frac{z^2}{\beta^2} \frac{Z}{A} \left(\ln \frac{2mc^2\beta^2}{1 - \beta^2} - \beta^2 - \ln I - \frac{C}{Z} - \frac{\delta}{2} \right) \qquad (3.17)$$

where all parameters have the same meaning as above, and C/Z represents shell corrections (47) that need not be discussed here. Equation (3.17) is similar to Eq. (3.12).

The material properties enter essentially only with Z/A (≈ 0.5) and $\ln I$, and the dependency on the particle parameters is approximately as β^{-2} (which is sim-

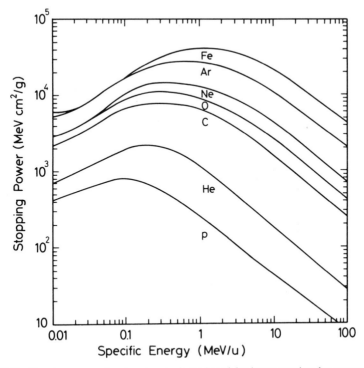

Figure 3.32. Stopping power of various heavy charged particles in water using the program of Ref. 74.

ilar to the case of electrons) and proportional to z^2 (Figure 3.32). Therefore, particles of the same velocity have different stopping powers according to their z^2, whereas particles with the same stopping power and different z have different velocities (with the consequences for their secondary electron cross sections mentioned in Section 3.5.2). Because of the similarity in the z and β dependences of cross sections and stopping power, an apparent correlation over a restricted parameter range of the size of a radiation effect with stopping power could in reality be due to an actual correlation with one or more important basic interactions since the cross sections for these have the same dependency.

Beyond the considerations outlined above, the applicability of stopping power in track structure research is rather limited for several reasons. First, the number of collisions and the energy transferred in them are random variables with certain

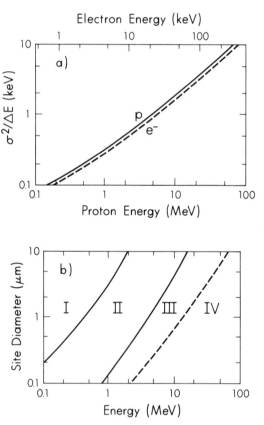

Figure 3.33. (*a*) Ratio of variance of energy loss spectrum to mean energy loss for protons and electrons in the free electron model. (*b*) Regions of site diameters and proton energies where the energy deposition is strongly influenced by their finite range (I), straggling (III + IV), and by the range of their delta rays (IV). In region II the energy deposition can be approximated by dE/dx. (79).

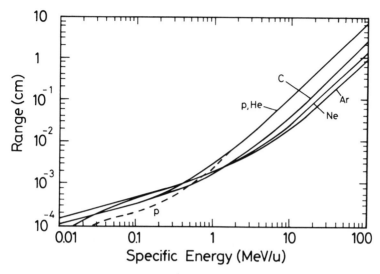

Figure 3.34. Ranges of protons and some heavier ions in water (calculated from Ref. 74).

probability densities. Therefore, the actual energy loss in a small region of interest can vary considerably (77), and the physical and chemical reactions do not "know" of average values but "feel" only actual values. This variation in actual energy loss is shown in Figure 3.33 (78, 79), which shows the variance σ^2 of energy loss in thin absorbers divided by the expected energy loss ΔE (stopping power times pathlength) for electrons and protons. For thin absorbers this straggling ratio is independent of absorber thickness. Second, secondary electrons can acquire rather high energy and form tracks of their own physicochemical importance. Third, a site of interest might be larger than the finite range of a particle track structure; this makes considerations of mean rates of energy loss useless in some cases. The regions of influence of these various aspects are illustrated in Figure 3.33 (76, 79).

Figure 3.34 gives data for the csda ranges of heavy charged particles in water, which can be used to estimate the longitudinal extensions of their track structures. The ranges of these particles are orders of magnitude shorter than those of electrons of the same kinetic energy.

3.6. SPATIAL ASPECTS OF TRACK STRUCTURES

3.6.1. Computational Methods of Track Structure Simulation

Single quantities like absorbed dose, energy imparted to a target site, and stopping power do not accurately predict physical, biological, or chemical radiation effects. Track structure theories give fuller information about spatial and temporal aspects of consequences of irradiation. Track structure calculations require adequate cross sections for the relevant processes in the material under consideration. The generation, synthesis, and testing of cross sections consumes most of the time spent on a problem. Once a satisfactory data base has been derived, one can use Monte Carlo techniques to simulate charged particle track structures with a computer, event by event.

The cross sections have to be processed in a way that later permits speedy sampling from the probability distributions. To this purpose, it is useful to derive and store the mean free paths between subsequent collisions for primary and secondary particles. For the same reason it is appropriate to precalculate for all relevant processes the ratios of their cross sections to the total cross section. The cumulative distribution function $F(i)$ of the discrete density of these ratios $f(i)$ can then be stored and used directly to sample from it (with the help of a random number R) the type $i(R)$ of interaction at the next collision (Figure 3.35) such that

$$\sum_{i=1}^{j-1} f_i = F_{j-1} \leq R < \sum_{i=1}^{j} f_i = F_j \tag{3.18}$$

with R uniform on $[0, 1)$, where the bracket term indicates $0 \leq R < 1$ (81). For a continuous variable x (distance to next event, energy of secondary electron, scattering angle, and so on) the following equation has to be solved for x:

$$R = \frac{\displaystyle\int_0^x f(x')\, dx'}{\displaystyle\int_0^\infty f(x')\, dx'} \tag{3.19}$$

In both cases random numbers are needed, the use of which are the main characteristics of Monte Carlo calculations. The generation of pseudorandom numbers on digital computers is discussed elsewhere (80); details of sampling techniques, acceleration techniques (importance sampling, biasing, and so on) also can be found elsewhere (81).

The probability density for the distance to the next collision along the line of flight between s and $s + ds$ is, for any particle,

$$p(s) = \frac{1}{\lambda} e^{-\lambda s}\, ds \tag{3.20}$$

where λ is the mean free path between collisions (expectation value of distance between two succeeding events). This leads to the sampling scheme of the distance s to the next event,

$$s = -\lambda \ln R \tag{3.21}$$

with R uniform in $[0, 1)$.

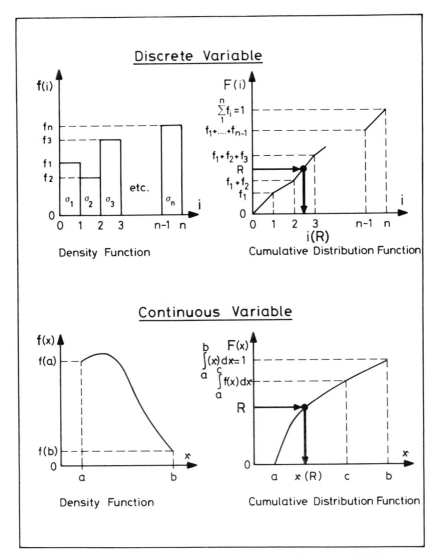

Figure 3.35. Principles of random sampling from discrete and continuous distributions.

During an interaction a photon, neutron, or charged particle can be scattered into a new direction. This can be taken into account by:

1. Using appropriate coordinate frames, namely one fixed Cartesian coordinate frame at a meaningful point of reference and local spherical coordinate frames with moving origins at the actual points of collisions and with their polar axis coincident with the path of flight (Figure 3.36).

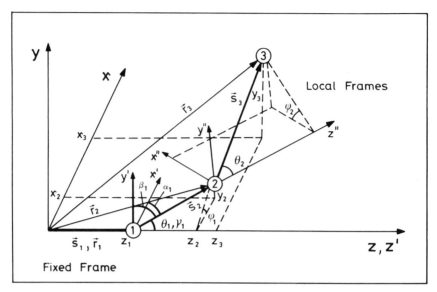

Figure 3.36. Coordinate systems used in Monte Carlo track structure calculations and relevant direction cosines.

2. Transformation by the Euler angle transformation of the direction cosines after scattering in the local frame into new direction cosines in the fixed frame (82, 83).

With the flight distance s to the next event determined by Eq. (3.21), this gives the Cartesian coordinates of the event in the fixed frame for bookkeeping and evaluation; then the next event can be calculated, and so on. These are the computational principles of Monte Carlo radiation track structure calculations.

This technique has been successfully applied to many aspects of radiation transport [photons (82), neutrons (83), charged particles (84), nucleon–meson cascades (85), and nonlinear radiation transport (86)] and to detailed track structure calculations (see below).

3.6.2. Characteristics of Tracks from Photons and Neutrons

Photons and neutrons in most cases produce primary events separated by distances that are large compared to the ranges of their secondary charged particles. This is illustrated in Figure 3.37 for water and air. It can also be seen by comparing the particle ranges given in Figures 3.12, 3.27, and 3.34. Therefore, the track structures produced by primary electrons, protons, and heavier ions are similar to those produced by photons and neutrons. However, with the latter two radiations more than one particle sometimes leaves the same affected atom, for example, because of Auger electron emission after inner shell ionization, or an $(n, 3\alpha)$ reaction.

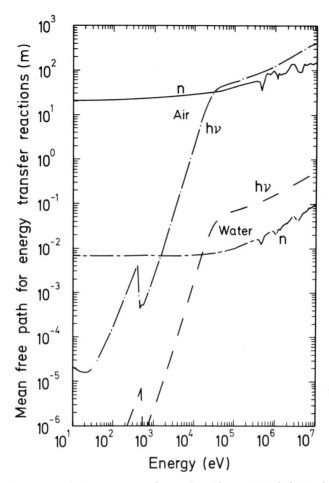

Figure 3.37. Mean free paths for energy transfer reactions of neutrons and photons in dry air and water (24, 87).

Figure 3.38 shows another characteristic of photon and neutron track structures (88). The ranges of secondary particles from neutrons (mainly protons and heavier charged particles) are much shorter than those from photons (electrons). Therefore, for the same absorbed dose neutrons have a smaller chance to affect a certain small site of interest in the irradiated matter. But if the small site happens to be affected, the amount of energy imparted to that site in this interaction is much larger (because of the higher stopping power of ions compared to that of electrons). This leads to a higher expectation value of the specific (imparted) energy $\bar{z}_n$ for neutrons than for photons $\bar{z}_\gamma$ per energy deposition event. At a fixed absorbed dose, for the same reason, the standard deviation σ of the distribution of z is much broader for neutrons than for photons (Figure 3.38). The determination of such distributions

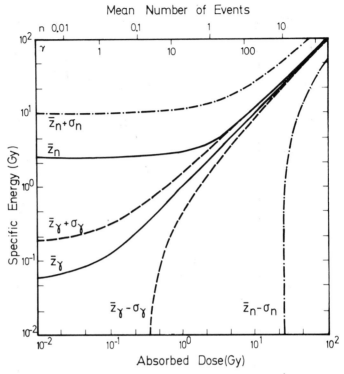

Figure 3.38. Range of specific energies (mean values $\bar{z} \pm$ standard deviation σ) and mean number of events as a function of absorbed dose for 14-MeV neutrons (n) and ^{60}Co γ rays in tissue spheres of 1 μm diameter (79).

is the main objective of microdosimetry. Further discussions of the properties of these characteristic functions can be found in the literature (79).

3.6.3. Structures of Electron Tracks

Electron track structures have been calculated for energies from about 10 eV to several MeV, and for gases, liquids, and solids (14, 40, 84a, 89–96). Here a few characteristic results are shown, using tracks of keV electrons in water vapor.

Figures 3.39–3.41 (and track structures in later sections) are two-dimensional projections of three-dimensional computed tracks, which originally contain full information about the location of each event and its physicochemical nature. For the decay mode of initially excited or ionized states, evaluated experimental data of Tan et al. (42d) were used. The figures show that (a) the number of events increases with electron energy, (b) with increasing electron energy the mean distance between inelastic collisions increases, (c) electrons suffer appreciable angular scattering, and (d) the event density is particularly high in the track ends of the primary and secondary electrons.

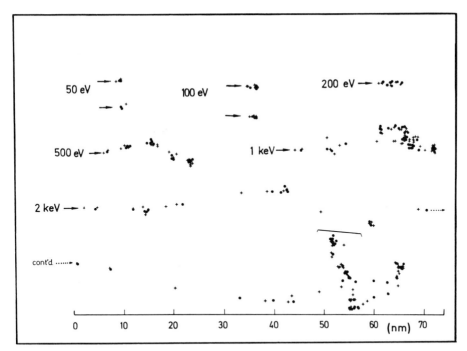

Figure 3.39. Two-dimensional projections of electron tracks in water vapor calculated with a Monte Carlo simulation program (MOCA-8). For graphical reasons the 2-keV e⁻ track is continued in the last line. For this plot all 20 types of activations are divided into two classes: +, excitation; ●, ionization.

The largest number of events with small distances to their neighbors can be found in tracks of electrons of around 500 eV; electrons with less energy produce less events per energy deposited, and those with higher energies produce them further apart on the average. This can be seen from the frequency distribution to the next neighbors (Figure 3.42) and to all other events in a track (Figure 3.43).

For electrons with energies in the keV range the probability to find less than 1, 2, 3, . . . additional events within a certain distance follows approximately a Gaussian function (Figure 3.44). However, this distribution describes only expected distances averaged over many events and tracks. Actual reactions of events in a track do not know of these averages and react according to their actual distances. If the reaction probabilities vary nonlinearly with reactant separation distances, for example, in a second-order reaction, the distance distributions must also be weighted nonlinearly. This is possible with Monte.Carlo simulations.

3.6.4. Structures of Proton Tracks

Proton track structures have been calculated for comparison with experimental proton data, for neutron track structure studies, and to provide basic data for ra-

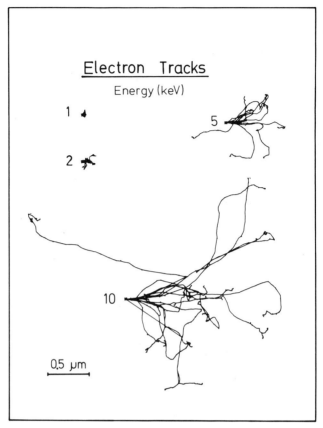

Figure 3.40. Ten tracks of 1-, 2-, 5-, and 10-keV electrons each in water, showing the stochastic nature of paths of identical projectiles and the widely different degrees of spatial concentrations of energy absorption events.

diation biology (97, 98). Figure 3.45 gives three tracks each for short segments of 0.3-, 1-, and 3-MeV protons in water vapor calculated with the simulation program MOCA-14 to demonstrate the similarities and differences between fast ion tracks of the same energy compared to those of different energy. In general, these tracks are straight because the heavier mass of fast ions prevents them from being scattered as much as electrons in elastic and inelastic collisions. With increasing ion energy the relative fraction of all events produced directly by the fast ion decreases from more than two-thirds for energies T around 0.3 MeV/u to around one-third at 5 MeV/u. The fraction of events on or close to the ion path also decreases. This is due to the $T^{-1} \ln T$ behavior of the secondary electron inelastic cross sections and the "hardening" of the secondary electron spectrum with increasing ion energy (the secondary electron maximum energy increases with T).

The linear density of fast secondary electrons along the ion path, however, is

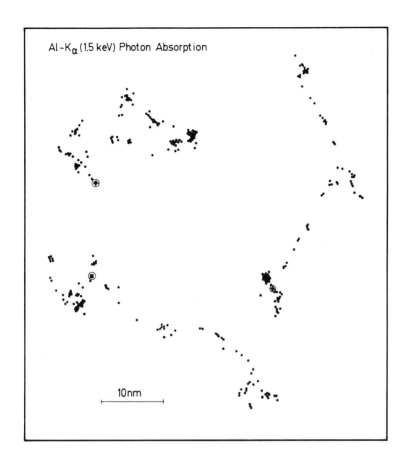

Al-K$_\alpha$ (1.5 keV) Photon Absorption

10nm

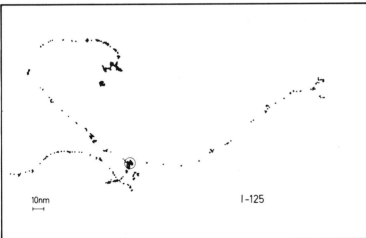

10nm   I-125

Figure 3.41. Simulated electron tracks from the absorption of Al-K$_\alpha$ X rays in water (leading typically to one photoelectron and one Auger electron) and from an ^{125}I decay (up to 22 electrons of widely different energies ejected per decay).

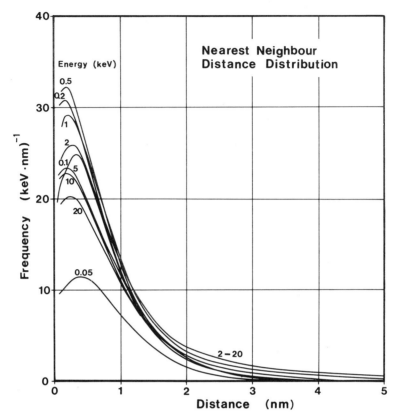

Figure 3.42. Frequency distributions of nearest neighbor distances between excitations and ionizations in tracks of electrons with starting energies from 0.1 to 20 keV.

rather small, and it decreases with increasing secondary electron energy ($\propto E^{-2}$). Therefore, the usefulness of a "radial dose concept" for classification of heavy-ion tracks (Section 3.4), which averages the energy deposited by secondary electrons at certain radial distances from the ion path along the path, is rather restricted (at least for protons and α particles).

The differences in the three track structures produced by protons of the same energy are completely due to the stochastic nature of all single-collision processes. It is still a challenging problem to derive a rigorous strategy for the classification of such track structures, regarding the similarities of tracks produced by identical particles and dissimilarities between tracks from different particles.

3.6.5. Structures of Heavy Charged Particle Tracks

For protons and α particles experimental cross section data and theory permit estimates of basic input data for track structure calculations with reasonable accuracy

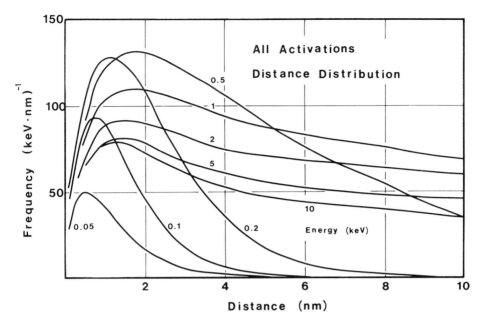

Figure 3.43. Frequency distributions of distances from each primary activation (excitation or ionization) to all others in the same track for electrons of various starting energies.

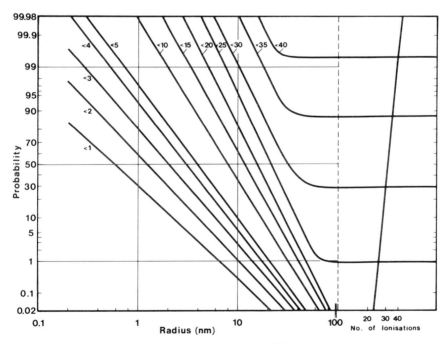

Figure 3.44. Probability (in percent) of finding less than *n* additional ionizations in a sphere of radius *r* around a randomly selected ionization in tracks of 1-keV electrons in water (left side of abscissa). Right side: cumulative probability of finding in total less than *N* ionizations in 1-keV e⁻ tracks in water vapor.

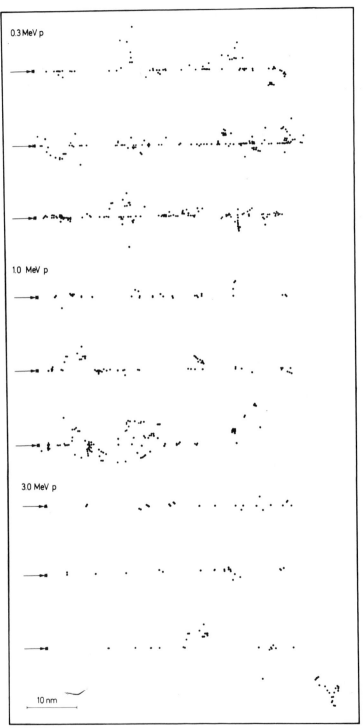

Figure 3.45. Proton track segments in water (0.3, 1.0, and 3.0 MeV, three tracks per energy) calculated with a Monte Carlo simulation program (MOCA-14).

142

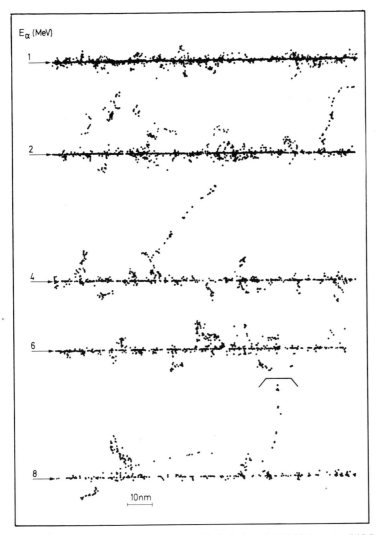

Figure 3.46. Calculated α particle track segments (1, 2, 4, 6, and 8 MeV) in water (MOCA-14).

for a few target media (e.g., water vapor) for a restricted energy range (say, 0.3–10 MeV/u). Figure 3.46 shows two-dimensional projections of track segments from 1 to 8 MeV α particles in water vapor, calculated with MOCA-14 (97). The higher density of events (all primary and secondary inelastic events are symbolized by a dot, though the full information on their chemical nature is available in the computer program) stems from the z^2 dependence of all ion cross sections. The electron transport, however, is independent of the charge of the ejecting ion. Because of the z^2 dependence of the primary cross section, more higher energy secondary electron tracks are also visible per unit track length for α particles than for protons (Figure 3.45).

In these computed α particle tracks one can see the transition from the "grain count" (single, separated events along the ion path) regime at high particle energies to the "track width" (overlapping events on and close to the ion path) regime at lower energies. This can lead to different types of radiation action (15) and can actually be seen microscopically in fast heavy-ion tracks registered in nuclear emulsions (5b, 16a).

Presently it is still impossible to perform reliable track structure calculations for ions heavier than α particles because of the lack of reliable or comprehensive experimental absolute cross sections and the lack of accurate enough theories. As mentioned above, the existence of loosely bound electrons in the projectile is the main reason for this problem. However, to give a semiquantitative example, track structures of fast carbon ions were derived from proton tracks by multiplying all primary ion cross sections with the velocity-dependent effective charge squared, as calculated from Barkas's formula (5a). Figure 3.47 gives an impression of such a heavy-ion track. The event density is very high along, and even at some distance

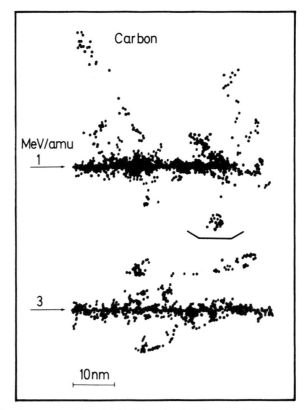

Figure 3.47. Track segments of 1- and 3-MeV/u carbon ions in water (MOCA-14) scaled from proton tracks of the same velocity by multiplication of primary electron ejection cross sections by the effective charge squared.

from, the ion path. Quite different chemical reactions and dynamics might be expected in the ''core'' of such tracks than in sparsely ionizing electron tracks (Chapter 4), such as are encountered at larger distances from an ion track or due to photon interactions.

3.7. CLASSIFICATIONS OF TRACK STRUCTURES

3.7.1. Concepts of Classification

As demonstrated in the preceding section, radiation track structures, due to their stochastic nature and the usually large number of events per track, contain much detailed information. For certain problems it is too much for easy evaluation. This fact introduces the problem of data reduction and classification (99). This is a general problem in science and various approaches to its solution can be found in different fields. It is a common objective in research to identify the essential parameters in the usually large number of descriptors that can be attached to an object, process, or phenomenon. Such an identification for data reduction is necessary for the derivation of underlying principles. The aim should first be to establish quantitative criteria for similarities and dissimilarities between the objects of study (here track structures) and then to derive strategies for classification and discrimination.

Classification of incident radiation fields (even of mixed fields) is an easy task. In most cases they can conveniently be described at any point in space by their spectral distributions of particle radiances p_i (i identifies the particle type) as a function of energy, solid angle, and time (100). These radiances are related to the particle fluence ϕ_i, which is the number of particles per unit area, by

$$\phi_i = \int_E \int_\Omega \int_t p_i(E, \Omega, t) \, dt \, d\Omega \, dE \qquad (3.22)$$

The classification of track structures produced in matter, however, is a difficult task and presently there is no rigorous solution. A few limited approaches are mentioned below, and ways are shown that might lead to improvements.

3.7.2. Macroscopic Dose, Kerma and Fluence

One possibility of classification is by the energy spent to produce all events of tracks, or of that spatial subgroup, which is in a certain site element of interest. Essentially this means that tracks are ordered according to the energy needed to produce them. The chemical nature of the events and their spatial arrangement (within a site element of interest) is ignored. This could be justified if the so-called optical approximation were valid throughout (9), namely, the relative spectrum of physicochemical initial species produced by interaction is independent of the parameters of the radiation and depends only on material properties (for example on its dipole oscillator strength distribution). However, this approximation is valid only for fast charged particles and not for slow electrons, which are very important for track structure formation. Furthermore, the primary interactions of photons, neutrons, and charged particles are quite different and the direct consequences of these primary collisions can also be quite different. Thus, the energy absorbed to produce a track is not a very good quantity for classification in radiation research in spite of its wide usage. Its frequent use, in fact, might be a major reason that insight into radiation action mechanisms is not more advanced.

For some applications, however, absorbed energy in mass elements might be an acceptable classification quantity, for example, in calorimetry, or if exactly the same radiation field and irradiation condition is used repeatedly to investigate the dose–effect relationships for this field. Absorbed energy can then be used as an implicit integral measure of the number of charged particle tracks produced. However, if the density in space and time of tracks from different primary particle interactions becomes so high that the track structures overlap and interfere with each other, even in this restricted sense classification by absorbed energy in a mass element can initiate misleading interpretations of observed radiation actions. For mixed fields or when comparing radiation effects produced by different irradiation modalities, even use of a ''radiation quality'' factor (accounting for different actions of different radiations at the same absorbed dose) cannot overcome the general shortcomings of the quantity ''energy absorbed in a mass element'' for classification of track structures and their effects. Therefore, it is doubtful whether dosimetric quantities actually ''provide a physical measure to correlate with actual or potential effects'' (100).

It is mainly for historical reasons and because of its usefulness in certain applications (see above) that absorbed dose is mentioned in this context. Absorbed dose D can be defined (100) as the quotient of the expectation value of the energy ϵ absorbed in an infinitesimally small element divided by its mass m:

$$D = \frac{d\bar{\epsilon}}{dm} = \lim_{m \to 0} \bar{z} \qquad (3.23)$$

where $z = \epsilon/m$ = specific energy
$\bar{z}$ = first moment of $f(z)$

Figure 3.48 gives a survey of classification schemes based essentially on dosimetry. By contrast, Figure 3.49 shows preliminary concepts for characterization of track structures based on the locations and types of events rather than on the total energy needed to produce them. Some of these concepts are discussed in more detail below.

Closely related to the quantity absorbed dose is the quantity *kerma*, which is the *k*inetic *e*nergy *r*eleased in *ma*tter by uncharged particles (photons, neutrons) divided by the mass of the scattering element (26, 101, 104):

Examples of Concepts of Macro- and Microscopic Dosimetry

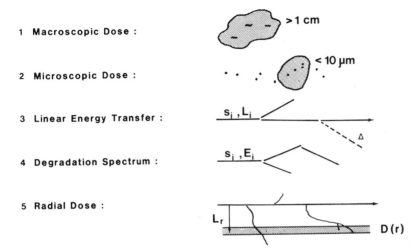

1 **Macroscopic Dose :**

2 **Microscopic Dose :**

3 **Linear Energy Transfer :**

4 **Degradation Spectrum :**

5 **Radial Dose :**

Figure 3.48. Some dosimetric classification concepts for charged particle tracks.

Examples of Concepts of Track Structure Analysis

$$\text{Track} = \left| S_1, \vec{x}_1; S_2, \vec{x}_2; \cdots S_n, \vec{x}_n \right|$$

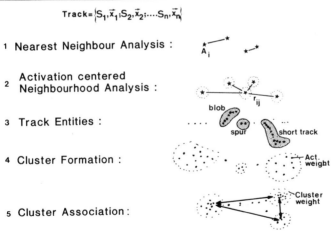

1 **Nearest Neighbour Analysis :**

2 **Activation centered Neighbourhood Analysis :**

3 **Track Entities :**

4 **Cluster Formation :**

5 **Cluster Association :**

Figure 3.49. Some classification concepts based on locations and types of new chemical species in charged particle tracks.

$$K = \frac{dE_{tr}}{dm} = \Phi \, E \, \frac{\mu_{tr}}{\rho} = \Psi \, \frac{\mu_{tr}}{\rho} \tag{3.24}$$

where Φ = particle fluence (particles/m^2)
$\quad\;\; \Psi$ = energy fluence (MeV/m^2)
$\quad\;\; E$ = particle energy
$\quad\;\; \mu_{tr}/\rho$ = mass energy transfer coefficient (100)

Kerma factors (kerma/fluence) for neutrons and photons in water and tissue are given in Figure 3.50. Both materials have low Z and thus the cross sections are almost identical, except for the additional nitrogen content of muscle, which influences the low energy neutron values through its (n, p) reaction. The partitioning of the kerma of neutrons in tissue in Figure 3.51 shows the large importance of proton recoils in neutron track structure, as mentioned in Section 3.3. At higher neutron energies nuclear (n, α) and $(n, 3\alpha)$ reactions in carbon and oxygen start to play a significant role.

Whereas absorbed dose is proportional to the energy actually absorbed in a mass element, kerma is the energy lost from a radiation field (into kinetic energy of secondary charged particles). They need not be numerically equal because of the potential production of bremsstrahlung, and some of the secondary electron tra-

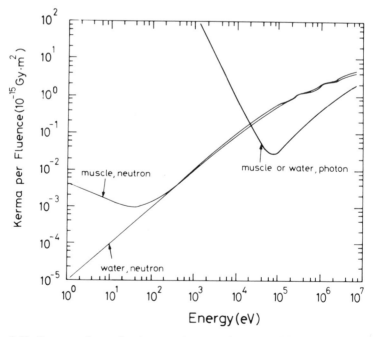

Figure 3.50. Kerma per fluence for photons and neutrons in water and tissue (muscle) (19, 101).

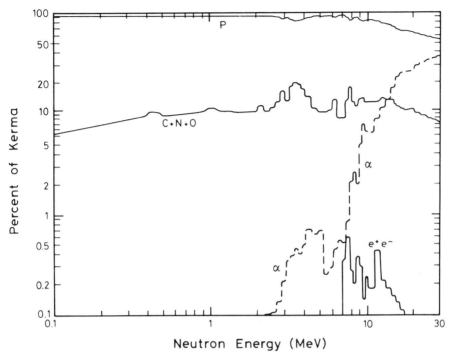

Figure 3.51. Partitioning of kerma in tissue-equivalent plastic (A 150) into the contributing charged particles (102).

jectories might extend beyond the small volumes where they began. Because of the finite ranges of secondary particles and their predominant forward emission, in extended irradiated bodies dose values will always be slightly larger than kerma values except for areas close to the surface where the tracks have not yet attained *secondary particle equilibrium*. This can be seen in Figure 3.52.

For the classification of track structures and their action, the differential particle radiance (100) is recommended to specify the local radiation field by direction, particle type, energy spectrum, time, and so on. At the next higher level of data reduction the fluence or fluence rate can be used [see Eq. (3.22)]. These primary quantities characterize the radiation field, and the corresponding track structures characterize its action on matter. A change in the energy spectrum of the radiation field can be used for classification of the interaction this field experiences but not for classification of the track structures produced in matter, where the new physicochemical species and their spatial arrangements are of interest and not the energy needed to produce them. If absorbed dose should be used at all as a derived quantity, the radiation modality should be stated in parentheses, for example D (X, 150 kV$_p$, 1 mm Cu), or D (^{60}Co, 2 cm steel), or D (n, 0.2–6 MeV, 2 mm Perspex).

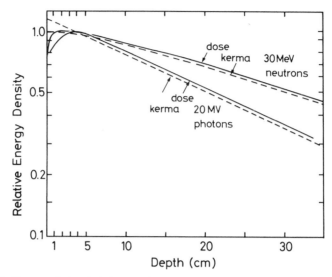

Figure 3.52. Kerma and absorbed dose on the central axis as a function of depth for high-energy photons and neutrons in tissue (derived from Refs. 26, 103, and 104 and normalized to the dose maximum).

3.7.3. Microdosimetric Concepts

Microdosimetry is a refinement in dose-based concepts (79, 105). Track structures are classified according to the energy needed to produce the events in a particular, small volume element of interest. Due to the stochastic nature of track formation, the imparted energy ϵ is also a stochastic quantity, and one can consider the frequency distribution $f(\epsilon)$ and its mean value $\bar{\epsilon}$. In microdosimetry the quantity ϵ itself plays only a minor role compared to the two derived quantities lineal energy y and specific energy z defined as:

$$y = \frac{\epsilon}{\bar{l}} \qquad z = \frac{\epsilon}{m} \qquad (3.25)$$

where $\bar{l}$ = mean chord length of volume element (for convex bodies $\bar{l}$ is four times the volume divided by the surface)

m = mass of volume element.

Distributions of y and z in a spherical volume element of tissue show large differences for sparsely ionizing radiations (photons) and densely ionizing radiations (neutrons) (Figure 3.53) and also differences among photon fields (Figure 3.54). This can be understood from the preceding discussion of the respective secondary particles and their track structure properties. Thus measured y or z spec-

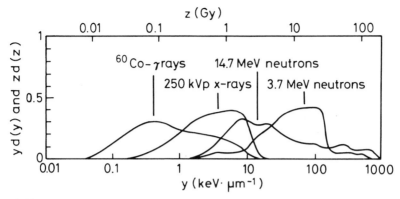

Figure 3.53. Distribution of dose in lineal energy y and specific energy z for a spherical tissue region of 1 μm diameter and various radiation fields, showing large differences in energy deposition properties of high- and low-LET radiation (105).

tra might be used to derive particle and energy radiances of the incident radiation and thus provide the basic information needed for a more appropriate classification.

The first moment, $\bar{y}_F$, of the distribution and the second moment divided by the first moment, $\bar{y}_D$, are also characteristics of the energy deposition by radiation in a particular site (Figure 3.55) and are used for classification in microdosimetry (79, 88, 105). However, all these distributions depend on the size and shape of the volume element of interest, and thus this information must be stated for each y or z spectrum and for their mean values.

Another possibility to classify track structures within the concept of microdosimetry is the $T(r)$ function. It gives the average energy deposited around initial species as a function of radial distance (79). For one track $T(r)$ increases with increasing distance. At higher doses it includes a contribution from an "average" single track (intratrack) and a component from track overlaps (intertrack) (Figure 3.56). The objections raised against dose and possibly wrong averages also apply to microdosimetry.

3.7.4. Restricted Stopping Power and Radial Dose

Higher densities of events in charged particle tracks are often more effective than lower densities in producing certain radiation effects (sometimes the opposite is true). Therefore, concepts have been put forward to classify radiation track structures according to their stopping power or their stopping power distribution (48). The average energy locally imparted to the medium is called *Linear Energy Transfer* (LET), which is essentially the same quantity as stopping power.

Energetic ionizing radiation can lead to the ejection of energetic secondary electrons. These electrons can travel considerable distances away from the path of the

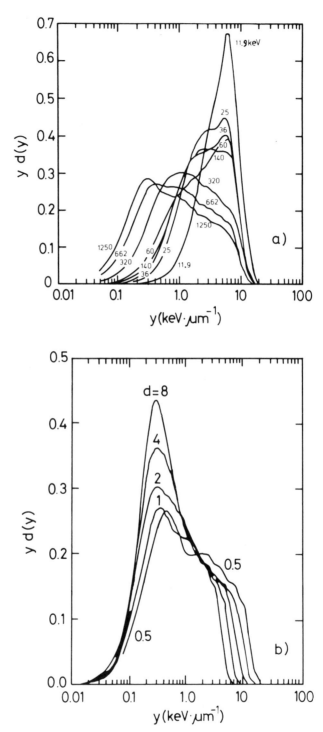

Figure 3.54. Spectra of lineal energy y for (a) a 1-μm-diameter site and various photon energies; (b) ^{60}Co γ rays and various site diameters (μm); adapted from Ref. 79, data from Ref. 106.

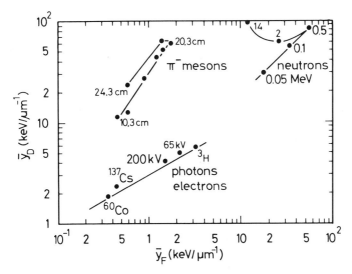

Figure 3.55. Second moment divided by first moment, $\bar{y}_D$, vs. first moment $\bar{y}_F$ of linear energy spectra in 1-μm-diameter sites for various radiation fields (79).

primary particle and virtually form sparse density tracks of their own. Therefore, modified classifications have been put forward (48) that neglect the energy expended in sparse tracks because it is almost "wasted." This neglect can be introduced either by an energy cutoff (L_q) in the energy transfer distribution or a radius cutoff (L_r) in the radial distance up to which energy is considered "locally imparted" (48). Figure 3.57 demonstrates the effect of extending the integral over the energy transfer distributions (essentially the secondary electron kinetic energy distributions) only up to the cutoff energy q (107, 108). Around 50% of the stop-

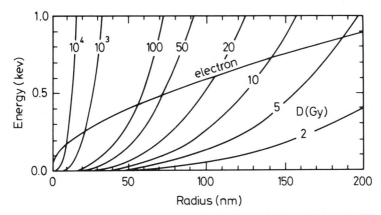

Figure 3.56. Comparison of inter- and intratrack contributions to the energy imparted within spheres around arbitrary activations in high-energy electron tracks: "electron" indicates the intratrack component, the parameterized lines show the intertrack part as a function of absorbed dose (79). The curve labelled electron corresponds to the proximity function T(r) for a single high-energy electron.

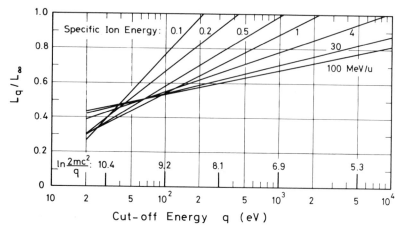

Figure 3.57. Ratio of the energy restricted stopping power L_q to the unrestricted stopping power L_∞ for various specific ion energies as a function of cutoff energy (107).

ping power is contained in energy transfers below 100 eV and thus remains rather close (say, within a few nanometers) to the point of interaction.

Whereas L_q can readily be calculated from basic differential cross sections by simple integration but not measured directly, L_r can be directly measured but not readily calculated (since full track structure calculations are needed). Even the radial derivative of L_r, which is the energy absorbed locally at a certain radial distance from the path of a primary particle, has been measured for several fast ions and used to check the reliability of track structure calculations (109–120) (Figure 3.58). In general, these radial dose profiles decrease approximately with $1/r^2$ up to the range of the maximum energy secondary electrons, which is proportional to the ion energy. They are also proportional to the square of the (effective) charge of the incident ion.

These general properties are depicted in Figure 3.59a, which also shows the contributions of secondary electrons ejected from inner and outer shells of target molecules. Based on such radial dose distributions averaged along the ion path, Katz has put forward an empirical classification scheme (15, 111) that agrees with many observations. However, this agreement still lacks a mechanistic model.

Figure 3.59 shows calculated L_r/L_∞ ratios (L_∞ is the unrestricted linear energy transfer and is thus essentially equal to the stopping power). As mentioned above, a large fraction of the energy lost by fast ions stays within the first nanometer around the path, but with increasing ion energy a large fraction can be transported by energetic electrons to large distances from that path. Also shown in this figure is the radius of the so-called track core as introduced by Mozumder and Magee (116).

Restricted LET concepts and Katz's radial dose concept for the classification of track structures depend on the validity of the averaging procedure. If an effect depends on the types and actual density of events on a microscopic scale, averages on the energy deposition level lead to doubtful conclusions. In many radiation

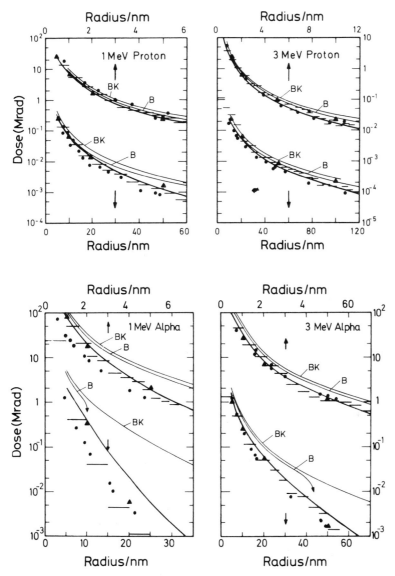

Figure 3.58. Calculated radial dose profiles around proton paths and α particle paths in tissue equivalent gas [csda (heavy line), mixed csda–Monte Carlo (triangles), and full Monte Carlo (histograms) models, for explanation see Ref. 69] compared to experimental data (●, Ref. 109), Baum's model (B, Ref. 110), and the model of Butts and Katz (BK, Ref. 111).

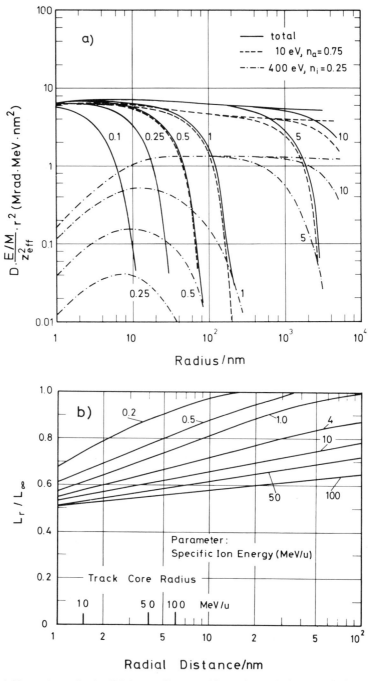

Figure 3.59. (*a*) Generalized radial dose profiles around heavy-ion tracks in a model substance (outer shell potential 10 eV, electron occupation fraction 0.75; inner shell potential 400 eV, electron occupation fraction 0.25). (*b*) Quotient of radius restricted stopping power L_r to unrestricted stopping power L_∞ as a function of radial distance from the ion path. For comparison the adiabatic limit for excitations is indicated as track core radius (116) for several specific ion energies.

effects averages are meaningful only for final yields, after the reactions are complete. Energy density in itself does not contain enough information.

3.7.5. Track Entities

Ninety-five percent of the energy loss events of fast electrons transfer less than 100 eV to matter. Each of the resulting low-energy electrons produces three or less additional ionizations during their own slowing down, which can be seen in the decrease of the yield curves (per 100 eV) in Figure 3.23. About 60% of them do not produce any further ionization. Therefore, in most cases there are only one or a few events in such isolated spatial areas, which are called "spurs" (10, 121), and the subsequent chemical reactions and their dynamics are quite different from those encountered in areas of higher ionization density. Areas of higher density are classified by Mozumder and Magee (10) as "blobs" (energy transfers between 100 and 500 eV) or "short tracks" (energy transfers between 0.5 and 5 keV). Secondary electrons produced in energy transfers above 5 keV are considered as "branch tracks," which are not likely to overlap with the other entities. The high-event density in blobs and short tracks favors recombination of radiogenic species over reaction with molecules of the material. The low-event density in spurs may favor reaction with molecules of the material (121, 122) (see also Chapters 2 and 4–6). The fraction of energy spent by fast electrons to form such entities is given in Figure 3.60. Figure 3.61 shows the yields of such track entities for fast and slow electrons, depicting the decrease of secondary blobs and spurs at low electron energies. Similar calculations have been performed for electrons by Berger (123)

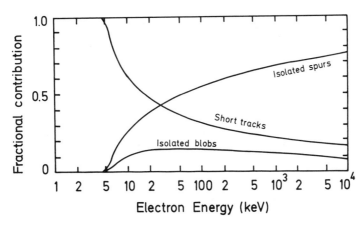

Figure 3.60. Track structure classification by so-called track entities spurs, blobs, and short tracks (10a).

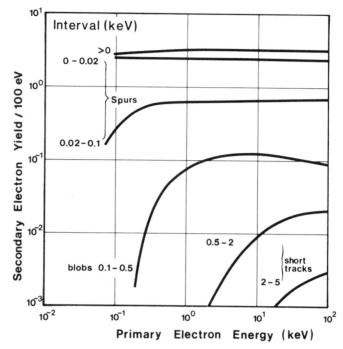

Figure 3.61. Yields in water vapor of secondary electrons in energy intervals relevant to track entity classification.

and for radiation chemical yields from heavy charged particles by Turner et al. (124) and Miller and Wilson (125).

The old concept of track entities is still a useful approach for the classification of track structures, since it takes into account the spatial arrangement of initial species, which affects their subsequent reactions. Refinement of this concept is needed in particular for heavy charged particle tracks (Chapter 4).

3.7.6. Outlook for Other Clustering Concepts for Track Structures

In 1937 Gray stated that "the most natural unit of radiation dosimetry is the absolute increase of energy of the absorbing medium" (127), and dosimetric concepts (for example, LET) have been used for classification of track structures for many decades. However, dosimetric concepts are inadequate to explain recently observed differences between results obtained with different radiation fields. Track structures are a better starting point for classification. This is illustrated in Figure 3.62, which compares energy-based evaluation concepts (left panel) with event-based evaluation (right panel). Such evaluation or similar approaches can follow

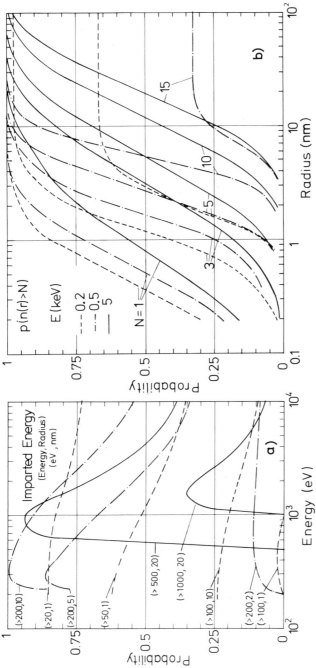

Figure 3.62. (*a*) Probability of finding more than a certain imparted energy within a certain radial distance around an excitation or ionization in electron tracks as a function of electron start energy. (*b*) Probability of finding more than *N* additional ionizations within a certain distance *r* around an arbitrary ionization in water for electrons of 0.2-, 0.5-, and 5-keV start energy (126).

a) simulation of particle track

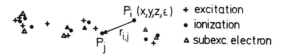

$P_i(x,y,z,\varepsilon)$ + excitation
  • ionization
$r_{i,j}$ ▲ subexc. electron

P_j

b) distances between events

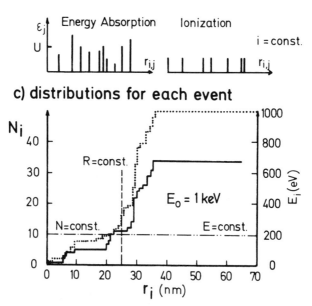

ε_j Energy Absorption   Ionization

U   i = const.

$r_{i,j}$ $r_{i,j}$

c) distributions for each event

N_i 40   1000

30   R=const.   800

20             $E_0 = 1\,keV$ 600 $E_i\,(eV)$

10   N=const.   E=const.   400

0   200

0   10   20   30   40   50   60   70   0

$r_i\,(nm)$

Figure 3.63. Some possibilities of track structure evaluation for the purpose of information reduction.

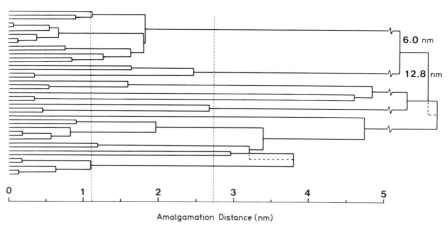

6.0 nm

12.8 nm

0   1   2   3   4   5

Amalgamation Distance (nm)

Figure 3.64. Dendrogram of 1-keV electron track using the k-means algorithm.

the principles outlined in Figure 3.63. First, track structures are measured [for example, in a low-pressure cloud chamber (2)] or calculated with an appropriate Monte Carlo program. Then the important distances and their distributions are derived from the structures. Finally, from these distributions of event densities reaction probabilities are computed and evaluated.

The evaluation can be made by using conventional Cluster algorithms (128, 129) designed for the analysis of more-dimensional data. Results from such types of classification can be displayed in the form of dendrograms (Figures 3.64 and 3.65). Here the merging to clusters of increasing sizes of closely located initial events is shown as a function of their increasing amalgamation distance (for example the square root of the sum of the distances of all members of a cluster to its virtual center) until the whole track forms one large cluster. The dendrograms of Figure 3.65 can be used to demonstrate that there are fewer events in electron tracks in water vapor than in liquid water (scaled for the density), they are further apart on a mass basis in the liquid (because of the assumed delocalization through

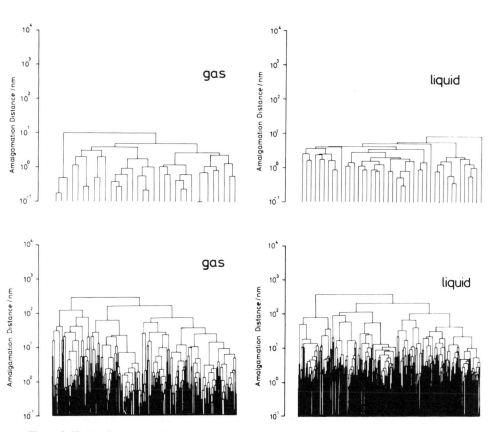

Figure 3.65. Dendrograms of single 0.5- and 5.0-keV electron tracks in water vapor and liquid water.

plasmon diffusion), but the total track lengths scaled for the density are very similar (as is also expressed in the phase state insensitivity of stopping power).

The classification of track structures requires further improvement. There are classification concepts used in other fields, for example, in fuzzy set theory (130), in pattern recognition, and in artificial intelligence (131), which should be explored for usefulness in this context.

REFERENCES

1. C. T. R. Wilson, *Proc. Roy. Soc. (London)*, **A85**, 285 (1911); **A87**, 277 (1912); **A104**, 192 (1923).

2. (a) H. J. Delafield and P. D. Holt, in *Charged Particle Tracks in Solids and Liquids*, G. E. Adams, D. K. Bewley, and J. W. Boag (Eds.), Institute of Physics, Conference Series No. 8, London, 1970, p. 35. (b) T. Budd and M. Marshall, *Radiat. Res.* **93**, 19 (1983). (c) T. Budd, C. S. Kwok, M. Marshall, and S. Lythe, *Radiat. Res.* **95**, 217 (1983).

3. (a) L. S. Schröder, *Nucl. Instr. Meth.* **162**, 395 (1979). (b) W. A. Wenzel, *Ann. Rev. Nucl. Sci.* **14**, 205 (1964).

4. (a) D. A. Glaser, *Phys. Rev.* **87**, 665 (1952); **91**, 496 (1953). (b) M. Dykes, P. Lecoq, et al., *Nucl. Instr. Meth.* **179**, 487 (1981).

5. (a) W. F. Barkas, *Nuclear Research Emulsions*, Academic Press, New York, 1963. (b) P. H. Fowler, in *Advances in Radiation Research*, Vol. 1, J. F. Duplan and A. Chapiro (Eds.), Gordon and Breach, New York, 1973, p. 39.

6. (a) R. L. Fleischer, P. B. Price, and R. M. Walker, *Nuclear Tracks in Solids: Principles and Applications*, University of California Press, Berkeley, 1975. (b) F. Granzer, H. G. Paretzke, and E. Schopper (Eds.), *Solid State Nuclear Track Detectors: Proceedings of the 9th International Conference, Neuherberg/Munich*, Pergamon Press, Oxford, 1978. (c) H. Francois, N. Kurtz, J.-P. Massue, M. Monnin, R. Schmitt, and S. A. Durrani (Eds.), *Solid State Nuclear Track Detectors: Proceedings of the 10th International Conference, Lyon*, Pergamon Press, Oxford, 1980. (d) P. H. Fowler and V. M. Clapham (Eds.), *Solid State Nuclear Track Detectors: Proceedings of the 11th International Conference, Bristol*, Pergamon Press, Oxford, 1982.

7. U. Fano, (a) *Ann. Rev. Nucl. Sci* **13**, 1 (1963); (b) in Ref. 2a, p. 1; (c) in *Radiation Biology*, Vol. 1, A. Hollander (Ed.), McGraw-Hill, New York, 1954, p. 1.

8. J. W. Boag, (a) in Ref. 5b, p. 9; (b) in *Progress and Problems in Contemporary Radiation Chemistry*, Vol. 1, *Proceedings of the 10th Czechoslovac Annual Meeting on Radiation Chemistry, Marianske Lazne, Czechoslovakia, June 22–25, 1970*, J. Teply et al. (Eds.), Academy of Science, Prague, 1971, p. 19.

9. (a) E. J. Hart and R. L. Platzman, in *Mechanisms in Radiobiology*, M. Errera and A. Forssberg (Eds.), Vol. 1, Academic Press, New York, 1961, p. 93. (b) R. L. Platzman, in *Radiation Research: Proceedings of the 3rd International Congress, Cortina d'Ampezzo, Italy*, G. Silini (Ed.), Wiley, New York, 1967, p. 20.

10. (a) A. Mozumder, in *Advances in Radiation Chemistry*, Vol. 1, M. Burton and J. L. Magee (Eds.), Wiley, 1969, p. 1. (b) A. Mozumder and J. L. Magee, *Radiat. Res.* **28**, 203 (1966).

11. R. Voltz, in Ref. 8b, pp. 45, 139.

12. W. Brandt and R. H. Ritchie, in *Physical Mechanisms in Radiation Biology*, R. D. Cooper and R. W. Wood (Eds.), Technical Information Center, US Atomic Energy Commission, Oak Ridge, TN, 1974, p. 20.

13. M. Inokuti, *Applied Atomic Collision Physics*, Vol. 4, Academic Press, New York, 1983, p. 179.

14. H. G. Paretzke, in *Advances in Radiation Protection and Dosimetry*, R. H. Thomas and V. Perez-Mendez (Eds.), Plenum Press, New York, 1980, p. 51.

15. D. E. Lea, *Actions of Radiations on Living Cells*, University Press, Cambridge, 1946.

16. (a) R. Katz, S. C. Sharma, and M. Homayoonfar, in *Topics in Radiation Dosimetry, Radiation Dosimetry Supplement 1*, F. H. Attix (Ed.), Academic Press, New York, 1972, p. 317. (b) R. Katz and E. J. Kobetich, *Phys. Rev.* **186**, 344 (1969). (c) R. Katz and E. J. Kobetich, in Ref. 2a, p. 109.

17. (a) R. D. Evans, in *Radiation Dosimetry*, Vol. I, F. H. Attix and W. C. Roesch (Eds.), Academic Press, New York, 1968, p. 93. (b) R. D. Evans, *The Atomic Nucleus*, McGraw-Hill, New York, 1955. (c) N. A. Dyson, *X-Rays in Atomic and Nuclear Physics*, Longmans Green, New York, 1973.

18. (a) D. A. Anderson, *Absorption of Ionizing Radiation*, University Park Press, Baltimore, 1984. (b) H. E. Johns and J. R. Cunningham, *The Physics of Radiology*, Charles Thomas Publisher, Springfield, IL, 1984.

19. E. F. Plechaty, D. E. Cullen, and R. J. Howerton, *Tables and Graphs of Photon-Interaction Cross Sections from 0.1 keV to 100 MeV derived from the LLL Evaluated-Nuclear Data Library*, Lawrence Livermore National Laboratory, Report UCRL-50400, Vol. 6, Rev. 3, Livermore, CA, 1981.

20. (a) J. H. Hubbell, *Photon Cross Sections, Attenuation Coefficients, and Energy Absorption Coefficients from 10 keV to 100 GeV*, U.S. National Bureau of Standards, Report NSRDS-NBS 29, 1969. (b) J. H. Hubbell, *Radiat. Res.* **70**, 58 (1977). (c) J. H. Hubbell, *Int. J. Appl. Radiat. Isot.* **33**, 1269 (1982). (d) J. H. Hubbell, W. J. Veigele, E. A. Briggs, R. T. Brown, D. T. Cromer, and R. J. Howerton, *J. Phys. Chem. Ref. Data* **4**, 471 (1975).

21. R. Blohm, Durchgang von Elektronen durch Strahlenempfindliche Bereiche des Zellkerns, Dissertation, Universität Göttingen, Göttingen, Federal Republic of Germany, 1983.

22. G. L. Brownell, W. H. Ellett, and A. R. Reddy, *J. Nucl. Med.*, *MIRD* **9** (Suppl. 1), 27 (1968).

23. P. Jacob, H. G. Paretzke, and J. Wölfel, *Nucl. Sci. Eng.* **87**, 113 (1984).

24. (a) K. H. Beckurts and K. Wirtz, *Neutron Physics*, Springer-Verlag, Berlin, 1964. (b) R. J. Howerton, D. E. Cullen, M. H. Mac Gregor, S. T. Perkins, and E. F. Plechaty, *The LLL Evaluated Nuclear Data Library (ENDL): Graphs of Cross Sections from the Library*, UCRL-50400, Vol. 15, Part B, University of California, Berkeley (1976). (c) S. F. Mughabghab and D. Garber (Eds.), *Neutron Cross Sections*, Vol. I, *Resonance Parameters*, BNL-325, Brookhaven National Laboratory, Upton, NY, 1973. (d) R. R. Kinsey and D. I. Garber (Eds.), *Neutron Cross Sections*, Vol. II, *Graphical Data*, BNL-325, Brookhaven National Laboratory, Upton, NY, 1979. (e) M. D. Goldberg, S. F. Mughabghab, S. N. Purohit, B. A. Magurno, and V. M. May (Eds.), *Neutron Cross Sections*, Vol. IIB, $Z = 41\text{-}60$, Report BNL-325,

Brookhaven National Laboratory, Upton, NY, 1966. (f) D. I. Garber, L. E. Stromberg, M. D. Goldberg, D. W. Cullen, and V. M. May (Eds.), *Angular Distributions in Neutron Induced Reactions*, Vol. I: $Z = 1$-20, Vol. II, $Z = 21$-94, BNL-400, Brookhaven National Laboratory, Upton, NY, 1970.

25. (a) R. S. Caswell and J. J. Coyne, *Radiat. Res.* **52**, 448 (1972). (b) R. S. Caswell, J. J. Coyne, and M. L. Randolph, *Radiat. Res.* **83**, 217 (1980).

26. International Commission on Radiation Units and Measurements, *Neutron Dosimetry for Biology and Medicine*, ICRU-Report 26, ICRU, Bethesda, MD, 1977.

27. (a) A. A. Edwards and J. A. Dennis, *Phys. Med. Biol.* **20**, 395 (1975). (b) H. G. Paretzke, F. Grünauer, E. Maier, and G. Burger, in *First Symposium on Neutron Dosimetry in Biology and Medicine*, G. Burger, H. Schraube and H. G. Ebert (Eds.), EUR 4896, Commission of the European Communities, Luxembourg, 1972, p. 73.

28. (a) U. Oldenburg and J. Booz, *Radiat. Res.* **51**, 551 (1972). (b) J. Booz and M. Coppola, in *Annual Report of the Joint Research Center*, EUR 5260, Commission of the European Communities, Luxembourg, 1974, p. 396.

29. (a) L. G. Christophorou *Atomic and Molecular Radiation Physics*, Wiley, New York, 1971. (b) L. G. Christophoru (Ed.), *Electron-Molecule Interactions and their Applications*, 2 vols., Academic Press, New York, 1984.

30. S. E. Schnatterly, in *Solid State Physics*, Vol. 34, H. Ehrenreich, F. Seitz, and D. Turnbull (Eds.), Academic Press, New York, 1979, p. 275.

31. H. Raether, *Excitations of Plasmons and Interband Transitions in Solids*, Springer Tracts in Modern Physics 88, Springer-Verlag, New York, 1980, p. 1.

32. C. E. Klots, in *Fundamental Processes in Radiation Chemistry*, P. Ausloos (Ed.), Wiley, New York, 1968, p. 1.

33. (a) N. F. Mott and H. S. W. Massey, *The Theory of Atomic Collisions*, Oxford University Press, London, 1949. (b) H. S. W. Massey and E. H. S. Burhop, *Electronic and Ionic Impact Phenomena*, Clarendon Press, Oxford, 1969.

34. (a) R. L. Platzman, *Vortex* **23**, 327 (1962). (b) R. L. Platzman, *Radiat. Res.* **17**, 419 (1962). (c) Y. Hatano, in *Proceedings of the Workshop on Electronic and Ionic Collision Cross Sections Needed in the Modelling of Radiation Interactions with Matter*, Argonne National Laboratory, Report ANL-84-28, 1984, p. 27.

35. (a) D. Rapp and P. Englander-Golden, *J. Chem. Phys.* **43**, 1464 (1965). (b) C. Ramsauer and R. Kollath, *Ann. Physik* **4**, 91 (1930). (c) A. V. Phelps, *Rev. Mod. Phys.* **40**, 399 (1968). (d) B. L. Schram, F. J. de Heer, M. J. van der Wiel, and J. Kistemaker, *Physica* **31**, 94 (1965). (e) H. Ehrhardt and F. Linder, *Phys. Rev. Lett.* **21**, 419 (1969). (f) H. J. Grosse and H.-K. Bothe, *Z. Naturforsch.* **25a**, 1970 (1970).

36. (a) L. J. Kieffer, *At. Data* **1**, 19 (1969); **2**, 293 (1971). (b) C. E. Brion, in Ref. 34c, p. 85. (c) F. J. de Heer and M. Inokuti, *Electron Impact Ionization*, in T. D. Märk and G. H. Dunn (Eds.), Springer-Verlag, Heidelberg, 1985. (d) D. F. Kyser, H. Niedrig, D. E. Newbury, R. Shimizu (Eds.), *Electron Beam Interactions with Solids for Microscopy, Microanalysis and Microlithography*, Scanning Electron Microscopy, POB 66507, AMF O'Hare, IL.

37. (a) A. E. S. Green and J. H. Miller, in Ref. 12, p. 68. (b) M. Inokuti, in Ref. 12, p. 51. (c) W. Lotz, *Z. Phys.* **232**, 101 (1970). (d) E. J. McGuire, *Phys. Rev.* **A3**, 267 (1971). (e) C. J. Powell, *Rev. Mod. Phys.* **48**, 33 (1976). (f) M. Gryzinsky, *Phys. Rev.* **138**, A336 (1965). (g) F. Trajmar, *Science* **208**, 247 (1980).

38. (a) H. A. Bethe, *Ann. Phys.* **5,** 325 (1930). (b) H. A. Bethe, in *Handbuch der Physik*, Vol. 24, Part I, H. Geiger and K. Scheel (Eds.), Springer-Verlag, Berlin, 1933, p. 273. (c) M. Inokuti, *Rev. Mod. Phys.* **43,** 297 (1971). (d) M. Inokuti, Y. Itikawa, and J. E. Turner, *Rev. Mod. Phys.* **50,** 23 (1978).

39. (a) J. Berkowitz, *Photoabsorption, Photoionization, and Photoelectron Spectroscopy*, Academic Press, New York, 1979. (b) Y.-K. Kim, *Radiat. Res.* **61,** 21 (1975); **64,** 96 (1975); **64,** 205 (1975). (c) Y.-K. Kim, in *Radiation Research: Biomedical, Chemical and Physical Perspectives*, O. F. Nygaard, H. I. Adler, and W. K. Sinclair (Eds.), Academic Press, New York, p. 219, 1975.

40. H. G. Paretzke and M. Berger, in *Sixth Symposium on Microdosimetry*, J. Booz and H. G. Ebert (Eds.), Report EUR 6064, Harwood Academic Publishers, London, 1978, p. 749.

41. (a) U. Fano, *Radiat. Res.* **64,** 217 (1975); *Phys. Rev.* **95,** 1198 (1954). (b) M. Inokuti, in *Electronic and Atomic Collisions*, N. Oda and K. Takayanagi (Eds.), North Holland Publishing, Amsterdam, 1980, p. 31. (c) M. Inokuti, Y.-K. Kim, and R. L. Platzman, *Phys. Rev.* **164,** 55 (1967).

42. (a) G. D. Zeiss, W. J. Meath, J. C. F. MacDonald, and D. J. Dawson, *Radiat. Res.* **70,** 284 (1977); *Mol. Phys.* **39,** 1055 (1980). (b) J. Schutten, A. J. de Heer, H. R. Mustafa, A. J. H. Boerboom, and J. Kistemaker, *J. Chem. Phys.* **44,** 3924 (1966). (c) T. D. Märk and F. Egger, *Int. J. Mass Spectrom. Ion Phys.* **20,** 89 (1976). (d) K. H. Tan, C. E. Brion, P. E. van der Leeuw, and M. J. Van der Wiel, *Chem. Phys.* **29,** 299 (1978).

43. C. B. Opal, E. C. Beaty, and W. K. Peterson, *At. Data* **4,** 209 (1972).

44. (a) M. A. Bolorizadeh and M. E. Rudd, in Ref. 34c, p. 52. (b) D. A. Vroom, *Energy Deposition Studies: Final Report*, Gulf Atomic, San Diego, CA, Report UC-34-Physics, 1976.

45. (a) N. Oda, *Radiat. Res.* **64,** 80 (1975). (b) N. Oda, F. Nishimura, and S. Tahira, *J. Phys. Soc. Japan* **33,** 462 (1972).

46. (a) E. Weigold, A. J. Dixon, I. E. McCarthy, and C. J. Noble, in *10th International Conference on the Physics of Electronic and Atomic Collisions*, Paris, 1977, p. 364. (b) H. Ehrhardt, K. H. Hesselbacher, K. Jung, M. Schulz, and K. Wittmann, *J. Phys. B.* **5,** 2107 (1972).

47. International Commission on Radiation Units and Measurements, *Stopping Power for Electrons and Positrons*, ICRU-Report 37, ICRU, Bethesda, MD, 1984.

48. International Commission on Radiation Units and Measurements, *Linear Energy Transfer*, ICRU-Report 16, ICRU, Bethesda, MD, 1970.

49. G. J. Kutcher and A. E. S. Green, *Radiat. Res.* **67,** 408 (1976).

50. M. Terrissol, J. P. Patau, and T. Eudaldo, in Ref. 40, p. 169.

51. (a) C. Møller, *Ann. Phys. Rev.* **57,** 485 (1940). (b) E. A. Uehling, *Ann. Rev. Nucl. Sci.* **4,** 315 (1954). (c) F. Rohrlich and B. C. Carlson, *Phys. Rev.* **93,** 38 (1953).

52. (a) E. Fermi, *Phys. Rev.* **57,** 485 (1940). (b) R. M. Sternheimer, *Phys. Rev.* **88,** 851 (1952). (c) R. M. Sternheimer, M. J. Berger, and S. M. Seltzer, *At. Data and Nucl. Data Tabl.* **30,** 261 (1984).

53. H. G. Paretzke, J. E. Turner, R. N. Hamm, H. A. Wright, and R. H. Ritchie, *J. Chem. Phys.* **84,** 3182 (1986).

54. (a) L. V. Spencer and U. Fano, *Phys. Rev.* **93,** 1172 (1954). (b) D. A. Douthat, *Radiat. Res.* **61,** 1 (1975). (c) U. Fano and L. V. Spencer, *Int. J. Radiat. Phys. Chem.* **7,** 63 (1975). (d) D. Combecher, *Radiat. Res.* **84,** 189 (1980).

55. (a) U. Fano, *Phys. Rev.* **72,** 26 (1947). (b) M. Inokuti, D. A. Douthat, and A. R. P. Rao, *Phys. Rev.* **A22,** 445 (1980). (c) D. G. Alkhazov, A. P. Komar, and A. A. Vorob'v, *Nucl. Instr. Meth.* **48,** 1 (1967). (d) W. Neumann, in *Seventh Symposium on Microdosimetry,* J. Booz, H. G. Ebert, and H. D. Hartfiel (Eds.), EUR 7147, Harwood Academic, London, 1981, p. 1067.

56. W. E. Wilson and H. G. Paretzke, in *Fourth Symposium on Microdosimetry,* J. Booz, H. G. Ebert, R. Eickel, and A. Waker (Eds.), Report EUR 5122, Commission of the European Communities, Luxembourg, 1974, p. 113.

57. W. E. Wilson and L. H. Toburen, *Phys. Rev.* **A7,** 1535 (1973).

58. (a) L. H. Toburen and W. E. Wilson, in *Radiation Research: Proceeding of the 6th Intern. Congress,* S. Okada, M. Imamura, T. Terashima, and H. Yamaguchi (Eds.), Japan Association of Radiation Research, Tokyo, 1979, p. 80. (b) L. H. Toburen, *Phys. Rev.* **A3,** (1971). (c) L. H. Toburen and W. E. Wilson, *J. Chem. Phys.* **60,** 5202 (1977).

59. (a) M. E. Rudd and J. H. Macek, in *Case Studies in Atomic Physics 3,* E. W. McDaniel and M. R. C. McDowell (Eds.), North-Holland, Amsterdam, 1974, p. 49. (b) M. E. Rudd, in Ref. 34c, p. 43. (c) M. E. Rudd, Y.-K. Kim, D. H. Madison, and J. W. Gallagher, *Rev. Mod. Phys.* **57** (1985). (d) M. E. Rudd, A. Itoh, and T. V. Goffe, *Phys. Rev. A* **32,** 2499 (1985). (e) M. E. Rudd, T. V. Goffe, and A. Itoh, *Phys. Rev. A* **32,** 2128 (1985).

60. (a) F. T. Smith and A. Salop, in Ref. 39c, p. 242. (b) U. Fano and W. Lichten, *Phys. Rev. Lett.* **14,** 627 (1965).

61. (a) D. H. Madison, *Phys. Rev. A* **8,** 2449 (1973). (b) J. Macek, *Phys. Rev. A* **1,** 235 (1970). (c) A. Salin, *J. Phys. B* **2,** 631 (1969). (d) J. S. Briggs, *J. Phys. B* **10,** 3075 (1977).

62. (a) M. E. Rudd, L. H. Toburen, and N. Stolterfoht, *At. Data Nucl. Data Tabl.* **23,** 405 (1979). (b) L. H. Toburen, N. Stolterfoht, Z. Ziem, D. Schneider, *Phys. Rev. A* **24,** 1741 (1981). (c) M. E. Rudd, R. D. DuBois, L. H. Toburen, C. A. Ratcliffe, and T. V. Goffe, *Phys. Rev. A* **28,** 3244 (1983). (d) R. D. DuBois, L. H. Toburen, and M. E. Rudd, *Phys. Rev. A* **29,** 70 (1984).

63. (a) N. Oda, F. Nishimura, K. Komatsu, and H. Shibata, in *13th Int. Conf. on the Physics of Electronic and Atomic Collisions,* J. Eichler, W. Fritsch, et al. (Eds.), ICPEAC e.V., Berlin, 1983, p. 369. (b) A. K. Kaminsky and M. I. Popova, *J. Phys. B,* **15,** 403 (1982).

64. (a) D. H. Jakubassa-Amundsen, in Ref. 63a, p. 377. (b) S. B. Berry, J. A. Sellin, et al., *IEEE Trans. Nucl. Sci.* **NS30**(2), 902 (1983).

65. (a) M. Breinig, S. Elston, et al., *Phys. Rev. Lett.* **45,** 1689 (1980). (b) W. Meckbach, I. B. Nemirovsky, and C. R. Garibotti, *Phys. Rev. A* **24,** 1795 (1981).

66. (a) H.-D. Betz, *Rev. Mod. Phys.* **44,** 465 (1972). (b) H.-D. Betz and A. B. Wittkower, *Phys. Rev. A* **6,** 1485 (1972).

67. (a) L. C. Northcliffe and R. F. Schilling, *Nucl. Data A* **7,** 233 (1970). (b) C. F. Williamson, J. P. Boujot, and J. Picard, *Tables of Range and Stopping Power of*

Chemical Elements for Charged Particles of Energy 0.5 to 500 MeV, Rapport CEA-R 3042, Fontenay-aux-Roses, 1966.

68. (a) I. S. Dimitriev and V. S. Nikolaev, *Soviet Phys. JETP* **20,** 409 (1965). (b) V. S. Nikolaev and I. S. Dimitriev, *Phys. Lett.* **28A,** 277 (1968).

69. H. G. Paretzke, in Ref. 56, p. 141.

70. (a) H. Bichsel, in *American Institute of Physics Handbook*, 3rd ed., D. E. Gray (Ed.), McGraw-Hill, New York, 1972, pp. 8.142–8.189. (b) H. Bichsel, in *Radiation Dosimetry*, Vol. 1, F. H. Attix and W. C. Roesch (Eds.), Academic Press, New York, 1968, p. 157.

71. L. C. Northcliffe, *Ann. Rev. Nucl. Sci.* **13,** 67 (1963).

72. (a) P. Sigmund, in *Radiation Damage Processes in Materials*, C. H. S. Dupuy (Eds.), Noordhoff, Leyden, 1975, p. 3. (b) J. Lindhard and M. Scharff, *K. Danske Vidensk. Selsk. Mat.-Fys. Medd.* **34**(15) (1953).

73. (a) S. P. Ahlen, *Rev. Mod. Phys.* **52,** 121 (1980). (b) International Commission on Radiation Units and Measurements, *Basic Aspects of High Energy Particle Interactions and Radiation Dosimetry*, Report 28, ICRU, Bethesda, MD, 1978. (c) R. H. Ritchie and W. Brandt, *Phys. Rev. A* **17,** 2102 (1978).

74. P. G. Steward, *Stopping Power and Range for any Nucleus in the Specific Energy Interval 0.01–500 MeV/amu in Any Nongaseous Material*, UCRL-18127, University of California, Berkeley, 1968.

75. (a) H. H. Andersen, *The Stopping and Ranges of Ions in Matter,* Vol. 2, *Bibliography and Index of Experimental Range and Stopping Power Data*, Pergamon Press, New York, 1977. (b) H. H. Andersen and J. F. Ziegler, Ref. 75a, Vol. 3, *Hydrogen: Stopping Powers and Ranges in all Elements*, 1977. (c) J. F. Ziegler, Ref. 75a, Vol. 4, *Helium: Stopping Powers and Ranges in all Elemental Matter*, 1977. (d) J. F. Ziegler, Ref. 75a, Vol. 5, *Handbook of Stopping Cross Sections for Energetic Ions in all Elements*, 1980.

76. (a) J. A. Dennis and D. Powers, in Ref. 40, p. 661. (b) R. B. J. Palmer and A. Akhavan-Rezayat, Ref. 40, p. 739.

77. (a) P. V. Vavilov, *Sov. Phys. JETP* **5,** 749 (1957). (b) C. Tschalär, *Nucl. Instr. Meth.* **61,** 141 (1968). (c) C. Tschalär and H. D. Maccabee, *Phys. Rev. B* **1,** 2863 (1970).

78. A. M. Kellerer and D. Chmelevsky, *Radiat. Res.* **63,** 226 (1975).

79. International Commission on Radiation Units and Measurements, *Microdosimetry*, ICRU-Report 36, ICRU, Bethesda, MD, 1983.

80. (a) T. E. Hull and A. R. Dobell, *SIAM Rev.* **4,** 230 (1962). (b) R. A. Kronmal and A. V. Peterson Jr., *Am. Stat.* **33,** 214 (1979). (c) N. Schmitz and F. Lehmann, *Monte Carlo-Methoden I. Erzeugen und Testen von Zufallszahlen*, Vol. 2. *Auflage*, Verlag A. Hain, Meisenheim, 1982. (d) H. E. Exner, *Stat. Software Newslett.* **9,** 71 (1983).

81. (a) Y. A. Shreider (Ed.), *The Monte Carlo Method*, Pergamon, New York, 1966. (b) L. L. Carter and E. D. Cashwell, *Particle-Transport Simulation with the Monte Carlo Method*, ERDA Critical Review Series, TID-26607, Oak Ridge, 1975. (c) E. D. Cashwell and C. J. Everett, *A Practical Manual on the Monte Carlo Method for Random Walk Problems*, Pergamon, New York, 1950.

82. (a) C. D. Zerby, in *Methods in Computational Physics*, Vol. 1, B. Alder, S. Fern-

bach, and M. Rotenberg (Eds.), Academic, New York, 1963, p. 89. (b) A. Razan and H. E. Hungerford, *Nucl. Sci. Eng.* **46**, 1 (1971). (c) E. D. Cashwell, J. R. Neergaard, et al., *Monte Carlo Codes: MCG and MCP*, USAEC Report LA-5157-MS, Los Alamos, 1973. (d) H. Lichtenstein, M. Cohen, H. Steinberg, E. Troubetz-koy, and M. Beer, *The SAM-CE Monte Carlo System for Radiation Transport and Criticality Calculations in Complex Configurations (Rev. 7.0)*, Mathematical Applications Group, New York, 1979.

83. (a) E. S. Troubetzkoy, *UNC-SAM2: A FORTRAN Monte Carlo Program Treating Time-Dependent Neutron and Photon Transport through Matter*, Report UNC-5157, United Nuclear Corporation, 1966. (b) P. A. Robinson Jr., *The Development and Application of a Coupled Monte Carlo Neutron–Photon Transport Code*, Report UCRL-51234, Lawrence Livermore Laboratory, 1972. (c) C. W. Drawbaugh, *Nucl. Sci. Eng.* **9**, 185 (1961). (d) E. D. Cashwell, J. R. Neergaard, W. M. Taylor, and G. D. Turner, *MCN: A Neutron Monte Carlo Code*, Report LA-4751, Los Alamos, 1972.

84. (a) M. J. Berger, in Ref. 82a, p. 135. (b) C. D. Zerby and F. L. Keller, *Nucl. Sci. Eng.* **27**, 190 (1967). (c) R. T. Giuli and T. A. Moss, *Astrophys. J.* **167**, 331 (1971).

85. (a) T. W. Armstrong and R. G. Alsmiller Jr., *Nucl. Sci. Eng.* **33**, 291 (1968). (b) T. W. Armstrong, R. G. Alsmiller Jr., K. C. Chandler, and B. L. Bishop, *Nucl. Sci. Eng.* **49**, 82 (1972).

86. (a) J. A. Fleck Jr. and J. D. Cummings, *J. Comput. Phys.* **8**, 313 (1971). (b) L. L. Carter and C. A. Forest, *Non-linear Radiation Transport Simulation with an Implicit Monte Carlo Method*, Report LA-5038, Los Alamos, 1973.

87. E. Storm and H. I. Israel, *Nucl. Data Tabl.* **A7**, 565 (1971).

88. J. Booz, in *Third Symposium on Neutron Dosimetry in Biology and Medicine*, G. Burger and H. G. Ebert (Eds.), EUR 5848, Commission of the European Communities, Luxembourg, 1978, p. 499.

89. (a) H. G. Paretzke, in Ref. 56, p. 141. (b) D. Combecher, J. Kollerbaur, G. Leuthold, H. G. Paretzke, and G. Burger, in Ref. 40, p. 295. (c) H. G. Paretzke, in Ref. 34c, p. 9. (d) H. G. Paretzke, in Ref. 58a, p. 157.

90. (a) M. J. Berger, in *Second Symposium on Microdosimetry*, H. G. Ebert (Ed.), Report EUR 4452, Commission of the European Communities, Brussels, 1969, p. 541. (b) M. J. Berger, in *Third Symposium on Microdosimetry*, H. G. Ebert (Ed.), EUR 4810, Commission of the European Communities, Luxembourg, 1972, p. 157. (c) M. J. Berger, S. M. Seltzer, and K. Maeda, *J. Atmos. Terr. Phys.* **32**, 1015 (1970).

91. (a) R. N. Hamm, H. A. Wright, R. H. Ritchie, J. E. Turner, and T. P. Turner, in *Fifth Symposium on Microdosimetry*, J. Booz, H. G. Ebert, and B. G. R. Smith, (Eds.), Report EUR 5452, Commission of the European Communities, Luxembourg, 1976, p. 1037. (b) J. E. Turner, H. G. Paretzke, R. N. Hamm, H. A. Wright, and R. H. Ritchie, *Radiat. Res.* **92**, 47 (1982). (c) H. A. Wright, J. E. Turner, R. N. Hamm, R. H. Ritchie, J. L. Magee, and A. Chatterjee, in *Eighth Symposium on Microdosimetry*, J. Booz and H. G. Ebert (Eds.), EUR 8395, Commission of the European Communities, Luxembourg, 1982, p. 101. (d) R. H. Ritchie, R. N. Hamm, J. E. Turner, and H. A. Wright, in Ref. 40, p. 345.

92. (a) B. Grosswendt, in Ref. 91c, p. 79. (b) B. Grosswendt and E. Waibel, *Nucl. Instr. Meth.* **155**, 145 (1978).

93. (a) M. Terrissol and J. P. Patau, in Ref. 55d, p. 414. (b) M. Terrissol, J. P. Patau, and T. Eudaldo, in Ref. 40, p. 169.

94. (a) R. P. Singhal, C. H. Jackman, and A. E. S. Green, *J. Geophys. Res.* **85,** 1246 (1980). (b) A. E. S. Green and D. E. Rio, in *Workshop on the Interface between Radiation Chemistry and Radiation Physics*, Argonne National Laboratory, Report ANL-82-88, 1982, p. 65.

95. (a) D. F. Kyser, in Ref. 36d, p. 119. (b) D. E. Newbury and R. L. Myklebust, in Ref. 36d, p. 153.

96. (a) L. Reimer, *Optik* **27,** 86 (1968). (b) R. Shimizu, Y. Kataoka, T. Ikuta, T. Koshikawa, and H. Hashimoto, *J. Phys. D* **9,** 101 (1976). (c) K. Murata, D. F. Kyser, and C. H. Ting, *J. Appl. Phys.* **52,** 4396 (1981). (d) G. Love, M. G. Cox, and V. D. Scott, *J. Phys. D* **10,** 7 (1977).

97. (a) H. G. Paretzke, G. Leuthold, G. Burger, and W. Jacobi, in Ref. 56, p. 123. (b) J. W. Baum, M. N. Varma, C. L. Wingate, H. G. Paretzke, and A. V. Kuehner, in Ref. 56, p. 93. (c) H. G. Paretzke and F. Schindel, in Ref. 55d, p. 387. (d) W. E. Wilson and H. G. Paretzke, *Radiat. Res.* **87,** 521 (1981). (e) M. Malbert, C. Carel, J. P. Patau, and M. Terrissol, in Ref. 55d, p. 359.

98. (a) D. E. Charlton, D. T. Goodhead, W. E. Wilson, and H. G. Paretzke, *Energy Deposition in Cylindrical Volumes: a) Protons, Energy 0.3–4.0 MeV, b) Alpha Particles, Energy 1.2–20.0 MeV*, Monograph 85/1, Medical Research Council, Radiobiology Unit, Chilton, 1985. (b) M. Zaider, D. J. Brenner, and W. E. Wilson, *Radiat. Res.* **96,** 231 (1983).

99. (a) H. G. Paretzke, in Ref. 6d, p. 3. (b) H. G. Paretzke, in Ref. 91c, p. 67. (c) H. G. Paretzke, in Ref. 40, p. 925. (d) G. Burger, D. Combecher, J. Kollerbaur, G. Leuthold, T. Ibach, and H. G. Paretzke, in Ref. 55d, p. 537.

100. International Commission on Radiation Units and Measurement, *Radiation Quantities and Units*, ICRU-Report 33, ICRU, Bethesda, MD, 1980.

101. International Commission on Radiation Units and Measurements, *Neutron Fluence, Neutron Spectra and Kerma*, ICRU-Report 13, ICRU, Washington, DC, 1969.

102. J. J. Coyne, in *Ion Chambers for Neutron Dosimetry*, J. J. Broerse (Ed.), Commission of the European Communities, EUR 6782, Harwood Academic Publishers, London, 1980, p. 195.

103. M. Cohen, D. E. A. Jones, and D. Greene, *Br. J. Radiol. Suppl.* **11,** 12 (1972); **11,** 81 (1972).

104. National Council on Radiation Protection and Measurement, *Protection against Neutron Radiation*, NCRP-Report 38, NCRP-Publications, Washington, DC, 1971.

105. A. M. Kellerer and H. H. Rossi, *Curr. Top. Radiat. Res. Q.* **8,** 85 (1972).

106. P. J. Kliauga and R. Dvorak, *Radiat. Res.* **73,** 1 (1978).

107. (a) H. G. Paretzke, in Ref. 6b, p. 87. (b) E. V. Benton and W. Nix, *Nucl. Instr. Meth.* **67,** 343 (1969).

108. F. Rohrlich and B. C. Carlson, *Phys. Rev.* **93,** 38 (1954).

109. C. L. Wingate and J. W. Baum, *Radiat. Res.* **65,** 1 (1976).

110. J. W. Baum, in Ref. 90a, p. 653.

111. J. J. Butts and R. Katz, *Radiat. Res.* **30,** 855 (1967).

112. (a) M. N. Varma, H. G. Paretzke, J. W. Baum, J. T. Lyman, and J. Howard, in

Ref. 91a, p. 75. (b) M. N. Varma, J. W. Baum, and A. V. Kuehner, *Radiat. Res.* **62**, 1 (1975). (c) M. N. Varma and J. W. Baum, *Nanometer Dosimetry of 337 MeV/ Nucleon* ^{20}*Ne-Ions*, Brookhaven National Laboratory, Report BNL-22756. (d) M. N. Varma, J. W. Baum, and A. V. Kuehner, *Phys. Med. Biol.* **25**, 651 (1980).

113. (a) M. N. Varma, J. W. Baum, and P. Kliauga, in Ref. 40, p. 227. (b) P. Kliauga and H. H. Rossi, in Ref. 91a, p. 127.

114. H. G. Menzel and J. Booz, in Ref. 91a, p. 61.

115. T. Ibach and D. Combecher, in Ref. 55d, p. 191.

116. A. Mozumder, *J. Chem. Phys.* **60**, 1145 (1974).

117. (a) A. Chatterjee, H. Maccabee, and C. Tobias, *Radiat. Res.* **54**, 479 (1973). (b) A. Chatterjee and H. J. Schäfer, *Radiat. Environ. Biophys.* **13**, 215 (1976).

118. (a) J. Fain, M. Monnin, and M. Montret, in Ref. 56, p. 169. (b) M. Monnin, in *International Topical Conference on Nuclear Track Registration in Insulating Solids and Applications*, Vol. 1, D. Isabelle and M. Monnin (Eds.), Universite de Clermont-Ferrand, 1969, p. II-73.

119. J. Kiefer and H. Straaten, *Phys. Med. Biol.* (submitted)

120. J. W. Hansen and K. J. Olsen, *Radiat. Res.* **97**, 1 (1984).

121. (a) J. L. Magee, *J. Chim. Phys.* **52**, 528 (1955). (b) A. H. Samuel and J. L. Magee, *J. Chem. Phys.* **21**, 1080 (1953). (c) A. Mozumder and J. L. Magee, *Radiat. Res.* **28**, 203 (1966). (d) A. Mozumder and J. L. Magee, *J. Chem. Phys.* **45**, 3332 (1966). (e) J. L. Magee and A. Chatterjee, in Ref. 58a, p. 166.

122. (a) A. Kuppermann, in Ref. 9b, p. 212. (b) G. R. Freeman, in Ref. 94b, p. 9. (c) W. G. Burns, in Ref. 55d, p. 471.

123. M. J. Berger, in Ref. 55d, p. 521.

124. J. E. Turner, J. L. Magee, R. N. Hamm, A. Chatterjee, H. A. Wright, and R. H. Ritchie, in Ref. 55d, p. 507.

125. J. H. Miller and W. E. Wilson, in Ref. 94b, p. 73.

126. (a) H. G. Paretzke, (a) in Ref. 40, p. 925; (b) in Ref. 91a, p. 41.

127. L. H. Gray, *Br. J. Radiol.* **10**, 600 (1937); 721 (1937).

128. H. Späth, *Cluster Analysis Algorithms for Data Reduction and Classification of Objects,* Wiley, New York, 1980.

129. (a) R. R. Sokal, in *Classification and Clustering*, J. van Ryzin (Ed.), Academic, New York, 1977, p. 1. (b) P. H. A. Sneath and R. R. Sokal, *Nature* **193**, 855 (1962). (c) R. R. Sokal and F. J. Rohlf, *Taxonomy* **11**, 33 (1962). (d) P. H. A. Sneath and R. R. Sokal, *Numerical Taxonomy,* Freeman, San Francisco, 1973.

130. (a) L. A. Zadeh, in Ref. 129a, p. 251. (b) J. C. Bezdek, *J. Math. Biol.* **1**, 57 (1974). (c) J. Kacprzyk and R. R. Yager (Eds.), *Management Decision Support Systems Using Fuzzy Sets and Possibility Theory,* Verlag TÜV Rheinland, Köln, 1985. (d) E. Gogala, *Probabilistic Sets in Decision Making and Control,* Verlag TÜV Rheinland, Köln, 1984.

131. (a) H. Haken (Ed.), *Pattern Formation by Dynamic Systems and Pattern Recognition,* Springer-Verlag, Berlin, 1979. (b) K. S. Fu, *Digital Pattern Recognition,* Springer-Verlag, Berlin, 1976. (c) C. H. Chen (Ed.), *Pattern Recognition and Artificial Intelligence,* Academic, New York, 1976.

4 TRACK REACTIONS OF RADIATION CHEMISTRY

John L. Magee and Aloke Chatterjee

Lawrence Berkeley Laboratory
University of California
Berkeley, California

4.1. INTRODUCTION

Radiation chemistry must always be considered in terms of track reactions. Energy is deposited by radiations in tracks and then follows a sequence of nonhomogeneous processes that create and transform reactive intermediates until final radiation chemical products are formed. Track models giving the initial spatial distribution of the energy deposits of charged particles are becoming quite elaborate (1), but the use of this information in the treatment of the subsequent radiation chemistry has not yet been made. In fact, track models oversimplified to the extent that they consist only of isolated radical pairs are widely used in radiation chemistry (2). In this chapter we discuss track reactions in a way that we hope will bring a perspective to the field. We introduce the notion of a general track and attempt to show how various simplified models can be used as approximations.

Most radiation chemical investigations, both experimental and theoretical, deal with energetic electrons as the source of radiation. This situation probably exists because electron accelerators, ^{60}Co sources, and X-ray machines are so generally available. It can be argued that the electron is the most important particle in radiation effects because secondary electrons are always created in an irradiated system. On the other hand, the tracks of heavy particles bring in additional characteristics, so there are qualitative differences between heavy particle and electron irradiations. Thus, the most general treatment of tracks in radiation chemistry must involve heavy-particle tracks.

Consider the track of a heavy particle in a low-atomic-number medium such as water. Figure 4.1 gives a schematic cross-sectional view of the track of a heavy particle going directly into the plane of the paper. About half the energy is deposited in the physical "core," which has a radius r_c; the other half is deposited in the tracks of knocked-out electrons in a region we call the "penumbra," which has a radius r_p (3). The radius r_c is determined by the Bohr adiabatic condition (see Section 4.2.2), and the radius r_p is determined by the penetration of secondary electrons. The chemical core radius r_{ch} is always intermediate between r_c and r_p; it depends on the competition of chemical reactions (as discussed in Section 4.6). The pattern of energy deposition implied by Figure 4.1 leads to a distribution of reactive species that initiate nonhomogeneous reactions.

The chain of events set off by the absorption of high-energy radiations in matter can be grouped in three characteristic temporal stages (4). During the first, or physical, stage of radiation action, lasting perhaps 10^{-14} s on a local time scale, energy is transferred from the radiation to matter with the creation of molecular excitations and ionizations (Chapter 2). The initially excited or ionized molecules are sometimes called primary species; they are usually extremely unstable and undergo secondary reactions, either spontaneously or in collisions with neighboring molecules. In the second, or physicochemical, stage of radiation action, lasting until 10^{-11} s or so after local energy deposit, some new chemical species are formed, and sequences of processes follow with the attainment of thermal equilibrium (or thermalization) but not chemical equilibrium. In the third, or chemical, stage, which starts at a local time of 10^{-11} s and lasts until all radiation chemical reactions are complete, diffusion and reaction of thermalized chemical species occur in the tracks. These reactions, which are extremely nonhomogeneous, are the track reactions of radiation chemistry. Their treatment is the most important objective of this chapter.

It is widely believed that thermalization requires a few picoseconds in a molecular material such as water (5). The radiation chemical processes that occur in a track can be divided into two categories: the early processes that precede thermalization and the nonhomogeneous track reactions of the intermediates (radicals)

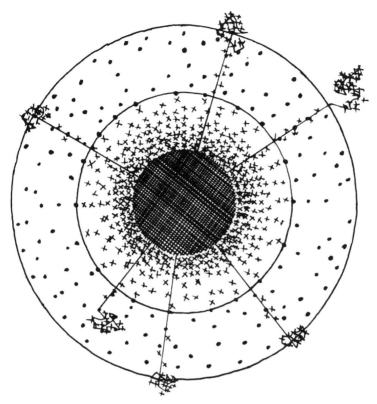

Figure 4.1. Schematic representation of the cross section of the track of a high-energy particle in water. The innermost region is the physical core with radius r_c; the next circle with radius r_{ch} marks the boundary of the chemical core; the outermost circle with radius r_p marks the boundary of the penumbra. The nonhomogeneous distribution of water radicals ($\times$) at the beginning of the thermal period is indicated along with a homogeneous distribution of scavenger molecules ($\circ$). Half of the radicals are near the physical core, and the other half are within the penumbra radius. The radius of the chemical core is determined by the competition of the radical recombination and reaction with scavenger. The dimensions are appropriate for a Ne particle with energy 10 MeV/nucleon and a radical scavenging time of 3×10^{-7} s if the radial scale is log r; $r_p \simeq 2100$ nm.

following thermalization. This classification involves lumping the physical and the physicochemical stages into one and calling it the *prethermal* stage. The earliest attempts at theoretical explanations featured the latter part of the problem (6), that is, the track reactions of the thermalized intermediates. It was thought that the identities of the transient species created by the radiation were known from their radiation chemical effects, although it was not actually understood how they were formed. For example, in the radiolysis of water (7, 8) radiation chemists believed that all effects were produced by the primary water radicals H and OH. Pulse radiolysis provided a method for observing processes at early times, but it was soon realized that the prethermal events were not accessible to direct observation

(9). The track reactions have now been developed to the point that they are actually quite well understood. At present, studies aimed at obtaining information about the prethermal period from theoretical considerations are underway.

We know from elementary considerations that the deposition of energy by a high-energy charged particle is a stochastic process (1). If we consider several tracks of any given type of particle with the same initial energy in a given medium, the tracks will all be different in detail because the energy will have been deposited in stochastic sequences of processes. The numbers of species of all kinds and also their relative distributions in space will vary from track to track. After tracks are formed, the physical and chemical processes that continue are also stochastic. Thus, the most logical way to consider the nonhomogeneous track reactions of radiation chemistry is through techniques that are also stochastic in nature [usually called *Monte Carlo* techniques (10)]. Until the recent past, however, the use of such methods was impractical, and ways of looking at the track reactions that attempted to model the *average track* were developed. These developments are the main subject of Chapter 4. They are important and will continue to have applications well into the future although we expect that stochastic methods will eventually dominate the radiation chemistry of tracks.

In radiation chemistry one is almost exclusively interested in the chemical effects produced by many tracks (i.e., a relatively large energy absorption or dose), and thus it is the average effect of a track that is important. The early development of a theory of track effects started with attempts to describe average tracks, and fluctuations of energy loss events were not considered (6, 11–13). Those models seem to bring out the essential features of radiation chemistry and thus have been useful. In Chapter 4 we do not consider fluctuations.

Section 4.2 considers in some detail a track model based on stopping power theory (which provides average energy loss), which can be applied to any heavy particle. It is a unified model in the sense that it deals with radiation "quality" effects in a uniform manner. Some of the largest uncertainties in track theory involve the early prethermal processes. In condensed phases such as liquid water most of our knowledge with regard to the early track processes is still based on theory and speculation; we look at these problems in Section 4.3. The central feature of radiation chemical models is the diffusion controlled track reactions of the early thermalized radicals. The principal models developed over the past 35 years have involved deterministic treatments using *prescribed diffusion* (2, 6, 14, 15). This type of treatment is considered in Section 4.4. The electron track model has special importance because in radiation interaction the secondary particles are always electrons. All low-LET (linear energy transfer) radiations (16) are composed of electron tracks and all heavy-particle tracks have half their energy deposit coming from generated electron tracks (3). Treatments of electron and heavy-particle tracks are given in Sections 4.5 and 4.6.

In this chapter we are primarily concerned with materials composed of low-atomic-number atoms, such as water. In fact, water is very much emphasized. The radiation chemistry of water has been studied extensively (7, 8), and a great many experimental results are available for comparison with model calculations.

4.2. PHYSICAL DESCRIPTION OF A PARTICLE TRACK

The stopping power of matter for high-energy particles is known very well (17, 18) and furnishes a good basis for a track model to be used in radiation chemistry (12, 13). Stopping power theory gives the average energy loss per unit distance and so it is directly applicable to an energy deposit pattern to be used in a deterministic model aimed at describing an average track.

We look for a model that applies to all tracks of a given heavy particle of energy E penetrating a medium. Consider the energy loss of such a particle with velocity v and charge ze in a material of atomic number Z. An approximate form of the energy loss expression (17, 18) is

$$-\frac{dE}{dx} = 2\kappa ZN \ln \frac{2mv^2}{I} \qquad (4.1)$$

where κ is the stopping parameter,

$$\kappa = \frac{2\pi z^2 e^4}{mv^2} \qquad (4.2)$$

e, m are the electronic charge and mass, N is the number of atoms per unit volume, and I is the mean excitation potential (4, 19, 20).

Equation (4.1) is valid only for nonrelativistic particles with velocity v greater than the classical velocities of the bound electrons of the atoms. The quantity I determines the energy loss of such particles in a given material. A theoretical expression relates I to the distribution of the oscillator strength of the atoms,

$$Z \ln I = \Sigma f_j \ln \epsilon_j + \int f'(\epsilon) \ln \epsilon \, d\epsilon \qquad (4.3)$$

where f_j is the oscillator strength for the jth discrete transition and $f'(\epsilon)$ is the oscillator strength per unit energy interval in the continuum at an energy loss ϵ. The sum extends over all discrete transitions, and the integral extends over the continuum.

To a good first approximation the stopping power of any molecular material is determined by the overall atomic composition and is, therefore, almost independent of specific chemical binding of the atoms. This fact is frequently called the *Bragg rule*. The approximate independence of stopping power on chemical binding has been interpreted to mean that the most important primary loss processes occur for energies that are high compared with energies of chemical binding, and at such excitation energies it is reasonable to assume that the density of oscillator strength depends only on atomic composition.

It is well known that equal contributions to the stopping power of a material for high-energy heavy particles come from glancing collisions, which have low energy loss per event, and head-on collisions, which have high energy loss per event. This fact has sometimes been called by the unfortunate expression "equipartition of energy," which confuses it with the heat capacity of the various degrees of freedom of a molecule. In any case, the principle means that of the total energy loss given by Eq. (4.1), $\kappa ZN \ln 2mv^2/I$ arises from glancing collisions and another $\kappa ZN \ln 2mv^2/I$ arises from knock-on collisions. We can see this as follows. Bethe (18) gives

$$\sigma_j = \kappa \frac{f_j}{\epsilon_j} \ln \left(\frac{2mv^2}{\epsilon_j} \right) \tag{4.4}$$

as the approximate cross section for excitation of a discrete level with frequency $\nu_j = \epsilon_j/h$ and oscillator strength f_j by a particle of velocity v. In our designation this excitation results from a "glancing" collision. We can also say that the differential cross section of an atom for glancing collisions in the continuum at the energy ϵ is given by

$$d\sigma_g(\epsilon) = \kappa \frac{f'(\epsilon)}{\epsilon} \ln \left(\frac{2mv^2}{\epsilon} \right) d\epsilon \tag{4.5}$$

where $f'(\epsilon)$ is $df/d\epsilon$. Although formulas (4.4) and (4.5) for glancing collisions are only approximately valid, they are useful because they allow the cross section to be estimated from the distribution of oscillator strength (4, 19, 20).

The differential cross section for the energy loss ϵ in the interval $d\epsilon$ by a free electron in collision with a heavy particle is given by classical considerations as

$$d\sigma = \kappa \frac{d\epsilon}{\epsilon^2} \tag{4.6}$$

In accordance with the customary assumptions of dispersion theory we take the number of electrons that receive such losses as equal to the oscillator strength of all transitions with energies less than ϵ. Thus, for head-on collisions, we get

$$d\sigma_h = n(\epsilon) \, \kappa \frac{d\epsilon}{\epsilon^2} \tag{4.7}$$

where

$$n(\epsilon) = \Sigma f_i + \int^\epsilon f'(\epsilon') \, d\epsilon' \tag{4.8}$$

The total contribution of each of these processes to the stopping power, $-dE/dx$, is the same; that is,

$$-\left(\frac{dE}{dx}\right)_g = -\left(\frac{dE}{dx}\right)_h = N \int \epsilon \, d\sigma_g(\epsilon) = N \int \epsilon \, d\sigma_h(\epsilon) = \kappa Z N \ln\left(\frac{2mv^2}{I}\right) \quad (4.9)$$

and the sum gives the result of Eq. (4.1).

4.2.1. Track Entities: Spurs, Blobs, Short Tracks, and Branch Tracks

The stopping power theory is based on an implicit assumption that is called the csda, or the continuous-slowing-down approximation. From this theory one gets quantitative information on the average rate of energy loss (i.e., the linear energy transfer, or LET). The usefulness of this information is extremely limited, and for an understanding of the radiation chemical processes it can actually be misleading if one assumes that all track segments receive an energy corresponding to the average energy deposit. This would be the situation if a continuous-slowing-down mechanism actually applied. The most profitable way to consider a track is in terms of the energy loss spectrum given implicitly in the equations of Section 4.2. Equations (4.4) and (4.5) show that the glancing collisions create excitations to energy levels that have oscillator strength for ordinary optical excitation. In low-atomic-number materials the oscillator strength distributions are essentially contained in the region 6–100 eV (4, 19, 20). Half of the energy loss of a heavy particle occurs in loss events along the trajectory, which individually involve less than 100 eV with a well-defined average in the region 30–40 eV (19, 20, 22, 23). These low-energy losses create track entities that are called "spurs." If the particle is fast and lightly ionizing, the spurs are widely spaced and develop independently, that is, without interference from the neighboring spurs. If the particle is slow enough to be heavily ionizing, the spurs overlap and form a continuous column.

Half of the energy deposit of a heavy particle is in the spurs formed in the core. If the particle is lightly ionizing (such as a proton with energy of about 100 MeV, or 10 pJ) the spurs are well separated and the isolated spur is roughly an approximation to the track. An average spur may contain two or three radical pairs, and so we see that even an approximation of isolated radical pairs (2) for a track may be reasonable.

The part of the spectrum arising from the knock-on electrons consists of electrons with energies from about 100 eV to a maximum value of $2mv^2(1 - \beta^2)^{-1}$. These recoil electrons are created in processes dependent on the Rutherford scattering cross section, which varies as ϵ^{-2}, and thus low energies are favored. The low-energy loss events (near 100 eV, or 10 aJ) are like spurs, whereas the very high energy loss events (say in the multikilovolt or femtojoule range) produce branch tracks. Thus, the behavior of individual loss events in the knock-on spectrum vary widely. The problem of describing the radiation chemistry of these events is the same as that of describing the radiation chemistry of the electron track itself (21, 24). The 100-eV upper limit for spurs is arbitrary. Track entities formed with energies a little above 100 eV are like spurs, but for some higher value of the energy loss a recoil electron energetic enough to create a series of spurs is formed and a short track results. Mozumder and Magee (21) considered these matters for electrons in water and decided that for an energy loss of 500 eV the track entity should be classified as a short track. The track entities in the range 100–500 eV are called "blobs." This noncommittal name is used because the shape of the entities varies from essentially spherical to rodlike as the energy increases from 100 to 500 eV.

The "short track" category of track entity also has an upper limit in energy. Mozumder and Magee (21) argued that at an energy of 5 keV a knocked-out electron forms several spurs before ending up in the continuous distribution called a short track. Thus 5 keV was chosen as the upper limit for the short track.

The radiation chemistry of electron tracks is considered in Section 4.5. The treatment depends on the system absorbing the energy, its reaction sequences, and so on. However, to a large extent the track entities are the same for all low atomic number materials. Mozumder and Magee suggested that all electron tracks can be considered as composed of three track entities:

Spur	Energy deposit of $\sim$ 6–100 eV
Blob	Energy deposit of $\sim$ 100–500 eV
Short track	Energy deposit of $\sim$ 500–5000 eV

4.2.2. Heavy-Particle Tracks: Core and Penumbra

The half of the energy deposited along the trajectory in the form of spurs is called the physical core. The other half of the energy lost in knock-on collisions that create recoil electrons with energies up to $2mv^2(1 - \beta^2)^{-1}$ is called the penumbra (13, 25) (see Figure 4.1).

It has been mentioned that energy deposition processes are stochastic in nature, but with respect to a heavy particle it may be a reasonable approximation to discuss its tracks in terms of average spurs. The validity of this approximation is based on the fact that the fluctuation in energy loss by a heavy particle is not very significant. For tracks generated by electrons this is not true.

In Section 4.6 we consider the radiation chemistry of heavy-particle tracks in a unified manner with a simple track model. In this section we discuss the physical parameters that characterize a heavy-particle track to help the reader to a better understanding of the model of Section 4.6.

The radius of the physical core is given by the Bohr adiabatic condition:

$$r_c = \frac{\beta c}{\Omega_p} \tag{4.10}$$

where $\beta = v/c$ and Ω_p is the plasma oscillation frequency* equal to $(4\pi ne^2/m)^{1/2}$, where n is the number density of the electrons in water, e and m are the charge

---

*Ω_p is the rate of change of phase of the plasma oscillation, more precisely called the angular velocity, with units radian/s. (Ed.)

TABLE 4.1. Track Parameters Independent of Particle Charge[a]

Energy (MeV/nucleon)[b]	β	r_c (nm)	r_p (nm)	ϵ_{max}(eV)[b]	λ
1000	0.876	9.016	2.7×10^5	3.4×10^6	23.42
800	0.843	8.675	2.4×10^5	2.5×10^6	22.46
600	0.794	8.169	2.1×10^5	1.7×10^6	22.31
400	0.715	7.354	1.6×10^5	1.1×10^6	21.98
200	0.568	5.843	8.4×10^4	4.9×10^5	21.15
100	0.430	4.421	3.9×10^4	2.3×10^5	20.17
80	0.390	4.012	3.0×10^4	1.8×10^5	19.84
60	0.343	3.526	2.1×10^4	1.4×10^5	19.38
40	0.284	2.923	1.3×10^4	0.9×10^5	18.80
20	0.204	2.099	5.3×10^3	4.4×10^4	17.67
10	0.145	1.496	2.1×10^3	2.2×10^4	16.49
8	0.130	1.340	1.6×10^3	1.8×10^4	16.17
6	0.113	1.163	1.1×10^3	1.3×10^4	15.70
4	0.092	0.951	6.2×10^2	8.8×10^3	14.96
2	0.065	0.673	2.4×10^2	4.4×10^3	13.75
1	0.046	0.477	9.6×10^1	2.2×10^3	12.61
0.8	0.041	0.426	7.1×10^1	1.8×10^3	12.23
0.6	0.036	0.369	4.8×10^1	1.3×10^3	11.74
0.4	0.029	0.302	2.8×10^1	8.8×10^2	11.06
0.2	0.021	0.213	1.1×10^1	4.4×10^2	9.89
0.1	0.015	0.151	4.3	2.2×10^2	8.70

[a] Reprinted with permission from *Journal of Physical Chemistry*, vol. 84. Copyright (1980) American Chemical Society.
[b] 1 eV = 0.16 aJ.

and mass of the electron, $n = 3 \times 10^{23}$ electrons/cm³. Values of r_c for various values of the specific energy of particles are given in Table 4.1. This radius is the approximate limiting distance from the particle trajectory at which an electronic excitation can be created. It has nothing to do with the subsequent processes that disperse the transient species produced. Values of radii at later times are also of interest. Another radius used in the heavy-particle track model involves the chemical competition of the recombination of transient species and their reaction with a scavenger in homogeneous distribution. This is called the radius of the chemical core r_{ch}.

Chatterjee et al. (25, 26) have calculated the energy density deposited in the penumbra by the knock-on electrons. For each velocity of a heavy particle there is a well-defined maximum distance that its knock-on electrons can go, and we call it r_p. Table 4.1 summarizes the track parameters that depend only on v and are independent of particle charge. The values of the average energy densities in the particle tracks are given by

$$\rho_{core} = \frac{LET}{2} [\pi r_c^2]^{-1} + \frac{LET}{2} \left[2\pi r_c^2 \ln \left(\frac{e^{1/2} r_p}{r_c} \right) \right]^{-1} \qquad r \leq r_c \quad (4.11)$$

$$\rho_{pen}(r) = \frac{LET}{2} \left[2\pi r^2 \ln \left(\frac{e^{1/2} r_p}{r_c} \right) \right]^{-1} \qquad r_c < r \le r_p \qquad (4.12)$$

Here the core refers to the physical core.

In Table 4.1 ϵ_{max} refers to the maximum value of energy a knock-on electron can have for an incident particle of a given velocity. The last column of Table 4.1 gives λ, the ratio of the *initial* energy density in the physical core to that in the penumbra near r_c, and it shows that concentrations of intermediates in the core are expected to be much larger than those in the penumbra.

Although we have given a formula [Eq. (4.12)] for *energy density* in the penumbra of a track, this region is actually comprised of electron tracks. These tracks are well separated from each other, and the concept of track entities is useful.

4.2.3. Formal Expression for Track Yield

In radiation chemistry we are always interested in radiation chemical yields, which are normally given in terms of G values, or the numbers of molecules formed per 100 eV of energy deposited by the radiation. Two kinds of G values are used: the integral yield G and the differential (or "local") yield G' (also called the "track segment" yield). The G and G' are functions of the same variables, such as the type of particle, the particle energy, the observed product, and so on; the notation we employ is flexible, and any number of the variables may appear as arguments; for example, the differential ferric yield for a particle of charge z and energy E per nucleon in a dilute solution with ferrous ion concentration c_s may be given as $G'(Fe^{3+}; E)$ when the other conditions are understood or $G'(Fe^{3+}; E, a, c_s)$ when the particle (a) and the concentration of solute (Fe^{2+}) need to be indicated explicitly. We are usually interested in the variation of a particular yield with heavy-particle energy and merely write $G'(E)$ or $G(E)$. The two kinds of G values are related as follows:

$$G'(E) = \frac{d}{dE} EG(E) \qquad (4.13)$$

From a theoretical point of view $G'(E)$ is the more fundamental quantity.

We are concerned in this chapter only with systems in which single-track effects dominate and radiation chemical yields are sums of yields from individual tracks. The simplest way to introduce the concept of a unified track model for heavy particles (12, 13) is to consider the special case of the track of a heavy particle with an LET below 2–3 eV/nm, which in practice limits us to protons, deuterons, or α particles with energy above 100 MeV/nucleon. At these LET values, to a good approximation, spurs formed by the main particle track can be considered to remain isolated throughout the radiation chemical reactions. For such a case the G' value of any radiation chemical product can be explicitly given if the $G_e(\epsilon)$ value of the same product is known for electron tracks. It is understood that we are always talking about the same radiation chemical product (not indicated explicitly), and we can write

$$G'(E) = (1 - f)G_{sp} + f \frac{\displaystyle\int_{\epsilon_0}^{\epsilon_{max}} G_e(\epsilon) w(E, \epsilon) \, d\epsilon}{\displaystyle\int_{\epsilon_0}^{\epsilon_{max}} w(E, \epsilon) \, d\epsilon} \qquad (4.14)$$

where $G'(E)$ is the differential yield of the product per 100 eV of energy deposited by the particle with energy E; G_{sp} is the yield in isolated spurs; $G_e(\epsilon)$ is the G value in an electron track that starts with energy ϵ; $w(E, \epsilon)$ is the fraction of energy lost by the particle of energy E in the creation of knock-on electrons per unit energy interval at ϵ; ϵ_0 is 100 eV, the borderline energy that is considered to separate glancing collisional losses from knock-on collisional losses; $\epsilon_{max} = 2mc^2\beta^2/(1 - \beta^2)$, the maximum knock-on electron energy; and f, the fraction of energy expended in the creation of knock-on electrons, is 0.5 in the case of heavy particles.

All spur properties, such as their distribution in energy between 6 and 100 eV, are expected to be independent of the particle that creates them. The first term on the right of Eq. (4.14) is the core contribution and the second term is the penumbra contribution. The validity of Eq. (4.14) is seen from elementary considerations: the total yield is made up of contributions from a finite number of track entities (spurs in the core and knock-on electron tracks in the penumbra) that develop and produce their yields independently of one another.

Equation (4.14) is important for heavy-particle track theory. It gives an explicit relationship between the differential yield of a heavy-particle track in the low-LET region in terms of the yields of electron tracks. Furthermore, a similar equation applies to electron tracks themselves (see Section 4.5).

Equation (4.14) is important for another reason; it is an integral relationship that the G values satisfy. Thus, one can set up a scheme to extend the knowledge of G values into regions that have not been investigated experimentally. In Section 4.5 an application of this method is made for electron tracks.

4.3. INITIAL PHASES OF TRACK DEVELOPMENT

In Section 4.2 we analyzed the stopping power formulas to obtain energy deposit patterns for tracks. Reactive chemical intermediates are formed in these tracks, and the nonhomogeneous track reactions follow. In this section we introduce chemical phenomena into the discussion; the radiolysis of water is used for illustration. Consider the following equation for the chemical action in the prethermal phase of track development:

$$H_2O \rightsquigarrow H_3O^+, \; OH, \; e_{aq}^-, \; H, \; H_2, \; H_2O_2 \qquad (4.15)$$

This equation is not balanced chemically and merely enumerates the species that are formed in the prethermal phase; the crooked arrow means radiation action is involved.

Equations like Eq. (4.15) indicate the products that have been formed under various conditions. Chemical mechanisms are implied, and there are always nonhomogeneous processes involved. Although Eq. (4.15) applies to the prethermal phase of the radiation action on water, similar equations can also be used for the thermal phase.

Pulse radiolysis, introduced as an experimental technique just before 1960, made possible the direct observation of species created by irradiation; the hydrated electron e_{aq}^- is the best known example. It was at first expected that experimentalists would be able to investigate successively earlier times as pulse radiolysis techniques improved, but later it was realized that a limitation exists with respect to direct observation.

The limit on the real-time measurement of early time phenomena may be imposed either by the time required to make the deposit or the transit time of the analyzing light. For the purpose of this consideration a minimal reaction cell can be taken as a cube 3 mm on a side. If the energy is deposited by a δ pulse of electrons traveling with the velocity c of light in a vacuum, the beam requires 10 ps to travel through the cube. The analyzing light requires a slightly longer time to traverse the sample (because the refractive index of the liquid is >1.0) and inevitably records effects arising from energy deposits with a time spread of 10 ps or more. It is unlikely that improvement of experimental techniques will allow time resolutions of less than a few picoseconds for irradiation with electrons. We refer to this limit as the *picosecond barrier* to early real-time observations. Information on the very earliest times must come from theory and interpretation of observations made at later times.

4.3.1. Prethermal Events

Let us consider the early events from a theoretical point of view to see if formation of the species on the right side of Eq. (4.15) appears to be reasonable. All the energy is initially utilized in the excitation of electrons, and the most probable event is an ionization. First consider the positive ion H_2O^+. In a condensed phase such as liquid water an ionization event cannot involve one molecule alone; an H_2O^+ ion interacts with its neighbors, and the charge must be transferred from one molecule to another. A more precise statement would say that the positive hole is created and moves initially as a wave packet. The water structure involves hydrogen bonding of each molecule to four others, and so there is a close coupling of the O–H stretching vibrations, which have frequencies of about 10^{14} s^{-1}. It has been conjectured that on the time scale of 10^{-14} s vibrational energy is transmitted to the OH vibrations. There is an irreversible process in this system involving the coupled OH vibrations; that is,

$$H_2O^+ + H_2O \longrightarrow H_3O^+ + OH \qquad (4.16)$$

which can occur by either a hydrogen atom or a proton moving from one molecule to a neighbor (34).

It is usually said that reaction (4.16) occurs in $1-2 \times 10^{-14}$ s. Thus, we expect that all H_2O^+ ions lead to (H_3O^+, OH) pairs on this time scale.

The ionized electrons quickly fall into the subexcitation range, that is, below 6 eV. How these electrons lose energy is not really understood, although there has been much speculation (27–29). It is generally agreed that the rate of loss is in excess of 10^{13} eV/s. For a long time the assumption that hydration could only occur after 10^{-11} s dominated the thinking on this question. This is the relaxation time for water stimulated by microwave radiation. Later it was realized that another type of relaxation, called *longitudinal relaxation*, occurs at a shorter time, 4×10^{-13} s (30), and it is now believed that electrons can be trapped and hydrated on this shorter time scale.

In addition to ionized states electronic excited states are formed. Some of these states (the Rydberg states) probably ionize immediately and some of them dissociate,

$$H_2O^* \longrightarrow H + OH \qquad (4.17)$$

$$H_2O^{**} \longrightarrow H_2 + O \qquad (4.18)$$

Reaction (4.18) is followed by the reaction of O with H_2O to form H_2O_2.

Thus all of the products of Eq. (4.15) can be accounted for, and they form in the prethermal time phase.

4.3.2. Prethermal Time Scale

The picosecond barrier prevents direct observation of prethermal events, but Hamill et al. (31–33) have used a technique involving the competition of reactions to obtain information about them. Ogura and Hamill (33) used concentrated chloride solutions to investigate the motion of the epithermal H_2O^+ ion. They reasoned that the positive charge jumps from molecule to molecule and that the reaction

$$H_2O^+ + Cl^- \longrightarrow H_2O + Cl \qquad (4.19)$$

occurs if the chloride ion is present to compete with reaction (4.16). In these solutions Cl atoms formed in (4.19) react with Cl^- according to

$$Cl + Cl^- \longrightarrow Cl_2^- \qquad (4.20)$$

and the yields of Cl_2^- can be measured as a function of Cl^- concentration. Ogura and Hamill (33) concluded from analysis of their experimental data that in pure water the H_2O^+ hole jumps 14 times on the average before reaction (4.19) occurs. Estimates of the positive-charge jump time are in the range of 10^{-15} s, and thus the ion molecule reaction (4.16) takes $\sim 1.4 \times 10^{-14}$ s in agreement with the conjecture of Section 4.3.1.

Experiments also give information on the prethermal electron. Hamill (31) suggested that in the earliest time scale the epithermal electrons (he called them "dry" electrons) can react with scavengers in competition with the hydration process. Sawai and Hamill (32) showed that large concentrations of scavengers (such as Cd^{2+}, Cu^{2+}, NO_2, H_2O_2, CH_3NO_2, etc.) can reduce the initial yields of hydrated electrons, and these authors interpreted the results in terms of dry electron reactions with the scavengers. Ogura and Hamill (33) also investigated concentrated solutions of scavengers in mixtures of water and alcohols. If the assumptions are made that the epithermal (or dry) electron hops from a molecule to neighboring molecules, that the trapping process that forms the hydrated electron can occur at any site, and that reaction with scavengers occurs on every encounter, a time scale can be established. This analysis sets the trapping time at about 10^{-13} s in both water and alcohol.

Hunt (35) studied dry electron reactions by making direct observations at 10 ps using pulse radiolysis. His results can be interpreted in the same framework of dry electron reactions in the prethermal period that Hamill uses. Czapski and Peled

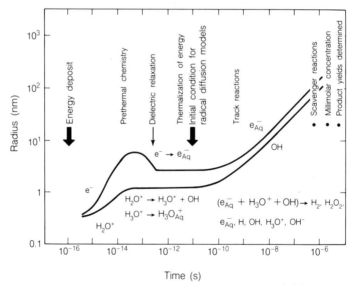

Figure 4.2. Space–time development of a small isolated energy deposit (spur) in a dilute aqueous system. In the prethermal time period ($t < 3 \times 10^{-12}$ s), phenomena involve electronic motion for the most part; the ion–molecule reaction creating H_3O^+ and the longitudinal relaxation of H_2O require only proton motion. In the postthermal period the species that have already been created undergo diffusion-controlled reactions with each other (indicated by the parenthesis on the left of the equation). (From *Radiation Research, Proceedings of Sixth International Congress of Radiation Research*, edited by S. Okada, M. Imamura, T. Terashima, and H. Yamaguchi, Copyright 1979. Reprinted with permission of Japanese Association for Radiation Research.)

(36), however, pointed out that another interpretation involving fast reaction in the early thermal period between reactants formed close together ("contact" reactions) is possible. The uncertainties in the prethermal mechanisms need not concern us in our considerations of the thermal track reactions.

4.3.3. Time Scale of Water Radiolysis

The prethermal processes discussed here are involved in the formation of a distribution of thermalized chemical intermediates that react in the track reactions of irradiated water. Consider a single energy deposit in the range 6–100 eV, that is, a spur. All processes that occur following the initial deposit increase the size of the affected region. Figure 4.2 shows how the radius of a spur in water is believed to change with time and the processes that are probably occurring at various times. The nonhomogeneous nature of the reactions of the thermalized intermediates is indicated by a parenthesis on the left side of the reaction in the figure. It would be desirable to describe all of the processes that occur in Figure 4.2 in a unified manner, and perhaps this will be possible in the near future using Monte Carlo techniques. At present, however, the method of the average track presented in the remainder of this chapter is the only available way to give the essentials of track effects.

4.4. RADICAL DIFFUSION MODEL

We assume that about 10 ps after the passage of a charged particle through a solution, transient chemical species have been formed in certain typical nonhomogeneous spatial distributions. The system is in local thermal equilibrium but not in chemical equilibrium, and hence reactions of these species will continue until such equilibrium is attained. The nature of the initial spatial nonhomogeneity of the chemical species is that their concentrations are high in the vicinity of the sites of energy deposition (spurs, blobs, short tracks, etc.). These species then proceed to react with or diffuse away from each other. The problem is to evaluate the yields of various products or radicals at any given time after the passage of the particle.

The radical diffusion model has been developed in most detail for the radiation chemical system in which each track is completely independent, which means that the intermediates formed in one track react completely with each other and constituents of the medium before they can diffuse far enough to encounter intermediates from another track (37). This situation is required for the yield to be independent of the dose rate, and this is found experimentally to be the most common case.

The complete set of chemical reactions that occur in the irradiated system must be obtained. This information can come from radiation chemistry or from general knowledge of the chemical nature of the system. Table 4.2 gives the reactions that occur in irradiated water. The reactions included are the ones that presumably

TABLE 4.2. Reactions in Irradiated Neutral Water

Reactions	Reaction Rate Constant[a] (1/mole·s)
A. Recombination of Primary Radicals[b]	
*1. $H + H \rightarrow H_2$	1×10^{10}
2. $e_{aq}^- + H \rightarrow H_2 + OH^-$	2.5×10^{10}
3. $e_{aq}^- + e_{aq}^- \rightarrow H_2 + 2\, OH^-$	6×10^{9}
4. $e_{aq}^- + OH \rightarrow OH^- + H_2O$	3×10^{10}
*5. $H + OH \rightarrow H_2O$	2.4×10^{10}
*6. $OH + OH \rightarrow H_2O_2$	4×10^{9}
7. $H_3O^+ + e_{aq}^- \rightarrow H + H_2O$	2.3×10^{10}
8. $H_3O^+ + OH^- \rightarrow H_2O$	1.4×10^{11}
B. Other Reactions of Radicals[b]	
*9. $H + H_2O_2 \rightarrow H_2O + OH$	1×10^{10}
10. $e_{aq}^- + H_2O_2 \rightarrow OH + OH^-$	1.2×10^{10}
*11. $OH + H_2O_2 \rightarrow H_2O + HO_2$	5×10^{7}
*12. $OH + H_2 \rightarrow H_2O + H$	6×10^{7}
*13. $HO_2 + H \rightarrow H_2O_2$	1×10^{10}
14. $e_{aq}^- + O_2 \rightarrow O_2^- + H_2O$	1.9×10^{10}
*15. $HO_2 + OH \rightarrow H_2O + O_2$	1×10^{10}
*16. $HO_2 + HO_2 \rightarrow H_2O_2 + O_2$	2×10^{6}
*17. $H + O_2 \rightarrow HO_2$	1×10^{10}
18. $O_2^- + H_3O^+ \rightarrow HO_2$	3×10^{10}
19. $H_3O^+ + HO_2^- \rightarrow H_2O_2$	3×10^{10}
20. $HO_2 + O_2^- \rightarrow HO_2^- + O_2$	5×10^{7}
21. $H_2O_2 + HO_2 \rightarrow H_2O + O_2 + OH$	530
22. $H_2O_2 + O_2^- \rightarrow OH + OH^- + O_2$	16
C. Dissociation Reactions[c]	
23. $H_2O \rightarrow H_3O^+ + OH^-$	2.5×10^{-5}
24. $HO_2 \rightarrow H_3O^+ + O_2^-$	1×10^{6}
25. $H_2O_2 \rightarrow H_3O^+ + HO_2^-$	3×10^{-2}

[a] Rate constants were selected from the collection published by Ross et al. (38–41).
[b] Equations preceded by an asterisk occur in acid water.
[c] Rate constants for Eqs. 23–25 are s^{-1}.

occur between thermalized species at times greater than 10^{-11} s after energy deposit; for neutral water there are 12 such species; they are H, e_{aq}^-, OH, H_3O^+, H_2O, H_2, H_2O_2, OH^-, HO_2, O_2, O_2^-, HO_2^-. In Eq. (4.15) there are six products formed in the prethermal period; five others are formed in track reactions, and H_2 is also included here. In 0.8 N acid (the medium for the Fricke dosimeter) hydrated electrons react in less than 10^{-10} s to form H atoms, and only seven species are present at longer times: H, OH, H_2O, H_2, H_2O_2, HO_2, O_2. Actually one can never be sure all reactions occurring in an irradiated system have been identified, and we must assume that truncated systems are inevitably used. An extreme case of truncation is the *one-radical model*, which has only a single radical intermediate (say, R) and a single radiation chemical product (say, R_2) arising from combination of the radicals.

The set of P species that react in an irradiated system undergo Q reactions. In the case of water $P = 12$ and $Q = 25$ (see Table 4.2). Let us write a system of differential equations for the P species based on the Q equations for water. It has generally been assumed that we have a set of partial differential equations (14, 42, 43)

$$\frac{\partial c_i}{\partial t} = D_i \, \nabla^2 c_i + \frac{dc_i}{dt} \quad i = 1, 2, 3, \ldots, P \tag{4.21}$$

to determine the track reactions; here dc_i/dt is to be obtained from Table 4.2. Equation (4.21) is the same as Eq. (1.5). The symbol c_i stands for $c_i(r, t)$ to designate the probability density of finding a species of type i at a time t and position r. For the sake of simplicity we will call this probability density a concentration. In Eq. (4.21) D_i is the diffusion coefficient of species of type i, and ∇^2 is the Laplacian operator. The physical interpretation of this equation is that the local change of concentration of species i in a small volume element of the system is equal to the net amount of the species that diffuses into that element divided by its volume (the ∇^2 term) plus the local rate of formation of the species due to chemical reactions involving other species minus its local rate of disappearance due to reaction with other species (the dc_i/dt term). This term involves the reaction rate expressions given in Table 4.3.

Given the initial radical distributions, $c_i(r, 0)$, the diffusion coefficients of species, the reaction mechanism, and the rate constants, the system of equations represented by (4.21) can be solved using numerical techniques. Kuppermann et al. (14, 42, 43) solved the set of equations for water using both one-radical and multiradical approximations. These authors assumed that the initial concentrations of species (at about 10^{-11} s) can be given in terms of continuous functions in space. Gaussian functions were used, and other theoretical radiation chemists have made the same approximations. Perhaps with the increasing development of Monte Carlo techniques the initial conditions will be known better in the near future.

The set of Eqs. (4.21) is nonlinear, and no general method of solution exists. The straightforward numerical procedure is to convert them to *difference* equations and solve by the method of the march of steps. In this process it must be remembered that the resulting *parabolic partial difference* equations may generate convergent solutions that are not the correct solutions of the corresponding differential equations. To safeguard against this difficulty, a certain relationship between increments in the spatial and temporal directions must be maintained. In some sense Eqs. (4.21) are like the self-consistent field equations of quantum chemistry, in that the mechanism, rate constants, initial conditions, and so on, can be varied

TABLE 4.3. Differential Equations for Transient Species and Radiation Products in Irradiated Neutral Water[a]

$$*\frac{d}{dt}(H) = -2k_1(H)^2 - k_2(e^-_{aq})(H) - k_3(H)(OH) + k_7(H_3O^+)(e^-_{aq}) - k_9(H)(H_2O_2) + k_{12}(OH)(H_2) - k_{13}(HO_2)(H) - k_{17}(H)(O_2)$$

$$\frac{d}{dt}(e^-_{aq}) = -k_2(e^-_{aq})(H) - 2k_3(e^-_{aq})^2 - k_4(e^-_{aq})(OH) - k_7(H_3O^+)(e^-_{aq}) - k_{10}(e^-_{aq})(H_2O_2) - k_{14}(e^-_{aq})(O_2)$$

$$*\frac{d}{dt}(OH) = -k_4(e^-_{aq})(OH) - k_5(H)(OH) - 2k_6(OH)^2 + k_9(H)(H_2O_2) + k_{10}(e^-_{aq})(H_2O_2) - k_{11}(OH)(H_2O_2) - k_{12}(OH)(H_2) - k_{15}\,(HO_2)(OH) + k_{21}(H_2O_2)(HO_2^-) + k_{22}(H_2O_2)(O_2^-)$$

$$\frac{d}{dt}(H_3O^+) = -k_7(H_3O^+)(e^-_{aq}) - k_8(H_3O^+)(OH^-) - k_{18}(H_3O^+)(O_2^-) + k_{23}(H_2O) + k_{24}(HO_2) - k_{19}(H_3O^+)(HO_2^-) + k_{25}(H_2O_2)$$

$$*\frac{d}{dt}(H_2O) = k_4(e^-_{aq})(OH) + k_5(H)(OH) + k_7(e^-_{aq})(H_3O^+) + k_8(H_3O^+)(OH^-) + k_9(H)(H_2O_2) + k_{11}(OH)(H_2O_2) + k_{12}(OH)(H_2) + k_{15}(HO_2)(OH) - k_{23}(H_2O) + k_{21}(H_2O_2)(HO_2)$$

$$*\frac{d}{dt}(H_2) = k_1(H)^2 + k_2(e^-_{aq})(H) + k_3(e^-_{aq})^2 - k_{12}(OH)(H_2)$$

$$*\frac{d}{dt}(H_2O_2) = k_6(OH)^2 - k_9(H)(H_2O_2) - k_{10}(e^-_{aq})(H_2O_2) - k_{11}(OH)(H_2O_2) + k_{13}(HO_2)(H) + k_{16}(HO_2)^2 + k_{19}(H_3O^+)(HO_2^-) - k_{21}(H_2O_2)(HO_2) - k_{22}(H_2O_2)(O_2^-) - k_{25}(H_2O_2)$$

$$\frac{d}{dt}(OH^-) = k_2(e^-_{aq})(H) + 2k_3(e^-_{aq})^2 + k_4(e^-_{aq})(OH) - k_8(H_3O^+)(OH^-) + k_{10}(e^-_{aq})(H_2O_2) + k_{23}\,(H_2O) + k_{22}(H_2O_2)(O_2^-)$$

$$*\frac{d}{dt}(HO_2) = k_{11}(OH)(H_2O_2) - k_{13}(HO_2)(H) - k_{15}(HO_2)(OH) + k_{17}(H)(O_2) + k_{18}(H_3O^+)(O_2^-) - 2k_{16}(HO_2)^2 - k_{24}\,(HO_2) - k_{20}(HO_2)(O_2^-) - k_{21}(HO_2)(H_2O_2)$$

$$*\frac{d}{dt}(O_2) = -k_{14}(e^-_{aq})(O_2) + k_{15}(HO_2)(OH) + k_{16}(HO_2)^2 - k_{17}(H)(O_2) + k_{20}(HO_2)(O_2^-) + k_{21}(H_2O_2)(HO_2) + k_{22}(H_2O_2)(O_2^-)$$

$$\frac{d}{dt}(O_2^-) = k_{14}(e^-_{aq})(O_2) - k_{18}(H_3O^+)(O_2^-) + k_{24}(HO_2) - k_{20}(HO_2)(O_2^-) - k_{22}(H_2O_2)(O_2^-)$$

$$\frac{d}{dt}(HO_2^-) = -k_{19}(H_3O^+)(HO_2^-) + k_{20}(HO_2)(O_2^-) + k_{25}\,(H_2O_2)$$

[a]Equations preceded by an asterisk apply to acidic water (see Equations 4.2 and 4.7).

until the results agree with experiment. No one has thought about the general problems in such a disciplined way that a satisfactory analysis can be made at this time. However, it is possible to get an indication of the validity of the entire scheme, and this has been done in the case of water, as we shall see later.

Once $c_i(r, t)$ is determined for every i, r, and t, one can integrate over space and time to obtain the total amounts of each reaction product formed. These are related in a simple manner to the G values of the several species for a spur or any other track entity.

4.4.1. Reactions in Spurs

A key concept in the understanding of track reactions in low-LET irradiations is the insensitivity of the yields of scavenger reactions to scavenger concentration when this concentration is low. In Chapters 2 and 5 of this volume it is shown that a pair of particles (either a charge pair or neutral particles) that starts at time $t = 0$ at a finite separation r_0 has a probability $p(r_0)$ that the particles will combine and a probability $1 - p(r_0)$ that they will escape from each other. Thus, an ensemble of such pairs of particles is divided into two categories: those that combine and those that escape. The same situation holds for clusters of particles generally: a finite fraction of such particles combine and the others escape. The radiation chemistry of systems that have scavengers to react with the uncombined radicals is simple if the tracks can be approximated as made up of strings of spurs. From each spur a fraction of the radicals escapes combination and ultimately reacts with the scavenger.

Consider a model system in which a single kind of radical R is formed that can react with another radical or with a scavenger S:

$$R + R \longrightarrow R_2 \tag{4.22}$$

$$R + S \longrightarrow RS \tag{4.23}$$

The track is a "string of beads," that is, it is composed only of widely spaced spurs. Let us compare this model track with the description of a track given in Section 4.2. There it was shown that about half the deposited energy is in the form of spurs along the particle trajectory. The average energy in a spur is about 40 eV, and so the average separation of spurs along the track is $\sim 40/(\text{LET}/2)$ or 80/LET. If the LET is small (for example, ~ 0.2 eV/nm as for a 100 MeV proton), the spurs are hundreds of nanometers apart, and of course the penumbra is even more widely spread out. This is a typical "low-LET" situation, and for it a string of beads track model is appropriate.

It is important to realize that the isolated spur is in some way a model for the track of a high-energy particle if the LET is small. And in the same spirit the model for the track of a higher LET particle is a string of spurs that overlap as they develop.

We now consider the treatment of an isolated spur. The set of Eq. (4.21) is reduced to four equations, one each for R, R_2, S, and RS. We consider in some detail only the equation for the radical R. In order to make any progress, we must introduce a further simplification, and this is done in the form of an approximation called *prescribed diffusion* (6, 14, 15). Prescribed diffusion has been used widely in studies of the track reactions of radiation chemistry. The approximation is largely intuitive and is justified only in that it gives results that are reasonable. According to this approximation, we assume that the radical concentration is determined by diffusion alone, except for depletion by recombination. Thus, the radical concentration in each spur is given by

$$c(r, t) = \frac{N(t) \exp (-r^2/4Dt)}{(4\pi Dt)^{3/2}}, \qquad t \geq t_0 \qquad (4.24)$$

where r is the distance from the center of the spur, $N(t)$ is the number of radicals that exist in the distribution at time t, and t_0 is a fictitious time that gives a reasonable initial concentration

$$r_0^2 = 4Dt_0 \qquad (4.25)$$

where r_0 is the appropriate parameter in the Gaussian required to give the desired initial spatial distribution. The radicals were created by processes that did not involve diffusion of thermalized species, and so the equation has no meaning for $t < t_0$.

In this model the track equation for the radical concentration becomes

$$\frac{\partial c}{\partial t} = D \nabla^2 c - kc^2 - k_s c_s c \qquad (4.26)$$

Another approximation, neglect of scavenger depletion, is introduced here. This means that the scavenger is assumed to remain at the same constant concentration at all times. These approximations allow us to solve Eq. (4.26). Integration is first carried out over space, yielding the following ordinary differential equation:

$$\frac{dN}{dt} = -\frac{kN^2}{(8\pi Dt)^{3/2}} - k_s c_s N \qquad (4.27)$$

If the variable $u = N^{-1}$ is introduced, Eq. (4.27) is transformed into a linear equation that can be solved analytically to give

$$N(x) = N_0 \, e^{-q(x-1)} \left[1 + \frac{kN_0 t_0}{(8\pi Dt_0)^{3/2}} J(x, 1) \right]^{-1} \qquad (4.28)$$

where $q = k_s c_s t_0$, $x = t/t_0$, and $J(x, 1) = \int_1^x e^{-qx}(dx/x^{3/2})$. The fraction of radicals that reacts with scavenger is given by

$$f_s = k_s c_s \int_1^\infty \frac{N(x)}{N_0} \, dx \tag{4.29}$$

Although the equation for f_s is a closed-form expression, it is not simple. The limit of zero scavenger concentration $c_s = 0$ is an interesting special case in which

$$N(x) = N_0 \left[1 + \frac{2kN_0 t_0}{(8\pi D t_0)^{3/2}} (1 - x^{-1/2}) \right]^{-1} \tag{4.30}$$

and at infinite time

$$N(\infty) = N_0 \left[1 + \frac{2kN_0 t_0}{(8\pi D t_0)^{3/2}} \right]^{-1} \tag{4.31}$$

which shows that only partial recombination of radicals occurs in a spur expanding into a three-dimensional scavenger-free region. If a small scavenger concentration exists, the uncombined radicals will react with the scavenger by default, having nothing else to do. At very small concentrations, therefore, the fraction reacting with the scavenger is

$$f_s^0 = \left[1 + \frac{2kN_0 t_0}{(8\pi D t_0)^{3/2}} \right]^{-1} \tag{4.32}$$

The near constancy of the fraction of intermediates that react with the scavenger is essentially explained by this example. Experimentally, there may be several decades of scavenger concentrations, say from 10^{-6} to 10^{-3} M, in which f_s is essentially the same for low-LET radiations.

Figure 4.3 shows the results of a sample calculation for this system. The number of radicals decreases at first by recombination, and we can see the number of R_2 molecules increasing and approaching an asymptote. The remaining radicals react with a scavenger with a lifetime of approximately 10^{-7} s in this example. If the scavenger reaction time had been set at some other value (10^{-6} s or longer) the amount of recombination would have been very nearly the same.

In order to get a simple estimate of the variation of f_s with c_s, let us note that the lifetime of a radical with respect to the scavenger reaction is

$$\tau = \frac{1}{k_s c_s} \tag{4.33}$$

We can argue that all the recombination that takes place must be over at this time. With the use of Eqs. (4.30) and (4.32) and

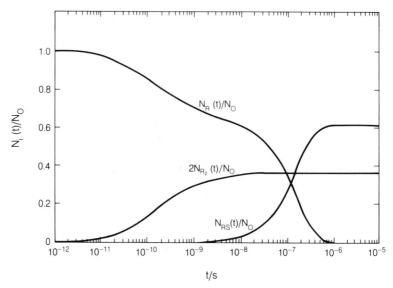

Figure 4.3. Variation of $N_R(t)/N_0$, $2N_{R_2}(t)/N_0$ and $N_{RS}(t)/N_0$ with time for a spur. Initial distribution is spherical Gaussian. $N_0 = 12$ radicals; $r_0 = 1.0$ nm; $D_R = 4 \times 10^{-5}$ cm²/s; $k_{RR} = k_{RS} = 6 \times 10^9$ M^{-1} s^{-1}; $c_{s_0} = 10^{-3}$ M. (From *Actions Chimiques et Biologiques des Radiations*, edited by M. Haissinsky, Copyright 1961. Reprinted with permission from Masson Editeurs.)

$$x = \frac{\tau}{t_0} = (k_s c_s t_0)^{-1} = q^{-1} \tag{4.34}$$

we get, for the fraction uncombined at the time τ,

$$f_s = f_s^0 [1 + (1 - f_s^0)q^{1/2} + \cdots] \tag{4.35}$$

and this is the fraction that reacts with the scavenger.

The variation of f_s with the square root of the scavenger concentration is also obtained in more rigorous treatments of the model. Balkas et al. (44) showed that their scavenging function introduced for hydrocarbons,

$$f_s = f_s^0 + (\alpha c_s)^{1/2}[1 + (\alpha c_s)^{1/2}]^{-1} \tag{4.36}$$

can also be applied in water.

The results of radiolysis using low-LET radiations are in apparent agreement with the string-of-beads model for the track. More sophisticated models take into account the knock-on electrons and penumbra effects.

4.4.2. Reactions in a String of Spurs

From the considerations above we see that information can be obtained from scavenger studies on the initial sizes of spurs. Changing the LET in the string-of-beads model for the track means changing the spur separations. As the spacing decreases, the radicals from neighboring spurs along the track can intermingle and react, increasing reaction (4.22) over (4.23) in our one-radical model. Tracks with variable interspur spacing were first discussed by Ganguly and Magee (11) and later by Chatterjee and Magee (12, 13). Let us consider a track with N spurs along the Z axis (the particle trajectory) with separations that can be varied. In one-radical prescribed diffusion the concentration of radicals in such a track of N spurs is given by

$$c(r, z, t) = \sum_{i=1}^{N} v_i \frac{\exp[(-r^2 - (z - z_i)^2)/(r_i^2 + 4Dt)]}{[\pi(r_i^2 + 4Dt)]^{3/2}} \tag{4.37}$$

The N spurs have centers located at positions on the z axis (along the track) at z_1, $z_2, \ldots, z_i, \ldots, z_N$, the ith spur has v_i radicals and a radius parameter r_i. If we use this concentration in the partial differential Eq. (4.26), take $c_s = 0$, and integrate over all space, we get Eq. (4.38), which is an equation summed over all the N spurs:

$$\frac{d}{dt} \sum_i v_i = -2k \sum_{i=1}^{N} \frac{v_i^2}{[2\pi(r_i^2 + 4Dt)]^{3/2}}$$

$$+ \sum_{\substack{i=1 \\ j \neq i}}^{N} \frac{v_i v_j \exp[-(z_i - z_j)^2/(r_i^2 + r_j^2 + 8Dt)]}{[\pi(r_i^2 + r_j^2 + 8Dt)]^{3/2}} \tag{4.38}$$

In prescribed diffusion the v_i values are considered as functions of time to be determined. Equation (4.38) is a single equation with N unknowns (the v_i). Equation (4.38) does not have enough information to allow a solution for the N v's, so we form a system of N equations in an intuitive manner to determine the unknown quantities. We write a system of N ordinary differential equations:

$$\frac{dv_i}{dt} = \frac{-2kv_i^2}{[2\pi(r_i^2 + 4Dt)]^{3/2}}$$

$$+ v_i \sum_{j \neq i} \frac{v_j \exp[-(z_j - z_i)^2/(r_i^2 + r_j^2 + 8Dt)]}{[\pi(r_i^2 + r_j^2 + 8Dt)]^{3/2}} \tag{4.39}$$

$$i = 1, 2, 3, \ldots, N$$

Although it is not a rigorous justification, we note that all possible spur–spur interaction terms are included, and summation of both sides of Eq. (4.39) over i leads to Eq. (4.38).

A rigorous treatment of this set of equations has not been made, but the following approximation leads to a result that has intuitive appeal. Consider the equation for an interior spur, the ith spur, and approximate the summation in Eq. (4.39) by an integration. We are interested in the average value of dv_i/dt and take the probability that the jth spur occurs in the interval dz_j as dz_j/Z_1, where Z_1 is a parameter (average interspur separation distance) that depends on the LET. Substitution of the appropriate integral gives

$$\frac{dv_i}{dt} = -2k\left(\frac{v_i^2}{[2\pi(r_i^2 + 4Dt)]^{3/2}} + v_i\left\langle\frac{v_j}{[\pi(r_i^2 + r_j^2 + 8Dt)]Z_1}\right\rangle\right) \quad (4.40)$$

The first term of Eq. (4.40) gives the ordinary prescribed diffusion result for a spur, and the second term gives its interaction with the other spurs; the angle brackets on the second term indicate an average over the spur distribution function. Spurs are created by resonant energy losses in the range 6–100 eV, and with 17 eV required to form a radical pair, v_i varies from 2 to 12; in order to get the track yield, this expression must be averaged over a spur distribution function giving the fraction of spurs having the various numbers of radicals from 2 to 12. Although such a distribution function for spurs has been proposed (23), we do not consider the average explicitly at this time but, for simplicity, propose that the following equation applies to the low-LET average spur that has v radicals:

$$\frac{dv}{dt} = \frac{-2kv^2}{[2\pi(r^2 + 4Dt)]^{3/2}}\left[1 + \frac{[2\pi(r^2 + 4Dt)]^{1/2}}{Z_1}\right] \quad (4.41)$$

where we choose v as the average number of radicals in a spur and the average r^2 parameter so that correct spur yields are obtained (for example, $G = 18.4$ for the Fe^{3+} yield of the Fricke system; see Section 4.5.1); the parameter Z_1 varies inversely with LET, and the proportionality constant can be adjusted for agreement with the experiment.

When the parameter Z_1 in Eq. (4.41) gets so small with the increase of LET that the second term in the bracket is much larger than unity, we get

$$\frac{dv}{dt} \simeq -\frac{2kv^2}{2\pi(r^2 + 4Dt)Z_1} \quad (4.42)$$

and we can write the equation in terms of the variable $v/Z_1 = n$, the number of radicals per unit distance

$$\frac{dn}{dt} \simeq -\frac{2kn^2}{2\pi(r^2 + 4Dt)} \quad (4.43)$$

which is the ordinary prescribed diffusion equation for cylindrical symmetry. Equation (4.41), therefore, has a convenient form that allows the low-LET core equation to remain valid as LET increases. The equation contains the essence of the Ganguly–Magee track treatment; here, however, we recognize that only part of the track is involved (the penumbra is not included). This treatment has also been extended to a multiradical case (13).

4.5. RADIATION CHEMISTRY OF ELECTRON TRACKS

Consider an electron with energy E that is completely absorbed in a system; it produces a yield $G_e(E)$ of some particular product per 100 eV absorbed. The only example we shall use is the Fricke system; so the $G_e(E)$ value will always be that for Fe^{3+}, although the results of this section are much more general than this. Here the subscript e refers to the electron. We are interested in a general relationship between the G_e values produced by electrons that have different energies. The concept of an integral equation relating the two yields is best introduced through a consideration of Eq. (4.14). This equation applies to a low-LET heavy particle, but it seems reasonable to expect that a similar equation should apply to a high-energy electron (which also has low LET).

The elementary arguments used in the derivation of Eq. (4.14) depend on the smallness of the individual energy losses with respect to the total energy of the particle under consideration. For all heavy particles the maximum energy loss is $\sim (2m/M)E$, where m/M is the mass ratio of the electron to the particle, is a very small fraction of the particle energy. Thus, the consideration of the right side of Eq. (4.14) as a differential effect is well justified.

The next step is to derive a similar equation for the electron as primary particle. In this case the maximum energy loss is one-half the incident energy, and the possibility for the use of such an equation is not at all clear. It can, however, be shown (24) that the similar equation

$$G'_e(E) = \frac{d}{dE} EG_e(E) = (1 - f)G_{sp} + f \frac{\int_{E_0}^{E/2} G_e(\epsilon)w_e(E, \epsilon)\, d\epsilon}{\int_{E_0}^{E/2} w_e(E, \epsilon)\, d\epsilon} \tag{4.44}$$

is valid for energies E that are sufficiently high, for example, $E \geq E_1$, where E_1 is an energy that is large enough so that the spurs formed by the electron are widely spaced and essentially finish recombination before they overlap. At $E = 20$ keV the LET is 1.35 eV/nm and spurs are expected to be, on the average, about 60 nm apart. Recombination of hydrogen atoms, with an initial radius of 1.5 nm or so, is expected to be 90% complete before overlapping. This value of E_1 would seem to be reasonable, at least in the radiolysis of water. Equation (4.44) is an asymptotic equation that becomes more valid as energy increases.

Equation (4.44) can be used to obtain $G_e(E)$. Suppose that $G_e(E)$ is known below $E = E_1$. Integrate Eq. (4.44) and obtain

$$EG_e(E) = E_1G_e(E_1) + (1 - f)(E - E_1)G_{sp} + f \int_{E_1}^{E} \Gamma(E')dE' \quad (4.45)$$

where

$$\Gamma(E') = \int_{E_0}^{E'/2} G_e(\epsilon)w_e(E', \epsilon) \, d\epsilon \left[\int_{E_0}^{E'/2} w_e(E', \epsilon) \, d\epsilon \right]^{-1} \quad (4.46)$$

Equations (4.45) and (4.46) determine the function $G_e(E)$ for energies above $E = E_1$.

The function $w_e(E, \epsilon)$ for the electron is complicated (45, 46), and full numerical procedures should be used in consideration of Eqs. (4.45) and (4.46). However, it is instructive to use a simple approximation arising from the Rutherford cross section for knock-on collisions and that leads to an analytical treatment. This approximation is

$$w_e(E, \epsilon) \simeq \frac{1}{\epsilon} \quad (4.47)$$

Insertion of Eq. (4.47) into Eq. (4.44) and transformation to the variable $\eta = \ln(E/2E_0)$ leads to the differential equation

$$\frac{d}{d\eta} \eta e^{-\eta} \frac{d}{d\eta} e^{\eta}[G_e(\eta)] = (1 - f)G_{sp} + fG_e(\eta - b) \quad (4.48)$$

where $b = \ln 2$. This is an equation valid for $E > E_1$, or $\eta > \ln(E_1/2E_0)$.

Equation (4.48) is an unusual equation in that the arguments of the function $G_e(\eta)$ are different on the left and right sides. It has a solution in the form of an infinite series. However, this solution is not convenient to use, and an approximate solution has been found:

$$G_e(\eta) = A + \frac{1}{\eta^\alpha}\left(a_0 + \frac{a_1}{\eta} + \frac{a_2}{\eta^2} + \frac{a_3}{\eta^3}\right) \quad (4.49)$$

Here A, α, a_0, a_1, a_2, and a_3 are constants to be determined. This solution can be shown to be correct to the order of η^{-4}.

If a Taylor series is used for $G_e(\eta - b)$, the following values are found for the constants in Eq. (4.49):

$$A = G_{sp}$$

$$\alpha = 1 - f$$

$$a_1 = \alpha(\alpha - bf)a_0$$

$$a_2 = \frac{\alpha(1 + \alpha)}{2}\left\{(1 + \alpha - bf)(\alpha - bf) - f\frac{b^2}{2}\right\}a_0 \qquad (4.50)$$

$$a_3 = \frac{\alpha(1 + \alpha)(2 + \alpha)}{6}\left[(\alpha + 2 - bf)\left\{(\alpha + 1 - bf)\right.\right.$$

$$\left.\left.\times (\alpha - bf) - f\frac{b^2}{2}\right\} - (\alpha - bf)fb^2 - f\frac{b^3}{3}\right]a_0$$

For heavy-particle tracks we have stressed that $f = 0.5$, but for high-energy electrons it is 0.4 (approximately 40% of the energy is spent in producing another generation of electrons). Substitution of this value leads to the explicit expression

$$G_e(\eta) = G_{sp} + \frac{a_0}{\eta^{0.6}}\left\{1 + \frac{0.1936}{\eta} + \frac{0.1587}{\eta^2} + \frac{0.2752}{\eta^3}\right\}$$

$$= G_{sp} + \frac{a_0}{\eta^{0.6}}F(\eta) \qquad (4.51)$$

where $F(\eta)$ has been substituted for the expression within the curly bracket.

Equation (4.51) tells us that all G values for electron tracks in the high-energy region have a common asymptotic form when written in terms of the variable η. There are two adjustable parameters in Eq. (4.51), G_{sp} and a_0. The first of these has a simple meaning, the G-value for the isolated spur for the particular reaction of interest; the second parameter, a_0, is more complicated. It depends on an integration over the G values for all of the track entities in the low-energy region, as we see from Eq. (4.44).

4.5.1. G Values for Fe^{2+} Oxidation in Electron Tracks

The Fricke reaction is the oxidation of Fe^{2+} in strongly acidic solution. Reaction (4.15) is followed by

$$e_{aq}^- + H_{aq}^+ \longrightarrow H$$

$$OH + Fe^{2+} \longrightarrow OH^- + Fe^{3+}$$

$$H_2O_2 + Fe^{2+} \longrightarrow OH^- + OH + Fe^{3+}$$

If oxygen is present, one has

$$H + O_2 \longrightarrow HO_2$$

$$HO_2 + Fe^{2+} \longrightarrow HO_2^- + Fe^{3+}$$

$$HO_2^- + H_{aq}^+ \longrightarrow H_2O_2$$

If oxygen is absent, one has

$$H + Fe^{2+} \longrightarrow Fe^{2+}(H)$$

$$Fe^{2+}(H) + H_{aq}^+ \longrightarrow Fe^{3+} + H_2$$

Ideally we would like to find a self-consistent solution for $G_e(E)$ that satisfies Eqs. (4.44) and (4.45) for all energies. In principle a scheme can be devised for the use of experimental data to find the best solution for $G_e(\epsilon)$ that satisfies both the data and Eqs. (4.44) and (4.45). However, experimental data are not known precisely enough for this method to be used effectively. The most straightforward procedure is to use a model calculation. Such a calculation must involve explicit consideration of electron tracks of all energies from 100 eV to large values at which the asymptotic Eq. (4.51) applies.

Electron tracks with energies below 5 keV are single entities. A model calculation to obtain their Fricke G values was carried out and reported (24). This model is very rough, but the results can be considered adequate because of the self-consistency of the problem. It is beyond the scope of this chapter to give the details of the model for which the reader can use Ref. 24.

The yield of Fe^{3+} is larger when oxygen is present in the solution than when it is absent. Given in terms of the "molecular" and "radical" yields of reaction (4.15), one has:

$$G(Fe^{3+}; O_2) = 3G_H + 2G_{H_2O_2} + G_{OH}$$

$$G(Fe^{3+}; \text{no } O_2) = G_H + 2G_{H_2O_2} + G_{OH}$$

The yields from isolated spurs estimated from a model calculation and satisfying the self-consistency requirements are:

$$G_{OH}^{sp} = 3.92 \qquad G_{H_2O_2}^{sp} = 0.52$$

$$G_H^{sp} = 4.48 \qquad G_{H_2}^{sp} = 0.24$$

Say that the radical and molecular yields for all track entities with less energy than 5 keV are known. We can calculate the Fricke G values for tracks with higher than 5 keV energy using Eqs. (4.45) and (4.46) repeatedly. We take a small in-crement in E and calculate a new value of G_e; this operation is repeated and the function $G_e(E)$ is built up. In this calculation we take into account the overlapping of the spurs for energies below 20 keV, and thus E_1 has a slightly different signif-icance than in Eq. (4.45).

Figure 4.4 shows the calculated $G_e(Fe^{3+}; E)$ curves for the Fricke dosimeter containing oxygen (aerated) or no oxygen (deaerated). The most accurate inelastic cross section, the Møller formula (45, 46), was used in the $w_e(E, \epsilon)$ function. It is found, however, that the analytical Eq. (4.51) is in excellent agreement if proper constants are chosen. This agreement with an approximate analytical expression obtained by the use of the Rutherford cross section for the electron knock-on spec-trum is not entirely understood. Perhaps it is largely a result of normalization [on the left side of Eq. (4.44) there is a ratio of integrals, both of which involve the

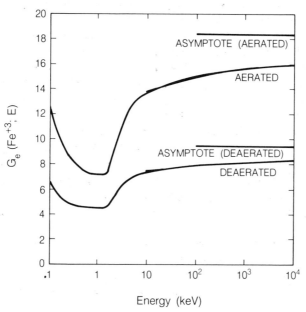

Figure 4.4 The $G_e(Fe^{3+}; E)$ curves calculated from the electron track model. Above 10 keV the curves calculated from the asymptotic equations are shown. Aerated:

$$G_e(Fe^{3+}; E) = 18.40 - \frac{9.7337}{\eta^{0.6}} F(\eta)$$

Deaerated:

$$G_e(Fe^{3+}; E) = 9.44 - \frac{4.3026}{\eta^{0.6}} F(\eta)$$

The asymptotes are indicated. (Reprinted with permission from *Journal of Physical Chemistry*, vol. 82. Copyright 1978, American Chemical Society.)

TABLE 4.4. Constants for $G_e(E)$ Curves and Values at 10^6 eV^a

G(product)	G_{sp}	a_0	G_e (10^6 eV)
G_{-H_2O}	4.96	-1.5870	4.51
G_{OH}	3.92	-3.4209	2.95
G_H	4.48	-2.7156	3.71
$G_{H_2O_2}$	0.52	0.9169	0.78
G_{H_2}	0.24	0.5643	0.40
$G(Fe^{3+}; O_2)$	18.40	-9.7337	15.64
$G(Fe^{3+}; no\ O_2)$	9.44	-4.3026	8.22

^{a}Reprinted with permission from *Journal of Physical Chemistry*, vol. 82. Copyright (1978) American Chemical Society.

cross section]. Table 4.4 summarizes the constants to be used in the high-energy regions of the curves in Figure 4.4.

The emphasis in this treatment has been on the general nature of the energy dependence of the yields rather than on the quantitative values, and the latter are still subject to a certain amount of revision. On the other hand, the authors believe that the $G_e(E)$ curves of Figure 4.4 are in good agreement with the experimental data of radiation chemistry. It has actually not been customary to report experimental yields in terms of $G_e(E)$ that correspond to values obtained with monoenergetic electron beams from accelerators. The G values obtained with X rays or γ radiations are combinations of such yields because the latter radiations produce their own characteristic spectra of recoil electrons. Table 4.5 contains yields calculated using Figure 4.4 and spectra assumed for several radiation sources, and a comparison is made with experiment. The spectra for ^{60}Co γ radiation and 220-kV X radiation were obtained from Cormack and Johns (47); the spectra for the β radiations were obtained from Nelms (48).

The Fricke dosimeter has been investigated extensively (49–52). Fregene (49) carried out an investigation spanning a larger range of energies than any other single investigation; β, X, and γ radiations were used. In Figure 4.5 eleven values measured by Fregene (49) are plotted along with the curve $G_e(E) = 18.40 -$

TABLE 4.5. Calculated $G(Fe^{3+})$ Compared with Experiment^a

Radiation	$G(Fe^{3+})$ Model Calculated	$G(Fe^{3+})$ Experimental	
^{60}Co γ ray	15.61	15.5 $\pm$ 0.3[49]	15.6 $\pm$ 0.3[50]
^{32}P β ray	15.34	15.35 $\pm$ 0.5[49]	15.4 $\pm$ 0.5[51]
220-kV X rays	14.70	14.6 $\pm$ 0.3[49]	15.0 $\pm$ 0.5[51]
Tritium β rays	12.94	12.8 $\pm$ 0.4[49]	12.9 $\pm$ 0.2[52]

^{a}Reprinted with permission from *Journal of Physical Chemistry* **82**, 2222. Copyright (1978) American Chemical Society.

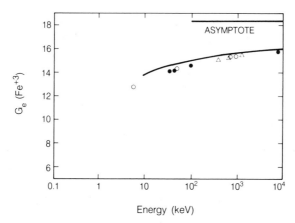

Figure 4.5. The $G_e(Fe^{3+})$ as a function of electron energy. Eleven experimental points obtained by Fregene (49) are shown along with the asymptotic curve (shown in Figure 4.4). The points are plotted at the mean electron energies given by Fregene. (Reprinted with permission from *Journal of Physical Chemistry*, vol. 82. Copyright 1978, American Chemical Society.)

$9.7337/\eta^{0.6}F(\eta)$, which we have chosen as the asymptotic yield curve for the aerated Fricke dosimeter. The points are plotted at the mean energies as listed by Fregene (Table I, Ref. 49). The error bars of the points (not shown in Figure 4.5) are large enough so that the solid curve is consistent with them.

The expression for the G value for the Fricke dosimeter is $G(Fe^{3+}, O_2) = 2G_{H_2O_2} + G_{OH} + 3G_H$, and we see that the larger the radical yields, the larger the G value. Low-LET radiations give larger G values than do high-LET radiations in this system. Traditionally we think of electrons with energies in the mega-electronvolt range to be typical low-LET radiations, and in fact most low-LET radiations transfer their energy to matter through the action of high-energy electrons. We notice that the Fricke G value for mega-electronvolt electrons is 15.6, and the G value for the spur is 18.4. If electron tracks were completely comprised of spurs, the Fricke G values for all electrons would be 18.4. It is an important fact that there are higher LET portions of electron tracks that have much lower Fricke G values. We see from Figure 4.4 that electrons in the energy range 0.5 and 1.5 keV have G values below 12. The G value 15.6 of the mega-electronvolt electron is an average of 18.4 for spurs and 10–12 for track ends.

4.6. RADIATION CHEMISTRY OF HEAVY-PARTICLE TRACKS

The most general problem of radiation chemistry is to describe the radiation chemistry of the heavy-particle track. It has been shown (Section 4.2.2) that such a track is initially comprised of a physical core and a penumbra, each of which contains about half of the deposited energy. If the core and penumbra were isolated from each other, the treatment would be simple. In that case the core would be considered by the methods of Section 4.4 and the penumbra (which is made up of electron tracks), by the methods of Section 4.5. Actually there are particles (say protons and α particles at energies of 10^3 MeV/nucleon) that have tracks in water at such low LET that every excitation produced directly by the particle initiates radiation chemical events that develop independently of all others. For these particles the differential G value for any product of interest in an aqueous solution containing appropriate scavengers is given by Eq. (4.14). This equation is very important in track theory because it relates heavy-particle tracks to electron tracks. Such a straightforward approach does not seem to be applicable for lower energy heavy charged particles (say, below 400 MeV/nucleon). Across the boundary between the core and the penumbra of the tracks of such particles there is a considerable interaction between the radicals formed in these two regions. This phenomenon must be considered carefully.

Figure 4.6 gives the specific energy dependence of the LET for some representative heavy particles in water (54). Accelerators exist that can give energies up to 10^3 MeV per nucleon to all nuclei up to uranium. The particles chosen for the figure include hydrogen and helium, which can be accelerated to these energies by many accelerators; carbon, neon, and argon, which are routinely accelerated in the BEVALAC (53) at Berkeley; and fermium, which is representative of the heaviest particles.

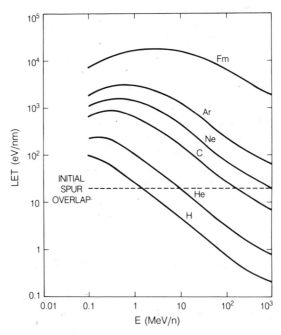

Figure 4.6. The stopping power (LET) for six selected particles in H_2O versus specific energy (in MeV/nucleon). Data are from Ziegler (54). The condition for initial spur overlap is indicated.

Consider the track of a proton with energy 10^3 MeV. Its LET (Figure 4.6) is 0.2 eV/nm and there is 0.1 eV/nm in spurs in the core. On the average each spur has 40 eV, and thus the average separation is $\sim 40/0.1 = 400$ nm. In a typical aqueous solution radicals from neighboring spurs will not begin to intermingle (and react) in the time required for a scavenger to react with them. For this situation the conditions of Eq. (4.14) are well fulfilled.

Now consider a higher-z nucleus at the same specific energy (1000 MeV/nucleon). Such a particle has the same velocity as the proton in the above example. If we take neon, the LET is 20 eV/nm and the interspur spacing is ~ 4 nm. If the spur radius is 2 nm, the spurs on the average are touching their neighbors, and we have a cylindrical distribution instead of a string of beads.

As particles of still higher charge at this specific energy are considered, we find that the cores contain more energy density and thus larger concentrations of radicals. We know from Eq. (4.12) that the energy density in the penumbra varies as LET r^{-2}, and thus as LET increases, the radial distance at which radicals can react with each other increases. In addition, as the core expands radially due to the diffusion of chemical species, it starts engulfing radicals from the penumbra region. Hence a part of the penumbra becomes involved in the core reaction. On the other hand, the parts of the penumbra sufficiently far from the core react as isolated electron tracks: they have a radical recombination phase followed by reaction with a scavenger without mutual interference.

This discussion suggests a model for the heavy-particle track. We can take a core that includes part of the penumbra (the part that it interacts with) and treat the rest of the penumbra as a collection of electron tracks. In order to obtain a method for setting a limit for the radius of the extended core (see Figure 4.1), consider the competition of radicals in recombination and reaction with a scavenger. We use a one-radical approximation here. At any point in a track, radicals recombine with a rate given by $2kc^2$ (the factor 2 appears because two radicals disappear) and react with the scavenger with a rate $k_s c_s c$, where c is the local radical concentration, c_s is the scavenger concentration, and k and k_s are rate constants. These two rates are equal when we have a radical concentration given by

$$c_1 = \frac{k_s c_s}{2k} \tag{4.52}$$

and we refer to this concentration, c_1, as a reference concentration. If the radical concentration is greater than c_1, recombination between water radical products is more important than scavenging, and if it is less than c_1, the reverse is true. The mean time for the scavenger reaction is given by Eq. (4.53), and we use t_1 as a reference time:

$$t_1 = (k_s c_s)^{-1} \tag{4.53}$$

The track reactions cannot last for times larger than t_1. The average distance, l_1, that a radical diffuses radially in time t_1 is given by

$$l_1 = \sqrt{4Dt_1} \qquad (4.54)$$

For a heavy-particle track the radical concentration decreases as we move radially out from the trajectory, except in the core, where it is considered to be uniform. Hence, for such a track the importance of the scavenging reaction depends on the radical concentration at a certain radius when compared with c_1 [Eq. (4.52)]. The values of the radii (r_1) for the six representative particles for which $c = c_1$ are given in Figure 4.7. Figure 4.7 shows that r_1 is equal to the physical core radius (r_c) at high particle energies; as particle energy decreases, r_1 increases and r_p decreases; at some energy r_1 becomes equal to r_p; for even smaller values of E, r_1 is taken to be equal to r_p. It is to be remembered that r_c and r_p are dependent on particle energies only, whereas r_1 depends also on the scavenger concentration.

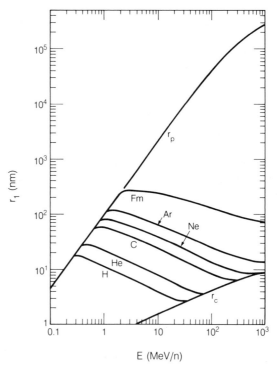

Figure 4.7. Chemical core radii r_1 for the six selected particles versus specific energy E (in MeV/nucleon). Scavenging conditions are appropriate for the Fricke dosimeter. At low energies, all values of r_1 are equal to the penumbra radius r_p; at high energies all values of r_1 become equal to the physical core radius r_c. The r_c and r_p are common for all particles. (Reprinted with permission from *Journal of Physical Chemistry*, vol. 84. Copyright 1980, American Chemical Society.)

The results of Figure 4.7 are for the Fricke system with ferrous ion concentration of 10^{-3} M (1 mol/m^3).

The fraction of the total energy of the tracks enclosed within the radius r_1 for this particular set of parameters is shown in Figure 4.8 for each of the six representative particles. This fraction is called F_{ch} because it is the fraction of energy in the chemical core. At values of E for which r_1 is equal to or less than the physical core radius (r_c), the fraction of energy contained in the chemical core is 0.52; at values of E for which r_1 is equal to r_p, the fraction becomes equal to 1.

4.6.1. Model Calculation for Heavy-Particle Tracks

In Section 4.2.3 it was pointed out that if $G_e(E)$, the G value of a radiation chemical product from electron tracks, is known, the G' values of the same product for low-LET particles can be obtained by elementary operations [see Eq. (4.14)]. The fundamental condition of validity of Eq. (4.14) is the nonoverlapping of the spurs along the track core as they expand and react with the scavenger. In reality, such a condition is valid at low LETs below 2–3 eV/nm or so. In other words, nonoverlapping spurs are produced only by protons, deuterons, and α particles or by higher atomic number particles of very high energy (greater than 1000 MeV/nucleon). Hence in order to include particles of all LETs, a more general relationship for the track yield must be written:

$$G'(E) = F_{ch}G'_{ch}(E) + [1 - F_{ch}(E)]\, G'_{pen}(E) \qquad (4.55)$$

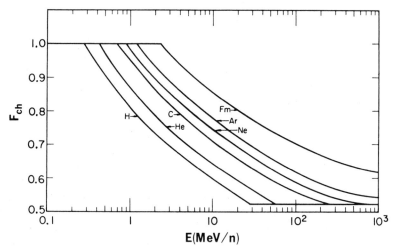

Figure 4.8. The fraction of track energy contained initially within the chemical core radius r_1 of Figure 4.7 for the six particles versus the specific energy E. (Reprinted with permission from *Journal of Physical Chemistry*, vol. 84. Copyright 1980, American Chemical Society.)

Here $G'_{ch}(E)$ is to be obtained according to appropriate models, and $G'_{pen}(E)$ is to be obtained by a modification of the integral expression of Eq. (4.14) for electron energies greater than 20 keV. For low-energy electrons the model calculations (see Section 4.5.1) can be considered. The basic electron yield curve as a function of energy must be used, such as the one in Figure 4.4 for the Fricke dosimeter system. All electrons in the spectrum of the ejected electrons that are absorbed in the region $r_1 < r < r_p$ must be used. Let us consider the following expression for the penumbra yield:

$$
G'_{pen}(E) = \frac{\displaystyle\int_{\epsilon_1}^{\epsilon_{max}} G_e(\epsilon')w(E,\,\epsilon)\,d\epsilon}{\displaystyle\int_{\epsilon_1}^{\epsilon_{max}} w(E,\,\epsilon)\,d\epsilon}
\tag{4.56}
$$

All of the knock-on electrons originate on the track axis and go into the chemical core region $r < r_p$; ϵ_1, the minimum energy that allows them to reach r_1, must be taken as the lower limit of the integral. The penetration of electrons in the penumbra is a statistical problem that has been considered by Chatterjee and Schaefer (25). These authors found that the value of r_p is essentially proportional to ϵ_{max}. The result means that on the average the penetration (not range) of an electron is linear in energy. Thus, we take

$$
\epsilon' = \epsilon - \epsilon_1
\tag{4.57}
$$

as the appropriate argument in the electron yield expression, that is, $G_e(\epsilon')$, because the electrons that go beyond r_1 leave this amount of energy in the penumbra.

In order to calculate $G'_{ch}(E)$ of Eq. (4.55), we have to consider several cases depending on the LET values. A detailed description of all these cases will not be given here. Instead the readers are referred to Refs. (12) and (13) for a full account with special considerations to the Fricke dosimeter system. Here we will briefly summarize the models for the various cases of LET values and provide results for the same system (the Fricke dosimeter).

There are three distinct situations: (i) low-LET chemical core, (ii) medium-LET chemical core, and (iii) high-LET chemical core. In the case of low-LET chemical core the individual spurs are initially separated, and the problem is to consider the overlapping of spurs along a track as expansion takes place with time. The treatment of this problem is very similar to that of the Ganguly–Magee model (11). The same model has also been applied to the case of the medium-LET chemical core. In this situation the spurs in the track core are initially merged, and the main problem for calculation of track yields is the interaction of the core with the penumbra. The Ganguly–Magee model is flexible in that the same equations apply to both low and high LET. They are the equations for a string of spurs at low LET and automatically go over to the appropriate cylindrical distribution at high LET.

The radicals of the penumbra that belong to the chemical core (i.e., in the region $r < r_1$) are included as they become engulfed by the physical core. In the third case (high-LET chemical core) recombination of radicals dominates so much that ideally one should calculate the track reactions for nondiffusing radicals. It is found, however, that recombination is so fast and so nearly complete in the Ganguly–Magee model that even in this situation we have applied the same equations. Thus, in all the three cases the Ganguly–Magee model has been applied and the track equations in prescribed diffusion have been used with a complete chemical mechanism.

Table 4.6 gives a set of spur expansion equations in the prescribed diffusion approximations for the limited set of reactions represented by the box in Table 4.7. In the actual calculations all reactions with asterisks in part B of Table 4.2 are included for the Fricke system. The complete set of differential equations actually used is not given in Table 4.6 because extension of the set shown to include all secondary reactions is trivial and can be done by inspection. An examination of the equations in Table 4.6 reveals that the first term in the large parentheses refers to non-overlapping spurs or isolated spurs, and the second term refers to the case of overlapping spurs.

The term Z_1 is the interspur separation distance and has been taken as a parameter. The value chosen is $50/(\text{LET}/2)$; we could guess that this parameter should

TABLE 4.6. Track Equations in Prescribed Diffusion[a]

$$\frac{dv_1}{dt} = -\frac{2k_1 v_1^2}{[2\pi(r_1^2 + 4D_1 t)]^{3/2}} \left(1 + \frac{[2\pi(r_1^2 + 4D_1 t)]^{1/2}}{Z_1}\right)$$

$$-\frac{k_2 v_1 v_2}{[\pi(r_1^2 + r_2^2 + 4D_1 t + 4D_2 t)]^{3/2}} \left(1 + \frac{[\pi(r_1^2 + r_2^2 + 4D_1 t + 4D_2 t)]^{1/2}}{Z_1}\right)$$

$$\frac{dv_2}{dt} = -\frac{k_2 v_1 v_2}{[\pi(r_1^2 + r_2^2 + 4D_1 t + 4D_2 t)]^{3/2}} \left(1 + \frac{[\pi (r_1^2 + r_2^2 + 4D_1 t + 4D_2 t)]^{1/2}}{Z_1}\right)$$

$$-\frac{2k_3 v_2^2}{[2\pi(r_2^2 + 4D_2 t)]^{3/2}} \left(1 + \frac{[2\pi(r_2^2 + 4D_2 t)]^{1/2}}{Z_1}\right)$$

$$\frac{dv_3}{dt} = \frac{k_2 v_1 v_2}{[\pi(r_1^2 + r_2^2 + 4D_1 t + 4D_2 t)]^{3/2}} \left(1 + \frac{[2\pi(r_2^2 + 4D_2 t)]^{1/2}}{Z_1}\right)$$

$$\frac{dv_4}{dt} = \frac{k_1 v_1^2}{[2\pi(r_1^2 + 4D_1 t)]^{3/2}} \left(1 + \frac{[2\pi(r_1^2 + 4D_1 t)]^{1/2}}{Z_1}\right)$$

$$\frac{dv_5}{dt} = \frac{k_3 v_2^2}{[2\pi(r_2^2 + 4D_2 t)]^{3/2}} \left(1 + \frac{[2\pi(r_2^2 + 4D_2 t)]^{1/2}}{Z_1}\right)$$

[a] Reprinted with permission from *Journal of Physical Chemistry*, vol. 84. Copyright (1980) American Chemical Society.

TABLE 4.7. Differential Equations for Transient Species and Radiation Products in Track[a]

$$\frac{d}{dt}(H) = -2k_1(H)^2 - k_2(H)(OH) \quad - \quad k_4(H)(H_2O_2) + k_6(OH)(H_2) - k_7(HO_2)(H)$$

$$\frac{d}{dt}(OH) = -k_2(H)(OH) - 2k_3(OH)^2 \quad + \quad k_4(H)(H_2O_2) - k_5(OH)(H_2O_2) - k_6(OH)(H_2) - k_8(HO_2)(OH)$$

$$\frac{d}{dt}(H_2O) = k_2(H)(OH) \quad + \quad k_4(H)(H_2O_2) + k_5(OH)(H_2O_2) + k_6(OH)(H_2) + k_8(HO_2)(OH)$$

$$\frac{d}{dt}(H_2) = k_1(H)^2 \quad - \quad k_6(OH)(H_2)$$

$$\frac{d}{dt}(H_2O_2) = k_3(OH)^2 \quad - \quad k_4(H)(H_2O_2) - k_5(OH)(H_2O_2) + k_7(HO_2)(H) + k_9(HO_2)^2$$

$$\frac{d}{dt}(HO_2) = k_5(OH)(H_2O_2) - k_7(HO_2)(H) \quad - \quad k_8(HO_2)(OH) - 2k_9(HO_2)^2$$

$$\frac{d}{dt}(O_2) = k_8(HO_2)(OH) + k_9(HO_2)^2$$

[a] Reprinted with permission from *Journal of Physical Chemistry*, vol. 84. Copyright (1980) American Chemical Society.

be 40/(LET/2) because 40 is the average energy of a spur, and half the LET of a heavy particle creates spurs. The larger value 50 gives results better in agreement with experiment. When Z_1 is very large (i.e., the spurs are very well separated from each other), the first term in the equation dominates. For very small Z_1 values (overlapping spurs) the second term dominates. The various r_j and D_j values that appear in Table 4.6 are tabulated in Table 4.8 and have been considered as param-

TABLE 4.8. Values of Initial Radii and Diffusion Constants[a]

Species	Initial Radius (nm)	Diffusion Constant $(10^{-5} \text{ cm}^2/\text{s})$
H	2.6	8
OH	1.3	2
H_2	2.6	8
HO_2	2.6	2
H_2O_2	1.3	2
O_2	2.6	2

[a] Reprinted with permission from *Journal of Physical Chemistry*, vol. 84. Copyright (1980) American Chemical Society.

eters. The radii for hydrogen and hydroxyl radicals are adjusted for very energetic protons in such a manner that the ferric yield in an isolated spur is 18.4 (24). Once adjusted for the protons as a heavy particle, no readjustments of the parameters have been made in calculations for all other particles.

The system of differential equations of Table 4.6 has been solved numerically from $t = 3 \times 10^{-12}$ s (thermalization time) to $t = 3 \times 10^{-7}$ s. The yields of H, OH, H_2O_2, and HO_2 at $t = 3 \times 10^{-7}$ s have been assumed to react with ferrous ions to produce the ferric yield.

The results of calculations based on these procedures are summarized in Figure 4.9 for proton, helium, carbon, neon, argon, and fermium particles. In this figure the differential yield, $G'(E)$, has been plotted against particle energy varying between 0.1 and 10^3 MeV/nucleon. In order to compare with experiments, some data points have also been plotted; at high energies $G'(E)$ can be measured directly. But at lower energies the data points are estimated from measured integral yields. The dashed lines a, b, and c are the experimental values for carbon, neon, and argon, respectively, as obtained by Christman, Jayko, and Appleby (55, 56). The experimental values are higher than the calculated curves by more than the estimated experimental error. The possible cause for such a discrepancy may be due

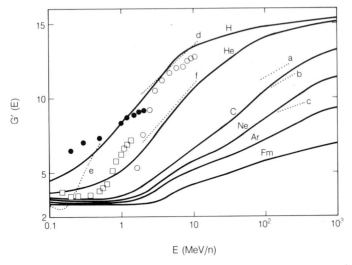

E (MeV/n)

Figure 4.9. Calculated differential ferric ion yields, $G'(E)$, for the six representative particles versus energy per nucleon, E, are given by the solid curves. Estimates of values for C, Ne, and Ar from direct experimental measurements by Appleby et al. are indicated by the dotted lines a, b, and c, respectively. In the low-energy region (below 10 MeV/nucleon), estimates have been made from analysis of integral G values by various authors. For H: (●) Hart, Ramler, and Rocklin (57); (dotted curve, e) Puchealt and Julien (58). For D: (○) Hart, Ramler, and Rocklin (57); (dotted curve, d) Schuler and Allen (59). For He: (□) Gordon and Hart (60); (dotted curve, f) Schuler and Allen (59). (Reprinted with permission from *Journal of Physical Chemistry*, vol. 84. Copyright 1980, American Chemical Society.)

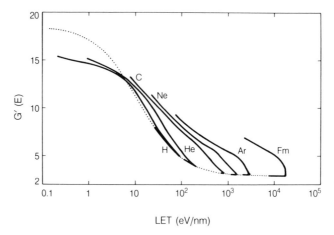

Figure 4.10. Calculated differential ferric ion yields, $G'(E)$, for the six representative particles versus total LET are given by the solid curves. The calculations for C, Ne, Ar, and Fm are in the range $0.1 \leq E \leq 100$; the calculations for H and He are in the range $0.001 \leq E \leq 100$. The dotted curve indicates the common curve obtained under conditions in which the penumbra vanishes. (Reprinted with permission from *Journal of Physical Chemistry*, vol. 84. Copyright 1980, American Chemical Society.)

to the phenomenon of nuclear fragmentation. According to this phenomenon, a fraction of a heavy-particle beam undergoes nuclear disintegration. This results in the dilution of the main beam with lower atomic number particles, and thus the overall ferric yield is higher. A careful examination of this process and its influence in chemistry is needed.

For proton and helium particles experimental data are available at the low-energy region, and there seems to be good agreement with the calculated results between 1 and 10 MeV/nucleon (57–64). Below 1 MeV/nucleon the experimental results themselves scatter greatly.

In Figure 4.10 the differential yields of Figure 4.9 have been replotted against the LET. This figure makes it quite clear that $G'(E)$ is not a unique function of LET. In the early development of experimental radiation chemistry it was generally believed that radiation chemical yields should be simple functions of LET, but analysis of the dependence of yield on track structure reveals a more intricate behavior.

4.7. SUMMARY

In this chapter the essential features of the track reactions of radiation chemistry have been set forth. These reactions are nonhomogeneous because the reactive species that form the radiation chemical products are created in tracks. The structure of the tracks is actually quite complicated, and a large part of the effort involves the devising of reasonable track models. Our treatment of the track reactions begins at about 3 ps, at which time the chemical species are more or less thermalized following the initial deposition of energy at about 10^{-16} s. A more desirable way to consider the problem of nonhomogeneous kinetics may be to follow each excitation or ionization in time and space with the progressive evolution of the chemical species until they are thermalized. From this point on the application of diffusion kinetics should be adequate. Efforts to fill in the early details are in their infancy and involve the application of Monte Carlo techniques. One of the problems associated with this approach is the lack of accurate knowledge of the cross sections for ionization and excitation in the liquid phase (see Chapter 3). Other difficulties are the assumptions that have to be made to initiate the Monte Carlo calculation, the validities of which have yet to be established. In addition, our present state of knowledge does not allow us to extend with confidence a calculation, based on accurate cross sections, to the formation of initial chemical species. Although we expect that progress in this direction will be made in the near future, at present an adequate Monte Carlo treatment is not available, and the unified approach to track reactions developed by the authors is chosen for presentation. In this treatment we calculate radiation chemical yields using a direct consideration of the behavior of the average track.

The problem we face is to obtain a realistic track model for high-energy particles that can be used in an ''average'' track treatment of the track reactions. We note that the energy loss of a high-energy heavy particle is divided into two equal parts: glancing collisions with energy losses confined essentially (in the case of water) between 6 and 100 eV and knock-on collisions with energy losses between 100 eV and $2m\beta^2c^2(1 - \beta^2)^{-1}$. The low-energy losses of the glancing collisions have well-defined average properties; we call the track entities that are formed by them spurs; and we call the region along the particle trajectory that they occupy the track core. The much larger region occupied by the knock-on electrons is called the penumbra.

We emphasize that all heavy-particle tracks are similar, each composed of a core and a penumbra, and we present a model for the radiation chemistry of such tracks in water. In the calculations various approximations have to be made. Reaction sequences obtained from theory and interpretation of experimental results are used; in the treatment of diffusion-controlled reactions we use prescribed diffusion. Parameters used in the calculations are chosen so that the results obtained for all particles and energies agree with experiment reasonably well.

A special technique is discussed for obtaining G values of monoenergetic electrons by consideration of a self-consistent equation involving the G values as a function of energy. The G values of monoenergetic electrons are important, and the method may have other important applications. Models for average electron tracks are also of interest because so much of radiation chemistry involves irradiation with electrons. The simplest model for the high-energy electron track is the spur. The isolated spur is also a good track model in many low-LET situations. It is particularly significant that high-energy protons and α particles have tracks with radiation chemical yields related precisely to those of high-energy electron tracks. Thus "radiation quality" effects can be treated with confidence.

The overall picture of the nonhomogeneous reactions of radiation chemistry that emerges from the treatment in this chapter is not expected to change significantly with further development of the Monte Carlo approach. The latter technique will allow more details to be filled in from time to time. Thus we believe that the principles outlined in this chapter form a good basis for understanding the track reactions of radiation chemistry.

REFERENCES

1. Chapter 3.

2. Chapter 5.

3. See Section 4.2.

4. R. L. Platzman, in *Radiation Research*, G. Silini (Ed.), North Holland, Amsterdam, 1967, p. 20.

5. Chapter 2.

6. A. H. Samuel and J. L. Magee, *J. Chem. Phys.* **21,** 1080 (1953).

7. A. O. Allen, *The Radiation Chemistry of Water and Aqueous Solutions*, D. Van Nostrand, Princeton, NJ, 1961.

8. I. G. Draganic and Z. D. Draganic, *The Radiation Chemistry of Water*, Academic, New York, 1971.

9. A. Mozumder and J. L. Magee, *Int. J. Radiat. Phys. Chem.* **7,** 83 (1975).

10. J. E. Turner, J. L. Magee, H. A. Wright, A. Chatterjee, R. N. Hamm, and R. H. Ritchie, *Radiat. Res.* **96,** 437 (1983).

11. A. K. Ganguly and J. L. Magee, *J. Chem. Phys.* **25,** 129 (1956).

12. J. L. Magee and A. Chatterjee, *J. Phys. Chem.* **84,** 3529 (1980).

13. A. Chatterjee and J. L. Magee, *J. Phys. Chem.* **84,** 3537 (1980).

14. A. Kuppermann, in *Radiation Research*, G. Silini (Ed.), North Holland, Amsterdam, 1967, p. 212.

15. G. Jaffé, *Ann. Phys. Leipzig IV* **42,** 303 (1913).

16. LET stands for linear energy transfer; it is the average value of $-dE/dx$.

17. S. P. Ahlen, *Rev. Mod. Phys.* **52,** 121 (1980).

18. H. A. Bethe, *Ann. Phys. (Leipzig)* **5,** 325 (1930).

19. G. D. Zeiss, W. J. Meath, J. C. F. MacDonald, and D. J. Dawson, *Radiat. Res.* **63,** 64 (1975).

20. G. D. Zeiss, W. J. Meath, J. C. F. MacDonald, and D. J. Dawson, *Radiat. Res.* **70,** 284 (1977).

21. A. Mozumder and J. L. Magee, *Radiat. Res.* **20,** 203 (1966).

22. J. L. Magee, *Ann. Rev. Phys. Chem.* **12,** 389 (1961).

23. J. L. Magee and A. Chatterjee, *Radiat. Phys. Chem.* **15,** 125 (1980).

24. J. L. Magee and A. Chatterjee, *J. Phys. Chem.* **82,** 2219 (1978).

25. A. Chatterjee and H. J. Schaefer, *Rad. Environ. Biophys.* **13,** 215 (1976).

26. A. Chatterjee, H. D. Maccabee, and C. A. Tobias, *Radiat. Res.* **54,** 479 (1973).

27. J. L. Magee, *Can. J. Chem.* **55,** 1847 (1977).

28. H. Frohlich and R. L. Platzman, *Phys. Rev.* **92,** 1152 (1953).

29. G. R. Freeman (Ed.), *Can. J. Chem.* **55,** 1797–2277 (1977).

30. A. Mozumder, *J. Chem. Phys.* **55,** 3020 (1971).

31. W. H. Hamill, *J. Phys. Chem.* **73,** 1341 (1969).

32. T. Sawai and W. H. Hamill, *J. Phys. Chem.* **74,** 3914 (1970).

33. H. Ogura and W. H. Hamill, (a) *J. Phys. Chem.* **77,** 2952 (1973); (b) *J. Phys. Chem.* **78,** 504 (1974).

34. S. Kalarickal, Ph.D. Dissertation, University of Notre Dame (1959).

35. J. W. Hunt, in *Advances in Radiation Chemistry*, Vol. 5, M. Burton and J. L. Magee (Eds.), Wiley-Interscience, New York, 1976, pp. 185–315.

36. G. Czapski and E. Peled, *J. Phys. Chem.* **77,** 893 (1973).

37. J. L. Magee, *J. Am. Chem. Soc.* **73,** 3270 (1951).

38. M. Anbar, M. Bambenek, and A. B. Ross, *Selected Specific Rates of Reactions of Transients from Water Aqueous Solution. I. Hydrated Electron*, Report NSRDS-NBS 43, U.S. Department of Commerce/National Bureau of Standards, Washington, DC, 1973.

39. A. B. Ross, *Selected Specific Rates of Reactions of Transients from Water in Aqueous Solution. I. Hydrated Electron, Supplemental Data*, Report NSRDA-NBS 43, U.S. Department of Commerce/National Bureau of Standards, Washington, DC, 1975.

40. M. Anbar, Farhataziz, and A. B. Ross, *Selected Specific Rates of Reaction of Transients from Water. II. Hydrogen Atom*, Report NSRDS-NBS 51, U.S. Department of Commerce/National Bureau of Standards, Washington, DC, 1975.

41. Farhataziz and A. B. Ross, *Selected Specific Rates of Reactions of Transients from Water in Aqueous Solutions. Hydroxyl Radical and Perhydroxl Radical and Their Radical Ions*, Report NSRDS-NBS 59, U.S. Department of Commerce/National Bureau of Standards, Washington, DC, 1977.

42. A. Kuppermann and G. G. Belford, *J. Chem. Phys.* **36,** 1412 (1962).

43. A. Kuppermann, *Act. Chim. Biol. Radiat.* **5,** 85 (1961).

44. T. I. Balkas, J. H. Fendler, and R. H. Schuler, *J. Phys. Chem.* **74,** 4497 (1970).

45. E. A. Uehling, *Ann. Rev. Nucl. Sci.* **4,** 315 (1954).

46. F. Rohrlich and B. C. Carlson, *Phys. Rev.* **93,** 38 (1954).

47. D. V. Cormack and H. E. Johns, *Br. J. Radiol.* **25,** 369 (1952).

48. A. T. Nelms, *Natl. Bur. Stand. Circ. no. 524*, 1953.

49. A. O. Fregene, *Radiat. Res.* **31,** 256 (1976).

50. H. Fricke and E. J. Hart, in *Radiation Dosimetry*, Vol. 2, F. H. Attix and W. C. Roesch (Eds.), Academic Press, New York, 1966, p. 107.

51. J. L. Haybittle, R. D. Saunders, and A. J. Swallow, *J. Chem. Phys.* **25,** 1213 (1956).

52. H. C. Sutton, *Phys. Med. Biol.* **1,** 153 (1956).

53. A. Ghiorso, H. Grunder, W. Hartsough, G. Lambertson, E. F. Lofgren, K. Lou, R. M. Mair, R. Mobley, R. Morgado, W. Salsig, and F. Selph, *IEEE Trans. Nucl. Sci.* **NS-20**(3), 155 (1973).

54. J. F. Ziegler, in *The Stopping Power and Ranges of Ions in Matter*, Vol. 4, J. F. Ziegler (Ed.), Pergamon, Elmsford, NY, 1977.

55. E. A. Christman, A. Appleby, and M. Jayko, *Radiat. Res.* **85,** 443 (1981).

56. M. E. Jayko, T. L. Tung, G. P. Welch, and W. M. Garrison, *Biochem. Biophys. Res. Commun.* **68,** 307 (1976).

57. E. J. Hart, W. J. Ramler, and S. R. Rocklin, *Radiat. Res.* **4,** 378 (1956).

58. J. Pucheault and R. Julien, *Proc. Tihany Symp. Rad. Chem.* **2,** 1191 (1972).

59. R. H. Schuler and A. O. Allen, *J. Am. Chem. Soc.* **79,** 1565 (1957).

60. S. Gordon and E. J. Hart, *Radiat. Res.* **15,** 440 (1961).

61. E. Collinson, F. S. Dainton, and J. Kroh, *Proc. R. Soc. London Ser. A* **265,** 407 (1962).

62. G. L. Kochanny, Jr., A. Timnick, C. J. Hochandadel, and C. D. Goodman, *Radiat. Res.* **19,** 462 (1963).

63. A. R. Anderson and E. J. Hart, *Radiat. Res.* **14,** 689 (1961).

64. N. E. Bibler, *J. Phys. Chem.* **79,** 1991 (1975).

5 SINGLE-PAIR DIFFUSION MODEL OF RADIOLYSIS OF HYDROCARBON LIQUIDS

Andries Hummel

Interuniversitair Reactor Instituut
Delft, The Netherlands

5.1. INTRODUCTION

The spatial correlation of the ions in the tracks of high-energy particles was already noted early in the century during the studies of the electrical conductance of irradiated gases and liquids. Immediately after the publication of the first cloud chamber pictures of particle tracks by Wilson in 1912 (1), Jaffé presented the first mathematical treatment of the diffusion and reaction of the positive and negative ions formed in α particle tracks (2) (see Chapter 1, Figure 1.4a, and Chapter 2, Section 2.4.1). He assumed freely diffusing ions and did not take the electrostatic interaction of the ions explicitly into consideration. The treatment was not very satisfactory for the case of ionization in dielectric gases and liquids, for which the theory was developed, since the coulombic interaction between the charged species in these fluids plays a most important role. Many years later, however, the Jaffé treatment turned out to be quite useful for the description of the tracks of neutral species and charged species in polar liquids, where the coulombic interaction has a much shorter range (3).

The primary energy losses along the track of a charged particle in a liquid are predominantly relatively small, and the events are separated along the track by an average distance that depends on the velocity of the primary particle. Each event produces a single ionization or a small cluster of ionizations. For fast electrons with an energy on the order of 1 MeV, the distance between events is very large, so the clusters of ionizations do not interact with one another (see Chapters 3 and 4). While no direct information is available about the distribution of the number of ionizations per group for the condensed phase, it is often assumed that this is the same as for the gas phase. In the gas phase this number distribution has been estimated from cloud chamber pictures, and it has been found that approximately 60% of the groups contains only one ionization, while the probability for occurrence of larger groups rapidly decreases with increasing number.

In a polar liquid like water, where the coulombic interaction has a comparatively short range, most of the low-energy electrons formed in the ionizations get thermalized outside the field of attraction of the parent positive ion. In nonpolar liquids, however, the range of the coulombic interaction is very large and usually considerably in excess of the separation between the thermalized electron and the parent positive ion, with the result that a large fraction of the pairs of positive ions and electrons often does not escape from the mutual field, and "recombines." Recombination of the positive ion with the electron usually results in formation of an electronically excited molecule that often dissociates. When solutes are present that react with the charged species during their lifetime, the course of the chemical reactions (and the final product formation) is changed.

In this chapter we are concerned with the behavior of the charged species after thermalization in the track of a high-energy electron in molecular dielectric liquids. This involves the study of the relative motion due to diffusion and drift in each other's field, of single pairs and small groups of pairs of oppositely charged species, which eventually leads to recombination or escape, and the reaction of the charged species with solutes. For single pairs of charged species that carry out a classical diffusion motion, the mathematical problem has been solved, partly analytically and partly numerically. In Section 5.2 several aspects of this problem are reviewed. Also in Section 5.2 some recently obtained results for groups with a small number of ion pairs are presented.

In Section 5.3 the track of the fast electron is dealt with. It is shown that a number of experimental observations can be satisfactorily explained using our knowledge about the behavior of classical ion pairs. Also the limitations of the single-ion-pair treatment are discussed. We have not attempted to present a complete review of the literature on the subject. We have rather aimed at providing the reader with an insight into the use of nonhomogeneous kinetics of ion pairs for obtaining information about the track of high-energy electrons, and only material that seemed most suited for this purpose has been presented.

5.2. DIFFUSION AND REACTION OF PAIRS OF OPPOSITELY CHARGED IONS

In this section we consider the diffusion and reaction of pairs of oppositely charged ions that initially are placed at some distance away from each other, but in each other's coulombic field. The ions undergo a diffusive movement, as well as a drift motion due to the mutual field, and eventually either react with each other or escape from each other's field. If suitable "scavenger" molecules are present as solutes, the ions during their lifetime may react with these solutes. For a single ion pair the evolution in time of the diffusion and reaction can be described by a differential equation, which can be solved. In the following we shall discuss some results, dealing with various aspects of the problem. In a concluding section we shall then briefly deal with the problem of two and three ion pairs initially in each other's field and present some results obtained by simulation of the diffusion and drift in the field by means of a Monte Carlo technique.

5.2.1. Diffusion

Before we turn to the discussion of the ion pair problem, we want to say a few words about diffusion in general. An example given by Chandrasekhar (4) is illustrative. He considers the random walk of a particle on equidistant positions with a spacing l along a straight line, with an equal probability of going forward and backward for each step. When the particle starts at the origin, the probability of finding the particle after N steps at position m, $W(m, N)$, is equal to

$$W(m, N) = \frac{N!}{[\frac{1}{2}(N + m)]![\frac{1}{2}(N - m)]!} \left(\frac{1}{2}\right)^N \tag{5.1}$$

which is called a Bernoulli distribution. The average displacement, $\overline{m}l$, is equal to zero, since the movement is symmetrical,

$$\overline{m} = 0 \tag{5.2}$$

and the mean square displacement is equal to

$$\overline{m^2 l^2} = N l^2 \qquad (5.3)$$

In Table 5.1 $W(m, N)$ is listed for $N = 10$. It can be shown that for large N and $m \ll N$, the probability of finding the particle at position m, at a distance ml from the origin, is

$$W(m, N) = \left(\frac{2}{\pi N}\right)^{1/2} \exp\left(-\frac{m^2}{2N}\right) \qquad (5.4)$$

In Table 5.1 we compare $W(m, n)$ as obtained with Eqs. (5.1) and (5.4) for $N = 10$. The probability of finding the particle between x and $x + \Delta x$ along the straight line where $\Delta x \gg l$ is now $W(m, N)\, \Delta x/2l$ (the factor 2 results from the fact that m and N are either even or odd), which can also be written as

$$c(x, N)\,\Delta x = \frac{1}{2l}\, W(m, N)\Delta x \qquad (5.5)$$

where $c(x, N)$ is the probability per unit distance, the probability density, with a dimension of length^{-1}. Since $x = ml$, with Eqs. (5.4) and (5.5) we have

$$c(x, N) = \frac{1}{(2\pi N l^2)^{1/2}} \exp\left(-\frac{x^2}{2N l^2}\right) \qquad (5.6)$$

which is the well-known Gaussian distribution. The average displacement is $\bar{x} = 0$ again, and the mean square displacement $\overline{x^2} = N l^2$. If we now define a rate ν as the number of steps that take place per unit time, we have $N = \nu t$, and we can write the probability density as a function of time,

$$c(x, t) = \frac{1}{(2\pi\nu l^2 t)^{1/2}} \exp\left(\frac{-x^2}{2\nu l^2 t}\right) \qquad (5.7)$$

TABLE 5.1. Random Walk in One Dimension[a]

m	$W(m, N)$ from Eq. (5.1)	$W(m, N)$ from Eq. (5.4)
0	0.24609	0.252
2	0.20508	0.207
4	0.11715	0.113
6	0.04374	0.042
8	0.00977	0.010.
10	0.00098	0.002

[a]The probability $W(m, N)$ to find a particle m steps away from the origin after $N = 10$ steps, as obtained from the Bernoulli distribution [Eq. (5.1)] and from Eq. (5.4).

We now define the diffusion coefficient for this random movement along a straight line,

$$D = \tfrac{1}{2} \nu l^2 \tag{5.8}$$

where D has a dimension of length^2 × time^{-1} and which can be seen to be equal to one-half of the mean square displacement of the particle per unit of time. Equation (5.7) can now be rewritten as

$$c(x, t) = \frac{1}{(4\pi Dt)^{1/2}} \exp\left(-\frac{x^2}{4Dt}\right) \tag{5.9}$$

By considering the development of $c(x, t)$ in a Taylor series, we can see that Eq. (5.9) is the solution of the equation

$$\frac{\partial c}{\partial t} = D \frac{\partial^2 c}{\partial x^2} \tag{5.10}$$

which is the familiar diffusion equation for one dimension. The widening of the probability density distribution with time, as seen from Eq. (5.9), causes a "flow" of probability (ϕ) away from $x = 0$, which can be described by

$$\frac{\partial c}{\partial t} = -\frac{\partial \phi}{\partial x} \tag{5.11}$$

because of continuity. From Eqs. (5.10) and (5.11) we have

$$\phi(x) = -D \frac{\partial c}{\partial x} \tag{5.12}$$

The development of the probability density distribution in three dimensions of a freely diffusing particle is described by

$$\frac{\partial c}{\partial t} = D \left(\frac{\partial^2 c}{\partial x^2} + \frac{\partial^2 c}{\partial y^2} + \frac{\partial^2 c}{\partial z^2} \right) \tag{5.13a}$$

which is also written as

$$\frac{\partial c}{\partial t} = D \operatorname{div} \operatorname{grad} c \tag{5.13b}$$

or

$$\frac{\partial c}{\partial t} = D\nabla^2 c \tag{5.13c}$$

For the flux of probability passing through a unit surface per unit time, we have now

$$\varphi = -D \text{ grad } c \tag{5.14a}$$

or

$$\varphi = -D \nabla c \tag{5.14b}$$

If, instead of one particle, we have a large number of identical particles that carry out a random movement independent of one another, the number of particles present in a small, but not too small, volume will be given by the sum of the probability densities of all particles, and in Eqs. (5.13) and (5.14) the concentration may be used instead of the probability density. It is important to realize that these equations describe the behavior of the probability density of one particle due to the random diffusive movement, and as a result of this, the equations also apply for distributions of several particles, as long as they do not interact with each other. A particle does not "know" that it is situated in a region of relatively high concentration and then moves to a region of lower concentration. Its probability to diffuse in different directions is completely independent of the concentration (presence of the other species). In fact, if this is not the case, the diffusion equations given above do not hold.

We consider the diffusion of a particle in three dimensions starting at the origin of a spherical coordinate system. The function describing the probability of finding the particle at a given time in a volume will be spherically symmetrical around the origin. The probability density $c(r, t)$ is the probability of finding the species in a shell between r and $r + dr$ divided by the volume of the shell, $4\pi r^2 dr$. The flow of probability through a sphere at r is now

$$J(r) = 4\pi r^2 \phi(r) = -4\pi r^2 D \frac{\partial c}{\partial r} \tag{5.15}$$

and for the rate of change of the probability in the shell between r and $r + dr$ we have

$$4\pi r^2 \frac{\partial c}{\partial t} = -\frac{\partial J}{\partial r} \tag{5.16}$$

which yields, with Eq. (5.15),

$$\frac{\partial c}{\partial t} = \frac{D}{r^2} \frac{\partial}{\partial r} (r^2 \frac{\partial c}{\partial r})$$

$$= D \left(\frac{\partial^2 c}{\partial r^2} + \frac{2}{r} \frac{\partial c}{\partial r} \right) \tag{5.17}$$

The solution of this equation is

$$c(r, t) = \frac{1}{(4\pi Dt)^{3/2}} \exp\left(-\frac{r^2}{4Dt}\right) \tag{5.18}$$

It is seen that the only difference between this expression and the solution for the one-dimensional case is the power $\frac{3}{2}$ in the denominator instead of $\frac{1}{2}$. We see that the probability density $c(r, t)$ has a maximum at $r = 0$ at all times. The probability of finding the particle in a spherical shell at r per unit thickness is $4\pi r^2 c(r, t)$. The most probable distance to find the particle is where $(\partial/\partial r)\, [4\pi r^2 c(r, t)] = 0$, and it can be found that this corresponds to $r = (4Dt)^{1/2}$. The mean square displacement from the origin is

$$\overline{r^2} = \int_0^\infty r^2 c(r, t) 4\pi r^2 \, dr = 6Dt \tag{5.19}$$

(The diffusion coefficient is seen to be equal to one-sixth of the mean square displacement during one unit of time.)

5.2.2. Isolated Pairs of Ions

5.2.2.1. The Smoluchowski Equation. In the following we shall consider the simultaneous diffusion of two particles. It can be shown that the diffusion of the two neutral particles can be described by the equation

$$\frac{\partial c}{\partial t} = D \text{ div grad } c \tag{5.20}$$

where D is now the sum of the diffusion coefficients of the particles and where the origin of the coordinate system is taken at one of the two particles. Since we shall not be interested in the angular dependence of the relative motion but only in the relative distance of the particles, spherically symmetrical initial conditions are chosen. For example, when we want to study the relative diffusion of two particles initially separated by r_0, we consider the average of the cases where the second particle is initially on all possible positions on a sphere at $r = r_0$ around the first particle; in other words, the second particle is "smeared out" over the sphere $r = r_0$. We may now use the spherically symmetrical diffusion equation

$$\frac{\partial c}{\partial t} = \frac{D}{r^2} \frac{\partial}{\partial r}\left(r^2 \frac{\partial c}{\partial r}\right) \tag{5.21}$$

where D is the sum of the diffusion coefficients of the two particles.

Thus far we have considered the diffusion in the absence of any force between the particles, which is an entirely random process. When the particles are ions, they exert a force on one another, and the relative motion is biased, which can be expressed in a flux of probability equal to the product of the relative drift velocity due to the field and the local probability density. For the spherically symmetrical case this gives rise to a flow through a sphere at r of

$$J_F = 4\pi r^2 v c \tag{5.22}$$

We now consider oppositely charged ions with a charge e. The relative drift velocity v is equal to the product of the sum of the mobilities of the ions (u) and the field due to the ion at the origin at the position of the other ion, or

$$v = -\frac{e}{4\pi\varepsilon_0\epsilon_r r^2} u \tag{5.23}$$

and since $u/D = e/k_B T$, with D the sum of the diffusion coefficients,

$$J_F = -\frac{e^2}{\varepsilon_0\epsilon_r k_B T} Dc \tag{5.24}$$

or

$$J_F = -4\pi r_c Dc \tag{5.25}$$

where we have written $r_c = e^2/4\pi\varepsilon_0\epsilon_r k_B T$, the distance at which the potential energy between the ions is equal to $k_B T$.

The total flow due to diffusion and the field is now

$$J = -4\pi r^2 D \frac{\partial c}{\partial r} - 4\pi r_c Dc \tag{5.26}$$

and for the change in time of the probability density we have, using Eq. (5.16),

$$\frac{\partial c}{\partial t} = \frac{-1}{4\pi r^2} \frac{\partial J}{\partial r} \tag{5.16}$$

together with Eq. (5.26),

$$\frac{\partial c}{\partial t} = \frac{D}{r^2} \frac{\partial}{\partial r}\left(r^2 \frac{\partial c}{\partial r} + r_c c \right) \tag{5.27a}$$

or

$$\frac{\partial c}{\partial t} = D\left[\frac{\partial^2 c}{\partial r^2} + \left(\frac{2}{r} + \frac{r_c}{r^2} \right) \frac{\partial c}{\partial r} \right] \tag{5.27b}$$

This equation is often referred to as the Smoluchowski or Debye–Smoluchowski equation for a pair of ions (5).

We consider a pair of ions initially separated by a distance r_0. If we take one of the particles at the origin, the probability density of the other at $t = 0$ can be written as

$$c(r, 0) = \frac{\delta(r - r_0)}{4\pi r_0^2} \tag{5.28}$$

The development of the probability density in space and time is described by Eq. (5.27). We have now a sink at $r = \infty$ representing the pairs that escape reaction and a sink close to the origin representing the neutralization or recombination reaction between the ions. The first sink is described by

$$c(\infty, t) = 0 \tag{5.29}$$

The second boundary condition describes the reaction of the ions on encounter. Reaction takes place when the particles have approached each other to a distance $r = R$, the reaction radius. The rate of reaction, or rather the probability of reaction per unit time, may be taken to be equal to the product of the probability density at $r = R$ and a specific rate k_R. This is equal to the flow due to diffusion and field toward $r = R$,

$$4\pi D\left(R^2 \frac{\partial c}{\partial r} + r_c c\right)_{r=R} = k_R c(R, t) \tag{5.30}$$

This is called the radiation boundary condition. When k_R is extremely large, we may take

$$c(R, t) = 0 \tag{5.31}$$

This is the Smoluchowski boundary condition.

Sometimes it is of interest to express the Smoluchowski equation in dimensionless variables,

$$\rho = \frac{r}{r_c} \quad \text{and} \quad \tau = \frac{Dt}{r_c^2}$$

Equation (5.27) then reads

$$\frac{\partial c}{\partial \tau} = \frac{\partial^2 c}{\partial \rho^2} + \left(\frac{2}{\rho} + \frac{1}{\rho^2}\right)\frac{\partial c}{\partial \rho} \tag{5.32}$$

with the initial condition

$$c(\rho, \tau = 0) = \frac{\delta(\rho - \rho_0)}{4\pi\rho_0^2} \tag{5.33}$$

The radiation boundary condition at the reaction radius $\rho_R = R/r_c$ is given by

$$\frac{\partial c}{\partial \rho} + \frac{c}{\rho^2} = \frac{k_R}{k_D} \frac{1}{\rho_R} \tag{5.34}$$

where $k_D = 4\pi RD$; the Smoluchowski boundary condition is

$$c(\rho_R, \tau) = 0 \tag{5.35}$$

5.2.2.2. Survival and Escape. We consider now the development in time of the diffusion and reaction with each other of a pair of oppositely charged ions initially separated by a distance $r = r_0$. The initial condition is given by $c(r, t = 0) = \delta(r - r_0)/4\pi r_0^2$. The probability that the ions have not reacted at a given time, or the survival probability, $W(t)$, is equal to the probability density $c(r, t)$ integrated over space,

$$W(t) = \int_R^\infty c(r, t) 4\pi r^2 \, dr \tag{5.36}$$

The probability that the ions have not survived at time t, $1 - W(t)$, is the probability that the ions have reacted with each other, and this is equal to the total flow across the reaction radius during t,

$$1 - W(t) = \int_0^t 4\pi D \left(R^2 \frac{\partial c}{\partial r} + r_c c \right)_{r=R} dt \tag{5.37}$$

or

$$1 - W(t) = \int_0^t k_R c(R, t) \, dt \tag{5.38}$$

as follows from Eq. (5.30), which expresses the radiation boundary condition. The probability for escape from neutralization W_{esc} is equal to $W(t = \infty)$ and is expressed by Eq. (5.37) or (5.38) for $t = \infty$.

It is interesting to note that one does not need to know explicitly the probability density in space and time in order to calculate the survival probability. Laplace transformation of Eq. (5.37) or (5.38) gives us an expression for the Laplace-transformed survival probability, $\overline{W}(s)$, in terms of the Laplace-transformed probability density, $\overline{c}(r, s)$ (6). If the latter one can be obtained from solution of the

Laplace-transformed Smoluchowski equation, $W(t)$ can be obtained by back transformation of $\overline{W}(s)$.

We see from Eq. (5.38) that the escape probability $W_{esc} = W(t = \infty)$ is simply given by $1 - W_{esc} = k_R \overline{c}(R, s = 0)$. In this way Monchick obtained an expression for the probability of escape for an arbitrary spherically symmetrical potential between the two species (7). For two oppositely charged species this reads

$$W_{esc} = 1 - \frac{1 - \exp(-r_c/r_0)}{1 - \exp(-r_c/R) + (4\pi Dr_c/k_R) \exp(-r_c/R)} \tag{5.39}$$

For $k_R \gg 4\pi Dr_c$ and $r_c \gg R$, we obtain

$$W_{esc} = \exp\left(\frac{-r_c}{r_0}\right) \tag{5.40}$$

which is the result obtained by Onsager (8). The expression for W_{esc} can also be obtained by assuming a continuous production of species at the sphere $r = r_0$, resulting in a steady flow toward the sink at the reaction radius and a flow toward infinity [Eq. (5.26)]. The fraction of the production going inward and outward, which is equal to the probability of reaction and escape, respectively, can be calculated.

We now return to the time dependence of the probability density and the probability of survival. While for a pair of neutral particles rather simple exact analytical solutions have been obtained (9), for ions the exact expressions for $c(r, t)$ and $W(t)$ are rather cumbersome and require considerable numerical treatment (10). The probability density $c(r, t)$ can also be obtained by direct numerical integration of the Smoluchowski equation (11) and by Monte Carlo simulation of the relative movement of the ions (12). The survival probability $W(t)$ has also been obtained numerically by the method outlined above, using the Laplace-transformed Smoluchowski equation (13). In Figure 5.1 we show the development in time and space of the probability density for two ions initially at $r = r_0$. In one case the rate of reaction on encounter (k_R) is very large so that we effectively have the Smoluchowski boundary condition [Eq. (5.31)]. In the other case the reaction rate k_R is not so large, so that accumulation at $r = R$ results, due to the fact that the ions are held together by the coulombic field, while it takes some time for reaction to take place. In Figure 5.2 we show results on the survival probability with time for different initial separations. For long times $W(t)$ can be written as (14)

$$\lim_{R \to 0} W(t) = \exp\left(\frac{-r_c}{r_0}\right)\left(1 + \frac{r_c}{(\pi Dt)^{1/2}}\right) \tag{5.41}$$

which shows that for large t the survival probability at a given time divided by the escape probability [$\exp(-r_r/r_0)$] is independent of the initial separation. It should

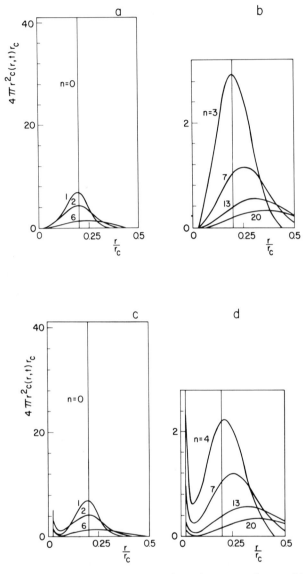

Figure 5.1. Probability that an ion with initial separation r_0 has a separation r at different times t for two values of k_R (11); $r_0/r_c = 0.2$, $R/r_c = 0.025$; $k_R/Dr_c = 103$ (a, b) and $k_R/Dr_c = 1.03$ (c, d); $Dt/r_c^2 = n \times 10^{-3}$.

be realized, however, that this equation holds only for values of $W(t)$ slightly above W_{esc}, as is shown in Figure 5.2. An approximate analytical solution obtained by Mozumder with the so-called prescribed diffusion approximation is (15)

$$W(t) = \exp\left[-\frac{r_c}{r_0}\left(1 - \text{erfc}\,\frac{r_0}{(4Dt)^{1/2}}\right)\right] \tag{5.42}$$

This equation has also been plotted in Figure 5.2 for comparison.

5.2.2.3. Reaction with a Scavenger.
We now consider the case where a solute (S) is present that may react with the ion that is not at the origin. This can be accounted for in the Smoluchowski Eq. (5.27) with an additional sink term, $-k_s C_s c$, where k_s is the specific rate of reaction with the scavenger and C_s the scavenger concentration,

$$\frac{\partial c}{\partial t} = \frac{D}{r^2}\frac{\partial}{\partial r}\left(r^2\frac{\partial c}{\partial r} + \frac{c}{k_B T}\frac{dV}{dr}\right) - k_s C_s c \tag{5.43}$$

It is assumed that k_s may be taken to be constant. (For diffusion-controlled reactions and at comparatively large concentrations this obviously is not correct. We shall return to this later.) The solution of this equation, $c(r, t)$, is now simply related to the case where no solute is present, $c_0(r, t)$, by

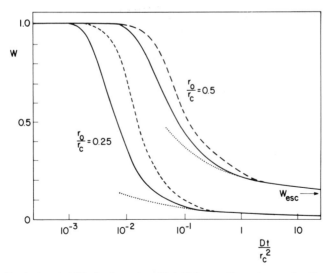

Figure 5.2. Survival probability for ion pairs with initial separation r_0/r_c as a function of time (14); also indicated are the results obtained with Eq. (5.41) for the long-time approximation (dotted lines) and with the prescribed diffusion approximation (15), Eq. (5.42) (dashed lines).

$$c(r, t) = c_0(r, t) \exp(-k_s C_s t) \qquad (5.44)$$

which is seen when we substitute Eq. (5.44) into Eq. (5.43) since we find the original Smoluchowski equation in c_0, without the scavenging term. The probability for reaction with the scavenger per unit time is equal to the local rate of reaction, $k_s C_s c(r, t)$, per unit volume and time integrated over space,

$$S(t) = \int_R^\infty k_s C_s c(r, t) 4\pi r^2 \, dr \qquad (5.45)$$

which on substitution of Eq. (5.44) gives

$$S(t) = \int_R^\infty k_s C_s c_0(r, t) \exp(-k_s C_s t) 4\pi r^2 \, dr \qquad (5.46)$$

which can be written as

$$S(t) = k_s C_s W_0(t) \exp(-k_s C_s t) \qquad (5.47)$$

where $W_0(t)$ is the probability of survival in the absence of S. The probability for the ion to have reacted with the solute at time $t = \infty$, $P_s(k_s C_s)$, is now obtained by integrating the rate of reaction $S(t)$ over time. Using Eq. (5.47) for $S(t)$ we find

$$P_s(k_s C_s) = \int_0^\infty k_s C_s W_0(t) \exp(-k_s C_s t) \, dt = k_s C_s \overline{W}_0(k_s C_s) \qquad (5.48)$$

This shows that the scavenging probability is related to the survival probability function by the Laplace transformation (provided k_s is constant).

Since the fraction that does not react with the solute reacts with the ion at the origin, $P_s(k_s C_s)$ can also be obtained from consideration of the flow of unreacted ions toward the origin. The rate of reaction with the ion at the origin at time t in the presence of scavenger, $R(t)$, is equal to the rate in the absence, $R_0(t)$, multiplied by the probability that during the time t scavenging has not taken place, $R(t) = R_0(t) \exp(-k_s C_s t)$ and $R_0(t) = -dW_0/dt$. The total probability for reaction with the ion at the origin is $P_R(k_s C_s) = \int_0^\infty R(t) \, dt$,

$$P_R(k_s C_s) = \int_0^\infty R_0(t) \exp(-k_s C_s t) \, dt = \overline{R}_0(k_s C_s) \qquad (5.49)$$

With the radiation boundary condition we have $R_0(t) = k_R c_0(R, t)$ [Eq. (5.30)], and therefore (16, 17)

$$P_R(k_s C_s) = k_R \overline{c}_0(R, k_s C_s) \qquad (5.50)$$

If we substitute $R_0(t) = -dW_0/dt$ in Eq. (5.49),

$$P_R(k_s C_s) = \int_0^\infty -\frac{dW_0}{dt} \exp(-k_s C_s t)\, dt \qquad (5.51)$$

and integrate by parts, after substitution of Eq. (5.48) we find that $P_R + P_s = 1$, as expected; the sum of the probability of reaction with the scavenger and with the ion at the origin must be equal to 1. It should be noted that when no scavenger is present, this equality does not hold, since in this case $P_s(0) = 0$, while $P_R(0) = 1 - W_{esc}$.

We see that the scavenging probability can be calculated from the Laplace-transformed probability density as well as from the survival probability in the case of no scavenger. When t and r are expressed as dimensionless variables in Eq. (5.43), $\tau = Dt/r_c^2$, $\rho = r/r_c$, the concentration variable is $\kappa = k_s C_s r_c^2/D$.

Several authors have presented calculations of the scavenging probability (18, 19). In Figure 5.3 we show some results. At very low concentrations the scavenging probability as a function of concentration approaches a $\sqrt{C_s}$ dependence, which is expressed as (19)

$$P_s(k_s C_s) = \exp\left(\frac{-r_c}{r_0}\right)\left[1 + \left(\frac{k_s C_s r_c^2}{D}\right)^{1/2}\right] \qquad (5.52)$$

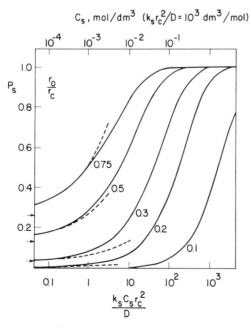

Figure 5.3. Scavenging probability for ion pairs with initial separation r_0/r_c as a function of $k_s C_s r_c^2/D$ (18a); dotted lines represent the limiting expression Eq. (5.52).

and is the counterpart of Eq. (5.41) for the survival at large t. For values of r_0/r_c below 0.5 the validity of this equation is limited to values of $P_s(k_s C_s)$ slightly in excess of the escape probability, as is shown in Figure 5.3.

If the specific rate of reaction with the scavenger cannot be assumed to be independent of time, Eq. (5.48) can be written as

$$P_s(k_s C_s) = \int_0^\infty k_s(t) C_s W_0(t) \exp\left[-\int_0^t k_s(t') C_s \, dt' \right] dt \qquad (5.53)$$

which no longer represents a simple Laplace transform. It is not obvious that for $k_s(t)$ the expression can be used as derived for the pairwise reaction between particles (in the absence of the drag force due to a third particle) (20). This problem is not treated here.

5.2.2.4. Escape with an External Field.

When an external field is present, the probability of escape is increased. It is clear that in this case we do not have spherical symmetry anymore. The problem has been solved by Onsager by means of a stationary flow treatment, and the solution has been expressed as an integral containing a Bessel function of zero order that has been expanded in a power series containing the external field E, r_c and r_0 (8).

Several authors have given different expansions of the original Onsager solution (21, 29). Examples of calculations of $W_{esc}(E)$ for different initial separations are given in Figure 5.4a. It can be shown that for relatively small fields and small values of r_0/r_c, the probability of escape is given by

$$W_{esc}(E) = \exp\left(\frac{-r_c}{r_0}\right)\left(1 + \frac{eEr_c}{2k_BT}\right) \qquad (5.54)$$

which shows that the ratio of the escape probability with and without the field is independent of the initial separation. In Figure 5.4b, $W_{esc}(E)/W_{esc}(E = 0)$ has been plotted for different values of r_0, which illustrates the limiting behavior represented by Eq. (5.54).

5.2.3. Groups of More Than One Ion Pair

In this section we briefly discuss some recently obtained results on the kinetics of neutralization in groups of more than one ion pair using Monte Carlo methods for simulating the movement of the ions (12). In these calculations the first positive ion is chosen at the origin, and the first negative ion is chosen at random from a spherically symmetrical distribution $f_1(r)$ around the first positive ion (which may be a delta function, a Gaussian, or some other function). Then the second positive ion is chosen with another distribution $f_{1,2}(r)$ around the first positive ion, the second negative ion around the second positive ion with $f_2(r)$, and so on. All dis-

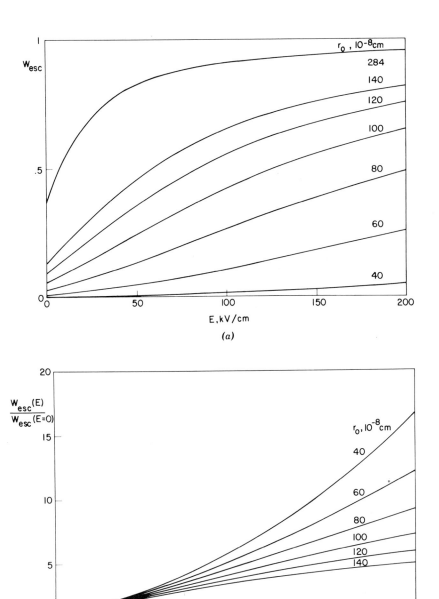

Figure 5.4. (*a*) Escape probability as a function of external field for different initial separations; $r_c = 284 \times 10^{-8}$ cm. (*b*) Escape probability in the presence of an external field divided by escape probability in the absence of a field as a function of the field for different initial separations; $r_c = 284 \times 10^{-8}$ cm.

tributions are spherically symmetrical. Then the movement of each ion is calculated for small time intervals. The movement consists of a drift due to the coulombic field resulting from the interaction with all the ions and a random (diffusive) movement with a distribution of jump lengths. When two (oppositely charged) ions approach each other within the reaction radius R, this pair is assumed to react. In this way the disappearance of the ions with time is calculated. The calculations are repeated for large numbers of groups.

In Figure 5.5 we show an example of results obtained for groups of two pairs initially. The fraction of groups of which no pair has recombined yet, $W_2(t)$, is shown to decay monotonically, giving rise to formation of groups of one pair. The fraction of groups containing one pair, $W_1(t)$, first increases, but then also the single (last) pairs get neutralized, so that eventually the fraction of one-pair groups starts to decrease, while the fraction $W_0(t)$ of groups of which both pairs have recombined grows. The sum of the fractions of the groups containing 2, 1, and 0 groups must equal unity,

$$W_2(t) + W_1(t) + W_0(t) = 1 \tag{5.55}$$

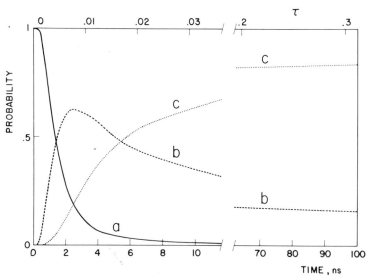

Figure 5.5. Probability of survival as a function of time for groups of two pairs of oppositely charged ions: (a) probability that group consists of two pairs (no recombination) $W_2(t)$; (b) probability that group consists of one pair, $W_1(t)$; (c) probability that both pairs of the group have recombined. For the separation distributions the function $\delta(r - r_0)/4\pi r_0^2$ is used with $r_0 = 20 \times 10^{-8}$ cm for +, + distribution and $r_0 = 84 \times 10^{-8}$ cm for the +, − distributions; $D_+ = D_- = 1.3 \times 10^{-9}$ m^2/s, $r_c = 284 \times 10^{-8}$ cm, $R = 15 \times 10^{-8}$ cm; $W_{esc} = 0.034$.

The number of ion pairs left over in a group at any time averaged over all groups is $2W_2(t) + W_1(t)$. The fraction of ion pairs left over of the total number present in the groups initially (2), again averaged over the groups, is equal to the survival probability of the ion pairs and is given by

$$W(t) = \tfrac{1}{2}[2W_2(t) + W_1(t)] \tag{5.56}$$

More generally, for a group of M_0 pairs initially this is written as

$$W(M_0; t) = \frac{1}{M_0} \sum_{n=1}^{M_0} nW_n(t) \tag{5.57}$$

In Figure 5.6 we show an example of the survival probability $W(M_0; t)$ as a function of time for clusters of two pairs initially, with the same initial separation distributions $f_1(r)$ and $f_2(r)$ together with the survival probability for a single pair with the same distribution of initial separations between the positive and negative species, $f(r) = f_1(r) = f_2(r)$. For these separation distributions a delta distribution $f(r) = \delta(r - r_0)/4\pi r_0^2$ was chosen. It is shown that the decay in the two-pair groups

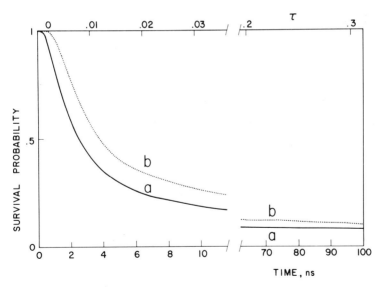

Figure 5.6. Survival probability as a function of time for ion pairs in groups of two pairs initially (curve a) compared with the survival probability for a single ion pair (curve b). For the groups of two pairs the initial conditions have been chosen as in Figure 5.5; also the same values for D_+, D_-, r_c, and R have been used. The separation distribution of the single pair is the same as that of the $+$, $-$ distribution in the two-pair group.

is somewhat faster but not drastically faster. It can be shown that the difference decreases when the distance between the positive charges, expressed in $f_{1,2}(r)$, is increased. The average escape probability for the groups of pairs can be calculated by determining the escape probability of the last pair in each group at the moment the last-but-one pair gets neutralized and averaging over all the cases. In Figure 5.7 we show an example of calculations of the probability of escape for groups of two ion pairs, using the same delta distribution for $f_1(r)$ and $f_2(r)$, for different initial separations between the positive charges. Comparison with the escape probability for a single ion pair shows that the average escape probability from the group is not drastically different from the escape probability of the constituting pairs with the same separation distribution considered independently. We shall return to this matter later, when dealing with the escape and survival in the track.

A few results have also been obtained by this method for the effect of an externally applied field on the probability of escape for the case of groups of two ion pairs. In Figure 5.8 the escape probability as a function of the external field is shown for two cases of groups of two ion pairs where the initial r_{++} distance is different. For the r_{+-} distribution the function $\delta(r - r_0)/4\pi r_0^2$ with $r_0 = 60 \times 10^{-8}$ cm is taken for both cases; the lower curve shows the results obtained with the same function for r_{++} with $r_0 = 10 \times 10^{-8}$ cm, the upper curve with $r_0 = 50 \times 10^{-8}$ cm ($r_c = 300 \times 10^{-8}$ cm, $T = 300$ K). For comparison the results for

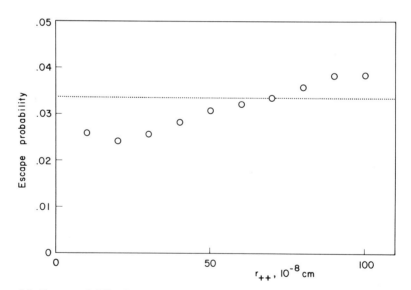

Figure 5.7. Escape probability for ions in groups of two pairs as a function of the initial separation of the positive ions. For the separation distribution the function $\delta(r - r_0)/4\pi r_0^2$ has been used, with $r_0 = 84 \times 10^{-8}$ cm for the $+, -$ distributions; the values for D_+, D_-, r_c, and R are as in Figure 5.5. The horizontal line indicates the escape probability for a single ion pair with $r_0 = 84 \times 10^{-8}$ cm.

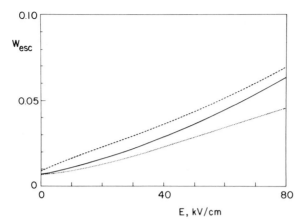

Figure 5.8. Escape probability of ion pairs in groups of two ion pairs initially as a function of the external field for two different distributions of the initial separation between the positive ions. For the separation distributions the function $\delta(r - r_0)/4\pi r_0^2$ has been used with $r_0 = 60 \times 10^{-8}$ cm for the r_{+-} distributions and 10×10^{-8} cm for the r_{++} distribution (lower curve) and 50×10^{-8} cm (upper curve) ($r_c = 300 \times 10^{-8}$ cm, T = 300 K). For comparison the field dependence for the single ion pair is given for the same initial separation (full curve).

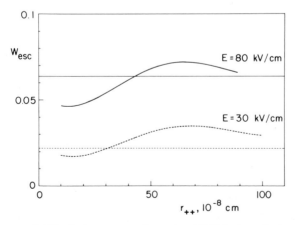

Figure 5.9. Escape probability in the presence of an external field for ion pairs in groups of two ion pairs initially as a function of the initial separation of the positive ions in the group, r_{++}. For the separation distributions the function $\delta(r - r_0)/4\pi r_0^2$ has been used, with $r_0 = 60 \times 10^{-8}$ cm for the r_{+-} distribution; $r_c = 300 \times 10^{-8}$ cm, T = 300 K. The horizontal lines indicate the escape probability for a single ion pair with $r_0 = 60 \times 10^{-8}$ cm.

a single ion pair with the same initial $+$, $-$ separation is shown. In Figure 5.9 the escape probability in the presence of an external field is shown as a function of the initial distance between the two positive ions (r_{++}), while the initial distance between the negative ions is kept constant ($r_0 = 60 \times 10^{-8}$ cm). For comparison the escape probability for single pairs with the same initial separation has been given. We see a somewhat similar dependence on r_{++} as observed in Fig. 5.7 for the escape probability in the absence of an external field.

Approximate analytical treatments of the multiple-ion-pair problem for various initial separation distributions have been given by several authors (25d, 29). It appears that in the near future it will be possible to check the accuracy of these methods by means of Monte Carlo calculations, at least for a moderate number of ion pairs.

5.3. CHARGE SEPARATION IN TRACK OF FAST ELECTRON

5.3.1. Energy Loss and Ionization Along Track

Fast-moving charged particles lose their energy predominantly by interaction with the electrons of the medium. We can distinguish two modes of energy loss, the hard or knock-on collisions and the soft, or resonant, collisions. In the hard collisions an amount of energy is transferred to an electron of the medium that is very large compared to the binding energy of the electron, and for the description of these processes the medium electrons can be considered free. The cross section for such processes decreases with increasing energy loss $W(\propto W^{-2})$. In the soft or distant collisions, where energy losses are involved on the order of magnitude of the binding energy, interaction with the whole molecule takes place, and the same transitions occur that are observed with light. The cross sections for the various resonant losses have been studied extensively in the gas phase. It is found that most of the losses involve ionization and that the most probable loss is around twice the ionization potential. The relative cross sections for the various resonant processes are only weakly dependent on the velocity of the primary particle, and therefore the energetic secondary electrons that are formed in the hard collisions may be considered to give rise to approximately the same spectrum of resonant transitions as the primary particle. As a result, the spectrum of these transitions is approximately independent of the energy of the primary particle. Therefore, the number of each of the transitions occurring during the slowing-down process divided by the total energy absorbed, or the yield of these transitions, will be approximately independent of the initial energy of the primary. This yield, when expressed in events occurring per 100 eV absorbed, is customarily indicated as the G value; when the unit of energy absorbed is not specified, we indicate the yield as N'.

It is illustrative to consider the energy losses that occur when a fast electron is slowing down. In Table 5.2 we show the energy losses W in intervals of W for a 1-MeV electron slowing down. This may be considered typical for a hydrocarbon

TABLE 5.2. The Number of Energy Losses in the Track of a 1-MeV Electron in a Hydrocarbon Liquid for Intervals of the Energy Loss W^a

Energy Interval ΔW (eV)	Number of Losses, $m(W) \Delta W$	Yield per 100 eV, G (per 100 eV)
<10	15000	1.5
10–20	18400	1.84
20–30	6100	0.61
30–50	3000	0.30
50–100	1500	0.15
100–300	370	0.037
300–500	74	0.0074
500–1000	55	0.0055
1000–5000	44	0.0044
(>5000)	(11)	—

[a] From Ref. 22.

liquid. In this case secondary electrons with an energy >5000 eV (a total number of 11) have been considered as primaries, and the losses of these secondaries (tertiaries, and so on) have been included in the numbers for the smaller losses. The reason for choosing the limit of 5000 eV will become clear later. We see the predominant occurrence of losses between 0 and 30 eV and the rapidly decreasing frequency of occurrence of larger losses. Losses above the ionization threshold may give rise to further ionizations and formation of next generation electrons, until finally all electrons lose their energy by neutral excitation and become thermalized at some distance away from the original positive ion.

We now turn to the spatial distribution of the losses along the track of the primary particle. For a 1-MeV electron in a medium with a density of 1000 kg/m^3 the total energy deposited per unit distance (or LET, linear energy transfer) is approximately 0.2 eV/nm (in a low-Z material). Using a value of 40 eV for the average loss of the primary, we find that the average spacing between the events, most of which are ionizations, is about 200 nm. This distance is considerably larger than the range of the coulombic field of a singly charged species, even in a nonpolar medium. The distance at which the coulombic interaction between two singly charged species is equal to $k_B T$, the Onsager distance $r_c = e^2/4\pi\epsilon_0\epsilon_r k_B T$, is about 30 nm in a nonpolar medium with $\epsilon_r = 2$, while in water ($\epsilon_r = 80$) it is about 0.75 nm.

For a 1-keV electron, however, the LET is approximately 10 eV/nm, the average spacing between the primary events is about 3 nm, and we see that while in water the distance between the primary ionizations is still larger than the Onsager radius, in the nonpolar material this is no longer the case. Therefore, in the latter case the 1-keV electron will give rise to a large group of ions that influence each other's movement because of the mutual coulombic field. We see now that the groups resulting from the secondary electrons with sufficiently large energy can be subdivided into separate groups and that below a certain energy, dependent on r_c, this is not possible. In Table 5.2 losses above 5000 eV have been subdivided. The LET of a 5000-eV electron is about 2 eV/nm, and therefore the average spacing between primary events is about 20 nm. However, the choice of 5000 eV, although somewhat arbitrary, is not very critical.

5.3.2. Separate Groups of Ion Pairs

We have seen that the primary ionizations along the track of a 1-MeV electron in a liquid take place on the average at such a large distance from each other that they are outside of each other's coulombic attraction. Most of the secondaries have only very little energy, which is dissipated in close proximity to the parent positive ion, and results in small groups of ions. Occasionally a secondary is formed with such a large energy that the spacing between the successively formed positive ions in its track is so large that this track can again be subdivided into separate groups. When we now study the behavior of the charged species in the track of a fast electron in a liquid, we can treat the track as a collection of groups of ionizations that may be considered independent of one another. We shall now consider the ionizations in these groups.

In order to account for the complete slowing down of the electrons to energies where they cannot cause any further ionization or excitation, we have to have detailed knowledge about the various energy loss processes that may take place as well as the cross sections for these processes as a function of the energy of the energetic electron. This information has been obtained from experiments with low-energy electrons in the gas phase for a number of small molecules. For these gases the slowing down of electrons can be adequately described, and the number of ionizations measured is in agreement with expectation. It is found that the average amount of energy deposited per ionization observed is approximately twice the ionization potential of the atom or molecule.

However, in the liquid phase the situation is more uncertain. No direct experimental knowledge is available about the nature of the ionization process and about the cross sections for the various loss processes for low-energy electrons. The total yield of ionizations is known experimentally only for a few liquids. In liquid noble gases, where total yields of ionization have been determined from ionization chamber measurements, values have been found that substantially exceed the gas phase values (23). In (liquid) water, where the yield of solvated electrons has been determined spectrophotometrically at a picosecond time scale after pulsed irradiation, a yield has been found that is 20% larger than that in the gas phase (24).

Since we are uncertain about the number of ionizations caused by electrons with energies approaching the ionization potential, it is difficult to convert the primary energy losses discussed above to numbers of ionizations. Based on assumed values of the total number of ionizations, distributions of numbers of ionizations per group have been estimated. These results may be compared with the distributions as obtained in dense gases from cloud chamber measurements. In Table 5.3 we give

TABLE 5.3. Fraction f_n of Groups of Ion Pairs Containing n Ion Pairs in Track of High-Energy Electron and Fraction of Ion Pairs in Groups with n Pairs, $F_n = nf_n/\Sigma\, nf_n$

n	f_n^a	f_n^b	F_n^b	f_n^c	F_n^c
1	0.62	0.66	0.23	0.78	0.52
2	0.20	0.19	0.13	0.10	0.13
3	0.09	0.06	0.06	0.03	0.062
4	0.04	0.031	0.044	0.013	0.034
>4	—	0.066	—	0.076	0.0254
5	0.02	—	—	—	—
6	0.01	—	—	—	—
>6	0.02	—	—	—	—
5–10	—	—	0.106	—	—
11–100	—	—	0.160	—	—
>100	—	—	0.256	—	—

[a] Obtained from cloud chamber measurements (25a,b).
[b] Calculated (25c).
[c] Calculated (25d).

cloud chamber data as well as examples of theoretical distributions (22, 25). It is shown that on the basis of these results a large contribution of single-ion-pair groups is expected, although the contribution of larger groups in the total yield of ion pairs is still quite considerable. It is important to realize however, that the distribution is rather uncertain for the liquid.

5.3.3. Escape from Initial Recombination

5.3.3.1. Last Pairs. As shown above, the slowing down of a fast electron in a liquid leads to the formation of a track of groups of pairs of positive ions and thermalized electrons. Subsequently, the charged species start to move with respect to each other both due to diffusion and drift in their mutual fields, and chemical reaction may take place with molecules of the medium. In nonpolar media the range of the coulombic interaction is mostly much larger than the thermalization distance of the electrons, so that only a small fraction of the ions and electrons escapes from the coulombic field. The charged species that do escape from recombination in their group eventually get neutralized in reactions with charged species from other groups of the same track or of a different track.

In the following we shall use the word *ion* for both ions and thermal electrons. The recombination of ions of the same group we shall denote as initial recombination, and the recombination of ions of different groups as homogeneous recombination. When a liquid is irradiated continuously, the concentration of escaped ions builds up to a stationary value depending on the homogeneous rate of recombination and on the radiation intensity. Under most experimental conditions the homogeneous concentration is much lower than the local concentration of the ions in the groups, the latter of which, of course, is independent of the intensity. As a result, the time scale at which the initial recombination takes place is much shorter than that of the homogeneous recombination, and the two processes can be separated experimentally.

The yield of ions escaping initial recombination, or the yield of escaped ions, or free ions, has been determined in different ways and for different liquids (26, 27). (When an external field is present during the irradiation, the yield of escape will be increased. This will be discussed in Section 5.3.3.3.) In Table 5.4 we show yields of escape for some liquids.

The value of the yield of escape contains information about the initial spatial distribution of the charged species in the groups of the track. If the number of charged species escaping on the average from the group resulting from an energy loss W is indicated as n_{esc}^w, for the number of species escaping from the track of the electron with initial energy E_0, we can write

$$N_{esc} = \int_0^{W_{max}} m(W)n_{esc}^w \, dW \tag{5.58}$$

where $m(W) \, dW$ is the number of losses in the interval W to $W + dW$ in the track. For the yield per unit energy absorbed, $N'_{esc} = N_{esc}/E_0$, we have

$$N'_{esc} = \int_0^{W_{max}} m'(W)n_{esc}^w \, dW \tag{5.59}$$

where $m'(w) = m(W)/E_0$, which is only weakly dependent on the initial energy of the electron E_0, as noted above.

TABLE 5.4. Yield of Escaped Ions at Zero Field in Some Liquids Together with Values of r_c and b (Gaussian Distribution)[a]

	G_{esc} (Per 100 eV)	r_c (10^{-8} cm)	b (10^{-8} cm)	b/r_c
n-Hexane	0.13	299	67	0.22
n-Tetradecane	0.12	277	61	0.22
2,2,4-Trimethylpentane	0.33	286	95	0.33
Neopentane	1.2	318	230	0.72
Cyclohexane	0.13	279	61	0.22
Benzene	0.053	248	42	0.17
Perfluoromethyl-cyclohexane	0.028	305	44	0.14
CCl$_4$	0.096	253	51	0.23
CS$_2$	0.31	314	72	0.34
Tetramethylsilane	0.76	313	159	0.51

[a]See Section 5.3.3.2 (27, 29).

For n-hexane the yield of escape is approximately 0.1/100 eV, or $N_{esc} = 10^3$ for $E_0 = 1$ MeV, and n_{esc}^w is approximately 1 at 5000 eV (28, 29, 32); from the yield of energy losses for the different intervals of W as shown in Table 5.2, we can see that losses considerably smaller than 100 eV must contribute to the escape. In fact, the contribution of the small losses dominates, as we shall see below.

The probability of escape for an ion pair with a given initial separation r_0 is known to be given by the simple expression $\exp(-r_c/r_0)$ [Eq. (5.40)]. We now consider the separations of the single pairs in the track and of the last pairs of oppositely charged species in each group of pairs after recombination of all the other pairs of that group. If the probability for occurrence of the separation r to $r + dr$ of the last pair of the group or of a single pair due to the loss W is $g^W(r)\, dr$, we can write, for the yield of escape from the track,

$$N'_{esc} = \int_0^\infty \int_0^{W_{max}} m'(W)\, g^w(r) \exp\left(-\frac{r_c}{r}\right) dr\, dW \qquad (5.60)$$

If the integration over the losses W is performed, we can write

$$N'_{esc} = \int_0^\infty g'(r) \exp\left(-\frac{r_c}{r}\right) dr \qquad (5.61)$$

where $g'(r)\, dr$ is the number of separations of the single and last pairs in the interval dr for the whole track divided by the energy of the primary, or the yield in the interval dr. Since r_c is dependent on temperature, Eq. (5.61) predicts a temperature dependence of N'_{esc}.

Estimates of the last pair separations as a function of energy W have been made that fit the experimentally observed yields of escape in n-hexane, using Eq. (5.60) together with calculated energy loss distributions $m'(W)$ as presented in Table 5.2 (29). Although there is some uncertainty about the probability that ionization occurs for the smallest losses W, the important conclusion may be drawn that apparently the thermalization length of the subexcitation electrons is so large that the contribution of the values of W, corresponding to only one or a few pairs, dominates in the yield of escape.

5.3.3.2. Independent Pairs.

We now consider the escape from recombination in the groups of ion pairs along the track in more detail. In Figure 5.10 we have pictured schematically the position of the charged species after thermalization for the case of a primary ionization giving rise to a secondary electron 1 that causes two additional ionizations and two tertiary electrons, 2 and 3. The ionizations are spaced rather closely together, as compared to the larger thermalization distance of the subionization electrons. The distribution function for distances between the electron after thermalization and its parent ion for the pair i may be indicated as $f_i(r)$ and for the distance between subsequently formed positive ions $f_{kl}(r)$ (see also Section 5.2.3).

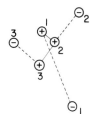

Figure 5.10. Group of three ion pairs resulting from a primary energy loss of a high-energy electron; the secondary electron formed in the first ionization, 1, causes two further ionizations, 2 and 3.

For comparatively small energy losses with only a few ion pairs the $+$, $-$ separation distribution functions will probably be approximately spherically symmetrical and not very different for the different pairs. For larger losses one might possibly expect a somewhat wider distribution for the energetic secondary 1.

In Section 5.2.3 we have shown results of calculations of the survival probability for groups of two ion pairs for spherically symmetrical distributions of separations between the oppositely charged species $f_i(r)$ (with a separation of 84×10^{-8} cm). Rather similar values of the escape probability were found for the groups of two pairs and for single pairs with the same $+$, $-$ separation.

We now assume that the average number of pairs escaping recombination from a group of M_0 pairs with $f_1(r)$, $f_2(r)$, . . . , is approximately equal to the sum of the escape probabilities of the pairs considered independently,

$$M_{\mathrm{esc}}(M_0; f_i, f_{kl}) = M_0 W_{\mathrm{esc}}(M_0; f_i, f_{kl}) \simeq \sum_{i=1}^{M_0} \int_0^\infty f_i(r) \exp\left(-\frac{r_c}{r}\right) dr \quad (5.62)$$

where f_i symbolizes the different separation distributions $f_1(r)$, $f_2(r)$, ... Equation (5.62) shows that the escape probability for the group is approximately equal to the escape probability of the independent pairs,

$$W_{\mathrm{esc}}(M_0; f_i, f_{kl}) \simeq \frac{1}{M_0} \sum_{i=1}^{M_0} \int f_i(r) \exp\left(-\frac{r_c}{r}\right) dr \qquad (5.63)$$

With

$$\frac{1}{M_0} \sum_{i=1}^{M_0} f_i(r) = f(M_0; r) \qquad (5.64)$$

we can also write

$$W_{\mathrm{esc}}(M_0; f_i) \simeq \int f(M_0; r) \exp\left(-\frac{r_c}{r}\right) dr \qquad (5.65)$$

where $f(M_0; r)$ is the average distribution of separations of all the independent pairs in the group normalized to 1, $\int f(M_0; r)\, dr = 1$.

If $p(M_0)$ is the number of groups with M_0 pairs during the thermalization of the primary and $p'(M_0)$ is the yield of these groups, we have, for the yield of escape for the track,

$$N'_{esc} \simeq \sum_{M_0} p'(M_0) M_0 W_{esc}(M_0; f_i, f_{kl}) \tag{5.66}$$

Using the independent pair approximation expressed in Eqs. (5.62) and (5.65), we find

$$N'_{esc} \simeq p'(M_0 = 1) \int f(M_0 = 1, r) \exp\left(-\frac{r_c}{r}\right) dr$$

$$+ \int \sum_{M_0 > 1} p'(M_0) M_0 f(M_0; r) \exp\left(-\frac{r_c}{r}\right) dr \tag{5.67}$$

where we have written the contribution of single pairs, which is exact, and that of the multiple ion pair groups, which is approximate, separately. We can also write

$$N'_{esc} \simeq \int_0^\infty f'(r) \exp\left(-\frac{r_c}{r}\right) dr \tag{5.68}$$

with

$$f'(r) = \sum_{M_0} p'(M_0) M_0 f(M_0; r) \tag{5.69}$$

where now $f'(r)\, dr$ represents the yield of pairs with separation between r and $r + dr$, including both single pairs and pairs in groups. The total yield of ion pairs initially is given by

$$N'_0 = \int_0^\infty f'(r)\, dr \tag{5.70}$$

and also by

$$N'_0 = \sum_{M_0} p'(M_0) M_0 \tag{5.71}$$

We may define a thermalization range distribution for the track, which is normalized to 1, as $f_T(r) = f'(r)/N'_0$ [see Eq. (5.70)] and write, for the yield of escape for the whole track, using Eq. (5.68),

$$N'_{esc} = N'_0 W_{esc} \simeq N'_0 \int_0^\infty f_T(r) \exp\left(-\frac{r_c}{r}\right) dr \tag{5.72}$$

The distribution $N_0' f_T(r)$ obtained from the experimental value of N_{esc}' and Eq. (5.72) is an approximation, due to the approximation involved in the second term on the right-hand side of Eq. (5.67).

We now want to study the thermalization range distribution $f_T(r)$ for some liquids, using the experimentally obtained yields of escape and employing Eq. (5.72). Unfortunately, the total yield of ions formed, N_0', is not accurately known for any dielectric molecular liquid. In liquid noble gases this yield is known from measurement of the current collected with an applied electric field. While in liquid noble gases a saturation current is observed for not too large external fields, in molecular liquids, even with very large fields, saturation is not achieved. In liquid argon a total yield of ions $N_0' = 4.4 \pm 0.4/100$ eV has been found, whereas gas phase values of 3.6–4.2/100 eV have been reported (23). Charge scavenging experiments, in which the product of the reaction of an added solute with the charged species is measured, do not give reliable information about the total yield of ion pairs. The scavenging yield at high concentration does not reach a plateau, and no theory exists that can be used to extrapolate to infinite concentration. Also, the problem exists that at larger scavenger concentration excited states may give rise to the same products as the ions. Estimates of the yield of ion pairs in *cis*-decalin have been made from the yield of solvent fluorescence and the probability of formation of a fluorescing excited state on recombination of an ion pair (30). However, these experiments have not provided very accurate results either. In water the yield of solvated electrons has been measured by means of pulse radiolysis with spectrophotometeric detection (24). The yield at approximately 30 ps after formation has been found to be 4.0/100 eV, considerably in excess of the total ion yield in the gas phase of 3.3/100 eV. To date, pulse radiolysis of dielectric liquids has not provided information about the total yields of ion pairs.

In order to investigate the range distributions with the aid of Eq. (5.72), it is often assumed that the total yield of ions in the liquid is equal to that in the gas. It should be borne in mind, however, that this is an approximation, and differences between the two values, in either direction, may be expected.

Various shapes of the range function $f_T(r)$ can be made plausible. We shall consider the Gaussian distribution

$$f_T(r) = \frac{4\pi r^2}{(\pi b^2)^{3/2}} \exp\left(-\frac{r^2}{b^2}\right) \tag{5.73}$$

which is based on a picture of the thermalizing electron as a classical particle undergoing a random walk, and an exponential distribution

$$f_T(r) = \frac{1}{b} \exp\left(-\frac{r - R}{b}\right) \tag{5.74}$$

which results from considering the thermalization of the electron as an outgoing wave undergoing multiple scattering (31). In Figure 5.11 we show examples of

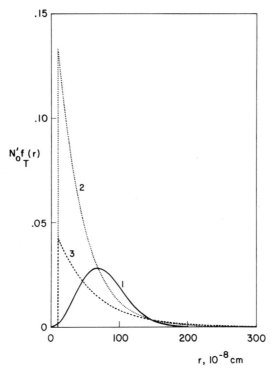

Figure 5.11. Yield of ion pairs with initial separation r per unit interval of r, $N'_0 f_T(r)$, as a function of r: (1) $N'_0 = 2.3/100$ eV, Gaussian distribution [Eq. (5.73)] with $b = 68 \times 10^{-8}$ cm; (2) $N'_0 = 5.0/100$ eV, exponential distribution [Eq. (5.74)] with $b = 38 \times 10^{-8}$ cm, $R = 10 \times 10^{-8}$ cm; (3) $N'_0 = 2.3/100$ eV, exponential distribution with $b = 54 \times 10^{-8}$ cm, $R = 10 \times 10^{-8}$ cm. All distributions have the same yield of escape, $N'_{esc} = 0.08/100$ eV, with $r_c = 289 \times 10^{-8}$ cm.

different distributions $f'(r) = N'_0 f_T(r)$ with the same yield of escaped ions, $N'_{esc} = 0.08/100$ eV, for $r_c = 289 \times 10^{-8}$ cm (CCl_4 at 20 °C). For the Gaussian distribution $N'_0 = 2.3/100$ eV was taken, and $b = 68 \times 10^{-8}$ cm; for the exponential ones $N'_0 = 5.0/100$ eV with $b = 38 \times 10^{-8}$ cm and $N'_0 = 2.3/100$ eV with $b = 54 \times 10^{-8}$ cm ($R = 10 \times 10^{-8}$ cm). This figure illustrates that the integral Eq. (5.72) is very insensitive to the contribution for small values of r.

To try to understand the thermalization process, yields of escape for numerous liquids with different molecular properties have been studied (27). In Table 5.4 we show free ion yields for various liquids and b values for a Gaussian distribution, assuming total ion yields are equal to the gas phase values [Eqs. (5.72) and (5.73)]. For some of the liquids this Gaussian distribution gives a satisfactory description of the temperature dependence of the yield of escape (n-hexane); for others, however, this is not the case (neopentane). No detailed information about the shape of the distributions can be obtained from these measurements alone, however. Whatever the true form of the distribution is, it is clear from Table 5.4 that considerable differences exist in the thermalization distance for the different liquids. It has also been observed that hydrocarbon liquids with a large thermalization length have large mobilities of the excess electron. It appears that the thermalization distance is determined by the interaction with the medium of electrons with energies close to thermal energy.

5.3.3.3. Escape with an External Field.
In the foregoing section, when considering the escape without applied field, we have assumed that the average number of pairs escaping recombination in a multipair group can be calculated approximately by assuming the geminate pairs in the group to be independent. This was based on results obtained with Monte Carlo calculations for groups of two and three pairs without external field. No results have been obtained yet for the case of an applied field, and we cannot predict which errors are introduced when we calculate the field dependence with the same assumption. If we do make this approximation and assume that the separations are not affected by the external field, we find, for the yield of escape, analogous to Eq. (5.67),

$$
N'_{esc}(E) \simeq p'(M_0 = 1) \int_0^\infty f(M_0 = 1; r) W_{esc}(r; E)\, dr
$$
$$
+ \sum_{M_0 > 1} p'(M_0) \int_0^\infty f(M_0; r) W_{esc}(r, E)\, dr
$$
(5.75)

or, analogous to Eqs. (5.68) and (5.72),

$$
N'_{esc}(E) \simeq \int_0^\infty f'(r) W_{esc}(r; E)\, dr
$$
(5.76)

$$
\simeq N'_0 \int_0^\infty f_T(r) W_{esc}(r; E)\, dr
$$
(5.77)

The probability of escape for an ion pair in the presence of a field $W_{esc}(r; E)$ can be written as a series in r and E, as we have seen in Section 5.2.2.4. For $r \ll r_c$ and small E this series reduces to

$$
W_{esc}(r; E) = \exp\left(-\frac{r_c}{r}\right)\left(1 + \frac{er_c E}{2k_B T}\right)
$$
$$
= W_{esc}(r; E = 0)\left(1 + \frac{er_c E}{2k_B T}\right)
$$
(5.78)

which means that the ratio of the escape in the presence and in the absence of a field is independent of the separation r and increases linearly with the field. If r

$\ll r_c$ may be assumed to hold for the whole distribution of separations $f'(r)$, we have

$$N'_{\text{esc}}(E) = N'_{\text{esc}}(E = 0) \left(1 + \frac{er_cE}{2k_BT} \right) \tag{5.79}$$

This behavior is indeed observed experimentally for a variety of liquids with a comparatively small yield at zero field for quite different temperatures. Figure 5.12a shows the experimental results for n-hexane (32). The value of the slope divided by the intercept of the straight line is 0.60×10^{-4} cm/V, while the value of $er_c/2k_BT$ in Eq. (5.79) is 0.58×10^{-4} cm/V. In Table 5.5 we show some more examples. The results reported in the table have all been obtained for irradiation with electrons of high energy. It is interesting to note that with primary electrons of 2.5 keV the limiting slope is not found (29). In the track of these electrons most of the ion pairs are correlated, and the independent pair treatment breaks down.

Liquids with larger yields of escape do not show the limiting behavior given by Eq. (5.79), probably because the requirement $r \ll r_c$ is not fulfilled. For these liquids, and for the low-yield liquids at high field strength, higher terms in the series representing $W_{\text{esc}}(r; E)$ should also be considered, which in turn means that the field dependence of the yield of escaped ions contains information about the spatial distribution of the charged species. Several authors have determined separation distributions that give best fits to the experimentally determined $N'_{\text{esc}}(E)$ using the full series for $W_{\text{esc}}(r; E)$. In Figure 5.12b we show experimental results as obtained with n-hexane for fields up to 190 V/cm (33, 34). Good fits are obtained for n-hexane for a Gaussian distribution with $b = 76 \times 10^{-8}$ cm together

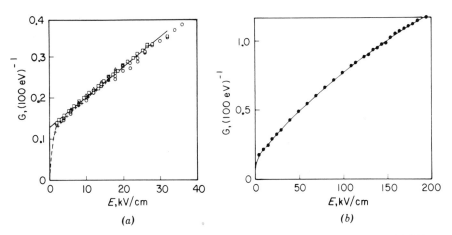

Figure 5.12. Yield of ion pairs escaping initial recombination as a function of the external field in n-hexane (32–34).

TABLE 5.5. Ratio of Slope to Intercept[a]

	T (K)	d (10^{-6} m/V)	$e^3/8\pi\varepsilon_0\epsilon_r k_B^2 T^2$ (10^{-6} m/V)
n-Hexane	298	0.60	0.58
	219	1.12	1.01
iso-Octane	300	0.40	0.58
	210	0.93	0.95
Toluene	236	0.49	0.47
CS_2	293	0.43	0.46
Liquid O_2	90	8.0	7.9
Liquid N_2	77	11	11.2

[a]Obtained from measurements of the yield of escape as a function of the electric field, when represented as $N'_{esc}(E) = N'_{esc}(E = 0)(1 + dE)$, together with calculated values of $er_c/2k_BT$ ($= e^3/8\pi\varepsilon_0\epsilon_r k_B^2 T^2$) for various liquids at various temperatures (29).

with a total yield of ions $N'_0 = 2.7/100$ eV and also for an exponential distribution with $b = 50 \times 10^{-8}$ cm ($R = 0$) and $N'_0 = 5.9/100$ eV (35). (Assuming for N'_0 the gas phase yield $4.3/100$ eV, we find from the zero field yield $N'_{esc} = 0.13/100$ eV a value of b of 67×10^{-8} cm for the Gaussian and 58×10^{-8} cm for the exponential distribution.) We do not gain much information about the separation distribution for small distances from the field dependence of the yield in this case, which is not surprising considering the limiting behavior that is independent of the separation distribution. In Figure 5.13 we have plotted results of the field depen- dence of the yield of escape in neopentane (36). Rather good fits have been ob- tained for an exponential function with $b = 250 \times 10^{-8}$ cm as well as with a Gaussian with $b = 230 \times 10^{-8}$ cm (37) with values of N'_0 of $\sim 5.4/100$ eV and $\sim 3.8/100$ eV, respectively (38). Improved fits can be obtained with somewhat different separation distributions (39). Application of the multiple-pair model of

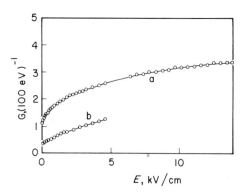

Figure 5.13. Yield of escaped ions as a function of the external field in neopentane (a) and in a solution of SF_6 in neopentane (b) (36b).

Dodelet and Freeman to the neopentane results leads to estimates of the separation distributions that are wider than the ones obtained from the independent pair treatment (39).

It appears that in this way we can get a rough estimate of the shape of the separation distributions for large separations. More information can be obtained from the field dependence at different temperatures provided it can be assumed that the range distribution does not change with temperature. It may be noted, however, that for a large-yield liquid like neopentane the assumption of independence of the range distribution from the external field may break down at the larger field strengths. The main problem at present is our insufficient knowledge of the effect of the field on the escape from multiple-ion-pair groups.

5.3.4. Initial Recombination and Reaction with Solutes

In the foregoing section we have considered the escape of ion pairs from initial recombination at infinite time. We shall now discuss the time evolution of the initial recombination and the reaction of the charged species with solutes during their lifetime. The probability for reaction with a solute is determined by the lifetime distribution of the charged species involved; in turn, the lifetime distribution is determined by the initial spatial distribution and therefore, indirectly, the scavenging probability also. When considering the yield of escape from recombination at infinite time for a group of ion pairs in Section 5.3.3.2, we have used the simplifying approximation [Eq. (5.62)]

$$M_{esc}(M_0; f_i, f_{kl}) \simeq \sum_{i=1}^{M_0} \int_0^\infty f_i(r) \exp\left(-\frac{r_c}{r}\right) dr \qquad (5.80)$$

which means that the average number of pairs expected to escape from the group is equal to the sum of the escape probabilities of the pairs treated independently. This approximation can also be written as [see Section 5.3.3.2, Eq. (5.63)]

$$W_{esc}(M_0; f_i, f_{kl}) \simeq \frac{1}{M_0} \sum_{i=1}^{M_0} \int_0^\infty f_i(r) \exp\left(-\frac{r_c}{r}\right) dr \qquad (5.81)$$

which shows that the escape probability for the group is equal to the average escape probability of the independent pairs.

The independent pair approximation for all times reads

$$W(M_0; f_i, f_{kl}; t) \simeq \frac{1}{M_0} \sum_{i=1}^{M_0} \int_0^\infty f_i(r) W(r; t) dr \qquad (5.82)$$

where $W(M_0; f_i, f_{kl}; t)$ is the survival probability for the pairs in the group with M_0 initial pairs with initial distributions $f_i(r)$ for the separations between the pairs of positive and negative ions and f_{kl} for the separations between positive ions in the cluster (see Section 5.3.3.2); $W(r, t)$ is the survival probability for a single ion pair with initial separation r. As we have seen in Section 5.2.3 for the case of two pairs, for short times, the decay in the multipair group is faster than predicted by the right side of Eq. (5.82). For the initial separation distributions studied, the differences amounted to a shift in time of about a factor 2. If, however, we accept Eq. (5.82) as an approximation, we can write, for the yield of ion pairs in the track,

$$N'(t) \simeq \int_0^\infty f'(r)W(r; t)\, dr \tag{5.83}$$

using again $f'(r)$ for the yield distribution of separations (as in Section 5.3.3.2) and where $N'(t = 0) = N'_0 = \int_0^\infty f'(r)\, dr$.

For the yield of reaction of one of the ions with a scavenger it is tempting to assume that

$$N'_s(C_s) = \int_0^\infty N'(t)\, k_s C_s\, \exp\, (-k_s C_s t)\, dt \tag{5.84}$$

neglecting the time dependence of k_s. Since scavenging of one ion in a multiple-pair group with an accompanying change in diffusion coefficient has an effect on the lifetimes of the other ions, deviations from Eq. (5.84) may be expected.

Although the treatment given above is attractive because of its simplicity, the physical significance of the approximation Eq. (5.82) is unclear. We therefore also want to consider the problem from a different point of view.

When the track is formed, we have a collection of groups of ion pairs characterized by the yields $p'(M_0)$ of groups of M_0 pairs. After a time t the probability that a group with initially M_0 pairs contains n pairs is $P_n(M_0; t)$, and the average number of pairs left over, $M(t)$ is

$$M(M_0; t) = \sum_{n=1}^{M_0} n P_n(M_0; t)$$

The yield of pairs in the track at time t is now

$$N'(t) = \sum_{M_0} p'(M_0) \sum_n n P_n(M_0; t) \tag{5.85}$$

If we write the contribution of single pairs separately, we get

$$N'(t) = p'(M_0 = 1)P_1\,(M_0 = 1; t) + \sum_{M_0 > 1} p'(M_0)\, P_1\,(M_0; t)$$

$$+ \sum_{M_0 > 1} p'(M_0) \sum_{n > 1} n P_n(M_0; t) \tag{5.86}$$

The first term on the right-hand side represents the contribution of the pairs that are formed initially as single pairs; the second term is due to the last pairs of the groups with initially more than one pair. The first term on the right side continuously decreases with time until a constant value is reached at $t = \infty$. The second term on the right-hand side is zero initially but then increases because in the groups with two pairs one recombines and finally decreases again due to recombination of the last pairs.

As we have seen in Section 5.2.3, it appears that the recombination of all pairs in the group but the last one often takes place relatively rapidly, before appreciable recombination of last pairs takes place. Therefore, it is of interest to designate a time t_L after which we consider only single ion pairs to be present. This means that for $t > t_L$ the last term in Eq. (5.86) is negligible, and

$$N'(t > t_L) = p'(M_0 = 1)P_1 (M_0 = 1; t)$$

$$+ \sum_{M_0 > 1} p'(M_0)P_1 (M_0; t) \qquad (5.87)$$

It may be noted that the total yield of last pairs of the groups and single pairs $\Sigma_{M_0} p'(M_0)$ represents a large fraction of the ions formed (see Table 5.3). Since at $t = t_L$ some fraction of the last pairs have recombined, $N'(t_L)/N_0'$ will be smaller than this value. In Figure 5.14 we have illustrated the contribution of the different groups to the decay. If the separation distribution at $t = t_L$ is $h'(t_L; r)$, we have, for the yield of survival at t,

$$N'(t > t_L) = \int_0^\infty h'(t_L; r) \, W(r; t - t_L) \, dr \qquad (5.88)$$

where $W(r; t - t_L)$ is the probability of survival for a single pair at t with a separation r at t_L. We see that from the experimentally obtained yield of survival as a function of time, by means of Eq. (5.88), we obtain information about the last pair separation distribution at $t = t_L$, $h'(t_L; r)$. This is obviously different from the distribution $f'(r)$ of all the ion pairs initially that we have used above, the difference being mainly that the last pair distribution contains a much smaller contribution of pairs with a small r. It is important to note that this expression is exact, although only applicable for $t > t_L$, while $N'(t)$ as obtained from the independent pair treatment [Eq. (5.83)] is an approximation at all times.

Returning now to the ion scavenging, we see that as long as the rate of scavenging is so low that all the scavenging may be considered to take place after $t = t_L$, the scavenging yield is obtained by

$$N_s'(C_s) = \int_0^\infty h'(t_L; r) P_s (r; k_s) \, dr \qquad (5.89)$$

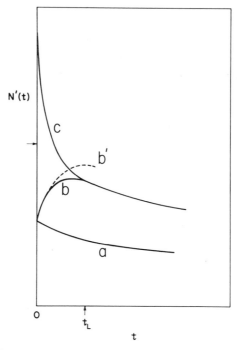

Figure 5.14. Schematic representation of the yield of survival as a function of time for (1) ion pairs formed as single ion pairs, $p'(M_0 = 1)P_1(M_0 = 1, t)$, (curve a); (2) last pairs remaining from the groups initially containing more than one pair, $\Sigma_{M_0 > 1} p'(M_0)P_1(M_0; t)$ (the region between curve b and curve a); (3) pairs in groups of more than one pair, $\Sigma_{M_0 > 1} p'(M_0) \Sigma_{n > 1} nP_n(M_0; t)$ (the region between curves c and b); (4) all ion pairs (curve c); (5) last pairs in the case where no recombination of the last pairs takes place during t_L, the time during which groups of more than one pair are present (between curves b' and a); the arrow indicates the total yield of single pairs and last pairs formed, $\Sigma_{M_0} p'(M_0)$.

where $P_s(r; k_s)$ is the scavenging probability for an ion pair with separation r. Also, for low concentrations we have

$$N'_s(C_s) = \int_0^\infty N'(t > t_L) k_s C_s \exp(-k_s C_s t) \, dt \tag{5.90}$$

neglecting the time dependence of k_s. For high concentrations Eqs. (5.89) and (5.90) do not apply anymore, since scavenging with pairs other than the last one will also take place.

We see that for relatively small yields of scavenging and yields of survival as a function of time, we can treat the track as a collection of single ion pairs. The fraction of the total ion yield that can be treated this way depends on the group size distribution and on the behavior of the survival with time of the multiple-pair groups in comparison to that of the single and last pairs [see Eq. (5.86)]. It would seem that the treatment can be applied safely to at least 30–40% of the ions. For shorter times and for larger yields the last pair treatment eventually breaks down, and the multiple-pair groups have to be considered.

5.3.5. Scavenging; Effect of Change of Mobility on Scavenging of Counter Ion

Studies have been made of the scavenging of both the positive and negative species in several liquids. The concentration dependence of the yield of scavenging may conveniently be expressed over a large range of concentration by an approximate function

$$G_s(C_s) = G_{esc} + B \frac{(\alpha C_s)^n}{1 + (\alpha C_s)^n} \tag{5.91}$$

where n is a power varying between 0.5 and 0.7. In Figure 5.15 we show the CH_4 yield as a function of the concentration of CH_3Br in cyclohexane as an example. In this case the excess electron reacts with CH_3Br to give Br^- ions and CH_3 radicals, and the latter form CH_4 by abstraction of a hydrogen atom. At lower concentrations the yield approaches the yield of escaped ions, and the concentration dependence in this range of concentrations is best represented by a linear function of $C_s^{0.6}$, as we shall see in the next section. The drawn curve in Figure 5.15 represents the function

$$G_s(C_s) = G_{esc} + B \frac{(\alpha C_s)^{0.5}}{1 + (\alpha C_s)^{0.5}} \tag{5.92}$$

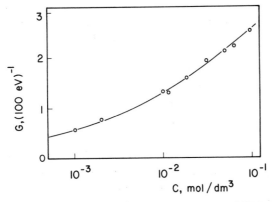

Figure 5.15. Yield of CH_4 as a function of concentration in solutions of CH_3Br in cyclohexane. The curve represents Eq. (5.92) with $G_{esc} = 0.13/100$ eV, $B = 5/100$ eV, and $\alpha = 9$ dm^3/mol (40).

which is often used in the literature and provides a good description over a large range of concentrations (41). It should be realized that Eqs. (5.91) and (5.92) are empirical functions that approximately describe the concentration dependence of the yield in a limited range of concentration, which a priori have no validity outside of that range of concentration. Therefore, the functions cannot be used to determine the total initial yield of ions by extrapolation to $C_s = \infty$. The total yield of ion pairs may be either smaller or larger than the parameter B in Eqs. (5.91) and (5.92).

If scavenging of only single ion pairs took place, the scavenging yield, expressed as a function of the product $k_s C_s$, would be approximately independent of the scavenger (see Section 5.2.3.3), α would be expected to be proportional to k_s for different scavengers, and B would be independent of the scavenger. In Table 5.6 we present values of B, α, and $B^2\alpha$ as obtained for charge scavenging with different scavengers in cyclohexane, using Eq. (5.92). Since B is obtained by plotting $(G_s - G_{esc})^{-1}$ against $C_s^{-0.5}$ and extrapolating to $C_s^{-0.5} = 0$, the value for B^{-1} obtained from the intercept is often not accurately known. Therefore, $B^2\alpha$, which can be determined more accurately, has been given. In this table we also present the specific rates k_s of reaction of the solutes with charged species, determined independently, and the value of $B^2\alpha/k_s$. The values of $B^2\alpha/k_s$ for the different scavenging reactions are very similar. The agreement for scavenging of the positive and the negative species is especially striking since the specific rates differ by more than an order of magnitude (46).

When more than one scavenger is present that can react with the same charged species, and considering the track as a collection of single ion pairs, the yield of scavenging by solute 1 in the presence of both solutes 1 and 2 can be described by

$$N'_{s1}(C_1, C_2) = \frac{k_1 C_1}{k_1 C_1 + k_2 C_2} N'_s \qquad k_s C_s = k_1 C_1 + k_2 C_2 \qquad (5.93)$$

where $N'_s(k_s C_s)$ is the yield of scavenging expressed as a function of $k_s C_s$, which is assumed to be independent of the scavenger. If Eq. (5.92) holds with B and α/k_s independent of the scavengers, we have for the yield of reaction with solute 1 in the presence of both solute 1 and 2,

$$G_{s1}(C_1, C_2) = \frac{\alpha_1 C_1}{\alpha_1 C_1 + \alpha_2 C_2} G_s \qquad \alpha_s C_s = \alpha_1 C_1 + \alpha_2 C_2 \qquad (5.94)$$

with $G_s(\alpha C_s)$ given by Eq. (5.92). By measuring the yield of scavenging by solute 1 in the presence of solutes 1 and 2, the ratio α_1/α_2 can be obtained, which is equal to the ratio of the rate constants, k_1/k_2 (40).

We now discuss the effect of scavenging of one of the charged species of a pair on the scavenging probability of the other species when the diffusion coefficient of the first species changes on scavenging. This effect has been studied in some

TABLE 5.6. Values of G_{esc}, α, B, and $B^2\alpha$ Obtained from Yield of Scavenging of Charged Species in Cyclohexane,[a] Specific Rates k_s of Reaction for the Different Scavengers and $B^2\alpha/k_s$

	G_{esc} ([100 eV]⁻¹)	α (dm³/mol)	B ([100 eV]⁻¹)	$B^2\alpha$[b] (dm³/mol [100 eV]²)	k_s[b] (10¹² dm³/mol s)	$B^2\alpha/k_s$ (10⁻¹² s/[100 eV]²)
CH₃Br + e⁻	0.12	16.2	3.8	234 (43a)	4.1 (45)	57
	0.12	11.2	4.9	274 (43b)	—	67
	0.13	9	5	225 (40)	—	54
CH₃Cl + e⁻	0.12	5.6	3.8	81 (43a,c)	1.4 (44a)	58
	0.13	3.5	5	86 (40)	—	61
	0.12	5	4.2	87 (43d)	—	—
ND₃ + ⊕	0.12	0.85	3.8	12.3 (43e)	0.26 (44b)	47
c-C₃H₆ + ⊕	0.10	0.4	3.8	5.8 (42)	0.10 (44b)	58
	0.13	0.18	5	4.5 (40)	—	45

[a] $G_s(C_s) = G_{esc} + B(\alpha C_s)^{1/2}/[1 + (\alpha C_s)^{1/2}]$
[b] Numbers in parentheses are reference numbers.

255

detail in cyclohexane (29, 42). In this liquid the mobility of the excess electron is 0.2 cm^2/V s, while molecular ions have a mobility on the order of 10^{-4} cm^2/V s, the mobility of the electron hole is 0.01 cm^2/V s. When considering pairs of electron holes and excess electrons, scavenging of electrons will increase the lifetime of the holes and therefore also the probability for being scavenged by a hole scavenger. The scavenging probability of the positive species of a pair in the presence of a scavenger reacting with the negative species can be found to be (29, 42)

$$P_p(k_p C_p, k_N C_N) = (-k_N C_N - k_p C_p + r_D k_p C_p)^{-1} \times [(-k_p C_p$$
$$+ r_D k_p C_p) P_s(k_s C_s = k_N C_N + k_p C_p)$$
$$- k_N C_N P_s(k_s C_s = r_D k_p C_p)] \qquad (5.95)$$

where $P_s(k_s C_s)$ is the scavenging probability when only one of the charged species is reacting; p and N refer to the positive and the negative species, and $r_D = [D(+) + D(-)]/[D(+) + D(S^-)]$, the ratio of the sum of the diffusion coefficients before and after scavenging of the electron. If the track may be considered as a collection of single ion pairs, $N'_p(k_p C_p, k_N C_N)$ can be expressed in $N'_s(k_s C_s)$ analogously to Eq. (5.95), and in terms of the function (5.92) for the 100-eV yield we get

$$G_p(C_p, C_N) = (-k_N C_N - k_p C_p + r_D k_p C_p)^{-1} \times [(-k_p C_p$$
$$+ r_D k_p C_p) G_s(\alpha C_s = \alpha_N C_N + \alpha_p C_p)$$
$$- k_N C_N G_s(\alpha C_s = r_D \alpha_p C_p)] \qquad (5.96)$$

For $C_N = 0$ Eq. (5.96) reduces to $G_p(C_p, C_N = 0) = G_s(\alpha C_s = \alpha_p C_p)$, as expected. For very large values of C_N, where $k_N C_N \gg r_D k_p C_p$, we find $G_p(C_p, C_N) = G_s(\alpha C_s = r_D \alpha_p C_p)$, which shows that the scavenging efficiency of the positive species has increased by a factor r_D due to the scavenging of the electrons. If the electrons are scavenged very early, the lifetime of the counterion is increased by r_D. Figure 5.16a illustrates the yield of positive-ion scavenging as a function of concentration of the positive-ion scavenger, cyclopropane, without the electron scavenger and in the presence of 1 mol/dm^3 CCl$_4$ as an electron scavenger. We see that the curves are shifted by approximately a factor of 17 in concentration, which is in agreement with expectation, since the value of r_D found from mobility measurements is 20. In Figure 5.16b the effect of increasing concentrations of CCl$_4$ on the yield of positive-ion scavenging is plotted for different concentrations of cyclopropane. The drawn curves have been calculated using Eq. (5.96) with experimentally determined values of α_N, α_p, taking $k_N/k_p = \alpha_N/\alpha_p$ and using $r_D = 17$. The agreement shows that the description of the track as a collection of single ion pairs is quite satisfactory.

In Section 5.3.7 we shall deal once more with the effect of scavenging on the lifetime of the charged species in cyclohexane when considering the yield of survival as a function of time in biphenyl solutions. It will be shown that in this case

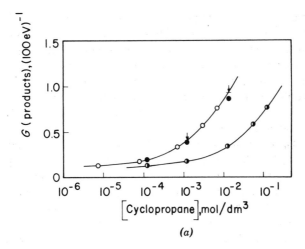

(a)

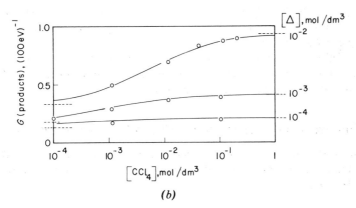

(b)

Figure 5.16. (a) Yield of positive-ion scavenging in cyclohexane by cyclopropane as a function of cyclopropane concentration; ◑, no electron scavenger; ●, with 1 mol/dm³ CCl₄ present as electron scavenger. The estimated effect of incomplete electron scavenging is indicated by the arrows. The open circles (○) have been obtained by shifting the results without electron scavenger to lower concentrations by a factor of 17. (b) Yield of positive-ion scavenging by cyclopropane in cyclohexane as a function of the concentration of the electron scavenger CCl₄ for different cyclopropane concentrations. The yields in the absence of CCl₄ are indicated at the left by the dashed lines; the limiting yields at high concentration are to the right. (Reprinted with permission from ref. 42b.)

as well the experimental results are in agreement with a single-ion-pair picture together with the assumption that the scavenger only affects the mobility of the charged species and not the initial separation distribution. It should be remarked that in both cases we are dealing with maximum observed yields of around 1/100 eV.

While the effect of the change in the mobility as a result of scavenging on the lifetime distribution of a pair is relatively easy to calculate, for multiple-pair groups this problem is not so simple and has not been solved. Therefore, although the agreement of the results of the cyclopropane scavenging experiments with the single-ion-pair treatment is rather striking, we cannot evaluate a possible contribution of multiple-pair effects on the basis of these experiments.

Similar positive-charge scavenging experiments have been carried out with other liquids. The results obtained with the decalins, where the mobilities of the excess electrons and the electrons holes have been measured independently, are also in agreement with expectation (30b).

In the calculation given above use has been made of the assumption that scavenging does not affect the initial separation distribution and that only the mobility changes by a fixed ratio. A complication may arise when (at large scavenger concentrations) scavenging occurs before the thermalization is completed, thus causing the initial separation distribution to be changed. This, of course, will also result in a decrease of the yield of escape, an effect that has been observed in liquids with large electron mobilities (and large yields of escape) (47). Also, deviations are expected when the original charged species do not obey the classical diffusion equation, as might be expected for high-mobility species in the strong coulombic field at small separations. However, for low-yield liquids these effects do not appear to present a problem.

In the following section we shall discuss the yield of scavenging at low concentration in relation to the initial separation distribution, and in Section 5.3.8 we shall deal with the scavenging at high concentration.

5.3.6. Survival at Long Times; Scavenging at Low Concentrations

As we have shown before, for low concentrations the concentration dependence of the yield of scavenging is related to the time dependence of the survival by the Laplace transformation

$$N_s'(C_s) = \int_0^\infty N'(t)k_sC_s \, \exp\,(-k_sC_st)\, dt \tag{5.97}$$

The time dependence of k_s is neglected again, which is justified when considering large values of t and small concentrations. We have also seen in Section 5.2.2.3 that for single ion pairs and vanishingly small concentrations the scavenging probability is given by

$$[P_s(C_s)]_{C_s \to 0} = \exp\left(-\frac{r_c}{r_0}\right)\left[1 + r_c\left(\frac{k_sC_s}{D}\right)^{1/2}\right] \tag{5.98}$$

For the probability of survival at very long times we have (Section 5.2.2.2)

$$[W(t)]_{t \to \infty} = \exp\left(-\frac{r_c}{r_0}\right)\left[1 + \frac{r_c}{(\pi Dt)^{1/2}}\right] \tag{5.99}$$

For the track we therefore expect

$$[N_s'(C_s)]_{C_s \to 0} = N_{esc}' \left[1 + r_c \left(\frac{k_s C_s}{D} \right)^{1/2} \right] \qquad (5.100)$$

and

$$[N'(t)]_{t \to \infty} = N_{esc}' \left[1 + \frac{r_c}{(\pi D t)^{1/2}} \right] \qquad (5.101)$$

It is seen from Eq. (5.100) that for small C_s the ratio of the yield of scavenging to the yield of escape, $N_s'(C_s)/N_{esc}'$, is expected to increase linearly with $C_s^{1/2}$ and the slope is equal to $r_c(k_s/D)^{1/2}$, independent of the initial separation distribution. Equation (5.101) shows that for large t, $N'(t)/N_{esc}'$ increases linearly with $t^{-1/2}$, with a slope $r_c/(\pi D)^{1/2}$.

The concentration dependence of the probability of scavenging was plotted for different initial separations in Figure 5.3. It was shown that for separations smaller than approximately $0.5r_c$ the limiting expression Eq. (5.100) is only applicable for scavenging probabilities slightly in excess of the escape probability. At higher concentrations the scavenging probability first increases somewhat faster than linearly with $C_s^{1/2}$; at larger concentrations the rate of increase with concentration, of course, gets smaller again, since the probability has to saturate at the value 1.

In Figure 5.17 we show experimentally determined yields of electron scavenging plotted against $C_s^{1/2}$ for small scavenger concentrations together with the linear dependence as predicted by Eq. (5.100) for the low-concentration limit. The experimental curve bends away from the line that represents the limiting behavior (49). As can be seen from Figure 5.18, the scavenging yield in this range of concentrations is approximately linear with $C_s^{0.6}$.

In Figure 5.19 we show the yield of survival for long times in solutions of cyclohexane, n-hexane, and isooctane plotted against $t^{-0.6}$, as obtained from measurements of the microwave absorption (45). For comparison, Figure 5.20 is a plot against $t^{-0.5}$ for the cyclohexane results together with the linear dependence given by Eq. (5.101). In Figure 5.21 results are shown for pure CCl_4 at various temperatures (46), all plotted against $t^{-0.6}$.

We now want to investigate the relation between the survival of the charged species and the initial separation distribution of the pairs. In Figure 5.22 we have plotted the ratio of the probability of survival at a given time to the probability of escape as a function of time for pairs with different initial separation distributions and calculated by the methods outlined in Section 2. The probability of survival is expressed as a function of the reduced time $\tau = Dt/r_c^2$. Curves are presented for the initial separations ($\rho_0 = r_0/r_c$) 0.25, 0.33, and 0.5, corresponding to values of W_{esc} of 0.018, 0.048, and 0.135, respectively. It is interesting to observe that for values of $\tau^{-0.5}$ up to approximately 3–4, the time dependence of $W(\tau)/W_{esc}$ is the same. The value of $\tau^{-0.5} = 3$ corresponds to a time of approximately 40 ns for a

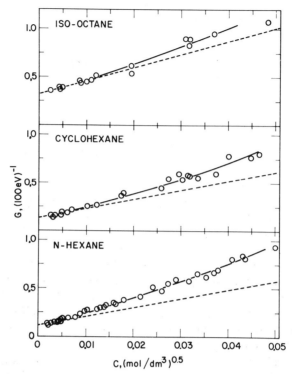

Figure 5.17. Yield of electron scavenging in solutions of CH_3Br in n-hexane (43c, 48), cyclohexane (43b, c, 48), and isooctane (43c, 48). The dashed lines represent the limiting square root relationship given by Eq. (5.100).

mobility of 1×10^{-7} m^2/Vs in a liquid with a low dielectric constant at room temperature ($r_c \simeq 30$ nm), which is typical for a molecular ion. Also, results are plotted as obtained with Gaussian distributions of initial separations,

$$f(\rho) = \frac{4\rho^2}{\pi^{1/2}\rho_b^3} \exp\left(-\frac{\rho^2}{\rho_b^2}\right) \qquad (5.102)$$

and with exponential distributions,

$$f(\rho) = \frac{1}{\rho_b} \exp\left(-\frac{\rho - \rho_R}{\rho_b}\right) \qquad (5.103)$$

The probability of escape is given by

$$W_{\text{esc}} = \int_{\rho_R}^{\infty} f(\rho) \exp\left(-\frac{1}{\rho}\right) d\rho \qquad (5.104)$$

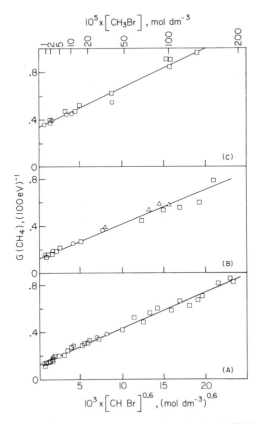

Figure 5.18. Yield of CH_4 formed as a result of electron scavenging by CH_3Br plotted as a function of $[CH_3Br]^{0.6}$ in liquid hydrocarbons: (*a*) *n*-Hexane; $\bigcirc$, Refs. 43c and 48; $\square$, Ref. 43b. (*b*) Cyclohexane; $\bigcirc$, Refs. 43c and 48; $\square$, Ref. 43b; $\triangle$, Ref. 43b. (*c*) Isooctane; $\bigcirc$, Ref. 43c and 48. (Reprinted with permission from ref. 45, Copyright Pergamon Press.)

We see from Figure 5.22 that the time dependence of $W(\tau)/W_{esc}$ for long times is the same for the three distributions.

The limiting expression for $t \to \infty$, [Eq. (5.99)] expressed in τ is

$$\frac{[W(\tau)]_{\tau \to \infty}}{W_{esc}} = 1 + (\pi\tau)^{-0.5} \tag{5.105}$$

This expression is also plotted in Figure 5.22.

It is found that for $0 < \tau^{-0.5} < 3$, $W(\tau)/W_{esc}$ can be rather well represented by (45, 53)

$$W(\tau)/W_{esc} = 1 + 0.6\tau^{-0.6} \tag{5.106}$$

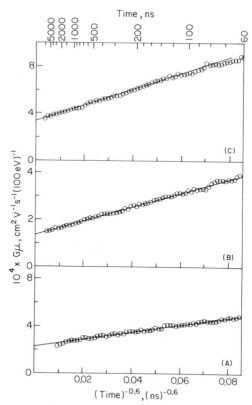

Figure 5.19. Product of yield ions surviving recombination after pulsed irradiation and the sum of the ion mobilities, $G\mu$, plotted against $t^{-0.6}$. (a) A solution of n-hexane containing 1×10^{-2} mol/dm³ biphenyl. (b) A solution of cyclohexane with 2.6×10^{-2} mol/dm³ benzene and 1.4×10^{-2} mol/dm³ CO_2. (c) A solution of isooctane with 1.0×10^{-2} mol/dm³ C_6F_6. (Reprinted with permission from ref. 45, Copyright 1984, Pergamon Press.)

For the range of values of τ mentioned, this expression is apparently applicable for a very wide range of initial separations. In liquids with relatively low yields of escape we now assume this expression to also hold for the track. Using Eq. (5.106), we have, for the yield of charged species at time t,

$$N'(t) = N'_{esc} \left[1 + 0.6 \left(\frac{r_c^2}{Dt} \right)^{0.6} \right] \tag{5.107}$$

which shows the $1/t^{0.6}$ dependence that we have observed experimentally in Figures 5.19 and 5.21. Values of the sum of the diffusion coefficients (D) and the yield of escape, as determined from these plots, have indeed been found to be in agreement with independently determined values (45, 46).

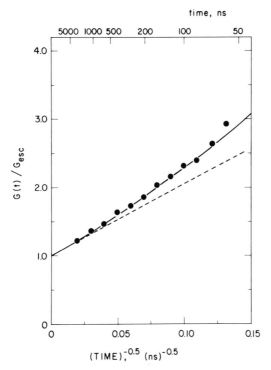

Figure 5.20. Yield of ion pairs at time t after pulsed irradiation divided by yield of escaped ions in a solution of 1.4×10^{-2} mol/dm³ CO_2 and 2.6×10^{-2} mol/dm³ benzene in cyclohexane (50). The full line gives the probability of survival divided by the escape probability, calculated with Eq. (5.106); the dashed line represents the limiting expression Eq. (5.105); $r_c = 279 \times 10^{-8}$ cm, $D = 2.2 \times 10^{-5}$ cm²/s.

Once the time dependence of the yield of survival has been obtained for large t, the concentration dependence of the scavenging yield for small concentrations can be found with the transformation Eq. (5.97). In this way, using Eq. (5.107) for $N'(t)$, we find, for $N'_s(C_s)$,

$$N'_s(C_s) = N'_{esc} \left[1 + 1.33 \left(\frac{er_c^2 k_s}{u k_B T} \right)^{0.6} C_s^{0.6} \right] \tag{5.108}$$

The values obtained for the slope divided by the intercept (S/I) from the plots of the yield of electron scavenging by CH_3Br against $C_s^{0.6}$ (as presented in Figure 5.18) are in excellent agreement with the values calculated with Eq. (5.108) using independently determined values of k_s and u (45).

The yield of scavenging as a function of concentration as well as the yield of survival against time for small yields is in good agreement with expectation on the basis of a single-ion-pair treatment. No information is obtained, however, about

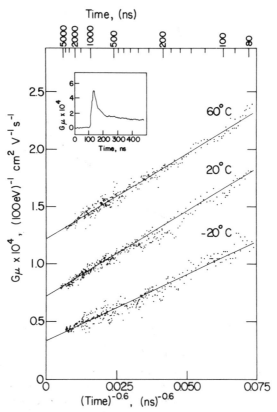

Figure 5.21. Product of yield of ions surviving recombination after pulsed irradiation and the sum of the mobilities, $G\mu$, plotted against $t^{-0.6}$, in CCl_4 at different temperatures. Insert: signal obtained at 20 °C. (Reprinted with permission from ref. 51, Copyright 1982, Pergamon Press.)

the separation distribution in addition to what had already been concluded from the value of the yield of escape.

5.3.7. Survival at Short Times

As we have seen from Figure 5.22 in the foregoing section, in order to obtain substantially more information about the initial separation of the ion pairs than we have from the yield of escape, we need to study the survival for values of $1/\tau^{1/2}$ considerably in excess of 3. For a value of 10^{-7} m^2/Vs for a molecular ion and $r_c = 3 \times 10^{-8}$ m, $1/\tau^{1/2} = 3$ corresponds to $t = 40$ ns; for excess electrons in cyclohexane, with a mobility $u = 0.2 \times 10^{-4}$ m^2/Vs, this time is 200 times smaller, or 200 ps. From Figure 5.22 we see that we must reach values of $1/\tau^{1/2} = 10$ or above to start to discriminate between different distributions, which means that we need a time resolution more than 10 times better than the values mentioned above.

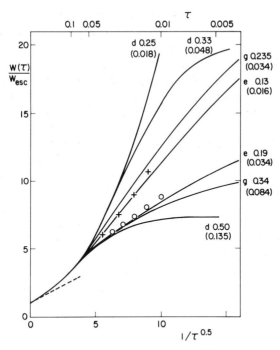

Figure 5.22. Probability $W(\tau)$ of survival at $\tau = Dt/r_c^2$, divided by the escape probability W_{esc}, plotted against $1/\tau^{0.5}$ for different initial separation distributions. The curves for the delta distributions are indicated with **d**, together with the value of the initial separation, expressed in the reduced variable $\rho_0 = r_0/r_c$. The Gaussian distributions are indicated with **g**, and the value of $\rho_b = b/r_c$ [see Eq. (5.102)], the exponential distributions with **e** and $\rho_b = b/r_c$ ($\rho_r = 0.035$) [see Eq. (5.103)]. The escape probability is given within brackets. The dashed line represents the limiting behavior given by Eq. (5.105) (51). Experimentally obtained values of the yield of ions divided by the yield of escaped ions are also plotted for CCl_4 at -20 °C (51) (+) and solutions of biphenyl in cyclohexane (52) ($\bigcirc$).

Direct observation of the charge carriers in this time domain has been achieved with CCl_4, by means of microwave absorption measurements (51), and with biphenyl solutions of cyclohexane, where the biphenyl ions are studied by means of optical absorption (50).

In CCl_4 the thermalized electron is captured efficiently by a CCl_4 molecule to give $CCl_3 + Cl^-$, while the CCl_4^+ ion initially formed may or may not dissociate to give $CCl_3^+ + Cl$, so we are dealing with pairs of Cl^- ions and either CCl_4^+ or CCl_3^+ ions. The results obtained for CCl_4 at -20 °C are plotted in Figure 5.22. (It is quite remarkable that the survival yield is approximately linear with $1/t^{0.6}$ up to about 10 times the escape yield, which is $G_{esc} = 0.08/100$ eV; since $u = 4.1 \times 10^{-8}$ m^2/Vs and $r_c = 2.89 \times 10^{-8}$m, $t = 10$ ns corresponds to $\tau = 1.08 \times 10^{-2}$.) Both a Gaussian distribution of separations with $b/r_c = 0.235$ and an exponential distribution with $b/r_c = 0.13$ (and $R/r_c = 0.035$) are rather satisfactory. With a 100-eV yield of escape $G_{esc} = 0.08/100$ eV, the value of $W_{esc} = 0.034$

for the probability of escape for the Gaussian distribution implies assumption of an initial yield of ion pairs of 2.3/100 eV; similarly for the exponential distribution, with $W_{esc} = 0.016$, a value of 5.0/100 eV is assumed. It is interesting that these two initial distributions with very different initial yields show approximately the same time dependence of the survival at long times. In Figure 5.23 we have plotted the yield $G(t)$ against time for $0 < t < 10$ ns. In this figure we have also pictured the distribution of the yield over the separations at $t = 0$ and $t = 3$ ns for both the Gaussian and exponential initial distributions. This shows that the large differences for short separations that are present initially have disappeared after 3 ns, and the distributions have become approximately the same.

The experimental results represented above have been obtained with 10-ns pulses, and with deconvolution techniques it is possible to get reliable information down to a few nanoseconds. From Figure 5.23 we see, however, that a time resolution substantially below 1 ns is needed. Attempts have been made to study the charge separation in liquid CCl_4 by means of spectrophotometric detection with a higher time resolution. An absorption is observed that is probably due to a product of the recombination of the ions and that shows a rapid growth at a time scale of a few hundred picoseconds. However, there still are some ambiguities about the reaction mechanism (51).

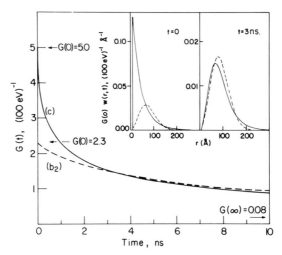

Figure 5.23. Yield of ion pairs that have not recombined at time t after formation, calculated for a Gaussian distribution of initial separations (**g**, $b/r_c = 0.235$, $W_{esc} = 0.034$) and an initial yield of ion pairs of $G_0 = 2.3/100$ eV and for an exponential distribution (**e**, $b/r_c = 0.13$, $R/r_c = 0.035$, $W_{esc} = 0.016$) with an initial yield $G_0 = 5.0/100$ eV; in both cases the yield of escape is the same, $G_{esc} = 0.08/100$ eV; $r_c = 289 \times 10^{-8}$ cm, $D = 0.9 \times 10^{-5}$ cm^2/s (see also Figure 5.22). The insert shows the probability density for an ion pair to be found with a separation r at time t, $c(r, t)$, multiplied by the initial yield of ion pairs, G_0, for the **g** and **e** distributions at $t = 0$ and $t = 3$ ns ($t = 10$ ns corresponds to $\tau = 1.08 \times 10^{-2}$). (Reprinted with permission from ref. 51, Copyright 1982, Pergamon Press.)

We now turn to the discussion of the survival of the ions in biphenyl solutions in cyclohexane as studied by measuring the optical absorption of the biphenyl ions (52). In Figure 5.24 we have collected data obtained by different authors through the years. When both charged species may be considered to be scavenged immediately after thermalization (and the thermalization distance is not affected by the scavenging), the recombination is expected to be retarded by the ratio of the diffusion coefficients before and after scavenging. For the biphenyl solutions this ratio, r_D, is approximately 300 (50). In order to efficiently scavenge electrons with a lifetime t, so that after scavenging their lifetime may be considered to be approximately $r_D t$, we require a scavenger concentration such that during a small fraction of t, scavenging is completed to a sufficiently high degree. It would seem that the leveling off observed at the shortest times in the experiments of Ref. 52d possibly reflects incomplete scavenging. A comparative study of the biphenyl solutions by means of optical detection and microwave absorption has shown that the electron transfer from biphenyl anion to the cyclohexyl radical takes place (50). It is estimated that in the results presented in Figure 5.25 this effect is negligible for $t < 50$ ns. The results for longer times therefore cannot be used to determine $G(t)G_{esc}$ [or rather $\epsilon G(t)/\epsilon G_{esc}$], in order to verify Eq. (5.107), which expresses the $1/\tau^{0.6}$ dependence, and we therefore assume Eq. (5.107), to hold for $t > 50$ ns.

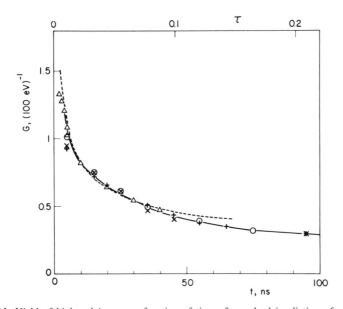

Figure 5.24. Yield of biphenyl ions as a function of time after pulsed irradiation of solutions of biphenyl in cyclohexane; $\bigcirc$, 0.1 mol/dm³ biphenyl, 10-ns pulse, Ref. 52a; ×, 0.1 mol/dm³, 10-ns pulse, Ref. 52b; +, 0.1 mol/dm³, 11-ns pulse, Ref. 52c; $\triangle$, 0.5 mol/dm³, 10-ps pulse, Ref. 52d; the middle of the pulse has been taken as $t = 0$ (54, 55). The dashed curve has been calculated using Eq. (5.112).

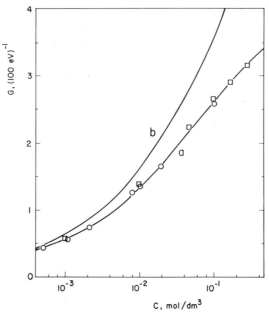

Figure 5.25. Yield of scavenging of excess electrons in solutions of CH_3Br in cyclohexane as a function of CH_3Br concentration. Curves a and b have been calculated with $G_0 = 4.6/100$ eV and an exponential distribution of separations with $b = 49 \times 10^{-8}$ cm ($R = 10 \times 10^{-8}$ cm); a, neglecting the time dependence of the specific rate; b, taking the time dependence into consideration (see text). Experimental data from Refs. 43b ($\square$) and 48 ($\bigcirc$).

Taking now the sum of the mobilities of the biphenyl ions at $u = 7 \times 10^{-8}$ m^2/ Vs [which we estimate is accurate within 10% (50)], for $t = 50$ ns we find $\tau = 0.11$ ($1/\tau^{1/2} = 3.0$), and $W(\tau)/W_{esc} = 3.26$. Taking from Figure 5.24 the value for the yield at 50 ns of 0.41/100 eV, with $W(\tau)/W_{esc} = 3.3$, we find $G_{esc} = 0.126/100$ eV. Using this assumption, we have plotted the results from Figure 5.24 as $G(t)/G_{esc}$ against $1/\tau^{0.5}$ in Figure 5.22 together with the results obtained for CCl$_4$. We see that in order to fit the results for the cyclohexane solutions, we need slightly wider separation distributions than for CCl$_4$.

The maximum yields observed in the experiments discussed above are approximately $G = 1.3/100$ eV. This yield is due to scavenging of the tail of the yield of survival as a function of time of the original charged species, where we expect predominantly single-pair scavenging. As pointed out above for single-pair scavenging, at moderate concentrations the Laplace transform Eq. (5.97) is applicable, which relates the yield of scavenging with the time dependence of the survival:

$$N'_s(C_s) = \int_0^\infty N'(t)k_sC_s \exp(-k_sC_st) \, dt \qquad (5.109)$$

In the foregoing section we also employed this equation for the study of $N'(t)/N_{esc}$ and $N'_s(C_s)/N_{esc}$ in relation to the separation distribution.

It is of interest to present the yield of survival as obtained from the expression for the scavenging yield [Eq. (5.92)]:

$$G_s(C_s) = G_{esc} + \frac{B(\alpha C_s)^{1/2}}{[1 + (\alpha C_s)^{1/2}]} \tag{5.110}$$

Using eq. (5.109), we find

$$G(t) = G_{esc} + B \exp\left(\frac{k_s t}{\alpha}\right) \text{erfc}\left[\left(\frac{k_s t}{\alpha}\right)^{1/2}\right] \tag{5.111}$$

where k_s and α apply to the same scavenging reaction (29, 56). If the recombination is retarded due to scavenging by the factor r_D, which is equal to the ratio of the sums of the diffusion coefficients before and after scavenging, we have

$$G(t) = G_{esc} + B \exp\left(\frac{k_s t}{\alpha r_D}\right) \text{erfc}\left[\left(\frac{k_s t}{\alpha r_D}\right)^{1/2}\right] \tag{5.112}$$

Using the values $B = 5/100$ eV, $\alpha = 9$ dm^3/mol, and $k = 4.1 \times 10^{12}$ dm^3/mol s for electron scavenging by CH$_3$Br, with $G_{esc} = 0.13/100$ eV and $r_D = 300$ for the biphenyl ions, we can now calculate $G(t)$ with Eq. (5.112). The results are also shown in Figure 5.24 (dashed line) (57). Considering the experimental errors in the various parameters used, the very good agreement is somewhat fortuitous; the agreement shows, however, that for yields up to $\sim 1/100$ eV [where we expect the Laplace Eq. (5.109) to be applicable] scavenging in cyclohexane by biphenyl apparently does not change the initial separation distribution appreciably.

Recently results have become available on the growth of the solvent fluorescence after a short pulse in *cis*- and *trans*-decalin on a subnanosecond time scale (30). Since the recombination of the holes and electrons in these liquids results in the formation of solvent fluorescent excited states, the growth kinetics of the fluorescence reflects the time dependence of the survival of the ions. It has been shown in this way that in *cis*-decalin $\sim 80\%$ of the fluorescence is formed within 100 ps (the time resolution of the experiments), which corresponds to $1/\sqrt{\tau} = 6$. In *trans*-decalin, where the mobility of the electron is substantially lower than in *cis*-decalin, the time resolution of 100 ps corresponds to $1/\sqrt{\tau} = 18$ at -27 °C and therefore a much larger fraction of the decay of the ions is expected to be observed, which indeed has been shown to be the case; 50% of the fluorescence is formed within ~ 100 ps. With respect to the analysis of the fluorescence growth kinetics at the shortest times, some uncertainty exists with respect to the possible contribution to the fluorescence from "direct" excitation by the fast electrons. Furthermore, the efficiency of formation of the solvent fluorescent excited state from ion recombination as determined for relatively long-lived ion pairs may not be representative for all the pairs. If it is assumed that neither of these effects plays a role, the total yield of ion pairs is simply equal to the yield of fluorescent excited states divided by the efficiency of formation. In this way the survival probability of the ions as a function of time can be determined and compared with calculated curves as presented in Figure 5.22. Although the uncertainty in the data is considerable, it appears that the separation distribution is strongly peaked toward small distances, a conclusion that is of some importance for our understanding of the ionization process in the liquid. Considering, however, the various assumptions involved, this conclusion should be considered with some reservation. Experimental results with a better time resolution are clearly needed.

5.3.8. Scavenging at High Concentrations

As we have seen in Section 5.3.4, with comparatively low scavenger concentrations, where the probability of scavenging is small during the time that multiple pairs are present (t_L), the scavenging is determined by the separation distribution of single pairs and last pairs,

$$N_s'(C_s) = \int_0^\infty h'(t_L; r)P_s(r; k_sC_s) \, dr \tag{5.113}$$

where $P_s(r; k_sC_s)$ is the scavenging probability for a single pair. At larger concentrations scavenging of other pairs than the last ones takes place, and this equation no longer applies. We have little information about the time it takes for the disappearance of the multiple-pair groups (t_L), and it is therefore difficult to predict when Eq. (5.113) ceases to be applicable. Since, however, the yield of the last pair and single pairs is probably somewhat more than half of the total initial yield, and the latter is probably around 4–5/100 eV, the last pair treatment will probably break down for scavenging yields around $G = 2/100$ eV.

In Sections 5.2.3 and 5.3.4 we have discussed the probability of survival for multiple-ion-pair groups. We have seen that the final probability of escape for the pairs in the group on the average did not differ greatly from the average of the probability of escape for the pairs independently from each other. We have also seen that the decay of the survival on the average for the group was faster than the average of the independent pairs. Therefore, if we calculate the yield of survival with [Eq. (5.83)]

$$N'(t) = \int_0^\infty f'(r)W(r; t) \, dr \tag{5.114}$$

where $f'(r)$ is the separation distribution of all the (independent) pairs, we calculate a survival yield $N'(t)$ that is decaying more slowly than in reality. Using this $N'(t)$ to calculate the scavenging yield by [Eq. (5.84)]

$$N'_s(C_s) = \int_0^\infty N'(t)k_sC_s \exp(-k_sC_st)\, dt \qquad (5.115)$$

will lead to too high values of $N'_s(C_s)$. On the other hand, when $N'(t)$ is known, using Eq. (5.114), we find a too narrow separation distribution $f'(r)$. As pointed out in Section 5.3.4, it should be realized that application of Eq. (5.115) is somewhat questionable for multiple-ion-pair groups because a large change in diffusion coefficient of one ion will affect the lifetime distribution of the others.

Another problem that arises with large concentrations of scavengers is that the specific rate of reaction k_s may no longer be considered to be independent of time, so that we should write

$$N'_s(C_s) = \int_0^\infty N'(t)k_s(t)C_s \exp\left[-\int_0^t k_s(t')C_s\, dt'\right] dt \qquad (5.116)$$

The expression for $k_s(t)$ for a homogeneous distribution of scavenger is (20)

$$k_s(t) = k_D \frac{k}{k + k_D}\left(1 + \frac{k}{k_D} \exp x^2 \operatorname{erfc} x\right) \qquad (5.117)$$

with $x = (1 + k/k_D)(Dt)^{1/2}/R$, and where $k_D = 4\pi RD$, with R the reaction radius, D the sum of the diffusion coefficients of the reactants, and k the specific rate at $t = 0$ (57). It may be questioned if this expression can be used in our case. However, it should give us an impression of the order of magnitude of the effect (58). For $k \gg k_D$, and $(Dt)^{1/2}/R > 1$, $k_s(t)$ reduces to the familiar expression

$$k_s(t) = k_D\left(1 + \frac{R}{(\pi Dt)^{1/2}}\right) \qquad (5.118)$$

As an illustration, we show in Figure 5.25 experimentally obtained yields of scavenging of electrons in cyclohexane with CH_3Br together with calculated curves. The calculated results have been obtained by first calculating $N'(t)$ with Eq. (5.114) for an exponential distribution with $b = 49 \times 10^{-10}$ m, $R = 10 \times 10^{-10}$ m, and $N'_0 = 4.6/100$ eV ($N'_{esc} = 0.14/100$ eV), and then calculating $N'_s(C_s)$ with Eq. (5.115), neglecting the time dependence of the specific rate, and with Eq. (5.116), which incorporates the time dependence. For k_D the experimentally determined value of 4.1×10^{12} $dm^3mol^{-1}s^{-1}$ has been taken (45). Using $u = 0.22 \times 10^{-4}$ m^2/V s for the mobility of the excess electron, we find the reaction radius $R = 10 \times 10^{-10}$ m. The results are rather insensitive to the value of k above $10k_D$ in this range of concentrations. We see that the effect of the time dependence is quite considerable, even at relatively low concentrations. In order to accommodate the increased specific rate of reaction in the calculation, a narrower separation distribution has to be used. However, as we have pointed out earlier in this section, as a result of the use of Eq. (5.114), the distribution found is probably already narrower than the real one. Considerable uncertainty therefore remains about the separation distribution at very short distances and about the initial yield of ion pairs.

A problem that we have touched upon earlier is the interference of the scavenging with the thermalization process. The decrease in the yield of escape due to the presence of electron scavengers in liquids with a large yield of escape is well established and has been illustrated in Figure 5.13, where the yield of escape in the presence of an external field is shown in neopentane with and without SF_6 present. The effect is probably at least partly due to a decrease of the thermalization ranges (47b). In liquids with a comparatively low yield of escape, this effect does not appear to play an important role, as judged from the lack of effect of electron scavengers on the yield of escape as well as from the results on the yield of survival with time in the biphenyl solutions (Section 5.3.7). However, as we have seen, these experiments are not very sensitive to the behavior of pairs with small separations, and it is conceivable that at high scavenger concentrations the scavenger does interfere with the early charge separation processes, even in liquids with a small yield of escape. This adds to the uncertainty of conclusions about the separation distributions for short distances and moreover about the nature of the initial charge separation processes caused by high-energy radiation.

The single-pair description appears to be successful in describing the diffusion, recombination, and reaction with solutes of an appreciable fraction of the ion pairs formed in various molecular dielectric liquids irradiated with high-energy electrons. Both development of the theory of the diffusion kinetics of multiple pair groups and of the experimental techniques for obtaining results in the picosecond domain are needed now to study the earliest charge separation processes caused by high-energy radiation.

REFERENCES

1. C. T. R. Wilson, *Proc. Roy. Soc. (London)* **A87,** 277 (1912).

2. G. Jaffé, *Ann. Phys.* **42,** 303 (1913).

3. H. A. Schwarz, *J. Phys. Chem.* **73,** 1928 (1969).

4. S. Chandrasekhar, *Rev. Mod. Phys.* **15,** 1 (1943).

5. For an arbitrary potential this equation is written as

$$\frac{\partial c}{\partial t} = \frac{D}{r^2}\frac{\partial}{\partial r}\left[r^2\left(\frac{\partial c}{\partial r} + \frac{c}{k_B T}\frac{dV}{dr}\right)\right]$$

or

$$\frac{\partial c}{\partial t} = \text{div } D(\text{grad } c + \frac{c}{k_b T}\text{ grad } V)$$

6. The Laplace transform of a function $F(t)$ is defined as

$$L\{F(t)\} = \int_0^\infty F(t) \exp(-st)dt$$

and written as $\bar{F}(s)$.

7. L. Monchick, *J. Chem. Phys.* **24**, 381 (1956).

8. L. Onsager, *Phys. Rev.* **54**, 554 (1938).

9. A. Hummel, in *Radiation Chemistry: Principles and Applications*, Farhataziz and M. A. J. Rodgers, (Eds.), Verslag Chemie, to be published.

10. K. M. Hong and J. Noolandi, *J. Chem. Phys.* **69**, 5026 (1978).

11. (a) A. Hummel and P. P. Infelta, *Chem. Phys. Lett.* **24**, 559 (1974). (b) W. de Zeeuw and A. Hummel, Report Interuniversitair Reactor Instituut, Delft, No. 133-75-09.

12. W. M. Bartczak and A. Hummel, *Radiat. Phys. Chem.* **27**, 71 (1986).

13. G. C. Abell, A. Mozumder, and J. L. Magee, *J. Chem. Phys.* **56**, 5422 (1972).

14. K. M. Hong and J. Noolandi, *J. Chem. Phys.* **68**, 5163 (1978).

15. A. Mozumder, *J. Chem. Phys.* **48**, 1659 (1968).

16. It has been shown that the Laplace-transformed rate of recombination $\bar{R}(s)$, as obtained with the radiation boundary condition, can be related in a simple way to the Laplace-transformed rate obtained with the Smoluchowski boundary condition, $\bar{R}_s(s)$, by

$$\bar{R}(s) = \frac{\bar{R}_s(s)}{1 - \dfrac{k_D R}{k_R}\left(\dfrac{\partial \bar{R}_s}{\partial r}\right)_{r=R}}$$

where $k_D = 4\pi R D$; see also Ref. 17.

17. J. B. Pedersen, *J. Chem. Phys.* **72**, 771, 3904 (1980).

18. (a) R. J. Friauf, J. Noolandi, and K. M. Hong, *J. Chem. Phys.* **71**, 143 (1979). (b) G. C. Abell, A. Mozumder, and J. L. Magee, *J. Chem. Phys.* **56**, 5422 (1972). (c) P. P. Infelta, *J. Chem. Phys.* **69**, 1526 (1972). (d) M. Tachiya, *J. Chem. Phys.* **70**, 238 (1979).

19. J. L. Magee and A. B. Tayler, *J. Chem. Phys.* **56**, 3061 (1972).

20. F. C. Collins and G. E. Kimball, *J. Coll. Sci.* **4**, 425 (1949).

21. (a) G. R. Freeman, *J. Chem. Phys.* **39**, 1580 (1963). (b) G. C. Abell and K. Funabashi, *J. Chem. Phys.* **58**, 1079 (1973). (c) A. Jahns and W. Jacobi, *Z. Naturf.* **21**, 1400 (1966). (d) J. Terlecki and J. Fiutak, *Int. J. Rad. Phys. Chem.* **4**, 469 (1972). (e) A. Mozumder, *J. Chem. Phys.* **60**, 4300, 4305 (1974).

22. A. Mozumder and J. L. Magee, *J. Chem. Phys.* **47**, 939 (1967).

23. (a) S. S. S. Huang and G. R. Freeman, *Can. J. Chem.* **55**, 1838 (1977). (b) N. V. Klassen and W. F. Schmidt, *Can. J. Chem,* **47**, 4286 (1969).

24. R. K. Wolff, M. J. Bronskill, J. E. Aldrich, and J. W. Hunt, *J. Phys. Chem.* **77**, 1350 (1973).

25. (a) W. J. Beekman, *Physica* **15**, 327 (1949). (b) A. Ore and A. Larsen, *Radiat. Res.* **21**, 331 (1964). (c) K. Kowari and S. Sato, *Bull. Chem. Soc. Jap.* **54**, 2878 (1981). (d) J. P. Dodelet and G. Freeman, *Int. J. Radiat. Phys. Chem.* **7**, 183 (1975).

26. A. Hummel and W. F. Schmidt, *Radiat. Res. Rev.* **5**, 199 (1974).

27. (a) A. O. Allen, NSRDS-NBS-57, U.S. Government Printing Office, Washington DC, 1976. (b) W. F. Schmidt and A. O. Allen, *J. Chem. Phys.* **52**, 2345 (1970).

28. Direct experimental knowledge about n_{esc}^W in the region of interest (below 5 keV) exists only for $W = 2.4$ keV in *n*-hexane, for which $n_{esc}^W = 1.3$ has been found. See also Refs. 29 and 32.

29. A. Hummel, in *Advances in Radiation Chemistry*, Vol. 4, M. Burton and J. L. Magee (Eds.), Wiley, New York, 1974, p. 1.

30. (a) L. H. Luthjens, H. C. de Leng, C. A. M. van den Ende, and A. Hummel, in *Proceedings of the 5th Tihany Symposium on Radiation Chemistry 1982*, J. Dobo, P. Hedvig, and R. Schiller (Eds.), Akademiai Kiado, Budapest, 1983, p. 471. (b) Unpublished results, Interuniversitair Reactor Instituut, Delft.

31. K. Funabashi, *Adv. Rad. Chem.* **4**, 103 (1974).

32. (a) A. Hummel, A. O. Allen, and F. H. Watson, Jr., *J. Chem. Phys.* **44**, 3431 (1966). (b) A. Hummel, Thesis Free University, Amsterdam, 1967.

33. (a) J. Mathieu, D. Blanc, P. Caminada, and J. P. Patau, *J. Chim. Phys.* **64**, 1679 (1967). (b) J. Casanovas, R. Grob, D. Blanc, G. Brunet, and J. Mathieu, *J. Chem. Phys.* **63**, 3673 (1975). (c) A. Hummel and A. O. Allen, *J. Chem. Phys.* **46**, 1602 (1967).

34. Since the yields at zero field presented by Casanovas in Ref. 33b are somewhat higher than the yields measured by several other authors, their yields have been adjusted to the results of Ref. 33c; also the results of Ref. 33a have been adjusted to the absolute height of the yields of Ref. 33c.

35. Abell and Funabashi (Ref. 37) mention that they get good agreement with experiment with both an exponential and a Gaussian distribution. For the exponential distribution they find $b = 50 \times 10^{-8}$ cm $(R = 0)$; it may be shown that for this distribution they implicitly assume $N_0' = 5.6$ $(100$ eV$)^{-1}$, in substantial agreement with the value given above. They do not mention parameters of their best fit for the Gaussian.

36. (a) J. P. Dodelet, P. G. Fuochi, and G. R. Freeman, *Can. J. Chem.* **50**, 1617 (1972). (b) W. F. Schmidt, *Radiat. Res.* **42**, 73 (1970).

37. G. C. Abell and K. Funabashi, *J. Chem. Phys.* **58**, 1079 (1973).

38. The values of N_0' can be obtained by comparison of the relative values of the yield of escape given in Ref. 37 with the absolute ones in Ref. 36b.

39. J. P. Dodelet and G. R. Freeman, *Int. J. Radiat. Chem. Phys.* **7**, 183 (1975).

40. E. L. Davids, J. M. Warman, and A. Hummel, *J. Chem. Soc. Faraday Trans. I* **71**, 1252 (1975).

41. J. M. Warman, K. D. Asmus, and R. H. Schuler, *J. Phys. Chem.* **73**, 931 (1969).

42. (a) A. Hummel, *J. Chem. Phys.* **49**, 4840 (1968). (b) S. J. Rzad, R. H. Schuler, and A. Hummel, *J. Chem. Phys.* **51**, 1369 (1969).

43. (a) J. M. Warman, K. D. Asmus, and R. H. Schuler, *J. Phys. Chem.* **73**, 931 (1969). (b) S. J. Rzad, private communication. (c) S. J. Rzad and J. M. Warman, *J. Chem. Phys.* **49**, 2861 (1968). (d) C. Klein and R. H. Schuler, *J. Phys. Chem.* **77**, 978 (1973). (e) K. D. Asmus, *Int. J. Radiat. Chem.* **3**, 419 (1971).

44. (a) A. O. Allen, T. E. Gangwer, and R. A. Holroyd, *J. Phys. Chem.* **79**, 25 (1975).

(b) J. M. Warman, P. P. Infelta, M. P. de Haas, and A. Hummel, *Chem. Phys. Lett.* **43,** 321 (1972).

45. C. A. M. van den Ende, J. M. Warman, and A. Hummel, *Radiat. Phys. Chem.* **23,** 55 (1984).

46. The values of α and B reported here were obtained by fitting the empirical Eq. (5.92) to results obtained with a large range of concentrations, varying from ca. 10^{-4} mol/dm^3 to a few 0.1 mol/dm^3. As shown in Section 5.3.8, at high concentrations of scavenger different effects may take place that cause differences in the behavior of the scavenging yield as a function of k_sC_s, $N_s'(k_sC_s)$. Therefore a priori the value of $B^2\alpha/k$ is not expected to be the same for all scavenging reactions. From Table 5.6 it appears that these problems are not too serious in this case.

47. (a) A. Mozumder, *J. Chem. Phys.* **60,** 4305 (1974). (b) A. Mozumder and M. Tachiya, *J. Chem. Phys.* **62,** 979 (1975).

48. S. J. Rzad and J. M. Warman, *J. Chem. Phys.* **52,** 485 (1970).

49. If the concentration dependence of the scavenging yield, determined experimentally, is described as linear with $C_s^{1/2}$, it should be realized that the slope is somewhat larger than the limiting slope given in Eq. (5.100).

50. C. A. M. van den Ende, L. Nyikos, J. M. Warman, and A. Hummel, *Radiat. Phys. Chem.* **15,** 273 (1980).

51. C. A. M. van den Ende, L. H. Luthjens, J. M. Warman, and A. Hummel, *Radiat. Phys. Chem.* **19,** 455 (1982).

52. (a) J. K. Thomas, K. Johnson, T. Klippert, and R. Lowers, *J. Chem. Phys.* **48,** 1608 (1968). (b) A. Hummel and L. H. Luthjens, *J. Chem. Phys.* **59,** 654 (1973). (c) E. Zador, J. M. Warman, and A. Hummel, *J. Chem. Phys.* **62,** 3897 (1975). (d) Y. Yoshida, S. Tagawa, and Y. Tabata, *Radiat. Phys. Chem.* **23,** 279 (1984).

53. Another good representation is

$$W(\tau)/W_{\text{esc}} = 1 + (\pi\tau)^{-0.5} + 0.05\ \tau^{-1}$$

54. Since no absolute values of the yields are given in Ref. 52d, the results have been normalized to the results obtained in Refs. 52a–c at $t = 30$ ns.

55. The results from Refs. 52a, b have been obtained with 10-ns pulses, and those from Ref. 52c with 11-ns pulses, and the middle of the pulse has been taken as $t = 0$ for the plot of the results. In Ref. 52d a pulse of 10 ps has been used.

56. S. J. Rzad, P. P. Infelta, J. M. Warman, and R. H. Schuler, *J. Chem. Phys.* **52,** 3971 (1970).

57. A good approximation (within a few percent) is

$$\exp x^2 \operatorname{erfc} x = \frac{2}{\pi^{1/2}} \frac{1}{x + (x^2 + 4/\pi)^{1/2}}$$

58. In Section 5.2.2.3 we mentioned that the application of Eq. (5.117) may be problematic for the single pair. With multiple pairs an additional problem arises due to the effect of depletion.

6 STOCHASTIC MODEL OF CHARGE SCAVENGING IN LIQUIDS UNDER IRRADIATION BY ELECTRONS OR PHOTONS

Gordon R. Freeman

Department of Chemistry
University of Alberta
Edmonton, Canada

The electrons and ions generated in a system under irradiation are highly reactive. They produce changes in the starting material. For example, irradiated ethane decomposes to hydrogen, ethene, and n-butane. A partial reaction mechanism is

$$C_2H_6 \rightsquigarrow C_2H_6^+ + e^- \tag{6.1}$$

$$C_2H_6^+ + e^- \longrightarrow C_2H_6^* \tag{6.2}$$

$$C_2H_6^* \longrightarrow C_2H_5 + H \tag{6.3}$$

$$H + C_2H_6 \longrightarrow H_2 + C_2H_5 \tag{6.4}$$

$$2C_2H_5 \longrightarrow C_2H_4 + C_2H_6 \tag{6.5}$$

$$2C_2H_5 \longrightarrow C_4H_{10} \tag{6.6}$$

If a suitable reactant is present at adequate concentration, the ion or electron can be intercepted before neutralization (6.2) occurs. For example, the $C_2H_6^+$ can be intercepted by ND_3. This results in the formation of HD in place of some of the H_2:

$$C_2H_6^+ + ND_3 \longrightarrow C_2H_5 + ND_3H^+ \tag{6.7}$$

$$ND_3H^+ + e^- \longrightarrow ND_2H + D \tag{6.8}$$

$$ND_3H^+ + e^- \longrightarrow ND_3 + H \tag{6.9}$$

$$D + C_2H_6 \longrightarrow HD + C_2H_5 \tag{6.10}$$

The hydrogen atom formed in reaction (6.9) produces H_2 in (6.4), so not all of the intercepted ions result in HD.

Interception of e^- by N_2O leads to the formation of N_2 and H_2O instead of some of the H_2.

$$e^- + N_2O \longrightarrow N_2 + O^- \tag{6.11}$$

$$C_2H_6^+ + O^- \longrightarrow C_2H_5 + OH \tag{6.12}$$

$$OH + C_2H_6 \longrightarrow H_2O + C_2H_5 \tag{6.13}$$

The fraction of the ions or electrons intercepted is determined by measuring the yield of a new product such as N_2. This chapter describes the calculation of the probability of interception as a function of reactant concentration. The model mechanism is an extension of that for ionization in liquids irradiated with high-energy photons or electrons (see Section 2.4.2).

6.1. MICROZONES OF REACTIVITY (SPURS)

The collision of a 100-fJ (0.6-MeV) photon with a molecule usually causes an electron to be ejected with about half the initial photon energy. The ~50-fJ electron moves through the liquid losing energy in small bits (a few attojoules, tens of electronvolts) and ionizes ~10^4 other molecules along its path. Thus, nearly all of the physical and chemical changes in the system are produced by the energetic electron and not by the initial photon. The kinetics of reactions induced by high-energy photons are therefore similar to those obtained if high-energy electrons are used as the primary radiation. The model described in this chapter applies to radiolysis by either high-energy photons or electrons.

As mentioned in Section 2.4.2, a high-energy electron passing through a liquid loses energy in bits of a few attojoules (tens of electronvolts) each, separated by distances of a few hundred nanometers. The few attojoules deposited at a spot cause an electron to be ejected from a molecule. In about 75% of the cases the ejected electron does not have enough energy to ionize another molecule (1–3), but it does have enough to move away from its sibling ion. As the electron moves away, it undergoes collisions with molecules, loses its excess energy, and becomes thermalized within a few nanometers of its sibling ion. Each ion–electron pair is a long distance from any other pair, so there is no interaction between these pairs. However, in about 15% of the cases the ejected electron has ~4 aJ excess energy, and it ionizes one more molecule. The average rate of energy loss of a 4-aJ electron in a compound composed of light atoms (4, 5), such as an organic material or water, is ~10 pJ·m^2/kg (Figure 6.1). This rate is normalized for the density of the absorber because it applies over a range of densities. An organic liquid or water has a density of ~10^3 kg/m^3, so the average rate of energy loss per unit pathlength is ~10^4 pJ/m. The second ionization therefore occurs at an average distance of ~0.4 nm from the first. The two cations are within a molecular diameter, or two to three CH$_2$ groups of each other. The two electrons scatter more or less randomly and become thermalized a few nanometers from the ions. A typical geometry of a two-pair microzone is displayed in Figure 2.7a.

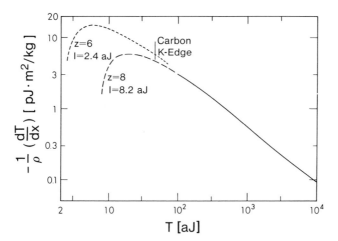

Figure 6.1. Density-normalized rate of energy loss of electrons in hydrocarbons (CH$_2$)$_n$ for $3 < T$ (aJ) $< 10^4$; ρ = density. The curves were calculated from the Bethe equation (4) using the values of Z and I indicated (5) (Z = number of electrons/CH$_2$ group that interact with the projectile electron; I = mean excitation energy of the electrons that interact with the projectile). The K shell electrons in carbon are deleted from Z and I when $T < 50$ aJ; 1 aJ = 6.2 eV.

In about 10% of the cases the ejected electron has enough energy to ionize two or more other molecules. The distribution of sizes of the bits of energy deposited along the high-energy electron path extends tenuously to quite large bits (1, 2). The relative number of microzones containing j ionizations decreases with increasing j; the overall average value of j is 2, but the average for microzones with $j \geq$ 3 might be 5–10.

The cations in a five-pair microzone would initially all be within about 1 nm of each other. With the strong scattering one would have an average distance between the first and fifth ionizations of about $0.4\sqrt{5} \approx 1$ nm, estimated from a random walk model. The electrostatic repulsion between the cations would be $\sim 10^{10}$ V/m, which would drive them apart. The diameter of the positive core of the microzone in a liquid like hexane would expand to ~ 2 nm within 1 ps.

As described in Section 2.4.2, the electrons that become thermalized closest to the ion core in a multipair microzone are rapidly drawn back to it by the coulombic attraction. The attraction of the electrons enhances the outward drift of the cations and makes the central core more diffuse.

The net effect of the electron and cation motions and neutralizations is that the multipair zones decay rapidly to single-pair zones. Larger multipair zones can decay to two or more single-pair zones. For the collection of ions and electrons at electrodes by application of an electric field, the rapid recombination of the inner pairs in the multipair zones reduces the effective value of the total ionization yield, or G_{tot} (Section 2.4.2). However, the reaction of a neutral solute with an electron or ion can occur nearly as easily in a multipair as in a single-pair microzone. In the scavenging model to be described, the effective value of G_{tot} is essentially the true value. In water and many simple organic liquids $G_{tot} \approx 5$–6 ionizations per 100 eV absorbed, or 5–6×10^{-7} mol of ionizations per joule absorbed.

The estimated distribution of 75% for $j = 1$, 15% for $j = 2$, and 10% for $j \geq$ 3 is based on a calculated distribution of energy depositions in liquids (2), $G_{tot} = 5$ in the liquid (3), and a comparison with a measured distribution in air (1), for which $G_{tot} = 3.0$ (6).

The reaction model to be described includes the approximation that there is only one ion–electron pair per microzone. The main errors caused by this approximation occur at high solute concentrations, where scavenging occurs in the inner parts of multipair microzones. The model tends to underestimate the minimum concentration of solute required to scavenge all of the electrons or ions, because the model does not account for the possibility of solute depletion in multipair microzones and overestimates the neutralization times in such zones. The shorter actual neutralization times reduce the random components of the motions of the reactive species and the probability that the neutral scavenger will encounter an ion or electron. These errors are compensated in the model by exaggerating the short distance end of the ion–electron separation distance distribution.

In the jargon of radiation chemistry, a reactive microzone is called a *spur*. However, more than one specific definition of the term spur exists in the literature.

General agreement on a single definition has not been achieved. As models become more precise, the term might slip into disuse.

6.2. CHARGE SCAVENGING BY SOLUTES

6.2.1. The Model and Its Application to Electron Scavenging in Nonpolar Liquids

In stochastic models the emphasis is usually on the calculation of reaction probabilities rather than on reaction rates. The rates are included implicitly, and may be obtained explicitly by separating out the appropriate dynamic terms.

The time scales of the reactions in an irradiated liquid divide naturally into two regimes: those that occur quickly within the individual reactive microzones and those of the species that diffuse away from the microzones, which occur at later times.

To describe the model, a schematic mechanism similar to those in Sections 2.2.3 and 2.4.2 is used.

In the absence of solute the mechanism is

$$M \rightsquigarrow [M^+ + e^-] \tag{6.14}$$

$$[M^+ + e^-] \longrightarrow P_1 + X_1 \qquad \text{(geminate neutralization)} \tag{6.15}$$

$$[M^+ + e^-] \longrightarrow M^+ + e^- \qquad \text{(free ions)} \tag{6.16}$$

$$M^+ + e^- \longrightarrow P_1 + X_1 \qquad \text{(random neutralization)} \tag{6.17}$$

where M is a molecule such as ethane, P_1 is a measured product such as hydrogen, and X_1 represents other products. The square brackets indicate that the species inside them exert an appreciable coulombic attraction on each other. The microzone is defined by the region around the cation where the energy of attraction of the electron is not small compared to the thermal energy $k_B T$. The coulombic attraction tends to draw the ion and electron back together to undergo geminate neutralization. Random thermal motion tends to cause the electron and ion to wander away from each other, to ultimately become free of each other's coulombic influence. The free ions, usually at a much later time, undergo neutralization with free ions that originated in other microzones.

When the liquid contains a small concentration of a solute S that can react with the electron, the free ion electrons can be captured:

$$e^- + S \longrightarrow S^- \longrightarrow P_2 + X_2^- \tag{6.18}$$

where P_2 is a different measured product, such as nitrogen when S is nitrous oxide. The solute and free ions are randomly distributed in space. They are a homogeneous system, and their reaction rates are described by conventional kinetics.

At sufficiently higher concentrations of S, electrons can also be captured inside the microzones as they drift back toward their sibling ions:

$$[e^- + S] \longrightarrow [S^-] \longrightarrow [P_2 + X_2^-] \qquad (6.19)$$

The square brackets in (6.19) indicate the coulombic influence of the sibling cation.

At concentrations where scavenging inside the microzones is appreciable, scavenging of the free ions is virtually complete, and reaction (6.18) has completely replaced (6.17). The stochastic model permits the calculation of the yield of P_2 in reaction (6.19). The total yield of P_2 is the sum of the yield in (6.18), taken as the entire free ion yield, and that of (6.19).

The geminate neutralization and free ion reactions now include

$$[M^+ + X_2^-] \longrightarrow P_3 \qquad \text{(geminate neutralization)} \qquad (6.20)$$

$$[M^+ + X_2^-] \longrightarrow M^+ + X_2^- \qquad \text{(free ions)} \qquad (6.21)$$

$$M^+ + X_2^- \longrightarrow P_3 \qquad \text{(random neutralization)} \qquad (6.22)$$

These reactions do not contribute to the yield of P_2.

We wish to calculate the yield of P_2 as a function of the concentration of S. An example is the yield of nitrogen as a function of nitrous oxide concentration (7), shown in Figure 6.2. We calculate the probability that the electron, while drifting toward the cation in the foreplay of reaction (6.15), encounters and reacts with an S before encountering and reacting with the cation. To do this, we estimate the total number of molecules that an electron encounters after its energy has been reduced to where it can become attached to a molecule and before it becomes attached to the cation (neutralization). It is assumed for simplicity that S can only capture the electron after it becomes thermalized. If S can capture the electron before it becomes thermalized, the scavenging probability is somewhat larger, as described below.

The total number of molecules encountered by an electron between thermalization and neutralization is the sum of the number of nearest neighbors it has at the instant of thermalization, b_0, and the number it encounters while drifting toward the cation. If attachment to S can occur only after the electron has been thermalized, the value of b_0 is taken as the number of nearest neighbors in a random close packed liquid, ~ 10. If attachment to S can occur before thermalization, the number of molecules encountered while the electron energy falls between the

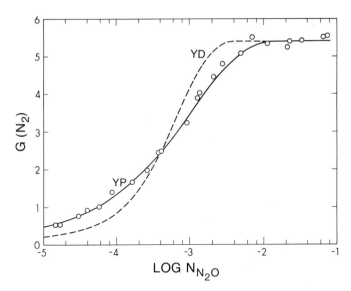

Figure 6.2. Yield of nitrogen from electron reaction with nitrous oxide in γ-irradiated liquid ethane at 183 K. $G(N_2)$ = number of nitrogen molecules formed per 100 eV absorbed by the ethane. N_{N_2O} is mol fraction of nitrous oxide. ○, experimental (7). The curves were calculated from Eqs. (6.28)–(6.34) using the indicated $F(y)$ distributions from Figure 6.3 and parameters from Table 6.1.

upper limit of attachment and thermal energy is included in b_0, in which case b_0 can be $\gg 10$ (see Section 6.3).

In most liquids thermal electrons spend much of their time in localized states (liquids with $\mu_e < 10^{-2}$ m²/V·s; Figure 2.1). In such liquids electron thermalization is assumed to be achieved at the instant a localized state forms. The thermal electron then diffuses in a series of discrete displacements. The number of molecules encountered by the electron after it leaves the initial localization site and before it becomes attached to the M^+ in reaction (6.15) is $n_- b_-$, where n_- is the number of diffusive displacements made by the electron and b_- is the average number of new molecules that it meets per displacement. The number of diffusive displacements is taken as the ratio of the geminate neutralization time t_{gn} to the average time t_{1-} required for one displacement:

$$n_- = \frac{t_{gn}}{t_{1-}} \qquad (6.23)$$

The geminate neutralization time is estimated in the manner described in Section 2.4.2:

$$t_{gn} \approx \frac{\epsilon(y^3 - r_0^3)}{4.32 \times 10^{-9}(\mu_+ + \mu_-)} \quad \text{(s)} \qquad (6.24)$$

where ϵ is the dielectric constant of the medium between the ion and electron, y is the initial distance (in meters) between the thermalized electron and the ion, r_0 is the distance of the final jump of the electron onto the cation, and $\mu_+ + \mu_-$ is the sum of the mobilities (m^2/V·s) of the ion and electron.

The average time for one diffusive displacement is (8)

$$t_{1-} \approx \frac{\lambda_-^2}{6D_-} \quad \text{(s)} \tag{6.25}$$

where λ_-^2 is the mean square displacement distance, $D_- = k_B T \mu_- / \xi$ is the diffusion coefficient of the electron, k_B is Boltzmann's constant, T is the temperature, and ξ is the protonic charge (in coulombs). Equation (6.25) is valid for our purposes because the number of displacements that the electron undergoes during the measurement of μ_- or D_- is large.

The average number b_- of new molecules that an electron encounters per displacement depends on the electron–molecule effective encounter radius, the displacement distance λ_-, the packing of the molecules in the medium, and the diffusion coefficients of the molecules in the medium.

The easiest method of calculating the probability that reaction (6.19) occurs instead of (6.15) is to estimate the probability that (6.19) does *not* happen and then subtract that from unity (9).

The probability that the electron at the instant of localization finds itself in a site that does not contain an S is $(1 - N_s)^{b_0}$, where N_s is the mol fraction of S in the solution. If S can efficiently attach the electron in an energy range prior to localization, the value of b_0 is increased to include the total number of molecules the electron encounters while passing through that energy range. The thermalized electron makes n_- diffusive jumps to reach the cation. The probability that it does not encounter an S molecule during those jumps is $(1 - N_s)^{n_- b_-}$. During the interval t_{gn} an S molecule also makes an average of $n_s = t_{gn}/t_{1s}$ displacements, where t_{1s} is the average time per S displacement, and it encounters $n_s b_s$ new molecules. The total probability that the electron *does not* encounter an S during the interval between dropping below the upper energy limit of attachment and capture by M$^+$ is $(1 - N_s)^{Z_-}$, where

$$Z_- \approx b_0 + n_- b_- + n_s b_s$$

$$\approx b_0 + \frac{6\epsilon(y^3 - r_0^3)}{4.32 \times 10^{-9}(\mu_+ + \mu_-)} \left(\frac{D_- b_-}{\lambda_-^2} + \frac{D_s b_s}{\lambda_s^2} \right) \tag{6.26}$$

Therefore, the probability that the electron *does* encounter at least one S during the interval is $1 - (1 - N_s)^{Z_-}$.

The electron might not attach to the S upon the first encounter. The probability f of attachment per encounter might be only 0.5, or 0.1, or some other fraction. In fact, both of the reactions (6.15) and (6.19) might have encounter efficiencies less than unity. The ratio of the encounter efficiencies is $f_- = f_{6.19}/f_{6.15}$. The individual efficiencies $f_{6.15}$ and $f_{6.19}$ are not known, so f_- is treated as an adjustable parameter. The probability $\phi_-(y)$ that reaction (6.19) occurs instead of (6.15) is approximately

$$\phi_-(y) = 1 - (1 - f_- N_s)^{Z_-} \tag{6.27}$$

Equation (6.27) is the probability of scavenging the electron in a thermalized pair that has an initial separation distance y and is headed for geminate neutralization. The electrons generated in reaction (6.14) do not all have the same energy, and there is a large random component in the scattering during thermalization. The thermalization distances y are not all the same. There is a distribution of y values, and Eq. (6.27) must be averaged over it:

$$\phi_- = 1 - \int_0^\infty (1 - f_- N_s)^{Z_-} F(y) \, dy \tag{6.28}$$

where $F(y) \, dy$ is the fraction of electrons that have thermalization distances in the interval y to $y + dy$.

At values of N_s where ϕ_- is appreciable, essentially all of the electron free ions are scavenged. The fraction ϕ_{fi}^0 of ions that becomes free ions is added to the scavenging probability of the recombining ions to obtain the total fraction Φ_- of electrons that is scavenged:

$$\Phi_- = \phi_{fi}^0 + (1 - \phi_{fi}^0) \, \phi_- \tag{6.29}$$

where, from Section 2.4.2,

$$\phi_{fi}^0 = \int_0^\infty e^{-r_c/y} F(y) \, dy \tag{6.30}$$

where $r_c = \xi^2/4\pi\varepsilon_0\epsilon k_B T = 1.67 \times 10^{-5}/\epsilon T$ is the distance (in meters) at which the coulombic attraction between the ion and electron equals the thermal agitation energy $k_B T$, and ε_0 is the permittivity of a vacuum.

Finally, the yield of product P_2 is

$$G(P_2) = \Phi_- G_{tot} \tag{6.31}$$

In practice, the nb terms in Z_- are split into two factors, so Eq. (6.26) becomes

$$Z_- = b_0 + \nu\beta_- \tag{6.32}$$

where ν contains the drift distance and inverse coulombic attractive force parameters,

$$\nu = \frac{6\epsilon(y^3 - r_0^3)}{4.32 \times 10^{-9}} \tag{6.33}$$

and β_- contains the transport parameters,

$$\beta_- = \frac{1}{\mu_+ + \mu_-} \left(\frac{D_- b_-}{\lambda_-^2} + \frac{D_s b_s}{\lambda_s^2} \right) \tag{6.34}$$

The first step in the calculations is to fit a distribution function $F(y)$ to the measured free ion yield using Eqs. (6.30) and (6.35):

$$G_{\mathrm{fi}}^0 = \phi_{\mathrm{fi}}^0 G_{\mathrm{tot}} \tag{6.35}$$

The function $F(y)$ and the value of ϵ determine the operational values of ν in Eq. (6.33) and hence in (6.28).

In Eq. (6.34) the values of μ_+, μ_-, D_-, and D_s can be measured or estimated. For example, if the mobility μ of a thermal energy electron or ion is known, the diffusion coefficient D can be calculated from the Einstein relation,

$$D = \frac{\mu k_B T}{\xi} \tag{6.36}$$

The values of b and λ in Eq. (6.34) are not known, but they are related to each other. Diffusion of molecules in liquids probably occurs by way of a large number of very small displacements rather than by a smaller number of jumps of about one molecular diameter each (10). The value of b would increase with λ until λ reached the unusually large value of two molecular diameters. The ratio b/λ^2 is less subject to error than are individually assumed values of b and λ. However, in practice, the value of β_- as a whole is used as an adjustable parameter in fitting Eqs. (6.28)–(6.31) to experimental yields of P_2 as a function of N_s.

Nitrogen yields from the γ radiolysis of solutions of nitrous oxide in liquid ethane at 183 K were fitted to Eq. (6.31), as shown in Figure 6.2. The shape of the calculated $G(N_2)$ curve is sensitive to the assumed shape of the $F(y)$ distribution. Information is therefore obtained about $F(y)$, within the limitations of the single-pair microzone model, by fitting calculated $G(N_2)$ curves to the experimental results.

The influence of the shape of $F(y)$ on the shape of the calculated scavenging yield curve can be demonstrated using an oversimplified distribution, such as the power function YP (Fig. 6.3):

$$\begin{aligned} F(y) &= 0 & y &\leq y_{\mathrm{min}} \\ F(y) &= x\, y_{\mathrm{min}}^x\, y^{-(x+1)} & y &\geq y_{\mathrm{min}} \end{aligned} \tag{6.37}$$

This function has two adjustable parameters, x and y_{min}.

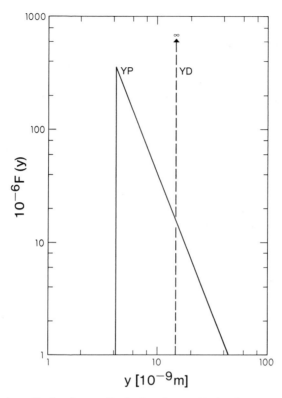

Figure 6.3. Two thermalization distance distributions that would give $G(N_2)_{fi} = 0.16$ in γ-irradiated liquid ethane at 183 K. YD represents Eq. (6.38). YP represents Eq. (6.39).

If all electrons generated by reaction (6.14) had the same thermalization distance, $F(y)$ would be a delta function. This corresponds to YP with $x = \infty$ and $F(y) = 0$ everywhere except at the single value of y. The delta function, designated YD, for the ethane solutions is

$$F(y)\, dy = 0 \qquad y \neq 1.45 \times 10^{-8}\ \text{m}$$
$$F(y)\, dy = 1 \qquad y = 1.45 \times 10^{-8}\ \text{m} \tag{6.38}$$

It gives the desired free ion yield $G(N_2)_{fi} = 0.16$. However, the use of YD in Eq. (6.28) gives too steep a slope to the calculated $G(N_2)$ curve in Figure 6.2.

The distribution that gave the best agreement with the measured yields in Figure 6.2 had $x = 1.5$:

$$F(y) = 0 \qquad\qquad\quad y \leq 4.2 \times 10^{-9}\ \text{m}$$
$$F(y) = 4.1 \times 10^{-13}\ y^{-2.5} \qquad y \geq 4.2 \times 10^{-9}\ \text{m} \tag{6.39}$$

The value of β_- was 1.3×10^{18} V/m^2, assuming $f_- = 1$, $r_0 = 1 \times 10^{-9}$ m, and $b_0 = 10$.

The distributions (6.38) and (6.39) are shown in Figure 6.3. A broader distribution of thermalization distances produces a gentler slope in the scavenging yield curve. The YP distribution that gives the best fit to the scavenging yields has a smaller slope than the one that best fits the electric field effect on the free ion yields collected at electrodes (Section 2.4.2). For propane the latter has $x = 2.5$ and $y_{min} = 6.0$ nm at 183 K (Figure 2.8). For electron scavenging in propane the best YP has $x = 1.5$ and $y_{min} = 2.9$ nm at 183 K (7). The values of y_{min} and the slope are smaller in the scavenging model because scavenging occurs successfully within multipair microzones, whereas separation of the inner ions and electrons by the application of an external electric field is not successful. The smaller values of y in the single-pair microzone model of scavenging represent the smaller values of t_{gn} for the inner pairs in the multipair microzones.

The value of the scavenging parameter $\beta_- = 1.3 \times 10^{18}$ V/m^2 obtained for the ethane solutions at 183 K is consistent with known transport parameters. The value of μ_- [5.2×10^{-5} m^2/V·s (11)] is 430 times that of μ_+ [1.2×10^{-7} m^2/V·s (12)]. The value of D_{N_2O} would be somewhat greater than D_+ because of the charge-induced dipole drag on the ion, but D_- would be $\geq 100\, D_{N_2O}$. Equation (6.34) for liquid ethane at 183 K therefore reduces, with Eq. (6.36), to

$$\beta_- = \frac{k_B T b_-}{\xi \lambda_-^2} \qquad (6.40)$$

The empirical value of β_- is 1.3×10^{18} V/m^2, so $\lambda_-^2/b_- = 1.2 \times 10^{-20}$ m^2. The relatively high mobility of the electron implies that the displacement distance λ_- is probably about a molecular diameter, $\sim 4 \times 10^{-10}$ m. This would give $b_- \approx 13$, which is a reasonable estimate of the number of new molecules encountered by the electron per jump, considering that the electron mean localization energy of ~ 0.02 aJ (~ 0.1 eV) is associated with a de Broglie *radianlength* (Section 2.1.1) $\lambdabar \approx 6 \times 10^{-10}$ m. The corresponding average jump time is $t_{1-} \approx 3 \times 10^{-14}$ s, which is the order of a molecular vibration time; it is six times greater than the average time required by quasifree electrons at 183 K to traverse 4×10^{-10} m. Thus, the electrons spend most of their time in localized states, and the random walk (Brownian motion) model is appropriate.

In liquids comprised of spherelike molecules, such as xenon, methane, and neopentane, the value of the product $f_-\beta_-$ is 10^{-5}–10^{-2} times smaller than that in ethane (13, 14). The small values are related to the fact that the electrons spend much or all of their time in the delocalized state. The value of μ_- is large, corresponding to a larger jump frequency and a larger λ_- than in ethane. The larger λ_- can produce a smaller b_-/λ_-^2 and hence a smaller β_- [Eq. (6.40)]. The value of f_- for capture by nitrous oxide is also apparently smaller for delocalized than for localized electrons (14). The model requires further development for these liquids.

6.2.2. Field-Dependent Mobility Near Center of Microzone

The electric field strength close to an ion is very large. At a distance r(m) from an ion of charge $z\xi$ (in coulombs) the field strength is

$$E = \frac{z\xi}{4\pi\varepsilon_0\epsilon r^2} \quad \text{(V/m)} \tag{6.41}$$

In liquid ethane at 183 K, which has $\epsilon = 1.8$, the field strength 3 nm from a singly charged ion is 1×10^8 V/m.

The mobilities of electrons in liquid ethane at 142–197 K increase with increasing electric field strength above 1×10^7 V/m (15). At 2×10^7 V/m the mobility is double that at fields below 1×10^7 V/m. Thus, the electron mobility increases as it drifts the last 7 nm toward the ion. This might decrease the probability of scavenging within 7 nm of the cations in ethane. The problem is worse in multipair microzones. The magnitude of the effect depends on the mechanism of the increased mobility. The increase was originally interpreted (15) in terms of the Bagley model (16), which pictures the electron as migrating on a lattice of localized states, with the field in the direction of one axis of the lattice and the other axes at right angles to it. The electron hops from site to site over energy barriers. The field lowers the barrier for hopping down-field and raises it for hopping up-field. The electrons are assumed to remain in thermal equilibrium with the lattice. This model yields estimates of λ_- about 10 times larger, and jump frequencies about 10^{-2} times smaller, than those estimated from the scavenging model β_-. If the reaction encounter efficiency ratio f_- were smaller than unity [Eq. (6.28)], the required value of β_- would be larger by the inverse of f_-. Reasonable values of b_- would then indicate a still smaller value of λ_-, and the disagreement with the Bagley model (15) would worsen.

Because the Bagley model would require unreasonable values of b_-, λ_-, and jump frequency, it is not favored for ethane. However, if it did apply, the increasing mobility of the electrons near the cations would decrease the scavenging efficiency simply by increasing the average value of μ_-, which decreases the geminate neutralization time and the value of β_- [Eq. (6.34)]. The value of D_- is less affected than is μ_- by high fields because the motions transverse to the field are assumed to be unchanged; Eq. (6.36) no longer applies.

A different model of electron transport in liquid ethane involves transitions between localized and delocalized states [see Eq. (2.26) and environs]:

$$e_{loc}^- \rightleftharpoons e_{deloc}^- \tag{6.42}$$

The temperature dependence of the mobility is attributed mainly to that of the step $e_{loc}^- \rightarrow e_{deloc}^-$, in which the electron absorbs energy from the medium. The field dependence might be attributed mainly to the increasing energy of the electron in the delocalized state, thereby decreasing the efficiency of $e_{deloc}^- \rightarrow e_{loc}^-$ and increasing λ_-. The localization step involves energy loss by the electron through inelastic scattering. The value of b_-/λ_-^2 probably decreases as λ_- increases, because b_- would not increase as rapidly as λ_-^2. This would tend to decrease the scavenging probability ϕ_- [Eq. (6.27)]. Details depend on the extent of elastic scattering in the delocalized state and whether capture occurs from that state. The relative efficiency $f_- = f_{6.19}/f_{6.15}$ could either increase or decrease with increasing electron energy, depending on the scavenger (17, 18). The net effect of the increased mobility should be studied by comparing the shapes of scavenging curves between $G(P_2)/G_{tot} = 0.7$ and 1.0, after superimposing the lower portions of the curves by horizontally shifting plots such as that in Figure 6.2.

In the application of the model, instead of decreasing the value of β_- by increasing μ_- near the cations, the effective value of ν [Eq. (6.33)] is decreased by decreasing the minimum value of the initial distance y. The distribution $F(y)$ obtained by fitting the model to the experimental yields has smaller values of y_{min} and x [Eq. (6.37)] than those that would approximate the real spatial distribution.

By contrast with ethane, in solvents comprised of spherelike molecules, such as methane, μ_- is large at low fields and decreases at high fields (19–22). At fields above a certain threshold the electrons gain more energy by acceleration in the field than they lose by collision with the molecules. The average energy of the electrons increases, and they are said to be heated by the field. The greater random speed of the hot electrons increases their collision rate, which interferes with their drift in the direction of the field [see Eq. (2.19) and environs]. However, the "heating" is not necessarily isotropic, so the enhancement of diffusive motion perpendicular to the field might be different from that in the direction of the field (23, 24). The change in electron energy might also increase or decrease the reaction efficiency ratio f_-. The field effect in these liquids requires much more study.

6.2.3. Diffusion Approximations

The model in Section 6.2.1 separates reactions into two time zones: that of geminate neutralization and that of the free ion reactions. The stochastic calculation applies to scavenging electrons or ions that undergo geminate neutralization. The scavenging of free ions is a homogeneous process that is calculated separately and added to the geminate part to obtain the total scavenging yield.

This model approximates the geminate neutralization time t_{gn} by Eq. (6.24), which gives a unique neutralization time for a given initial separation distance y. Actually, during the time t_{gn} the electron and ion undergo more diffusive jumps in random directions (Brownian motion) than in the direction of the field. The preceding results for ethane provide an example. The 14.5 nm represented by Eq. (6.38) could be traversed by 36 jumps of 0.4 nm each straight back to the cation. However, Eq. (6.24) gives $t_{gn} = 2.3 \times 10^{-11}$ s for $y = 14.5 \times 10^{-9}$ m and $r_0 = 1 \times 10^{-9}$ m, while Eq. (6.25) gives $t_{1-} \approx 3 \times 10^{-14}$ s, so by Eq. (6.23) the electron makes 760 nearly random jumps during the drift across the 14.5 nm. The random jumps would generate a relatively broad distribution of arrival times for a large sample of electrons traversing 14.5 nm under the double influence of the coulombic field and thermal agitation. Equation (6.24) ignores the broadening and is therefore called the forced diffusion (FD) approximation (25).

An alternative is the prescribed diffusion (PD) approximation. Beginning with a population of single-pair microzones that all have the same initial ion–electron separation y, it is assumed that the random motions rapidly generate a Gaussian distribution of separation distances. The distribution is assumed to remain Gaussian under the double influence of the coulombic field and the Brownian motion (26). The random motions cause some pairs to be neutralized sooner than predicted by Eq. (6.24) and some later. The PD method estimates the probability of neutralization as a function of time.

An electron–ion pair begins Brownian motion and drift at $t = 0$ and separation y. The probability F_y that the pair remains unneutralized at time t is given by (25, 26)

$$-dF_y = \frac{\xi^2 F_y}{4\pi\varepsilon_0 \epsilon k_B T (4\pi D)^{0.5} t^{1.5}} \exp\left(\frac{-y^2}{4Dt}\right) dt \tag{6.43}$$

where $-dF_y$ is the fraction of pairs that undergoes neutralization between the times t and $t + dt$, and D is the sum of the diffusion coefficients of the positive and negative species.

In the PD method Eqs. (6.30), (6.31), and (6.35) remain the same, but the total probability $\Phi_-(y)$ that an electron with a given y value will be scavenged is changed to

$$\Phi_-(y) = 1 + \int_{t=0}^{\infty} (1 - f_- N_s)^{b_0 + \alpha t} \, dF_y \tag{6.44}$$

where $\alpha = b_-/t_{1-}$ is the number of new molecules an electron meets per unit time. The sign in front of the integral is positive because dF_y is negative [Eq. (6.43)].

The scavenging probability $\Phi_-(y)$ is averaged over the y distribution to obtain the total fraction Φ_- of electrons that is scavenged:

$$\Phi_- = \int_{y=0}^{\infty} \Phi_-(y) F(y) \, dy \tag{6.45}$$

As in the FD treatment, the first step in the calculation is to pick a distribution function $F(y)$ that gives the correct free ion yield with Eqs. (6.30) and (6.35). The $F(y)$ is then used with Eqs. (6.44), (6.45), and (6.31) to calculate scavenger product yields at different concentrations N_s. The value of α is constant for all N_s. It is selected to minimize the error of the curve calculated with that particular $F(y)$.

The procedure is repeated with different forms of $F(y)$, for example, YP with different values of x, to find the best fit to the plot of experimental yields against N_s.

The fitting is illustrated with the nitrogen yields from the γ radiolysis of solutions of nitrous oxide in cyclohexane at 298 K (Figure 6.4 and Ref. 27). Calculated curves obtained with the y distribution YD and the best-fitting YP are shown for both the FD and PD approximations (25). The parameter values are listed in Table 6.1. The curves calculated with the PD approximation are less sensitive to the shape of the $F(y)$ distribution than are those obtained with the FD approximation. In fact, the unrealistically narrow distribution YD gives almost as good a fit as the best YP in the PD approximation. The PD method overestimates the spread of neutralization times caused by Brownian motion of the electron and ion.

Another example is the scavenging of electrons in neopentane (tetramethylmethane). The diffusion coefficient of electrons in neopentane at 295 K is ~ 160 times greater than that in ethane at 183 K or cyclohexane at 298 K (Table 6.1). The free ion yield in neopentane is ~ 7 times larger than those in the other liquids. Nitrogen yields from the scavenging of electrons by nitrous oxide in neopentane under γ irradiation (13) are displayed against the calculated curves in Figure 6.5. The best-fit curve from the FD method was obtained with the distance distribution YP with $x = 1.4$, which is relatively broad. By contrast, the best-fit curve from the PD method was obtained with the delta function distribution YD. The narrow

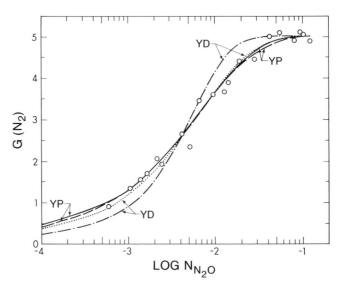

Figure 6.4. Nitrogen yields from electron reaction with nitrous oxide in γ-irradiated cyclohexane at 298 K, corrected for the direct radiolysis of N_2O at high concentrations (27). N_{N_2O} is mol fraction of N_2O. Curves were calculated with the $F(y)$ distributions indicated. ——— and –·–·, using the FD approximation, Eqs. (6.28)–(6.34); – – – – and · · · · · , using the PD approximation, Eqs. (6.43)–(6.45) and (6.31). Parameter values are listed in Table 6.1.

TABLE 6.1. Parameters for the Calculated Curves in Figures 6.2, 6.4, and 6.5[a]

	Solvent			
	C_2H_6	$c\text{-}C_6H_{12}$	$C(CH_3)_4$	CH_3OH[b]
T (K)	183	298	295	298
ϵ	1.80	2.02	1.82	33
d (kg/m^3)	551	774	590	788
G_{fi}^0	0.16	0.13	1.0	2.0
G_{tot}	5.5	5.0	4.7	4.6
D_- (10^{-7} m^2/s)	8	9	1400	0.016
D_+ (10^{-10} m^2/s)	19	10	30	38
	FD Model Parameters			
YD, y (10^{-10} m)	145	76	195	22[c]
β_- (10^{17} V/m^2)	1.9	1.4	0.035	1.8
YP, x	1.5	1.9	1.4	1.5
$y_{\min}$ (10^{-10} m)	42	31	96	12
β_- (10^{17} V/m^2)	13	6.4	0.083	7.5
	PD Model Parameters			
YD, y (10^{-10} m)		76	195	20
α (10^{12} s^{-1})		7.7	40	0.32
YP, x		2.5	∞	∞
$y_{\min}$ (10^{-10} m)		39	195	20
α (10^{12} s^{-1})		19	40	0.32

[a] Assume $f_- = 1$, $b_0 = 10$, and for FD $r_0 = 1 \times 10^{-9}$ m.
[b] Curves not shown, see text.
[c] The time-dependent dielectric constant was averaged using Eq. (6.49) and $\epsilon_s = 33$, $\epsilon_\infty = 5.6$, and $\tau = 4.7 \times 10^{-11}$ s (36). The static dielectric constant was used with the other functions to simplify the calculations. The distance $y = 20 \times 10^{-10}$ m was obtained for YD when ϵ_s was used instead of ϵ_{av}.

initial distribution function is required to compensate for the excessive broadening of the distribution that is generated within the model by the prescribed diffusion approximation (28).

6.2.4. Yields as Functions of t^{-x} and C_s^x, $x \approx \frac{1}{2}$

The yield of scavenger product P_2 increases approximately linearly with the square root of the scavenger concentration at low concentrations (29, 30). The approximate square root dependence applies to about one-third of the electrons or ions that react within the microzones, after which scavenging becomes progressively more difficult.

Nitrogen yields from the reaction of electrons with nitrous oxide in liquid ethane during γ radiolysis at 183 K are plotted against the square root of the nitrous oxide concentration, $C_{N_2O}^{0.5}$ (mol/m^3)$^{0.5}$, in Figure 6.6. The dashed straight line fits mod-

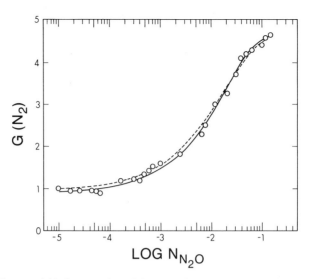

Figure 6.5. Nitrogen yields from reaction of electrons with nitrous oxide in γ-irradiated neopentane at 295 K (13). N_{N_2O} = mol fraction of N_2O. Curves represent the best fits obtainable from the FD (———) and PD ($\cdot$ $\cdot$ $\cdot$ $\cdot$) methods. Parameter values are listed in Table 6.1.

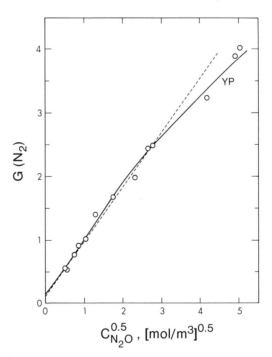

Figure 6.6. Nitrogen yields from reaction of electrons with nitrous oxide in ethane during γ radiolysis at 183 K plotted against the square root of the concentration (mol/m³) of nitrous oxide. Experimental points are those from Figure 6.2 up to $N_{N_2O} = 1.4 \times 10^{-3}$. Full curve is the corresponding part of the YP curve (FD method) from Figure 6.2 calculated down to $N_{N_2O} = 1 \times 10^{-7}$. Dashed straight line is drawn for comparison.

erately well up to $G(N_2) = 2.5$, which is nearly half of the total electrons. However, the YP ($x = 1.5$) curve from Figure 6.2 fits the overall yields quite well, and in the square root plot it has a slight curvature that indicates $G(N_2) \propto C^m$, with m slightly greater than 0.5 for the capture of the first 20% of the electrons. The value $m = 0.6$ gives a more nearly linear plot up to $G(N_2) = 0.9$, as shown in Figure 6.7 (see also Section 5.3.6). The intercept corresponds to G_{fi}. Hence, for about the first 20% of scavenging within the microzones the scavenging yield is approximately given by

$$G(P_2) - G(P_2)_{fi} \approx AC_s^{0.6} \qquad (6.46)$$

where A is a constant the value of which depends on the solute, solvent and temperature.

The prescribed diffusion method, used with the thermalization distance delta function YD, generates yields that follow the relation (6.46) for nearly half of the scavenging within the microzones in cyclohexane (31).

The best agreement with experimental yields in a number of liquids is provided by YP with $x = 1.5$ in the FD model and by YD in the PD model (25, 28, 31). A simple Gaussian distribution of thermalization distances, such as Eq. (2.35), does not provide an appropriate concentration dependence of $G(P_2)$ in either the FD or PD model.

It follows implicitly from relation (6.46) that the longest lived 20% of the electrons and ions that undergo geminate neutralization decay approximately as $t^{-0.6}$:

$$G(e^-)_t - G(e^-)_{fi} \propto t^{-0.6} \qquad (6.47)$$

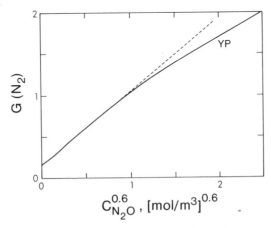

Figure 6.7. Low concentration end of calculated yield curve YP in Figure 6.6 plotted against $C^{0.6}$ Dashed straight line is drawn for comparison.

where $G(e^-)_t$ is the yield of electrons that remains unneutralized at time t in the absence of scavenger; $t = 0$ is the instant of formation of the thermalized electron–ion pairs. This time dependence is also discussed in Section 5.3.6.

6.2.5. Application to Polar Liquids

The single-pair approximation is less appropriate for highly polar liquids that have large dielectric constants, such as methanol or water. In these liquids most of the actual single-pair microzones produce free ions. The large dielectric constant greatly decreases the coulombic attraction that tends to pull the charges back together, against the randomizing Brownian motion. Most of the geminate neutralization occurs in multipair microzones, in which the several cations near the center exert a stronger attractive force on the electrons than occurs in the single-cation microzones. However, the FD model works quite well for alcohols (28, 32) and even for water (28, 33). The reason is that the electron–ion separation distance distribution in the model is used as an adjustable parameter. A larger number of cations near the center of the microzone is compensated in the single-pair model by using smaller y values.

Another factor is that in alcohols the dielectric relaxation time τ is similar in magnitude to t_{gn}. This means that the dielectric constant of the liquid between the electron and ion increases during a significant portion of the time that the charges are drifting together. At the instant the deenergized electron forms a localized state, the dielectric constant has its high-frequency (short-time) value ϵ_∞. As the molecular dipoles reorient in the field of the ion and electron, the dielectric constant $\epsilon(t)$ increases toward its static (long-time) value ϵ_s (34):

$$\epsilon(t) = \frac{\epsilon_s}{1 + [(\epsilon_s - \epsilon_\infty)/\epsilon_\infty]\exp(-\epsilon_s t/\epsilon_\infty \tau)} \tag{6.48}$$

When τ is similar in magnitude to t_{gn}, an average value ϵ_{av} is used in Eqs. (6.24), (6.26), and (6.33) (28):

$$\epsilon_{av} = t_{gn}^{-1} \int_0^{t_{gn}} \epsilon(t)\, dt$$

$$= \epsilon_s + \frac{\epsilon_\infty \tau}{t_{gn}} \ln\left\{\frac{1 + [(\epsilon_s - \epsilon_\infty)/\epsilon_\infty]\exp(-\epsilon_s t_{gn}/\epsilon_\infty \tau)}{\epsilon_s/\epsilon_\infty}\right\} \tag{6.49}$$

The strong charge-dipole torque that acts on polar molecules near the ion or electron probably reduces τ below the normal value that is measured with the weak fields in microwave beams (35).

In nonpolar liquids, where only electronic and atomic polarization are appreciable, $\tau < 10^{-13}$ s $< t_{gn}$, so ϵ_{av} equals ϵ_s.

*

In water at 296 K, $\epsilon_\infty \tau / \epsilon_s$ is $\leq 6 \times 10^{-13}$ s (36). The estimated values of t_{gn} are greater than 10^{-12} s, so $\epsilon_{av} = \epsilon_s$ has been used in the calculations for water (28, 33). However, further examination might reveal that some geminate neutralization occurs at $t < 10^{-12}$ s, in which case the time dependence of ϵ will be important.

In methanol at 298 K, $\epsilon_\infty \tau / \epsilon_s$ is $\leq 8 \times 10^{-12}$ s (36). The time dependence of the dielectric constant therefore makes a contribution (Table 6.1). The effect is especially important in alcohols at low temperatures (32, 37). The calculations should be renewed because most of those published (32, 37) were done with an $F(y)$ distribution that is now outdated. The conclusion about the participation of the time dependence of ϵ would not be altered, but the details of the curve fitting would be. For this reason the calculated yield curves for water (28) and alcohols (32, 37) are not shown here.

6.2.6. Scavenging of Positive Ions

For a positive-ion scavenger that forms a product P_4, the equations are the same as those for electron scavenging, with two exceptions. Instead of Eq. (6.31), one has

$$G(P_4) = \Phi_+ G_{tot} \tag{6.50}$$

and instead of (6.34), one has

$$\beta_+ = \frac{1}{\mu_+ + \mu_-} \left(\frac{D_+ b_+}{\lambda_+^2} + \frac{D_s b_s}{\lambda_s^2} \right) \tag{6.51}$$

In the other equations the minus signs on n_-, b_-, t_{1-}, λ_-, D_-, Z_-, ϕ_-, and f_- are simply changed to plus signs.

An example scavenger is ND_3, for which P_4 is HD [Eqs. (6.7)–(6.10)] (9, 29).

The cationic D_+ is much smaller than the electronic D_- in most liquid hydrocarbons, so β_+ is usually smaller than β_-. Thus, a larger concentration N_s of scavenger is required to scavenge cations than to scavenge the same fraction of electrons (9).

6.3. SCAVENGING ELECTRONS PRIOR TO SOLVATION

The reactions of electrons with solutes in alcohols under γ irradiation can be separated into four time zones. The times correlate inversely with regions of solute concentration where the reactions are measured.

1. At low solute concentrations and long times, electrons that are free-ions react. The electrons have relaxed into the solvated state, e_s^-, and are in thermal

equilibrium with the medium. The species are randomly distributed in space, forming a homogeneous system, and the kinetics are of the conventional first order when the concentration of S is much greater than that of e_s^-.

Electrons that undergo geminate neutralization make up the other three categories. They have progressively shorter lifetimes, and to be scavenged, they require progressively higher concentrations of solute.

2. The geminate electrons that have the largest y values live long enough to relax into the solvated state, e_s^-. To scavenge them, higher concentrations of solute are required than in 1.

3. At still higher concentrations of solute, electrons can be captured before they have relaxed into the solvated state. Incompletely relaxed states are designated e_{ir}^- (Section 2.2.2.2). The energy E of e_{ir}^- is greater than that of e_s^-, and the relative reaction efficiency f_- of e_{ir}^- is greater than that of e_s^- for some solutes such as benzene and toluene (38, 39).

4. The earliest state that can be captured is quasifree electrons e_{qf}^- that have not yet become localized, but have fallen low enough in energy that they can attach to an appropriate molecule. The term *quasifree* simply implies that the electron is not localized and is not in free space. Quasifree electrons are captured either by localization traps in the liquid (potential wells, Figure 2.3) or by scavenger molecules when they are present in high enough concentration (39–41).

Rates of reactions of e_s^- in categories (1) and (2) can be strongly affected by the solvation energy of the electrons (39, 42, 43). If the electron affinity of the solute molecule is small, it is easier for an electron to jump onto the molecule from a shallower potential well in the solvent than from a deeper well. The well depth, or solvation energy, depends on the nature of the solvent and is roughly measured by the optical absorption energy of e_s^-.

Rates of reactions of e_{ir}^- and e_{qf}^- in categories (3) and (4) show relatively little solvent effect. The energies of the states e_{ir}^- and e_{qf}^- are higher than that of e_s^- and are less sensitive to details of solvent structure.

The reactions being considered are

$$[e_{qf}^-] \longrightarrow [e_{ir}^-] \longrightarrow [e_s^-] \longrightarrow \text{normal products}$$
$$\quad {}_{(4)}\downarrow {}^S \qquad\qquad {}_{(3)}\downarrow {}^S \qquad\qquad {}_{(2)}\downarrow {}^S \qquad\qquad\qquad\qquad (6.52)$$
$$\quad S^- \qquad\qquad\quad S^- \qquad\qquad\quad S^-$$

The state e_{qf}^- does not absorb light appreciably at any wavelength. The state e_{ir}^- absorbs at longer wavelengths (lower energies) than does e_s^-. In 2-butanol at 162 K the optical absorption spectra were measured after a 100 ns pulse of 1.7 MeV electrons (44). At 200 ns after the beginning of the pulse the spectrum has an absorption maximum at 950 nm (Figure 6.8). As time progresses, the absorbance decreases at 950 nm and grows at 550 nm. The change of the spectrum is caused by relaxation of the e_{ir}^- states toward e_s^-, by geminate neutralization, and reaction of e_{ir}^- with other substances. After 45 μs all the remaining electrons are e_s^- (Figure 6.8), which decay at much longer times.

In the presence of about 1 mol % of an electron scavenger such as benzene, the initial optical absorbance is smaller and the decay pattern is different. Spectra at three times, in the absence and presence of 2 mol % benzene, are shown in Figure 6.9. The curves at 0.14 μs show that scavenging at short times preferentially decreases the absorbance at long wavelengths. The absorbance at 500 nm grows by a factor of 2 between 0.14 and 9 μs in the pure alcohol but does not grow in the presence of 2 mol % benzene.

The fractional decrease in absorbance in the benzene solution compared to that in pure 2-butanol is shown as a function of benzene concentration at two wavelengths, 1000 and 500 nm, and for several times in Figure 6.10.

Such data are explained by the following reaction probability sequence (39).

An energetic electron being deenergized by collisions in the liquid becomes scavengeable after the energy has been decreased below about 1 eV. Scavenging is possible during the last train of collisions that the quasifree electron undergoes prior to localization. If reaction does not occur during these collisions, it might occur if the initial localization site contains a scavenger molecule. If scavenging still has not occurred, electrons initially in shallow traps might migrate and react before settling into the equilibrium solvated state. The reaction encounter efficiencies f_{qf} and f_{ir} of electrons in the two preceding states are probably different.

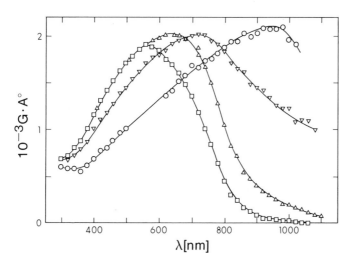

Figure 6.8. Optical absorption spectrum of localized electrons in 2-butanol at 162 K at several times after the start of a 100-ns pulse of 1.7-MeV electrons. The times are (μs): $\bigcirc$, 0.20; $\triangledown$, 1.0; $\triangle$, 5.0; $\square$, 45. Data from Ref. 44. $G \cdot A^0$ is the product of the electron yield G (electron/100 eV absorbed) and the decadic optical absorbance coefficient A^0 (m^2/mol). The non-SI molar absorbance unit dm^3/mol·cm is equivalent to 0.1 m^2/mol.

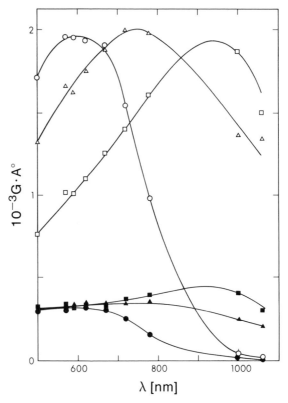

Figure 6.9. Optical absorption spectra of localized electrons in 2-butanol at 163 K at several times after the start of a 100-ns pulse of 1.7-MeV electrons. $\square$, $\blacksquare$, 0.14 μs; $\triangle$, $\blacktriangle$, 0.54 μs; $\bigcirc$, $\bullet$, 9 μs. Open points, pure alcohol; closed points, 2.0 mol % benzene. Data from Ref. 39.

The absorbance ratio A_i/A_{i0} expresses the fraction of electrons that have not been scavenged up to time i.

For the quasifree regime let p represent the probability that an electron does not become localized at a given collision. The probability per collision that it does not react with the scavenger present at mol fraction N_s is $1 - f_{qf}N_s$. The fraction of electrons that survive the quasifree period is

$$\left(\frac{A_i}{A_{i0}}\right)_{qf} = \sum_{m=1}^{\infty} (1 - p)[p(1 - f_{qf}N_s)]^{m-1} \tag{6.53}$$

where m is the number of collisions. Upon inversion, Eq. (6.53) reduces to

$$\left(\frac{A_{i0}}{A_i}\right)_{qf} \approx 1 + \frac{p}{1 - p} f_{qf}N_s \tag{6.54}$$

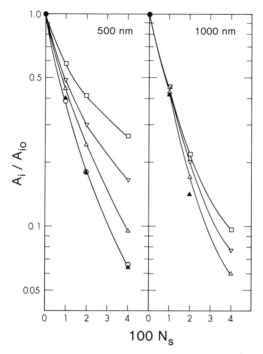

Figure 6.10. Effect of benzene concentration on the fractional absorbance A_i/A_{i0} of localized electrons in 2-butanol at 163 K at different times and wavelengths. A_{i0} is the absorbance in pure 2-butanol; A_i is that in the solution. $\square$, 0.14 μs; $\triangledown$, 0.24 μs; $\triangle$, 0.54 μs; $\blacktriangle$, 2.0 μs; $\bigcirc$, 9 μs. Data from Ref. 39.

Equation (6.54) is only an approximation because both p and f_{qf} are approximated as average values, although they both are energy dependent.

The probability of scavenging after localization, but prior to reaching the equilibrium solvated state, includes the probability that the initial localization site contains a scavenger molecule. There is also a probability that the electron will migrate or tunnel to a scavenger while relaxing to the equilibrium state. We now speak of the probability f_{ir} of reaction per encounter, rather than per collision, because a localized electron has an extended interaction with each of the molecules in its localization site. The probability that a localized electron survives the e_{ir}^- stage, unscavenged through encounters with n molecules, to reach the e_s^- state is

$$\left(\frac{A_i}{A_{i0}}\right)_{ir} \approx (1 - f_{ir}N_s)^n \tag{6.55}$$

When $f_{ir}N_s \ll 1$, Eq. (6.55) reduces to*

---

*Different stochastic mechanisms can lead to "survival fraction" equations of exponential form. See Eq. (7.1) and Section 7.2.1.

$$\left(\frac{A_i}{A_{i0}}\right)_{ir} \approx \exp\left(-nf_{ir}N_s\right) \tag{6.56}$$

The probability that electrons survive both periods is given by

$$\frac{A_i}{A_{i0}} \approx \frac{(1 - f_{ir}N_s)^n}{\left\{1 + [p/(1-p)]f_{qf}N_s\right\}}$$

$$\approx \frac{\exp\left(-nf_{ir}N_s\right)}{\left\{1 + [p/(1-p)]f_{qf}N_s\right\}} \tag{6.57}$$

Experimentally, $fN_s \ll 1$, under which circumstances many mathematical functions are approximately equivalent to (6.54)–(6.57). Furthermore, depending on the scavenger and solvent, either Eq. (6.54) or (6.56) may dominate in (6.57). For this reason, different authors have arrived at various expressions that are approximations of (6.57) (41, 45–47).* The integral form of Eq. (6.53) (38) leads to

$$\frac{A_i}{A_{i0}} \approx \frac{(1 - f_{ir}N_s)^n \ln p}{\ln\left[p(1 - f_{qf}N_s)\right]} \tag{6.58}$$

which is equivalent to (6.57) in that the values of p and $1 - f_{qf}N_s$ are each only slightly less than unity, in which case $\ln x \approx x - 1$.

For benzene in 2-butanol at 163 K, at 0.14 μs, and $\lambda < 600$ nm, Eq. (6.54) is a reasonable approximation (39). Optical absorption at these short wavelengths involves electrons in deep traps from which migration is negligible in short times. The decrease of A_i/A_{i0} is therefore mainly caused by scavenging from the quasifree state. The quantity $f_{qf}p/(1 - p)$ is ≤ 70 at 500 nm. The value of nf_{ir} increases with time, and at 5 μs $nf_{ir} \geq 35$. By this time relaxation toward the solvated state is well advanced (Figure 6.8), and negligible further scavenging occurs from the e_{ir}^- state (39). This early scavenging apparently occurs from the infrared band, not from the band as it relaxes further in the visible. Thus, nf_{ir} is negligible for electrons in the visible band, when benzene is the scavenger.

The ratio of amounts of scavenging from the quasifree and incompletely relaxed states, S_{qf}/S_{ir}, for benzene in 2-butanol at 163 K is 1.4 at 1 mol % and 1.0 at 4%.

The parameters $f_{qf}p/(1 - p)$ and nf_{ir} each vary with temperature approximately as T^{-4}. This is equivalent to a net Arrhenius temperature coefficient of about −8 kJ/mol (39). Thus, the Arrhenius coefficients for the rates of localization and relaxation of the electron states are about 8 kJ/mol greater than those for scavenging from the quasifree and incompletely relaxed states, respectively. The temperature

---

*See also Chapter 7 for modified exponential survival functions.

dependences are similar for benzene and toluene in several primary and secondary alcohols, including methanol (39).

REFERENCES

1. W. J. Beekman, *Physica* **15**, 327 (1949).
2. A. Mozumder and J. L. Magee, *J. Chem. Phys.* **47**, 939 (1967).
3. J.-P. Dodelet and G. R. Freeman, *Int. J. Radiat. Phys. Chem.* **7**, 183 (1975).
4. H. A. Bethe and J. Ashkin, in *Experimental Physics,* Vol. 1, E. Segre (Ed.), Wiley, New York, 1953, pp. 166–357.
5. G. R. Freeman, *Quaderni dell'Area di Ricerca dell'Emilia-Romagna* **2**, 55 (1972).
6. J. Booz and H. G. Ebert, *Strahlentherapie* **120**, 7 (1963).
7. M. G. Robinson and G. R. Freeman, *J. Chem. Phys.* **55**, 5644 (1971).
8. S. Chandrasekhar, *Rev. Mod. Phys.* **15**, 1 (1943).
9. G. R. Freeman, *J. Chem. Phys.* **46**, 2822 (1967).
10. B. J. Alder and T. Einwohner, *J. Chem. Phys.* **43**, 3399 (1965).
11. N. Gee and G. R. Freeman, *Phys. Rev. A* **22**, 301 (1980).
12. N. Gee and G. R. Freeman, *Can. J. Chem.* **58**, 1490 (1980).
13. K. Horacek and G. R. Freeman, *J. Chem. Phys.* **53**, 4486 (1970).
14. M. G. Robinson and G. R. Freeman, *Can. J. Chem.* **51**, 650 (1973).
15. W. F. Schmidt, G. Bakale, and U. Sowada, *J. Chem. Phys.* **61**, 5275 (1974).
16. B. G. Bagley, *Solid State Commun.* **8**, 345 (1970).
17. A. O. Allen, T. E. Gangwer, and R. A. Holroyd, *J. Phys. Chem.* **79**, 25 (1975).
18. G. Bakale and W. F. Schmidt, *Z. Naturforsch.* **36a**, 802 (1981).
19. L. S. Miller, S. Howe, and W. E. Spear, *Phys. Rev.* **166**, 871 (1968).
20. W. F. Schmidt, *Can. J. Chem.* **55**, 2197 (1977).
21. J.-P. Dodelet and G. R. Freeman, *Can. J. Chem.* **55**, 2264 (1977).
22. N. Gee and G. R. Freeman, *Phys. Rev. A* **20**, 1152 (1979).
23. P. Kleban and H. T. Davis, *J. Chem. Phys.* **68**, 2999 (1978).
24. S. L. Lin, R. E. Robson, and E. A. Mason, *J. Chem. Phys.* **71**, 3483 (1979).
25. J.-P. Dodelet and G. R. Freeman, *Can. J. Chem.* **49**, 2643 (1971).
26. A. Mozumder, *J. Chem. Phys.* **50**, 3153, 3162 (1969).
27. M. G. Robinson and G. R. Freeman, *J. Chem. Phys.* **48**, 983 (1968).
28. J.-P. Dodelet and G. R. Freeman, *Can. J. Chem.* **50**, 2729 (1972).
29. F. Williams, *J. Am. Chem. Soc.* **86**, 3954 (1964).
30. J. W. Buchanan and F. Williams, *J. Chem. Phys.* **44**, 4377 (1966).
31. G. R. Freeman, *Int. J. Radiat. Phys. Chem.* **4**, 237 (1972).
32. K. N. Jha and G. R. Freeman, *J. Chem. Phys.* **51**, 2839, 2846 (1969).
33. J. C. Russell and G. R. Freeman, *J. Chem. Phys.* **48**, 90 (1968).
34. H. Fröhlich, *Theory of Dielectrics,* 2nd ed., Clarendon, Oxford, 1958, p. 72.
35. G. R. Freeman, *Adv. Chem. Ser.* **82**, 339 (1968).

36. From data in F. Buckley and A. A. Maryott, *Tables of Dielectric Dispersion Data,* National Bureau of Standards U.S. Circ. No. 589, Washington, DC, 1958.

37. K. N. Jha and G. R. Freeman, *J. Chem. Phys.* **48,** 5480 (1968).

38. G. L. Bolton, K. N. Jha, and G. R. Freeman, *Can. J. Chem.* **54,** 1497 (1976).

39. K. Okazaki and G. R. Freeman, *Can. J. Chem.* **56,** 2313 (1978).

40. W. H. Hamill, *J. Chem. Phys.* **49,** 2446 (1968).

41. D. Razem and W. H. Hamill, *J. Phys. Chem.* **81,** 1625 (1977).

42. A. M. Afanassiev, K. Okazaki, and G. R. Freeman, *J. Phys. Chem.* **83,** 1244 (1979).

43. A. M. Afanassiev, K. Okazaki, and G. R. Freeman, *Can. J. Chem.* **57,** 839 (1979).

44. K. Okazaki and G. R. Freeman, *Can. J. Chem.* **56,** 2305 (1978).

45. G. Czapski and E. Peled, *J. Phys. Chem.* **77,** 893 (1973).

46. K. Y. Lam and J. W. Hunt, *Int. J. Radiat. Phys. Chem.* **7,** 317 (1975), and earlier work.

47. J. R. Miller, *J. Phys. Chem.* **79,** 1070 (1975).

7 MODELS OF CELLULAR RADIATION ACTION

Albrecht M. Kellerer

Institut für Medizinische Strahlenkunde
der Universität Würzburg
Würzburg, Federal Republic of Germany

7.1. INTRODUCTION

Living cells can be affected by exposure to chemicals or to ionizing radiation. The resulting effects are not inherently different. However, it is a singular feature of ionizing radiation that energy is transmitted to the exposed media in discrete packages and that the microdistribution of imparted energy and of subsequent radiation products is highly nonuniform. This nonuniformity determines the relative biological effectiveness of different types of ionizing radiation, and its role is particularly important at small doses. Nonhomogeneous kinetics is, therefore, a central issue in radiation biophysics.

The physics of the interaction of ionizing radiation with matter and the spatial distributions of energy in charged particle tracks are treated in detail in Chapter 3. Figure 7.1 is a simplified diagram of tracks of sparsely and densely ionizing charged particles in relation to the superimposed micrograph of part of a mammalian cell. The microdistribution of energy has always been a principal topic of quantitative radiobiology. It is also the subject of microdosimetry, a new branch of dosimetry and radiation physics created by H. H. Rossi. Despite the more rigorous concepts of microdosimety, semiquantitative treatments—for example, in terms of the concept of linear energy transfer (LET)—are still common. To put the different approaches into perspective, it is necessary to deal first with simple approaches and with the theories of radiation action that preceded microdosimetry. Most of the models treated in Section 7.2 are too crude to be of pragmatic value in themselves. The basic probabilistic notions are, however, essential, and to understand the simplified approaches and their limitations is a condition for the development and utilization of more sophisticated treatments.

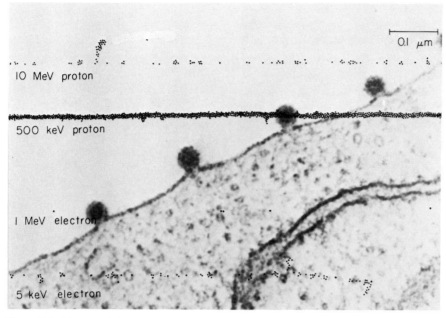

Figure 7.1. Diagram of charged particle tracks superimposed on a micrograph (adapted from Ref. 1) of part of a mammalian cell. The viruses budding from the outer cell membrane permit an added comparison of size. In the projected track segments the dots represent ionizations. The lateral extension of the track core is somewhat enlarged in order to resolve the individual energy transfers. For more accurate diagrams of particle tracks see Chapter 3.

7.2. ESSENTIALS OF TARGET THEORY

Quantitative radiobiology began with investigations of the inactivation of bacteria, viruses, or certain enzymes by X rays. In such experiments dose–effect relations were found that differ characteristically from those familiar in cytotoxicology. The differences are fundamental to an understanding of the action of ionizing radiations. An initial, general consideration of dose–effect relations is, therefore, required.

The simplest dose dependence would result if the exposed organism, for example, a bacterium, tolerated doses up to a certain threshold but were inactivated if this threshold were exceeded. The dose–effect relation would then be a step function. In reality, one can never attain entirely homogeneous populations in microbiological studies. Furthermore, one must note that biological processes are inherently stochastic; the complexity even of simple cells is such that the slightest differences in initial conditions can lead to unpredictable fluctuations in the response to various factors. Although such systems are in principle deterministic,

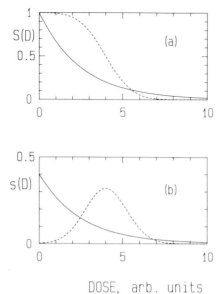

Figure 7.2. Exponential and sigmoid dose–effect relations and their derivatives.

their reactions can be described only in stochastic terms. Instead of the threshold reaction, one expects a response curve such as the dashed line in Figure 7.2. Dependences of this type are termed *sigmoid curves* or *shoulder curves*.

A naive explanation of a sigmoid dose–effect relation invokes the notion of a distribution of sensitivities within a population. The stochastic response of the cell is not considered. Instead it is assumed that individuals of the population have different critical thresholds of dose. If, for example, one postulates the Gaussian distribution of critical doses symbolized in Figure 7.2*b*, one obtains the integral of this distribution, that is, the sigmoid curve in Figure 7.2*a*, as a response function. The example shows that the derivative, $-dS(D)/dD$, of the survival curve can, in the simplest interpretation, be considered as the probability distribution of critical doses; one could also speak of the probability distribution of resistance within the population. However, this interpretation in terms of biological variability is only one among other possibilities. It disregards the potential influence of other stochastic factors that may codetermine the dose–response relation.

In early radiobiological experiments, when enzymes, viruses, or certain bacteria were exposed to X rays, entirely different dose–effect relations were obtained (2). The fraction of viable units, $S(D)$, decreases—as exemplified in Figure 7.3 for the DNA phage T7—exponentially with the absorbed dose D:*

$$S(D) = \exp(-aD) \tag{7.1}$$

*The form of Eq. (7.1) is similar to that of Eq. (6.56), where the parameter N_s is analogous to D and nf_{ir} is analogous to a. Both equations are based in stochastics. (Ed.)

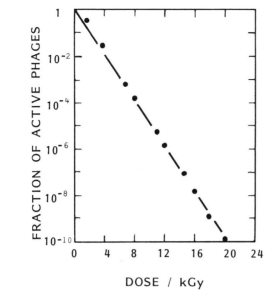

Figure 7.3. Exponential dose dependence for the inactivation of T1 phages by Co γ rays (redrawn from Ref. 3).

The solid line in Figure 7.2a corresponds to this relation. The distribution of re-sistance that could explain such a curve would also be an exponential, as shown in Figure 7.2b (solid line):

$$\frac{-dS(D)}{dD} = s(D) = a \exp(-aD) \tag{7.2}$$

This distribution has its maximum at $D = 0$, that is, a relatively large fraction of the exposed entities would have to be highly sensitive. The broad tail would imply, on the other hand, that a substantial fraction of the microbiological entities is highly resistant. Biological variability cannot, in general, be the reason for such relations. However, analogy to radioactive decay led to a different interpretation that was to become the basis of target theory (2, 4–6).

7.2.1. Exponential Dose–Effect Relation and Single-Hit Process

The decay of a radioactive substance is characterized by the fact that equal frac-tions of the remaining atoms disintegrate in equal time intervals:

$$\frac{dN(t)}{N(t)} = -a\, dt \quad \text{or} \quad \frac{d \ln N(t)}{dt} = -a \tag{7.3}$$

Therefore,

$$N(t) = N_0 \exp(-at) \tag{7.4}$$

where $N(t)$ is the number of atoms still present at time t. The probability of an atom to disintegrate is, accordingly, independent of its age. This independence reflects the fact that the decay process is a spontaneous random event rather than the result of gradual deterioration.

The interpretation of the exponential survival curves in radiobiology is analogous. Dose takes the place of time; with constant dose rate it is proportional to time. Regardless of the dose already applied, a constant dose increment reduces the number, $N(D)$, of survivors by a constant fraction:

$$\frac{dN(D)}{N(D)} = -a \, dD \quad \text{or} \quad \frac{d \ln N(D)}{dD} = -a \tag{7.5}$$

Therefore,

$$N(D) = N_0 \exp(-aD) \tag{7.6}$$

Normalized to the surviving fraction, $S(D) = N(D)/N_0$, one has

$$S(D) = \exp(-aD) \tag{7.7}$$

The exponential survival curves can, therefore, be understood in terms of individual random events. Such random events have been called hits because they had to be discrete acts of energy transfer from the radiation field to sensitive structures of the exposed organism. Dessauer introduced the notion of *point heat* to characterize the hit process (7). Crowther, who developed the formalism of target theory independently (8), postulated somewhat more pragmatically that the hits were individual ionizations in the sensitive structures. His assumption has been verified in many radiobiological investigations on enzymes, or single-strand viruses. It was demonstrated that such comparatively simple systems can, indeed, be inactivated by the detachment of individual electrons, with subsequent damage caused directly, or induced indirectly, through the formation and action of free radicals.

From Eq. (7.5) one concludes that $a \, dD$ is the probability of a hit per dose increment, dD. Accordingly, aD is the mean number of hits per exposed unit at dose D. Assuming a certain magnitude E of energy deposition in individual events (ionization or cluster of ionizations), one can utilize this relation to deduce formally a *critical mass* or a corresponding *critical volume*. The example of Figure 7.3 may explain the method. The T1 phage is inactivated according to Eq. (7.7) with $a = 0.0011 \text{ Gy}^{-1}$. The mean number of hits in the assumed target region is then aD, and the mean energy per target region is aDE. The mean energy is also equal to the dose times the mass m of the target region:

$$aDE = Dm \qquad (7.8)$$

Therefore,

$$m = Ea \qquad (7.9)$$

For example, if one primary ionization corresponds to the mean energy transfer E = 80 eV, one has (with 1 eV = 1.602×10^{-19} J and 1 Gy = 1 J/kg):

$$m = Ea = 80 \text{ eV} \times 0.0011 \ Gy^{-1} = 80 \times 1.6 \times 10^{-19} \times 0.0011 \text{ kg}$$
$$= 1.4 \times 10^{-17} \text{ g} \qquad (7.10)$$

The actual mass of the DNA double-strand molecule of the T1 phage is 5×10^{-17} g. The inactivation probability due to a single collision is thus substantially less than 1. It is likely that the critical events are those that produce a double-strand break in the phage DNA.

7.2.1.1. Geometric Illustration of Single-Hit Process.

The single-hit mechanism is a comparatively simple random process. Other schemes are considered in subsequent sections of this chapter. To elucidate the relations between the processes of different complexity, simplified two-dimensional schemes are employed. Such schemes can also be used to illustrate the single-hit process and to bring out certain factual deviations from the model.

The three panels in Figure 7.4 illustrate schematically the occurrence of absorption events (hits) in individual cells, which are represented as squares and rectangles. The dots symbolize energy deposits (ionizations) randomly occurring within the collective of cells.

Panel A depicts the simplest case where all cells are of equal sizes and consequently have equal probabilities to be hit. Panel B gives cells of varying sizes,

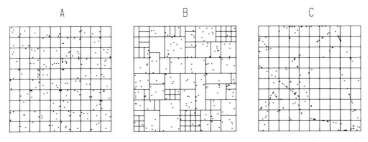

A   B   C

Figure 7.4. Diagrams of different Poisson processes with an average number of 1.8 hits/cell. Panel A: Simple Poisson process of independent hits on cells of equal size. Panel B: Simple Poisson process of independent hits on cells of varying size. Panel C: Poisson process of spatially correlated hit events, also termed compound Poisson process.

with corresponding variation of the probabilities to be hit. Panel C represents, again, an array of equal cells. However, the energy deposits occur in clusters, which are simplified two-dimensional analogs of charged particle tracks.

The panels give the random configurations at a specified "dose". Dose is here measured in terms of the mean number of events per exposed unit, the value being 1.8 in the example of Figure 7.4. The graphs of Figure 7.5 represent results of simulated exposures. The decreasing number of undamaged cells (that is, cells still without a hit) is plotted versus dose. The somewhat irregular functions result from the relatively small size of the samples of only 100 cells. If the number of cells were vastly increased, or if many repeated simulations were performed, one would obtain the dashed lines in the graphs.

The simplest case of equal cells in panel A (that is, the case with no biological variability) results in an exponential survival curve. One speaks of a pure Poisson process.

The more complex case of panel B (that is, the case that corresponds to cells with different sensitivities) results in a curve that is not straight in the semilogarithmic plot but is concave upward. This is understood from the fact that the smaller units, which represent the more resistant cells, tend to survive to higher doses, so that the average sensitivity (that is, the probability to be hit) of the survivors declines with increasing dose. Accordingly, the slope of the survival curve in the semilogarithmic plot decreases at higher doses.

Panel C represents the case of spatial correlation of some of the energy deposits. Such spatial correlation—the central aspect of microdosimetry—occurs in charged particle tracks; it has the general effect that there are fewer energy deposition events

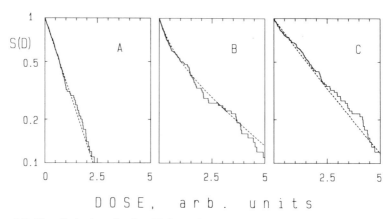

Figure 7.5. The elimination of cells with increasing dose in the samples of Figure 7.4. The step functions are the results of random trials; the broken curves apply to large samples. For the assumed one-hit process the dependences are exponential if the cells are of equal size (case A and C). The spatial correlation of hits (case C) reduces the slope of the dose dependence. The differences in sensitivity (case B) cause a nonexponential dependence.

(passages of charged particles) per cell but there can be multiple ionizations per event. Due to the statistical independence of events, one obtains an exponential dose dependence, but the reduced event frequency causes a reduced slope of the survival curve; in other words, the close spatial correlation of energy transfers creates some waste of energy in the single-hit process. A radiation of higher ionization density has in general less biological effectiveness. In a multihit process, that is, in a case where damage accumulation is required for the biological effect, the situation can be reversed (Section 7.2.2).

7.2.2. Interpretation of Sigmoid Dose–Effect Relations

When target theory was conceived for the interpretation and quantitative analysis of single-hit processes, it opened up an intriguing new field of study to the biophysicist. The term *quantum biology* was coined (4). It referred to the fact that the detachment of single electrons could have substantial effects on complex cellular systems containing billions of atoms. It was natural, in view of the fascination with a novel field of research, that attempts were made to explore also broader implications beyond the immediate scientific issues, and such considerations extended to general philosophical discussions of determinism or indeterminism in living objects (9). The formalism of target theory was also extended, and attempts were made to explain all survival curves—not merely the exponential ones—in terms of the statistics of energy deposition.

Sigmoid dose dependences are obtained when certain bacteria or, as in the example of Figure 7.6, higher cells are exposed to X or γ rays. The assumption was made that with these dose relations also, the deviations from a step function are caused by the statistics of energy deposition. Later it was recognized that this must be an oversimplification, and the deviations are the composite result of several factors, among which the fluctuations of energy deposition need not always be the dominant one. It is nevertheless useful to review the classical multihit or multitarget models, even if they have little pragmatic importance. The underlying mathematics is, in modified form, required in any treatment of the statistics of energy deposition by ionizing radiations. The subsequent considerations serve as a simple introduction to necessary elements of probability theory. In particular, the Poisson distribution and the closely related Γ distribution will be referred to.

7.2.2.1. Multihit Process and Poisson Distribution.

The slope of the survival curve in the semilogarithmic representation, $-d(\ln S(D))/dD$, is constant for the exponential relation. For sigmoid curves it increases with increasing dose, and this is the expression of a gradual accumulation of damage. The slope determines the fraction of the surviving cells that is inactivated by an additional dose increment. The slope increases with absorbed dose in those cases where successive hit events accumulate damage up to a critical level.

The simplest model of a multihit process results from the assumption of a definite threshold, that is, from the postulate that the cell survives with less than n

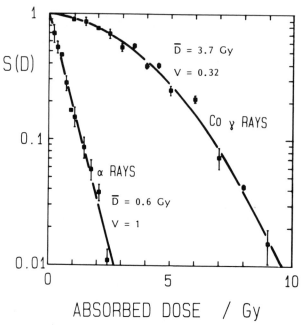

Figure 7.6. Loss of proliferative ability of mouse fibroblast cells exposed to Co γ rays and americium α rays. Data are from experiments by Lücke-Huhle et al. (10). Parameters: mean inactivation dose $\bar{D}$ and relative variance V are indicated for the two survival functions; for definitions see Section 7.2.4.

hits but is inactivated if there are n or more hits. If the hits are statistically independent, and if all exposed units have the same critical threshold n, one obtains comparatively simple equations.

The Poisson equation gives the probability for exactly ν events in a trial, when the number of events averaged over many trials is x:

$$p(\nu) = \exp(-x)\,\frac{x^{\nu}}{\nu!} \qquad (7.11)$$

This relation can be illustrated in terms of panel A in Figure 7.4. The average number of events per field in the diagram is $N = 1.8$. Figure 7.7 gives for this expectation value the Poisson distribution and a corresponding sum distribution:

$$P(\nu) = \sum_{k=0}^{\nu-1} \exp(-x)\,\frac{x^{k}}{k!} \qquad (7.12)$$

Inserted in Figure 7.7 as dashed lines are also the graphs that correspond to the particular trial represented in panel A of Fig. 7.4 for the sample of 100 cells.

For the Poisson distribution the variance σ^2 is equal to the mean x, and the

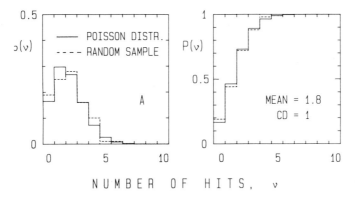

Figure 7.7. Poisson distribution of hits for the expectation value 1.8 and its sum distribution (solid lines). Broken lines are frequencies resulting from a random trial with the sample of 100 cells in panel A of Figure 7.4. The value 1 of the coefficient of dispersion (CD) applies to Poisson distributions regardless of their mean value.

dispersion coefficient, $CD = \sigma^2/x$, is equal to unity. The dispersion coefficient for the data from panel A (Figure 7.4) is

$$CD = \sum_{i=1}^{I} \frac{(v_i - x)^2}{Ix} = 0.99 \approx 1 \tag{7.13}$$

where v_i is the number of events in cell $i (i = 1, \ldots, I)$, and $x = 1.8$ the mean number of events per cell, which is estimated as $x = \Sigma\, v_i/I$.

Figure 7.8 gives, for comparison, the distribution of the number of hits per cell

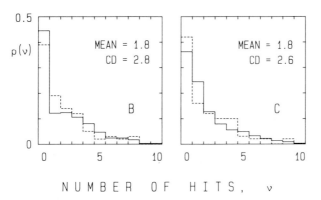

Figure 7.8. Relative frequencies of the number of hits at a mean value of 1.8 hits/cell for the processes B and C in Figure 7.4. Broken lines are the results of random trials for the samples of 100 cells. The solid lines and the coefficients of dispersion apply to large samples.

for the two random trials in panels B and C of Figure 7.4. The distributions are substantially broader, and the coefficients of dispersion are 2.8 and 2.6. Biological variability of sensitivity enhances, in the same way as the statistics of energy deposition, the fluctuations of the effect on individual cells.

According to the multihit model, all those cells survive that have less than n hits. With $x = aD$, the survival function is

$$S(D) = \sum_{\nu=0}^{n-1} \exp(-aD) \frac{(aD)^\nu}{\nu!} \tag{7.14}$$

Figure 7.9 represents the multihit relations (i.e., the survival functions according to this equation) for selected values of the parameter n. As expected, the shoulder of the curves is most pronounced for large n; the stochastic character of the response is less marked when many random events must be accumulated to reach the critical level of damage.

The steplike function in Figure 7.9 gives the result of the random simulation of a three-hit process corresponding to the simple Poisson process represented in panel A of Figure 7.4. Although the result is obtained with the comparatively small sample of only 100 cells, it is in general accord with the theoretical curve for $n = 3$.

The more complex conditions represented in Figure 7.4 can be utilized to illustrate fundamental inadequacies of the simple multihit model. Figure 7.10 com-

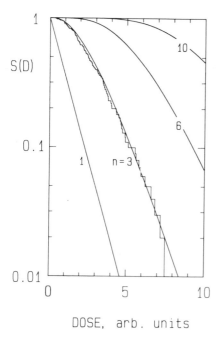

Figure 7.9. Multihit survival curves according to Eq. (7.14). Values aD are given on the abscissa. The random curve results from a trial of a three-hit process on the sample of 100 equal cells in panel A of Figure 7.4.

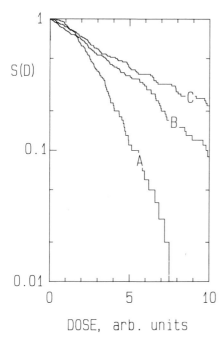

Figure 7.10. Comparison of three-hit survival curves obtained from random trials of the processes A, B, and C in Figure 7.4. Deviations from the simple Poisson process (A), such as varying size of the cells (B) or spatial correlation of hits (C), lead to a reduction of the shoulder, that is, to an increased relative variance of the survival functions (see Section 7.2.4).

pares the survival relations obtained from random trials for the samples of 100 cells for the three cases and an assumed three-hit process. The result for case *B* illustrates the general effect that any variations of sensitivity decrease the shoulder of the curve. This fact has frequently been overlooked in applications of multihit theories to observed survival curves where it was usually assumed that the combined influence of all factors other than the statistics of energy deposition would somehow "average out." Instead, the general effect is one of reducing the apparent hit numbers. Such numbers may, therefore, be misleading and quite meaningless as estimates of the multiplicity of events involved in the cellular action of ionizing radiations.

Curve *C* illustrates the influence of the spatial correlation of energy deposits within charged particle tracks. Spatial correlation of energy deposits, too, leads to a reduction of the shoulder of the survival curve, that is, to smaller apparent hit numbers. This is in agreement with the general observation that survival functions, as well as other dose–effect relations, for sparsely ionizing radiations can have pronounced shoulders while relations obtained with densely ionizing radiations tend to be exponential (see Figure 7.6). The explanation is that if more energy is imparted to the cell in one event, fewer events are required on the average to reach the critical level of damage. In Section 7.2.4 the relative variance V of the dose–effect relation is defined, a parameter that quantifies the extent of the deviations from a threshold response. For the survival curves of Figure 7.6 one obtains the value $V = 0.32$ for γ rays and $V = 1$ for α rays.

7.2.2.2. Time Factor. The simple multihit model postulates accumulation of damage due to random events that are statistically independent. The model disregards the additional factors that codetermine the response of the irradiated cells. One factor, beyond those that have been considered, is the temporal distribution of dose. DNA is the major target of radiation damage in the cell, and there are various repair mechanisms that eliminate or reduce damage to DNA. The time constants of the repair processes are seconds to hours. Due to the repair processes, a short-term irradiation is, in general, more effective than the protracted or fractionated application of the same dose. When there is sufficient time during the irradiation for partial reversion of sublethal damage, the accumulation of damage and the subsequent radiation effects are reduced.

Because of the oversimplified nature of the multihit model, there is little justification for formal modifications of the equations that would account for the dose-rate effect. But even without a quantitative treatment, it is helpful to illustrate the general influence of the time factor. One can postulate that the sublesions produced by individual hits are restituted randomly with constant probability b per unit time. There is then an exponential distribution of lifetimes of the lesions. It is easy to simulate such a model. The results in Figure 7.11A are two-hit curves for the simple Poisson process (i.e., the random process represented in panel A of Figure 7.4); they are derived with the assumed repair process under the condition of constant dose rate, which is most common in practice. Random trials for the sample

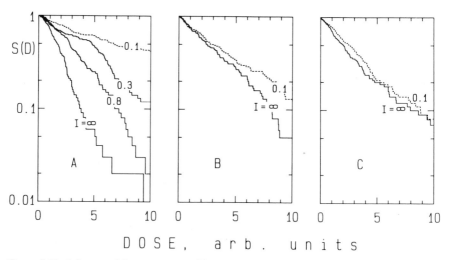

Figure 7.11. Influence of dose rate on two-hit curves resulting from random trials for the processes in panels A, B, and C in Figure 7.4. Parameter I corresponds to the dose rate; it equals the mean number of hits per cell during the mean restitution time. The influence is largest for the pure Poisson process (A). It is substantially reduced for the process with different cell sizes (B), and it is nearly absent in the process with spatially correlated hits (C).

of 100 cells are given for different dose rates. The dose rate I is here expressed as the ratio of the mean number of hit events per unit time and the repair rate b:

$$I = \frac{a\,(dD/dt)}{b} \tag{7.15}$$

The general result is that the survival curves have substantially reduced shoulders when the dose rate is sufficiently small for the influence of repair to become important. This is in agreement with a wide range of experimental results.

According to the simple model, the probability of an observable effect can be vanishingly small at very small dose rates. In most experimental systems such a complete reduction is not observed, and there are two main reasons. The first is that part of the cellular damage can be irreversible. The second reason is that the statistical fluctuations of energy deposition are such that even with sparsely ionizing radiations single charged particles can, with small probability, bring about effects that usually require the accumulation of several energy deposition events. Figure 7.11C shows that, in fact, the dose–rate dependence is substantially reduced for a model that includes spatial correlations of energy transfers (panel C of Figure 7.4). For densely ionizing radiations this microdosimetric aspect, the instantaneous local energy accumulation, is the predominant characteristic. Dose–rate dependences of cellular effects are then largely absent. On the tissue level dose–rate effects can occur even with densely ionizing radiations, but they do not always consist in a reduction of the effect with protraction or fractionation of an exposure (11).

7.2.3. Limited Validity of Target Theory Models

As pointed out in the preceding sections, the multihit theory is a gross oversimplification because it attempts to explain the dose–effect relation merely in terms of the statistics of energy deposition, disregarding other factors that codetermine the dose dependences. It is, nevertheless, possible to derive from observed dose–effect relations certain rigorous statements on the accumulation of damage in independent events of energy deposition. This is considered in Section 7.2.4.

7.2.3.1. A More General Treatment in Terms of Markov Processes. The multihit model is the simplest description of a process of random accumulations of damage. The same mathematical relations have been postulated in multistep theories of cancer. There are, nevertheless, a variety of alternative models.

One familiar model is the multitarget postulate. It is assumed that there are m targets in the cell and that all targets have to be damaged if the effect is to occur. Making the simplest assumption of independent single-hit inactivation of the individual targets with equal probabilities, one obtains the survival function

$$S(D) = 1 - [1 - \exp(-aD)]^{m} \tag{7.16}$$

This equation can describe adequately the inactivation of clumps of m cells if the survival relation of the individual cell is exponential. But apart from this nearly trivial example, there have never been experiments where a response function has been fitted by the equation and where such a fit has then led to the identification of the corresponding number of actual critical structures.

With a modification to account for the finite initial slope of dose–effect relations at small doses and with the admission of noninteger values for the exponent m, the equation

$$S(D) = \exp(-cD) \{1 - [1 - \exp(-aD)]^m\} \qquad (7.17)$$

is frequently used to describe observed survival functions, for example, of mammalian cells, exposed to sparsely ionizing radiations. However, it is realized that the applicability of the equation provides no verification of the underlying model; the fit is merely empirical.

The multitarget equation differs from the multihit equation chiefly by the fact that the curves have asymptotic tangents in the semilogarithmic representation. For both Eqs. (7.16) and (7.17) these tangents intersect the ordinate at the value m. One speaks of an "extrapolation number" rather than a "target number" in order to avoid the identification of the equation with the underlying target theory model.

A vast variety of other models could be postulated, but it makes little sense to employ equations that contain sufficient free parameters to fit nearly any dose–response relation. To put the target theory models into a more general context, it is nevertheless helpful to consider a broader class of stochastic models that contains the multihit or multitarget conditions as special cases. These are linear stochastic processes, which can be depicted as Markov chains (multicompartment models). Figure 7.12 gives diagrams that represent the multihit process, the multitarget process, and an example of a somewhat more general two-hit process. The latter model includes repair and also a single-hit component that may correspond to the action of densely ionizing radiation. In the diagrams the dots represent successive states of damage. The lowest dot represents the undamaged state, and the highest dot represents cell death or another specified irreversible effect. The coefficients of transition into states of higher damage are proportional to the dose rate I if time is chosen as the independent variable. The coefficients for the restitution processes are taken to be constant in the simple treatment referred to in the preceding paragraph. However, it is evident that there could be further complexities, such as a dependence of the repair rates on dose or on dose rate.

The purpose of the present discussion is not a detailed formal treatment but the critical assessment of somewhat arbitrary assumptions underlying all simplified models. More complex linear models could be constructed, and nonlinearities would lead to further complexities.

The linear models can always be represented by the simple equation

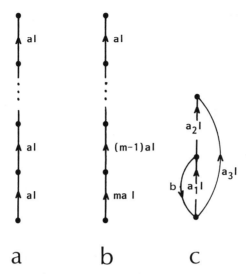

Figure 7.12. Schematic diagram of Markov processes that correspond to the multihit model [Eq. (7.14)], the multitarget model [Eq. (7.16)], and a two-hit process with repair [Eq. (7.19)].

$$\frac{d}{dD} X = AX \tag{7.18}$$

where X is a column vector that contains as components the probabilities for the individual states, say $0, \ldots, n - 1$; the last state need not appear in the equation, since its probability is the complement of the sum of all other probabilities. The matrix A contains the transition probabilities. The solution $X(D)$ can be given in terms of the eigenvectors and eigenvalues of the transition matrix A (6, 12).

The matrix of transition coefficients characterizes a particular model, and the solution for any such model is then straightforward. For example, Eqs. (7.14) and (7.16) are derived from Eq. (7.18) with the transition matrices that correspond to Figs. 7.12a, b. The transition matrix for Fig. 7.12c but with dose as independent variable is

$$A = \begin{pmatrix} -a_1 - a_3 & \dfrac{b}{I} \\ a_1 & -a_2 - \dfrac{b}{I} \end{pmatrix} \tag{7.19}$$

For the dose dependence of the survival $S(D)$, that is, the sum of the probabilities of the two states 0 and 1, one obtains

$$S(D) = c \exp(-\lambda_1 D) + (1 - c)\exp(-\lambda_2 D) \tag{7.20}$$

with the eigenvalues:

$$\lambda_{1,2} = \frac{1}{2}\left(a_1 + a_2 + a_3 + \frac{b}{I} \pm \sqrt{\left(-a_1 + a_2 - a_3 + \frac{b}{I}\right)^2 + \frac{4a_1 b}{I}}\right)$$

and

$$c = \frac{\lambda_2 - a_3}{\lambda_2 - \lambda_1}$$

Somewhat different expressions result if the two eigenvalues are equal.

The comparatively simple two-event process has special interest because, as pointed out in later sections, cellular effects of ionizing radiations appear to result largely from second-order processes.

7.2.4. Generalized Characterization of Dose–Effect Relations

It has been shown in the preceding sections that the shoulder of a dose–response curve can be reduced by a variety of factors. The statistics of energy deposition is one of these, but this need not always be the dominant influence. To quantify statements on the shape of the dose–effect relation, one can utilize basic parameters that apply if the dose–response relation is treated formally as a probability distribution function. This notion was invoked at the beginning of this chapter in a specific explanation of a dose–effect relation as the sum distribution of the resistance of microorganisms (Section 7.2, Figure 7.2). As a general way to look upon dose–effect relations, the concept is, however, not sufficiently familiar, and it is helpful to illustrate it first in terms of the elementary Poisson process postulated by the multihit model.

Figure 7.13 represents several random paths that correspond to the Poisson process of equal independent events occurring with constant probability per unit dose. The vertical dotted line indicates a specified dose. The random lines intersect this line at integer ordinate values equal to the number of hits that have occurred. The distribution of the points of intersection follows the Poisson distribution. With a slight change in notation, Eq. (7.11) for the Poisson density (that is, the probability for exactly n events) is written as

$$p(n;x) = \exp(-x)\frac{x^n}{n!} \tag{7.21}$$

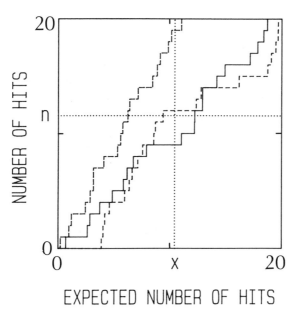

EXPECTED NUMBER OF HITS

Figure 7.13. Three random paths for the simple Poisson process. The points of intersection with the vertical dotted line are subject to a Poisson distribution [Eq. (7.20)] with mean value $x = 10.4$. The points of intersection with the horizontal dotted line are subject to a gamma distribution [Eq. (7.23)] of order $n = 12$.

Equation (7.12) for the sum distribution takes the form

$$P(n; x) = \sum_{\nu=0}^{n-1} \exp(-x) \frac{x^\nu}{\nu!} = 1 - \int_0^x \exp(-z) \frac{z^{n-1}}{(n-1)!} dz \qquad (7.22)$$

where the last equality is obtained by taking the derivative with respect to x of the sum.

In its interpretation as the Poisson sum distribution $P(n; x)$ is the probability that a random path traverses the vertical line below n. An equivalent condition is that the random path intersects the horizontal dotted line at a value in excess of x. Therefore, $G(x; n) = 1 - P(n; x)$ is the sum distribution of values x to reach n events. The corresponding density is designated by $g(x; n)$; this is the differential distribution of the values x required to reach n events:

$$G(x; n) = \int \exp(-z) \frac{z^{n-1}}{(n-1)!} dz \qquad (7.23)$$

$$g(x; n) = \exp(-x) \frac{x^{n-1}}{(n-1)!} \qquad (7.24)$$

The distribution function $G(x; n)$ and the density of the random variable x are called Γ distributions of order n. For different values of the parameter n, the distributions are depicted in Figure 7.14.

The connection between the sum distributions $P(n; x)$ and $G(x; n)$ remains valid for a more general process with steps of variable size that correspond to the highly variable energy deposition by charged particles (see Section 7.4). The expressions for the densities relative to n or x are, however, different in this case. The subsequent relations for the moments are also specific to the pure Poisson process.

The mean value of the Γ distribution and its second moment are

$$\bar{x} = \int_0^\infty x \exp(-x) \frac{x^{n-1}}{(n-1)!} \, dx = n \tag{7.25}$$

$$\overline{x^2} = \int_0^\infty x^2 \exp(-x) \frac{x^{n-1}}{(n-1)!} \, dx = n(n+1) \tag{7.26}$$

As with the Poisson distribution the variance is equal to the mean:

$$\sigma^2 = \overline{x^2} - \bar{x}^2 = n \tag{7.27}$$

and the relative variance V is the inverse of the mean:

$$V = \frac{\sigma^2}{\bar{x}^2} = \frac{1}{n} \tag{7.28}$$

According to these considerations, the multihit survival functions of Eq. (7.14) correspond to the Γ distribution:

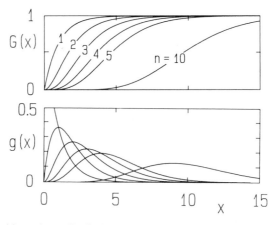

Figure 7.14. Densities and sum distributions for gamma distributions of order 1, 2, 3, 4, 5, and 10.

$$S(D) = 1 - G(x; n) = P(n; x) \tag{7.29}$$

where

$$x = aD \tag{7.30}$$

The survival function can be understood as a probability distribution of the inactivation dose D. The mean of the distribution is the mean inactivation dose $\overline{D}$ [see Eq. (7.2)]:

$$\overline{D} = \int_0^\infty D\, s(D)\, dD = \int_0^\infty S(D)\, dD \tag{7.31}$$

The variance of the inactivation dose is

$$\sigma^2 = \int_0^\infty (D - \overline{D})^2\, s(D)\, dD = 2 \int_0^\infty DS(D)\, dD - \overline{D}^2 \tag{7.32}$$

The relative variance is

$$V = \frac{\sigma^2}{\overline{D}^2} \tag{7.33}$$

In the specific case of Eq. (7.14) one obtains, with the mean and the variance of the Γ distribution,

$$\overline{D} = \frac{n}{a} \quad \text{and} \quad \sigma^2 = \frac{n}{a^2} \tag{7.34}$$

and therefore,

$$V = \frac{\sigma^2}{\overline{D}^2} = \frac{1}{n} \tag{7.35}$$

The mean inactivation dose (or mean effect dose) $\overline{D}$ and the relative variance V of the dose–response relation have here been explained for the particular example of the n-hit process, that is, for the pure Poisson process. However, the considerations have more general validity. One can utilize the parameters D and V for any dose–effect relation.

The mean inactivation dose is not an unfamiliar concept for particular dose dependences. It equals the 50% survival dose D_{50} for a symmetrical dose–response relation; for exponential dose dependences it is identical with D_{37}. However, it is a universal parameter that applies to any dose–effect relation provided the relation can be extrapolated to doses where the effect probability approaches unity.

The relative variance of a dose–effect relation, or specifically a survival function, is an even more important parameter. It can serve as a quantitative measure for the deviation of the dependence from a threshold reaction. For a pure threshold reaction (that is, a step function) the relative variance V would be zero. For the multihit curves the relative variance equals the inverse of the hit number n [Eq. (7.35)]. The relative variance or its inverse, which is called relative steepness (6, 12), is therefore a generalization of the familiar concept of the hit number, which gives this concept general validity by making it independent of any particular model. Figure 7.15 exemplifies the general applicability of the parameters $\overline{D}$ and V. It is striking that the two most fundamental parameters of a distribution, the mean and the variance, are not very familiar concepts in the analysis of radiobiological dose–response relations. The reason is probably the convenience of using the conventional parameters [see for example (13)], which can be read off a graph and can be roughly estimated, even from crude data.

The relative variance of a dose–response relation is jointly determined by the statistical factors that are relevant to the response of the cells. The main factors are statistics of energy deposition, biological variability within the population, and the inherently stochastic response of the cell. It is in general not possible to separate the individual factors. An exception is the case of the simple exponential relation, for which the dominant factor can be the statistics of energy deposition.

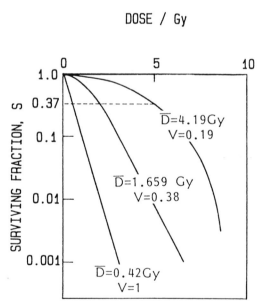

Figure 7.15. Schematic examples of different types of survival functions (adapted from Ref. 13). Parameters mean inactivation dose $\overline{D}$ and relative variance V are inserted. For comparison with observed survival functions for mammalian cells see Figure 7.6.

To explain a response function with small relative variance, one must invoke at least $1/V$ independent acts of energy deposition within the cell. No model involving a smaller mean hit number can lead to a response with this or a smaller value of the relative variance (6, 12, 14). The inferred number of events, the inverse of the relative variance, is merely a lower limit. The actual number of events relevant to the process will in general be larger because sigmoid dose–effect relations are codetermined by other statistical factors. Only an unknown part of the observed relative variance can then be ascribed to the statistics of energy deposition, and the number of energy deposition events must be larger than the theoretical limit. In subsequent sections it will be seen that the varying magnitude of the energy deposition events has an additional influence that works in the same direction.

The dose–effect relations in Figure 7.10 for the processes of Figure 7.4 illustrate the contribution of the statistics of energy deposition—including the spatial correlation of energy depositions—and the role of biological variability. It is evident that the nonuniformity of the reaction is dependent on the Poisson statistics, the "track structure," and the biological variability. Although the examples are highly schematic, they serve to demonstrate the complexity of the interplay of random factors that determines the dose–response relations. Simple relations are obtained only when the statistics of energy deposition is entirely dominant as a random factor; an example is the exponential survival functions obtained with very small objects (see Figure 7.3) or with densely ionizing radiation (see Figure 7.6).

In the subsequent sections the emphasis will be on the quantitative analysis of the one factor that is most characteristic for ionizing radiations, the statistics of energy deposition. It is difficult to assess the role of this factor in terms of complicated multihit processes. There have, nevertheless, been attempts to formulate threshold models in analogy to the target theory models. Such attempts will be surveyed after a section (7.3) that deals with fundamentals of microdosimetry. In a subsequent section simpler models will be considered that describe second-order processes of radiation action in terms of the actual microscopic distribution of energy deposition. There is good evidence from a variety of radiobiological investigations that ionizing radiations work on higher cells in a way that can broadly be termed a quadratic reaction. One can, accordingly, consider models that are sufficiently simple to remain tractable even if the highly complex patterns of energy deposition with ionizing radiations are taken into account.

7.3. CONCEPTS AND QUANTITIES OF MICRODOSIMETRY

7.3.1. Objective of Microdosimetry

The stochastic models of target theory postulate independent and equal random events of energy deposition. In reality, ionizing radiations impart energy to the exposed media in complex microscopic random patterns (Chapter 3). An approximate characterization of such patterns is afforded by the concept of linear energy transfer (LET). One pictures the charged particle tracks as straight lines with constant rate of energy loss or as strings of ionizations regularly or randomly spaced along the trajectory of the particle with a density that corresponds to the LET, that is, to the mean energy loss of the particle per unit pathlength. This concept is widely used, but it has limited validity (see also Chapter 4). A more sophisticated approach was first developed by Lea (15). Later, when Rossi and his co-workers laid the theoretical and experimental foundations of microdosimetry (16–20), a coherent framework of concepts and quantities was created to quantify the statistics of energy deposition and the characteristic differences between various types of ionizing radiations.

The microscopic patterns of energy imparted by ionizing radiations to the exposed media can be determined by measurement or by computation. The most ambitious approach would be to determine all individual electronic alterations and their spatial coordinates. Although such an approach is not feasible, one can, as explained in Chapter 3, achieve the complex task of simulating the inchoate (initial) distribution of electronic changes produced by ionizing radiations within a specified volume. Any such distribution is a random configuration. Two identical exposures or random simulations would never result in equal spatial distributions of energy. Suitable parameters are required to characterize the random patterns. Ideally one would construct a hierarchy of characteristic parameters to express, with increasing degrees of sophistication, those properties of the distributions that determine the biological effectiveness of different radiations. Microdosimetry, in its general sense, is concerned with such characterization. In a conventional, more restricted sense it deals merely with certain quantities that are linked to the notion of energy concentration. These quantities are the specific energy z and the lineal energy y. They are considered first because they are suitable for the more elementary stochastic models of radiation biophysics.

When certain molecules are uniformly distributed in a solution, one can quantify their reactions in terms of equations that depend on their concentrations. The notion of concentration is not directly applicable when one considers small volumes that may contain only few of the molecules (see also Chapter 11). But even then the reaction kinetics remains comparatively simple because one can utilize the Poisson statistics of independently distributed molecules. The concentration remains the sole parameter that determines the distribution.

With ionizing radiations the situation is characteristically different. The energy imparted, and the subsequent radiation products such as free radicals, are not distributed in simple, uniform random patterns. Instead, they occur in clusters along the trajectories of charged particles (Chapters 3 and 4). Depending on absorbed dose and on the type and energy of the charged particles, the resulting inhomogeneity of the microdistribution can be very substantial. It is the reason that one cannot apply the notion of concentration directly. Measurements in randomly selected microscopic volumes will yield energy concentrations or concentrations of subsequent radiation products, which deviate considerably from their expectation values, and these variations depend in intricate ways on the size of the reference volumes, the magnitude of the doses, and the types of ionizing radiations. Any quantitative description of radiation quality must account for the fluctuations of energy concentration, and it must therefore utilize concepts of probability theory.

The conventional quantities of microdosimetry are defined as concentrations in microscopic volumes. To account for the inapplicability of the simple concept of concentration, they must be treated as random variables.

7.3.2 Definition of Quantities and Distributions

7.3.2.1. Quantities

1. The energy imparted, ϵ, to a specified volume is equal to the energy of ionizing radiation incident on the volume minus the energy of ionizing radiation emerging from the volume.

A rigorous formulation (21) accounts also for possible changes of rest mass that are here disregarded.

2. The specific energy z is equal to the energy imparted divided by the mass of the reference volume.

The specific energy is the random analog of absorbed dose. Its expectation value is the absorbed dose in the specified volume. Its actual values can deviate substantially from the expectation value.

Specific energy has sometimes been called local dose. In those cases where one deals with a cell or a cell nucleus as reference volume, the term *cell dose* has been utilized. Such terms may be illustrative, but they are here avoided in order to exclude any confusion between the random variable, specific energy, and its mean value, the absorbed dose.

3. The lineal energy y is equal to the energy imparted divided by the mean diameter of the reference volume.

The term *mean diameter* stands for mean chord length under uniform, isotropic randomness. It is the average length of the straight-line segments that result when the reference volume is randomly traversed by straight particle tracks from a uniform, isotropic field. For a sphere the mean chord length is equal to two-thirds of the diameter. For any convex region the mean chord length is equal to four times the ratio of volume and surface (22, 23).

The quantity lineal energy has been conceived as a random analog to LET, and it is conventionally expressed in the same unit, keV/μm, as LET. In view of the analogy to LET, lineal energy is utilized with reference to single events only, that is, it refers to energy deposition due to one charged particle and/or its secondaries.

It is important for many considerations in microdosimetry to define an ''energy deposition event'' as energy deposition due to correlated particles. Energy deposition in a volume due to an α particle and its δ rays is one event. When an α particle misses a microscopic volume and injects several δ rays, the combined energy imparted belongs also to the same event, because the δ rays are associated particles. In principle, energy deposition by two or more charged particles belongs also to the same event if these particles are released by the same uncharged particle; however, this case is rarely of importance because uncharged particles do not tend to produce charged secondaries in close proximity. An exception is electrons of an Auger cascade.

The notion of energy deposition events is important in microdosimetry because, according to the definition, two energy deposition events are statistically independent. This is the condition for the application of the Poisson statistics; the occurrence of one event must neither increase nor decrease the probability for further events.

The three quantities ϵ, z, and y are closely related and largely equivalent. Subsequent considerations that are phrased in one of these variables can therefore be readily translated into another.

7.3.2.2. Dose-Dependent Distributions.

When a reference volume is repeatedly exposed to the same dose of a radiation, different values of specific energy z occur. With a sufficiently large number of exposures one obtains a probability distribution of the values of specific energy. Here $f(z; D)\, dz$ is the probability that a value of specific energy between z and $z + dz$ occurs in an exposure with absorbed dose D:

$$f(z; D)\, dz = \text{Prob}\{z \leq \underline{z} < z + dz \mid D\} \qquad (7.36)$$

The function $f(z; D)$ is the probability density (differential distribution) of specific energy z at absorbed dose D. The corresponding sum distribution is linked to the density:

$$F(z; D) = \int_0^z f(z; D)\, dz \qquad (7.37)$$

The distribution $f(z; D)$ contains a delta function at $z = 0$, that is, there is always a probability $F(0; D)$ for no energy deposition. This probability is vanishingly small at sufficiently high doses where many charged particles are expected to traverse the reference volume.

The probability distributions of specific energy or of the related microdosimetric quantities are fundamental concepts of microdosimetry, although their explicit shape is not often required. In pragmatic applications it is usually sufficient to utilize basic parameters of the distributions, which will be considered subsequently. In later sections certain diagrams will be employed to illustrate the distributions for different reference volumes, radiations, and absorbed doses.

7.3.2.3. Single-Event Distributions. Energy deposition in a specified microscopic volume occurs by independent charged particles and their correlated secondaries. The increments of specific energy in individual events can vary greatly, and the probability distribution of these event sizes in a specified volume is characteristic for a type of ionizing radiation.

The function $f_1(z)$ is the probability density of specific energy produced by individual events; that is, $f_1(z) \, dz$ is the probability that an energy deposition event produces a specific energy between z and $z + dz$.

The sum distribution of the event sizes is defined in analogy to Eq. (7.37):

$$F_1(z) = \int_0^z f_1(z) \, dz \qquad (7.38)$$

In contrast to the dose-dependent distributions, there is no delta function at $z = 0$ in the single-event distribution $f_1(z)$.

7.3.3. Relations between Distributions

Microdosimetric distributions can be determined with detectors called Rossi counters. These are spherical devices with tissue-equivalent walls and an interior sensitive volume filled with tissue-equivalent gas. A central multiplication wire inside a suitable helix defines a multiplication region for the electrons that are liberated by ionizing radiations and are then collected. For the technical aspects of microdosimetry one may refer to the literature (18–20). In the initial development of microdosimetry, the dose-dependent distributions of specific energy were measured by multiple exposures of the counter to the same dose of a radiation. However, it was then realized that there is no need to measure the dose-dependent distributions, because they can be computed from distributions of the increments of specific energy produced by single events. Microdosimetric measurements need therefore be performed merely to obtain the single-event spectra. Figure 7.16 shows examples of such spectra for a spherical tissue region of diameter 1 μm and for different radiations.

In this diagram the dose-weighted distributions and their sum distributions are presented. They are defined as

$$d(y) = \frac{yf(y)}{\bar{y}_f} \qquad \bar{y}_f = \int_0^\infty yf(y) \, dy \qquad (7.39)$$

with analogous formulas for the dose-weighted distributions of specific energy per event. The subscript f distinguishes the frequency average from the average of the dose-weighted distribution.

A relatively simple connection between the single-event distributions and the dose-dependent distributions exists because the individual events are statistically independent. Due to this statistical independence, the number of events at a spec-

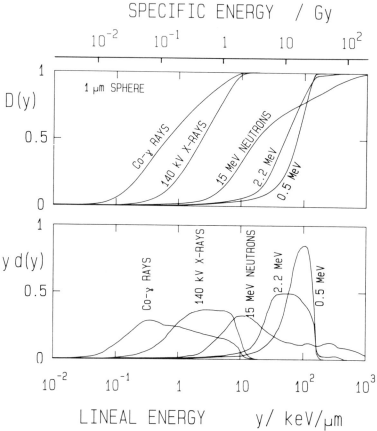

Figure 7.16. Distributions of lineal energy in spherical tissue regions 1 μm diameter exposed to various radiations. In the lower panel the distributions are represented as dose-weighted densities, $yd(y)$, relative to a logarithmic scale of lineal energy y. These spectra determine the fraction of absorbed dose delivered per unit logarithmic interval of lineal energy. In the upper panel the corresponding sum distributions $D(y)$ are given; they specify the fraction of energy delivered by events up to lineal energy y. On top of the upper panel an additional abscissa is given for the specific energy z. Relative to this scale the curves in the lower panel are the weighted densities $zd_1(z)$ of specific energy in single events; the curves in the upper panel are the sum distributions $D_1(z)$ of specific energy in single events. Data from Refs. 14, 27, 28.

ified dose follows the Poisson distribution (Section 7.2). The probability for exactly ν events is

$$p(\nu) = \exp(-n)\frac{n^{\nu}}{\nu!} \qquad (7.40)$$

where n equals $D/\bar{z}_f$, and $\bar{z}_f$ is the mean specific energy produced per event:

$$\bar{z}_f = \int_0^\infty z f_1(z) \, dz \tag{7.41}$$

The formulas for the distributions of specific energy are more complicated than Eq. (7.40) because they do not deal with a simple Poisson process that is characterized by the identity of events. Individual energy deposition events can impart greatly different energies to the target volume. For this reason one speaks of a mixed (or compound) Poisson process. If a random variable results from a mixed Poisson process, its value depends on the number of events that have taken place and on the randomly varying size of the events. The equation for the specific energy at a specified dose is

$$f(z; D) = \sum_{\nu=0}^\infty p(\nu) f_\nu(z) \tag{7.42}$$

where $p(\nu)$ is the probability for exactly ν events, and $f_\nu(z)$ is the probability distribution of specific energy that results when exactly ν events occur. The latter distribution is obtained by the operation of convolution, which is fundamental in probability theory.

When a random variable is the sum of two independent random variables, its distribution is the convolution of their distributions. The specific energy from two events is the sum of the two statistically independent increments produced in these events, and the probability distribution of specific energy for exactly two events is therefore

$$f_2(z) = \int_0^z f_1(z - s) f_1(s) \, ds \tag{7.43}$$

A straightforward extension gives the recursion formula that links the distribution for exactly ν events with that for $\nu - 1$ events:

$$f_\nu(z) = \int_0^z f_{\nu-1}(z - s) f_1(s) \, ds \tag{7.44}$$

In view of the importance of the operation of convolution, one uses an abbreviated notation instead of the explicit integral. Equations (7.43) and (7.44) read, using this notation,

$$\begin{aligned} f_2(z) &= f_1(z) * f_1(z) = f_1^{*2}(z) \\ f_\nu(z) &= f_{\nu-1}(z) * f_1(z) = f_1^{*\nu}(z) \end{aligned} \tag{7.45}$$

Instead of using the recursion formula (7.44), it is economical to compute first the convolutions that correspond to powers of 2:

$$f_2(z) = f_1(z) * f_1(z) \tag{7.46}$$

$$f_4(z) = f_2(z) * f_2(z)$$

$$\vdots \tag{7.47}$$

and then to perform appropriate convolutions of these distributions to reach any desired number. For example, the distribution for exactly 100 events is obtained as

$$f_{100}(z) = f_{64}(z) * f_{32}(z) * f_4(z) \tag{7.48}$$

In actual computations a further refinement is utilized (24–26). Its consideration will facilitate some of the developments in subsequent sections.

If two exposures are applied, each with dose D, the resulting specific energy is the sum of the independent contributions of the first and the second exposure. In other words, the distribution of specific energy for the dose $2D$ is the convolution of the distributions for dose D:

$$f(z; 2D) = f(z; D) * f(z; D) \tag{7.49}$$

Repeating the process, one can readily reach the distribution for the fourfold dose, the eightfold dose, and so on. This implies that a distribution of specific energy for high doses can be obtained from the distribution for low doses. For very low doses, however, one can give the distribution directly in terms of the single-event distribution and of a delta function at $z = 0$. If a dose d is small compared to $\bar{z}_f$, the average specific energy in one event, it is very likely that no event takes place, that is, the distribution contains a delta function at $z = 0$ with a coefficient close to 1. With small probability, $d/\bar{z}_f$, one event occurs, and this corresponds to the single-event distribution with coefficient $d/\bar{z}_f$. The probability for two or more events is proportional to the square of $d/\bar{z}_f$, that is, to a small number that can be disregarded. For a small dose d one can thus approximate the distribution of specific energy by the equation

$$f(z; d) = \left(1 - \frac{d}{\bar{z}_f}\right) \delta(z) + \frac{d}{\bar{z}_f} f_1(z) \tag{7.50}$$

To compute the distribution for a larger dose D, one chooses a small submultiple, $d = D/2^N$, and derives the desired distribution by N successive convolutions. In this way one can obtain the distribution for a mean value of 100 events by 14 convolutions starting with the dose that corresponds to only 0.0061 events on the average. By this method one generates sets of distributions for doses increasing by factors of 2. Figures 7.17 and 7.18 give such sets of distributions and sum distributions for sparsely ionizing radiations and densely ionizing radiations.

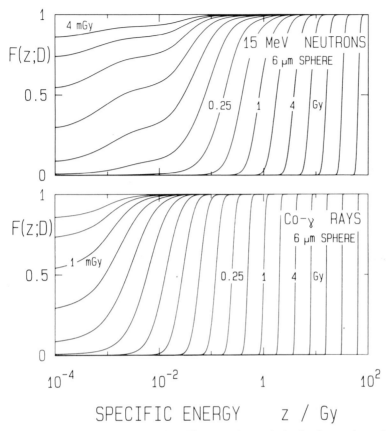

Figure 7.17. Sum distributions $F(z; D)$ of specific energy in a unit density tissue sphere of 6 μm diameter exposed to different doses of Co γ rays and 15-MeV neutrons. The distributions are calculated by the algorithm of successive convolutions (26).

A single-event distribution can extend over an extremely broad range of energy imparted. As seen, for example, in the spectrum for 15-MeV neutrons (Figure 7.16), possible values extend from a few electronvolts to several hundred kiloelectronvolts. Such distributions can be displayed on a logarithmic scale. Fast algorithms have been formulated that perform the convolution directly on the logarithmic scale (24, 26); they are more practical than the use of the fast Fourier transform, which necessitates linear scales.

7.3.4. Moments of Microdosimetric Distributions

In applications of microdosimetry one rarely requires the explicit distributions of specific energy. It is frequently sufficient to use certain parameters that characterize these distributions. Most important among such parameters are the moments of the

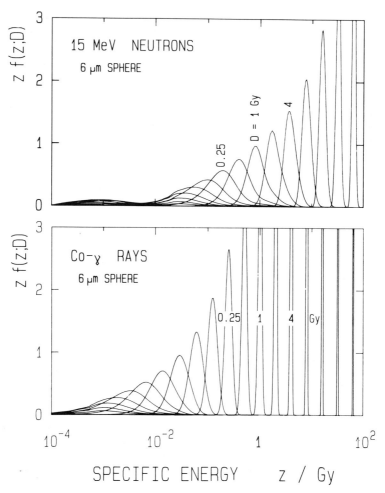

Figure 7.18. Densities $z f(z; D)$ of specific energy that correspond to the sum distributions in Figure 7.17.

distribution. The moments are the expectation values of the integer powers of the random variable. The first moment is the mean value; the second moment is closely related to the variance. The moments can be derived either for the single-event spectra or for the dose-dependent spectra, and it will be seen that they are inter-related.

The moments for the single-event spectra are

$$\overline{z_1^n} = \int_0^\infty z^n f_1(z) \, dz \tag{7.51}$$

The index 1 indicates that the quantities correspond to the occurrence of exactly one energy deposition event in the reference region. To adhere to the conventional notation, the symbol $\bar{z}_f$ is used instead of $\bar{z}_1$.

In contrast to the moments of the single-event distributions, the moments of the distributions $f(z; D)$ depend on absorbed dose. They are defined as

$$\overline{z^n} = \int_0^\infty z^n f(z; D) \, dz \tag{7.52}$$

The moments of the dose-dependent distributions can be expressed in terms of the moments of the single-event distribution (24–26):

$$\bar{z} = \int z f(z; D) \, dz = D$$

$$\overline{z^2} = \int z^2 f(z; D) \, dz = \frac{\overline{z_1^2}}{\bar{z}_f} D + D^2 \tag{7.53}$$

$$\overline{z^3} = \int z^3 f(z; D) \, dz = \frac{\overline{z_1^3}}{\bar{z}_f} D + \frac{3\overline{z_1^2}}{\bar{z}_f} D^2 + D^3$$

. . .

$D/\bar{z}_f$ is the mean number of events at absorbed dose D. The mean specific energy per event, $\bar{z}_f$, is the inverse of the event frequency per unit absorbed dose.

The relation for the second moment of the dose-dependent distribution is of special importance in biophysical considerations of second-order processes, that is, of reactions that depend on the square of the specific energy so that their yield is proportional to the expectation value of the square of specific energy. Because of its special pragmatic importance, the relation for the second moment is here derived, although reference is also made to the more general derivation of the relations (7.53) for all moments (24–26).

When a random variable such as z is the sum of two statistically independent random variables, say x and y, its variance is the sum of the variances of the individual variables. This fundamental relation is readily derived. For any random variable one has

$$\sigma^2 = \overline{(z - \bar{z})^2} = \overline{z^2} - \overline{2 z \bar{z}} + \bar{z}^2 = \overline{z^2} - \bar{z}^2 \tag{7.54}$$

For the second moment of the sum of two independent random variables one obtains (with $\overline{xy} = \bar{x}\,\bar{y}$, due to the statistical independence)

$$\overline{z^2} = \overline{(x + y)^2} = \overline{x^2} + \overline{2xy} + \overline{y^2} = \overline{x^2} + \overline{y^2} + 2\bar{x}\,\bar{y} \tag{7.55}$$

The square of the mean is

$$\bar{z}^2 = (\overline{x + y})^2 = \bar{x}^2 + \bar{y}^2 + 2\bar{x}\bar{y} \qquad (7.56)$$

Therefore,

$$\sigma_z^2 = \overline{z^2} - \bar{z}^2 = \overline{x^2} - \bar{x}^2 + \overline{y^2} - \bar{y}^2 = \sigma_x^2 + \sigma_y^2 \qquad (7.57)$$

The variances of specific energy due to two dose increments applied successively are additive, and the variance of specific energy is, therefore, proportional to absorbed dose:

$$\sigma_z^2(D) = kD \qquad (7.58)$$

The constant k can be obtained from the approximation [Eq. (7.50)] for a small dose d:

$$\bar{z} = d \quad \text{and} \quad \overline{z^2} = \frac{d}{\bar{z}_f}\overline{z_1^2} = \frac{\overline{z_1^2}}{\bar{z}_f}d \qquad (7.59)$$

Utilizing Eq. (7.54) and the symbol $\bar{z}_d$ for the mean of the dose-weighted distribution [see Eq. (7.39)], one has

$$\sigma_z^2 = \bar{z}_d d - d^2 \quad \text{where} \quad \bar{z}_d = \frac{\overline{z_1^2}}{\bar{z}_f} \qquad (7.60)$$

The quadratic term d^2 is an inaccuracy due to the omission of multiple events in Eq. (7.50); the resulting fractional error $d/\bar{z}_d$ of the variance vanishes for sufficiently small values of d. The important formula for the variance of z agrees with Eq. (7.53):

$$\sigma_z^2 = \bar{z}_d D \qquad (7.61)$$

It is essential for models of cellular radiation action that postulate a dependence of the effect on the square of the energy concentration (see Section 7.4.2).

7.3.5. Illustration of Microdosimetric Data

In Chapter 3 examples and graphic representations of charged particle tracks are given. Such descriptions provide the information that is essential for an understanding of microdosimetric spectra and for an appreciation of the substantial differences of local energy densities produced by sparsely and densely ionizing ra-

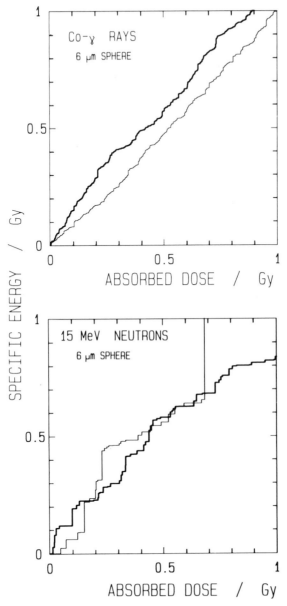

Figure 7.19. Two random paths that represent the stochastic sequence of events of energy deposition in a 6-μm tissue sphere exposed to Co γ rays and 15-MeV neutrons.

diations. When one considers distributions of specific energy or of lineal energy, one replaces the detailed information on the spatial positions of energy transfers by a description that contains some of the information implicitly. It is useful to illustrate the distributions by a series of diagrams.

Figure 7.19 refers to spherical tissue volumes with the approximate dimensions of the nucleus of a mammalian cell. Measured single-event spectra of specific energy produced by ^{60}Co γ rays and by 15-MeV neutrons in such spheres are utilized to create random simulations of the accumulation of specific energy in the sites. Each panel gives two sequences of the random process from dose zero up to 1 Gy. The statistical fluctuations are smaller for the sparsely ionizing γ rays; nevertheless, one notes that at least in one of the two random sequences (heavy line), the deviation of specific energy from its mean, the absorbed dose, happens to be sizable even at the dose of 1 Gy. Therefore, one cannot disregard the microdosimetric fluctuations of energy deposition entirely even if one deals with sparsely ionizing radiation and with the entire nucleus of the cell at the relatively high dose of 1 Gy.

The random steps of energy deposition are far larger and correspondingly less frequent with 15-MeV neutrons. For example, one of the two random sequences contains at its end (only partially represented in the diagram) an increment of spe-

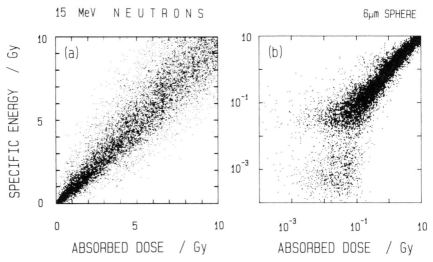

Figure 7.20. Scatter diagrams of the distribution of specific energy at specified absorbed doses in spherical tissue regions 6 μm diameter exposed to 15-MeV neutrons. In analogy to Figure 7.19 a linear scale of absorbed dose and specific energy is used in the left panel. The right panel uses logarithmic scales; it demonstrates that the relative fluctuations of specific energy decrease at larger doses. In each diagram a large number of absorbed dose values uniformly distributed on the abscissa scale is used. Each dot represents the value of specific energy from a random simulation of the exposure with the chosen value of absorbed dose.

cific energy of about 0.5 Gy. A carbon recoil of 1 MeV can produce in the nucleus of the cell an increment of specific energy in excess of 20 Gy; although this is a rare event, it illustrates the enormous range of increments of specific energy caused by energetic neutrons.

A small number of random paths can give a rough indication of the process of energy accumulation in a microscopic structure. For a quantitative evaluation one needs a large number of simulations. A superposition of many random paths would not provide a readable diagram. To obtain a better representation, one can consider varying values of absorbed dose and derive for each of these a random value of

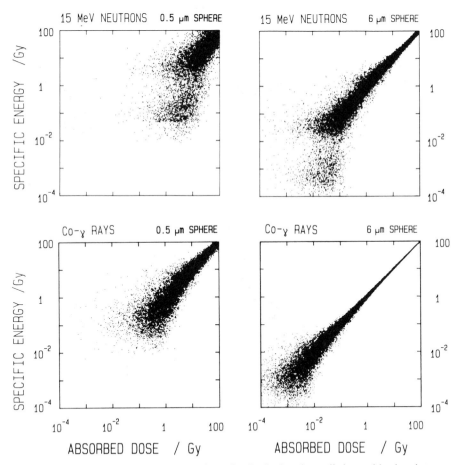

Figure 7.21. Scatter diagrams for a comparison of z distributions in small sites and in sites that correspond roughly to the diameter of the nucleus of a cell (6 μm). Here and in Figure 7.22 each panel contains 4000 simulations per decade of D. The number of plotted points is considerably less at low doses because the events with $z = 0$ are not visible.

the specific energy. In this way one obtains a cloud of points in a D–z diagram, as represented in Figure 7.20 for the 15-MeV neutrons. In the comparison with the right panel of Figure 7.19, one must note that the diagram is extended to an absorbed dose of 10 Gy. In the simulation 10,000 points are chosen with values of absorbed dose equally distributed between 0 and 10 Gy.

The diagrams of Figure 7.19 and the left panel of Figure 7.20 show that the fluctuations are largest at the highest doses. However, the absolute magnitude of the fluctuations is not the relevant parameter. In most applications the relative fluctuations are more meaningful. To judge such deviations and, at the same time, to obtain a diagram that covers a vastly larger dose range with sufficient accuracy, one can utilize a logarithmic representation as in the right panel of Figure 7.20. Such representations are also utilized in Figures 7.21 and 7.22 where the distri-

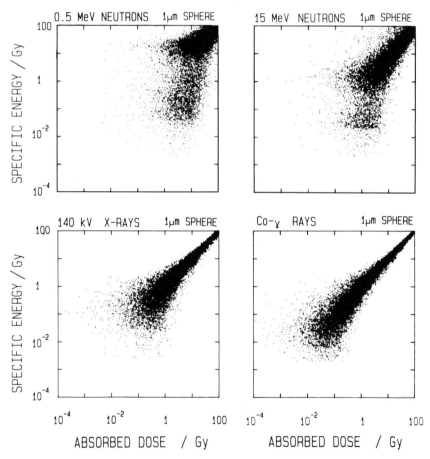

Figure 7.22. Scatter diagrams as in Figure 7.21 but for a comparison of 140-kV X rays, Co γ rays, 0.55-MeV neutrons, and 15-MeV neutrons for a spherical tissue site 1 μm diameter.

TABLE 7.1. Frequency Average $\bar{z}_f$ and Dose Average $\bar{z}_d$ of Specific Energy z (Gy) Produced by Single Events in Spherical Tissue Regions[a]

Diameter of Spherical Region, d (μm)	Type of Radiation		
		Neutrons	
	^{60}Co γ Rays	0.43 MeV	15 MeV
12	$\bar{z}_f$, 0.0004	0.018	0.016
	$\bar{z}_d$, 0.0011	0.033	0.092
5	$\bar{z}_f$, 0.003	0.24	0.10
	$\bar{z}_d$, 0.008	0.32	0.75
2	$\bar{z}_f$, 0.02	2.0	0.63
	$\bar{z}_d$, 0.07	2.8	4.9
1	$\bar{z}_f$, 0.09	10	2.6
	$\bar{z}_d$, 0.37	14	20
0.5	$\bar{z}_f$, 0.56	48	11
	$\bar{z}_d$, 2.2	62	82

[a] Values obtained by interpolation from a variety of measured and calculated data (20). With the units Gy, keV, and μm one has for spherical regions the following relations between specific energy z, lineal energy y, and energy imparted ϵ: $z = 0.204\,y/d^2$; $y = 4.9zd^2$; $y = 1.5\epsilon/d$; $z = 0.306\epsilon/d^3$. The event frequency per unit dose is $\bar{z}_f^{-1}$.

butions are compared for different radiation qualities and for different site sizes. To create these diagrams, the chosen values of absorbed dose were equally distributed on the logarithmic scale. There are few points at the small doses where the value 0 of specific energy becomes very probable; points that correspond to the absence of any absorption event do not appear on the diagram.

Although the scatter diagrams can visualize the magnitude of the fluctuations of specific energy for different radiation qualities, for different site sizes, and for different doses, they contain more information than is usually required in applications of microdosimetric data. Event frequencies and mean event sizes are more frequently utilized. Table 7.1 illustrates the magnitude of these quantities in typical cases.

7.4. MICRODOSIMETRIC MODELS OF CELLULAR RADIATION ACTION

7.4.1. Threshold Models and Their Extension

When a new technique has been developed, it is natural to seek improvements of earlier approaches. Accordingly, certain notions of target theory were revived and extended after the concepts of microdosimetry had been introduced and microdosimetric data had become available.

The target theory approaches, such as the multihit and multitarget models, were unrealistic in postulating events of equal energy deposition in certain assumed target regions. Microdosimetric data permit a more accurate description. If one wishes to retain the general concepts of the multihit models, one can replace them by threshold models in terms of microdosimetry. Such threshold models were considered during the earlier stages of the development of microdosimetry and were then replaced by models that were more meaningful and in better agreement with radiobiological findings. These more pragmatic models are dealt with in the subsequent section. A brief consideration of the threshold models is, nevertheless, required to put into perspective the instances where earlier approaches are retraced, usually with somewhat different and not always with clearly defined terminology.

The essence of the threshold models is the postulate of a sensitive site in the cell with a critical threshold of energy deposition for the cellular effect (for example, loss of proliferative ability) to occur. Although a model of this type is an abstraction, it can lead to certain firm conclusions.

In the simplest approach one assumes that the sensitive site is a sphere of specified diameter and that the critical threshold of specific energy in the site is z_c. The survival relation is then

$$S(D) = F(z_c; D) \tag{7.62}$$

where $F(z; D)$ is the sum distribution of specific energy for the radiation in question and for the specified diameter [see Eq. (7.37)]. While Eq. (7.62) is the analog of Eq. (7.29), it applies to a compound Poisson process, that is, to events of energy deposition of randomly varying magnitude. The solutions are therefore more complicated than those for the multihit models; however, as explained in the preceding section, they are readily computed from single-event spectra. For a diameter of 6 μm and for 15-MeV neutrons and Co γ rays, the functions are given in Figure 7.17. Replotting these data versus absorbed dose rather than specific energy, one obtains the dependences in Figure 7.23. They would be the dose–effect relations if the cell nucleus were the critical site and if it were to react at a sharp threshold of specific energy.

The relative variance of these dose dependences is substantially less than typical values observed in cell inactivation studies with γ rays or fast neutrons (see Figure 7.6). Either the critical targets in the cell must therefore be smaller, so that the fluctuations of energy deposition are larger, or other stochastic factors (that is, biological variability and the inherent stochastic response of the cell) must contribute largely to the observed variance of the response. Probably both conditions apply.

Considering the statistics of energy deposition as the only relevant factor, one could determine the site diameter, which yields, for a postulated threshold reaction, the relative variance of an observed cell survival function. Figure 7.24 gives the solutions for a target diameter of 1 μm. Cellular survival data either fit into the set of these curves (see Figure 7.6) or they have larger relative variances. For a

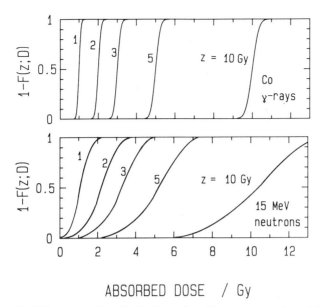

Figure 7.23. Probabilities to exceed a critical value z of specific energy in a spherical tissue site 6 μm diameter. The data correspond to those of Figure 7.17.

smaller target size the relative variances would increase; that is, they would be inconsistent with experimental observations. Any model based only on the reaction of a single target smaller than 1 μm must therefore be rejected. Equally excluded are models that invoke a multiplicity of independent targets of smaller size. On the other hand, one may postulate a larger target or a multiplicity of interdependent small targets spread over a larger region, and one can then obtain dose depen-

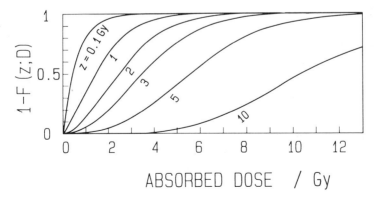

Figure 7.24. The probabilities to exceed a critical value z of specific energy in spherical tissue sites 1 μm diameter exposed to X rays.

dences with small relative variances. If the model yields a dose dependence with too little relative variance, it is still consistent with the observations since part of the observed variance can be ascribed to additional random factors not linked to the statistics of energy deposition. This argument is analogous to the earlier considerations of the multihit model. The main conclusion is that a threshold-type argument in terms of microdosimetry can provide a lower bound for the size of the gross sensitive region; the actual dimension will usually be larger.

The argument is often misinterpreted. The preceding considerations cannot exclude models that postulate smaller targets—for example, localized damage in the DNA—as the lethal event in the cell. However, to explain a nonexponential dose dependence (relative variance < 1), the model must invoke additional mechanisms that extend over longer ranges. Thus, one could postulate a dose-dependent inhibition or overload of repair. But there is, at present, no experimental evidence for a reduction of the repair capacity or the rate of repair at doses of a few gray which are relevant to cellular radiation effects. Reduced efficiency of repair or enhanced misrepair are apparent at elevated doses of sparsely ionizing radiations and at all doses of densely ionizing radiations, but they can be understood in terms of the greater proximity of sublesions of DNA and the resultant failure of DNA repair. A simple example would be two neighboring single-strand breaks on opposing strands of DNA, which interfere with excision repair. Such interference with repair due to spatial proximity of sublesions is, in a somewhat loose terminology, included in the general notion of the "interaction" of sublesions. Microdosimetric analysis provides information on the magnitude of distances involved; it cannot, by itself, identify the molecular nature of the processes.

Threshold considerations provide general conclusions without actual commitment to the reality of a threshold. Biological objects do not exhibit defined thresholds in terms of energy deposition in certain target structures. More complex models have therefore been considered in the past, and they have been revived from time to time.

These approaches postulate a spherical target region, sometimes the cell nucleus, sometimes a smaller structure. The effect probability at a certain dose is the integral of the effect probabilities for all possible values of specific energy in the target volume. Formally this is written as

$$E(D) = \int_0^\infty E(z)f(z; D)\, dz \tag{7.63}$$

If one postulates this relation and assumes that one knows, with sufficient precision, not only the microdosimetric distribution $f(z; D)$ but also the dose–effect relation $E(D)$, one can invert Eq. (7.63) and deduce the response function $E(z)$. The knowledge of this function is then expected to elucidate underlying mechanisms or, at least, to make possible a prediction of the efficiency of other types of ionizing radiations for which the microdosimetric distributions are known.

The approach may appear attractive. However, there are hidden assumptions and hidden difficulties. First, the selection of the reference target is uncertain. The only reasonable choice may be the entire nucleus of the cell. However, for sparsely ionizing radiations it is apparent from Figure 7.18 that the distributions $f(z; D)$ are narrow and that, accordingly, the function $E(z)$ is not substantially different from the observed dose–effect relation $E(D)$. The additional insight is then minimal. The fluctuations of specific energy in the entire nucleus of the cell are simply too small for sparsely ionizing radiations to play a major role at doses of one or several grays. A second weakness of the approach is an inherent assumption that underlies Eq. (7.63). It is assumed that the effect probability depends merely on the specific energy in the reference region; no account is taken of the fact that equal values of specific energy may be associated with different distributions of energy imparted within the target region. In certain extreme cases the differences can be very substantial; it is then not justified to expect the same efficiency of the radiations at equal values of z in the nucleus of the cell. The standard example is that of electrons of very short range. A 1-keV electron deposits its energy very locally in the nucleus of the cell and can thus be quite effective. This has been substantiated by various comparative studies of the effects of soft X rays and of energetic sparsely ionizing radiations (30, 31). The cellular effects of the low-energy electrons are found to be comparable to or even larger than those of higher energy electrons, although the latter produce larger values of specific energy in the cell nucleus as a whole. The effect on the cell can, therefore, not be a mere function of the specific energy in the nucleus; it depends in an insufficiently understood way on the spatial microdistribution of energy.

Further conclusions are obtained from experiments by Rossi et al. (32–34) with correlated pairs of protons or deuterons. In these experiments pairs of particles traverse mammalian cells with small lateral separations of variable magnitude. By comparison to single particles of twice the stopping power, it is found that a lateral separation of only 0.1 μm causes an appreciable reduction of the effect. On the other hand, nearly the same energy is deposited in the nucleus under the two conditions (left and central panel of Figure 7.25). Therefore, the energy in the nucleus of the cell cannot be the parameter that determines the probability of the effect on the cell.

Even for the same radiation Eq. (7.63) need not be valid. At a larger value of absorbed dose a specified value of z will more likely arise from several events than from one event. If it is produced by several events, the energy will tend to be more loosely spaced (as indicated in the diagram of Figure 7.26), and the effectiveness will then be reduced. The temporal distribution of events is an additional factor that can determine the extent of DNA repair and thus the magnitude of the observed cellular effect.

The main weakness of the approaches considered in the present section is the multiplicity of assumptions and parameters. Radiobiological data are not usually of sufficient precision to permit complex numerical analysis. Simpler approaches are required to test basic principles of the action of ionizing radiations on the cell.

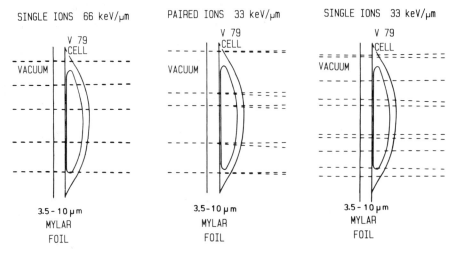

Figure 7.25. Schematic diagram to indicate the result of the molecular-ion beam experiments (adapted from Ref. 33). Protons with linear energy transfer 66 keV/μm (left panel) are substantially more effective at a given dose than pairs of protons of half the stopping power traversing the nucleus of the cell simultaneously with lateral distance on the order of 0.1 μm (center panel). Uncorrelated protons of 33 keV/μm (right panel) are still somewhat less effective.

The subsequent section deals with one particular approach which, in spite of its restricted validity, has led to tangible insights.

7.4.2. Site Model of Dual Radiation Action

7.4.2.1. General Considerations. In the preceding section notions of target theory have been reconsidered within the framework of microdosimetry. This section deals in more detail with one particular approach, the application of microdosimetry to a second-order process. The objective of the treatment is to bring out essentials.

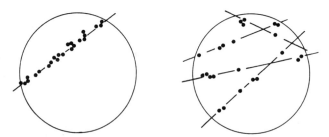

Figure 7.26. Schematic diagram to indicate that the distances between energy transfers tend to be smaller for one track segment in the site than for several track segments in the site with the same total energy transfer.

When a charged particle traverses the cell, it distributes its energy, with a degree of randomness, to an array of sensitive structures. Even if one were to assume a multiplicity of cells subjected to the same microscopic configuration of energy deposition, one could not expect equal effects. The cell responds somewhat randomly to the initial electronic alterations produced by ionizing particles, and any quantitative description of the action of ionizing radiations must therefore be of statistical nature (see also Chapter 6).

The immediate consequence is the rejection of any threshold model of radiation action. As indicated in the preceding section, threshold considerations may nevertheless be useful in the derivation of general conclusions. They can serve to exclude certain modes of radiation action. As a pragmatic description, however, a threshold model must fail. In fact, the models have been tried and rejected in earlier attempts to apply microdosimetric data—for example, to the analysis of cataracts (35) caused by neutrons or X rays or to the analysis of mutations in maize (36) produced by these radiations.

As also pointed out in the preceding section, one can postulate general empirical response functions for specific energy in certain targets [see Eq. (7.63)] and then try to derive from a set of observations, with different radiation qualities and different doses, those parameters that appear most consistent with the experiments. The approach has so many degrees of freedom that it will always yield solutions; in fact, it will usually give a wide choice of solutions that fit the data equally well. Meaningful conclusions will therefore be minimal.

A more pragmatic procedure is to start with narrow assumptions and to test them against available experimental evidence. By a stepwise process of modifications better descriptions may be attained. Unlike other models, the dual-action model was originally (29) based on assumptions narrow enough to permit conflicts with observations.

Many dose–effect relations for sparsely ionizing radiations can be adequately described by a linear-quadratic function in absorbed dose or by an exponential function that contains a linear-quadratic term in absorbed dose. The quadratic term tends to predominate with sparsely ionizing radiation, and the linear term becomes more important for densely ionizing radiations. Even in certain experimental investigations [for example, radiation-induced cataracts (37)] where the dose–effect relation, in itself, is hardly amenable to a numerical description, the relative biological effectiveness–dose relation is indicative of an underlying process that appears to be quadratic for sparsely ionizing radiations and linear for densely ionizing radiations. Various radiobiological observations are consistent with a proportionality of the relative biological effectiveness (RBE) of neutrons to the inverse square root of the neutron dose (14, 29). Some of the early observations have motivated a quantitative analysis of second-order processes in terms of microdosimetry. Results of subsequent studies have been the most tangible result of insights obtained or of experiments suggested by the microdosimetric analyses.

Lea (15) may have been the first to attempt a general treatment of second-order processes in radiation biology. He was concerned with the production of chromosome aberrations by pairs of sublesions. His essential arguments were later rephrased in terms of microdosimetry.

7.4.2.2. Notion of Concentration Applied to Nonhomogeneous Distributions.

A second-order process due to chemical agents is a simple matter when one deals with homogeneous kinetics. The yield is proportional to the square of the concentration of the reactant or to the product of the concentrations of two reactants. A condition for the simple relation is that effects of saturation or depletion can be disregarded.

With ionizing radiations a second-order process cannot result in a purely quadratic dose–response relation. The reason is the failure of the notion of concentration of energy or of any subsequent radiation products. The irregular microscopic pattern of energy deposition produced by the random appearance and random behavior of charged particles and their secondaries precludes the naive application of the notion of concentration. Conventional microdosimetry is nevertheless an attempt to apply the notion of concentration, but in a more general sense. One measures concentrations of energy over specified microscopic regions, the reference volumes of microdosimetry (see also Chapter 4). Choosing a certain scale (that is, the diameter of a spatial probe, the reference region), one samples the exposed medium, and a probability distribution of concentrations in the probe is obtained. The random variable specific energy takes the place of concentration. Probability distributions are used instead of single-valued parameters.

A second-order process results from the pairwise combination of reaction partners. If the reaction partners can diffuse sufficiently fast and if they live sufficiently long, they may react over large distances. If they are short-lived or have fixed positions, they react only over small distances. A simple example mentioned in the last section is that of two adjacent single-strand breaks on opposite strands of the DNA; with excision repair they can lead to misrepair. In general, reaction partners may have a high probability to interact if they are created a short distance apart; they have a reduced reaction probability if they are created further apart. A mathematical description is given in Section 7.5. The present section deals with a simple approximation that is not unlike the arguments utilized by Lea.

The essence of the approach is to postulate initial radiation products (sublesions) whose yield is proportional to energy imparted and that can react pairwise. A fixed reaction probability is assumed between sublesions less than a certain distance apart. Interactions beyond this distance are disregarded. On the basis of microdosimetric data for different radiations, such assumptions can be translated into dose–response relations. One can then try to identify those sizes of the reference region that agree best with experimental observations for substantially different radiation qualities. This approach is chosen in the first simple version of dual-action theory.

7.4.2.3. Solution for Second-Order Process.

Due to the irregular energy deposition throughout an irradiated medium, one must replace the notion of energy concentration by the specific energy in a reference sphere. Within the approximation described in the preceding section, one can then assume that the yield E of radiation products from a second-order process will be proportional to the square of the specific energy z:

$$E(z) = cz^2 \tag{7.64}$$

The term *sensitive site* (or *target region*) has occasionally been used. However, this can be misleading. The image of reaction partners moving in an extended region and of sampling with a spherical probe may be more pertinent than that of geometrically defined sites that contain sublesions.

Equation (7.64) applies to one value of the random variable. The observable effect is the average over the distribution of the random variable. As for any random variable, the expectation of the square is equal to the square of the mean of the variable plus its variance [see Eq. (7.54)]:

$$E(D) = c\overline{z^2} = c(\sigma_z^2 + \bar{z}^2) \tag{7.65}$$

The relations for the mean and the variance of the specific energy have been obtained in Section 7.3 [Eqs. (7.53) and (7.61)]:

$$\bar{z} = D \tag{7.66}$$

$$\sigma_z^2 = \zeta D \tag{7.67}$$

where the more usual notion ζ is used, instead of $\bar{z}_d$, for the weighted mean event size:

$$\zeta = \frac{\displaystyle\int_0^\infty z^2 f_1(z)\,dz}{\displaystyle\int_0^\infty z f_1(z)\,dz} = \frac{\overline{z_1^2}}{\overline{z_f}} \tag{7.68}$$

The resulting dose dependence is

$$E(D) = c(\zeta D + D^2) \tag{7.69}$$

This is the main result of the original version of the dual-action model. A purely quadratic dependence on specific energy is transformed into a dose dependence that contains an additional linear term. The linear term is due to the intratrack reactions that occur even at very small doses. The quadratic term is due to the intertrack reactions, which predominate, at least for sparsely ionizing radiations, at larger doses.

The essence of the result is that energy concentrations in the cell or in smaller subcellular sites cannot be arbitrarily low even at very low doses of ionizing radiations. Within the track of charged particles there are always finite energy concentrations. When the dose is sufficiently low, the tracks tend to be separated, and all energy transfers have then in their proximity energy transfers only from the same track. Their statistical frequencies depend on the ionization density within the tracks but not on absorbed dose. The coefficient for the linear term in Eq. (7.69) is the weighted average of the values of specific energy produced in the reference site by individual events, that is, by individual charged particles.

To replace the formal derivation in Section 7.3.4 by an intuitive explanation, one can utilize the diagram of Figure 7.27. In a second-order process a potential reaction partner (an ionization, or a subsequent radiation product) has an efficiency that is proportional to the average number of reaction partners within the sphere of possible interaction. Potential reaction partners can be on the same particle track or on other tracks. The expected number of partners on the same track is proportional to the average event size ζ. This quantity needs to be a weighted average because an energy transfer chosen at random is more likely to be found in those events that contain more ionizations. The average number of reaction partners not on the same track is independent of radiation quality and is merely proportional to absorbed dose. The expected yield is proportional to the product of energy transfers and their expected reaction partners. The first quantity is proportional to absorbed dose, and the second quantity is proportional to the sum of ζ and the absorbed dose:

$$E(D) \propto D(\zeta + D) \tag{7.70}$$

It may appear paradoxical that the average energy density in the vicinity of an energy transfer should be larger than the absorbed dose. Naively one may feel that the contribution of independent tracks had to be reduced to compensate for the presence of the reference track. However, this is a misconception that arises when one disregards the difference between weighted and unweighted sampling. The

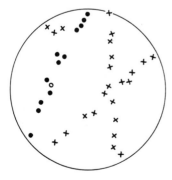

Figure 7.27. Schematic diagram to indicate the intratrack contribution (dots) and the intertrack contribution (crosses) in the same site as a randomly selected transfer (open dot). The contribution of the intratrack term to the specific energy is ζ; the contribution of the intertrack term is D.

point can be explained by the example of a Poisson process of pairs of points. Consider pairs of energy transfers placed with uniform randomness on the plane. For a random point in the plane the circumscribed circle of radius r will contain $kr^2\pi$ dots on the average, where the constant k is the mean number of dots per unit area. This process is called unweighted sampling. One can also perform weighted sampling and select not a random point of the plane but one of the dots (energy transfers). According to the definition of a Poisson process, the presence or absence of other events (pairs of energy transfers) remains uninfluenced by the presence of the selected event. The circle centered at a dot (or a circle positioned randomly over a dot) will always contain the selected dot and, with a probability, p, which depends on the spacing of the dots in a pair and on the radius r of the circle, it will also contain the associated second transfer. Jointly the term $1 + p$ corresponds to the intratrack term ζ. Due to the statistical independence, the expected contribution from other tracks, that is, the analog of the intertrack term, will be $kr^2\pi$. With weighted sampling the average number of transfers contained in a disk is $1 + p + kr^2\pi$.

The analogous argument applies also to a more complicated compound Poisson process. The essential point is that the intratrack term ζ in Eq. (7.70) does not reduce or influence the intertrack term D.

7.4.2.4. Dose–Effect Relation and RBE–Dose Relation. The linear-quadratic dependence on absorbed dose need not always be the direct result of experimental observations. In cell survival studies, for example, $E(D)$ is not simply the probability for cellular inactivation. Instead, one equates $E(D)$ with the negative logarithm of the survival probability. The resulting relation,

$$S(D) = \exp\left[-E(D)\right] = \exp\left(-aD - bD^2\right) \tag{7.71}$$

is in good agreement with many experimental studies. In other systems the dose dependences are more complicated. For example, in lens opacification studies the severity of effect is measured on an arbitrary numerical scale that is reproducible but provides not more than a ranking of the level of effect. In studies on radiation tumorigenesis similar problems arise, as there are different possibilities to quantify the increase of tumor rates after irradiation. Generally the dose–effect relation need not reflect directly the dose dependence of the underlying cellular damage. However, one can study RBE–dose dependences rather than dose–effect relations. This approach, introduced by Rossi (38), has been applied in many radiobiological studies. It has become an important tool in radiation biophysics for the comparison of the effectiveness of different types of ionizing radiations at specifically low doses.

The RBE of a type of radiation equals the ratio of the dose of a reference radiation (usually γ rays or X rays) to the dose of the specified radiation that produces the same level of effect. In studying RBE one assumes that the observed severity H of the effect, say, for X rays and neutrons, depends on the underlying and not directly observable cellular damage that is a linear-quadratic function of absorbed dose:

$$H_x(D_x) = H(E(D_x)) = H(\zeta_x D_x + D_x^2)$$
$$H_n(D_n) = H(E(D_n)) = H(\zeta_n D_n + D_n^2) \qquad (7.72)$$

Equality of the observed effect implies equality of the arguments in these equations, and one therefore has the relation between the equivalent doses:

$$\zeta_n D_n + D_n^2 = \zeta_x D_x + D_x^2 \qquad (7.73)$$

From these equations one obtains the functional dependence of the neutron RBE R_n on neutron dose D_n:

$$R_n = \frac{2(\zeta_n + D_n)}{\zeta_x + [\zeta_x^2 + 4(\zeta_n + D_n) D_n]^{1/2}} \qquad (7.74)$$

In the present context it is not necessary to deal with particularities of this dependence (14, 29). An important simplification is, however, the relation that results in the dose range where the underlying dose dependence for densely ionizing radiations is still linear, while it is effectively quadratic for sparsely ionizing radiations. One obtains

$$R_n = \sqrt{\frac{\zeta_n}{D_n}} \qquad (7.75)$$

This relation, the inverse proportionality of neutron RBE to the square root of neutron dose, has been a guiding principle in experimental studies not only of cellular but especially of tissue effects, such as radiation tumorigenesis. The investigations have uncovered high values of the RBE of neutrons that agree with the microdosimetric considerations, while they exceed substantially the value 10 of the quality factor utilized for purposes of radiation protection (14). Figure 7.28 represents the RBE–dose dependence according to Eq. (7.74) for the parameters $\zeta_n = 20$ Gy and $\zeta_x = 0$, together with observed relations for lens opacification in the mouse (37) and mammary tumors in the rat (39, 40). (For additional data see Ref. 14.)

In spite of the broad experimental evidence for Eq. (7.75), one cannot exclude the possibility that certain radiation effects have a more complex dependence on specific energy. Although this case may be of less pragmatic importance, it is useful to give the solutions. Utilizing the relations for the moments in Section 7.3.4 and assuming the dependence of the effect probability on specific energy,

$$E(z) = a_1 z + a_2 z^2 + a_3 z^3 \qquad (7.76)$$

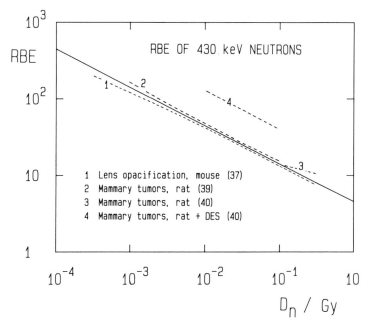

Figure 7.28. Relative biological effectiveness of 430-keV neutrons as a function of neutron dose for lens opacification in mice (37) and for the induction for mammary tumors in Sprague–Dawley (39) and in ACl rats (40). The solid line corresponds to Eq. (7.74) with $\zeta_n = 20$ Gy and $\zeta_x = 0$.

one obtains the dose dependence

$$E(D) = (a_1 + a_2\zeta + a_3\eta)D + (a_2 + 3a_3\zeta)D^2 + a_3D^3$$

where

$$(7.77)$$

$$\zeta = \frac{\overline{z_1^2}}{\overline{z}_f} \quad \text{and} \quad \eta = \frac{\overline{z_1^3}}{\overline{z}_f}$$

Higher-order terms could be dealt with similarly. In the considerations of this chapter the general solution will not be required. However, the possible existence of a linear term in specific energy needs to be taken into account in the interpretation of observed linear-quadratic dependences.

7.4.2.5. Site Model and Proximity Model. When an energy transfer or, for simplicity, an ionization happens to be contained in the sampling sphere of diameter d, and if another energy transfer occurs at distance x from it, the latter has a certain probability $U(x)$ to be contained in the same sampling sphere. Here $U(x)$ tends to unity if x is much smaller than d. For larger values of x the function $U(x)$ decreases, and it is zero for x larger than d. The function $U(x)$ is called the geometric reduction factor; it is an important concept that applies not only to spheres but also to other configurations. The term $U(x)$ is the probability that a random shift in random direction from a random point in the specified object S leads to a point that is still contained in S. For a sphere the solution is

$$U(x) = 1 - \frac{3x}{2d} + \frac{x^3}{2d^3} \qquad 0 \le x \le d \qquad (7.78)$$

This function is given in Figure 7.29. The function $U(x)$ is closely related to the distribution of distances of two random points in the sphere, and this is dealt with in more detail in Section 7.5. The essential point in the present context is that the dual-action model in its simplest form can also be interpreted in terms of a proximity model, that is, a model that applies to a homogeneous site and that postulates not certain critical sites but merely a distance-dependent interaction probability. The simple case of a spherical site corresponds to a distance dependence that is proportional to the function $U(x)$ given in Eq. (7.78) and represented in Figure 7.29. From the standpoint of a proximity model this particular dependence is arbitrary; however, it is not more arbitrary and certainly less unlikely than an assumed threshold dependence (broken line in Figure 7.29). The notion of a threshold dependence was inherent in the treatment by Lea (15), who assumed that biological sublesions (chromosome breaks) can combine if they are formed at a distance closer than a critical value.

There is no reason to assume that either of the two functions in Figure 7.29 equals actual interaction or combination probabilities of sublesions in the cell. The analysis in terms of Lea's proximity model or in terms of the site model is merely an approximation to bring out general characteristics of dose–response relations that may result with different radiation qualities. The simplest proximity model was a convenient choice for the considerations Lea formulated in terms of the LET concept. The site model, on the other hand, is the natural choice for a treatment

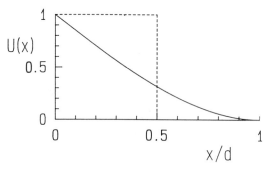

Figure 7.29. Geometric reduction factor $U(x)$ for a sphere of unit diameter. Broken line indicates the distance dependence of the interaction probability invoked in the proximity model of Lea (15).

that utilizes microdosimetric data obtained experimentally with spherical detectors. In spite of their differences, the two models are largely equivalent first approximations. More sophisticated analyses based on different distance dependences are considered in Section 7.5.

7.4.2.6. Limitations of Site Model.

The approximations inherent in the simple form of the dual-action model were noted even in the initial formulation. The most significant simplification is the assumption of a constant interaction probability of energy transfers or sublesions within the assumed site. The assumption is arbitrary. One may, nevertheless, expect that deviations from actual distance dependences should be minor if the compared radiations produce charged particles with ranges exceeding cellular dimensions. Gross inadequacies of the model arise when such radiations are compared to soft X rays, which produce electrons of very short ranges. Nevertheless, even for conventional radiations the comparison of the intratrack and intertrack mechanisms can be biased by the disregard of an increased interaction of energy transfers in close proximity. The reason is that close proximity of transfers within the site is more likely for events on the same track (Figure 7.26). The site model is therefore likely to underestimate the linear intratrack component of the effect.

A further feature of the simple model is the neglect of processes of depletion or competition. The quadratic dependence on specific energy implies the absence of such factors. The simple formulation is valid if there are many potential reaction partners and if only a few of them interact pairwise toward the observed effect. These conditions may largely be fulfilled. For radiation effects such as chromosome aberrations the average lifetime of sublesions, governed by repair, is short enough that only a minor fraction combines. Enzymatic repair processes are highly efficient, at least with sparsely ionizing radiations. Even with densely ionizing radiations a majority of initial lesions is likely to be repaired.

In principle, one can attempt more complex formulations that account for competition of reaction partners. Such approaches would be indicated if one dealt with well-known specific molecular or structural mechanisms. At present the necessary knowledge has not been reached in radiobiology. It is therefore not justified to add minor modifications to the theory, while one retains the crude assumption of spherical sites or of similarly simple geometries, which do not correspond to the reality of the cellular organization.

There is a restriction in the interpretation of the linear component of the dose-effect relation. Deducing a value $\zeta = a/b$ from a linear-quadratic dependence $aD + bD$ and equating it with the weighted mean event size, one disregards a possible component of the effect probability that is inherently linear. The relation

$$E(z) = a_1 z + a_2 z^2 \tag{7.79}$$

corresponds, in agreement with Eq. (7.77), to the dose dependence

$$E(D) = (a_1 + a_2 \zeta)D + a_2 D^2 = aD + bD^2 \qquad (7.80)$$

where

$$a = a_1 + a_2 \zeta \ , \ b = a_2 \ , \ \frac{a}{b} = \frac{a_1}{a_2} + \zeta$$

The term a/b is therefore merely an upper bound for ζ. Smaller values of ζ corresponding to larger site diameters cannot be excluded. In analogy to considerations in the preceding section, one concludes that any formally derived target size or effective interaction distance is a lower limit. Actual distances will usually be larger.

7.4.2.7. General Considerations on Dual Action.

The dual-action model describes a possible mechanism that is consistent with many observed linear-quadratic dose–effect relations and, specifically, with observed dose dependences of RBE of densely ionizing radiations. However, agreement between predicted dose and RBE–dose dependences and radiobiological observations cannot be taken as proof for specific postulates of the model.

The dual-action model is broad. The formalism applies to any second-order process. Such processes can either be due to the combination or accumulation of two lesions or to the combined reaction of two radiation products. The term *interaction* includes reactions of free radicals, energy transport along macromolecules and the resultant production of closely spaced DNA lesions, and the migration and reaction of sublesions, such as chromosome instabilities or breaks. There is as yet no firm evidence which of these mechanisms or which combination of the mechanisms is most relevant. It has also been concluded that the deduced formal interaction distances have only qualitative meaning, and that one deals with interaction or damage accumulation distances that range from nanometers to micrometers (see Section 7.5.4).

Furthermore, the assumed mechanisms and the corresponding equations are of such a general type that they are open to an even wider range of interpretation. In particular, one cannot reject the possibility that one deals not with the pairwise reaction of radiation products or sublesions, but instead with a combined process of highly localized potential lesions and dose-dependent probabilities for repair and/or fixation. Such an interplay of two processes can also have the properties of a second-order reaction.

A dose dependent slowdown or impairment of repair ability has occcasionally been invoked as an alternative explanation of the shoulder of the survival curve or other dose–response relations. However, there is little or no evidence for an impairment of enzymatic repair processes at doses of a few grays. Studies, for example by Virsik et al. on chromosome aberrations (41), have established characteristic repair times that are substantially constant up to 10 Gy, that is, up to the highest doses investigated. Similar observations have been obtained in various cell survival studies. Most of the enzymatic DNA repair processes that are known are of the catalytic type. The enzymes are not used up in the repair process, and under usual conditions it is safe to assume that the concentration of enzymes is sufficient to maintain constant repair efficiency at the concentration of lesions produced by doses of several grays. An exception is the repair enzyme for alkylated DNA bases; it works stoichiometrically, that is, by a suicide reaction. With alkylating agents one obtains a shoulder of the response curve due to the gradual disappearance of the repair enzymes. However, it is doubtful whether similar processes are relevant with ionizing radiations.

Another factor of possible influence on the shoulder of a dose–response curve could be the dose-dependent temporary suppression of DNA synthesis or of cell division. It appears that potentially lethal radiation damage in mammalian cells is partly fixed in an S phase and partly in mitosis. By radiation-induced delay of DNA synthesis or of cell division, a protective effect is achieved, as more time becomes available for repair. If this mechanism were relevant at small and moderate doses but less effective at high doses, it could codetermine the shoulder of the survival curve. Similar effects have been seen, although their role is still uncertain, in studies with Ataxia cells. It has been surmised that the increased sensitivity of these cells to ionizing radiation may not be exclusively due to decreased DNA repair ability, but may also be related to the absence or partial absence of the radiation-induced delay of DNA synthesis.

A detailed analysis would be required to determine the possible influence of such mechanisms, specifically of cell kinetic changes, due to irradiation and any concomitant changes of the proportion of repaired DNA damage. One of the relevant questions in the analysis of such mechanisms is whether the induced delay of DNA synthesis is a dose-dependent process that affects the nucleus of the cell as a whole or whether it is a spatially nonuniform process dependent on the microscopic distribution of energy. There are other potential mechanisms of similar complexity. Present knowledge of the molecular mechanisms is insufficient to verify or reject them. However, any quantitative treatment that may ultimately be reached will have to account for the microdosimetric distributions of energy that determine the biological effectiveness of ionizing radiations.

7.5. PROXIMITY FUNCTION AND ITS APPLICABILITY

7.5.1. Rationale for Distance Model

In the initial formulation of dual radiation action a weak dependence of the interaction probability on distance has been postulated. This assumption has subsequently been rejected on the basis of two groups of experiments. One group consists, as pointed out in Section 7.4, of studies with low-energy photons, which produce short-range electrons with energies between a few hundred electronvolts to several kilo-electronvolts (30, 31). These particles are substantially more efficient than Eq. (7.69) would predict with the parameter ζ for site sizes near 1 μm.

One concludes that energy transfers in close proximity are particularly effective. In other words, the interaction probabilities over small distances are substantially larger than those for larger distances. Interaction over larger distances, with greatly reduced probability, must nevertheless be assumed to explain the curvature of dose–response relations obtained with sparsely ionizing radiations.

A combination of pairs of reactants or of sublesions would not need to be postulated if one could invoke other factors (such as the ones mentioned in Section 7.4.2) to explain the curvature of the dose relation. But the latter processes would also be inherently quadratic, and the interplay of the two modes of damage would have to be quantified in terms of spatial and temporal interrelations.

The second group of experiments, namely the investigations of Rossi et al. (32–34), with spatially correlated heavy ions also showed that the interaction probabilities are greatly increased for energy transfers in close proximity, that is, at distances less than 100 nm. On the other hand, one found, both for cell killing and for chromosome aberrations, that the correlated protons, although separated by distances on the order of 100 nm, were more efficient than the same particles with no spatial correlation (see Figure 7.25). In this way the experiments have not only demonstrated the increase of the interaction probability with increasing spatial proximity of energy transfers but they have also shown the presence of a small interaction probability extending to larger distances. This latter finding is in line with the conclusions drawn earlier (6, 12, 15) from the curvature in dose–effect relations for cell survival or for the production of chromosome aberrations. But the experimental evidence is somewhat more specific. It proves that the effect is not merely a gross dose-dependent process (for example, reduction of repair) but is also a result of the spatial proximity of two charged particle tracks.

The evidence from the soft X ray experiments and the correlated ion studies has motivated a further development of the model of dual radiation action and a formal treatment that takes into account a distance-dependent interaction probability between energy transfers or sublesions (42). A precondition for the modified treatment was the earlier development of new microdosimetric concepts that are more closely linked to the computational approach than to microdosimetric measurements (44, 45). The biophysical considerations are related to the problem of the random intercept of geometric objects. Essential results are, therefore, introduced first in purely geometric terms. They can then be applied readily to the energy deposition problem.

Energy deposition in a cellular structure by a charged particle is a stochastic process that can be seen as the random intersection of two geometric objects, the site S and the particle track T. The site is assumed to be part of a uniform medium and, furthermore, the radiation field is taken to be uniform. A particle track is the random configuration of energy transfers produced by a charged particle and/or its secondaries. Figure 7.30 gives a diagram that explains the notion of energy transfers ϵ, that is, the individual energy deposits, which may either be ionizations or excitations. The term *particle track* denotes the set of all transfers produced by a charged particle and its secondaries. Each transfer point is ascribed a value of

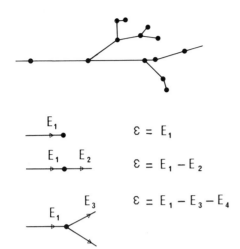

$$\varepsilon = E_1$$

$$\varepsilon = E_1 - E_2$$

$$\varepsilon = E_1 - E_3 - E_4$$

Figure 7.30. Schematic diagram of a segment of a charged particle track with indication ($\bullet$) of the transfer points and the corresponding energy transfers ϵ.

energy that equals the energy of the incident ionizing particle minus the energy of the emerging ionizing particle(s). These notions are utilized in Section 7.5.3. For the present geometric considerations it is sufficient to note that the site is a certain domain and that the track is also a geometric object. Frequently, as in the preceding sections, a simple geometry is postulated for the site. However, the subsequent considerations apply generally to domains that may be of complicated shape and need not be simply connected. The particle tracks are subdimensional structures, that is, sets of points. Nevertheless, one could visualize each transfer point as a small sphere with volume proportional to the corresponding energy transfer. Thus, a simple analogy between energy and volume is established, and the geometric theorems can readily be applied to the biophysical problem of the interception of cellular structures by charged particle tracks.

7.5.2. Mean and Weighted Mean of Random Overlap of Geometric Objects

7.5.2.1. Random Intercept of Two Domains. Assume that two geometric objects S and T are randomly superimposed in the sense of isotropic, uniform randomness, This type of randomness, also termed μ randomness, corresponds to the situation where one of the objects is in a fixed position while the other object has a uniformly and isotropically distributed random position. A different concept termed weighted randomness, or ν randomness, will also be considered. The two types of randomness were originally introduced in the context of random processes of straight lines (43). A more general use of the concepts in microdosimetric calculations (44, 45) led to a definition for arbitrary geometries (46).

Figure 7.31 illustrates the two types of randomness. Here S is represented by a circular site and T by linear tracks of variable size to indicate the general case. For

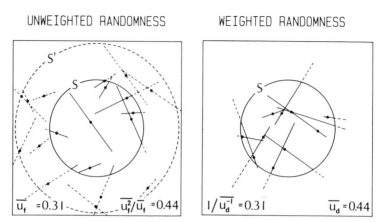

Figure 7.31. Linear tracks of length 0.25 and 0.75 intercepting a unit diameter circular site S. The intercepts with S are drawn as solid line segments. The dots are the random points that determine the position of the tracks according to the definition of the two types of randomness (see text). The mean intercept, 0.31, and the weighted mean intercept, 0.44, result with the two types of sampling as indicated in the diagrams.

μ randomness the object T is assumed to have a reference point C taken to be the center in the tracks of Figure 7.31 (left panel). Consider a region S' that contains S and all potential positions of C that permit an overlap between S and T. A random point R uniformly distributed in S' is chosen, and T—or, in the general case, an element of the set of objects T—is positioned with random orientation so that its reference point C coincides with R. Repeating this procedure and disregarding all events with no intercept of S and T, one obtains a probability distribution $f(u)$ of size u of the intercepts $U = T \cap S$. In the example of Figure 7.31 (left panel) there are 10 intercepts.

Weighted randomness (ν randomness) is obtained as follows. Two independent random points P_S and P_T are selected that are distributed uniformly and independently in S and T. If there is a set of objects T, each element has a selection probability proportional to its measure. The relative frequency of the larger objects is therefore enhanced in weighted sampling. Although short and long tracks are taken to be equally frequent in the example, there are far more long tracks in the right panel of Figure 7.31, which corresponds to weighted sampling. The selected object T is positioned with random orientation so that the points P_S and P_T coincide. With this procedure [originally introduced for sampling procedures in microdosimetric computations (44)] there is always a positive intercept u. In an extension of the earlier result of Kingman (43) for straight lines, it has been shown (46) that the distribution $d(u)$ of intercepts under weighted randomness is related to the distribution of intercepts for uniform, isotropic randomness:

$$d(u) = \frac{u f(u)}{\displaystyle\int_0^{u_{\max}} u f(u)\,du} = \frac{u f(u)}{\overline{u}_f} \tag{7.81}$$

The relation holds for arbitrary shapes of the geometric objects S and T. Weighted sampling can therefore be used as an efficient procedure to obtain not only the distribution $d(u)$ but also the distribution $f(u)$ and its moments. In Figure 7.31 two important identities are indicated for the unweighted and the weighted mean intercept:

$$\bar{u}_f = \frac{1}{\overline{u_d^{-1}}} \quad \text{and} \quad \frac{\overline{u_f^2}}{\bar{u}_f} = \bar{u}_d \tag{7.82}$$

The weighted average $\bar{u}_d$ is always larger than the unweighted average $\bar{u}_f$. The difference determines the variance of the intercept under μ randomness:

$$\sigma^2 = \overline{u_f^2} - \bar{u}_f^2 = (\bar{u}_d - \bar{u}_f)\bar{u}_f \tag{7.83}$$

Blaschke (47) and Santalo (48) have derived an equation, the fundamental kinematic formula, which determines, among other quantities, the mean intercept $\bar{u}_f$:

$$\bar{u}_f = \frac{V_S V_T}{A} \tag{7.84}$$

where V_S is the volume of S and V_T the measure of T, while A is the volume of the Minkowski product [also called dilatation (49)] of S and T. In the present case, where isotropic randomness is considered, the volume of the Minkowski product is averaged over all directions (for an elementary treatment of these matters see Ref. 50). If S is a spherical region, A is equal to the associated volume of T, one of the important notions introduced by Lea (15) to radiation biophysics.

A further result is of great generality and of special importance in the context of the present chapter. This is the relation for the mean overlap $\bar{u}_d$ of S and T under weighted randomness (46):

$$\bar{u}_d = \int_0^{x_{\max}} \frac{t(x)\, s(x)}{4\pi x^2}\, dx \tag{7.85}$$

where $x_{\max}$ is the maximum point pair distance in S or T, whichever is smaller. The two functions $s(x)$ and $t(x)$ are called proximity functions of S and T; they could also be called spatial autocorrelation functions. The proximity functions are equal to the probability distributions of distances between pairs of independent random points in the objects multiplied by the volume of the objects. Thus, $s(x)$ dx/V is the probability that two random points in S are separated by a distance between x and $x + dx$. If a geometric object is a random configuration, its proximity function is the volume-weighted average of the proximity functions for all possible configurations.

Equation (7.85) is the reason that the proximity functions are important characteristics of geometric objects. The functions have been derived for simple geometries such as spheres, spheroids, spherical shells, cylinders, cubes, or slabs (e.g., Ref. 43). For more complicated geometries such as charged particle tracks Monte Carlo computations may be required (Section 7.5.3).

If a domain is not subdimensional, it is often convenient to utilize a related quantity, the geometric reduction factor, a concept introduced originally by Berger (51) in the context of dosimetric computations:

$$U(x) = \frac{s(x)}{4\pi x^2} \tag{7.86}$$

As stated in Section 7.4, $U(x)$ is equal to the probability that a random point in S remains in S after it is shifted by the distance x in a random direction.

7.5.2.2. Intercept of Site by Poisson Process of Geometric Objects.

In many applications one deals with the intercept of a domain S by a Poisson process of geometric objects, T. The Poisson process, also called a Boolean scheme (52), results from a uniform, isotropic field of objects, T, with a density of λ objects per unit volume. Nominal coverage is $\psi = V\lambda$, where V is the mean volume of T. One will realize in Section 7.5.4 that ψ is the analog of absorbed dose D.

The volume u of the intercept of the Poisson process with S can be defined either as the sum of the volumes of all intercepts or as the volume of the union of all intercepts. The former case is considered first and is referred to in the subsequent sections. It corresponds to a compound Poisson process. The expected intercept is

$$\bar{u}(\psi) = V_S\psi \tag{7.87}$$

and with Eq. (7.53) for the compound Poisson process one obtains the second moment:

$$\overline{u^2}(\psi) = [\bar{u}_d + \bar{u}(\psi)]\bar{u}(\psi) = (\bar{u}_d + V_S\psi)V_S\psi \tag{7.88}$$

This is the geometric analog of the basic result for dual radiation action.

A more complicated result is obtained (46) when u is the measure of the union of all intercepts, that is, when multiple overlap is not weighted with the corresponding multiplicity:

$$\bar{u}(\psi) = V_S[1 - \exp(-\psi)]$$
$$\overline{u^2}(\psi) = [u* + \bar{u}(\psi)]\bar{u}(\psi) \tag{7.89}$$

where

$$u* = \frac{\displaystyle\int_0^{x_{max}} s(x)[\exp(\psi t(x)/4\pi x^2) - 1]\, dx}{\exp(2\psi) - \exp(\psi)}$$

Although this result is related to possible formulations of the saturation problem in cellular radiation action, too little is yet known about actual mechanisms to make quantitative formulations worthwhile. The result is here cited in view of its generality and the potential applications to a broad range of problems of stochastic geometry. It is also given to elucidate the point that, in the presence of saturation, a second-order process with nonhomogeneous kinetics need not lead to a linear-quadratic dependence on dose.

7.5.3. Application to Energy Deposition Problem

7.5.3.1. Proximity Function of Energy Transfers. The results of the preceding section can be applied to the energy deposition problem to obtain a generalized formulation of dual radiation action. Furthermore, apart from any particular application, the proximity function for a radiation field affords a fundamental characterization of radiation quality that extends the LET concept and links it to the conventional microdosimetric quantities.

The concept of the proximity function $t(x)$ of particle tracks is analogous to that of the geometric proximity function, with the modification that energy replaces volume. In analogy to the volume proximity function, the energy proximity function $t(x)$ can be understood as the distribution of pair distances of energy transfers multiplied by the total energy of the track. The proximity function includes a delta function at $x = 0$ that is proportional to the weighted mean of individual energy transfers ϵ_i. More formal definitions have been given elsewhere (14, 20, 45).

The actual computation is best explained in terms of the integral proximity function

$$T(x) = \int_0^x t(s) \, ds \tag{7.90}$$

From a simulated track $T(x)$ is obtained by considering all pairs of transfer points that are separated by a distance less than x:

$$T(x) = \frac{\sum_{i,k} \epsilon_i \epsilon_k}{\sum_i \epsilon_i} \tag{7.91}$$

where the summation runs over all i and over all transfer points k separated by distance up to x from the transfer point i. The value $T(0)$ determines the delta function $T(0) \, \delta(x)$ in $t(x)$. It results from the pairs with $i = k$:

$$T(0) = \frac{\sum \epsilon_i^2}{\sum \epsilon_i} \tag{7.92}$$

Equation (7.91) implies that $T(x)$ equals the expected energy on the particle track in a sphere centered on a randomly selected energy transfer. Accordingly, $t(x)\,dx$ is the expected energy within the distance interval x to $x + dx$ from a randomly selected energy transfer.

One can also consider the proximity function that contains the trivial dose-dependent intertrack term from unrelated particles:

$$t(x;\,D) = t(x) + 4\pi x^2 \rho D \tag{7.93}$$

With the density $\rho = 1$ g/cm^3 and with the units electronvolts, nanometers, and grays one obtains the following relation, which permits a convenient comparison of the relative magnitude of the intratrack and intertrack contributions:

$$t(x;\,D) = t(x) + 7.82 \times 10^{-5}\,x^2 D \tag{7.94}$$

In computations the proximity function is derived by sampling all pairs of energy transfers in a sufficiently large number of simulated particle tracks. If the particle tracks have different initial energies, the proximity function is the energy-weighted average for the different tracks. Proximity functions have been computed for electrons and for track segments of heavy ions (44, 53–56). Figure 7.32 represents the differential proximity function $t(x)$ for electrons of energy up to 10 keV. The upper panel contains also the dose-dependent intertrack term for doses of 10 and 100 Gy. The comparison shows clearly that damage accumulation over short distances is entirely determined by the intratrack term even at high doses.

In the simplest LET approximation one pictures the tracks as straight lines with an average value of LET, which ought to be not the track average $\bar{L}_f$ but the dose average $\bar{L}_d$. The proximity function is then a constant:

$$t(x) = 2\bar{L}_d \tag{7.95}$$

This is not a good approximation. The approximation in terms of linear tracks with LET varying according to the continuous slowing-down (CSD) approximation is also of limited applicability, as shown in Figure 7.33. At small distances the LET approximation indicates energy concentrations that are too low because no account is taken of energy loss straggling, that is, the discontinuous structure of the electron tracks. The fluctuations of energy loss must therefore be taken into account if the LET concept is to be applied to electrons. On the other hand, one sees from the figures that the correlated energy transfers at larger distances are overestimated with the LET concept and linear tracks. This is due to the neglect of angular scattering, which leads to curled electron tracks. However, one obtains a fairly good approximation to the proximity functions when one depicts the electron tracks as linear segments with continuous, constant energy loss and two-thirds of their CSD range.

For track segments of heavy ions the influence of straggling and radial energy distribution can be separated (44, 45). The proximity function is then the sum of

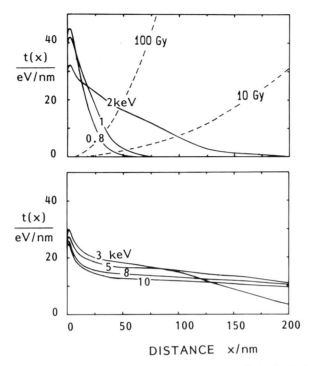

Figure 7.32. Proximity functions $t(x)$ for low-energy electrons (from Ref. 53). Broken lines in the upper panel give the contribution due to unrelated tracks, that is, the intertrack contribution, at doses of 10 and 100 Gy.

a term $t_\delta(x)$, which is the weighted average of the proximity functions for all δ rays, and a term that is equal to the proximity function of a continuous track with the same radial energy distribution as the actual track:

$$t(x) = t_\delta(x) + Lt_a(x) \tag{7.96}$$

where $t_a(x)$ refers to an amorphous track that averages out the structure of the δ rays. The term contains the stopping power L as a factor; $t_a(x)$ depends only on the radial distribution of dose and can be calculated from it. Figure 7.34 gives integral proximity functions for track segments of heavy ions with an energy of 20 MeV/nucleon and indicates their separation into the term $t_\delta(x)$ and the term for the continuous track. Where the δ term (dotted line) is a minor part of $T(x)$, the random occurrence of the δ rays can be disregarded, although the lateral energy transport by δ rays may still be an important factor.

The graphs of the differential proximity function in Figure 7.32 can be understood rather directly. Here $t(x)$ is a spatial autocorrelation function of the charged particle tracks, that is, it gives the distribution in distance of the potential reaction partners around a randomly chosen energy transfer. Distance distributions for sub-

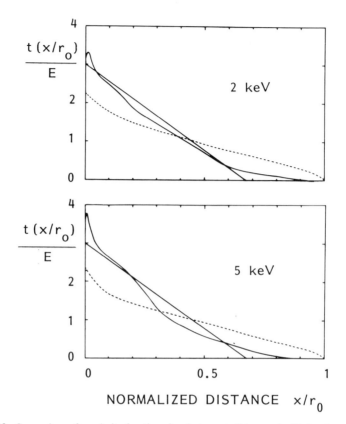

Figure 7.33. Comparison of proximity functions for electrons (solid curves) with functions that result from simplified models (from Ref. 53). The dashed curves are obtained if the electron tracks are treated as straight lines with LET from the CSD approximation. The solid straight lines result if the tracks are pictured as straight lines with constant energy loss rate and ranges equal to two-thirds of the CSD range. The ranges and LET values are for water. The curves show that the LET approximation fails to account for energy loss straggling and angular scattering.

sequent radiation products, such as free radicals, are modified due to diffusion or energy transport processes. An example for proximity functions subjected to diffusion is given in Figure 7.35. Zaider and Brenner (57) have applied proximity functions and distance distributions to radiation chemistry. Such applications may be expected to play an increasing role in problems of nonhomogeneous kinetics.

7.5.3.2. Relation between Proximity Function and Weighted Mean Event Size.
Equation (7.85) implies one of the central theorems of microdosimetry. The weighted average of energy imparted per event to a site S is

$$\bar{\epsilon}_d = \int_0^{x_{max}} \frac{t(x)s(x)}{4\pi x^2}\, dx = \int_0^{x_{max}} t(x)U(x)\, dx \qquad (7.97)$$

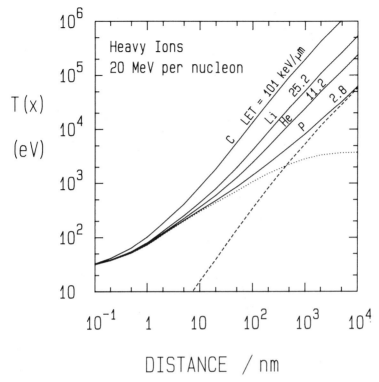

Figure 7.34. Integral proximity functions $T(x)$ for different heavy ions of kinetic energy 20 MeV/ nucleon (from Ref. 44). The dotted line represents the contribution, $T_\delta(x)$, due to δ rays. The dashed line corresponds to the LET-dependent term for protons. For the heavier ions this latter term is multiplied by a factor corresponding to the increased LET [see Eq. (7.96)].

where $t(x)$ is the proximity function of the radiation, and $s(x)$ and $U(x)$ are the proximity function and the geometric reduction factor of the site. For the sphere of diameter d one obtains, with Equation (7.78), the important equation

$$\bar{\epsilon}_d = \int_0^d \left(1 - \frac{3x}{2d} + \frac{x^3}{2d^3}\right) t(x)\, dx \qquad (7.98)$$

The weighted mean linear energy and mean specific energy per event are obtained from the relations

$$\bar{y}_d = \frac{\bar{\epsilon}_d}{\bar{l}} \quad \text{and} \quad \zeta = \frac{\bar{\epsilon}_d}{m} \qquad (7.99)$$

where the mean chord length $\bar{l}$ equals $4V/S$ for a convex site of volume V and surface area S, and m is the mass of the site.

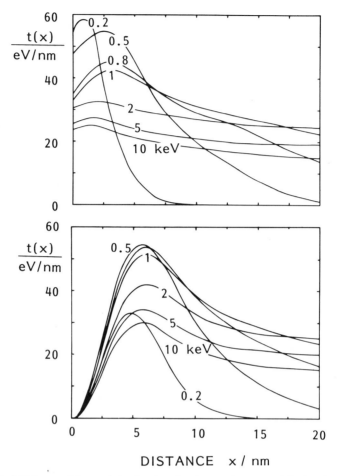

Figure 7.35. Initial parts of the proximity functions for electrons in water and the functions that result if the inchoate pattern of energy transfers has diffused by a characteristic distance of 5 nm. The characteristic distance is the mean separation that results due to the diffusion for two points initially coinciding (from Ref. 53).

Equation (7.97) permits the derivation of the weighted mean event size and therefore also of the variance ζD of the specific energy in any structure, even one of complex shape. This gives broad applicability to Eq. (7.69), that is, to the result for the site model of dual radiation action. In Section 7.5.4 it will be shown that Eq. (7.97) can also be modified to account for a distance model.

The function $t(x)$ cannot be measured directly with present experimental methods, but it can, at least in principle, be derived from measured values of $\bar{y}_d$, ζ, or $\bar{\epsilon}_d$ (58). It can be computed readily from simulated charged particle tracks. The use of the proximity function obviates the need to measure the quantities $\bar{y}_d$ or ζ for different geometries. If the function $t(x)$ for a radiation is known, one merely needs the geometric proximity function $s(x)$ for the site to compute the microdosimetric parameters. The geometric proximity function can be expressed analytically for simple geometries. In other cases it can be obtained from Monte Carlo simulations. Although one may not often find it necessary to compute microdosimetric parameters for irregular geometries, the use of Eq. (7.97) can be essential in general considerations. Even complicated sites can have fairly simple proximity functions. For example, small targets uniformly distributed over a spherical region have, except for small distances comparable to the targets, a proximity function equal to that of the sphere. Accordingly, one expects, even for the complicated structure, the linear-quadratic relation of Eq. (7.69) with the parameter ζ for the spherical site. The general relation is also important because it implies that microdosimetric parameters for spherical and cylindrical regions can be nearly equal (59). This reduces substantially the need for complicated detectors, particularly in charged particle fields where wall-less microdosimetric instruments are required (60).

In the context of models of cellular radiation action proximity functions can be employed to account more rigorously for spatial correlations of energy transfers and subsequent sublesions than did the simple site model or the earlier distance model of Lea.

7.5.4. General Formulation of Dual Radiation Action

In the following brief treatment *interaction of pairs of energy transfers* is again used as a convenient shorthand expression. It refers to a mechanism whereby sublesions are formed by two energy transfers and where they interact to form a lesion. The sublesions could be molecular alterations in DNA, for example, single-strand breaks in close proximity; the "interaction" would then be misrepair due to the interference of excision repair of the two single-strand breaks. The sublesions could also be alterations on a more complex level, for example, pairs of double-strand breaks in chromosomal structures that may form chromosome aberrations by misrepair.

The probability of an energy transfer to be transformed into a lesion is assumed to be proportional to the sum of its neighboring transfers, the contribution of each transfer being weighted by a distance-dependent interaction probability $\gamma(x)$. Accordingly, one obtains the probability of interaction of an energy transfer ϵ_i:

$$P_i = \epsilon_i \int_0^\infty t(x; D) \, \gamma(x) \, dx \qquad (7.100)$$

where $t(x; D)$ is the proximity function that includes the trivial dose-dependent term due to independent particle tracks [Eq. (7.93)]. To simplify the resulting formulas, it is practical to normalize $\gamma(x)$ so that its spatial integral is unity:

$$\int_0^\infty \gamma(x) 4\pi x^2 \rho \, dx = 1 \qquad (7.101)$$

The interaction probability is therefore

$$P_i = k\epsilon_i \left[\int_0^\infty t(x)\gamma(x)\ dx + D \right] \tag{7.102}$$

The total yield, $E(D)$, of lesions, arbitrarily normalized to unit mass, is then

$$E(D) = kD \left[\int_0^\infty t(x)\gamma(x)\ dx + D \right] = k(\xi D + D^2) \tag{7.103}$$

The main result is that one obtains, as in the simple site model, a linear-quadratic dose dependence. The coefficient ξ of the linear term is, as with the site concept, a measure of an effective local concentration of energy transfers in the individual charged particle tracks of the radiation. One can consider the result for the site model [that is, Eq. (7.69)] as a special case of the present formula. If one assumes that an energy transfer has constant interaction probability with all transfers within the same site and zero interaction probability with transfers outside the site, the function $\gamma(x)$ will be equal to the geometric reduction factor $U(x)$ divided by m. The dose dependence is then

$$E(D) = kD \left[\frac{\int_0^\infty t(x)U(x)\ dx}{m} + D \right] = k(\zeta D + D^2) \tag{7.104}$$

This is the earlier result for the site model.

In actuality, the site and proximity aspects play joint roles. The function $\gamma(x)$ is the product of two terms representing the inherent dependence on distance and the influence of the geometry of the site, $\gamma(x) = g(x)U(x)$. Usually it will be difficult in an experiment to separate the two factors. The aim of the biophysical investigations of cellular radiation action is to determine the compound function $\gamma(x)$.

These considerations describe the general approach, but additional factors play a role. There may be, as pointed out in Section 7.4.2, an inherently linear component in the dependence of the effect on z. Furthermore, the quadratic term in absorbed dose is, unlike the linear intratrack term, dependent on dose rate. In spite of such added complexities, the formulation in terms of the proximity function has led to definite conclusions when applied to experiments with correlated heavy particles or to the experiments with soft X rays (30–34). It is found that $\gamma(x)$ decreases sharply at small distances but reaches out to distances on the order of several micrometers (33, 34). Due to the abundance of neighboring energy transfers in close proximity, the linear term results mostly from short-range interactions. Due to the relatively large number of more distant energy transfers from independent tracks, the quadratic component results largely from the interplay of radiation products formed in independent particle tracks. This difference and the influence of the time factor on the quadratic component account for a different effect of various dose-modifying factors on the linear and the quadratic component in dose dependences of cellular radiation effects.

Models of cellular radiation action are still tentative. More needs to be known about the molecular mechanisms, but it follows from the different effectiveness of sparsely and densely ionizing radiations that the mechanisms are greatly affected by the spatial correlation of energy in cellular and subcellular structures. Microdosimetric concepts and data will, therefore, remain essential in any stochastic model of the effects of radiation on cells.

REFERENCES

1. B. Alberts, D. Bray, J. Lewis, M. Raff, K. Roberts, and J. D. Watson, *Molecular Biology of the Cell*, Garland, New York and London, 1983, Chapter 1, p. 4.

2. N. V. Timofeeff-Ressowsky and K. G. Zimmer, *Das Trefferprinzip in der Biologie*, Hirzel, Leipzig, 1947.

3. G. Hotz, *Z. Naturforschg.* **21b,** 148–152 (1966).

4. F. Dessauer, *Z. Physik* **84,** 218 (1933).

5. K. G. Zimmer, *Studies on Quantitative Radiation Biology*, Oliver and Boyd, London, 1961.

6. O. Hug and A. M. Kellerer, *Stochastik der Strahlenwirkung*, Springer-Verlag, Berlin, 1966.

7. F. Dessauer, *Z. Physik* **12,** 38–47 (1923).

8. J. A. Crowther, *Proc. Roy. Soc. (London)* **B96,** 207 (1924).

9. P. Jordan, *Naturwissenschaften* **26,** 537 (1938).

10. C. Lücke-Huhle, W. Comper, L. Hieber, and M. Pech, *Radiat. Environ. Biophys.* **20,** 171 (1982).

11. H. H. Rossi, *Radiat. Environ. Biophys.* **20,** 1 (1981).

12. A. M. Kellerer and O. Hug, *Biophysik* **1,** 33 (1963).

13. *Quantitative Concepts and Dosimetry in Radiobiology*, ICRU Report 30, International Commission on Radiation Units and Measurements, Washington, DC, 1979.

14. A. M. Kellerer and H. H. Rossi, in *Cancer, A Comprehensive Treatise*, Vol. 1, 2nd. ed., F. F. Becker (Ed.), Plenum, New York, 1982, pp. 569–616.

15. D. E. Lea, *Actions of Radiations on Living Cells*, 2nd ed., Cambridge University Press, London, 1955.

16. H. H. Rossi, *Radiat. Res.* **10,** 522 (1959).

17. H. H. Rossi, *Radiat. Res.* Suppl. 2, 290 (1960).

18. H. H. Rossi, in *Advances in Biological and Medical Physics*, Vol. 11, J. H. Lawrence and J. W. Gofman (Eds.), Academic, New York, 1967, pp. 27–85.

19. H. H. Rossi, *Radiation Dosimetry*, Vol. 1, *Fundamentals*, F. H. Attix and W. C. Roesch (Eds.), Academic, New York, 1968, pp. 43–92.

20. *Microdosimetry*, ICRU Report 36, International Commisssion on Radiation Units and Measurements, Bethesda, MD, 1983.

21. *Radiation Quantities and Units*, ICRU Report 33, International Commission on Radiation Units and Measurements, Washington, DC, 1980.

22. M. G. Kendall and P. A. P. Moran, *Geometrical Probability*, Griffin, London, 1963.

23. A. M. Kellerer, *Radiat. Res.* **98**, 425 (1984).

24. A. M. Kellerer, *Mikrodosimetrie, Grundlagen einer Theorie der Strahlenqualität*, Series of Monographs B1, Gesellschaft für Strahlenforschung, Munich, 1968, pp. 1–157.

25. A. M. Kellerer, in *Proceedings of the 2nd Symposium on Microdosimetry*, Euratom 4452 d-f-e, H. G. Ebert (Ed.), Brussels, 1970, pp. 107–134.

26. A. M. Kellerer, in *Techniques in Radiation Dosimetry*, K. R. Kase and B. E. Bjärngard (Eds.), Academic, in press.

27. J. Booz and M. Coppola, *Proceedings of the 4th Symposium on Microdosimetry*, Vol. II, J. Booz et al. (Eds.), Euratom 5122 d-e-f, Luxembourg, 1974, pp. 983–998.

28. H. H. Rossi, M. H. Biavati, and W. Gross, *Radiat. Res.* **15,** 431 (1961).

29. A. M. Kellerer and H. H. Rossi, *Curr. Top. Radiat. Res. Quarterly* **8,** 85 (1972).

30. D. T. Goodhead and D. J. Brenner, *Proceedings of the 8th Symposium on Microdosimetry*, J. Booz and H. G. Ebert (Eds.), Euratom 8395, Luxembourg, 1983, pp. 597–609.

31. R. P. Virsik, D. T. Goodhead, R. Cox, J. Thacker, Ch. Schäfer, and D. Harder, *Int. J. Radiat. Biol.* **38,** 545 (1980).

32. H. H. Rossi, *Radiat. Res.* **78,** 185 (1979).

33. A. M. Kellerer, Y.-M. P. Lam, and H. H. Rossi, *Radiat. Res.* **83,** 511 (1980).

34. M. Zaider and D. J. Brenner, *Radiat. Res.* **100,** 213 (1984).

35. G. R. Merriam, Jr., B. J. Biavati, J. L. Bateman, H. H. Rossi, V. P. Bond, L. Goodman, and E. F. Focht, *Radiat. Res.* **25,** 123 (1965).

36. H. H. Smith and H. H. Rossi, *Radiat. Res.* **28,** 302 (1966).

37. J. L. Bateman, H. H. Rossi, A. M. Kellerer, C. V. Robinson, and V. P. Bond, *Radiat. Res.* **51,** 381 (1972).

38. H. H. Rossi, *Phys. Med. Biol.* **15,** 255 (1970).

39. C. J. Shellabarger, D. Chmelevsky, and A. M. Kellerer, *J. Nat. Cancer Inst.* **64,** 821 (1980).

40. C. J. Shellabarger, D. Chmelevsky, A. M. Kellerer, and J. P. Stone, *J. Nat. Cancer Inst.* **69,** 1135 (1982).

41. R. P. Virsik, R. Blohm, K.-P. Herman, H. Modler, and D. Harder, *Proceedings of the 8th Symposium on Microdosimetry*, J. Booz and H. G. Ebert (Eds.), Euratom 8395, Luxembourg, 1982, pp. 409–422.

42. A. M. Kellerer and H. H. Rossi, *Radiat. Res.* **75,** 471 (1978).

43. J. F. C. Kingman, *J. Appl. Prob.* **2,** 162 (1965).

44. D. Chmelevsky, *Distributions et Moyennes des Grandeurs Microdosimetriques a l'echelle du Nanometre*, Rapport CEA-R-4785, Service de Documentation, C. E. N. Saclay. B.P.2—F-91190 Gif s. Yvette, 1976.

45. A. M. Kellerer and D. Chmelevsky, *Radiat. Environ. Biophys.* **12,** 321 (1975).

46. A. M. Kellerer, *J. Appl. Prob.* **23,** 307 (1986).

47. W. Blaschke, *Math. Z.* **42,** 399 (1937).

48. L. A. Santalo, *Integral Geometry and Geometric Probability*, Addison-Wesley, London, 1976.

49. J. Serra, *Image Analysis and Mathematical Morphology*, Academic, London, 1982.

50. A. M. Kellerer, *Proceedings of the 7th Symposium on Microdosimetry*, Vol. II, J. Booz, H. G. Ebert, and H. D. Hartfiel (Eds.), Euratom 7147 DE-EN-FR, Harwood Academic Publishers, London and New York, 1981, pp. 1049–1062.

51. M. J. Berger, in *Medical Radionuclides: Radiation Dose and Effects*, R. J. Cloutier, C. L. Edwards, and W. S. Snyder (Eds.), USAEC Report CONF-691212, National Technical Information Service, Springfield, VA, 1970, pp. 63–86.

52. G. Matheron, *Random Sets and Integral Geometry*, Wiley, New York, London, Sydney, and Toronto, 1974.

53. D. Chmelevsky, A. M. Kellerer, M. Terrissol, and J. P. Patau, *Radiat. Res.* **84,** 219 (1974).

54. H. G. Paretzke and F. Schindel, *Proceedings of the 7th Symposium on Microdosimetry*, Vol. I, J. Booz, H. G. Ebert, and H. D. Hartfiel (Eds.), Euratom 7147 DE-EN-FR, Harwood Academic Publishers, London and New York, 1981, pp. 387–398.

55. B. Grosswendt, *Proceedings of the 7th Symposium on Microdosimetry*, Vol. I, J. Booz, H. G. Ebert, and H. D. Hartfiel (Eds.), Euratom 7147 DE-EN-FR, Harwood Academic Publishers, London and New York, 1981, pp. 319–328.

56. F. Zilker, U. Mäder, and H. Friede, *Proceedings of the 7th Symposium on Microdosimetry*, Vol. I, J. Booz, H. G. Ebert, and H. D. Hartfiel (Eds.), Euratom 7147 DE-EN-FR, Harwood Academic Publishers, London and New York, 1981, pp. 449–456.

57. M. Zaider and D. J. Brenner, *Radiat. Res.* **100,** 245 (1984).

58. M. Zaider, D. J. Brenner, K. Hanson, and G. N. Minerbo, *Radiat. Res.* **91,** 95 (1982).

59. A. M. Kellerer, *Radiat. Res.* **86,** 277 (1981).

60. W. A. Glass and W. A. Gross, in *Topics in Radiation Dosimetry, Supplement 1*, F. H. Attix (Ed.), Academic, New York, 1972, pp. 221–260.

8 MECHANISMS AND KINETICS OF RADIATION EFFECTS IN METALS AND ALLOYS

Louis K. Mansur

Metals and Ceramics Division
Oak Ridge National Laboratory
Oak Ridge, Tennessee 37831

Research sponsored by the Division of Materials Sciences, U.S. Department of Energy under contract DE-AC05-840R21400 with Martin Marietta Energy Systems, Inc.

8.1. INTRODUCTION

In energy-transforming devices based on nuclear reactions, such as fission reactors and planned fusion reactors, materials are subjected to intense irradiation with energetic neutrons. A metal or alloy possesses a crystalline lattice on which atoms are arranged. Irradiation disrupts the perfection of this structure. Atoms are displaced, creating pairs of point defects, vacant lattice sites (vacancies), and atoms occupying interstitial positions (interstitials). Neutron irradiation also induces transmutation reactions that lead to impurity production. An important transmutation product is helium produced by (n, α) reactions. Helium is a highly insoluble inert gas that degrades the properties of materials. The diffusion and clustering of the point defects and the concomitant elemental redistribution of alloying constituents and transmutation products made possible by this diffusion lead to changes in the microstructure and microcomposition of the material. These local changes in turn are responsible for gross changes in macroscopic properties. The primary disruption of the crystal lattice is termed radiation damage. The resulting changes in microstructure, microcomposition, and physical and mechanical properties are sometimes identified as radiation damage but more often as radiation effects. The most serious radiation effects in structural materials are dimensional instability and degradation of mechanical properties.

Dimensional instability mainly results from induced swelling and creep. Swelling occurs by the formation of cavities, a new phase in the crystal, consisting of empty space. The cavity diameters are known to range from about 1 nm, the smallest resolvable in the modern transmission electron microscope, up to hundreds of nanometers. Irradiation creep is a gradual plastic deformation that can be many orders of magnitude larger than the more familiar thermal creep of materials subjected to stress at elevated temperatures. A shape change in compliance with the applied stress results. The practical importance of dimensional instability may be imagined when it is realized that swelling levels of tens of percent have been observed in some highly irradiated stainless steels.

Irradiation also degrades mechanical properties. Loss of ductility, or irradiation embrittlement, is a problem. Stainless steels, for example, possess high ductility (the ability to deform under stress without fracture) typically in the range of tens of percent. After high dose irradiation, however, ductility may be decreased to tenths of a percent.

8.2. BACKGROUND

Radiation effects is an important branch of materials science. In addition to the obvious technological need, research is stimulated by a very active effort on basic scientific questions. Because irradiation can produce point defects, defect clusters, and wholesale changes in microstructure and composition in a controlled manner, in a way that cannot be achieved by conventional thermomechanical treatments, it provides a unique tool and new avenues for the study of the basic physics of defects and materials.

Fortunately, radiation effects are sensitive to many variables including composition, preirradiation microstructure, and thermomechanical treatment during fabrication, as well as to irradiation conditions. It is therefore possible to control the response of materials and to tailor materials for specific environments. To take maximum advantage of these sensitivities requires that we understand the physical processes occurring during irradiation. Major theoretical and experimental efforts have gone into developing the principles and characterizing the details of radiation effects processes (1, 2). This work ranges from the properties of the point defects to the complex kinetic processes and cooperative phenomena leading to the observed dose, temperature, and compositional dependencies of property changes.

8.2.1. Historical Sketch

Effects of radiation on solids have been studied for over 150 years. In the last century some minerals, which were later shown to contain small amounts of thorium and uranium, were found to exhibit anomalous physical and mechanical properties. The term *metamict* was used to describe such minerals. These possess regular external crystalline habits but showed properties characteristic of amorphous or glassy materials. However, the properties could be restored to those of freshly prepared specimens by heating; this led to speculation that the crystal structures were inherently unstable or were disordered by some unidentified force. The discovery of radioactivity suggested that the unidentified force was bombardment over geological times by α particles emitted by uranium and thorium atoms.

Of much more interest to the present discussion, however, are the intense radiation fields in reactors. That the success of reactor technology would depend critically on the choice of high-temperature materials with satisfactory neutronic properties was pointed out by Fermi in 1946 (3). Wigner is generally credited with the first suggestions that fluxes of energetic neutrons and fission fragments in neutron chain reactors would displace significant numbers of atoms and thus alter physical and mechanical properties (4). It is an understatement to say that this has indeed turned out to be the case.

Swelling was discovered in 1967 as a result of electron microscopic observations of a stainless steel that had been irradiated in the Dounreay Fast Reactor in Scotland (5). Cavities were found ranging in size from the (then) resolution limits of the electron microscope of a few nanometers up to more than 150 nm. Macroscopic swelling was confirmed by immersion density measurements. Two early conferences were convened to bring together researchers to exchange a rapidly expanding volume of data and theoretical analyses on the void swelling problem (6, 7). Since that time there has been a great deal of further work on the problem. Swelling is sometimes said to be the most important effect of irradiation, although this statement should be qualified by reference to temperature range, type of reactor, alloy type, dose, the application intended, and so forth. Other effects can be more important in certain regimes.

Irradiation creep was known somewhat earlier. The first work was on uranium metal, which exhibited irradiation creep under neutron irradiation (8, 9). In 1959 (10) and 1962 (11) investigators reported irradiation creep in a stainless steel. Later work was predominantly either on stainless steels, of interest for fast reactors, or zirconium alloys, of interest for thermal reactors. The first results of more systematic studies were reported in the late 1960s (12, 13). Irradiation creep strains are generally smaller than swelling strains. Nevertheless, this is an important phenomenon because, on the one hand, it can relax stresses associated with differential swelling across large components; on the other hand, if it is too great, it can lead to bulging and bowing of components.

Irradiation embrittlement of austenitic [denoting the face-centered-cubic (fcc) structure] stainless steels and Inconel was first reported in 1959 (14), when it was noted that lower ductility was observed in tensile and stress-rupture tests after irradiation. Systematic results later demonstrated that small amounts of helium produced by neutron-induced transmutation of boron led to significant reductions in the total elongation at fracture (15). The somewhat earlier work on radiation-induced loss of ductility in other alloy types is conveniently summarized in the volume containing Ref. 14. Significant efforts were mounted at Oak Ridge National Laboratory in the United States and the AERE Harwell Laboratory in England. In 1972 a major international conference was convened that was largely devoted to this topic (16). In the intervening period work on irradiation embrittlement was somewhat overshadowed by more intensive efforts on swelling and creep. In recent years, however, irradiation embrittlement has again come to the fore because of the realization that it may be the lifetime limiting factor in the first wall of a fusion reactor (17).

Irradiation was early perceived as a useful tool for studying basic aspects of materials. Many of the solid-state physicists who were prominent in the Manhattan Project recognized that radiation offered an exciting new way to introduce controlled populations of defects into solids. This opened opportunities for detailed studies of the defects or of the solids themselves. These studies developed rapidly throughout the world in the 1950s and led to our present understanding of particle solid interactions and point defect behavior. Several treatises were written later in this period that detail the rapid early progress in this area (18–20).

At the present time there is an active worldwide community carrying out research in the area of radiation effects on materials. Since 1969, roughly the time that major impacts on fast reactor technology were perceived, a large number of international conferences have addressed the area (6, 7, 21–37). The United States, Canada, Japan, France, England, Germany, the USSR, and other countries devote substantial levels of resources to research on radiation effects. For example, in recent years the United States spent approximately $25 million annually on research, in addition to large sums on materials testing reactors and accelerator-based irradiation facilities.

8.2.2. Radiation Effects

In this section certain basic changes produced by irradiation are described. Much of the information available is for iron–nickel–chromium alloys (various types of steels and model alloys), especially with respect to high-dose effects. Because of this and their technological relevance, our discussion of experimental data will be weighted toward these materials.

8.2.2.1. Swelling. In stainless steels not designed to be resistant, swelling can reach tens of percent at the high doses required of these materials in advanced reactors. Figure 8.1 shows an analytic fit to the measured dose and temperature dependence in type 316 stainless steel irradiated in the Experimental Breeder Reactor II near Idaho Falls, Idaho (38). The increase in swelling with dose can be expressed in terms of a delay period followed by an increase according to a power

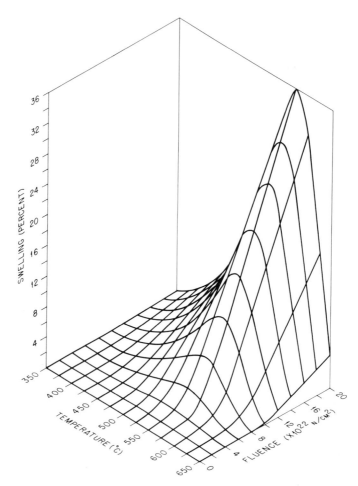

Figure 8.1. Dose and temperature dependences of swelling in type 316 stainless steel irradiated in the EBR-II reactor. Prepared by J. O. Stiegler, Oak Ridge National Laboratory, based on the empirical relations given in Ref. 38.

of dose. At low doses the power is greater than unity and approaches unity at higher doses. At very high exposures or under special conditions, the power may also be less than unity. The unit of exposure is the fluence in neutrons per unit area—the total linear pathlength per unit volume traveled by the particles.

Swelling depends on composition. Both major alloying constituents and minor elements or impurities are important. Large variations in swelling in the pure iron–nickel–chromium system have been demonstrated (39). Figure 8.2 shows the swelling after ion bombardment in a range of alloy compositions. The unit of dose is the displacement per atom (dpa). In a material irradiated to 1 dpa, every atom has, on the average, been displaced from its lattice position once. (The use of charged

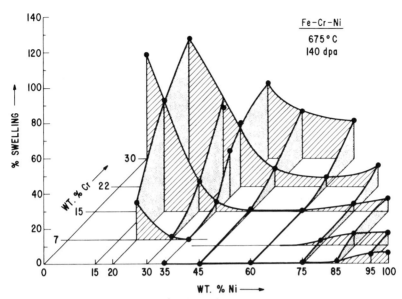

Figure 8.2. Swelling after nickel ion bombardment to 140 displacements per atom (dpa) at 675 °C in various iron–nickel–chromium alloys. From W. G. Johnston, T. Lauritzen, J. H. Rosolowski, and A. M. Turkalo, in N. L. Peterson and S. D. Harkness (Eds.), *Radiation Damage in Metals*, American Society for Metals, 1976, p. 242. With permission.

particle irradiation and the conversion of fluence to displacement per atom are described in later sections.) More recent work has confirmed this behavior for neutron irradiation (40). How strongly swelling depends on minor alloying additions is shown in Figure 8.3. Swelling is shown for several alloys in the nominal range for type 316 stainless steel. The alloys contain different amounts of alloying additions at levels of about 1% or less (41). The lowest swelling alloys were designed to be swelling resistant and are remarkably better even at high doses. Recent compilations of experimental data for swelling in the stainless steels contain data for a variety of conditions (42).

Helium is an impurity that must be singled out. It is produced in neutron irradiations by (n, α) transmutation reactions. Figure 8.4 is a set of micrographs showing the effect of helium on the size and number of cavities (43) in the alloy Fe–17Ni–17Cr–2.5Mo. The experiment was performed on a dual accelerator system where nickel ions (self-ions of the target material) produce the atomic displacements, and the second accelerator simultaneously injects helium ions into the damaged region. Differences in swelling are pronounced. In general, it is found that helium may increase or decrease swelling, depending on conditions. In a titanium-modified stainless steel it was found that no swelling occurred during ion irradiation unless helium was also introduced (44). The effects of helium on cavity swelling and the evolution of microstructure under irradiation in aluminum alloys,

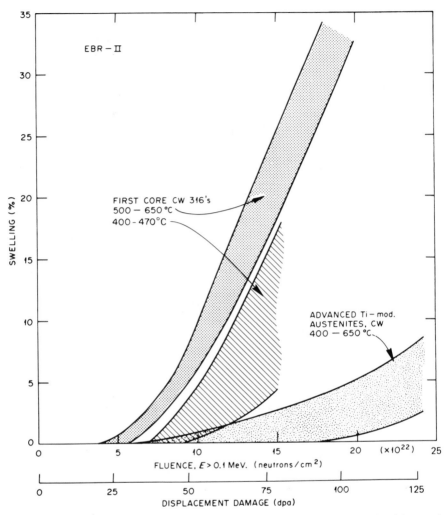

Figure 8.3. Swelling, up to high doses in alloys in the nominal range, for type 316 stainless steel irradiated in the EBR-II in the temperature range 400–650°C. Reprinted, with permission, from Ref. 41.

iron–nickel–chromium alloys, and other alloys have been reviewed recently (45, 46).

How crystal structure affects swelling is not fully understood. For example, ferritic materials [body-centered-cubic (bcc) crystal structure] are generally more resistant than the austenitic materials (fcc crystal structure) described above. Ferritic alloys also tend to exhibit their maximum swelling at temperatures substantially lower than the austenitic alloys (47, 48). Zirconium, with the hexagonal-

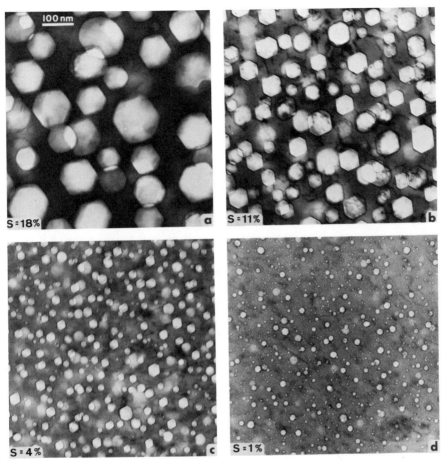

Figure 8.4. Micrographs showing cavity distributions produced by ion irradiations of the alloy Fe–17Ni–17Cr–2.5Mo at 625°C and a dose of 70 dpa. Results for no helium and simultaneous injection of helium are shown together with those for preinjection of helium at room temperature and at the irradiation temperature. The helium content is 1400 atom parts per million (appm). After Ref. 43.

close-packed (hcp) crystal structure, is highly resistant to swelling (49, 50). Magnesium, however, with the same structure swells readily (51).

8.2.2.2. Irradiation Creep. Irradiation creep is a time-dependent plastic deformation in the direction of an applied stress that occurs in excess of ordinary thermal creep. It is generally much larger than thermal creep at low and moderate temperatures. It also proceeds with quite different temperature and stress dependences than thermal creep. For example, thermal creep is not of concern in typical reactor structural materials below about 500°C, while irradiation creep occurs well below this temperature. As the temperature is increased, the irradiation creep rate generally experiences a mild increase. As shown in Figure 8.5, the creep rate in type 316 stainless steel pressurized tubes increases by a factor of about 2.5 from 375 to 600°C (52). This comparatively mild increase is to be compared with that of thermal creep, where the dependence is generally exponential in temperature (53).

The stress dependence of irradiation creep is another distinguishing feature. If we denote the stress dependence as n in the relation (creep rate) $\propto$ (stress)^n, we find that n is usually between 1 and 2 (54, 55). Stress dependences for thermal creep are much higher. The irradiation creep rate per unit of dose decreases with increasing dose rate (56).

8.2.2.3. Ductility Loss.
Ductility loss characteristics caused by irradiation are different in different classes of materials. Two classes of great technological significance are the austenitic stainless steels and the ferritic steels. The austenitic stainless steels have application in fast reactor fuel elements and in the first wall designs of fusion reactors, while the ferritic steels have application elsewhere in fast reactor cores, as alternate materials in the fusion reactor first wall, and in the pressure vessels of light-water reactors.

The austenitic stainless steels undergo significant loss of ductility after high-dose irradiation. Figure 8.6 (bottom) shows how the ductility decreases in type 316 stainless steel with increasing dose (57). Loss of ductility is generally related to increases in the strength of the material. Figure 8.6 (top) shows the increases in strength observed after high-dose irradiation of the same material. The data described in this figure are from postirradiation tests of materials irradiated at 370°C. Under some conditions ductility may be higher during irradiation than in postirradiation tests because of dynamic recovery processes. Also, irradiation creep tends to relax stresses, and this mitigates the consequences of loss of ductility.

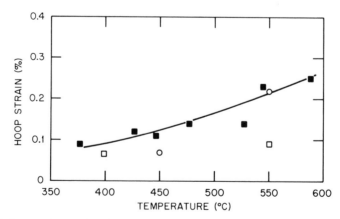

Figure 8.5. Creep strain as a function of temperature for type 316 stainless steel irradiated as pressurized tubes in the EBR-II reactor. After Ref. 52.

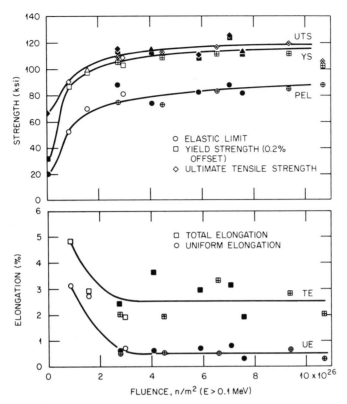

Figure 8.6. Effects of neutron fluence on the strength (top) and ductility (bottom) of type 316 stainless steel irradiated in EBR-II at 370°C. Three measures of strength are shown: proportional elastic limit (PEL), yield strength (YS), and ultimate tensile strength (UTS). Two measures of elongation are shown: uniform elongation (UE) and total elongation (TE). After Ref. 57.

In the ferritic steels there is a well-known ductile-to-brittle transition. Above the ductile-to-brittle transition temperature (DBTT) the material fails in a ductile mode, and the fracture toughness, a measure of the energy required to cause fracture, is high. Below the DBTT the energy required to cause fracture is low, and the material is termed brittle. Irradiation of ferritic steels causes the DBTT to increase and also reduces the energy for fracture in the ductile region. Figure 8.7 shows an example of these effects on SA533B steel, a steel used for light-water reactor pressure vessels (58). The test is a Charpy-V notch test, in which a pendulum is used to fracture a notched specimen, and the energy absorption is measured. In this example irradiation to a fluence of about 6×10^{23} n/m^2 with neutrons of energies greater than 1 MeV at 288°C—the approximate end-of-life fluence and operating temperature for a light-water reactor pressure vessel operating for 30 years—raised the DBTT by about 117 °C and reduced the upper shelf energy by about 20% (top figure).

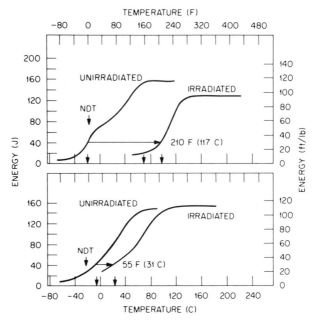

Figure 8.7. Results of Charpy V-notch ductility tests on A533-B pressure vessel plate steel. Material was irradiated to about 6×10^{23} n/m² (neutron energies > 1 MeV) at 288°C. Upper plot shows material in which copper and phosphorus contents were not controlled, and lower plot shows improved material in which copper and phosphorus are reduced to low levels. After Ref. 58.

8.2.3. Microstructure and Microcomposition

Underlying the macroscopic property changes described in the previous section are pronounced changes in structure and composition at the microscopic level. Figure 8.8 is a transmission electron micrograph of a 300 series stainless steel irradiated at about 500°C to a dose of 10 dpa (59). In this micrograph are shown many of the extended defects that will be of interest to us in the remainder of this chapter. The striped elliptical images are termed interstitial dislocation loops. These are formed by the planar condensation of interstitials to form an extra plane of atoms. Each platelet is bounded by a dislocation line and hence the term *dislocation loop*. The tangled spaghettilike line images are dislocation lines. Similar to dislocation loops, these are the boundaries of extra planes of atoms in the crystal. The light images are cavities. These are essentially empty spaces in the crystal like the holes in Swiss cheese. They are a manifestation of the macroscopic swelling of the crystal since the empty atomic lattice sites in the cavities record an equivalent enlargement of the crystal by the creation of new atomic lattice sites for the atoms so removed. The dark images are second-phase precipitates of structure and composition different than the matrix.

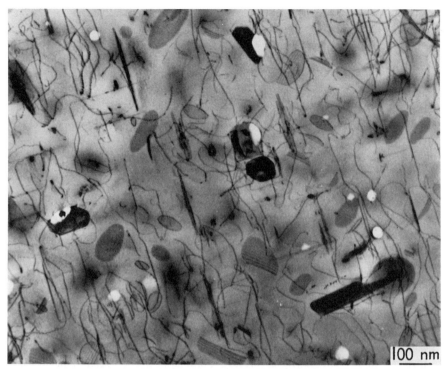

100 nm

Figure 8.8. Micrograph showing various extended defects produced by EBR-II neutron irradiation of commercial 300 series stainless steel to a fluence of 1.9×10^{26} n/m^2 at about 580°C. Imaged are dislocation loops, dislocations, cavities, and precipitates. After Ref. 59.

Most of these defects are present solely as a result of irradiation. A micrograph taken prior to irradiation showed a gray background corresponding to a defect-free solid solution on the scale of the micrograph (of course, there are dislocation lines and second-phase precipitates formed by the thermomechanical processing of the alloy). These extended defects are responsible for the observed drastic changes in dimensional stability and mechanical properties. For example, the dislocation loops, cavities, precipitates, and dislocations themselves are obstacles to the mechanical deformation of the material that arises from the glide of dislocations. This leads to increases in tensile strength and to loss of ductility. The cavities, in addition to representing swelling, provide indirect evidence of the presence of transmuted helium and residual gases since, as will be discussed later, it is very difficult for cavities to form without the initial stabilizing effect of gases even though cavities at the large sizes necessary for significant swelling are nearly unpressurized, that is, they are nearly empty.

There is another major effect of irradiation that is not apparent in Figure 8.8.

Until now we have noted only changes in microstructure, although, of course, precipitates and cavities have compositions different than the host matrix. However, drastic changes in the composition of microscopic regions may occur without triggering phase transformation. Near "sinks" for radiation-produced point defects, the composition of an alloy may be substantially altered. Such sinks include cavities, dislocation loops, grain boundaries, precipitates, and dislocations. The effect is designated radiation-induced segregation (RIS). Phase transformations may occur as a result of segregation (that is, enrichment or depletion of certain alloying elements) or as a result of other radiation-induced processes. The latter changes are termed radiation-induced precipitation (RIP).

RIS was anticipated to be a result of loss of radiation-induced vacancies at sinks, based on earlier work that showed segregation during thermal annealing studies of vacancies retained by quenching from higher temperatures (60). As is now known, however, both vacancies and interstitials contribute to RIS. Early experimental evidence for the phenomenon was reported stemming from unusual strain contrast effects in the TEM observation of cavity images of an Fe–18Cr–8Ni–1Si alloy irradiated to low dose with 1-MeV electrons. At higher doses the segregated regions giving rise to the strain contrast transformed to a new phase on the cavity surface (61). This type of phenomenon was also observed in aluminum alloys irradiated to high dose. Silicon resulting from a neutron transmutation reaction with aluminum was found to be thickly coated on cavities (62), as shown in Figure 8.9.

Cavities are not unique, however. Observations of solute segregation to dislocation loops were reported in 1976. In an early application of analytical electron microscopy (AEM) to segregation in irradiated materials, it was shown that the region around a loop in an ion-irradiated titanium- and silicon-modified iron–nickel–chromium alloy was substantially enriched in silicon (63). In thermal equilibrium grain boundaries often exhibit solute segregation. Grain boundaries are also known to be good sinks for radiation-induced point defects. It is therefore no surprise that strong segregation effects have been observed near grain boundaries. Figure 8.10 shows profiles measured by AEM of the chromium–iron ratio in an Fe–13% Cr alloy irradiated with carbon ions (64).

Similar effects have been observed at free surfaces after irradiation. Because this is a more readily accessible point defect sink, a more extensive body of data has been generated on segregation to free surfaces. The measurements are usually made by Auger electron spectroscopy (AES), secondary ion mass spectroscopy (SIMS), and Rutherford back scattering (RBS) or other nuclear microanalysis techniques. Figure 8.11 shows the solute concentration profiles determined by AES and sputter profiling near the free surfaces of four Ni–1% solute alloys irradiated with nickel ions of 3.5 MeV. The solutes were molybdenum, titanium, aluminum, and silicon (65). It can be seen that one solute is enriched while the others are depleted near the free surface.

Figure 8.9. Micrograph showing cavities surrounded by a silicon-rich coating in aluminum irradiated with neutrons in the HFIR at 55°C to a dose of 1.4×10^{27} n/m². The spheres in the border between the dark and light areas are actually silicon-coated cavities or "hollow beads" protruding out of the edge of a thin aluminum foil into vacuum. After Ref. 62.

In addition to the technological significance of microstructural and microcompositional evolution during irradiation, there is another significant aspect that should be emphasized. The highly nonequilibrium processes occurring during irradiation offer means to generate unique microstructures as well as materials consisting of composites of different alloys on a microscale.

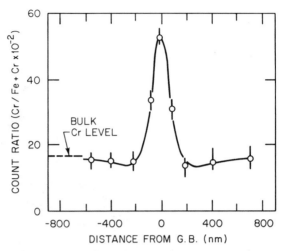

Figure 8.10. Concentration profile of chromium near a grain boundary in Fe–13% Cr alloy irradiated with carbon ions at 525°C to a dose of 57 dpa. After Ref. 64.

8.2.4. Mechanisms

Radiation-induced processes in metals and alloys can be thought of in two broad divisions. The first, termed *damage production* or *radiation damage*, concerns the initial events in the processes embracing the displacement of atoms. The second,

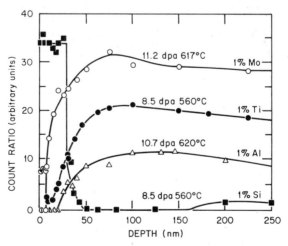

Figure 8.11. Solute segregation profiles determined by AES in four Ni–1% X alloys, where X denotes molybdenum, titanium, aluminum, or silicon. Irradiations were with Ni$^+$ ions of 3.5 MeV to the doses and at the temperatures shown. After Ref. 65.

termed *radiation effects*, encompasses the kinetics of processes resulting from the atomic defects so produced and the concomitant evolution of microscopic structure, composition, and macroscopic properties these induce. In metals and alloys the possibility of radiation effects stems from both the atomic unit and the crystalline structure. Random parts of the crystal are continually disassembled into atomic units, making possible a gradual rearrangement under prevailing conditions. Atomic rearrangement of this type is not ordinarily important without radiation except at elevated temperatures where diffusion via thermal lattice vacancies takes place.

Generally, the important atom-displacing reactions that occur in fission and fusion reactor structural materials are induced by neutrons. Elastic scattering of the neutron by the host nucleus dominates below about 1 MeV. Reactions where the neutron is absorbed by the nucleus, which then emits a gamma ray [(n, γ) reaction], causing the nucleus to recoil, also contribute. At higher energies other nonelastic processes can play a significant role. These include (n, n'), (n, p), (n, α) and $(n, 2n)$ reactions, where n' denotes a neutron of a different energy, p denotes proton, and α denotes the helium nucleus.

Irradiations using charged particles, such as electrons and the ions of hydrogen, helium, or self-ions of the alloy under study, also produce atomic displacements and are used extensively in radiation effects research.

When kinetic energy above a certain threshold level is imparted to a nucleus by any of the above processes, the atom is displaced. It then becomes a projectile that displaces other atoms in a cascading process until each projectile's energy falls below the threshold. The atom initially struck is termed the *primary knock-on atom* (pka or pko). Over a wide range of pka energies the number of atoms displaced is roughly proportional to the fraction of the initial pka energy called the damage energy. The damage energy does not include energy consumed in producing ionization and electronic excitation.

Following the displacement events the energy of the disturbed region is degraded, ultimately to thermal energies. The point defects and small-defect clusters relax and diffuse. Some are mechanically unstable to recombination, for example, a vacancy and interstitial created as near neighbors. Typically, during diffusion most vacancies and interstitials are annihilated by mutual recombination. Some avoid recombination, however, and it is these that are responsible for radiation effects. These defects are either absorbed at preexisting sinks such as grain boundaries or dislocations or aggregate with like defects to produce extended defect clusters. Most of the defects absorbed at sinks ultimately recombine there. However, some are left to accumulate. As we saw in the previous section, when the defect clusters become large enough, they are often observed as cavities or dislocation loops.

At the same time, by diffusing, the point defects can induce preferential flows of alloying elements toward or away from sinks. These flows arise from preferential binding of certain alloying elements to point defects and from preferential diffusion jumps of defects with certain alloying elements. Demixing of the alloy results in local regions being highly enriched or depleted in alloying constituents. Often phase transformations occur when the local solubility limit for precipitation is exceeded. Phase stability is also affected by the presence of defects and defect clusters themselves and by related processes.

As indicated above, the accumulation of defect clusters on a vast scale, the high fluxes of defects to preexisting sinks, and the segregation and phase instability can lead to profound changes in properties.

Figure 8.12 gives the characteristic time and energy scales of radiation-induced processes. Part of the energy of a neutron, typically in the mega-electron-volt range, is absorbed by the crystal. The energy is progressively degraded to the thermal range, where ordinary atomic transport plays a major role. The initial defect production events are completed in less than 10^{-11} s. Recovery from much of the damage and the accumulation of more permanent radiation effects takes place on the much longer time scales of defect diffusion and cluster evolution. Figure 8.13 is a schematic flow diagram summarizing the fates of point defects and the ultimate effects on properties. In the end they may either recombine by various processes, producing little ultimate change, or they may agglomerate at sinks, thereby leading to higher order or extended microstructural defects and variations in microcomposition.

In the next section damage production is described in more detail. Following this, in Section 8.4, the kinetics of radiation effects are described. Emphasis is placed on the theory, especially the kinetics, as a focus of this chapter. In this section experimental data have been discussed to give an overall characterization of radiation-induced processes. In the following sections other experimental data are detailed mainly to enable comparisons of theory and experiment to be made.

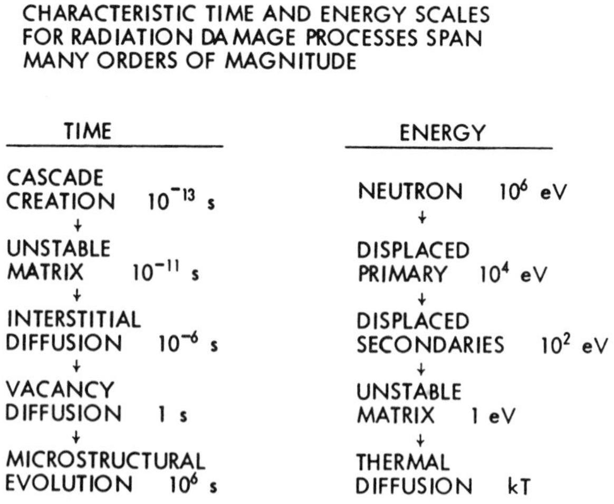

CHARACTERISTIC TIME AND ENERGY SCALES
FOR RADIATION DAMAGE PROCESSES SPAN
MANY ORDERS OF MAGNITUDE

TIME		ENERGY	
CASCADE CREATION	10^{-13} s	NEUTRON	10^6 eV
↓		↓	
UNSTABLE MATRIX	10^{-11} s	DISPLACED PRIMARY	10^4 eV
↓		↓	
INTERSTITIAL DIFFUSION	10^{-6} s	DISPLACED SECONDARIES	10^2 eV
↓		↓	
VACANCY DIFFUSION	1 s	UNSTABLE MATRIX	1 eV
↓		↓	
MICROSTRUCTURAL EVOLUTION	10^6 s	THERMAL DIFFUSION	kT

Figure 8.12. Characteristic time and energy scales of radiation-induced processes in metals and alloys.

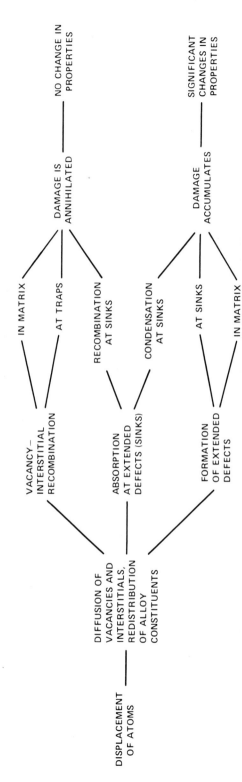

Figure 8.13. Flow diagram showing the various possible fates of point defects generated by radiation.

8.3. DAMAGE PRODUCTION

Our discussion of damage production begins with types and sources of irradiating projectiles and ends with the defects produced as a result of a chain of interactions and energy transfers. Several reviews are available that cover this general area or more specialized aspects of it (66–69).

8.3.1. Irradiating Particles

Of the fission reactors, the fast spectrum reactors experience by far the most severe radiation effects. In these reactors the neutrons are not deliberately moderated or slowed down to thermal energies. The liquid-metal-cooled fast breeder reactor (LMFBR) has been the mainstay of fast reactor concepts. The core is typically cooled with liquid sodium which, because of its moderate atomic mass, is not very effective at slowing neutrons down. The Phenix and Superphenix in France, the Prototype Fast Reactor (PFR) in Great Britain, the JOYO in Japan, the BN-600 in the Soviet Union, the Experimental Breeder Reactor-II (EBR-II), and the Fast Flux Test Facility in the United States have reactor cores that are prototypes of LMFBRs. They also serve as materials testing reactors where the effects of radiation on fuel and structural materials of interest are measured. Figure 8.14 shows the neutron energy spectrum of the EBR-II (70). The maximum energy is in excess of 10 MeV, corresponding to the highest energy neutrons from fission of plutonium and uranium atoms. Most of the neutrons are concentrated in a range surrounding 0.6 MeV.

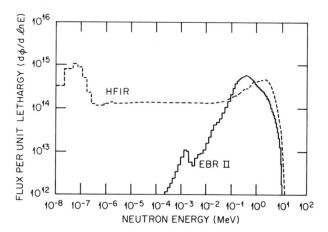

Figure 8.14. Neutron energy spectra for the fast spectrum reactor EBR-II and the mixed spectrum reactor HFIR. The unit on the ordinate is the neutron flux per unit lethargy, where the lethargy is related to the energy logarithmically, $\mu = \ln E$. After Ref. 70.

Fusion reactors, currently the object of aggressive scientific and technological efforts in many countries (71), will have spectra with a large component of much higher energy neutrons. Most near-term conceptual designs use a plasma of deuterium and tritium leading to the reaction

$$d + t \rightarrow He + n + 17.6 \text{ MeV},$$

which yields a neutron of approximately 14.1 MeV. Figure 8.15 shows the energy spectrum calculated for a fusion reactor that has been designed recently, the STARFIRE reactor (72). Although the details of the spectrum depend on the specific design, the main features are generic. There is a large peak near 14 MeV, with a significant flux also at lower energies because of neutron moderation by collisions and reactions.

Much radiation effects research has also been carried out using water-cooled and water-moderated mixed spectrum reactors such as the High Flux Isotope Reactor (HFIR) and the Oak Ridge Research Reactor (ORR). The water moderates the fission spectrum extremely effectively by collisions of the neutrons with hydrogen nuclei. However, these reactors are undermoderated so that there is both a large fast-flux component and a large slow-flux component in the neutron spectrum. Figure 8.14 shows the neutron energy spectrum of the HFIR (70). The fast-

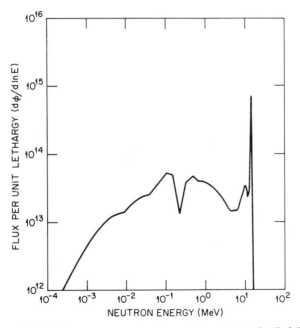

Figure 8.15. Neutron energy spectrum for a fusion reactor. After Ref. 70.

flux component is comparable to those of the EBR-II and FFTF so that damage can be accumulated rapidly. The thermal flux component is employed to advantage in producing those transmutation reactions whose cross sections are high at thermal energies. This combination has proven extremely useful in fusion reactor materials research. The 14-MeV neutrons in a fusion reactor will lead to high helium transmutation rates. As will become evident in Section 8.4, helium, an insoluble rare gas, is extremely potent in altering radiation effects resulting from displacements alone. However, fast reactor irradiations produce only much smaller amounts of helium. Since fusion reactors do not yet exist, how then does one obtain data on helium effects during neutron irradiation? Mixed spectrum reactor irradiations provide a partial solution to this problem. There is high helium generation by the two-step reaction $^{58}Ni + n \rightarrow {}^{59}Ni$ and $^{59}Ni + n \rightarrow {}^{56}Fe + {}^{4}He$. Thus in nickel-bearing alloys, such as the stainless steels that are near-term candidates for fusion reactor structural materials, high levels of helium can be produced. The study of interactions between helium and radiation effects is thereby greatly facilitated.

The above should not be construed to imply that commercially operating light-water-cooled and light-water-moderated reactors (LWR) do not experience any radiation effects in their structural materials. However, neutron energy spectra in these reactors are much poorer in fast neutrons, and the effects are commensurately less. For example, over the operating lifetime (about 30 years) of a LWR a small displacement dose is accumulated in the pressure vessel encasing the reactor core. As discussed in the previous section, this low dose can lead to embrittlement of the vessel material. However, in newer alloys, designed to be embrittlement resistant, this is not a significant problem.

Although neutron irradiations are the premier research tool of technological relevance, they are time-consuming and expensive. Qualitatively similar effects can be produced by charged particle irradiations. Stimulated in part by the applications but also because radiation effects is an area of materials science to be explored, extensive use has been made of charged particle accelerators. The irradiating particles range from heavy ions such as nickel and iron (self-ions of many of the alloys of interest), through light ions of the hydrogen and helium isotopes, to electrons. Energies employed by various research groups range from the low kilo-electron-volt range to hundreds of mega-electron-volts. Most of the work has been done in the range of 1 to a few tens of mega-electron-volts.

Because of coulombic interactions, charged particles lose their energy much more rapidly per unit distance in a solid than do neutrons. Two effects that arise from this are that much higher displacement rates can be achieved than with neutrons and that the particles are stopped at relatively shallow depths. The collision mean free path of a 1-MeV neutron in iron is several centimeters, while a 5-MeV iron ion is brought to rest in about 1 μm.

The characteristics with which the irradiating projectile transfers its energy to the target material are important in determining the material's response. The initially struck target atoms have a characteristic energy spectrum determined by the type and energy spectrum of the irradiating particles.

8.3.2. Primary Recoil Spectra

The next step in understanding damage production is to determine the primary recoil energy spectra produced by irradiating particles. The upper limit of recoil energy is useful as a benchmark. For projectiles whose rest energy is large compared to their kinetic energy, such as neutrons or ions in the mega-electron-volt range, classical collision theory is adequate to describe the energy transfers in elastic collisions. Conservation of energy and momentum give the result

$$T_m = \frac{4E_1 M_1 M_2}{(M_1 + M_2)^2} \tag{8.1}$$

where T_m is the maximum possible energy transfer from a projectile of mass M_1 and kinetic energy E_1 to a target atom of mass M_2. The corresponding relativistic expression for the maximum energy transfer to a target atom by an electron of kinetic energy E_1 and rest mass M_1 is

$$T_m = \frac{2E_1(E_1 + 2M_1 c^2)}{M_2 c^2} \tag{8.2}$$

where c is the speed of light. Figure 8.16 shows the maximum energy transfer by electrons to target atoms of several atomic masses for a range of incident electron kinetic energies.

The maximum energy transfer can only occur in head-on collisions. Other collisions lead to lower energy transfers. Irradiating a target with monoenergetic projectiles generates a complete range of elastic (and inelastic) collisions, resulting in a corresponding distribution of recoil energies. This probability distribution has a characteristic shape for each type of projectile and is a fundamental concept in the calculation of damage production. The probability is described by a cross section. The cross section has units of area, so that when taken in a product with projectile flux and number of target atoms per unit volume, it gives the reaction rate producing the outcome for which the cross section selects. In detail, the primary recoil spectrum is therefore often described by a differential cross section, this being the derivative of the cross section with respect to energy. It describes the probability that an incident projectile produces a recoil in a given differential energy interval.

For electrons the primary recoil spectrum is given by the Mott differential scattering cross section (69). In practice, this is cumbersome and approximations have been developed. A widely used expression is based on that given by McKinley and Feshbach (73, 67). Since electrons are of limited interest in the present context, this result will not be reproduced here. (Electrons in the energy range in common use do not generate collision cascades, while the neutrons and heavy ions do.)

The differential scattering cross section for ions is given over a wide energy range by the Rutherford cross section

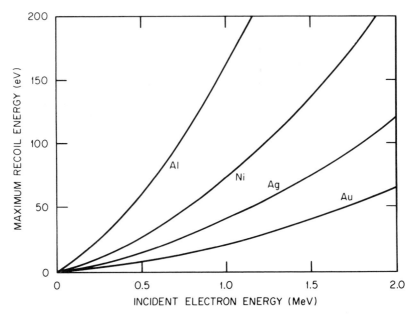

Figure 8.16. Maximum energy transfer by an incident electron of kinetic energy shown on the abscissa to a target nucleus, for a range of target atomic masses.

$$\frac{d\sigma}{dT} = \frac{\pi M_1 Z_2^2 Z_1^2 e^4}{M_2 E_1 T^2} \qquad 0 < T \le T_m \qquad (8.3)$$

Here σ is the cross section, Z_1 and Z_2 denote the atomic numbers of projectile and target, e denotes the charge of the electron, and T denotes the energy transferred to the target atom. Note the inverse square dependence on T, thus strongly favoring low-energy transfers. At very high energies, a few mega-electron-volts per projectile nucleon, this coulombic-interaction-based cross section must be corrected both for nuclear forces and for inelastic nuclear collisions. At the opposite extreme, at energies below about

$$Q_2 = (24.8 \text{ keV}) M_1 Z_1^{4/3} \qquad (8.4)$$

capture of electrons by the projectile in the target becomes significant (66). The atomic interaction again then differs significantly from the coulomb-based interaction between projectile and nucleus, Eq. (8.3). Thus, for much of the heavy ion irradiation work in progress, the cross section of Eq. (8.3) is not generally applicable, but for typical light-ion irradiation experiments, it is quite good.

For heavy ions, below Q_2, the theory of Lindhard, Scharff, and Schiott (74) is used. This is an energy partition model where energy loss is divided into electronic and nuclear components. The differential scattering cross section of this model gives a pka spectrum that obeys a dependence on pka energy similar to that of Eq. (8.3)—it decreases with increasing T only slightly less rapidly than T^{-2}.

For neutrons in the range from the lowest energy that can produce displacement to 1 MeV, elastic scattering dominates. A simple approximation is the isotropic elastic scattering expression

$$\frac{d\sigma}{dT} = \frac{\sigma_e}{T_m} \quad \text{for } T < T_m \tag{8.5}$$

where σ_e is the (energy-dependent) total elastic scattering cross section. This model is fairly accurate up to about 0.1 MeV, but it is inadequate at energies of 1 MeV and above. The elastic scattering distribution at these higher energies is complex. Details are given elsewhere (66). Low-energy transfers are favored. Above about 1 MeV nonelastic processes begin to contribute where the nucleus is excited by the incident neutron and emits a particle or particles. These events include inelastic neutron scattering (n, n') as well as reactions like $(n, 2n)$, (n, p), (n, t), and (n, α). The nuclear recoil from these events contributes characteristic components to the pka spectrum.

Figure 8.17 shows the pka spectra of several types of irradiating particles on nickel and iron at energies of interest (75). Results are shown for neutrons, 5-MeV protons, and 5-MeV nickel ions (after a penetration depth of 0.9 μm into an iron target). The neutron result is based not on monoenergetic particles but on an energy spectrum for a fast reactor of the type whose neutron energy spectrum is shown in Figure 8.14. It is clear that neutron irradiation produces far fewer low-energy recoils than does ion irradiation.

8.3.3. Energy Loss and Defect Production

Loss of energy by the pka takes place by both electronic and nuclear processes, according to the LSS theory (74). In metals and alloys it is usually assumed (for good reason) that only the energy lost in nuclear collisions can displace atoms. This energy is denoted as T_d. A standard procedure for the calculation of T_d, given T, has been recommended (66, 67). It is taken as valid for $0.01 < \epsilon < 500$, where

$$\epsilon = \frac{T}{0.08693Z^7/3} \tag{8.6}$$

for T in kilo-electron-volts. Then T_d is given by

$$T_d = T[1 + kg(\epsilon)]^{-1} \tag{8.7}$$

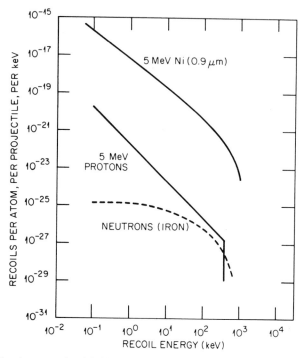

Figure 8.17. The pka spectra for nickel ion and proton irradiation of nickel and for fast reactor neutron irradiation of iron. After Ref. 75.

where

$$g(\epsilon) = \epsilon + 0.40244\epsilon^{3/4} + 3.4008\epsilon^{1/6} \tag{8.8}$$

and

$$k = 0.1337Z^{2/3}(M_1/M_2)^{1/2}. \tag{8.9}$$

The energy transferred in nuclear collisions, the damage energy, is translated into displaced atoms via a secondary displacement model. Secondary displacements are all atoms that are displaced in the stopping of a pka. The following model based on the initial work of Kinchin and Pease (76) is in general use:

$$
\begin{align}
N_d &= 0 & T_d &< E_d \\
N_d &= 1 & E_d &\leq 2T_d < 2E_d \tag{8.10} \\
N_d &= \frac{\beta T_d}{2E_d} & T_d &\geq 2E_d
\end{align}
$$

Here N_d is the number of displaced atoms, or vacancy–interstitial pairs, and E_d is a threshold energy, generally in the range of tens of electron-volts, that is required to displace an atom. The recommended value of β is 0.8, and this accounts for the deviation from a hard sphere ($\beta = 1$) of the interatomic potential (77). The factor of 2 in the denominator arises because on the average the pka energy is shared equally between two atoms after the first collision and between 2^n atoms after the nth generation (78). When the energy of each atom falls below $2E_d$, it does not create more displacements. Thus $N_d = 2^n = \beta T_d/2E_d$.

The form of Eq. (8.10) assumes a sharp threshold E_d. This is, of course, a weighted average. Direction in a crystal is highly structured, and the threshold varies with direction. The assumption that the threshold is sharp also has been the subject of investigation. It has been found that these two assumptions can lead to significant errors in N_d if many of the knock-ons have energies near E_d. A more detailed form of N_d, the damage function, is then required. Figure 8.18 is an example of the threshold energy surface $N_d(T)$ derived for copper and shown on the crystallographic unit triangle [as determined by electron irradiation and resistivity measurements (79)]. The range is broad, covering from ~ 19 to ~ 79 eV. When translated into a damage function, the result differs substantially from Eq. (8.10).

Similarly, at high energies there is evidence that Eq. (8.10) gives a factor of 2–3 too many displaced atoms (80). With the prescribed adjustment of E_d in Eq. (8.10), these relations are appropriate to compare damage production of particles, like neutrons, that emphasize high-energy transfers. For particles that emphasize low energy transfers, like protons, the actual $N_d(T)$ versus T damage functions (80) are to be used rather than Eq. (8.10).

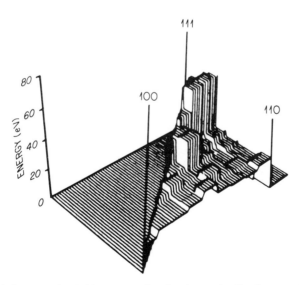

Figure 8.18. Displacement threshold energy surface for electron-irradiated copper. Reprinted, with permission, from Ref. 79.

An example of how incident ion energy is degraded as the projectile travels through a solid is shown in Figure 8.19. It applies to 5-MeV copper ion irradiation of a stainless steel target. Also shown are the electronic and nuclear (damage) energy loss components (81). The damage energy curve also gives the displaced atom or defect production profile by Eq. (8.10) or more detailed damage functions.

A link with common experimental situations is provided by the following sketch. The rate of displaced atom production is approximated by

$$G = \sigma_d \phi N_d \rho_A \qquad (8.11)$$

where σ_d is the spectrum-averaged displacement cross section, ϕ is the neutron flux, and ρ_A is the density of atoms per unit volume. In a liquid metal fast breeder reactor (LMFBR) the mean neutron energy is about 0.6 MeV. The mass M_2 of the major constituent atoms in stainless steel is about 60 in units of neutron mass. The maximum energy transfer from a neutron of mean energy is then about 40 keV from Eq. (8.1). Thus, N_d, the number of atoms displaced by typical pka's resulting from fast reactor neutron irradiation is of order 100. Fast fission reactors typically have neutron fluxes, ϕ, of 10^{19} m^{-2} s^{-1}, and the displacement-producing collision cross section, σ_d, of the nuclei of the structural materials is about 10^{-27} m^2 (10 barns). The result is $\sim 10^{23}$ displacements/(m^3 s). In fusion reactors the neutron fluxes are somewhat lower, but the number of atoms displaced by typical pka's is higher due to their higher energy. The resulting displacement rate is similar to that of fast reactors. Normalized to the atomic density, this gives $\sim 10^{-6}$ displacements/(atom s). The meaning of this is that after 10^6 s, roughly 12 days, every atom in the structural material of a reactor has on average been displaced once by

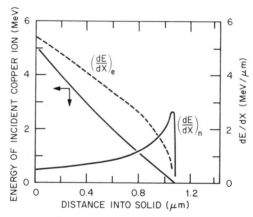

Figure 8.19. Energy of an incident 5-MeV copper ion as a function of distance into a stainless steel target. Also shown are the electronic (e) and nuclear (n) (damage) components of the energy loss per unit distance (dE/dX). After Ref. 81.

irradiation! The unit *displacement per atom* (dpa) is a convenient measure of the dose of displacement-producing radiation.

8.3.4. Cascade Characteristics

The characteristics and structure of the cascade in which point defects and point defect clusters are born is important. A displacement cascade is the branching sequence of collisions initiated by a pka. As discussed in the previous section, neutron and ion irradiations typically produce pka's with damage energy in the range of many kilo-electron-volts and more. Thus, cascades may consist of numbers of displaced atoms ranging into the hundreds. The interstitials may be produced by simple events where an atom is knocked from its site and reaches thermal energies in an interstitial position. Interstitials are also produced by replacement collision sequences. Here the vacancy member of the defect pair is created by placing the corresponding atom on the next lattice site. The atom on that lattice site is nudged onto the next site and so on. When the sequence stops, the last atom nudged from its lattice site becomes the interstitial. This is a mechanism for the long-range separation of a vacancy from its interstitial. It accounts in part for the form of the cascade—vacancy-rich "core" surrounded by a "shell" of interstitials. Figure 8.20 is a three-dimensional representation of a cascade generated by computer calculation. The result was produced by a pka of 5 keV in copper (82).

Atomic level observations of the structure of cascades have been made primarily in tungsten and platinum using the field ion microscope (83, 84). In this technique a very sharp needle-shaped metal specimen tip is bombarded with ions. The tip is then imaged using a gas that is ionized and projected onto the imaging screen. After the image is recorded, an atomic layer is peeled from the tip by electric field evaporation. The new surface is then imaged and so on. Vacancy and interstitial defects can be seen in successive layers and a three-dimensional picture of cascades built up. Figure 8.21 shows the observed vacancy structure of a cascade in tungsten irradiated with 30-keV chromium ions, where the experiment was carried out at a temperature of 10 K. Note that the cascade consists of both small and large clusters as well as monovacancies.

Cascade size increases with pka energy. Above a certain energy, however, the cascade morphology begins to change. It breaks up into subcascades that are separated from each other. Each subcascade has the usual form (85). In materials of interest for technological applications, subcascade formation occurs when the pka energy exceeds 30–50 keV.

Both the field ion microscope studies and the electron microscope studies reflect the state of a cascade after passage of a time that is very long compared with that necessary to produce the initial displacements (Figure 8.12). After the initial displacement events mechanical relaxations, or collapse of unstable configurations, occur as well as subthreshold energy transfers and thermal migration. A large number of observations of surviving cascade structure have been made using the elec-

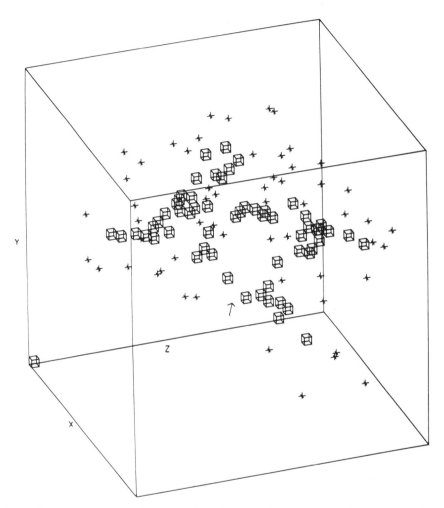

Figure 8.20. Three-dimensional representation of a cascade in copper produced by computer simulation for a 5-keV pka. The vacancies are shown as small cubes, and the interstitials are shown as the intersections of three orthogonal axes. Undisplaced atoms are not shown. The large cube edge length is 20 lattice parameters. The arrow indicates the pka direction. After Ref. 82.

tron microscope (85–92). These are mainly based on imaging separate or mingled defect clusters. As a result of these observations, it has been found that interstitials form clusters usually as dislocation loops, while vacancies may exhibit more variety in their clustering morphologies. Cascade cores often collapse into vacancy loops (86). Vacancy clustering to create stacking fault tetrahedra is observed in fcc materials such as copper, gold, nickel, Cu_3Au, and stainless steel (Fe–Ni–Cr) (87, 90). Other three-dimensional morphologies also result and are the precursors of much larger cavities. A significant observation is that vacancy clusters form in

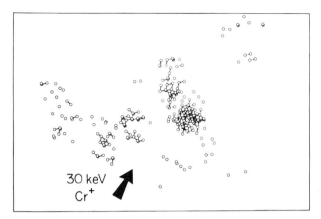

Figure 8.21. Computer-reconstructed images of the vacancy structure of a cascade observed by field ion microscopy for 30-keV chromium ion irradiation of tungsten. The rod connecting two vacancies represents the first-nearest-neighbor distance of 0.274 nm. Reprinted, with permission, from Ref. 84.

cascades at temperatures both higher and lower than where they would be expected to form by homogeneous diffusion and aggregation of thermal vacancies and divacancies. The efficiency of clustering processes also appears to be strongly affected by the presence of nearby sinks. Experiments have been carried out that show enhanced formation of clusters near dislocations and cavities that preexisted in the material.

8.4. KINETICS OF RADIATION EFFECTS

Point defect production and initial clustering is the first stage of larger scale and longer term defect agglomeration. We are thus led into a discussion of the kinetics of these larger scale processes that ultimately determine the radiation-driven macroscopic changes in materials properties.

8.4.1. Point Defects

Describing the kinetics of radiation effects subsequent to cascade formation rests largely on understanding the reactions of vacancies and interstitials as well as higher order clusters of these self-point defects. Indeed, understanding mass transport in general rests on knowledge of defect mechanisms. The study of point defect structure and properties is a large field. (See for example Refs. 2 and 93 and many papers in Refs. 24 and 94–99.)

Defects produced in cascades that are not so close as to be mechanically unstable cannot be annihilated with the opposite type defect or at sinks without first diffusing. The possible fates of the defects during diffusion lead to macroscopic changes in physical and mechanical properties, as shown schematically in Figure 8.13. For the kinetics of radiation effects we are therefore primarily interested in the diffusion parameters, including the types of jumps, activation energies, and entropies, and related aspects of the defect structure. In this section some basic aspects of self-point defects (vacancies and interstitials) are discussed from this point of view.

Other species are also referred to as point defects. Isolated transmutation products are an example. Perhaps the most important from our present perspective is helium produced by (n, α) reactions. Its diffusion and interaction with other defects and its role in the formation of clusters is very important. Much of the available information on diffusion coefficients, binding energies, and similar parameters has been developed recently. An indication of the scope of this field is contained in the literature (100, 101). Direct manifestations in the theory of radiation effects are both the effective diffusion coefficient of helium during irradiation and the effect of helium in pressurizing cavities. A theoretical analysis of the former is described elsewhere (102) while the latter is reviewed in some detail in a subsequent section.

8.4.1.1. Vacancies. A most important diffusion parameter for vacancies is the migration energy, E_v^m, the energy involved in the exchange of lattice sites between a vacancy and a neighboring atom. Table 8.1 gives the migration energy ranges for vacancies in selected metals (93). The diffusion coefficient of vacancies is given by

$$D_v = D_v^0 \exp\left(\frac{-E_v^m}{k_B T}\right) \qquad (8.12)$$

where D_v^0 is a constant that depends on the material, and $k_B T$ has its usual meaning.

As discussed previously and described in subsequent sections, vacancies and interstitials are formed in enormous numbers by irradiation.

TABLE 8.1. Point Defect Migration and Formation Energies in Selected Metals[a]

Metal	E_v^m (eV)	E_v^f (eV)	E_i^m (eV)	E_i^f (eV)
Ag	0.58–0.88	1.09–1.19	0.088–0.100	—
Al	0.58–0.65	0.60–0.77	0.112–0.115	3.0–3.6
Au	0.62–0.94	0.89–1.00	—	—
Cu	0.67–1.06	1.04–1.31	0.117	1.7–4.3
α-Fe	0.53–1.24	1.4–1.6	0.300	4.7–12
Mo	1.25–1.62	3.00–3.24	0.083	—
Ni	0.92–1.46	1.45–1.8	0.140–0.150	—

[a] Ranges shown correspond to extremes in reported measurements by various techniques and workers. From Ref. 93.

They are also formed at elevated temperatures by thermal disorder. The thermodynamic equilibrium concentration of vacancies is given by

$$C_v^0 = \Omega^{-1} \exp\left(\frac{S_v^f}{k_B}\right)\exp\left(\frac{-E_v^f}{k_BT}\right) \tag{8.13}$$

where Ω denotes atomic volume, S_v^f is the entropy of vacancy formation and E_v^f is the energy (enthalpy) of vacancy formation. Here S_v^f arises mainly from the changes in the phonon spectrum of the crystal on forming a defect. For typical metals, it is in the range 0.5–3 in units of k_B. The E_v^f measures the work required to remove an atom from a lattice position and place it at the crystal surface. Values for E_v^f are given in Table 8.1.

In cascades higher order defect clusters can be formed directly. Higher order clusters also can be formed simply by diffusional encounters of lower order mobile defects. For the thermal equilibrium concentrations of divacancies, trivacancies, or n-vacancies, in general, we may write

$$C_{nv}^0 = \eta(C_v^0)^n \exp\left(\frac{g_{nv}^b}{k_BT}\right) \tag{8.14}$$

Here n is 2 for divacancies or 3 for trivacancies and so on, η is the number of different orientations of the stable configuration of the polyvacancy, and g_{nv}^b is the binding Gibbs free energy of the polyvacancy, $g_{nv}^b = E_{nv}^b - TS_{nv}^b$. A typical value for E_{2v}^b is 0.3 eV. The stable configurations of n-vacancy clusters are discussed elsewhere (2, 93).

8.4.1.2. Interstitials. The migration energy for the interstitial is typically much less than that for the vacancy. The reason for this is structural. In the metals aluminum, copper, nickel, molybdenum, iron, and zinc, it has been confirmed that interstitials exist in a split configuration. Two atoms share the same lattice site. Figure 8.22 shows the split interstitial configuration. In the fcc lattice the split interstitial axis is in the $\langle 100 \rangle$ direction, in the bcc lattice it is in the $\langle 110 \rangle$ direction, and in the hcp structure it is parallel to the c axis.

Figure 8.22 also shows how the split interstitial migrates. Transport occurs by a combined rotation and translation. This requires several-fold less energy than translation or rotation alone because of the particularly easy path. For example, the splitting distance of the interstitial is about 20% smaller than the normal atomic separation in nickel. This leads to a strong force which leads to low-frequency resonant vibration modes. The "compressed spring" between the two atoms of the interstitial is unstable to displacements perpendicular to the spring axis—the spring tends to enlarge the perpendicular displacement. Table 8.1 shows the values of E_i^m for selected metals. The values are all 0.3 eV or less.

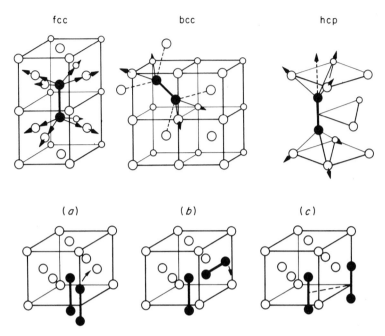

Figure 8.22. Configuration of the split interstitial in fcc, bcc, and hcp metals. The lower part of the figure shows, for the case of the fcc structure, how interstitial migration is accomplished by combined translational and rotational motions: (*a*) stable position, (*b*) metastable position during a jump process, and (*c*) new stable position after a jump. After Ref. 2.

Although interstitials are formed profusely by irradiation, their concentrations are usually entirely negligible in thermal equilibrium. The thermal interstitial concentration is given by Eq. (8.13) with the subscript i substituted for v. However, E_i^f is generally several times larger than E_v^f and this leads to extremely low thermal equilibrium interstitial concentrations. We note here that E_i^f plus E_v^f is generally much lower than the threshold energies noted in the previous section for producing a vacancy–interstitial pair by collisional displacements. The difference is due to the fact that the thermal pair is predominantly formed by the lowest energy *reversible* path. Several values of interstitial formation energy are given in Table 8.1.

As with vacancies, multiple interstitial clusters may be formed thermally once interstitials are available. Equation (8.14) describes the concentrations of multiple interstitial clusters when the subscript i is substituted for v. Multiple interstitial clusters generally have high binding energies and are quite stable against thermal dissociation even at fairly high temperatures. The binding energy of a di-interstitial is about 1 eV. As the cluster size increases, the binding becomes larger—in the range of the formation energy of the interstitial in a perfect lattice. Multiple interstitials of large order become interstitial dislocation loops.

8.4.2. Theory of Point Defect Clustering

Here we outline the theory of defect clustering. Various disciplines of the theory are based on systematic approximations to the general clustering equations. The discussion mainly is restricted to the clustering of vacancies to form cavities, with some mention of point defect clustering to form dislocation loops.

In a material undergoing irradiation, point defects are generated in spatially and temporally discrete cascades and are lost to spatially localized sinks or in atomically discrete recombination events. However, the mainstream of the kinetic theory follows continuum descriptions, wherein the real system is replaced with an effective medium in which all processes occur continuously in time and throughout the medium. By construction, the effective medium is given the correct overall source and sink properties. Each point in the continuum has both source and sink character and contributes infinitesimal increments to point defect generation and point defect loss corresponding to each sink and to recombination. This approach was adopted in the earliest work (20). A rigorous substantiation of effective medium theory has been provided recently (103). Most of our subsequent discussion will be in terms of this *(quasi-) chemical rate theory* picture. Recently a theory that departs from the continuum defect production aspect of this picture has been developed. It is termed the *cascade diffusion theory*. Its main feature is that it explicitly includes the discreteness of point defect production in cascades (104–106). Its salient features will be described in a subsequent section.

8.4.2.1. Clustering Equations. The most general description can be represented in terms of the discrete master equations (107)

$$\frac{dC_j}{dt} = \sum_k w(k; j)C_k - \sum_k w(j; k)C_j + G_j - L_j. \qquad (8.15)$$

Here C_j is the concentration of clusters of type j, for example, containing j vacancies. The transition rate $w(k; j)$ is the rate per unit concentration of cluster type k of transitions to a cluster of type j. The usual initial conditions on Eq. (8.15) at the beginning of an irradiation would be that $C_k = 0$. Just after the initiation, point defects would be present but no higher order clusters, that is, for $k > 1$, $C_k = 0$. At longer times clusters of larger size would evolve. The summations denote all possible reactions leading to a cluster of type j and all possible reactions removing a cluster of type j, respectively. The term G_j is the rate at which clusters of type j are formed directly, excluding the diffusive encounter reactions included in the first summation. This includes direct formation in cascades as well as thermal emission of defects from sinks. The term L_j is the loss rate of clusters of type j by diffusion to sinks such as large cavities and dislocation lines that are not counted as clusters in the second summation.

Equation (8.15), together with its time integral and initial conditions, can be viewed as a general continuity equation for the defect agglomeration process. The physics of each process is expressed in the various reaction parameters. Knowledge of many mechanisms is required, such as the details of diffusion of each species, species interaction parameters, point defect generation characteristics, and interaction with sinks.

8.4.2.2. Direct Solutions.

One approach to solving Eq. (8.15) is by direct numerical methods. No mathematical analysis beyond writing down the equations is required in principle. However, detailed atomistic calculations are needed first to find the configurations and energies of small clusters. This is sometimes termed the hierarchy-of-equations method. In practice, this approach is limited to a relatively small number of equations that can be solved in reasonable times for any forseeable level of computational power. Large clusters and long irradiation times cannot be treated. However, this approach provides the most detailed analysis of the very early stages of clustering, and it has been developed by a number of workers (108–112). Applications have been made to formation of both vacancy and interstitial clusters. In this approach the transition rates and reaction parameters required in the equations are obtained from separate analyses involving solution of problems in combinatorial analysis, continuum elasticity, discrete lattice calculations at small sizes, diffusion, and thermodynamics. These analyses are also used in other approaches to solving Eq. (8.15) described subsequently.

Figure 8.23 shows the computed evolution of loop size distributions from the beginning of an irradiation (110). The development of a loop size distribution is a feature in qualitative agreement with experiments at low doses, where loop size distributions are observed. However, observed distributions are always at larger sizes than shown in Figure 8.23 because of the finite resolution limits of electron microscopy ($\gtrsim 1$ nm for this application).

This approach is limited by the presence in the description of one equation for each type of cluster. In a high-dose irradiation cavities and dislocation loops of diameters larger than 100 nm can be generated. This would require discrete cluster equations running into hundreds of millions. One approach to circumvent this difficulty and extend the description to cover larger clusters and longer times is to devise grouping schemes as cluster sizes increase; groups centered about a certain size are all ascribed the characteristics of one type of cluster. Very small clusters are treated discretely while larger clusters are lumped into progressively larger groups. These methods have extended the reach of the direct hierarchy method (108, 113, 114) but must be applied with care. Otherwise, conservation rules can be violated, leading to incorrect results.

8.4.2.3. Fokker–Planck Descriptions.

Most approaches employ further analysis prior to solving Eq. (8.15). The models usually assume that all clusters are the result of stepwise additions of monomers.

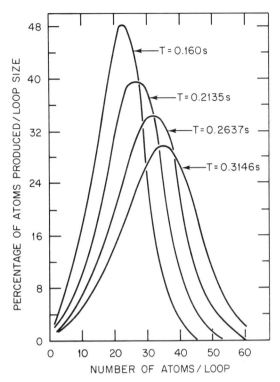

Figure 8.23. Interstitial dislocation loop size distributions computed using a hierarchy-of-equations clustering method. The material is nickel at a temperature of 450°C and a dose rate of 10^{-6} dpa/s. After Ref. 110.

Under these assumptions, Eq. (8.15) may be rewritten as

$$\frac{\partial C_j(t)}{\partial t} = w(j-1; j)C_{j-1}(t) + w(j+1; j)C_{j+1}(t)$$

$$- [w(j; j+1) + w(j; j-1)]C_j(t); \quad j \geq 2. \tag{8.16}$$

The equations for $j = 1$ are different than Eq. (8.16) because they contain the sink loss term L, the generation term G, and the recombination term annihilating interstitials with vacancies. These equations will be described in a subsequent section. If we imagine j to be a continuous variable and use a Taylor series expansion in Eq. (8.16) to relate all functions to their values at size j, we simplify the description to a continuum diffusional approximation in phase space, known as a Fokker–Planck equation:

$$\frac{\partial C_j(t)}{\partial t} = -\frac{\partial}{\partial j}\{C_j(t)[w(j; j+1) - w(j; j-1)]\}$$

$$+ \frac{1}{2}\frac{\partial^2}{\partial j^2}\{C_j(t)[w(j; j-1) + w(j; j+1)]\}. \tag{8.17}$$

This is a well-known equation for the evolution of clusters (115, 116). It has been adapted recently for the radiation effects problem by several workers (117–121).

The first term in Eq. (8.16) describes a "drift" in size space driven by the "force" of excess condensation of one point defect type over the other, in analogy with the physical drift of diffusing species in real space due to elastic or coulombic forces, for example. The second term describes a "diffusion" in size space by a similar analogy. The first term ensures that a large cluster will grow inexorably in a radiation field, while the second term accounts for the fact that two different clusters introduced at the same time and the same size may differ in size at a later time due to random encounters with point defects.

Hybrid approaches have also been developed, where discrete cluster equations are used at small sizes and the Fokker–Planck equation is used at larger sizes (121, 122).

"Nucleation" and "growth" regimes of cluster formation can be demarcated according to whether the first term in Eq. (8.17) is negative or positive, respectively. The first term is equivalent to the common distribution evolution function in the continuity equation, consisting of the sum of a concentration times the derivative of "velocity" with respect to j, plus a velocity times the derivative of concentration with respect to j. The second term is the curvature in the transition rate. In the nucleation regime, stochastic processes are dominant. Nucleation theory is primarily concerned with predicting the rates and time evolution of densities of viable clusters. Growth theory is mainly concerned with the growth kinetics of cluster size distributions in a dynamically changing environment. Often, time scale is a convenient demarcation in irradiation experiments. Prior to a certain time, cluster number densities are established. Afterward, the number density remains relatively stationary and the clusters grow in size.

8.4.2.4. Nucleation Theory. The nucleation rate can be defined as the rate at which clusters achieve a critical size above which the first term in brackets in Eq. (8.17) becomes positive and remains so for all larger sizes. We first find the cluster size distribution, C_j, below that critical size. For simplicity the analysis is restricted to steady state, where both the number flowing past the critical size and the distribution C_j are stationary in time.

To proceed, we must characterize physically the transition rates w. The method (123, 124) is based on a modified approach to the "classical" theory of precipitate nucleation of solute clusters in a solvent (125). The modification accounts for the fact that in an irradiated material precipitation of both the "solute" (vacancy) and the "antisolute" (interstitial) determines the nucleation rate.

We label the arrival rate of an interstitial or vacancy at a cluster containing j vacancies as $\beta_{i,v}(j)$, and the emission rate as $\alpha_{i,v}(j)$. Thus, the first term in Eq. (8.16) becomes $\beta_v(j - 1)C_{j-1}$, the second term becomes $[\beta_i(j + 1) + \alpha_v(j + 1)]C_{j+1}$, and the third becomes $[\beta_v(j) + \beta_i(j) + \alpha_v(j)]C_j$. As discussed in Section 8.4.1, we neglect terms of the type $\alpha_i(j)$. If the net number arriving at size $j + 1$ from j is denoted as $M(j)$, we may write (126)

$$M(j) = \beta_v(j)C_j - [\beta_i(j + 1) + \alpha_v(j + 1)]C_{j+1} \qquad (8.18)$$

The general steady-state condition on Eq. (8.18) is that for which $M(j) = M(j - 1) = M$, the steady-state nucleation rate. This rate can be determined by deriving the concentration at the critical size. Let the critical size be denoted by j^c and the corresponding concentration by C_{jc}.

Nucleation theory employs an ingenious device to find M. The argument is based on the special solution to Eq. (8.18), $M(j) = M(j - 1) = 0$. It is plausible that if the forward reaction bringing clusters of size j^c to $j^c + 1$ is not allowed to proceed, this would zero the nucleation rate but would not significantly affect the cluster size distribution C_j below j^c. This is because the net forward reaction rate M is but a small fraction of the gross forward or backward reaction rates given in Eq. (8.18). The size distribution C_j^0 is obtained for this so-called constrained distribution and the problem is solved. With $M(j) = 0$ in Eq. (8.18) we employ the equation recursively to find the concentration of j-mers in terms of monomers,

$$C_j^0 = \prod_{k=1}^{j-1} \frac{\beta_v(k)C^0 1}{\beta_i(k + 1) + \alpha_v(k + 1)} \qquad (8.19)$$

The superscript 0 denotes quantities in the constrained distribution. Equation (8.19) may be rewritten as

$$C_j^0 = C_1^0 \exp\left[\frac{-\Delta G(j)}{k_B T}\right] \qquad (8.20)$$

where

$$\Delta G(j) = -k_B T \sum_{k=1}^{j-1} \ln \frac{\beta_v(k)}{\beta_i(k + 1) + \alpha_v(k + 1)} \qquad (8.21)$$

The quantity ΔG may be described as a kinetic analog of the Gibbs free energy of j-mer formation under the nonequilibrium conditions of irradiation.

Assuming that defects go to sinks at their steady-state rates, α and β can be expressed in terms of point defect concentrations as (126)

$$\alpha_v(j) = \frac{4\pi\Omega^{1/3}j^{1/3} \, Z_v^c(j)D_vC_v^e(j)}{(4\pi/3)^{1/3}} \tag{8.22}$$

and

$$\beta_{i,v}(j) = \frac{4\pi\Omega^{1/3}j^{1/3} \, Z_{i,v}^c(j)D_{i,v}C_{i,v}}{(4\pi/3)^{1/3}} \tag{8.23}$$

The new symbols Ω, and $Z_{i,v}^c$, and $C_v^e(j)$ denote atomic volume, capture efficiency of a vacancy cluster (cavity) for interstitials or vacancies, and the vacancy concentration in thermal equilibrium with a cluster of size j, respectively.

We may now find the nucleation rate M from Eq. (8.18) as

$$M = \beta_v(j)C_j^0 \left(\frac{C_j}{C_j^0} - \frac{C_{j+1}}{C_{j+1}^0}\right) \tag{8.24}$$

The repeated summation of Eq. (8.24) gives

$$\sum_{k=1}^{\infty} \left(\frac{C_k}{C_k^0} - \frac{C_{k+1}}{C_{k+1}^0}\right) = \frac{C_1}{C_1^0} - \frac{C_\infty}{C_\infty^0} = \sum_{k=1}^{\infty} \frac{M}{\beta_v(k)C_k^0} \tag{8.25}$$

We know, however, that $C_\infty/C_\infty^0 \to 0$ and $C_1/C_1^0 \to 1$ in analogy with classical nucleation theory (125). Using Eq. (8.21) and noting that $C_1^0 = C_v$, the vacancy concentration, we obtain

$$M = C_v \bigg/ \sum_{k=1}^{\infty} \frac{\exp[\Delta G(k)/k_BT]}{\beta_v(k)} \tag{8.26}$$

Substituting for $\beta_v(k)$ from Eq. (8.23), Eq. (8.26) may also be written as

$$M = \frac{4\pi\Omega^{1/3}C_v^2D_v}{(4\pi/3)^{1/3}} \bigg/ \sum_{k=1}^{\infty} \frac{\exp[\Delta G(k)/k_BT]}{k^{1/3}Z_v^c(k)} \tag{8.27}$$

The last expression shows that the nucleation rate is proportional to the square of the vacancy concentration or more precisely to the vacancy flux D_vC_v times the vacancy concentration C_v. This product measures the probability that two vacancies encounter each other, and this probability governs the nucleation rate.

Figure 8.24 shows the theoretically calculated ΔG function for typical conditions where nickel is irradiated in a fast reactor.

8.4.2.5. Growth Theory. A major share of the work in cavity swelling and dislocation loop formation has been in growth kinetics (127–130). Here it is assumed that there are only two basic kinds of entities—point defects and point defect sinks. The latter may be large point defect clusters as well as objects such as dislocation

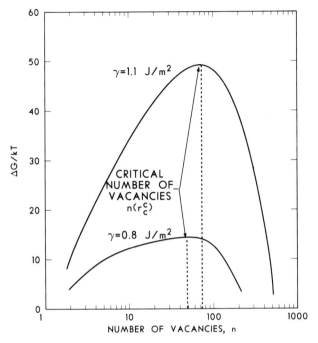

Figure 8.24. Theoretically derived ΔG function for fast reactor irradiation of nickel at $T = 500°C$ and a dose rate of 10^{-6} dpa/s. After Ref. 126.

lines and grain boundaries. The clusters grow and are characterized at any time by the dimension j, indicating the number of contained point defects, or equivalently the dimension r, indicating the cluster radius. Thus, the complete cluster size distribution of Eq. (8.15) is simplified drastically to only single-point defects and one type of large growing cluster. Often, to achieve higher accuracy, the large clusters are characterized by a size distribution instead of a uniform size (130), thus maintaining more generality in the description.

In terms of Eq. (8.17) only the first term in curly braces is retained in growth theory. Rewriting this in terms of the variable r, characterizing cluster radius, gives

$$\frac{\partial N(r, t)}{\partial t} = - \frac{\partial [\dot{r}_c N(r, t)]}{\partial r} \tag{8.28}$$

where now r replaces j and $N(r)\, dr$ replaces C_j as the number density of clusters of size r. The crucial term in Eq. (8.28) describing how clusters grow and therefore how a cluster size distribution evolves is $\dot{r}_c$, denoting the time derivative of the cavity radius. Much of cluster growth theory centers on understanding the physics underlying $\dot{r}_c$.

The growth rate of a cavity of radius r_c is determined by the net flux of vacancy volume per unit cavity area per unit time,

$$\frac{dr_c}{dt} = \frac{\Omega}{r_c} [Z_v^c D_v C_v - Z_i^c D_i C_i - Z_v^c D_v C_v^e(r_c)] \tag{8.29}$$

The symbol $C_v^e(r_c)$ denotes the vacancy concentration in thermal equilibrium with a cavity of radius r_c. Similarly, the growth rate of an interstitial loop of radius r_l is given by

$$\frac{dr_l}{dt} = a_l [Z_i^l D_i C_i - Z_v^l D_v C_v + Z_v^l D_v C_v^e(r_l)] \tag{8.30}$$

The new symbols a_l, $Z_{i,v}^l$ and $C_v^e(r_l)$ denote the area per atom enclosed in a loop, the loop capture efficiency for interstitials and vacancies, and the vacancy concentration in thermal equilibrium with a loop of radius r_l, respectively.

8.4.2.6. Point Defect Concentrations. In the previous two sections the concentrations of vacancies and interstitials are crucial in determining cluster nucleation and growth rates. The concentrations are obtained by solving continuity equations describing point defect generation and loss. The equations can be written as

$$\nabla \cdot \left(D_v \nabla C_v + \frac{D_v C_v}{k_B T} \nabla U_v \right) + G_v - RC_v C_i - K_v C_v = \frac{\partial C_v}{\partial t} \tag{8.31}$$

and

$$\nabla \cdot \left(D_i \nabla C_i + \frac{D_i C_i}{k_B T} \nabla U_i \right) + G_i - RC_v C_i - K_i C_i = \frac{\partial C_i}{\partial t} \tag{8.32}$$

Here G denotes vacancy or interstitial generation both by displacement and by thermal emission from sinks. The symbol R is the recombination coefficient given by $4\pi r_r(D_i + D_v)$, where r_r is the radius of the recombination volume. The remaining terms on the left sides describe losses to sinks. These are modeled in two ways. The terms involving spatial derivatives describe leakage and drift to the particular type of sink that is modeled as discrete. The symbol U is interaction energy with the discrete sink, and the K is the reaction rate constant that describes the loss rates per unit point defect concentration to sinks that are modeled as continuum sinks. Of these sinks there may be n types such as dislocations, cavities, dislocation loops, and so on, and $K_{v,i} = \sum_n K_{v,i}^n$, where $K_{v,i}^n$ is the vacancy or interstitial loss rate to the particular sink of type n. Here $K_{v,i}^n$ is defined as the product of a sink strength, $S_{v,i}^n$, and the appropriate diffusion coefficient, $D_{v,i}$. The sink strength reflects the physics of the point defect loss processes and is one of the most important quantities in the theory. The form of the sink strength is discussed in the next subsection.

The right sides of Eqs. (8.31) and (8.32) express changes in defect concentration with time. However, the relaxation times for changes in the concentrations are generally much shorter than characteristic times for changes in K by microstructural evolution. Therefore, except for a short time after the beginning of an irradiation (this time is given below), the right sides of Eqs. (8.31) and (8.32) can be set to zero. Similarly, when we treat all sinks according to their continuum descriptions, the spatial derivatives can be neglected. In that case Eqs. (8.31) and (8.32) become algebraic equations that are easily solved for C in terms of the processes governing production and loss,

$$C_v = \frac{[K_iK_v + R(G_i - G_v)]}{2RK_v} \left[\left(1 + \frac{4RG_vK_iK_v}{[K_iK_v + R(G_i - G_v)]^2} \right)^{1/2} - 1 \right] \quad (8.33)$$

and

$$C_i = \frac{[K_iK_v + R(G_v - G_i)]}{2RK_i} \left[\left(1 + \frac{4RG_iK_iK_v}{[K_iK_v + R(G_v - G_i)]^2} \right)^{1/2} - 1 \right] \quad (8.34)$$

The time until these steady-state concentrations prevail is given roughly by K_v^{-1}. This is the mean time it takes a vacancy to diffuse to a sink from where it is created by displacement. The interstitial typically diffuses much faster than the vacancy, and so it might be thought that it tends to come into steady state earlier. However, because of the recombination reaction the interstitial concentration does not reach steady state until the vacancy concentration does. Prior to this the interstitial concentration maintains a quasi-steady state with respect to the instantaneous value of the evolving vacancy concentration. We also note that what we refer to here as steady state is really also a quasi-steady state. Over long times clustering leads to microstructural evolution and the point defect concentrations gradually vary to maintain a steady state with respect to the instantaneous sink strength. Typical sink strengths in materials of interest are in the range 10^{12}–10^{16} m^{-2}.

8.4.3. Sink Strengths and Sink Efficiencies

The sink strengths in Eqs. (8.33) and (8.34) are essential in determining the point defect concentrations. The sink strength is that quantity that, when taken in a product with the volume-averaged point defect concentration and the point defect diffusion coefficient, gives the loss rate to all sinks of that specific type. The sink strength is obtained by solving a full spatial diffusion problem for the type of sink in question, such as a cavity, dislocation, or grain boundary. This solution always shows that the point defect loss rate is given by an operator in the sink characteristics and geometry times the bulk or far-field point defect concentration. The theory of sink strengths has been discussed in many papers. The most complete treatment is given in Ref. 131, where other work is also mentioned. As a specific example, we illustrate a derivation of sink strength for the cavity. Results for other sink types are quoted.

Consider an isolated cavity in an infinite medium with smeared-out sinks of strength S. For simplicity, neglect recombination and thermal vacancy generation by sinks other than cavities. In spherical coordinates we may write, for either type defect,

$$\frac{D}{\rho^2}\frac{d}{d\rho}\rho^2\frac{dc}{d\rho} - DSc + G = 0 \tag{8.35}$$

where r_c is the cavity radius, and the boundary condition at the cavity surface is given by

$$D\frac{dc}{d\rho}\bigg|_{\rho=r_c} = w[c(r_c) - C^e(r_c)] \tag{8.36}$$

Here ρ is the distance from the cavity center, and w is the point defect transfer velocity at the cavity surface. Equation (8.36) specifies that the flux due to the gradient equals the net forward reaction rate. The solution is

$$c = C - \frac{[C - C^e(r_c)]wr_c/D}{1 + S^{1/2}r_c + wr_c/D}\frac{r_c}{\rho}\exp\left[-S^{1/2}(\rho - r_c)\right] \tag{8.37}$$

where C is the bulk concentration, Eqs. (8.33) and (8.34). The current of point defects to the cavity is

$$4\pi r^2 D\frac{dc}{d\rho}\bigg|_{\rho=r_c} = \frac{4\pi[C - C^e(r_c)]wr_c^2(1 + S^{1/2}r_c)}{1 + S^{1/2}r_c + wr_c/D} \tag{8.38}$$

By the definition of sink strength, this gives the sink strength of a cavity as

$$S^c = 4\pi r_c(1 + S^{1/2}r_c)\frac{wr_c/D}{1 + S^{1/2}r_c + wr_c/D} \tag{8.39}$$

The first part $(4\pi r_c)$ is a simple geometric term, the second part $(1 + S^{1/2}r_c)$ is often designated as a multiple sink correction or interactive term, and the third part is the sink capture efficiency, designated as Z^c. The latter measures how good an absorber the sink is. Alternatively, the second (interactive) and third terms can be lumped together as a generalized capture efficiency (132). The capture efficiency provides a convenient device to maintain the expression for loss rate in terms of $C - C^e(r_c)$ even though the concentration at the surface $C(r_c) > C^e(r_c)$. An equivalent definition of capture efficiency is (130)

$$Z^c = \frac{C - C(r_c)}{C - C^e(r_c)} \tag{8.40}$$

The sink strengths of straight dislocations, dislocation loops, grain boundaries, free surfaces, and precipitates can also be derived by analogous methods. Here we record the results for dislocations and grain boundaries. Results for dislocation loops (133–135) and precipitates (136) are given elsewhere.

The standard result for the straight dislocation is

$$S^d = Z^d L \tag{8.41}$$

where

$$Z^d = \frac{2\pi}{\ln (r_L/r_d)} \tag{8.42}$$

Here L is the dislocation density, $r_L = (\pi L)^{-1/2}$ is a measure of dislocation spacing and r_d is the dislocation capture radius for the point defect. The quantity r_d is assigned the further subscript v or i to denote the different radii for vacancies and interstitials. Generally, $Z_i^d > Z_v^d$ because $r_{di} > r_{dv}$. This difference leads to the bias that drives cavity swelling, as will be discussed later. When cavities are included in the medium, Eq. (8.41) takes on a more complex form. Results for both random and periodic arrays are given elsewhere (131).

The sink strength of a grain boundary is obtained in a similar way. A spherical grain is either embedded in a continuum or modeled as part of a grain boundary lattice. Expressions for S^{gb} are obtained in a straight-forward way. These generally involve hyperbolic functions. Limiting expressions that are useful are

$$S^{gb} = \frac{60}{d^2} \qquad S^{1/2}d \ll 1 \tag{8.43}$$

and

$$S^{gb} = \frac{6S^{1/2}}{d} \qquad S^{1/2}d \gg 1 \tag{8.44}$$

where d is the grain diameter. Note that when other sinks are important, Eq. (8.44), their sink strength is an inherent part of the grain boundary sink strength. In typical irradiated microstructures Eq. (8.44) is a good approximation.

8.4.4. Cascade Diffusion Theory

The above discussion of kinetics has been in terms of a continuum reaction rate theory wherein both point defect generation and loss are modeled as occurring continuously in time and throughout the medium. Most of the theoretical work in the field has been in this approximation. We know, however, that point defects are generated in discrete bursts (the cascades described in Section 8.3) and that they are lost to spatially discrete sinks. Recently we have developed a generalization (104–106) that accounts for the spatially and temporally discrete production of point defects.

The continuum theory gives rise to steady point defect concentration, Eqs. (8.33) and (8.34) after buildup times as described in Section 8.4.2.6. The cascade diffusion theory is more descriptive of the actual situation in that it describes sporadically fluctuating concentrations at any point in the medium because of the superposition of contributions from random discrete cascades occurring throughout the material. These fluctuations can be crucial in understanding certain processes in an irradiated material. For example, the theoretical discovery of cascade-induced creep, a topic to be discussed in a later section, was made possible (106). On the other hand, the continuum rate theory can be derived through approximations and limits of the cascade diffusion theory (104).

The approach is to introduce a cascade as a discrete unit of point defect production modeled as a delta function. The cascade is allowed to diffuse in the lossy medium where the solution for point defect concentration is described by the equation

$$\frac{\partial c(\rho, t)}{\partial t} = D(\nabla^2 - S)c(\rho, t) + G(\rho, t) \tag{8.45}$$

The solution is given by

$$c(\rho, t) = \frac{1}{(4\pi Dt)^{3/2}} \exp\left(-\frac{\rho^2}{4Dt}\right) \exp\left(-SDt\right), \tag{8.46}$$

where ρ is the distance from the cascade center and t is the time since its introduction. This solution is applicable far from a discrete sink. However, the presence of a sink modeled as discrete, such as a cavity or dislocation, perturbs the concentration given by Eq. (8.46). The solutions in the presence of these two types of discrete sinks have also been derived (104). Multiple solutions of this type corresponding to the occurrence of cascades anywhere in space for all previous time are superimposed on each other to give the concentration at the point of observation.

Figure 8.25 shows the point defect concentrations at an arbitrary point calculated for a typical neutron irradiation of nickel at 500 °C and 10^{-6} dpa/s. On the left is shown the vacancy concentration and on the right is shown the interstitial concentration. As can be seen, the vacancy concentration appears to have a fairly steady background level punctuated by large spikes. The background is contributed mainly by the very numerous cascades at large distances, while the spikes are contributed by the relatively few nearby cascades and by arrival coincidences of defects from different cascades. For these conditions it is found that $\sim 10^3$ cascades in various stages of decay make nonnegligible contributions to the concentration. By contrast, because of its much more rapid diffusion, the interstitial picture is different. At any instant there is at most one cascade making a nonnegligible con-

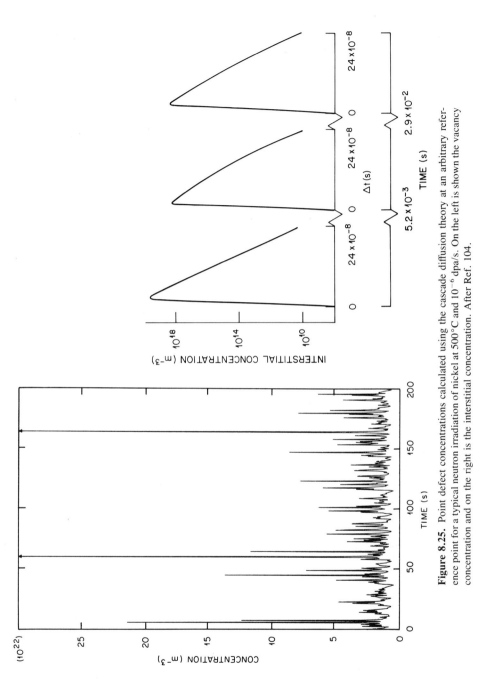

Figure 8.25. Point defect concentrations calculated using the cascade diffusion theory at an arbitrary reference point for a typical neutron irradiation of nickel at 500°C and 10^{-6} dpa/s. On the left is shown the vacancy concentration and on the right is the interstitial concentration. After Ref. 104.

423

tribution to the concentration at a point. There is no background. The interstitial concentration profile, Figure 8.25 (right), consists of extremely large spikes whose decay time is less than a microsecond separated by a few milliseconds. For most of the time there is negligible interstitial concentration.

By contrast, the continuum rate theory picture is an average of these results. The equivalent concentration for the vacancy, for example, would be a horizontal line at 2×10^{22} vacancies/m^3 in Figure 8.25.

Similarly, microscopic information on where in space the point defects come from can be obtained from the theory. Figure 8.26 shows differential and integral

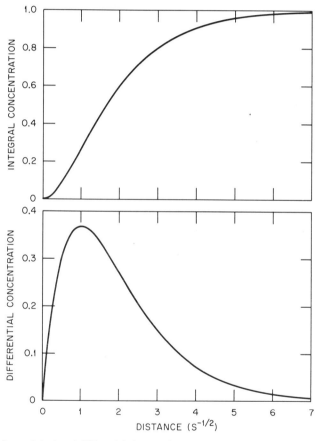

Figure 8.26. Integral (top) and differential (bottom) importance functions for point defect concentration calculated using the cascade diffusion theory. These represent concentrations normalized to the concentration obtained by considering contributions from cascades out to infinite distance. The differential function shows the fraction of the average concentration at a reference point contributed by the cascades in the material at any radial distance. The integral function shows the fraction of the average concentration contributed by cascades within a sphere of any radius. The distance on the abscissa is in units of the inverse square root of the sink strength. After Ref. 104.

importance functions versus distance from the observation point. These results show, on the average, how important is each shell of volume corresponding to a radial distance increment in contributing point defects at the point of observation. For example, from the lower (differential) figure we see that the most important infinitesimal radial increment is at one absorption length, defined as the inverse square root of the sink strength ($S^{-1/2}$). From the upper figure (integral) we see that the region of space within three absorption lengths contributes more than 80% of the total average defect concentration observed at the reference point.

An interesting application is the investigation of cascade coincidence probabilities. For example, within a small volume, cascades occur at an average rate defining an average time interval for n cascades to occur (average n-fold interval). However, for any time interval, even one much shorter than the average one, there is a finite probability that n cascades will occur. In particular, when the volume is of radius of the order of the mean point defect absorption length, $S^{-1/2}$, and the time interval is of the order of the mean point defect diffusion time, the probability of multiple cascade occurrence is of interest. Figure 8.27 shows the fraction of n-fold intervals of duration t or less as a function of t. Although the fraction of n-fold intervals for times of very short duration is small, it nevertheless may be significant when it is realized that typically 10^{21}–10^{24} cascades/s m^3 occur during

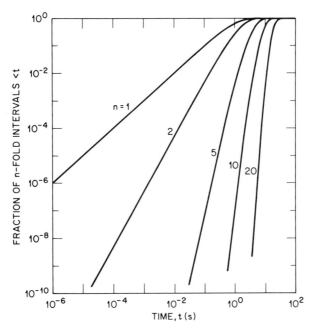

Figure 8.27. Coincidence probabilities for n cascades occurring within a sphere of radius $S^{-1/2}$ and time interval t as a function of t. An n-fold interval is a time interval containing n events. After Ref. 104.

irradiation. Very often within intervals of the order of the vacancy lifetime ($\geq 10^{-3}$ s) and even within intervals of the order of the interstitial lifetime ($< 10^{-6}$ s) for materials and temperatures of interest, several cascades regularly occur in a volume whose radius is of the order of the point defect absorption length. The "coincidence" of n cascades in this physically relevant volume may lead to intense interaction, possibly resulting in enhanced agglomeration of significant numbers of point defects and to other effects.

8.5. THEORETICAL MODELS

In this section a number of specific theoretical models for radiation effects phenomena are described.

8.5.1. Swelling

8.5.1.1. Bias. When the expressions for the point defect concentrations, Eqs. (8.33) and (8.34), are substituted into Eq. (8.29), the result can be written (130) as

$$\frac{dr_c}{dt} = \frac{\Omega}{r_c} (Z_i^d Z_v^c - Z_v^d Z_i^c) A \qquad (8.47)$$

Here A is a function of total dislocation and cavity sink strengths and their ratio, of dose rate, and of point defect recombination. The quantity $Z_i^d Z_v^c - Z_v^d Z_i^c$ is termed the bias. Its value determines the potential of a material to swell; it must be positive in order for any swelling to occur. It expresses the degree of preferential partitioning of vacancies to cavities and interstitials to dislocations. Values of this quantity necessary to explain observed swelling range from about 0.01 to 1 depending on the material and on values assigned to other parameters in Eq. (8.47).

Frequently, when one is interested only in swelling rate rather than the radial growth rate of individual cavities, Eq. (8.47) is expressed directly in terms of swelling rate, $\dot{V} = 4\pi r_c^2 (dr_c/dt)$, where V is the cavity volume fraction.

8.5.1.2. Dose and Dose Rate Dependence. Equation (8.47) can be written in two limiting forms:

$$\frac{dr_c}{dt} = \frac{\Omega}{r_c} \left(\frac{D_i D_v G}{Z_i^d Z_v^d R}\right)^{1/2} \frac{Q_i^{1/2} Q_v^{1/2}(Z_i^d Z_v^c - Z_v^d Z_i^c)}{(1 + Q_i)^{1/2}(1 + Q_v)^{1/2}} \quad \text{(recombination dominated)}$$

$$(8.48)$$

and

$$\frac{dr_c}{dt} = \frac{\Omega G Q_i Q_v (Z_i^d Z_v^c - Z_v^d Z_i^c)}{r_c Z_i^d Z_v^d L (1 + Q_i)(1 + Q_v)} \quad \text{(sink dominated)} \qquad (8.49)$$

Equation (8.48) applies where most point defects are lost by bulk recombination, and Eq. (8.49) applies where most are absorbed at sinks.* These are the two extremes in modes of point defect loss. As can be seen, in the former case the cavity growth rate depends parabolically on the dose rate, while in the latter case the cavity growth rate depends linearly on the dose rate. In these equations $Q_{i,v}$ denotes the ratio of dislocation to cavity sink strength,

$$Q_{i,v} = \frac{Z_{i,v}^d L}{4\pi r_c N_c Z_{i,v}^c} \tag{8.50}$$

where L is the dislocation density and N_c is the cavity density.

To integrate Eq. (8.48) or (8.49) to obtain a swelling-versus-dose prediction, the dose dependence of dislocation density must be known. Equations (8.48) and (8.49) are multiplied by $4\pi r_c^2 N_c$ so that the left side becomes $\dot{V}$ if N_c is constant ($V = \frac{4}{3}\pi r_c^3 N_c$). Modeling the evolution of dislocation density from the interstitial loop stage through interactions and annihilation is a complex problem that is of current interest (137, 138). We have examined a wide variety of limiting cases by noting that the behavior in large classes of experimental observations can be expressed as simple phenomenological relations (130):

1. At the early stages every vacancy in a cavity is matched by an interstitial in a dislocation loop (139). In this case

$$L \propto V^{1/2} \tag{8.51}$$

2. At higher doses dislocation loops give way to a dislocation line network whose density is maintained constant by competing loop formation and segment annihilation processes,

$$L = \text{constant} \tag{8.52}$$

These results are inserted into Eqs. (8.48) and (8.49) together with expressions for Z^c, which determine whether point defect absorption is diffusion or reaction controlled at the cavity surface. The results show that swelling may be expressed as proportional to (dose)^n, where n may take on values from $\frac{3}{5}$ to 3. A more complete discussion and a table of exact values for many cases are provided elsewhere (130). In general, during an extended irradiation several regimes of characteristic exponents will be traversed. Figure 8.28 is a schematic depiction of the predicted dose dependence of swelling. These theoretical predictions for dose dependence of swelling are generally in agreement with results observed experimentally.

8.5.1.3. Temperature Dependence. The temperature dependence of swelling implied by Eq. (8.29) can be obtained by computation. The typical result is that

---

*Of course, most point defects absorbed at sinks recombine there because of their absorption in nearly equal numbers. Only a small excess accumulates at each sink dictated by the bias as discussed.

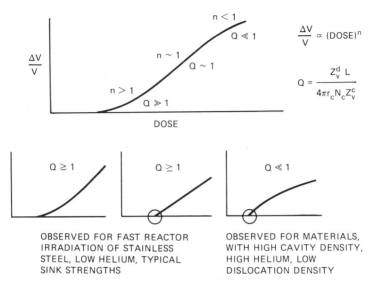

Figure 8.28. Schematic summary of theoretically predicted dose dependences of swelling. When the ratio of dislocation to cavity sink strength is much greater than unity, the swelling rate is low and the dose exponent is high; when the ratio of sink strengths is reversed, the swelling rate is low and the dose exponent is low. When the sink strengths are approximately equal, the swelling rate is maximum and the dose exponent is approximately unity. The circles in the plots in the lower part indicate that the early (high-dose exponent) stages are compressed into small intervals on the scale of the figures. After Ref. 130.

swelling exhibits a peak at intermediate temperatures and is negligible at much lower and higher temperatures. The swelling is low at low temperatures because the vacancy is practically immobile. Its concentration in the matrix therefore builds up and more vacancies and interstitials are lost by mutual recombination. At high temperatures another process diminishes swelling. Cavities emit thermal vacancies rapidly at high temperatures, and this counterbalances the net vacancy influx driven by irradiation. At intermediate temperatures swelling is maximized—both thermal emission and mutual recombination are less important, and the excess flow of vacancies to cavities is maximized. "High" and "low" temperatures are dependent on the properties of the material and in particular on how the overall sink strength changes with temperature. In some materials, over a certain temperature range, lowering the temperature increases the sink strength so strongly that recombination is actually less at the lower temperature. Theoretical predictions for temperature dependence of swelling are generally in good agreement with the results of experimental observations.

The location of the peak swelling temperature depends on dose rate, sink strength, and the predominant mode of defect loss. Of particular interest for comparisons of ion irradiation results ($\sim 10^{-3}$ dpa/s) with neutron irradiation results ($\sim 10^{-6}$ dpa/s), there is a strong shift in the peak with dose rate. The reason for the shift is that when dose rate is increased, more point defects are created, but their diffusion velocities remain fixed. To remove defects at the higher rate in the steady state requires the point defect concentrations to be higher. This results in relatively more recombination. However, by increasing temperature, the same relative ratio of recombination rate to absorption rate at sinks may be restored.

A general equation for temperature shift of the swelling rate has been derived (140):

$$T^{(2)} - T^{(1)} = \frac{[kT^{(1)2}/(E_v^m + nE_v^f)] \ln [(G^{(2)}/G^{(1)}) M]}{1 - [kT^{(1)}/(E_v^m + nE_v^f)] \ln [(G^{(2)}/G^{(1)}) M]} \qquad (8.53)$$

Here $T^{(2)}$ is the peak swelling temperature corresponding to dose rate $G^{(2)}$, and $T^{(1)}$ is that corresponding to $G^{(1)}$. The number n is 1 or 2 depending on whether loss is dominated by sinks or by recombination, respectively. The quantity M contains microstructural information. Figure 8.29 is a compilation showing observed peak swelling temperatures as a function of dose rate used in various experiments (141). The line labeled "theory" shows the predicted behavior (140).

8.5.1.4. Gas Effects. The effects of gaseous impurities on swelling are of particular interest in the theory because of very significant effects observed experimentally. The insoluble rare gas helium is one of primary concern. It is produced at

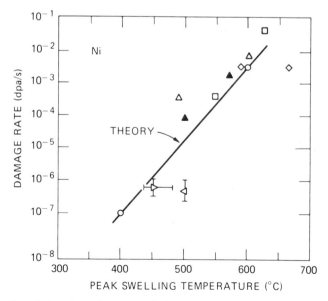

Figure 8.29. Compilation of experimental results for peak swelling temperature as a function of dose rate. Theoretically predicted trend is shown as the line. After Refs. 140 and 141.

levels of tens of parts per million by (n, α) transmutation reactions in fast reactors. In future fusion reactors helium is expected to be produced to levels of thousands of parts per million. Gases may affect swelling through several reaction pathways. These include cavity pressurization, induced changes in microstructural sink strength, induced changes in microcomposition, and alterations in atomic transport as discussed in our recent review (142).

One of the most important effects is cavity pressurization. Gas pressure suppresses thermal vacancy emission from cavities, as can be seen from the expression for $C_v^e(r_c)$, the thermal vacancy concentration near a cavity of radius r_c,

$$C_v^e(r_c) = C_v^0 \exp\left[\left(\frac{2\gamma}{r_c} - P\right)\frac{\Omega}{kT}\right] \tag{8.54}$$

When Eq. (8.54) is substituted into Eq. (8.29), the resulting cavity growth rate versus radius is as shown in Figure 8.30. For a given number of contained gas atoms, there are two roots, one at a stable radius, r_c^s, and one at a critical radius r_c^c. The cavity must achieve a size beyond r_c^c to grow by bias-driven growth. (The point on the figure denoted r_c^{c0} is the critical radius with no gas.) Gas makes this easier. With more contained gas r_c^c is reduced. At the same time more gas brings r_c^s closer to r_c^c. As more gas is added, r_c^c is decreased and r_c^s is increased until at a critical number of gas atoms, n_g^*, they meet at r_c^*. These quantities are also shown in Figure 8.30. If any more gas is added, a critical radius no longer exists, and bias-driven cavity growth is inevitable. There are thus two qualitatively different

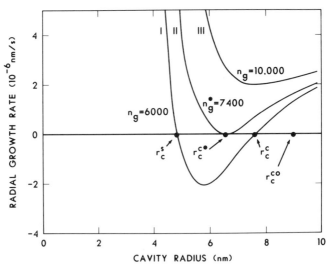

Figure 8.30. Cavity growth versus cavity radius for several different levels of contained gas in the cavity, as obtained from the theory. After Ref. 142.

paths to achieve bias-driven growth. The cavity may depart from its stable size by stochastic fluctuations, on occasion exceeding r_c^c. Alternatively, the cavity may accumulate more than the critical number of gas atoms, whereby bias-driven growth is insured; no fluctuations are required. These two possibilities merge continuously into each other as gas is added.

These critical quantities have been found to be extremely useful in understanding a number of aspects of swelling (142–155). The main ideas arising from effects of gas on swelling can be summarized as follows:

1. An increase in swelling at high temperatures.
2. A related theoretical analysis to calculate temperature shifts with changes in dose rate and microstructures [e.g., resulting in Eq. (8.53)].
3. The prediction of bimodal cavity size distributions.
4. A method for assessing fundamental parameters such as vacancy migration energy and bias.
5. Principles for the design of radiation-resistant alloys.

Item 1 is synthesized from the forms of the equation above and Eq. (8.29). Thermal emission is decreased by gas at any temperature. Item 2 is a method developed based on the critical radius and used in deriving Eq. (8.53). It is described in the literature (140, 130).

Items 3 and 4 are also important results of the theory. When gas is present, it is predicted that cavities containing less than n_g^* gas atoms remain stationary at or below r_c^* but those containing more than n_g^* grow. Thus the distribution should separate into two parts. As dose increases, the gap between growing and stationary cavities should increase. Figure 8.31 is an observed bimodal distribution in Fe–10% Cr after ion irradiation to a dose of 30 dpa at a temperature of 850 K (154). Theoretical expressions have been derived to predict r_c^* as a function of parameters of the material and irradiation conditions. Having measured r_c^* by reference to Figure 8.31, we may reverse the process and determine reasonable values of the unknown parameters that could give the result (154). In this way the relatively poorly known parameters for Fe–10% Cr, an important model material for structural alloys, have been estimated.

The last item designates the uncovering of two important principles for the design of swelling-resistant alloys. For moderate and high temperatures the critical radius is large. Thus, there is not a significant possibility that it is achieved by fluctuations. The cavity must wait until it acquires (nearly) n_g^* gas atoms before rapid swelling can begin. Two possibilities suggest themselves. By manipulation of materials compositions and treatments, we might be able to maximize n_g^*. Increasing the nickel content in iron–chromium–nickel alloys has been shown to increase n_g^* (155) by decreasing the bias. Also, we can delay the time to achieving any given value of n_g^* by increasing the number of cavities in the system, thus diluting the fixed generation rate of gas among more sites. As an example, this

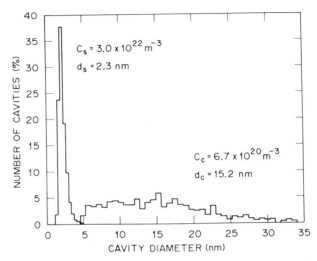

Figure 8.31. Histogram showing the bimodal cavity size distribution observed after irradiation of Fe–10% Cr with 4-MeV iron ions to 30 dpa at 850 K. The material was preinjected with 300 appm helium. After Ref. 154.

approach has been applied successfully by increasing the internal interfacial area, and thus the number of cavities, by means of a fine distribution of phosphide precipitates in a 300 series stainless steel (156). The onset of swelling was delayed drastically.

8.5.1.5. Impurity and Alloying Effects on Swelling.

As described in Section 8.2, impurities and alloying additions can have large effects on swelling. Correspondingly, there has been extensive work on the theoretical bases of impurity action. The possibilities can be appreciated when it is realized that any process that changes the fluxes of point defects to cavities, dislocations, or other sinks can alter swelling and, as we shall describe later, irradiation creep.

Several modes by which impurities may affect fluxes have been identified. First, impurities may change the overall point defect concentrations. Since fluxes are proportional to concentrations, this leads to changes in swelling. Point defect trapping at immobile impurity clusters is one example of such a process. By binding the defects for a time depending exponentially on the binding energy, the overall concentration, free plus trapped, of that type of defect is increased. This increases the probability of recombination with the opposite free defect. The free concentrations of defects of both types are thereby decreased by trapping and correspondingly so are the fluxes. These results are reviewed elsewhere (157).

The effect can be cast also in terms of effective point defect diffusion coefficients. The overall concentration, free plus impurity-associated, of point defects of one type is characterized by an effective diffusion coefficient. It is defined mathematically as a weighted composite of free and impurity-associated diffusion coefficients. Theoretical derivations of effective diffusion coefficients are given in the literature (158). In general, impurities may either lower or raise D^{ef}, depending on whether the point defect–impurity complex is immobile or highly mobile. In the case of enhanced vacancy diffusion, effects on swelling may also result (159).

A second effect of impurities and alloying additions on point defect fluxes is caused by segregation. When the layer of material near a sink becomes enriched or depleted in certain elements, changes in point defect capture efficiency result. Thus the Z's of Eq. (8.47), for example, may be changed. This may occur, for example, by changes in the diffusion coefficients within the layer compared to the matrix or by changes in elastic interactions of point defects with sinks (160–165). These effects have been reviewed recently (157).

Finally, under certain conditions phase changes occur as a result of impurity or alloying additions. Second phases may affect the fluxes through several mechanisms (166). They may change the overall point defect concentrations by acting as sinks for point defects or by themselves (their interfaces) acting as point defect traps. Layers of a second phase may form around a cavity leading to similar effects on capture efficiencies as from solute segregation discussed above. Precipitates may also form as particles attached to cavities. This larger composite sink (precipitate–cavity pair) can lead to higher point defect fluxes to the attached cavity and to its enhanced growth as compared to a cavity in the matrix (167, 168).

To describe point defect trapping, the continuity equations for point defects must be generalized. Equations (8.31) and (8.32) (dropping the spatial derivatives) are generalized to a set of four equations as folows:

Free vacancies

$$G_v + \tau_v^{-1}C_v' - RC_iC_v - C_vR_iC_i' - C_v\kappa_v(C^t - C_v' - C_i') - K_vC_v = 0 \quad (8.55)$$

Free interstitials

$$G_i + \tau_i^{-1}C_i' - RC_iC_v - C_iR_vC_v' - C_i\kappa_i(C^t - C_v' - C_i') - K_iC_i = 0 \quad (8.56)$$

Trapped vacancies

$$C_v\kappa_v(C^t - C_v' - C_i') - \tau_v^{-1}C_v' - C_iR_vC_v' = 0 \quad (8.57)$$

Trapped interstitials

$$C_i\kappa_i(C^t - C_v' - C_i') - \tau_i^{-1}C_i' - C_vR_iC_i' = 0 \quad (8.58)$$

The C with a prime denotes the population of point defects trapped at impurities. The symbol τ denotes the mean time a defect is trapped, given by $\tau_{i,v} = (b^2/D_{i,v}^0)\exp[(E_{i,v}^b + E_{i,v}^m)/kT]$. The symbols E_i^b and E_v^b denote the interstitial and vacancy binding energy, and R_i and R_v denote the recombination coefficients of free vacancies with trapped interstitials and of free interstitials with trapped vacancies, given by $4\pi r_iD_v$ and $4\pi r_vD_i$, respectively (r_i and r_v are the radii of the corresponding recombination volumes). Finally, κ_i and κ_v denote capture coefficients describing capture of free interstitials and free vacancies at traps. They are given by $4\pi r_i' D_i$ and $4\pi r_v' D_v$, respectively, where r_i' and r_v' are the radii of the corresponding capture volumes.

Figure 8.32 shows the fraction of point defects recombining versus temperature with and without point defect trapping. Trapping increases the fraction recombining [(1-fraction recombining) is the fraction lost to sinks]. The figure shows the interesting result that vacancy trapping with binding energy $E_v^b = 0.5$ eV produces the same effect as interstitial trapping with $E_i^b = 1.69$ eV. This is more than a coincidence. It has been shown that vacancy trapping with binding energy E_v^b is mathematically equivalent (157,170) to interstitial trapping with binding energy

$$E_i^b = E_v^m - E_i^m + E_v^b + kT \ln \left(\frac{D_i^0 r_v'}{D_v^0 r_i'} \right)$$

Figure 8.33 shows the theoretically calculated effects of vacancy trapping at solute impurities as a function of binding energy and solute concentration. It is clear that swelling can be strongly suppressed by point defect trapping. The minimum binding energy to produce significant effects is greater than 0.1 eV. Even at 10^{-5} atomic fraction solute, a swelling reduction can be produced when the binding energy is as high as ~ 1 eV.

As another example of recent progress, we show the effects of point defect collection at the precipitate–matrix interface and the channeled diffusion of the

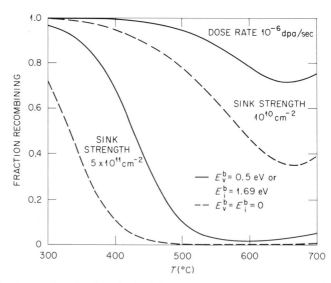

Figure 8.32. The calculated fraction of point defects recombining versus temperature for cases of no point-defect trapping and for either vacancy or interstitial trapping. The material modeled in the calculations is nickel. After Ref. 169.

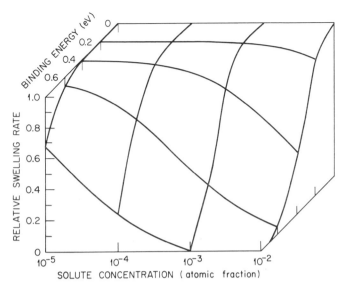

Figure 8.33. Three-dimensional plot illustrating the theoretically calculated effects on swelling rate of vacancy trapping as functions of binding energy and solute concentration. After Ref. 157.

defects along the interface to the attached cavity. The mechanism was developed and explored quantitatively elsewhere (167). Basically, the cavity–precipitate pair is a larger sink for point defects than is a cavity isolated in the matrix. Thus, the growth rate of the precipitate-attached cavity is larger than that of a cavity in the matrix. The ratio of growth rates dr_{cp}/dr_c is a convenient way to express this. Here r_{cp} is the radius of the attached cavity and r_c is the radius of the cavity in the matrix. The integral result r_{cp}/r_c is most convenient to compare with experiments. The solid lines in Figure 8.34 give the theoretically predicted attached cavity radius r_{cp} as a function of the matrix cavity radius r_c. Both quantities are normalized to r_0, their initial size. The parameter r_p is the precipitate size, given in relation to the initial cavity size. It can be seen that when a dose has been accumulated such that the matrix cavity increases by a factor of 3 in radius, for example, the precipitate-attached cavity is predicted to increase in radius by a factor of 6 when attached to a precipitate that is 15 times the size of the starting cavities.

A critical experiment was carried out (168). It employed nickel ion irradiation of a modified stainless steel, initially without gas injection to 70 dpa. Large precipitates formed, but no swelling occurred. Next, 400 atom parts per million (appm) helium was injected, resulting in many small bubbles, some on precipitate–matrix interfaces and some in the matrix. After the injection was completed, the nickel ion irradiation was continued. The cavities on the precipitate–matrix interface grew to larger sizes than those in the matrix. The cross-hatched area in Figure 8.34 represents the measured results. The agreement between theory and experiment is good.

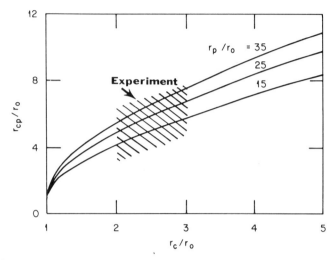

Figure 8.34. Plot of theoretically predicted and measured enhancement of growth rate of cavities attached to precipitates, reflecting the operation of the point defect collector mechanism. After Refs. 167 and 168.

8.5.2. Irradiation Creep

The point defect concentrations produced by irradiation drive many other radiation effects in addition to swelling. The fundamental processes underlying irradiation creep are dislocation climb and/or glide. Climb occurs when dislocations absorb or emit point defects. Glide occurs when a dislocation line segment moves by pure slip. The processes are easy to visualize for a simple edge dislocation, which marks the terminal row of an extra plane of atoms within a crystal. Climb consists of adding or removing atoms to the extra plane. Glide consists of the translation of the dislocation perpendicular to the extra plane. For example, by adding interstitials to dislocations whose extra planes are perpendicular to an applied stress and adding vacancies (removing atoms) to dislocations whose extra planes are parallel to the applied stress, the material will extend in the direction of the applied stress and contract in perpendicular directions. In other words, it will creep. Dislocation glide also can lead to creep. Glide can lead to the translation of planes inclined to the stress axis, resulting in a net extension of the material in the stress direction.

A rate equation approach is used that is similar to that for swelling. To produce climb, point defects must be partitioned to dislocations asymmetrically so that a net flux of interstitials or vacancies arrives at individual dislocation lines to cause climb. The glide mechanisms are more accurately termed *climb-enabled glide mechanisms* since dislocations are normally rapidly pinned by obstacles and, at lower temperatures at least, remain so until climb causes them to clear the obstacle and glide again. Four primary mechanisms of irradiation creep are discussed below. These mechanisms have been shown to reproduce some aspects of the experimental data. The mechanisms depend on several types of asymmetrical partitioning.

Point defects may be partitioned differently to dislocations that are oriented differently with respect to the applied stress. Dislocations whose extra plane is oriented perpendicular to the applied stress have a larger capture efficiency for interstitials than those that are orthogonal to it. This effect is responsible for the *stress induced preferred absorption mechanism, SIPA, or PA, creep* (171–173). This effect also leads to the climb-enabled glide mechanism termed *preferred absorption glide, PAG, creep* (174).

Defects may also be partitioned between dislocations and another sink. The outstandinng example is swelling, where excess interstitials condense at dislocations and excess vacancies precipitate at cavities. The resulting climb of dislocations causes them to overcome obstacles and to glide, leading to creep. This effect can be termed *swelling-driven creep* (175–178). Other mechanisms of this type have been suggested. For example, the ''other'' sink may be dislocations in a different local environment. An isolated dislocation has a larger capture efficiency for interstitials than does a dislocation in a dense group of dislocations (179). In a group the long-range dislocation stress fields tend to cancel, leading to a reduction in capture efficiency. The ''other'' type of sink also may be grain boundaries.

A fourth mechanism of irradiation creep results from climb excursions of dislocations. This process is caused by the large fluctuations in point defect fluxes produced by cascades, as discussed in Section 8.4.4, and is termed *cascade-induced creep*, or simply cascade creep (180). It operates even under conditions where there is no cumulative or long-term climb of dislocations resulting from stress-induced or other sink differences in capture efficiency. In response to the large and out-of-phase fluctuations in point defect arrival rates, as depicted in Figure 8.25, a dislocation segment pinned at an obstacle makes climb excursions. There is a finite probability that the segment will make a large enough excursion or series of excursions to become unpinned. The resulting glide leads to an increment of irradiation creep.

The relationships of these processes to swelling also have been developed recently (181).

8.5.2.1. Climb Creep.

By writing equations for the extension rate in the direction of an applied tensile stress, it is found that the creep rate by dislocation climb can be expressed as

$$
\dot{\epsilon}_c = \frac{2}{9}\Omega L \left\{ [\Delta Z_i^d D_i C_i - \Delta Z_v^d D_v C_v] \right.
$$

$$
\left. + D_v C_v^e \left[Z_v^{dA} \exp\left(\frac{\sigma\Omega}{kT}\right) - Z_v^{dN} \right] \right\} \tag{8.59}
$$

Here $\Delta Z_{i,v}^d = Z_{i,v}^{dA} - Z_{i,v}^{dN}$, where Z^{dA} denotes the capture efficiency of dislocations whose Burgers vectors† are aligned and Z^{dN} denotes the corresponding quantity for nonaligned dislocations. The quantity σ denotes applied stress. The first set of terms in square brackets is the creep rate by stress-induced preferred absorption of interstitials over vacancies on aligned dislocations. The second set of square brackets denotes the creep rate by the stress-induced preferred emission of vacancies from aligned dislocations and their absorption at nonaligned dislocations. This latter (174) is similar to a well-known mechanism of thermally induced creep termed Herring–Nabarro creep (182).

Figure 8.35 shows the behavior predicted by Eq. (8.59) for the creep rate as a function of temperature for several dose rates for a typical dislocation density of 5×10^{14} m^{-2}. It can be seen that the thermal creep rate is predicted to become important at a dose rate of $\sim 10^{-6}$ dpa/s at $\sim 525°C$, but it does not become important at 10^{-3} dpa/s until nearly 700°C. Equation (8.59) also predicts a linear dependence on dose rate, $\dot{\epsilon}_c \propto G$, for sink-dominated cases, high temperture, or low dose rate, for example. It predicts a parabolic dependence, $\dot{\epsilon}_c \propto G^{1/2}$, for recombination-dominated cases (183). It is also important to point out that the stress-induced preferred absorption part of Eq. (8.59) is linear in stress because ΔZ^d, the stress-induced difference in capture efficiency, is linear in stress (171–173).

8.5.2.2. Climb-Enabled Glide Creep by Cumulative Point Defect Absorption. In these processes an individual dislocation segment pinned on an obstacle climbs

†The Burgers vector is perpendicular to the extra plane of atoms for the edge dislocation.

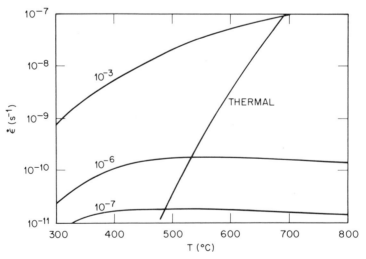

Figure 8.35. Dislocation climb creep rate versus temperature for typical reactor dose rates (10^{-7}–10^{-6} dpa/s) and charged particle dose rates (10^{-3} dpa/s). The thermal creep rate and the radiation induced creep rate are shown separately. After Ref. 183.

inexorably in one direction driven by one of the forms of point defect bias mentioned above. In preferred absorption glide the climb comes from the preferred absorption process discussed in Section 8.5.2.1. In swelling-driven creep the climb comes from swelling. In general, climb components from each process will be superimposed. This more general case has been treated recently (181). Here we discuss each component separately for simplicity.

The climb glide creep rate by these mechanisms can be written as

$$\dot{\epsilon}_{cg} = vf, \tag{8.60}$$

where v is the net dislocation climb velocity and f is a physically defined function that transduces climb to glide. Two simple models for f have been proposed: the dislocation bowing model (178), $f = (\pi L)^{1/2} \sigma/E$, where E is Young's modulus, and the obstacle model (184), $f = \alpha dbL/h$, where α is a factor less than 1, d is the obstacle spacing, b is Burgers vector magnitude, and h is the obstacle height.

Expressions for v can be determined for preferred absorption and for swelling. For preferred absorption the result is (174)

$$v = \frac{4}{9} \frac{\Omega}{b} \Delta Z^d D_i C_i \tag{8.61}$$

Thus we see immediately that PA and PAG creep have many similar dependences on parameters.

When swelling is responsible for dislocation climb, it is easy to show that the climb velocity is given by

$$v = \frac{\Omega}{b} (Z_i^d D_i C_i - Z_v^d D_v C_v) \tag{8.62}$$

It can be seen that swelling-driven creep has similar parametric dependences to those of swelling.

8.5.2.3. Cascade-Induced Creep.
Large locally unbalanced point defect fluxes at dislocation segments arise from fluctuations in point defect concentrations produced by cascades. Above, the quantity of interest was v, the climb velocity. Actually, we only employ v to calculate the release frequency of dislocations from obstacles. In cascade creep we can calculate the release frequency directly (106). The release frequency ω is given by

$$\omega = \sum_j F_j R_j, \tag{8.63}$$

where F_j is the frequency with which a dislocation segment experiences a climb event of height jb or greater, where j is distance in units of the Burgers vector.

The spectrum F_j is calculated based on cascade size, frequency, and materials properties. Here R_j is the probability that a dislocation segment resides a distance jb from an unpinning position. The creep rate from this process, using the obstacle model for f, is given by

$$\dot{\epsilon}_{cas} = \alpha bdL \sum_j F_j R_j \qquad (8.64)$$

8.5.2.4. General Remarks. In a theoretical comparison of the magnitudes of cascade-induced creep, swelling-driven creep, PAG creep and PA creep, it was found that cascade-induced creep gave the largest value for small obstacle sizes (up to several tens of b in length) (180). Otherwise, swelling-driven creep is larger, PAG creep gave a smaller magnitude, and PA creep gave the smallest.

While complete and detailed comparisons of irradiation creep models with all available experimental data have not been undertaken, several general trends are in agreement. For example, the theoretical ratio of creep to swelling when the dislocation and cavity densities have stabilized gives either a linear or two-thirds (or a combination) power of creep strain on cavity volume fraction for the models above (181). This agrees with a variety of experimental results. The models predict a gentle dependence on temperature, as is experimentally observed. The above models lead to a low power (*1 to 2*) dependence of creep rate on stress and a linear-to-parabolic dependence on dose rate. These results are in agreement with the data. It may be that more than one or even all four models described above are operating, with the mechanism giving the largest contribution dependent on the parametric regime.

8.5.3. Solute Segregation

Theoretical models of solute segregation describe the fluxes of alloying components induced by point defect fluxes and the resulting solute buildup. The most extensively developed work treats solute segregation to point defect sinks (185–190). Segregation may also occur by local perturbations in point defect recombination rate leading to annihilation of point defects and deposition of the associated solutes (191).

The differential equations governing solute segregation are a generalization of those for point defect trapping, Eqs. (8.55–8.58), to account for the mobility of solute and solute complexes. For example, Eqs. (8.55) and (8.57) become for dilute binary alloys with mobile solute and solute point defect complexes,

$$\frac{\partial C_v}{\partial t} = (1 + \sigma_v C')D_v \nabla^2 C_v + \sigma_v D_v \nabla C_v \nabla C' + G + \tau_v^{-1} C_v'$$
$$- RC_i C_v - C_v R_i C_i' - C_v \kappa_v (C' - C_v' - C_i') - K_v C_v \qquad (8.65)$$

and

$$\frac{\partial C_v'}{\partial t} = D_v' \nabla^2 C_v' + C_v \kappa_v (C' - C_v' - C_i') - \tau_v^{-1} C_v' - C_i R_v C_v', \qquad (8.66)$$

respectively. The interstitial equations are analogous. The coefficients R_i and R_v now contain a sum of free and trapped defect complex diffusivities (158) rather than the coefficient of the free defect only. In addition, there is another analogous equation describing continuity of the free solute concentration C'. Here the equations are written in their general time-dependent forms. The spatial derivatives describe the influence of a nearby sink, usually an interface, in inducing concentration gradients. The σ_v in Eq. (8.65) is a phenomenological coupling coefficient between vacancies and solute. The first term containing σ_v is the component of the divergence in free vacancy flux arising from the presence of solute, and the second term containing σ_v is the divergence arising through the coupling of the vacancy and solute gradients. The new symbol in Eq. (8.66), D'_v, is the diffusion coefficient of bound vacancy–solute complexes. The D'_i utilized in Eq. (8.67) is defined analogously.

The physical effects driving segregation are perhaps easiest to visualize by writing the equation for the solute current rather than its continuity equation for concentration as above. An approximation to this equation, valid when recombination of free defects with bound complexes containing the opposite type defect is not predominant (188), is given by

$$J' = -\left(\frac{D'_i C'_i}{D_i C_i} + \sigma_i C'\right) D_i \,\nabla C_i - \left(\frac{D'_v C'_v}{D_v C_v} - \sigma_v C'\right) D_v \,\nabla C_v. \quad (8.67)$$

Here the term in $D'_i C'_i$ gives the solute current carried by solute–interstitial complexes. The term in $\sigma_i C'$ gives the net solute current caused by diffusional encounters between solute and interstitial defects. In this form a direct comparison of the magnitudes of the two driving forces can be made. Analogous terms appear for vacancies. The terms in $\sigma_i C'$ and $\sigma_v C'$ are described as *inverse Kirkendall* terms, in inverse analogy with the Kirkendall effect in thermal diffusion where preferential alloying element exchanges with vacancies result in vacancy segregation (192). A feature of Eq. (8.67) is that it shows that solute segregation due to interstitials is always in the same direction as the interstitial flux driven by the interstitial gradient. For the solute segregation due to vacancies, however, the inverse Kirkendall term drives segregation opposite to the direction of vacancy flow induced by the vacancy gradient (that is, in competition with the flow of vacancy–solute complexes).

Similar equations have been developed as a first approximation to describe segregation in concentrated alloys based on a picture using partial diffusion coefficients (189).

Figure 8.36 gives an example of the model predictions for irradiation of a thin foil (185) using the set of defect parameters described in the reference. Segregation is pronounced at all temperatures. The lowest temperature shows the steepest profile because thermal back diffusion, which tends to relax the irradiation-induced segregation, is weakest at low temperatures. The temperature dependence of the overall segregation shows an interesting behavior that is qualitatively similar to

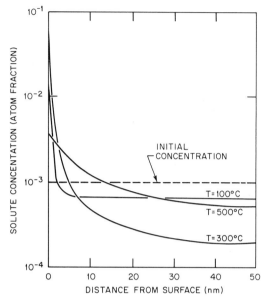

Figure 8.36. Calculated steady-state solute impurity concentration as a function of distance from a free surface at several temperatures for a dose rate of 10^{-3} dpa/s. After Ref. 185.

that of swelling. It is peaked at intermediate temperatures and diminished at both low and high temperatures.

8.6. OTHER PHENOMENA

Several additional phenomena that occur during irradiation are of interest. The most prominent among these are radiation-induced phase instability and radiation-induced embrittlement. Both phenomena depend on processes additional to diffusion and clustering, the main focus of the preceding sections. For example, analyzing phase instability involves nonequilibrium thermodynamics and the effects of mechanical, electrical, and magnetic interactions. Understanding embrittlement requires analyses of the dynamics of flow and fracture.

Mechanistic understanding in both fields is in an earlier stage of development than for those described above. In addition, because the analysis required departs from the kinetics of primary interest here, these phenomena lie somewhat outside the scope of this chapter. Nevertheless, for completeness, we shall describe some main features in the phenomenology. Brief comments are made on the underlying mechanisms. In both areas there are review papers, and these are cited to guide the reader who wishes to explore further.

8.6.1. Radiation-Induced Phase Instability

Phase transformations associated with irradiation have been observed for many years. Especially after swelling was discovered (itself an unusual kind of phase instability), many studies of microstructure were made. Often precipitates were seen.

More recently phase instability during irradiation has become an important field of research. This is for several reasons. For example, the formation of a phase with composition different than the matrix may deplete or enrich the matrix in certain solutes, thus altering mechanical and physical properties. The precipitates themselves may be responsible for other changes. They may interact with dislocations, hardening the material and changing fracture resistance and fracture modes. Similarly, precipitates may form on and interact with grain boundaries, again affecting fracture behavior. As we saw earlier, precipitates serve as sites for preferential cavity growth, and this effect can lead to greatly increased swelling.

A number of conferences, books, and reviews have appeared recently covering a wide range of work in this area. The first meeting dedicated to solute segregation and phase stability was the Workshop on Solute Segregation and Phase Stability (the proceedings of which constitute Ref. 193). This was followed by the Symposium on Phase Stability during Irradiation (32). A recent book containing chapters by authors active in the field is available (194). A recent extensive review (195) covers both theoretical concepts and experiments on many alloy systems.

Phase instability under irradiation includes acceleration of the decay of metastable phases, the destabilization of thermodynamically stable phases, and the induction of phases that are not observed under thermal conditions at the prevailing bulk alloy composition. Several mechanisms can be identified that may underlie these changes. Some of these are outlined below:

1. *Solute segregation.* When the *local* concentration of an alloying element or elements exceeds a solubility limit for transformation, a new phase may form. The local concentration may be highly enriched or depleted in comparison to the bulk, as we have discussed earlier. Such segregation usually occurs at a point defect sink. However, instabilities triggered by fluctuations in recombination rate may lead to a phenomenon reminiscent of spinodal decomposition (196).

2. *Radiation-enhanced diffusion.* A characteristic of irradiated materials is that they typically have instantaneous concentrations of vacancies and interstitials that are many orders of magnitude higher than those at thermal equilibrium. Diffusional processes that depend on interactions with point defects are accelerated greatly.

3. *Radiation disordering.* The displacement of atoms is a powerful process that will completely disorder an ordered structure in the absence of thermally activated restorative processes.

4. *Recoil mixing.* Similar to item 3, this process operates by the energetic displacement-induced removal of atoms from a precipitate particle into the neighboring matrix and the analogous process of energetic injection of nearby matrix atoms into the particle.

5. *Nucleation sites.* Precipitation is enhanced by the availability of sites for heterogeneous nucleation. Defects such as the clusters in cascade cores, dislocation loops, transmutation products, and other types of defects resulting from irradiation are abundant seeds for precipitation.

6. *Vacancies and interstitials.* In addition to item 2, point defects may assist the precipitation reaction itself. For example, for the continued growth of a strongly oversize phase (on the basis of volume per atom), vacancies may be required. To decrease the misfit strain energy, it may be necessary to regularly incorporate a vacancy as a species (''chemical vacancy''). A similar process can be visualized for interstitials.

7. *Radiation amorphization.* Similar to item 3, this process not only destroys the order with respect to the lattice but may eradicate the periodicity of the lattice itself.

The array of examples of radiation-induced phase instability is impressive. Many alloy systems have been studied, and the reader is directed to the references for a fuller exposure. Here we mention the observations in stainless steels and similar alloys. This class represents by far the most work in the field and is of great technological interest. It is a system that exhibits a very large variety of phenomena. Recent works by Lee et al. (197), Maziasz (198), Yang et al. (199), and Williams et al. (200) are sources of up-to-date and quantitative information.

In AISI type 316 stainless steel and titanium-modified variants in both the solution-annealed and cold-worked conditions, some 13 phases have been identified (197). In addition to the parent phase 3 phases were identified that were observed during alloy melting and fabrication. Nine additional phases were identified that resulted from either irradiation or thermal environments experienced by the alloys. Table 8.2 gives the crystallographic data determined for these phases.

Many of the precipitates were enriched in nickel and silicon, a feature that is commonly observed. We recall that nickel and silicon are enriched at point defect sinks such as loops and cavities by point defect flux-associated solute segregation. The phases were classified into three groups: radiation induced, radiation modified and radiation enhanced. The first are phases that do not appear in these alloys under thermal conditions, although they do appear in remote regions of the phase diagram corresponding to much different bulk alloy compositions. In post-irradiation annealing experiments these phases dissolve. The second group consists of phases that occur under normal conditions, but under irradiation their compositions are modified, especially by nickel and silicon enrichment. The temperature regime of the existence of the various phases is also altered by irradiation. The third group consists of phases that develop at significantly lower temperatures during neutron irradiation and/or whose volume fraction is enhanced by irradiation.

TABLE 8.2. Phases Observed in Austenitic Stainless Steels after Thermal Processing and Irradiations with Neutrons and Heavy Ions[a]

Phases	Crystal Structure	Lattice Parameter (nm)	Solute Atoms per Unit Cell	Typical Morphology	Orientation to γ Matrix
γ	Cubic, A1, Fm3m	$a_0 = 0.36$	4	Matrix	—
γ'	Cubic, L1$_2$, Fm3m	$a_0 = 0.35$	4	Small sphere	Cube on cube
G	Cubic, A1, Fm3m	$a_0 = 1.12$	116	Small rod	Random
Fe$_2$P	Hexagonal, C$_{22}$, P321	$a_0 = 0.0604$	6	Thin lath	$(1\overline{2}10)_{ppt}||(0.11)\gamma$
η	Cubic, E9$_3$, Fd3m	$a_0 = 1.08$	96	Rhombohedral	Cube on cube or twin
Laves	Hexagonal, C14, P6$_3$/mmc	$a_0 = 0.47$	12	Faulted Lath	Many variants
M$_{23}$C$_6$	Cubic, D8$_4$, Fm3m	$a_0 = 1.06$	92	Rhombohedral platelet	Cube on cube or twin
MC	Cubic, B1, Fm3m	$a_0 = 0.433$	4	Small sphere	Cube on cube
σ	Tetragonal, D8$_b$, P4/mnm	$a_0 = 0.88$	30	Various	Many variants
χ	Cubic, A12, I$\overline{4}$3m	$a_0 = 0.89$	58	Various	Many variants
TiN	Cubic, B1, Fm3m	$a_0 = 0.425$	4	Large cuboid	—
Zr$_4$C$_2$S$_2$	Hexagonal, P6$_3$/mmc	$a_0 = 0.34$ $c_0 = 1.21$		Globular	—
ZrO	Cubic, B1, Fm3m	$a_0 = 0.46$	4	Globular	—

[a] After Ref. 197.

Referring to the table, the principal phases observed in the neutron-irradiated alloys may be classified as follows: radiation induced, γ' and G; radiation enhanced, η and MC; radiation enhanced and modified, Laves and probably σ and Fe$_2$P; and neither enhanced nor modified, M$_{23}$C$_6$; the χ-phase was observed to be present only in minor amounts.

The above may give some indication of the scope of phenomena and results. These alloys are complex multicomponent alloys typically containing five alloying elements at concentrations greater than 1% and numerous other minor alloying elements and residual impurities. For this reason and those alluded to earlier, it is difficult to foresee first principles and quantitative understanding of phase instability in these alloys. However, work on simpler or model systems shows promise for quantitative modeling studies (201).

8.6.2. Ductility Loss

Macroscopic changes in mechanical properties have been shown to be dictated by the microstructural and microcompositional changes brought about by irradiation. However, the situation is not as simple as for swelling, for instance, where diffusion and clustering directly translate to volume changes. In embrittlement, irradiation alters the material microscopically such that flow and fracture behavior are changed, but the microscopic changes alone are not sufficient to understand how the dynamics are affected. Deformation and fracture are the integrated behavior of all components of the microstructure in the irradiation-altered material.

8.6.2.1. Ferritic Steels.

8.6.2.1. Ferritic Steels. As outlined in Section 8.2, irradiation increases the ductile-to-brittle transition temperature and decreases the fracture toughness above the transition temperature. This is a problem of current concern in light-water reactor pressure vessel steels. In steels not tailored to be radiation resistant, this effect can be quite large, as shown in Figure 8.7 (top). Of most concern is the possibility that with such large DBTT shifts, the brittle behavior may be encountered in the temperature regime of normal reactor operation or of reactor shutdown.

Metallurgical research has resulted in substantial improvements in this area. The poor behavior of some steels has been correlated with the level of alloying, especially of copper, phosphorus, and nickel. Steels with more than 0.15% copper generally exhibit high sensitivity to irradiation. Those with levels less than 0.10% are much improved. Figure 8.7 (bottom) shows results for A533-B type steel containing controlled amounts of copper (0.10%) and phosphorus (0.012%) (58). For comparison, also shown (top) are the results of tests on the same type of steels where the copper and phosphorus contents were not controlled. The shift in the DBTT is reduced by nearly a factor of 4. Nickel content has also been correlated with poor irradiation response. Figure 8.37 shows the results of Charpy-V notch ductility tests on weld metals containing similar copper (0.10%) and phosphorus (0.017%) contents but differing in nickel content by an order of magnitude (202). The higher-nickel alloys exhibit substantially higher DBTT.

The present status of the field, including its historical development and numerous irradiation data, is well described in a number of recent publications (203–206).

A similar phenomenon occurs in the ferritic stainless steels (207). These steels contain higher chromium levels than the pressure vessel steels and are possible candidates for the much higher dose applications of fusion reactor first walls and coolant ducts in sodium-cooled fast breeder reactors. Although there is a significant shift in the DBTT ($>120°C$ in alloy HT-9 containing 12% Cr) and a lowering of the upper shelf energy, the preliminary assessment is that the levels to which mechanical properties are degraded still appear acceptable for the intended application.

A simple qualitative explanation of the DBTT shift is as follows (208). For each steel there is a characteristic stress at which the material exhibits flow or yielding (yield stress). There is also a stress at which the material will fracture

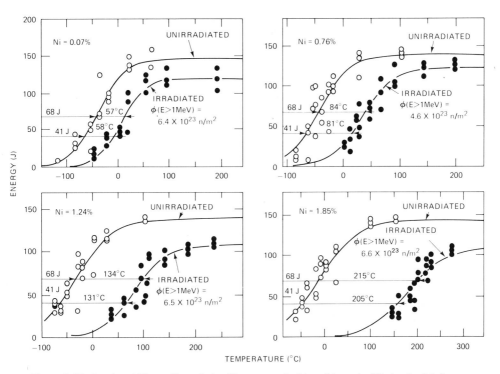

Figure 8.37. Results of Charpy V-notch ductility tests on ferritic weld metals differing in nickel content by an order of magnitude. The radiation-induced temperature shifts are indicated at both 41 and 68 J. Irradiations were done at 288°C. After Ref. 202.

(fracture stress). Below the DBTT the yield stress exceeds the fracture stress, and the alloy fails by cleavage with low fracture toughness. The yield stress is temperature dependent, while the fracture stress is much less so. As temperature is increased, the yield stress drops until, when it falls below the fracture stress, the DBTT is reached. Irradiation increases the DBTT by increasing the yield stress. The yield stress is increased because the various types of defects and clusters introduced by irradiation impede the motion of slip dislocations.

Microstructural studies using field ion microscopy (209, 210), electron microscopy, and small-angle X-ray scattering (211–213) have given some information on the microscopic hardening mechanisms. These studies indicate that several species at very high densities may be responsible. These include microvoids, possibly stabilized by copper and phosphorus, precipitates or "regions" rich in copper and phosphorus, and metal carbide precipitates.

Recent phenomenological models of hardening and ductility loss have had some successes in explaining effects of material and irradiation variables (214, 215). Several distinct types of extended defects are invoked with different characteristics regarding formation and annihilation kinetics.

8.6.2.2. Austenitic Stainless Steels. Austenitic stainless steels and similar alloys do not exhibit a DBTT. However, they do suffer embrittlement based on generally similar microstructural mechanisms. This type of embrittlement is well covered in a number of recent reviews (216–220).

Embrittlement is studied by two other types of mechanical tests in addition to those for fracture toughness described above. These are the tensile test and the creep rupture test. The tensile test is a rapid test where a high uniaxial tensile stress is imposed on a specimen and its yield strength, ultimate tensile strength, and total and uniform elongations are measured. The creep rupture test is a test of typically much longer duration where the stresses and strain rates are lower. The time to failure and the elongation are measured. This test is usually more realistic than the tensile test in the sense that it more closely approximates the loading conditions experienced by materials in service. Ideally, to learn about properties for the application, the tests should be carried out during irradiation. However, in-reactor testing is extremely difficult, time-consuming, and expensive, and the available reactor space is severely limited. By and large, therefore, most mechanical testing is done after irradiation.

Figure 8.38 shows both the yield stress and uniform strain of irradiated and unirradiated type 304 stainless steel observed in tensile tests (216). The irradiated material is significantly stronger and shows lower uniform strain, especially at lower temperatures. Figure 8.39 indicates the improvement possible by tailoring alloys for radiation resistance. Depicted are reference type 316 stainless steel and a similar alloy with minor changes, especially the addition of titanium. The modified alloy shows significantly better behavior.

Generally, creep rupture ductilities are worse than tensile ductilities because of the time factor, which allows for defect and solute diffusion and for microstructural evolution. Figure 8.40 is clear on this point, where the effects of strain rate on elongation to fracture, encompassing both tensile and creep rupture test regimes, are shown. The behavior of unirradiated as well as irradiated material is worse at the lower strain rates (221).

Helium, a transmutation product that we are acquainted with from our above discussion of swelling, can have a particularly deleterious effect on ductility. In fast reactor irradiations helium is produced to levels of tens of appm by the end of life of reactor core structural alloys. In fusion reactors it will be produced to levels of thousands of appm. Figure 8.41 shows a drastic effect of high helium levels on rupture life and elongation (223). There is very little radiation damage here. The helium was injected using an accelerator and the test performed after the implantation.

Embrittlement based on microstructural changes can be classified into three temperature regimes (216). At temperatures $\leq 0.35T_m$ dislocation loops and other small clusters dominate the defect structure, producing obstacles to dislocation glide and strengthening the grains. This strengthening reduces overall deformation and results in the concentration of deformation in localized channels. In the range $0.35T_m \sim 0.60T_m$, cavities, dislocation lines, dislocation loops, and precipitates

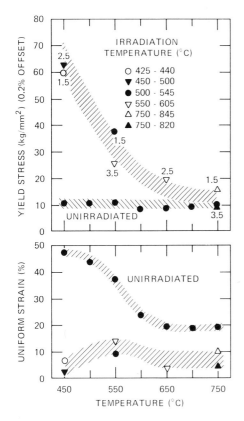

Figure 8.38. The yield stress and uniform strain of type 304 austenitic stainless steel observed in tensile tests. After Ref. 216.

are all prevalent. These act as barriers to gliding dislocations and again produce strengthening. Strengthening is always accompanied by ductility loss. Here the strengthening is less, however, because the spacing of the extended defects is larger than in the lower temperature regime. Above $\sim 0.60 T_m$ less pronounced alterations in microstructure and microcomposition are produced because of the high level of thermal activation supporting diffusional recovery processes. Bubbles are produced by helium accumulation, especially in grain boundaries. Some precipitation also occurs. Grain boundary bubbles are extremely effective in inducing intergranular failure and by their growth under stress may severely reduce ductility. Figure 8.42 shows typical microstructures in 300 series stainless steel in each of the three temperature regimes described above (226).

The mode of failure in both irradiated and unirradiated alloys varies as a function of stress and temperature (216). For example, at high stresses in irradiated alloys, channel fracture, a new mode of failure not encountered in unirradiated materials, is found (224). This takes place in highly localized regions of deformation. At lower stresses and strain rates there is a transition from transgranular to intergranular fracture. At the lower temperatures grains are strong, and unstable

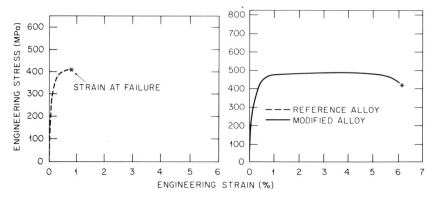

Figure 8.39. Tensile test results for reference type 316 stainless steel and a similar alloy tailored for embrittlement resistance. The alloys were irradiated at 600°C to approximately 3.5 dpa and 80 appm helium in the HFIR reactor. After Ref. 222.

grain boundary crack propagation is increased. At higher temperatures grain boundary fracture occurs as a result of the growth and coalescence of cavities on grain boundaries.

Changes in local composition are also correlated with failure. Recent work shows that the concentrations of several important alloying elements vary signifi-

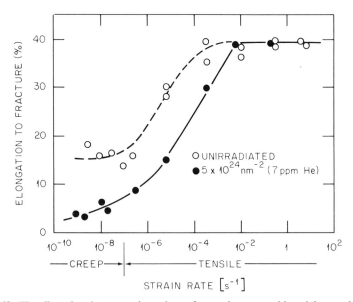

Figure 8.40. The effect of strain rate on elongation to fracture in an austenitic stainless steel at 550°C showing both tensile test and creep rupture regimes. After Ref. 221.

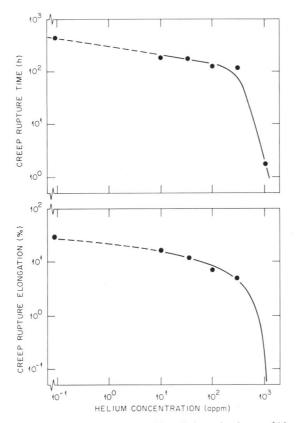

Figure 8.41. Effect of helium content on rupture life and elongation in type 316 austenitic stainless steel at 750°C and a tensile stress of 90 MPa. After Ref. 223.

cantly near the grain boundaries in highly irradiated high-nickel alloys (225). These observations are reminiscent of those described in Section 8.5.3 in the vicinity of other extended defects such as free surfaces, dislocation loops, and cavities.

As indicated earlier, embrittlement is a widely encompassing phenomenon. Theoretical understanding and modeling work is compartmentalized and fragmentary. Some success has been achieved in the following areas: (1) changes in deformation and crack propagation based on the buildup of hardening defects, (2) kinetics of solute segregation at grain boundaries, and (3) high-temperature grain boundary cavitation initiated by helium in the boundaries.

Changes in flow behavior are based on well-known principles for hardening against deformation by the introduction of obstacles to dislocation glide. Crack propagation is further enhanced by reduction in the effective fracture energy. The free energy required to create new surface area may be reduced by solute segregation. Another component of the required effective fracture energy, that expended by deformation in the material adjacent to the crack, may also be reduced by hard-

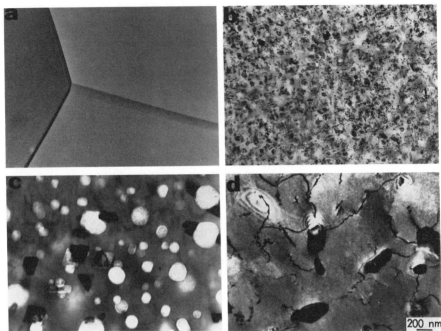

Figure 8.42. Typical microstructures in highly irradiated austenitic stainless steels in three temperature regimes: (a) unirradiated, (b) $<0.35T_m$, (c) $\sim 0.5T_m$, and (d) $>0.60T_m$. Micrographs courtesy of Dr. E. H. Lee, Oak Ridge National Laboratory, Oak Ridge, Tennessee.

ening resulting from solute segregation and precipitation. These considerations are summarized elsewhere (216).

We strongly suspect that radiation-induced segregation of solutes at grain boundaries may also weaken them in several ways. In extreme cases such segregation may lead to precipitation of a new phase with altogether different properties at the grain boundaries. A quantitative theory connecting solute segregation to embrittlement does not yet exist. One crucial ingredient, however, the kinetics of segregation, has been discussed in Section 8.5.3. It remains to understand how segregation may affect deformation and fracture. This last statement also applies to the effects of segregation even in the absence of irradiation (thermal segregation).

The third area, grain boundary cavitation, is more amenable to theoretical modeling, and significant advances have been made here (226–232). Aspects of it are similar to swelling. It involves diffusion and absorption of point defects and helium atoms. An analogy to swelling that is mathematically rigorous is based on the critical radius for unstable growth. Cavities of small radius or containing a small number of gas atoms grow slowly in response to the addition of transmutation-produced helium atoms arriving by diffusion. Just as for swelling, there is a critical radius and a critical number of gas atoms beyond which rapid unstable cavity growth occurs. In this case the rapid growth is driven by the applied stress, while in the case of cavity swelling the rapid growth rate is driven by the "chemical stress" of bias-induced excess vacancy condensation. Here it is found that the growth rates in the unstable regime are so much faster than those in the stable regime that the time to the achievement of the critical quantities can be identified with the time to rupture in a creep rupture experiment (218). In the rapid growth regime the stress-supporting grain boundary area rapidly disappears and the material fails. By recalling that the critical number of gas atoms is inversely proportional to the chemical or applied stress (142), and simply expressing the time to the accumulation of the critical number of gas atoms through the helium current to grain boundaries per cavity, the time to rupture is shown to be inversely proportional to the square of the stress (232). This dependence is in reasonable agreement with experimental results found during creep rupture tests under irradiation (it is weaker than the stress dependence found experimentally in postirradiation creep rupture tests).

With the current high level of interest in fusion reactor technology, where embrittlement is anticipated to be a significant problem, more work can be expected in this field. In view of recent high levels of interest in the scientific community in solute segregation and helium in metals, this should lead to significant further progress in radiation-induced embrittlement.

8.7. SUMMARY AND OUTLOOK

Radiation effects in structural materials have been investigated intensively over the past 20 years. Prior to this, numerous investigations were carried out on a more limited scale. The scope of possible phenomena has been explored, and many experimental characterizations have been carried out in specialized areas. In terms of macroscopic changes the area that has received most attention is that of dimensional instability—radiation-induced swelling and creep. Next in terms of research effort has been the area of changes of mechanical properties—strengthening, hardening, and loss of ductility.

Profound changes in the microstructure and microcomposition due to irradiation of materials have been discovered and studied. At lower doses point defects and defect clusters are detected. At higher doses the effects include the formation of dislocation loops, the evolution of a dense network of dislocations, the formation of cavities, the appearance of unexpected precipitates, and the demixing of initially homogeneous alloys to form segregated or microregion alloys whose compositions and properties differ greatly. Collectively, these effects are referred to as radiation-induced microstructural and microcompositional evolution. These effects are either absent entirely or occur on a much lower scale during ordinary thermal or mechanical treatments.

Much of the knowledge at the present high level of detail depends on sophisticated tools of solid-state physics and materials science. At the lowest doses for point defects and the smallest clusters, much has been learned from changes in electrical resistivity, positron annihilation spectroscopy, X-ray scattering and field ion microscopy, and other techniques. At higher doses, where defects with a dimension larger than the order of a nanometer are produced, analytical electron microscopy has been the premier tool. The defect structures are imaged, and regions of differing structure (precipitates) and local composition (segregation) are readily identified and quantitatively characterized. Tools for surface or near-surface analysis such as Auger electron spectroscopy, ion beam microanalysis, and secondary ion mass spectroscopy have been useful in probing segregation and diffusional behavior.

Owing in large part to the circumstance that virtually all work on radiation effects has taken place when such atomic and microstructural level tools were available, the approach in the field has been perhaps more mechanistically oriented than more traditional areas in metallurgy. Theoretical research has been closely coupled with experiments designed to uncover mechanisms and to improve basic understanding.

The theory of radiation effects has been developed extensively in a number of areas. The basics of defect production by displacement reactions are understood. The theory of microstructural and microcompositional evolution, based on the kinetics of point defect and extended defect interactions, has led to a basic understanding of swelling, irradiation creep, and solute segregation. The physics of many important phenomena have been incorporated into this picture, including the strong effects of transmutation-produced gases, the effects of alloying additions, and the effects of radiation-induced precipitation on dimensional instability. Areas of theoretical research that are receiving more emphasis and in which significant progress is expected in the future are those of radiation-induced phase instability and radiation-induced embrittlement.

Two rewards for this work accrue to researchers in the field, materials scientists interested in the mechanisms of defect behavior and materials engineers engaged in fission and fusion reactor development. The first benefit is an increased understanding of the physics of materials behavior. This applies especially to defect processes, where great progress in understanding has been achieved by grappling with some phenomena that have no counterpart in nonirradiation environments. The technological benefit is the discovery and application of principles for the design and selection of radiation-resistant materials. The engineering of fast-spectrum fission reactors and fusion devices would not be viable without the use of such radiation-resistant materials.

ACKNOWLEDGMENT

I thank Professor Freeman, the tireless editor of this volume, for persuading me to write this chapter and for providing the excellent forum. I am grateful to a number of colleagues for detailed reviews, especially Dr. A. D. Brailsford of Ford Motor Company and Drs. K. Farrell and E. H. Lee of Oak Ridge National Laboratory. My thanks to Wanda Gilliam for typing the manuscript.

REFERENCES

1. J. O. Stiegler and L. K. Mansur, *Ann. Rev. Mat. Sci.* **9**, 405 (1979).

2. H. Ullmaier and W. Schilling, in *Physics of Modern Materials*, Vol. 1, International Atomic Energy Agency, IAEA-SMR-46/105, Vienna, 1980, p. 301.

3. E. Fermi, *The Future of Atomic Energy*, U.S. Atomic Energy Commission Report MDDC-1, Technical Information Division, Oak Ridge, TN, May 27, 1946.

4. E. P. Wigner, *J. Appl. Phys.* **17**, 857 (1946).

5. C. Cawthorne and E. J. Fulton, *Nature* **216**, 575 (1967).

6. S. F. Pugh, M. H. Loretto, D. I. R. Norris (Eds.), *Voids Formed by Irradiation of Reactor Materials*, Brit. Nucl. Energy Soc., Reading 1971.

7. J. W. Corbett and L. C. Ianniello (Eds.), *Radiation-Induced Voids in Metals*, U.S. Atomic Energy Comm., Albany, 1971.

8. S. T. Konobeevsky, N. F. Pravdyuk, and V. I. Kutaitsev, 1955 Geneva Conference on the Peaceful Uses of Atomic Energy, Paper 681.

9. A. C. Roberts and A. H. Cottrell, *Philos. Mag.* **1**, 711 (1958).

10. J. W. Joseph, Jr., USAEC Report DP-369, E. I. Dupont de Nemours and Company, 1959.

11. R. E. Schreiber, USAEC Report DP-660, E. I. Dupont de Nemours and Company, 1962.

12. G. W. Lewthwaite, D. Mosedale, and I. R. Ward, *Nature* **216**, 472 (1967).

13. D. Mosedale, G. W. Lewthwaite, G. O. Leet, and W. Sloss, *Nature* **224**, 1301 (1969).

14. R. G. Berggren, W. E. Brundage, W. W. Davis, N. E. Hinkle, and J. C. Zukas in Proceedings of the USAEC Conference on the Status of Radiation Effects Research on Structural Materials, USAEC Report TID-7588, U.S. Atomic Energy Comm., Chicago, 1959, p. 205.

15. J. O. Stiegler and J. R. Weir, Jr., *Ductility*, American Society of Metals, Metals Park, OH, 1967, pp. 311–342.

16. J. S. Davis (Ed.), *Irradiation Embrittlement and Creep in Fuel Cladding and Core Components*, British Nuclear Energy Society, London, 1972.

17. S. D. Harkness, and B. Cramer, *J. Nucl. Mat.* **85/86**, 135 (1979).

18. G. J. Dienes, G. H. Vineyard, *Radiation Effects in Solids*, Vol. II, Interscience Publishers, New York, 1957.

19. D. S. Billington and J. H. Crawford, *Radiation Damage in Solids*, Princeton University Press, Princeton, NJ, 1961.

20. A. C. Damask, G. J. Dienes, *Point Defects in Metals*, Gordon & Breach, New York, 1963.

21. *Radiation Damage in Reactor Materials*, International Atomic Energy Agency, Vienna, 1969.

22. R. S. Nelson (Ed.), *The Physics of Irradiation Produced Voids*, Her Majesty's Stationery Office, London, 1974.

23. J. S. Watson and F. W. Wiffen (Eds.), *Radiation Effects and Tritium Technology for Fusion Reactors*, U.S. Energy Research & Development Admin., CONF-750989, Gatlinburg, TN, 1975, in 4 vols.

24. M. T. Robinson and F. W. Young (Eds.), *Fundamental Aspects of Radiation Damage in Metals*, U.S. Energy Research & Development Admin., CONF-751006, Gatlinburg, TN, 1975, in 2 vols.

25. S. D. Harkness and N. L. Peterson (Eds), *Radiation Damage in Metals*, American Society of Metals, Metals Park, OH, 1976.

26. F. R. Shober (Ed.), *Irradiation Effects on the Microstructure and Properties of Metals,* ASTM STP 611, American Society for Testing and Materials, Philadelphia, 1976.

27. J. O. Stiegler (Ed.), *Proceedings of the Workshop on Correlation of Neutron and Charged Particle Damage*, U.S. ERDA, Oak Ridge, TN, 1976, CONF-760673, 1976.

28. N. L. Peterson and R. W. Siegel (Eds.), 1976. *Proceedings of the International Conference on Properties of Atomic Defects in Metals*, Argonne, IL, 1976, in *J. Nucl. Mater.* **69** & **70** (1978).

29. M. L. Bleiberg and J. W. Bennett (Eds.), 1977. *Proceedings of the International Conference on Radiation Effects in Breeder Reactor Structural Materials*, Scottsdale, AZ, June 19–23, 1977, Metallurgical Society of American Institute of Mining, Metallurgical and Petroleum Engineers, New York, 1977.

30. F. W. Wiffen, J. H. DeVan, and J. O. Stiegler (Eds.), *Proceedings of the First Topical Meeting on Fusion Reactor Materials*, Miami Beach, FL, January 29–31, 1979. *J. Nucl. Mater.* **85** & **86** (1979).

31. J. Poirier and J. M. Dupuoy (Eds.), *Proceedings of the Conference on Irradiation Behavior of Metallic Materials for Fast Reactor Core Components*, 2 vols., Ajaccio, Corsica, France (1979), Commissariat à l'Energie Atomique, Gif-sur-Yvette. (1979).

32. J. R. Holland, L. K. Mansur, and D. I. Potter (Eds). *Proceedings of the Conference on Phase Stability During Irradiation*, Pittsburgh, PA, October 5–9, 1980, 32. Metallurgical Society of American Institute of Mining, Metallurgical and Petroleum Engineers, New York, 1980.

33. M. A. Kirk (Ed.), *Proceedings of the International Conference on Neutron Irradiation Effects in Solids*, Argonne, IL, Nov. 9–12, 1981. *J. Nucl. Mater.*, vols. **108** & **109** 1–755 (1982).

34. R. E. Nygren, R. E. Gold, and R. H. Jones (Eds.), *Proceedings of the Second Topical Meeting on Fusion Reactor Materials*, Seattle, WA, August 9–12, 1981. *J. Nucl. Mater.* **103** & **104** (1981).

35. H. R. Brager and J. S. Perrin (Eds.), *Proceedings of the Symposium on the Effects of Radiation on Materials*, Scottsdale, AZ, June 28–30, 1982, ASTM STP 782, American Society for Testing and Materials, Philadelphia.

36. D. R. Harries (Ed.), *Proceedings of the Conference on Dimensional Stability and Mechanical Behavior of Irradiated Metals and Alloys*, Brighton, England, April 11–13, 1983, 2 vols., British Nuclear Energy Society, London, 1984.

37. J. B. Whitley, K. L. Wilson, and F. W. Clinard, Jr. (Eds.), *Proceedings of the Third Topical Meeting on Fusion Reactor Materials*, Albuquerque, NM, September 19–22, 1983. *J. Nucl. Mater.* **122** & **123** (1984).

38. J. F. Bates and M. K. Korenko, *Nucl. Technol.* **48,** 303 (1979).

39. W. G. Johnston, T. Lauritzen, J. H. Rosolowski and A. M. Turkalo, p. 227 in Ref. 25.

40. J. F. Bates and W. G. Johnston, p. 625 in Ref. 29.

41. P. J. Maziasz, *J. Nucl. Mater.* **133** & **134**, 134(1985).

42. F. A. Garner, p. 459 in Ref. 37.

43. N. H. Packan and K. Farrell, *Nucl. Technol./Fusion* **3**, 392 (1983).

44. E. H. Lee, N. H. Packan, and L. K. Mansur, *J. Nucl. Mater.* **117**, 123 (1983).

45. K. Farrell, *Rad. Eff.* **53**, 175 (1980).

46. K. Farrell, P. J. Maziasz, E. H. Lee, and L. K. Mansur, *Rad. Eff.* **78**, 277 (1983).

47. R. W. Powell, D. T. Peterson, M. K. Zimmerschied, and J. F. Bates, *J. Nucl. Mater.* **103** & **104** 969 (1981).

48. L. L. Horton, Oak Ridge National Laboratory Report, ORNL/TM-8303, July 1982.

49. R. W. Gilbert, K. Farrell, and C. E. Coleman, *J. Nucl. Mater.* **84**, 137 (1979).

50. D. O. Northwood, *At. Energy Rev.* **15**, 547 (1977).

51. A. Jostsons and K. Farrell, *Rad. Eff.* **15**, 217 (1972).

52. J. L. Straalsund, *Radiation Effects in Breeder Reactor Structural Materials*, Metallurgical Society of American Institute of Mining, Metallurgical and Petroleum Engineers, New York, 1977.

53. J. Gittus, *Creep, Viscoelasticity and Creep Fracture in Solids*, Wiley, New York, 1975.

54. Chr. Schwaiger, P. Jung, and H. Ullmaier, *J. Nucl. Mater.* **90**, 268 (1980).

55. L. C. Walters, G. L. McVay, and G. D. Hudman, USAEC Report DP-660, E. I. Dupont de Nemours, 1962, p. 277.

56. G. W. Lewthwaite and D. Mosedale, *J. Nucl. Mater.* **90**, 205 (1980).

57. R. L. Fish, J. L. Straalsund, C. W. Hunter, and J. J. Holmes, *Effects of Radiation on Substructure and Mechanical Properties of Metals and Alloys*, ASTM STP 529, American Society of Testing and Materials, Philadelphia, 1973, p. 149.

58. J. R. Hawthorne, J. J. Koziol, and R. C. Groescher, *Properties of Reactor Structural Alloys after Neutron or Particle Irradiation*, ASTM STP 570, C. J. Baroch and F. R. Shober, (Eds.), American Society for Testing Materials, Philadelphia, 1975, p. 83.

59. See Ref. 1.

60. T. R. Anthony, p. 630 in Ref. 7.

61. P. R. Okamoto and H. Weidersich, *J. Nucl. Mater.* **53**, 336 (1974).

62. K. Farrell, J. Bentley, and D. N. Braski, *Scripta Met.* **11**, 243 (1977).

63. E. A. Kenik, *Scripta Met.* **10**, 733 (1976).

64. S. Ohnuki, H. Takahashi, and T. Takeyama, *J. Nucl. Mater.* **103** & **104**, 1121 (1981).

65. L. E. Rehn, P. R. Okamoto, D. I. Potter, and H. Wiedersich, *J. Nucl. Mater.* **74**, 242 (1978).

66. M. T. Robinson, p. 1 in Ref. 25.

67. D. G. Doran, J. R. Beeler, Jr., N. D. Dudey, and M. J. Fluss, Report of the Working Group on Displacement Models and Procedures for Damage Calculations, Hanford Engineering Development Laboratory Report, HEDL-TME 73-76 (1973).

68. J. R. Beeler, Jr., *Radiation Effects Computer Experiments*, North-Holland Publishing, Amsterdam, 1983.

69. J. W. Corbett, *Electron Radiation Damage in Semiconductors and Metals*, Academic Press, New York, 1966.
70. L. R. Greenwood, *J. Nucl. Mater.* **103 & 104,** 1433 (1981).
71. R. W. Conn, *Scie. Am.* **249,** 60 (1983).
72. R. E. Gold, E. E. Bloom, F. W. Clinard, D. L. Smith, R. D. Stevenson, and W. G. Wolfer, *Nucl. Technol. Fusion* **1,** 169 (1981).
73. W. A. McKinley, Jr., and H. Feshbach, *Phys. Rev.* **74,** 1759 (1948).
74. J. Lindhard, M. Scharff, and H. E. Schiott, *Mat. Fys. Medd. Dan. Vid. Selsk.* **33** (14) (1963).
75. A. D. Marwick, *J. Nucl. Mater.* **55,** 259 (1975).
76. G. H. Kinchin and R. S. Pease, *Rep. Prog. Phys.* **18,** 1 (1955).
77. P. Sigmund, *Appl. Phys. Lett.* **14,** 114 (1969).
78. F. A. Nichols, *Ann. Rev. Mat. Sci.* **2,** 463 (1972).
79. W. E. King and K. L. Merkle, *J. Nucl. Mater.* **117,** 12 (1983).
80. P. Jung, B. R. Neilsen, H. H. Andersen, J. F. Bak, H. Knudsen, R. R. Coltman, C. E. Klabunde, J. M. Williams, M. W. Guinan, and C. E. Violet, p. 963 in Ref. 35.
81. G. L. Kulcinski, J. J. Laidler, and D. G. Doran, *Rad. Eff.* **7,** 195 (1971).
82. H. L. Heinisch, Hanford Engineering Development Laboratory Report, HEDL-TME 83-17, June 1983.
83. D. N. Seidman, M. I. Current, D. Pramanik, and C. Y. Wei, *Nucl. Instr. Meth.* **182 & 183,** 477 (1981).
84. C. Y. Wei, M. I. Current, and D. N. Seidman, *Philos. Mag. A* **44,** 459 (1981).
85. K. L. Merkle, p. 58 in Ref. 25.
86. B. L. Eyre, p. 729 in Ref. 24.
87. M. Kiritani, in *Proceedings of Yamada Conference V on Point Defects and Defect Interactions in Metals*, J.-I. Takamura, M. Doyama, and M. Kiritani (Eds.), University of Tokyo Press, Tokyo, 1982.
88. J. B. Mitchell, R. A. Van Konynenberg, M. W. Guinan, and C. J. Echer, *Philos. Mag.* **31,** 919 (1975).
89. M. L. Jenkins, K. H. Katerban, and M. Wilkens, *Philos. Mag.* **34,** 1155 (1976).
90. S. Ishino, K. Fukuya, T. Muroga, N. Sekimura, and H. Kawanishi, *J. Nucl. Mater.* **122 & 123,** 597 (1984).
91. J. B. Roberto and J. Narayan, p. 120 in Ref. 24.
92. J. Narayan and S. M. Ohr, *J. Nucl. Mater.* **63,** 454 (1976).
93. H. J. Wollenberger, in *Physical Metallurgy*, R. W. Cahn and P. Haasen, (Eds.), Elsevier Science Publishers, New York, 1983, Chapter 17.
94. J.-I. Takamura, M. Doyama, and M. Kiritani, (Eds.) *Proceedings of Yamada Conference V on Point Defects and Defect Interactions in Metals*, University of Tokyo Press, Tokyo, 1982.
95. N. L. Peterson and R. W. Siegel, (Eds.), *J. Nucl. Mater.* vols. **69 & 70** (1978).
96. A. Seeger, D. Schumacher, W. Schilling, and J. Diehl (Eds.), *Proceedings of the International Conference on Vacancies and Interstitials in Metals*, Julich, Germany, September, 1968, North-Holland Publishing, Amsterdam, 1970.

97. R. R. Hasiguti (Eds.), *Lattice Defects and Their Interactions*, Gordon & Breach, New York, 1967.

98. M. Doyama and S. Yoshida (Eds.), *Progress in the Study of Point Defects*, University of Tokyo Press, Tokyo, 1977.

99. P. B. Hirsch (Ed.), *The Physics of Metals*, Part 2, *Defects*, Cambridge University Press, Cambridge, 1975.

100. D. J. Reed, Rad. Eff. **31**, 129 (1977).

101. H. Ullmaier (Ed.), Proc. Int'l. Symposium on Fundamental Aspects of Helium in Metals, *Rad. Eff.* **78** (1983).

102. N. M. Ghoniem, S. Sharafat, J. M. Williams, and L. K. Mansur, *J. Nucl. Mater.* **117**, 96 (1983).

103. A. D. Brailsford, *J. Nucl. Mater.* **60**, 257 (1976).

104. L. K. Mansur, A. D. Brailsford, and W. A. Coghlan, *Acta Met.* **33**, 1407 (1985).

105. L. K. Mansur, W. A. Coghlan, and A. D. Brailsford, *J. Nucl. Mater.* **85** & **86**, 591 (1979).

106. L. K. Mansur, W. A. Coghlan, T. C. Reiley, and W. G. Wolfer, *J. Nucl. Mater.* **103** & **104**, 1257 (1981).

107. H. Wiedersich, p. 147 in Ref. 22.

108. M. Kiritani, *J. Phys. Soc. Japan* **35**, 95 (1973).

109. R. A. Johnson, *J. Nucl. Mater.* **75**, 77 (1978).

110. M. R. Hayns, *J. Nucl. Mater.* **56**, 267 (1975).

111. B. O. Hall, *J. Nucl. Mater.* **85** & **86**, 565 (1979).

112. N. M. Ghoniem, *J. Nucl. Mater.* **89**, 359 (1980).

113. M. Koiwa, *J. Phys. Soc. Japan* **37**, 1532 (1974).

114. M. R. Hayns, *J. Nucl. Mater.* **59**, 175 (1976).

115. J. B. Zeldovich, *Acta Phys. Chim. USSR* **18**, 1 (1943).

116. J. Frenkel, *Kinetic Theory of Liquids*, Oxford University Press, Oxford, 1946.

117. W. G. Wolfer, L. K. Mansur, and J. A. Sprague, p. 841 in Ref. 29.

118. A. I. Bondarenko and Yu V. Konobeev, *Phys. Stat. Solidi A* **34**, 195 (1976).

119. C. F. Clement and M. H. Wood, AERE Harwell Report P 774, Oxfordshire, England, 1979.

120. H. Gurol, *Trans. Am. Nucl. Inc. TANSAO* **27**, 264 (1977).

121. N. M. Ghoniem and S. Sharafat, *J. Nucl. Mater.* **92**, 121 (1980).

122. B. O. Hall, *J. Nucl. Mater.* **91**, 63 (1980).

123. J. L. Katz and H. Wiedersich, *J. Chem. Phys.* **55**, 1414 (1971).

124. K. C. Russell, *Acta Met.* **19**, 753 (1971).

125. J. W. Christian, *The Theory of Transformations in Metals and Alloys*, Part I, 2nd ed., Pergamon, Elmsford, NY, 1975.

126. L. K. Mansur and W. G. Wolfer, Oak Ridge National Laboratory Report, ORNL/TM-5670, September 1977.

127. S. D. Harkness and C. Y. Li, *Met. Trans.* **2**, 1457 (1971).

128. H. Wiedersich, *Rad. Eff.* **12**, 111 (1972).

129. A. D. Brailsford and R. Bullough, *J. Nucl. Mater.* **44,** 121 (1972).

130. L. K. Mansur, *Nucl. Technol.* **40,** 5 (1978).

131. A. D. Brailsford and R. Bullough, *Philos. Trans. Roy. Soc. (Lond.)* **302,** 87 (1981).

132. L. K. Mansur, A. D. Brailsford, and W. G. Wolfer, *J. Nucl. Mater.* **105,** 36 (1982).

133. A. Si-Ahmed and W. G. Wolfer, in *International Conference on Dislocation Modeling of Physical Systems, Gainesville, Fla., June 1980,* M. F. Ashby, R. Bullough, C. S. Hartley, and J. P. Hirth (Eds.), Pergamon Press, New York, p. 142.

134. W. A. Coghlan and M. H. Yoo, p. 152 in Ref. 133.

135. R. Bullough, M. R. Hayns, and C. H. Woo, *J. Nucl. Mater.* **84,** 93 (1979).

136. A. D. Brailsford and L. K. Mansur, *J. Nucl. Mater.* **103 & 104** 1403 (1981).

137. F. A. Garner and W. G. Wolfer, p. 1073 in Ref. 35.

138. R. E. Stoller, Ph.D. Thesis, University of California at Santa Barbara, in preparation.

139. A. D. Brailsford, *Met. Trans.* **7A,** 333 (1976).

140. L. K. Mansur, *J. Nucl. Mater.* **78,** 156 (1978).

141. N. H. Packan, K. Farrell, and J. O. Stiegler, *J. Nucl. Mater.* **78,** 143 (1978).

142. L. K. Mansur and W. A. Coghlan, *J. Nucl. Mater.* **119,** 1 (1983).

143. V. F. Sears, *J. Nucl. Mater.* **39,** 18 (1971).

144. G. R. Odette and S. C. Langley, in Ref. 23, vol. 1, p. 395.

145. M. R. Hayns, M. H. Wood, and R. Bullough, *J. Nucl. Mater.* **75,** 241 (1978).

146. S. B. Fisher, R. J. White, and J. E. Harbottle, *Rad. Eff.* **40,** 87 (1979).

147. N. M. Ghoniem and H. Gurol, *Rad. Eff.* **55,** 209 (1981).

148. J. R. Townsend, *J. Nucl. Mater.* **108 & 109,** 544 (1982).

149. J. A. Spitznagel, W. J. Choyke, N. J. Doyle, R. B. Irwin, J. R. Townsend and J. N. McGruer, *J. Nucl. Mater.* **108 & 109,** 537 (1982).

150. A. Hishinuma and L. K. Mansur, *J. Nucl. Mater.* **118,** 91 (1983).

151. R. E. Stoller and G. R. Odette, *Effects of Radiation on Materials, ASTM STP 782,* H. R. Brager and J. S. Perrin (Eds.), American Society for Testing and Materials, Philadelphia, 1982, p. 275.

152. W. A. Coghlan and L. K. Mansur, *J. Nucl. Mater.* **122 & 123,** 495 (1984).

153. W. A. Coghlan and L. K. Mansur, in *Proceedings of the 12th International Symposium on The Effects of Radiation on Materials, Williamsburg, VA, June, 1984,* American Society of Testing and Materials, ASTM STP 870, F. A. Garner and J. S. Perrin (Eds.) Philadelphia, 1985, p. 481.

154. L. L. Horton and L. K. Mansur, p. 344 in Ref. 153.

155. E. H. Lee and L. K. Mansur, *Philos. Mag.* **52,** 993 (1985).

156. E. H. Lee, A. F. Rowcliffe, and L. K. Mansur, *J. Nucl. Mater.* **122 & 123,** 299 (1984).

157. L. K. Mansur, *J. Nucl. Mater.* **83,** 109 (1979).

158. L. K. Mansur, *Acta Metal.* **29,** 375 (1980).

159. F. A. Garner and W. G. Wolfer, *J. Nucl. Mater.* **102,** 143 (1981).

160. A. D. Brailsford, *J. Nucl. Mater.* **56,** 7 (1975).

161. M. Baron, A. Chang, and M. L. Bleiberg, *Trans. Amer. Nucl. Soc.* **28,** 148 (1978).

162. J. Weertman and W. V. Green, p. 256 in Ref. 26.

163. L. K. Mansur and W. G. Wolfer, *J. Nucl. Mater.* **69** & **70**, 825 (1978).

164. W. G. Wolfer and L. K. Mansur, *J. Nucl. Mater.* **91**, 265 (1980).

165. W. G. Wolfer, F. A. Garner, and L. E. Thomas, p. 1023 in Ref. 35.

166. L. K. Mansur, M. R. Hayns, and E. H. Lee, p. 359 in Ref. 32.

167. L. K. Mansur, *Philos. Mag.* **A44**, 867 (1981).

168. E. H. Lee, A. F. Rowcliffe, and L. K. Mansur, *J. Nucl. Mater.* **103** & **104**, 1475 (1981).

169. L. K. Mansur and M. H. Yoo, *J. Nucl. Mater.* **74**, 228 (1978).

170. A. D. Brailsford, *J. Nucl. Mater.* **78**, 354 (1978).

171. P. T. Heald and M. V. Speight, *Philos. Mag.* **29**, 1075 (1974).

172. R. Bullough and J. R. Willis, *Philos. Mag.* **31**, 855 (1975).

173. W. G. Wolfer and M. Ashkin, *J. Appl. Phys.* **47**, 791 (1976).

174. L. K. Mansur, *Philos. Mag.* **A39**, 497 (1979).

175. S. D. Harkness, J. A. Tesk, and Che-Yu Li, *Nucl. Appl. Technol.* **9**, 24 (1970).

176. G. Martin and J. P. Poirier, *J. Nucl. Mater.* **39**, 93 (1971).

177. W. G. Wolfer, J. P. Foster, and F. A. Garner, *Nucl. Technol.* **16**, 55 (1972).

178. J. H. Gittus, *Philos. Mag.* **25**, 345 (1972).

179. W. G. Wolfer, M. Ashkin, and A. Boltax, p. 233, in Ref. 26.

180. See Ref. 106.

181. L. K. Mansur and W. A. Coghlan, p. 65 in Ref. 26, Vol. 2.

182. F. R. N. Nabarro, *Philos. Mag.* **16**, 231 (1967).

183. L. K. Mansur and W. G. Wolfer, *Effects of Radiation on Structural Materials*, ASTM STP 683, J. A. Sprague and D. Kramer (Eds.), American Society for Testing and Materials, Philadelphia, 1979, p. 624.

184. D. J. Michel, P. L. Hendrick, and A. G. Pieper, *J. Nucl. Mater.* **75**, 1 (1978).

185. R. A. Johnson and N. Q. Lam, *Phys. Rev. B* **13**, 4364 (1976).

186. N. Q. Lam, A. Kumar, and H. Wiedersich, p. 985 in Ref. 35.

187. A. D. Marwick, *Nucl. Inst. Meth.* **182/183**, 827 (1981).

188. P. R. Okamoto, L. E. Rehn, and R. S. Averback, *J. Nucl. Mater.* **108**, **109**, 319 (1982).

189. H. Wiedersich and N. Q. Lam, p. 1 in Ref. 194.

190. A. R. Alnatt, A. Barber, A. D. Franklin, and A. B. Lidiard, *Acta Met.* **31**, 1307 (1983).

191. R. Cauvin and G. Martin, *Phys. Rev. B* **23**, 3322 (1981).

192. E. O. Kirkendall, *Trans. A.I.M.E.* **147**, 104 (1942).

193. J. O. Stiegler, *J. Nucl Mater.* **83**, (1979).

194. F. V. Nolfi, Jr., *Phase Transformations During Irradiation*, Applied Science Publishers, New York, 1983.

195. K. C. Russell, in *Progress in Materials Science*, J. W. Christian, P. Haasen, and T. B. Massalski (Ed.), Pergamon, Elmsford, NY (in press).

196. G. Martin, *Phys. Rev. B* **21**, 2122 (1980).

197. E. H. Lee, P. J. Maziasz, and A. F. Rowcliffe, p. 191 in Ref. 32.

198. P. J. Maziasz, "Effects of Helium Content on Microstructural Development in Type 316 Stainless Steel Under Neutron Irradiation," Ph.D. Dissertation, The University of Tennessee, Knoxville, 1984.

199. W. J. S. Yang, H. R. Brager, and F. A. Garner, p. 257 in Ref. 32.

200. T. M. Williams, p. 166 in Ref. 35.

201. E. H. Lee and L. K. Mansur, in preparation.

202. C. Guionnet, B. Houssin, D. Brasseur, A. Lefort, D. Gros, and R. Perdreau, p. 392 in Ref. 35.

203. J. R. Hawthorne, "Evaluation of IAEA Coordinated Program Steels and Welds for 288°C Radiation Embrittlement Resistance," U.S. Nuclear Regulatory Report NUREG/CR-2487, February 1982.

204. J. A. Spitznagel, private communication.

205. L. E. Steele, Technical Report Series No. 163, IAEA, Vienna, 1975.

206. E. A. Little, p. 141 in Ref. 35.

207. J. R. Hawthorne and F. A. Smidt, Jr., *J. Nucl. Mater.* **103, 104,** 883 (1981).

208. See Ref. 205.

209. S. S. Brenner, R. Wagner, and J. A. Spitznagel, *Met. Trans. A* **9,** 1761 (1978).

210. M. K. Miller and S. S. Brenner, *Res. Mechan.* **10,** 161 (1984).

211. S. B. Fisher, J. E. Harbottle, and N. B. Aldridge, p. 87 in Ref. 36, Vol. 2.

212. R. B. Jones and J. T. Buswell, p. 105 in Ref. 36, Vol. 2.

213. J. T. Buswell, "IAEA Coordinated Program . . . Examination of the Materials by Electron Microscopy," Central Electricity Generating Board Report, TPRD/B/0083/N82, July 1982, Berkeley Nuclear Laboratories, Berkeley, England.

214. R. Lott, in *Proceedings of the Topical Conference on Ferritic Alloys for Use in Nuclear Energy Technologies, Snowbird, Utah, June, 1983,* J. W. Davis and D. J. Michel (Eds.), The Metallurgical Society of AIME, p. 37. Warrendale, PA, 1984.

215. G. R. Odette, *Scripta Met.* **17,** 1183 (1983).

216. E. E. Bloom, p. 295 in Ref. 25.

217. D. R. Harries, *J. Nucl. Mater.* **82,** 2 (1979).

218. H. Ullmaier, *Nucl. Fusion* **24,** 1039 (1984).

219. M. L. Grossbeck, J. O. Stiegler, and J. J. Holmes, p. 95 in Ref. 29.

220. M. S. Wechsler, p. 991 in Ref. 24, Vol. 2.

221. B. Van Der Schaaf, M. I. DeVries, and J. D. Elen, p. 307 in Ref. 29.

222. P. J. Maziasz and E. E. Bloom, U.S. Department of Energy Quarterly Progress Report DOE/ET-0058/1, August 1978.

223. P. Batfalsky, H. Schroeder, and H. Ullmaier, to be published.

224. R. L. Fish, J. L. Straalsund, and C. W. Hunter, in *Effects of Irradiation on Substructure and Mechanical Properties of Metals and Alloys,* ASTM STP 529, American Society for Testing and Materials, Philadelphia, 1973, p. 149.

225. K. Farrell and J. Lehman, Oak Ridge National Laboratory, in preparation.

226. D. Hull and D. E. Rimmer, *Philos. Mag.* **8,** 673 (1958).

227. R. W. Balluffi and L. L. Siegle, *Acta Met.* **5,** 449 (1957).

228. R. S. Barnes, *Nature* **206,** 1307 (1965).

229. J. E. Harris, *Met. Scie.* **12,** 321 (1978).

230. M. V. Speight and W. Beere, *Met. Scie.* **9,** 190 (1975).

231. W. D. Nix, *Scripta Met.* **17,** 1 (1983).

232. H. Trinkaus, *J. Nucl. Mater.* **118,** 39 (1983).

9 Stochastic Theory of Electron–Hole Transport and Recombination in Amorphous Materials

Jaan Noolandi

Xerox Research Center of Canada,
Mississauga, Ontario, Canada

9.1. INTRODUCTION

In this chapter we look in detail at two examples of nonhomogeneous processes, namely the photogeneration and subsequent motion of charge in amorphous materials. The materials include solid photoconductors (such as amorphous selenium) and liquid dielectrics (such as cyclohexane). In the solid state the sequence of microscopic events that gives rise to light-induced charge patterns on the surfaces of photosensitive materials has been exploited in imaging science and has spawned a rapidly growing multibillion dollar industry. Studies of the liquid state have been less important practically but have provided a number of insights into fundamental processes, some of which have been helpful in developing an understanding of xerography. In this chapter we do not deal with the applied aspect of electron–hole recombination and transport in amorphous photoconductors since a number of excellent review articles are available in this field (1). Rather we restrict ourselves to a more fundamental description of charge carrier transport and recombination.

The theoretical models are oversimplified in order to be applicable to diverse materials. For example, the time evolution of electron–hole pairs in an amorphous medium will be described in entirely classical terms using a continuity equation, which is expected to be valid when the scattering length is comparable to the interatomic spacing. Quantum effects associated with the primary photogeneration event or the ultimate recombination process are buried in phenomenological parameters that are part of classical boundary conditions. Nevertheless, the classical diffusion of an electron–hole pair in the presence of their mutual coulombic interaction and an external electric field is sufficiently complicated that it merits separate study. A thorough understanding of this process is a necessary precursor to any discussion of quantum phenomena. One of the useful results to come out of the classical theory is the analytic dependence of the time-dependent pair distribution function on the boundary conditions, which gives some idea of how the analysis would be affected by a quantum mechanical treatment.

Our analysis of "stochastic" transport is also based on a classical continuum model describing the motion of charge through a transport band in the presence of traps. The traps (localized states) are assumed to be distributed homogeneously throughout the material and are described by classical capture and release rate parameters. For simplicity we assume discrete trapping levels so that traps with a range of microscopic rate parameters are lumped together and included in one level. In this way a few trapping levels properly distributed can adequately describe the behavior of charge moving through the material. Although the traps have a homogeneous distribution, the spatial distribution of charge in the medium can be highly nonhomogeneous. A typical example is the localized distribution of charge created by photoexcitation near the surface of a photoconductor. The time evolution of the nonhomogeneous distribution can give rise to a nonequilibrium process, namely anomalous dispersion, in which charges try to come to thermal equilibrium with the trap distribution but do not succeed before recombination or removal from the system by an absorbing boundary. This process is understood on the basis of the continuum multiple-trapping model.

Both of the theoretical models are exactly solvable by conventional analytic techniques. The more complicated of the two is the recombination problem. At first sight the analytic solution of this problem seems of limited value because it involves an infinite series of Bessel functions, which in any case have to be computed numerically. Why not solve the diffusion equation numerically in the first place? For the case where the external electric field vanishes, this can be done on a hand calculator. However, several points in favor of the analytic approach can be made. First, from a theoreticians point of view, it is more elegant to obtain an analytic solution. As a practical matter it is also preferable to use an analytic expression when computing the solution to some predetermined degree of accuracy. This problem is particularly important when dealing with the long-time behavior. For the case of a finite external electric field the numerical solution of the diffusion equation has not even been attempted and presumably would not give reliable results because of the particular nature of the solution. Second, analytic solutions allow one to predict new physical effects, such as the transition from three- to one-dimensional behavior for diffusion over the coulombic barrier when an external field is present. Third, analytic solutions allow for asymptotic expansions, treatment of limiting cases, and so on, which are physically interesting in their own right and can be used to test the accuracy of approximate methods and trial solutions. Finally, and perhaps most important, an understanding of the nature of the analytic solution can help one to incorporate other physical effects into the model in a systematic way. A good example of this is demonstrated in this chapter, where we include the effects of multiple trapping in the recombination model and obtain a new exact solution. Computer experiments, such as the Monte Carlo technique, can also be used to obtain useful insights into complicated "coupled" phenomena, but generally the results are difficult to generalize, and the effects cannot be studied in a systematic way. Exact analytic results, when they can be obtained, are interesting theoretically and useful practically and often form the basis of further advances in the field.

In Section 9.2 we discuss the recombination model based on the classical time-dependent diffusion equation with a coulombic potential and an external electric field. We review the general procedure for deriving the time-dependent Green's function for this problem and examine the nature of the general solution, emphasizing the transition from diffusive to nondiffusive behavior at a critical value of the electric field. For the zero-field case we then use the eigenfunction method to obtain expressions for the reaction rate and the survival probability as a function of time. The computed survival probabilities based on the analytic expression are given in Appendix A for different values of the initial geminate pair separation r_0. The effect of the electric field on the escape probability is then discussed, and applications of the theory to recent time-resolved measurements of geminate recombination in liquids and solids are reviewed. The theory of charge scavenging in liquids is covered, and some recent work on the detection of geminate ions by dc conductivity measurements is mentioned.

Section 9.3 deals with the stochastic theory of electron–hole transport. Beginning with a discussion of traditional time-of-flight experiments, we follow a historical approach and comment on the Scher–Montroll theory of dispersive transport. The modern theory of anomalous dispersion, based on the multiple-trapping model, is then reviewed in detail, and the concept of thermalization of charge carriers within a distribution of localized states is presented. A computer program for analyzing transient photocurrent data as well as calculating theoretical curves for various trap parameters is given in Appendix B.

In Section 9.4 we treat electron–hole recombination and anomalous dispersion on an equal footing and obtain an exact solution for the appropriate Green's function when both effects are combined. The combination of the two models is made possible by the simple way in which the multiple-trapping model describes the retarded response of the free current density to the total charge density, taking into account the presence of temporarily trapped carriers. Some results showing the influence of multiple trapping on the recombination rate and survival probability are presented. The conclusions are given in Section 9.5.

9.2. ELECTRON–HOLE RECOMBINATION

9.2.1. General Principles

In an amorphous material, whether it be liquid or solid, a moving test charge encounters potential fluctuations as it passes the atoms or molecules, as shown in the top panel of Figure 9.1. If an electron–hole pair is created somewhere by the absorption of a photon, the fluctuations shown in the upper panel of Figure 9.1 will have a coulombic potential superimposed upon it, as shown in the lower panel. One or both members of the charge pair move rapidly through the medium until they are at thermal equilibrium with their surroundings. With the hole or positive charge taken as the reference point in Figure 9.1, the negative charge with the greater excitation energy travels a greater distance before thermalization. A ma-

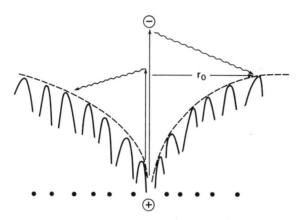

Figure 9.1. Diagram of one-dimensional potential profile before (top) and after (bottom) creation of a geminate electron–hole pair in an amorphous photoconducting medium.

terial will have a characteristic range of thermalization lengths for a given excitation energy, and one of the goals of the theory is to determine the distribution of these lengths, which can be used to obtain information about the thermalization process.

Figure 9.1 shows a one-dimensional coulombic potential well. In reality, the potential is three dimensional, and its shape along with a superimposed external electric field is shown in Figure 9.2. Although for purposes of illustration cross sections of this potential surface are shown in schematic diagrams, it is important to bear in mind that the three-dimensional potential offers many more possibilities for the relative motion of the electron–hole pair. It should also be noted that the static potential surface shown in Figure 9.2 represents an average over the motion of all the atoms or molecules of the medium. The motion of a particle along this potential surface is governed by thermal fluctuations, which give rise to diffusive or Brownian motion of the electron–hole pair under the influence of their mutual coulomb attraction and the external electric field.

A newly created electron–hole pair has available to it two possibilities for its evolutionary history. The pair may separate and escape from its mutual coulombic attraction with the assistance of the external electric field. The separated pairs can be measured to determine the quantum efficiency for the system (defined as the amount of free charge collected per photon absorbed). The other possibility is that the pair may recombine and give rise to fluorescence or increase the thermal energy of the medium. These two alternatives are shown in Figure 9.3 in a schematic

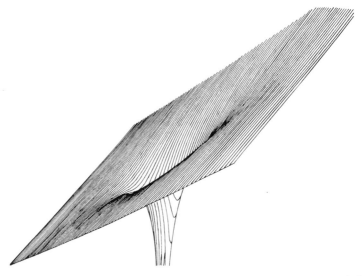

Figure 9.2. Plot of three-dimensional coulombic well with a superimposed linear electric field, corresponding to the interaction potential in the time-dependent Onsager problem.

energy level diagram. Excitation from a neutral state S_0 to a highly excited state S^* is followed by thermalization in the medium, giving rise to a bound pair that will either dissociate to free carriers or recombine to the first excited singlet state S_1. The ultimate (long-time) probabilities for thermalization and dissociation to free carriers are given by η_∞ and Ω, respectively. The recombination rate is $R = 1 - \Omega$. Some of the excited electron–hole pairs decay to the first excited singlet

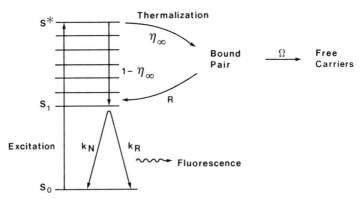

Figure 9.3. Schematic representation of charge carrier photogeneration in an inorganic photoconductor: Ω, escape probability for long times; R, corresponding reaction rate; η_∞ thermalization probability, with k_N and k_R the nonradiative and radiative rate constants for transitions to the ground state S_0 from the first excited singlet state S_1. Reprinted, with permission, from Ref. 6.

state with a probability $1 - \eta_{\infty}$. The first excited singlet state may in turn decay radiatively or nonradiatively to the ground state with the rate constants k_R and k_N respectively.

Greater excitation energies S^* lead to longer thermalization lengths, which in turn increase the probability of obtaining free carriers, so light quanta of longer wavelengths give rise to smaller quantum efficiencies for a given electric field strength. Figure 9.4 shows some quantum efficiency data for the photoconductor amorphous selenium as a function of applied electric field for different photon excitation energies (2). The dots indicate the experimental points, and the solid lines represent the results of a theoretical calculation based on the Onsager recombination model (3, 4) for various initial thermalization lengths.

9.2.2. Smoluchowski Equation

9.2.2.1. Formulation. To analyze the trends shown in Figure 9.4 quantitatively, we turn to a theoretical model of the recombination process. We follow the clas-

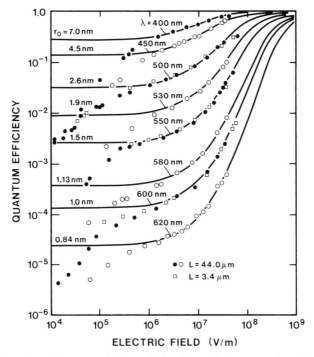

Figure 9.4. Experimental quantum efficiency of photoinjected holes versus the applied electric field for different values of the wavelength of exciting radiation in amorphous selenium. Circles and squares show the experimental quantum efficiencies, and the data on films of two different thicknesses are plotted. The solid lines are the theoretical Onsager dissociation efficiencies for various initial separations r_0 as indicated on the figure. (After Ref. 2.)

sical description based on the Smoluchowski equation. This model has the virtue that it can be solved analytically to obtain a number of useful insights into the recombination process. A classical model is appropriate for many amorphous materials because the mean scattering length of a moving particle is of the order of the interatomic spacing, thereby reducing quantum coherence effects. The continuity (Smoluchowski) equation for the two-particle distribution function, $\rho(r, t)$ is given by

$$\frac{\partial \rho}{\partial t} + div \, \mathbf{j} = 0 \tag{9.1}$$

where the current density $\mathbf{j}$ is

$$\mathbf{j} = -De^{-W} \nabla(e^{W} \rho) \tag{9.2}$$

The physical significance of Eq. (9.2) will be discussed shortly. Here D is the sum of the diffusion coefficients of the electron and hole, with one of the particles at the origin as shown in Figure 9.5. The potential energy of the particles consists of a coulombic term and the interaction with the external electric field,

$$W = -\frac{\xi^2}{4\pi\varepsilon_0\epsilon \, rk_B T} - \frac{\xi Er \cos \theta}{k_B T} \tag{9.3}$$

where the potential energy has been divided by $k_B T$ in order to make it dimensionless. Here ξ is the charge on a proton, ε_0 is the permittivity of the vacuum, ϵ is the dielectric constant, E is the magnitude of the external electric field, k_B is Boltzmann's constant, and T the absolute temperature. The radial distance r and the angle θ with respect to the external electric field E are indicated in Figure 9.5. It is useful to rewrite the expression for the potential energy in terms of a more

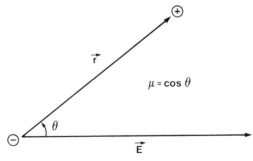

Figure 9.5. Frame of reference for diffusing geminate electron–hole pair described by the time-dependent Smoluchowski equation. The z axis is in the direction of the external electric field.

compact form involving the so-called Onsager radius, r_c, and a dimensionless parameter, F:

$$W = -\frac{r_c}{r} - \frac{2Fr}{r_c} \cos \theta \qquad (9.4)$$

where $r_c = \xi^2/4\pi\varepsilon_0\epsilon k_B T$, and $F = \xi E r_c/2k_B T$. The Onsager radius (sometimes also called the coulombic radius) is the distance at which the coulombic energy of the geminate pair has dropped $k_B T$ below the value at infinite separation. Roughly speaking, particles inside a sphere of radius r_c do not have enough thermal energy to escape recombination, whereas particles outside tend to dissociate.

The compact expression (9.2) for the current density $\mathbf{j}$, can be better understood by writing out the components,

$$\mathbf{j} = -D\,\nabla\rho + \rho\mathbf{v} \qquad (9.5)$$

with the first part involving only the diffusive motion of the two particles in the absence of any interaction energy, and the second part arising from the external electric field and the mutual coulombic interaction. With the help of the Einstein relation, $\mu/D = \xi/k_B T$, where $\mu = \mu_e + \mu_+$ is the sum of the electron and hole mobilities, the drift velocity $\mathbf{v}$ of the particles can be written as

$$\mathbf{v} = \mu\left(E - \frac{\xi}{4\pi\varepsilon_0\epsilon}\frac{\hat{r}}{r^2}\right) \qquad (9.6)$$

where $\hat{r}$ is the unit vector in the radial direction. This result shows the opposing forces of the external electric field and the coulombic interaction when the negative charge is fixed at the origin and the positive charge is mobile. We note that the only material parameters appearing in the model so far are the sum of the diffusion constants of the particles, D, and the dielectric constant and temperature of the medium.

A boundary condition must be imposed at the recombination center (origin). We choose the so-called radiation boundary condition, in analogy with the problem of heat conduction, describing a partly absorbing and partly reflecting sphere of radius $r = a$ at the origin. In this model the radial component of the current density at the surface of the sphere is given by

$$j_r = -\kappa\rho \quad \text{at } r = a \qquad (9.7)$$

where κ is the particle recombination velocity, which can have any value ranging from zero to infinity. When κ tends to infinity, the sphere at the origin becomes a perfectly absorbing black body, and the distribution function on the surface must vanish. However, the product of κ and ρ tends to some limiting value as determined by the solution of the Smoluchowski equation. When κ has a finite value, the sphere at the origin is partly absorbing, partly reflecting, and we assume that in some average way this boundary condition reflects the microscopic situation in which some electron-hole pairs approach one another and then separate again. When κ vanishes, no recombination takes place, and the quantum efficiency tends to unity. In Onsager's original calculation of the escape probability (4), the radius of the recombination sphere was assumed to be zero. However, there is little complication in keeping the radius of the sphere finite, and we leave this quantity as a parameter unless noted otherwise.

9.2.2.2. General Solution.
We now turn to the solution of the time-dependent Smoluchowski equation (also generally known as the Onsager problem). We wish to obtain the Green's function for this equation for a delta-function initial condition, which will then allow us to determine the general solution for any arbitrary initial distribution function. The small number of parameters in this model makes it an attractive one, and it is hoped that some of the insights we obtain will guide us toward better understanding of real materials. We solve the time-dependent Onsager problem for the general initial condition,

$$\rho(r, \mu, 0|r_0, \mu_0) = \frac{1}{2\pi r_0^2} \delta(r - r_0)\delta(\mu - \mu_0) \tag{9.8}$$

where at $t = 0$ the particles are separated by a distance r_0, and the line joining them makes an angle $\theta_0(\mu_0 = \cos\theta_0)$ with the polar axis, which is in the direction of the external electric field E, as shown in Figure 9.5. After obtaining the Green's function for this initial condition, we may calculate the time-dependent distribution function for any other *axially* symmetric initial distribution by integrating over r_0 and μ_0. Rather than going through the technical details of the solution, which can be found in the published papers (5-7), we will discuss the various steps in the general method and comment on the results. Looking back at Eq. (9.2) and remembering the vector identity,

$$\nabla \cdot (\theta \mathbf{a}) = \mathbf{a} \cdot \nabla\theta + \theta\nabla \cdot \mathbf{a} \tag{9.9}$$

where θ is a scalar function, we simplify the Smoluchowski equation by introducing a new function h according to the transformation

$$\rho(r, \mu, t|r_0, \mu_0) = \frac{h(r, u, t|r_0, \mu_0)}{2\pi\sqrt{r_0 r}} \exp\left\{(1/2)\left[W(r_0, \mu_0) - W(r, \mu)\right]\right\}$$

$$\tag{9.10}$$

so that the new function satisfies the equation

$$
\frac{\partial h}{\partial t} = \frac{\partial^2 h}{\partial r^2} + \frac{1}{r}\frac{\partial h}{\partial r} - \left[\left(\frac{F}{2}\right)^2 + \frac{1}{4r^2} + \frac{1}{r^4} \right] h
$$
$$
+ \frac{1}{r^2}\left\{ \frac{\partial}{\partial \mu}\left[(1 - \mu^2)\frac{\partial h}{\partial \mu} \right] + F\mu h \right\} \tag{9.11}
$$

with the corresponding initial condition now written as

$$
h(r, \mu, 0 | r_0, \mu_0) = \frac{1}{r_0}\delta(r - r_0)\delta(\mu - \mu_0) \tag{9.12}
$$

In these formulas we have introduced $r_c^2/4D = (r_c/2)^2/D$ as the unit of time, leading to the choice of $r_c/2$ as the unit of length. What has been gained by the transformation Eq. (9.10)? First, as observed by Onsager (4), the resulting equation is separable and its solution may be written as

$$
h(r, \mu, t | r_0, \mu_0) = \sum_{l=0}^{\infty} R_l(r, t | r_0) T_l(\mu) T_l(\mu_0) \tag{9.13}
$$

where $T_l(\mu)$ is a generalized Legendre function that satisfies the equation

$$
\frac{d}{d\mu}\left[(1 - \mu^2)\frac{dT_l}{d\mu} \right] + (F\mu + \lambda_l)T_l = 0 \tag{9.14}
$$

and λ_l are the eigenvalues for the angular components of the solution. Second, Eq. (9.11) has the property that it is formally the same as the (imaginary) time-dependent Schrödinger equation for a particle moving in a dipole-type potential field with an inverse fourth-power repulsive core. It should be noted that although this similarity is entirely formal, it is nevertheless very useful since it allows us to use certain results from well-known problems in quantum mechanics.

The associated equation for the radial function R_l is

$$
\frac{\partial R_l}{\partial t} = \frac{\partial^2 R_l}{\partial r^2} + \frac{1}{r}\frac{\partial R_l}{\partial r} - \left[\left(\frac{F}{2}\right)^2 + \frac{\lambda_l + 1/4}{r^2} + \frac{1}{r^4} \right] R_l \tag{9.15}
$$

with the initial condition

$$
R_l(r, 0 | r_0) = \frac{1}{r_0}\delta(r - r_0) \tag{9.16}
$$

Equation (9.15) can be solved by taking the Laplace transform,

$$\tilde{R}_l(r, s|r_0) = \int_0^\infty dt \, e^{-st} R_l(r, t|r_0)$$

$$= LR_l(r, t|r_0) \tag{9.17}$$

and writing down the complete solution in terms of the two linearly independent solutions to the corresponding homogeneous equation. As discussed in our earlier papers (5–7), the two independent solutions can be expressed in terms of a series of modified Bessel function products,

$$y_{1l}(r) = y_{2l}\left(\frac{1}{\sqrt{s_F r}}\right)$$

$$y_{2l}(r) = \sum_{n=-\infty}^{\infty} (-1)^n c_n I_n\left(\frac{1}{r}\right) K_{n+\nu}(\sqrt{s_F} r) \tag{9.18}$$

where the coefficients c_n are determined by the recursion relation

$$[(2n + \nu)^2 - \lambda_l - \tfrac{1}{4}]c_n = \sqrt{s_F}\,(c_{n+1} + c_{n-1}) \tag{9.19}$$

and $s_F = s + (F/2)^2$.

9.2.2.3. Analytic Properties.

It is helpful to describe the analytic properties of the Laplace transform of the solution to obtain insight into these formulas. Concerning the Laplace transform variable s, recall that small s corresponds to large $t (s = 0$ gives the steady-state limit) and large s corresponds to small t (8). Figure 9.6 shows the singularity structure of the Laplace transform for the dimensionless parameter $F = \xi E r_c / 2k_B T$ greater than the critical value $F_c \simeq 1.28$, to be discussed later (9). The first feature to note is that there is a branch cut along the negative real axis. This is characteristic of a pure diffusion solution without any interaction potential. A table of Laplace integrals shows that the free diffusion solution

$$\rho \simeq \frac{e^{-r^2/4Dt}}{t^{3/2}} \tag{9.20}$$

has the transform

$$\rho \simeq e^{-r\sqrt{s/D}} \tag{9.21}$$

which possesses a branch cut along the negative real axis starting from the origin, since it depends on $\sqrt{s}$. In these equations we have ignored numerical prefactors.

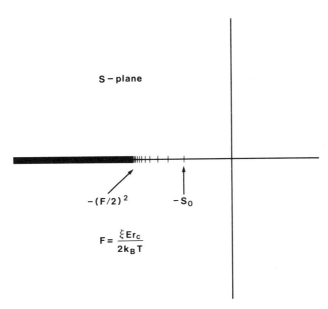

Figure 9.6. Singularity structure of the Laplace transform of the time-dependent electron–hole pair distribution function. For a given F, the branch point is always at $-(F/2)^2$. In addition, For $F > F_c$ $\simeq 1.28$, an infinite number of poles appear on the negative real axis. The poles merge into the branch cut and the largest pole, denoted by s_0, dominates the long-time behavior.

For a zero external electric field the solution of the time-dependent Onsager problem has the same singularity structure, which is not surprising since the solution also depends on the square root of the Laplace transform variable s, even though the full expression is more complicated. For no coulombic interaction but with a finite electric field, we expect a branch cut that is shifted away from the origin according to the theorem (8)

$$L^{-1}f(s + a) = e^{-at}F(t) \qquad (9.22)$$

where

$$L^{-1}f(s) = F(t) \qquad (9.23)$$

This reflects the superimposed drift velocity of the distribution function due to the external electric field (assumed to be in the z direction),

$$\rho \simeq t^{-3/2} \exp\left\{-\frac{[x^2 + y^2 + + (z - vt)^2]}{4Dt}\right\} \qquad (9.24)$$

where $v = \mu E$ is the drift velocity.

The surprising result in the solution of the Onsager problem with a finite electric field and a coulombic interaction is that the singularity structure changes, as shown in Figure 9.6. In addition to the branch cut that has separated from the origin, there now appear an infinite number of poles terminating at the end of the branch cut as a limit point. This new singularity structure has an important effect on the long-time behavior of the solution. For $t \to \infty$ we have

$$\rho \simeq t^{-a} \exp \left[-(\tfrac{1}{2}F)^2 \, t \right] \tag{9.25}$$

for $F < F_c$, where the exponent a varies from $\tfrac{3}{2}$ at $F = 0$ to 1 at $F = F_c$. For $F > F_c$ and $t \to \infty$, the long-time behavior of the distribution function is purely exponential,

$$\rho \simeq \exp \left(-|s_0|t \right) \tag{9.26}$$

where s_0 is the largest pole (closest to the origin) on the negative real axis.

9.2.2.4. Critical Field.

The change in the long-time behavior of the distribution function signals the onset of a transition from three-dimensional to one-dimensional diffusion. Similar behavior has been seen in the numerical simulation work of Silver et al. (10, 11). In addition, our earlier work on a model quadratic one-dimensional potential gave a singularity structure consisting of simple poles along the negative real axis, indicating a rate-limited Kramers type of behavior for diffusion over a potential barrier (12). For liquid hydrocarbons the critical field is on the order of 20 kV/cm, and its effects should be observable in a properly designed experiment. A simple physical interpretation of these unusual mathematical results may be given by referring to Figure 9.7. Here we show a schematic diagram comparing the drift time along the external electric field with the diffusion time perpendicular to the external field. Considering the drift velocity of a particle in a direction parallel to the field and neglecting the retarding effect of the coulombic attraction, we estimate that the time taken to reach the Onsager radius r_c is approximately

$$t_{\text{drift}} \simeq \frac{r_c}{\mu E} \tag{9.27}$$

The corresponding time required to reach the Onsager radius perpendicular to the electric field is

$$t_{\text{diffusion}} \simeq \frac{r_c^2}{2D} \tag{9.28}$$

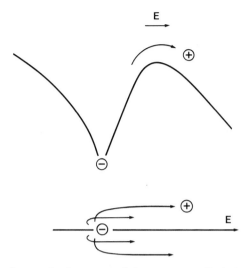

Figure 9.7. Schematic diagram showing escape of charge over an effectively one-dimensional cou-lombic barrier for large electric fields. The holes, which are assumed to thermalize in a roughly spher-ical distribution about the negative charge, are swept along the electric field lines as shown in the bottom panel of the figure.

Setting these two times equal gives a measure of the electric field strength re-quired to force the diffusing particles to move along the one-dimensional field lines. Using the Einstein relation, we obtain

$$\frac{\xi E r_c}{k_B T} \simeq 2 \tag{9.29}$$

This estimate gives us some idea of the electric field strength at which we may ignore the complications of the full three-dimensional Onsager dissociation process and study the simpler problem of diffusion over a one-dimensional potential well.

The result of a rigorous mathematical analysis shows that the critical value of the electric field is given by

$$F_c = \frac{\xi E_c r_c}{2 k_B T} \simeq 1.28 \tag{9.30}$$

above which the time-dependent behavior of all physical qualities changes from a power law (with a superimposed drift velocity) to exponential. The critical value of the electric field is obtained by referring back to Eq. (9.15) and noting that this equation becomes (after taking the Laplace transform) identical to the radial Schrö-dinger equation for a particle moving in the hard-core potential,

$$V(r) = \begin{cases} \dfrac{\lambda_l}{r^2} + r^{-4} & r > a \\ \infty & r \le a \end{cases} \tag{9.31}$$

From quantum mechanics we know that a potential of this type has no bound states for $\lambda_l > -\frac{1}{4}$, which means that the eigenvalue spectrum is continuous (13). Since these eigenvalues also give the singularities of $\tilde{R}_l(r, s \mid r_0)$, we conclude that the transform of the solution has no poles. However, as the electric field ($\propto F$) increases, the first eigenvalue, λ_0, for the angular eigenfunction $T_0 (\mu)$ becomes increasingly negative and reaches $-\frac{1}{4}$ for $F_c \simeq 1.28$. This critical value of F is determined once and for all from Eq. (9.14) using an expansion for $T_0(\mu)$ in terms of ordinary Legendre polynomials. For larger values of F the higher eigenvalues also become negative and less than $-\frac{1}{4}$, but this is not relevant for the long-time behavior of the solution, which is already dominated by the first eigenvalue.

9.2.2.5. Reaction Rate. We now discuss some of the other analytic results of the solution of the Onsager problem. The reaction rate $R(t)$ is defined as the rate at which the particles disappear into the sink at the origin,

$$R(t) = -2\pi a^2 \int_{-1}^{1} d\mu \, j_r \, (a, \mu, t) \tag{9.32}$$

$$= -4\pi a^2 \, j \, (a, t) \text{ (for the spherically symmetric case when } E = 0)$$

and the survival probability $\Omega(t)$ is given by the fraction of particles left in the system,

$$\Omega(t) = 4\pi \int_{a}^{\infty} dr \, r^2 \, \rho(r, t \mid r_0)$$

$$= 1 - \int_{0}^{t} dt' \, R(t') \tag{9.33}$$

Both the survival probability and the reaction rate can be determined once the two-particle distribution function ρ is known. We illustrate the determination of these quantities for the case when the electric field vanishes. The spherically symmetric Smoluchowski equation (9.1) is

$$\frac{\partial \rho}{\partial t} = D \frac{1}{r^2} \frac{\partial}{\partial r} \left(r^2 \, e^{r_c/r} \frac{\partial}{\partial r} (e^{-r_c/r} \, \rho) \right) \tag{9.34}$$

With $r_c^2/4D = (r_c/2)^2 \, /D$ as the unit of time and $r_c/2$ as the unit of length, Eq. (9.34) may be written in dimensionless form as

$$\frac{\partial \rho}{\partial t} = \frac{1}{r^2} \frac{\partial}{\partial r} \left(r^2 e^{2/r} \frac{\partial}{\partial r} (e^{-2/r} \rho) \right) \tag{9.35}$$

and the current density, given by Eq. (9.2), becomes

$$j(r, t) = -e^{2/r} \frac{\partial}{\partial r} (e^{-2/r} \rho) \tag{9.36}$$

Equation (9.32) for the reaction rate $R(t)$ is then given by

$$R(t) = 4\pi a^2 \frac{\partial \rho}{\partial r} \bigg|_{r=a} + 8\pi \rho(a, t | r_0) \tag{9.37}$$

and for $a \to 0$ only the second term survives since ρ is finite at the origin, so that

$$\lim_{a \to 0} R(t) = 8\pi \rho(0, t | r_0) \tag{9.38}$$

The expression for the survival probability $\Omega(t)$ is 1 minus the time integral of the density at the origin. Note that the particle recombination velocity κ does not appear explicitly in these expressions, although for finite a the solution for ρ depends on κ through the boundary condition. In the limit of vanishing a the same solution is obtained for any finite value of κ, so that in this limit the recombination center acts as a perfect black body.

9.2.2.6. Eigenfunction Expansion. A way to obtain ρ for $F = 0$ is to use an eigenfunction expansion, although the general solution outlined earlier could also be used. Hence we write

$$\rho(r, t | r_0) = \int_0^\infty \frac{dk}{2\pi} a_k \, g_k(r) \, e^{-k^2 t} \tag{9.39}$$

where $g_k(r)$ is an eigenfunction that satisfies the equation

$$\frac{1}{r^2} \frac{\partial}{\partial r} \left(r^2 e^{2/r} \frac{\partial}{\partial r} (e^{-2/r} g_k) \right) = -k^2 g_k \tag{9.40}$$

and the coefficients a_k are determined by the initial spherically symmetric condition

$$\rho(r, 0 | r_0) = \frac{1}{4\pi r_0^2} \delta(r - r_0) \tag{9.41}$$

The eigenfunctions $g_k(r)$ are orthogonal and have some normalization factor, which we denote by $2\pi W_k$, so that integrating over all space gives

$$4\pi \int_0^\infty dr \, r^2 e^{-2/r} \, g_k(r) g_{k'}(r) = 2\pi W_k \, \delta(k - k') \tag{9.42}$$

The weight factor $e^{-2/r}$ in this generalized orthogonality relation derives from the eigenvalue equation (9.40). Using the initial condition Eq. (9.41) in Eq. (9.39), multiplying both sides of the equation by $e^{-2/r} g_{k'}(r)$, and integrating over all space then gives the expression for a_k. The Green's function solution for ρ becomes

$$\rho(r, t | r_0) = e^{-2/r_0} \int_0^\infty \frac{dk}{2\pi} \frac{g_k(r_0) \, g_k(r) \, e^{-k^2 t}}{W_k} \tag{9.43}$$

The above method is very general, and a similar procedure can be used to obtain the exact solution for the Smoluchowski equation involving only diffusion and tunneling (without the coulombic interaction) (14). It remains to solve Eq. (9.40) for the eigenfunctions $g_k(r)$. With the transformation

$$g_k(r) = \frac{e^{1/r}}{\sqrt{r}} f_k(r) \tag{9.44}$$

the resulting equation for $f_k(r)$ becomes the radial Schrödinger equation with the potential given by Eq. (9.31) ($\lambda_0 = 0$ for $F = 0$), and the solutions are given in terms of a series of modified Bessel function products, Eqs. (9.18) and (9.19) (for $F = 0$). The normalization factor W_k can also be evaluated explicitly from Eq. (9.42). After changing the variable of integration from k to k^2, the final result for ρ is

$$\rho(r, t | r_0) = \frac{1}{\pi^2 \sqrt{r_0 r}} e^{(1/r) - (1/r_0)} \int_0^\infty \frac{du}{2\pi} \frac{f(r, u) f(r_0, u) e^{-ut}}{|M(u)|^2} \tag{9.45}$$

where $M(u)$ is related to the Wronskian of the Bessel function solutions. For $r \to 0$, we find, from the asymptotic properties of the Bessel functions (7),

$$f(r, u) \simeq (\tfrac{1}{2} \pi r)^{1/2} e^{-1/r} \tag{9.46}$$

independent of u. According to Eq. (9.38), the expression for the reaction rate, which involves $\rho(0, t | r_0)$, becomes

$$R(t) = 4(2/\pi r_0)^{1/2} e^{-1/r_0} \int_0^\infty \frac{du}{2\pi} \frac{f(r_0, u) e^{-ut}}{|M(u)|^2} \tag{9.47}$$

The corresponding survival probability is obtained by integration of this result, giving

$$\Omega(t) = \text{const} + 4(2/\pi r_0)^{1/2} e^{-1/r_0} \int_0^\infty \frac{du}{2\pi} \frac{f(r_0, u)}{|M(u)|^2} \frac{e^{-ut}}{u} \tag{9.48}$$

9.2.2.7. Escape Probability. The constant term is the infinite-time limit, $\Omega(\infty)$, or the probability for a geminate pair in zero external electric field and with an initial separation r_0 to dissociate completely. It can be calculated directly from the differential equation (9.35), which becomes, after Laplace transformation,

$$\frac{1}{r^2} \frac{\partial}{\partial r} \left(r^2 e^{2/r} \frac{\partial}{\partial r} (e^{-2/r} \tilde{\rho}) \right) = s\tilde{\rho} - \rho(t = 0)$$
$$= s\tilde{\rho} - \frac{1}{4\pi r_0^2} \delta(r - r_0) \tag{9.49}$$

making use of the initial condition Eq. (9.41). The steady-state solution is obtained by setting $s = 0$, and it can be verified by substitution that in this case the density is given by

$$\tilde{\rho}(r, s = 0 \mid r_0) = (1/8\pi) e^{-2/r_0}(e^{2/r} - 1) \tag{9.50}$$

for $r > r_0$ and is a constant for $r < r_0$,

$$\rho(r, s = 0 \mid r_0) = (1/8\pi) e^{-2/r_0} (e^{2/r_0} - 1) \tag{9.51}$$

so that the density is continuous at $r = r_0$.

Using the expression for the current density, Eq. (9.36), we find that the number of particles per unit time flowing outward across the surface of any sphere with radius $r > r_0$ is

$$4\pi r^2 j(r) = e^{-r_c/r_0}, \quad (r < r_0) \tag{9.52}$$

per unit particle created at $r = r_0$. Similarly, the number of particles per unit time flowing toward the recombination center for $r < r_0$ is

$$-4\pi r^2 j(r) = 1 - e^{-2/r_0} = 1 - e^{-r_c/r_0} \quad (r < r_0) \tag{9.53}$$

The steady-state escape probability, which we identify with $\Omega(\infty)$, is therefore

$$\Omega(\infty) = e^{-r_c/r_0} \tag{9.54}$$

which is the Boltzmann factor, $\exp W(r)$, with the coulombic potential $W(r)$ given by Eq. (9.4) with $F = 0$.

Another way of obtaining the above result is to use the theorem relating to the asymptotic behavior of the Laplace transform (8)

$$\lim_{t \to \infty} F(t) = \lim_{s \to 0} sf(s) , \tag{9.55}$$

where $f(s)$ is the transform of $F(t)$. In order to make use of this theorem, we take the transform of Eq. (9.33),

$$\tilde{\Omega}(s) = \frac{1 - \tilde{R}(s)}{s} \tag{9.56}$$

and find, from the analytic solution,

$$\lim_{s \to 0} s\tilde{\Omega}(s) = 1 - \tilde{R}(0) = e^{-2/r_0} = e^{-r_c/r_0} \tag{9.57}$$

which again is the Onsager escape probability for the zero field (3).

9.2.3. Experiments on Geminate Recombination

9.2.3.1. Zero-Field Measurements. Figure 9.8 shows the master curves for the escape probability given by Eq. (9.48) for a given initial separation r_0 plotted in

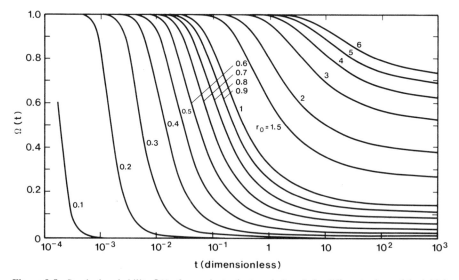

Figure 9.8. Survival probability $\Omega(t)$ of a geminate electron–hole pair for different values of the initial separation r_0 with no electric field. The unit of length is $r_c/2$, and the unit of time is $r_c^2/4D$. See Appendix A for tabulated numerical values corresponding to this plot.

terms of the dimensionless time and length parameters. The tabulated numerical values corresponding to the plot are given in Appendix A. A knowledge of the zero-field master curves for $\Omega(t)$ has been shown to be useful for an analysis of measurements of geminate recombination in liquids. In particular, Braun and Scott (15, 16) have explored the recombination of geminate cation–anion pairs of anthracene in liquid hexane by combining picosecond optical excitation with photoconductivity measurements. In these experiments the generation of free charge carriers by photoassisted geminate pair dissociation was studied, and the survival probability was obtained from the degree of photoconductivity as a function of the time delay between photoionization and infrared photodissociation. Comparison of the experimental results with the decay curves given in Appendix A allowed for the determination of the initial thermalized distribution of charge-pair separations. At present, the observed decay kinetics are in good agreement with the time-dependent diffusion theory. There is also no evidence that intermediate quantum mechanical states are important in the recombination process. Ultimately, the time-resolved measurements of geminate pair recombination kinetics should determine whether the classical diffusion theory gives an adequate description of the time evolution of geminate pairs.

Figure 9.9 shows two master curves for the zero-field reaction rate $R(t)$. The

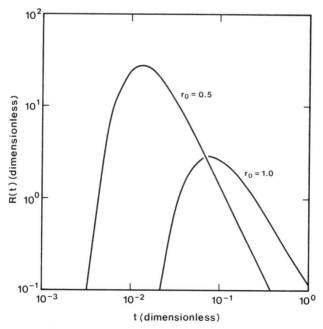

Figure 9.9. Reaction rate $R(t)$ for different values of r_0 calculated from the time-dependent Smoluchowski equation with no electric field. The units of length and time are $r_c/2$ and $r_c^2/4D$, respectively. Reprinted, with permission, from Ref. 5.

whole family of master curves has been used to analyze the photoluminescence decay in plasma-deposited amorphous hydrogenated silicon; for this case the effect of tunneling recombination had to be included (14). The low-temperature ($T = 8$ K) luminescence decay curve was used to determine the radial distribution function of thermalized electron–hole pairs, assuming purely tunneling recombination, and the high-temperature ($T = 150$ K) data were fitted by introducing the diffusion coefficient D as a parameter. The experimental data and the theoretical fits are shown in Figure 9.10. At high temperature the diffusion rate was found to be sufficiently large to quench the radiative recombination by tunneling, and the long-time luminescence decay showed a $t^{-3/2}$ decay characteristic of diffusion. The theory also led to good agreement with the overall quantum efficiency of the material, as measured by xerographic techniques, and gave an estimate of the microscopic electron mobility, which could be related to the time-of-flight transient photoconductivity experiments described in the next section.

9.2.3.2. Effect of Electric Field. We turn now to a discussion of electric field effects. Using the special properties of the generalized Legendre function, $T_l(\mu)$, Onsager originally showed that (3, 4)

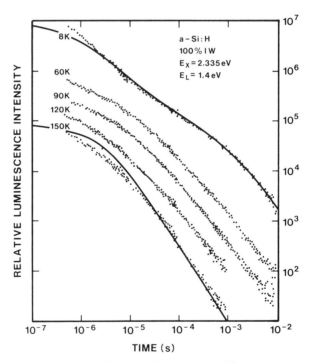

Figure 9.10. Log-log plot of photoluminescence decay curves at different temperatures in response to an optical excitation pulse of short duration. The solid lines show the fit to the theory. The excitation energy is E_X, and the luminescence energy is E_L. The sample was prepared from undiluted SiH_4, using low-rf power of 1 W.

$$\Omega(\infty, F) = \frac{1}{r_0} \exp\left[-\frac{Fr_0}{2}(1 + \mu_0)\right]$$

$$\int_2^\infty du \exp\left(-\frac{u_0}{r_0}\right) I_0 \left\{[2\, Fu\, (1 + \mu_0)]^{1/2}\right\} \qquad (9.58)$$

for the general nonspherical initial condition characterized by r_0 and μ_0 [Eq. (9.8)].

For a spherically symmetric one-parameter (r_0) delta-function initial condition [Eq. (9.41)], we must average over the angular dependence,

$$\Omega(\infty, F) = \frac{1}{2r_0} \int_0^2 dv \int_2^\infty du \exp\left[-\frac{Fr_0 v}{2} - \frac{u}{r_0}\right] I_0 (\sqrt{2Fuv}) \qquad (9.59)$$

and this expression is evaluated numerically for different values of r_0 to obtain the results plotted in Figure 9.4. For small F, Eq. (5.59) results in the low-field expansion,

$$\Omega(\infty, F) \simeq e^{-2/r_0} (1 + F) \qquad (F \ll 1), \qquad (9.60)$$

which gives a slope-to-intercept ratio independent of r_0 for low electric field. This allows for a unique test of the applicability of the Onsager theory provided that one can overcome charge trapping effects and accurately measure free carrier yields at low electric fields (17). There is a vast literature on the convolution of the above expression for $\Omega(\infty, F)$ with various empirical distribution functions to explain the results of radiolysis experiments (18–24). Some of the discussion in Chapters 2, 5, and 6 relates to this area of interest, and we do not pursue it further here.

In the photoconductor field many workers have fitted the Onsager theoretical prediction to the electric field dependence of the experimental photogeneration yield, with varying degrees of success. This "test" of the Onsager theory is not straightforward since the effects of surface recombination and bulk trapping must be accounted for (17). In a material such as amorphous selenium, however, there appears to be good quantitative agreement with Onsager theory using a delta-function initial distribution over a wide range of electric fields and thermalization distances (or photon energies) (2). However, a puzzle in the photoconductor field has been the large electron–hole "thermalization" lengths (2–3 nm) obtained for electron donor–acceptor crystals or polymers when the Onsager theory was used to fit the photogeneration yield. These results were inconsistent with spectroscopic evidence that a charge-transfer excited state involved electron transfer to mainly nearest neighbors (~0.5 nm). Braun (25) has shown that this problem may be solved by assuming that the lowest charge-transfer excited state is a reasonable precursor of free charge carriers. If the electric field dependence of the dissociation rate constant is assumed to be given by the theory for ion-pair dissociation of a weak electrolyte (3), a consistent description of photogeneration in donor–acceptor materials is obtained. Figure 9.11 shows a schematic diagram of carrier generation

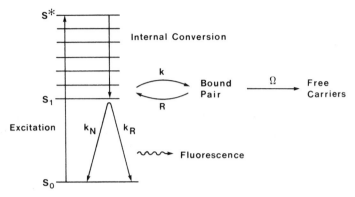

Figure 9.11. Schematic diagram of charge carrier generation in an organic (donor–acceptor) material, corresponding to the model described in the text. Reprinted, with permission, from Ref. 26.

for this case, when the lowest charge-transfer excited state is assumed to be a singlet. The charge carrier thermalizes by internal conversion, and this situation is to be contrasted with the ballistic model for thermalization shown in Figure 9.3. A virtue of the model is that thermalization lengths of 2–3 nm for the transferred electron do not need to be introduced. The picture here is a special case of a more general theory of photogeneration and fluorescence quenching in organic solids introduced by the author (26).

9.2.4. Scavenging

9.2.4.1. Theory. Finally we turn to the interesting area of charge scavenging. A great deal of information is available for the scavenging yield of electrons produced by the radiolysis of liquid hydrocarbons for many different solutes (scavengers) and solvents. It turns out that it is possible to extend the theory of geminate recombination to take into account the effect of a scavenger. We note that the scavenger competes with the cations of the geminate pairs for the electrons temporarily liberated by radiolysis. If the scavenger is present with a uniform concentration c_s, this process can be described by the addition of a term $-k_s c_s \rho_s(\mathbf{r}, t \,|\, r_0)$ to the left-hand side of Eq. (9.1), where k_s is the bimolecular rate constant for the scavenging reaction, and ρ_s is the distribution function for geminate pairs with the scavenger present. The solution of this modified continuity equation is related to the solution without a scavenger by (27)

$$\rho_s(\mathbf{r}, t \,|\, r_0) = \exp(-\lambda t)\, \rho(\mathbf{r}, t \,|\, r_0) \tag{9.61}$$

where $\lambda = k_s c_s$. It is important to remember that this result rests on the assumption that the distribution of the scavenger is completely uniform. If this is valid, the distribution function of geminate pairs retains its original spatial distribution (without a scavenger) but is gradually reduced in magnitude as time progresses.

The total scavenging yield at infinite time, when all electrons have either recombined or been scavenged, is given by the integral of $k_s c_s \rho_s(r, t|r_0)$ over all space and time. Denoting this quantity by $g_{sc}(c_s r_0)$, and remembering that the spatial integral of $\rho(\mathbf{r}, t|r_0)$ alone is the survival probability, we find

$$g_{sc}(c_s, r_0) = \lambda \, \tilde{\Omega}(\lambda) \, |_{\lambda = k_s c_s} \qquad (9.62)$$

The scavenging yield is therefore directly related to the Laplace transform of the survival probability. This remarkable result means that an experimental technique exists for measuring an abstract mathematical quantity provided that the assumptions inherent in the model indeed describe real scavenging experiments.

The Laplace transform of the survivial probability has been obtained earlier as an intermediate step in the solution of the time-dependent Smoluchowski equation. In order to make use of our analytical results, we note that the dimensionless Laplace transform variable s is related to $\lambda = k_s c_s$ by (28)

$$k_s c_s = \lambda = \frac{4D}{r_c^2} s \qquad (9.63)$$

For low scavenger concentration (small s) and $F \ll 1$, we obtain the following expansion from the exact solution (7):

$$s\tilde{\Omega}(s) = e^{-2/r_0}[1 + (4s + F^2)^{1/2} - F] \qquad (9.64)$$

for an isotropic initial distribution, which reduces to the zero-field value

$$s\tilde{\Omega}(s) = e^{-2/r_0}(1 + 2\sqrt{s}) \qquad (9.65)$$

and we have neglected the higher order terms. Based on Eq. (9.65), the scavenging yield is sometimes written

$$g_{sc}(c_s, r_0) = g_{fi} + g_{fi} \sqrt{4s} \qquad (9.66)$$

$$= g_{fi} + g_{fi} \sqrt{\frac{r_c^2}{D} k_s c_s}$$

where $g_{fi} = \exp(-r_c/r_0)$ is the normalized free ion yield, that is, $g_{fi} = G_{fi}/(G_{fi} + G_{gi})$, and G_{fi}, G_{gi} are the free and geminate ion yields, respectively. Equation (9.66) is expected to hold for extremely low scavenger concentrations, which, however, cannot be probed with current experimental techniques (29). It is known that the yield of charge scavenging in many liquids can be described very well over the entire range of concentration by the Warman–Asmus–Schuler (30) empirical formula,

$$g_{sc}(c_s) = g_{fi} + g_{gi} \frac{\sqrt{a_s c_s}}{1 + \sqrt{a_s c_s}} \tag{9.67}$$

where a_s is a parameter that measures the reactivity of the electrons for scavenging relative to their mobility in the solvent. Expanding this expression for small c_s gives agreement with Eq. (9.66) and leads to the identification

$$a_s = \left(\frac{g_{fi}}{g_{gi}}\right)^2 r_c^2 \frac{k_s}{D} \tag{9.68}$$

and this relation has been confirmed by an analysis of independent experimental data by Crumb and Baird (31).

In general, however, the empirical relationship Eq. (9.67) should be regarded as a good approximation to a complicated theoretical expression for $s\tilde{\Omega}(s)$. As has been shown elsewhere (5), the small s expansion, Eq. (9.65), is not a good approximation to $s\tilde{\Omega}(s)$ unless s is much less than unity.

9.2.4.2. Microwave and Optical Absorption.

More recently the geminate recombination of the ions formed in liquid CCl_4 by high-energy radiation has been studied by measuring the microwave and optical absorption, and the empirical expression (29)

$$\Omega(t) = \Omega(\infty)\left[1 + 0.6\left(\frac{r_c^2}{Dt}\right)^{0.6}\right] \tag{9.69}$$

has been shown to give a good representation of the results in the experimental time range, although for very long times (which are at present inaccessible experimentally) this formula does not reduce to the theoretically derived expression (5),

$$\Omega(t) = \Omega(\infty)\left[1 + \left(\frac{r_c^2}{\pi Dt}\right)^{1/2}\right] \tag{9.70}$$

Using the theorem for the asymptotic behavior of a Laplace transform (8),

$$\begin{array}{ll} \text{if } F(t) \propto t^a & \text{for } t \to \infty, \\ \text{then } f(s) \propto \Gamma(a+1)s^{-(a+1)} & \text{for } s \to 0 \end{array} \tag{9.71}$$

the small s limit of the transform of the survival probability corresponding to Eq. (9.69) gives the following expression for the scavenging yield:

$$g_{sc}(c_s, r_0) = g_{fi}\left[1 + \Gamma(0.4)\, 0.6\left(\frac{r_c^2 k_s c_s}{D}\right)^{0.6}\right] \tag{9.72}$$

where $\Gamma(0.4) = 2.22$ is the gamma function, and $g_{\mathrm{fi}} = \exp(-r_c/r_0)$. Equation (9.72) also gives a good representation of the scavenging data. This formula is an empirical fit over the experimental range of interest to the exact expression for the transform of the survival probability and should be regarded in the same spirit as Eq. (9.67). It should again be emphasized that the parameters in empirical formulas such as Eqs. (9.67) and (9.69) can be very sensitive to the fit to the available data and do not necessarily have any physical significance (32).

9.2.4.3. Scavenging in an Electric Field. Little has been done on the effects of scavenging in the presence of an electric field (33). For small scavenger concentrations the modified square-root dependence predicted by Eq. (9.64) seems to hold for measurements of the scavenging yield (34), although more experimental work is necessary. Studies of quenching of the recombination fluorescence of geminate pairs (35) by a scavenger (36) in the presence of an electric field should prove interesting, particularly since the Laplace transform of the reaction rate (or the associated survival probability) shown in Figure 9.12 could be compared with the results of these experiments. Measurements of the field dependence of the photoionization current in the presence of a scavenger show a reduction in the spatial extent of the distribution function of initial separation distances of the geminate

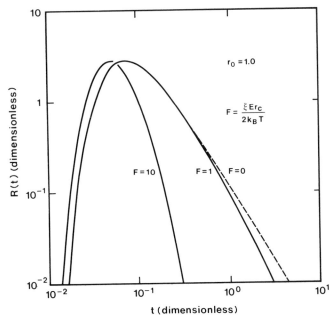

Figure 9.12. Reaction rate $R(t)$ calculated for different values of F according to the time-dependent Smoluchowski equation with the initial separation $r_0 = 1$. The dotted line shows the results for $F = 0$. The units of length and time are $r_c/2$ and $r_c^2/4D$, respectively.

pair (37, 38). An explanation of these results involves a theoretical model of epithermal electron scavenging (39), which is beyond the scope of this article.

9.2.5. Other Experiments

An exciting new development is the detection of geminate ions by dc conductivity in nanosecond pulse irradiated CCl_4 (40). It is found that the form of the dc transient is different from that obtained using microwaves. This is not surprising since the presence of the constant electric field results in a much higher yield of escaped ions. The effect can be seen in Figure 9.12, which shows the reaction rate, $R(t)$, evaluated for different values of the electric field parameter. As discussed earlier, the integral over time of the reaction rate is related to the survival probability. Figure 9.12 shows that the integral of $R(t)$ is smaller for $F = 10$ than for $F = 0$, corresponding to a higher survival probability for the higher electric field. The quenching of fluorescence due to geminate recombination by the electric field can be explained in a similar way (35). At present the form of the dc transient curves is in qualitative agreement with the predictions of the field strength effect on the survival probability as obtained from the solution of the time-dependent Onsager problem. More quantitative comparison with the theoretical results will prove interesting.

9.3. STOCHASTIC THEORY OF ELECTRON–HOLE TRANSPORT

9.3.1. Background

During the past decade a number of important advances have been made in the understanding of electron–hole transport through amorphous materials. While it has been known for some time through the work of Anderson (41) and others (42) that highly localized states can exist in amorphous materials, their effects on transport properties have only recently become understood. Although a variety of more sophisticated experimental techniques, such as induced photoabsorption as a function of bias illumination (43), are currently used to study electron–hole transport and recombination in amorphous materials, we will focus on the original time-of-flight transient photoconductivity experiment shown in Figure 9.13 (44). In this experiment a photoconducting sample is inserted between two electrodes, one of which is semitransparent. Sample thicknesses are typically 50–100 μm, and bias fields of 10–15 V/μm are commonly used. A light flash of short duration directed through the semitransparent electrode creates electron–hole pairs close to the surface of the material. Charge carriers of one polarity are swept through the bulk by the applied electric field, while charges of the opposite polarity are drawn into the semitransparent electrode. If the light flash is of sufficiently low intensity, space charge effects may be neglected, and we may treat the sheet of charge as a collection of individual charges. Experiments are usually carried out by monitoring the current decay for a fixed voltage drop across the sample.

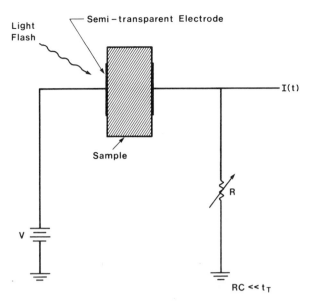

Figure 9.13. Schematic representation of a transient photoconductivity experiment. A short, strongly absorbed light pulse creates electron–hole pairs in the sample close to the semitransparent electrode. Charges of one sign are driven across the sample to the collecting electrode by a constant electric field. The change in photocurrent is monitored for a constant voltage drop.

The observed current transients as displayed on a linear plot are found to be quite featureless and uninteresting. Scharfe (45), however, showed that a log-log plot of the transient photocurrent against time displayed two straight lines of negative slope with a well-defined break, or knee, in the curve. This indicates a power law behavior of the transient photocurrent. Discussion of the exponents before and after the knee, which is called the apparent transit time t_T, has generated hundreds of papers. Figure 9.14 shows a typical experimental curve for a time-of-flight experiment on As$_2$Se$_3$ (46), where the plotted values have been normalized with respect to the value of the photocurrent at $t = t_T$. For many materials the normalized plots are identical for varying values of t_T obtained for different electric fields or different sample thicknesses, giving rise to the so-called universal behavior of the transient photocurrent decay.

9.3.2. Continuous-Time Random Walk (CTRW)

In 1975 the first theoretical model of the above phenomenon (often called anomalous dispersion) was put forward by Scher and Montroll (SM) (47). The Scher–Montroll model was based on a continuous-time random-walk (CTRW) theory on a regular lattice. It was assumed that the material could be partitioned into cells, each of which contained a large number of microscopic hopping sites for the electrons or holes. The transfer of charge from one cell to the next was described by

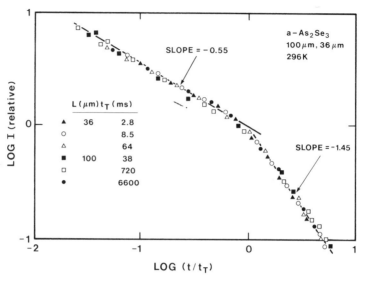

Figure 9.14. Master plot for transient hole currents in As$_2$Se$_3$, showing universal behavior for different values of the transit time. The plot was obtained by parallel shifting along the logarithmic axis of the individual traces recorded at different electric fields. The different transit times and sample thicknesses are listed in the legend. (After Ref. 46.)

a master equation with transfer rates into and out of the cells. The novel feature of the theory was the introduction of a waiting-time distribution function $\psi(t)$, which was introduced to describe the amount of time a charge waited in a particular cell before moving on. All cells were assumed to be equivalent. The motion of charge through the entire system was determined by the initial condition of creating a localized packet near one electrode, as shown in Figure 9.15, as well as the time evolution of the individual cell occupation probabilities. The figure also shows some dispersion in the carrier packet taking place as it moves across the sample. The exact nature of this dispersion and the associated instantaneous photocurrent can be determined from the theory, as we will now discuss.

The master equation describing the transport process is given by

$$\frac{dP(l, t)}{dt} = \int_0^t \Phi(t - t') \sum_{l'} [\eta(l - l') P(l', t') - \eta(l' - l) P(l, t')] \, dt' \quad (9.73)$$

where $P(l, t)$ is the probability that cell l is occupied by a carrier at time t, $\Phi(t)$ is a relaxation function, and $\eta(l - l')$ are transition probabilities. The relaxation function $\Phi(t)$ is in turn related to the distribution of waiting times, $\psi(t)$ (48), as we discuss later.

Solution of Eq. (9.73) using the Laplace transform technique gave, after some algebra, the transient photocurrent,

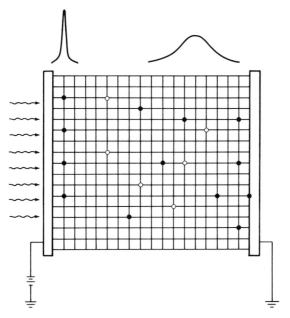

Figure 9.15. Schematic diagram for continuous-time random-walk model of transient photoconductivity. Charge carriers are created close to the semitransparent electrode on the left (solid circles), move across the sample on a regular lattice, where the lattice points represent quasimicroscopic cells in the material, and are absorbed at the right-hand boundary. The open circles signify localized trapping sites that are present at each lattice point. The charge packet undergoes either normal or anomalous dispersion during transit. (After Ref. 47.)

$$\tilde{I}(s) \simeq \frac{1 - \exp[-a(s)t_0]}{a(s)} \tag{9.74}$$

where t_0 is a constant, and the quantity $a(s)$ in the SM theory depends directly on the waiting-time distribution function $\psi(t)$. SM assumed a particular mathematical form for $\psi(t)$,

$$\psi(t) \sim t^{-(1+a)} \tag{9.75}$$

and gave an expression for the exponent a that, however, was highly model dependent. As we shall see later, the multiple-trapping theory gives an expression for $\psi(t)$ directly in terms of the microscopic parameters of the system. For the time being, however, we pursue the historical line of development. With the assumption of a power law form for the waiting-time distribution function, a number of interesting predictions immediately emerged from the theory. The average position of the carrier packet also evolved as a power law, $\langle x \rangle \simeq t^a$, and the rms deviation from the mean position, σ, was found to obey the same power law. The only feature in the transient photocurrent curve, the knee, was identified with the time taken for the leading edge of the carrier sheet to reach the collecting electrode. The time of the knee was therefore taken as the apparent transient time. The peculiar time dependence of the dispersion σ and the mean position of the carrier packet $\langle x \rangle$ derived from the power law dependence of the waiting-time distribution function. In particular, the ratio of the dispersion to the mean position was found to be a constant, in agreement with the observed universality of the experimental photocurrent traces when normalized according to t_T. The expression for the current is obtained from the time rate of change of the mean of the charge carrier packet, giving

$$I(t) \simeq \frac{d \langle x \rangle}{dt} \simeq \begin{cases} t^{-(1-a)} & \text{for } t < t_T \\ t^{-(1+a)} & \text{for } t > t_T \end{cases} \tag{9.76}$$

The change in the power law at $t = t_T$ is a consequence of the absorbing boundary, which is the collecting electrode. These results are shown schematically in Figure 9.16, which shows that many charge carriers are held back during transit, giving rise to the anomalous dispersion. The holding back of the charge carriers is due to a number of microscopic mechanisms, which we shall generally call trapping. The trapped charge carriers spend some time in localized states and then rejoin the crowd of untrapped charges on their way to the collecting electrode.

9.3.3. Anomalous Versus Normal Dispersion

The anomalous dispersion shown in Figure 9.16 contrasts with the case of Gaussian, or normal, dispersion shown in Figure 9.17. In the absence of any trapping, normal dispersion arises from the statistical variations in the microscopic scattering events of the charges during their drift along the electric field. In this case the center of the carrier packet is displaced linearly in time from the illuminated electrode, while the rms deviation σ varies as the square root of time. The ratio of the dispersion to the mean position then depends on the value of the transit time, as shown in Figure 9.17, and gives rise to nonuniversal behavior.

The nature of anomalous dispersion is determined by the form of the waiting-time distribution function. As we will show later, an exponential decay of $\psi(t)$ corresponds to normal, or Gaussian, statistics, whereas a superposition of many Gaussians gives an apparent power law decay in time and corresponds to the case of anomalous, or non-Gaussian, statistics. All of the interesting physics is contained in the shape of $\psi(t)$, and the calculation of the actual transient photocurrent shape from the master equation with an absorbing boundary is a matter of bookkeeping.

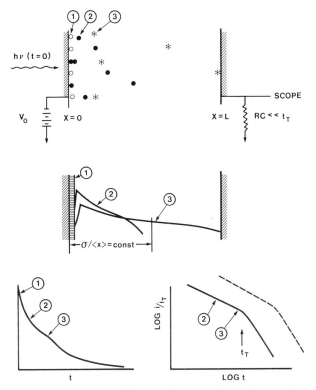

Figure 9.16. Schematic representation of charge carrier propagation with anomalous dispersion. The top panel shows the position of typical charge carriers in the bulk of the sample at (1) $t = 0$, (2) $t < t_T$, (3) $t \approx t_T$. The middle panel shows the charge distribution in the bulk of the sample at the different times, and the bottom panel shows the measured transient current. The plot at the bottom left is linear, while the plot at the bottom right is on a logarithmic scale. The dashed line shows the transient current trace for a lower electric field, corresponding to a longer transit time. (After Ref. 46.)

9.3.4. Multiple-Trapping Model

9.3.4.1. Continuity Equations. We now describe the more general multiple-trapping model of stochastic charge transport (49–51). In this model we introduce explicitly the microscopic parameters that determine the random stopping of the charge as it moves through the bulk of the sample. This is a continuum model and hence does not make use of the artificial division of the material into cells as in the SM description. Figure 9.18 shows the continuum electron and hole transport states in the energy level diagram, with electron and hole traps included, in the presence of an applied electric field. The charge density at any point in the material is the sum of the charge density moving in the extended states plus the charge density trapped in localized states,

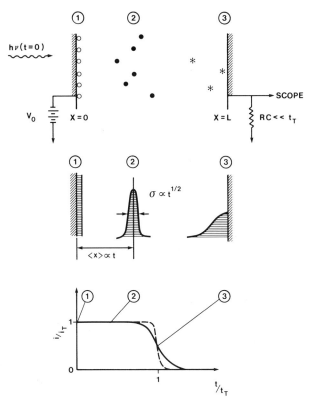

Figure 9.17. Schematic diagram of charge carrier propagation under ideal Gaussian conditions. The top panel shows the position of the typical carriers in the bulk of the sample at (1) $t = 0$, (2) $t < t_T$, (3) $t \approx t_T$. The middle panel shows the charge distribution in the bulk of the sample for the different times indicated above. The bottom panel shows the current pulse in the external circuit induced by the charge displacement. A linear plot is shown normalized to the transit time and the value of the current at the transit time. The dashed line represents the transient current for a lower electric field corresponding to a longer transit time. (After Ref. 46.)

$$n(x, t) = p(x, t) + \sum_i p_i(x, t) \tag{9.77}$$

where $p(x, t)$ is the local concentration of carriers in transport states, and $p_i(x, t)$ is the local concentration of carriers in trap i. The time rate of change of the total charge density is given by the continuity equation

$$\frac{\partial n}{\partial t} = g(x, t) - \frac{\partial f}{\partial x} \tag{9.78}$$

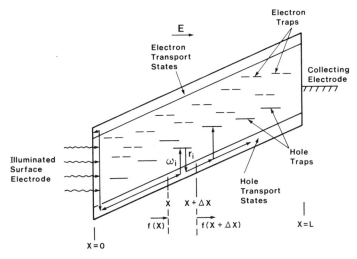

Figure 9.18. Schematic energy level diagram for multiple-trapping model. The electron and hole transport states are shown, and the holes are taken to be the mobile species. The trapping levels for both electrons and holes are shown, and $f(x)$ is the flux of holes.

where f is the flux of mobile carriers and g is the generation term. The continuity equation for the local concentration of carriers in trap i, say, is given in terms of the amount of charge captured by the trap less the amount released from the trap,

$$\frac{\partial p_i}{\partial t} = p\omega_i - p_i r_i \tag{9.79}$$

In this last equation the trap index i is general and could refer to traps of different energy. The main assumptions in writing the rate equation for the trap i are the spatial independence of the trap capture rate ω_i and the release rate r_i. The distribution of traps is taken to be completely homogeneous in this model. In order to solve Eq. (9.78) for the space and time dependence of the total charge density, we observe that the flux of free carriers is related to the free carrier concentration by $f(x, t) = \mu_0 E p(x, t)$, where μ_0 is the free carrier mobility. The total charge carrier density can be related to $p(x, t)$ using the trap equation, whose Laplace transform is

$$s\tilde{p}_i = \tilde{p}\omega_i - \tilde{p}_i r_i \tag{9.80}$$

and the total charge concentration can be written as

$$\tilde{n} = \tilde{p} + \sum_i \tilde{p}_i = \left(1 + \sum_i \frac{\omega_i}{s + r_i}\right) \tilde{p} \tag{9.81}$$

Inverting the Laplace transform for the local concentration of free carriers gives the following time convolution integral:

$$p(x, t) = \int_0^t Q(t - t') \, n(x, t') \, dt \qquad (9.82)$$

where $Q(t)$ is a relaxation function related to the amount of time carriers spend immobilized in the traps. Inserting this relation in Eq. (9.78) gives the final result for the transport equation,

$$\frac{\partial n}{\partial t} = g(x, t) - \mu_0 E \int_0^t Q(t - t') \frac{\partial n(x, t')}{\partial x} \, dt' \qquad (9.83)$$

9.3.4.2. Relation to CTRW. Equation (9.83) corresponds to the continuum limit of the Scher–Montroll master equation (47), as may be shown by letting

$$P(l, t) \rightarrow n(x, t) \qquad P(l \pm 1, t) \rightarrow n(x, t) \pm a_0 \frac{\partial n}{\partial x} \qquad (9.84)$$

$$\eta(\pm 1, 0, 0) = \tfrac{1}{6} \pm \tfrac{1}{2}\Delta \qquad \eta(0, \pm 1, 0) = \eta(0, 0, \pm 1) = \tfrac{1}{6}$$

for a cubic lattice, where a_0 is the SM lattice parameter, and Δ corresponds to the asymmetry in the transition probability between cells due to the electric field.

Aside from a proportionality factor, the relaxation functions $\tilde{\Phi}(s)$ and $\tilde{Q}(s)$ are directly related,

$$\tilde{\Phi}(s) = \tilde{Q}(s)/\tau \qquad (9.85)$$

where $\tau = a_0\Delta/\mu_0 E$. The above discussion shows that henceforth we do not need to consider the CTRW description of anomalous dispersion, and we can concentrate on the mathematically simpler and physically more transparent multiple-trapping model. By using the Laplace transformation, Eq. (9.83) for the total charge density may be solved for the initial condition,

$$g(x, t) = \gamma \, \delta(x) \, \delta(t), \qquad (9.86)$$

where γ is the photogeneration efficiency, to obtain

$$\tilde{p}(x, s) = \frac{\gamma}{\mu_0 E} \exp \left\{ - \left[a(s) t_0 x / L \right] \right\} \qquad (9.87)$$

where $t_0 = L/\mu_0 E$ is the transit time for untrapped carriers, and L is the sample thickness. The transform of the transient photocurrent is

$$\tilde{I}(s) = \xi\gamma \frac{\mu_0 E}{L} \int_0^L \tilde{p}(x, s)\, dx$$

$$= \xi\gamma \frac{1 - \exp[-a(s)t_0]}{a(s)t_0} \tag{9.88}$$

where hole transport has been assumed.

Now it can be verified numerically that

$$a(s) = s \left(1 + \sum_i \frac{\omega_i}{s + r_i} \right) \tag{9.89}$$

is well approximated by a power law,

$$a(s) \approx s^a \qquad (0 < a < 1) \tag{9.90}$$

for the case where charge carriers are likely to be trapped once or twice during transit ($M_i = \omega_i t_0 \sim 1$) and the release time from traps r_i^{-1} is roughly equal to the transit time t_T. In this case we get anomalous dispersion. Although there are many trapping events for the carriers, the small number of events when the carriers are delayed for times comparable to t_T give rise to non-Gaussian statistics. Thus we see that the original SM guess of a power law for the waiting-time distribution function gave a good approximation to the $a(s)$ function for the anomalous dispersion case.

9.3.4.3. Transient Photocurrent.

Inversion of the Laplace transform of the transient photocurrent Eq. (9.88) gives

$$I(t) = G_1(t) - G_2(t - t_0) \tag{9.91}$$

where G_1 is of the form

$$G_1(t) = \sum_i A_i e^{-a_i t} \tag{9.92}$$

The constants a_i are determined by the poles of the function $a(s)^{-1}$ and the coefficients A_i by the corresponding residues. The values of a_i fall in the spectrum of release rates r_i for the original traps and may in turn be regarded as the release rates from a set of renormalized traps. The function $G_1(t)$ represents the contribution to the current from traps that continually release carriers into the transport states.

The function $G_2(t - t_0)$ represents the decrease in the current due to charge carriers being collected at the downstream electrode. It may be calculated by noting that

$$a(s) = s \left(1 + \sum_i \frac{\omega_i}{s + r_i} \right) = s + \sum_i \omega_i - \sum_i \frac{r_i \omega_i}{s + r_i} \tag{9.93}$$

and looking up the inverse Laplace transform (52)

$$L^{-1} \exp \left(\frac{\beta_i}{s + r_i} \right) = \left(\frac{\beta_i}{t} \right)^{1/2} I_1 (2 \sqrt{\beta_i} t) e^{-r_i t} + \delta(t) \tag{9.94}$$

where $\beta_i = M_i r_i (x/L)$. The final result for $p(x, t)$ involves a convolution of these functions. A similar procedure gives the expression for $I(t)$ after averaging over $\bar{p}(x, s)$ according to Eq. (9.88). Appendix B gives a computer code for calculating $I(t)$ corresponding to a given set of trap parameters $\{\omega_i, r_i\}$ using a more efficient routine that involves contour integration in the complex s plane. Use of this program for various choices of $\{\omega_i, r_i\}$ illustrates the power of the multiple-trapping model, which, unlike the CRTW formalism, can describe both anomalous and normal dispersion by changing the values of the microscopic trapping parameters.

9.3.4.4. Waiting-Time Distribution Function. Turning now to the evaluation of the waiting-time distribution function $\Psi(t)$, we use the relation between this distribution function and the relaxation function $\Phi(t)$ (48) to get

$$\tilde{\Psi}(s) = [1 + s\tilde{\Phi}(s)^{-1}]^{-1}$$
$$= [1 + \bar{s}\tilde{Q}(\bar{s})^{-1}]^{-1} \tag{9.95}$$

where we have defined $\bar{s} = s\tau$, $\bar{\omega}_i = \omega_i \tau$, and $\bar{r}_i = r_i \tau$. We find that for m traps $\tilde{\Psi}(\bar{s})$ has $m + 1$ real poles, which we denote by

$$\bar{s}_i = -a_i, \qquad i = 1, \ldots, m + 1 \tag{9.96}$$

Rewriting Eq. (9.95) in terms of the a_i, we get

$$\tilde{\Psi}(\bar{s}) = \frac{\displaystyle\prod_{i=1}^{m} (\bar{s} + \bar{r}_i)}{\displaystyle\prod_{i=1}^{m+1} (\bar{s} + a_i)}$$

$$= \sum_{i=1}^{m+1} \frac{A_i}{\bar{s} + a_i} \tag{9.97}$$

where

$$A_i = \frac{\prod\limits_{j=1}^{m} (\bar{r}_j - a_i)}{\prod\limits_{j \neq i}^{m+1} (a_j - a_i)} \tag{9.98}$$

and all the r_i must be different for this expression for A_i to be valid. Using the Laplace transform theorem (8)

$$L^{-1}f(cs) = \frac{1}{c} F\left(\frac{t}{c}\right) \qquad (c > 0) \tag{9.99}$$

if

$$L^{-1}f(s) = F(t) \tag{9.100}$$

inversion of Eq. (9.97) gives

$$\tau\Psi(t) = \sum_{i=1}^{m+1} A_i \exp\left(\frac{-a_i t}{\tau}\right) \tag{9.101}$$

which is the full expression for $\Psi(t)$ in terms of the trap parameters. In practice, however, the long-time expression for $\Psi(t)$ has been shown to be adequate. To derive this form of the distribution function, we carry out the small s expansion of Eq. (9.95) to give

$$\tilde{\Psi}(\bar{s}) \simeq 1 - \bar{s}\left(1 + \sum_{i=1}^{m} \frac{\bar{\omega}_i}{\bar{s} + \bar{r}_i}\right)$$

$$\simeq 1 - \sum_{i=1}^{m} \bar{\omega}_i + \sum_{i=1}^{m} \frac{\bar{\omega}_i \bar{r}_i}{\bar{s} + \bar{r}_i} \tag{9.102}$$

where the last step is valid if the quantity ω_i/r_i, which corresponds to the total resting time in a trap, is much greater than unity for at least one of the traps.

On a physical basis we equate the constant τ with the lifetime for capture by any of the traps (49),

$$\tau^{-1} = \sum_{i}^{m} \omega_i \tag{9.103}$$

With this choice Eq. (9.102) becomes

$$\tilde{\Psi}(\bar{s}) = \sum_{i=1}^{m} \frac{\bar{\omega}_i \bar{r}_i}{\bar{s} + \bar{r}_i} \tag{9.104}$$

and the inverse is given by

$$\Psi(t) = \sum_{i=1}^{m} (\omega_i \tau) r_i \exp(-r_i t) \tag{9.105}$$

This function, besides being a very good approximation to the full expression for $\Psi(t)$, given by Eq. (9.101), also has a direct physical interpretation; $\Psi(t) \, dt$ is the probability that a charge carrier will be released during the time interval $(t, t + dt)$. The above expression shows that this quantity is the sum of the probabilities per unit time for the charge carrier to be released from each trap weighted by the fraction of times the carrier is captured in that trap. The waiting-time distribution function is translationally invariant, with any location in the sample being equivalent to any other as far as the probability of trapping and detrapping is concerned. Note that for all $\omega_i = 0$, the derivation of Eq. (9.105) is invalid since the condition $\omega_i/r_i \gg 1$ for at least one trap does not hold. In this case $\tilde{Q}(s) = 1$, and the waiting-time distribution function reduces to a single exponential. For the case of no multiple trapping the proportionality factor τ cannot be defined by Eq. (9.103). A detailed discussion relating to this point is given in the literature (50).

The computer code given in Appendix B can be used to determine the trap parameters for a given transient photocurrent trace, and using these values, the waiting-time distribution function can be calculated to a good approximation from Eq. (9.105). The code can accommodate up to five traps, although in practice three traps are sufficient for an analysis of most experimental data. Figures 9.19 and 9.20 show the results of an analysis of transient photocurrent traces for amorphous selenium. The use of the $\Psi(t)$ function to determine the density of trap states directly is discussed elsewhere (53).

9.3.4.5. Thermalization of Charge Carriers. In the final part of this section we discuss the recent interpretation of the multiple-trapping model based on the concept of thermalization of charge carriers within a distribution of localized states (54, 55). Here a thermalization energy E_c is assumed as a demarcation between shallow traps in thermal equilibrium with free carriers from deeper traps whose occupation probability remains almost constant. As time progresses, the photoexcited carriers, which initially occupy all trap states almost uniformly, drift across the sample and thermalize to successively lower energies, as shown in Figure 9.21 for a continuum of states. This occurs because carriers in the shallow traps are released preferentially into the transport states, while the random trapping process ensures that some fraction of these carriers always winds up in the deeper states. The net effect on the distribution of charge carriers is to cause a "sinking" in

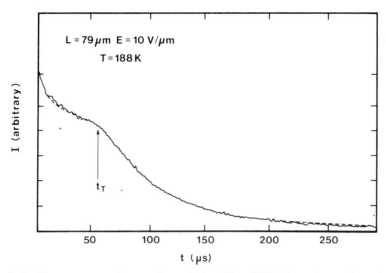

Figure 9.19. Photocurrent trace for amorphous selenium. The solid line shows the experimental curve, and the dashed line shows the best fit using the three-level multiple-trapping model. The photocurrent axis has a linear scale.

energy with increasing time, as shown in Figure 9.21. The picture is similar to a man running into a pond of quicksand at great speed and then slowing down as he sinks lower and lower.

The thermalization energy E_c, which is a function of time, is defined as the energy at which the trap release time is equal to the time delay t after the photoexcitation pulse,

$$t = v^{-1} \exp\left(\frac{E_c}{k_B T}\right) \tag{9.106}$$

which v is a prefactor, and the trap energy is measured relative to the bottom of the transport band. Equation (9.106) gives

$$E_c(t) = k_B T \ln(vt) \tag{9.107}$$

The waiting-time distribution function $\psi(t)$, which is the probability per unit time that a charge carrier is released during the interval $(t, t + dt)$, is then

$$\psi(t)\, dt = \frac{N_t(E_c)}{N_0} \frac{dE_c}{dt}\, dt \tag{9.108}$$

where $N_t(E)$ is the distribution in energy of a continuum of trap states, and N_0 is the total concentration of trap states,

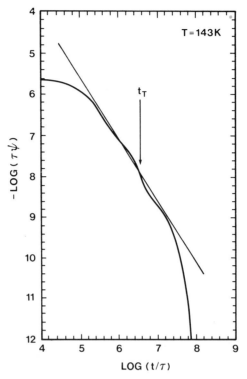

Figure 9.20. The solid curve shows the waiting-time distribution function $\tau\Psi(t)$ calculated for the amorphous photoconductor selenium using a three-level multiple-trapping model. The straight line indicates the power law behavior of the distribution function for the middle two time decades of the plot. The proportionality constant τ is defined as the lifetime for capture by any of the traps. The arrow indicates the position of the apparent transit time t_T.

$$N_0 = \int N_t(E)\, dE \tag{9.109}$$

Equation (9.108) follows from the assumption that all of the fraction of trapped charge within an energy range dE_c of E_c, that is, $N_t(E_c)\, dE_c/N_0$, is released after a time t given by Eq. (9.106). Using the corresponding expression for $E_c(t)$ then gives

$$\Psi(t)\, dt = \frac{N_t[E_c(t)]}{N_0} \frac{k_B T}{t}\, dt \tag{9.110}$$

For an arbitrary density of states function $N_t(E)$ the expression for $\Psi(t)$ must be evaluated numerically. However, for an exponential distribution

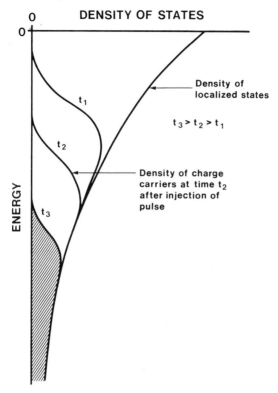

Figure 9.21. Diagram of density of localized states that trap charge carriers after photoexcitation. The carriers drifting across the sample thermalize to successively greater depths for longer times after the initial light pulse.

$$N_t(E) = N_A \exp\left(\frac{-E}{k_B T_c}\right) \qquad (9.111)$$

the result may be evaluated analytically to give

$$\Psi(t) \sim t^{-(1+a)} \qquad (9.112)$$

where $a = T/T_c$.

The above interpretation of the multiple-trapping model is attractive from a conceptual point of view; however, some caution must be used when applying the intuitive thermalization version of the model to an analysis of real materials. First, the thermalization assumption can be checked against the numerical results of the multiple-trapping model, as described earlier. This has been done by Marshall and Main (56), who have determined that the thermalization process is a complex function of the capture and release rates for the traps and that the deep traps prevent the establishment of quasi-thermal equilibrium for the shallower states. In particular, taking the thermalization assumption too literally can lead to erroneous results for the energy distribution of trapping states.

There are also other problems with the intuitive thermalization assumption. As we showed earlier, a *discrete* set of trap levels will give rise to an apparent power law behavior of the photocurrent over a broad time interval. It follows that the observation of a power law does not in itself prove the existence of any particular continuum distribution of localized states. Specifically, one cannot conclude the existence of an exponential distribution from the experimental observation of a power law for the photocurrent. Finally, in some materials, such as *a*-Se, the anomalous dispersion disappears as the temperature is raised, while the observed activation energy of the mobility remains constant (57). This observation is in contradiction with the prediction of the multiple-trapping model when the quasi-thermal equilibrium assumption is used. A lot remains to be done to obtain a complete physical understanding of anomalous dispersion and associated effects. Marshall (58) has recently written an excellent critical review of the latest advances in this field, and the reader is urged to consult this article for a thorough discussion of the main issues.

9.4. UNIFIED THEORY OF ELECTRON–HOLE RECOMBINATION AND ANOMALOUS DISPERSION

9.4.1. Constitutive Relation

An extension of the work outlined in the first two sections is to include the effects of trapping on the electron–hole recombination process. Shallow traps are taken into account by averaging over the potential fluctuations, as discussed (64) at the beginning of Section 9.2. Figure 9.22 shows the profile of the coulombic potential in which electrons diffuse in the presence of deep traps. In the following we outline one scheme in which recombination and multiple trapping can be treated on an equal footing. Butcher (59, 60) has shown that both normal and anomalous dispersion may be described by a single macroscopic constitutive relation between the current density and the total carrier number density. He dealt with the case in which a planar pulse of charge carriers moves across a bulk amorphous material subject to a constant electric field. For the case of normal dispersion, as discussed earlier, the charge carriers drift in the electric field and undergo Gaussian broadening due to diffusion. In this case the current density shows an instantaneous response to the *total* number density. For many experiments, however, anomalous dispersion is observed, and this occurs because the current density responds only to the moving or free charge density, which is a fraction of the total density. The effect of the temporarily trapped charge carriers leads to the retarded nature of the response of the current density to the total charge density. The extent of the retardation is described by the constitutive relation

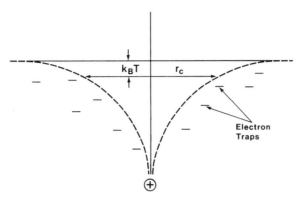

Figure 9.22. Profile of coulombic potential with homogeneously distributed traps, which forms the basis of a unified treatment of recombination and multiple trapping. The Onsager radius is denoted by r_c.

$$\mathbf{j}(\mathbf{r},\, t) = -\nabla \int_0^\infty D(t - \tau) n(\mathbf{r},\, \tau)\, d\tau + \mathbf{E} \int_0^\infty \mu(t - \tau) n(\mathbf{r},\, \tau)\, d\tau, \quad (9.113)$$

where the charge carriers are assumed to be holes; $D(t)$ and $\mu(t)$ are generalized diffusion and mobility functions that take into account the retarded nature of the response of $j(\mathbf{r},\, t)$ to the total number density $n(\mathbf{r},\, t)$. The second term in Eq. (9.113) has already been discussed earlier in connection with the multiple-trapping model alone. Since we now wish to include the effects of diffusion, we also retain the first term in Eq. (9.113). The Laplace transforms of the response functions may be written as

$$\tilde{D}(s) = D_0 \tilde{Q}(s) \qquad \tilde{\mu}(s) = \mu_0 \tilde{Q}(s) \qquad\qquad (9.114)$$

where D_0, μ_0 are the diffusion constant and mobility of "free" (untrapped) carriers and are connected by the Einstein relation, $\mu_0/D_0 = \xi/k_B T$. The "conversion" operator, $\tilde{Q}(s)$, defined earlier as

$$\tilde{Q}(s) = \left(1 + \sum_i \frac{\omega_i}{s + r_i}\right)^{-1} \qquad\qquad (9.115)$$

acts on the total (trapped and untrapped) charge density $n(r,\, t)$ to give the amount of free or untrapped charge alone. As has been pointed out by Butcher (59), the constitutive relation rests on the assumptions that the system is isotropic, the electric field E is small and constant, the charge carriers obey Boltzmann statistics, and the relation is spatially local. The first term in Eq. (9.113) is a generalization of Fick's diffusion law, to which it reduces if $D(\tau) = D_0\, \delta(\tau)$. The second term may be obtained from a more general expression for the current density involving the ac conductivity $\sigma(\omega)$ for a homogeneous system.

9.4.2. Geminate Recombination and Multiple Trapping

In order to make use of the constitutive relation for geminate recombination, we write Eq. (9.113) in terms of its Laplace transform,

$$\tilde{\mathbf{j}}(\mathbf{r}, s) = -\tilde{D}(s)\, \nabla \tilde{n}(\mathbf{r}, s) + \tilde{\mu}(s)\mathbf{E}n(\mathbf{r}, s), \tag{9.116}$$

which is similar in structure to the instantaneous response relation given by Eq. (9.5) if we identify $\tilde{n}(\mathbf{r}, s)$ with the "particle density" given by the geminate pair distribution function $\tilde{\rho}(\mathbf{r}, s)$. In order to include the effects of retardation due to multiple trapping in the time-dependent geminate recombination problem with external *and* coulombic electric fields, we make the substitution

$$\mathbf{E} \rightarrow \mathbf{E} - \frac{\xi}{4\pi\varepsilon_0\epsilon}\frac{\hat{r}}{r^2} \tag{9.117}$$

in Eq. (9.116) and use the resulting expression for the current density in solving the continuity relation Eq. (9.1). As may be verified from the formulas in Section 9.1, all of the mathematical analysis for the solution of the time-dependent Onsager problem then proceeds in the same way. The radial equation for the Laplace transform of $R_l(r, t)$ was previously

$$\frac{d^2\tilde{R}_l(r, s)}{dr^2} + \frac{1}{r}\frac{d\tilde{R}_l(r, s)}{dr} - \left[s + \left(\frac{F}{2}\right)^2 + \frac{\lambda_l + 1/4}{r^2} + \frac{1}{r^4}\right]\tilde{R}_l(r, s)$$
$$= -\frac{1}{r_0}\delta(r - r_0) \tag{9.118}$$

as may be checked using Eq. (9.15) and the initial condition Eq. (9.16). The corresponding equation for the transform of the "free" (untrapped) radial part of the charge density $\tilde{R}_l\tilde{Q}$ becomes

$$\frac{d^2(\tilde{R}_l\tilde{Q})}{dr^2} + \frac{1}{r}\frac{d(\tilde{R}_l\tilde{Q})}{dr} - \left[s\tilde{Q}^{-1} + \left(\frac{F}{2}\right)^2 + \frac{\lambda_l + \frac{1}{4}}{r^2} + \frac{1}{r^4}\right]\tilde{R}_l\tilde{Q} = -\frac{1}{r_0}\delta(r - r_0) \tag{9.119}$$

This result is obtained by *first* taking the Laplace transform of the continuity equation (9.1) and then noting that the divergence operator has no effect on the quantities $\tilde{D}(s)$ and $\tilde{\mu}(s)$ in Eq. (9.116) [with the substitution given by Eq. (9.117)] since they are assumed to have no spatial dependence. The quantity $s\tilde{Q}^{-1}$ is recognized from our earlier discussion as

$$s\tilde{Q}^{-1} = s\left(1 + \sum_i \frac{\omega_i}{s + r_i}\right) = a(s) \tag{9.120}$$

where we have used the notation $s_F = s + (F/s)^2$, so that when the electric field is present, we denote

$$a_F(s) = a(s) + \left(\frac{F}{2}\right)^2$$

$$= s_F + \sum_i \frac{s\omega_i}{s + r_i} \tag{9.121}$$

In the nondimensional form of these equations the unit of time is now $r_c^2/4D_0$, while the unit of length is still $r_c/2$.

The solution to Eq. (9.119) may again be written in terms of the two linearly independent solutions to the corresponding homogeneous equation, with the formal difference that s_F is to be replaced by $a_F(s)$. From Eq. (9.121) we see that we regain the original solution when all the trapping rates ω_i are set equal to zero. We denote the solution of Eq. (9.119) for the amount of untrapped charge by $\tilde{R}_l^0[r, a_F(s)]$, that is,

$$\tilde{R}_l \tilde{Q} = \tilde{R}_l^0[r, a_F(s)] \tag{9.122}$$

The *total* amount of charge, trapped and untrapped, is obtained using the inverse of the conversion operator,

$$\tilde{R}_l(r, s) = \tilde{Q}^{-1}(s) \, \tilde{R}_l^0[r, a_F(s)]$$

$$= \left(1 + \sum_i \frac{\omega_i}{s + r_i}\right) \tilde{R}_l^0 \tag{9.123}$$

and the inverse Laplace transform gives

$$R_l(r, t) = R_l^0(r, t) + \sum_i \omega_i \int_0^t e^{-(t-t')r_i} R_l^0(r, t') \, dt' \tag{9.124}$$

where the second part of this expression corresponds to the net amount of trapped charge at time t. The complete result for the charge density $\rho(r, t)$ is obtained by multiplying each $R_l(r, t)$ by the corresponding generalized Legendre function $T_l(\mu)$ (which only depends on F as a parameter, even for the case of multiple trapping) and summing over all the partial waves l.

9.4.3. Numerical Results

The analytic properties of $\tilde{R}_l^0[r, a_F(s)]$ may be obtained from the known analytical properties of $\tilde{R}_l^0(r, s_F)$, which satisfies a Schrödinger eigenvalue equation. However, the discussion of these properties is beyond the scope of this article. Here it

is more instructive to give a few examples of the effect of various trapping parameters on the reaction rate and the survival probability using the method of inverting
the Laplace transform along the positive real axis (61). Figure 9.23 shows the
results for the reaction rate, $R(t)$, for vanishing external electric field ($F = 0$) and
$r_0 = 1.0$ (that is, $r_0 = r_c/2$ in conventional units). A single trapping level is
assumed, and the unit of time is $r_c^2/4D_0$, which is the free diffusion time (without
coulombic interaction and no multiple trapping) taken to reach the origin from half
the Onsager radius. The expression for the reaction rate is given by Eq. (9.38),
where $\rho(0, t|r_0)$ is now the free charge density.

The solid line in Figure 9.23 shows the reaction rate for no multiple trapping
($\omega_i = 0$) and is identical to the curve shown in Figure 9.9 for $r_0 = 1.0$. The dashed
line shows the results for $\omega_i(r_c^2/4D_0) = 1.0$, $r_i(r_c^2/4D_0) = 1.0$ in conventional
units. In this case the charge carrier gets trapped once on its way from $r_c/2$ to the
origin, and the release time is equal to the free diffusion time to reach the origin.
From Section 9.3 we recall that anomalous dispersion occurs when a charge carrier
is captured approximately once in a trap whose mean release time is equal to the
empirical transit time. The dashed line in Figure 9.23 shows the decrease in $R(t)$
from the no trapping case and the beginning of an exponential tail that contrasts

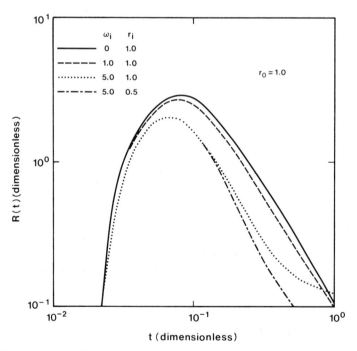

Figure 9.23. Reaction rate $R(t)$ for $r_0 = 1.0$ calculated from the time-dependent Smoluchowski equation with no external electric field and including the effects of multiple trapping. A single trapping
level is assumed, and the curves corresponding to different values of the capture rate ω_i and release
rate r_i are identified on the diagram. The unit of length is $r_c/2$, and the unit of time is $r_c^2/4D_0$, where
D_0 is the microscopic diffusion constant with no multiple trapping.

with the expected power law behavior for pure diffusion. The dotted curve corresponds to the case where the carrier gets trapped five times before recombination, and the exponential tail is evident. The effective diffusion time for trapped carriers to recombine in this case is approximately the free diffusion time plus the total resting time in the trap, given by ω_i/r_i. A bigger effect on the reaction rate can be obtained by reducing the release rate by half (the dot-dashed curve in Figure 9.23), thereby doubling the resting time in the trap. The calculated curve for this case drops considerably below the other curves for long times, and we have not shown the long tail for small values of the reaction rate.

Figure 9.24 displays the survival probabilities corresponding to the curves shown in Figure 9.23. These results are obtained by integrating the area under the reaction rate curves and subtracting the results from unity. The survival probabilities (for both free and trapped charges) are higher than the values for no trapping, and the features on these curves line up with the corresponding features in the reaction rate curves. In particular, the dotted curve for the trap with the long resting time ω_i/r_i = 5.0 shows a pronounced bump due to multiple trapping, and doubling the resting time pushes the survival probability curve out by a factor of 2, as expected. An interesting feature of Figure 9.24 is that all the survival probability curves tend to the same asymptotic value for long times. This occurs because multiple trapping

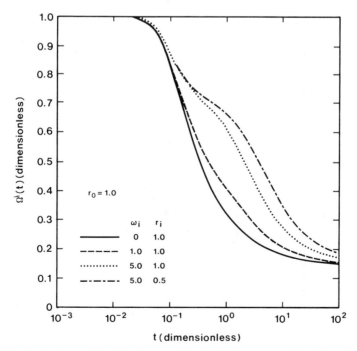

Figure 9.24. Survival probability $\Omega(t)$ for $r_0 = 1.0$ calculated from the exact solution of the unified theory of recombination and multiple trapping. The curves corresponding to different values of the capture rate ω_i and release rate r_i are shown on the diagram and a single trapping level is assumed. The unit of length is $r_c/2$, and the unit of time is $r_c^2/4D_0$.

does not affect the ultimate survival probability of a charge carrier but delays the recombination event by introducing random stopping and resting into the diffusion process. The exponential tails shown in Figure 9.23 for the reaction rate must therefore eventually revert to power law behavior, although this effect is not likely to be seen experimentally because of the long times involved. Any system for which multiple trapping is important is likely to have a distribution of trapping levels, and the long-time behavior of the reaction rate will be determined by the cumulative effect of the decays from the various traps.

Figure 9.25 shows the Laplace transform of the escape probability for the trap parameters used in Figures 9.23 and 9.24. We display these curves because the Laplace transform $\Omega(s)$ is directly measurable in a scavenging experiment, as explained in Section 9.2. The curves in Figure 9.25 are almost mirror images of those shown in Figure 9.24. In all cases the value of the transform is larger than expected for no trapping and deviates from the $s^{1/2}$ behavior predicted for extremely long times (corresponding to small s).

Our discussion of the effects of multiple trapping on geminate recombination has been restricted to a few examples that illustrate trends to be expected in a more complete treatment. These already appears to be some evidence that multiple trapping should be included in a general analysis of recombination (43, 62, 63). The

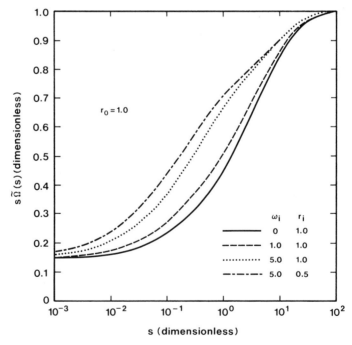

Figure 9.25. Laplace transform of the survival probability $\tilde{\Omega}(s)$ [$s\,\tilde{\Omega}(s) = 1 - \bar{R}(s)$] for $r_0 = 1.0$, calculated as a function of the transform variable s from the unified theory of recombination and multiple trapping. A single trapping level is taken into account, and the curves for different values of the capture rate ω_i and release rate r_i are indicated on the diagram. The unit of length is $r_c/2$, and the unit of time is $r_c^2/4D_0$.

design of experiments to test the theory, particularly in the liquid state with dopants that have only a few trapping levels, should prove valuable.

9.5. CONCLUSION AND SUMMARY

In Section 9.2 we have given a discussion of the time-dependent Smoluchowski equation with a coulombic potential and an external electric field. This model, which exploits some of Onsager's original ideas on ion recombination, forms the basis of our understanding of electron–hole recombination. We have analyzed the physical significance of the Green's function solution to the time-dependent Onsager problem and pointed out some applications to time-resolved measurements of geminate recombination in liquids and solids as well as in scavenging and radiolysis experiments. At present there is no evidence for the inadequacy of the continuum Onsager model in describing the observed phenomena, and even the photogeneration process can be understood with the help of the classical description. It seems safe to say that the continuum model of geminate recombination is solidly established and that future theoretical work in this area will be mostly concerned with the specific details of the recombination process when the particles approach one another within atomic distances. It is not likely, however, that many experiments will be sensitive to these details.

Considerable advances have also taken place in the area of dispersive transport, as reviewed in Section 9.3. The historically important Scher–Montroll model has now been superceded by the more physically appealing multiple-trapping description of charge transport. The Scher–Montroll model involves concepts that appear totally different from conventional trapping models. This occurs because of the mathematical structure of the continuous-time random-walk (CTRW) formalism on a lattice. The multiple-trapping model deals with separate generation, displacement, and stopping (capture) processes, while the CTRW description deals only with displacement events between artificially constructed pairs of sites. Hence the CTRW formalism cannot describe the kinetics of charge capture and release. The concept of a waiting-time distribution function, $\Psi(t)$, introduced by Scher and Montroll is useful, however, and we have shown how to construct this quantity in terms of trap-parameters. Other workers have shown how to use $\Psi(t)$ to determine the density of trap states directly (53).

In the multiple-trapping model the term *trap* is used in a general sense for any localized state that stops and holds a charge carrier for an observable length of time. By comparison, we call any fundamental state that determines the microscopic mobility of a charge carrier a *transport state*. Trap states are assumed to be spatially isolated from each other so that direct transitions between them do not take place, whereas transport states have sufficient spatial overlap to give rise to a measurable drift velocity of the carriers. An interesting discussion of an operational method of distinguishing between transport sites and traps has been given by Schmidlin (64).

We have also described a unified theory of anomalous dispersion and electron–hole recombination. The unified model is exactly solvable in the same way as the time-dependent Onsager model and involves the same kinds of special functions. The only formal change required to obtain the physically important quantities such as the survival probability and the reaction rate in the presence of multiple trapping is to replace the Laplace transform variable by the same variable times a conversion function, which relates the total (trapped and untrapped) charge density to the untrapped charge density. Using this theory, we gave a few examples of the effects of different trap parameters on the recombination process and ended with a suggestion for a novel experiment that uses the time-resolved technique to measure geminate recombination in the presence of dopant molecules with just a few trapping levels.

Appendix A gives tables of the survival probability of a geminate electron–hole pair for various values of the initial separation r_0 with an isotropic distribution of pairs. The time-dependent survival probability $\Omega(t)$ was computed on the basis of Eq. (9.48) in the text. Appendix B documents a FORTRAN program to estimate parameters for the multiple-trapping model from transient photocurrent data. This program has the feature that if one of the variables (LIM) is set equal to zero, it provides the model photocurrent function for various values of the capture and release rates for up to five different traps. In this case no fitting to data takes place. In the numerical inversion of the Laplace transform, the term st_0 in the theoretical expressions for the photocurrent, Eqs. (9.88) and (9.89), is ignored since it gives rise to an unmeasurable short-time contribution to the photocurrent.

We have reviewed the progress that has taken place over the past 10 years in the field of charge transport and recombination in amorphous materials. The past decade has seen the development of new experimental techniques that have been used on a wide range of liquids and solids, and several major advances in our theoretical understanding of these systems have been made. At present the field is expanding more rapidly than ever, and the approach to "steady state" seems a long way off.

ACKNOWLEDGMENT

I would like to thank L. Marks for evaluating the time-dependent survival probability tabulated in Appendix A as well as the calculations of the reaction rate and survival probability shown in Figures 9.23 and 9.25. The computer program used to calculate the transient photocurrent for the multiple-trapping model (Appendix B) is also due to L. Marks.

APPENDIX A: SURVIVAL PROBABILITY AS A FUNCTION OF TIME

The following table gives survival probability $\Omega(t)$ of a geminate electron–hole pair for different values of the initial separation r_0 with a spherically symmetric distribution and no electric field (see Figure 9.5). Note that the unit of length is $r_c/2$ and the unit of time is $r_c^2/4D$.

r_0(unit of length $r_c/2$)

Log$_{10}$t	.1	.12	.14	.16	.18	.20	.22	.24	.26	.28
-4.00	.9600	1.000	1.000	1.000	1.000	1.000	1.000	1.000	1.000	1.000
-3.90	.9370	.9998	.9999	1.000	1.000	1.000	1.000	1.000	1.000	1.000
-3.80	.6150	.9722	.9997	1.000	1.000	1.000	1.000	1.000	1.000	1.000
-3.70	.3735	.8802	.8994	1.000	1.000	1.000	1.000	1.000	1.000	1.000
-3.60	.1913	.6989	.9640	.9989	.9999	.9999	1.000	1.000	1.000	1.000
-3.50	.0861	.4746	.8685	.9870	.9994	.9999	1.000	1.000	1.000	1.000
-3.40	.0355	.2789	.6960	.9372	.9937	.9996	1.000	1.000	1.000	1.000
-3.30	.0139	.1461	.4887	.8201	.9642	.9959	.9997	.9999	.9999	1.000
-3.20	.0053	.0706	.3045	.6428	.8822	.9752	.9966	.9996	.9999	1.000
-3.10	.0020	.0325	.1727	.4505	.7377	.9122	.9793	.9965	.9995	.9997
-3.00	.0008	.0146	.0918	.2869	.5580	.7914	.9256	.9798	.9957	.9991
-2.90	.0003	.0066	.0470	.1701	.3853	.6281	.8192	.9295	9778	.9942
-2.80	.0001	.0030	.0236	.0962	.2475	.4580	.6694	.8305	.9269	.9733
-2.70	.0001	.0014	.0119	.0530	.1511	.3117	.5062	.6901	.8304	.9189
-2.60	.0000	.0007	.0060	.0289	.0895	.2017	.3588	.5343	.6959	.8217
-2.50	.0000	.0002	.0031	.0159	.0523	.1264	.2425	.3899	.5468	.6905
-2.40	.0000	.0001	.0017	.0088	.0305	.0780	.1589	.2722	.4072	.5473
-2.30	.0000	.0001	.0009	.0050	.0180	.0479	.1024	.1847	.2914	.4135
-2.20	.0000	.0000	.0005	.0029	.0108	.0297	.0657	.1234	.2033	.3017
-2.10	.0000	.0000	.0003	.0017	.0066	.0186	.0424	.0821	.1399	.2152
-2.00	.0000	.0000	.0002	.0011	.0041	.0119	.0276	.0548	.0959	.1519
-1.90	.0000	.0000	.0001	.0007	.0027	.0077	.0183	.0370	.0661	.1069
-1.80	.0000	.0000	.0001	.0004	.0017	.0052	.0124	.0253	.0459	.0756
-1.70	.0000	.0000	.0001	.0003	.0012	.0035	.0085	.0176	.0324	.0540
-1.60	.0000	.0000	.0000	.0002	.0008	.0025	.0060	.0125	.0232	.0391
-1.50	.0000	.0000	.0000	.0001	.0006	.0018	.0043	.0091	.0169	.0287
-1.40	.0000	.0000	.0000	.0001	.0004	.0013	.0032	.0067	.0126	.0214
-1.30	.0000	.0000	.0000	.0001	.0003	.0010	.0024	.0051	.0095	.0163
-1.20	.0000	.0000	.0000	.0001	.0003	.0007	.0018	.0039	.0073	.0126
-1.10	.0000	.0000	.0000	.0000	.0002	.0006	.0014	.0030	.0057	.0099
-1.00	.0000	.0000	.0000	.0000	.0002	.0005	.0011	.0024	.0046	.0079
-.90	.0000	.0000	.0000	.0000	.0001	.0004	.0009	.0020	.0037	.0064
-.80	.0000	.0000	.0000	.0000	.0001	.0003	.0008	.0016	.0031	.0053
-.70	.0000	.0000	.0000	.0000	.0001	.0003	.0006	.0013	.0025	.0044
-.60	.0000	.0000	.0000	.0000	.0001	.0002	.0005	.0011	.0022	.0037
-.50	.0000	.0000	.0000	.0000	.0001	.0002	.0005	.0010	.0018	.0032
-.40	.0000	.0000	.0000	.0000	.0001	.0002	.0004	.0008	.0016	.0028
-.30	.0000	.0000	.0000	.0000	.0000	.0001	.0003	.0007	.0014	.0024
-.20	.0000	.0000	.0000	.0000	.0000	.0001	.0003	.0007	.0012	.0021
-.10	.0000	.0000	.0000	.0000	.0000	.0001	.0003	.0006	.0011	.0019
.00	.0000	.0000	.0000	.0000	.0000	.0001	.0002	.0005	.0010	.0017
.10	.0000	.0000	.0000	.0000	.0000	.0001	.0002	.0005	.0009	.0016
.20	.0000	.0000	.0000	.0000	.0000	.0001	.0002	.0004	.0008	.0014
.30	.0000	.0000	.0000	.0000	.0000	.0001	.0002	.0004	.0008	.0013
.40	.0000	.0000	.0000	.0000	.0000	.0001	.0002	.0004	.0007	.0012
.50	.0000	.0000	.0000	.0000	.0000	.0001	.0002	.0003	.0006	.0011
.60	.0000	.0000	.0000	.0000	.0000	.0001	.0001	.0003	.0006	.0010
.70	.0000	.0000	.0000	.0000	.0000	.0001	.0001	.0003	.0005	.0009
.80	.0000	.0000	.0000	.0000	.0000	.0000	.0001	.0003	.0005	.0009
.90	.0000	.0000	.0000	.0000	.0000	.0000	.0001	.0002	.0005	.0008
1.00	.0000	.0000	.0000	.0000	.0000	.0000	.0001	.0002	.0005	.0008

r_0 (unit of length $r_c/2$)

$Log_{10}t$	.30	.32	.34	.36	.38	.40	.42	.44	.46	.48
-3.00	1.000	1.000	1.000	1.000	1.000	1.000	1.000	1.000	1.000	1.000
-2.90	.9988	.9997	.9999	.9999	1.000	1.000	1.000	1.000	1.000	1.000
-2.80	.9918	.9978	.9994	.9998	.9998	.9998	1.000	1.000	1.000	1.000
-2.70	.9662	.9876	.9959	.9987	.9996	.9997	.9997	.9997	1.000	1.000
-2.60	.9063	.9556	.9810	.9926	.9973	.9990	.9995	.9996	.9996	1.000
-2.50	.8061	.8883	.9408	.9710	.9868	.9944	.9977	.9989	.9994	.9994
-2.40	.6769	.7843	.8652	.9211	.9567	.9776	.9890	.9948	.9975	.9987
-2.30	.5392	.6564	.7570	.8369	.8961	.9370	.9636	.9799	.9893	.9944
-2.20	.4117	.5241	.6307	.7251	.8037	.8654	.9113	.9438	.9656	.9797
-2.10	.3051	.4034	.5042	.6012	.6896	.7662	.8294	.8794	.9174	.9450
-2.00	.2219	.3028	.3906	.4809	.5690	.6514	.7252	.7888	.8417	.8842
-1.90	.1600	.2239	.2964	.3747	.4555	.5354	.6116	.6818	.7446	.7990
-1.80	.1154	.1646	.2225	.2873	.3570	.4291	.5013	.5713	.6373	.6981
-1.70	.0836	.1211	.1664	.2186	.2764	.3384	.4027	4676	.5314	.5928
-1.60	.0612	.0897	.1248	.1661	.2130	.2646	.3196	.3768	.4350	.4929
-1.50	.0454	.0671	.0943	.1268	.1644	.2064	.2523	.3012	.3521	.4041
-1.40	.0342	.0509	.0720	.0976	.1275	.1616	.1993	.2402	.2836	.3288
-1.30	.0262	.0391	.0557	.0759	.0998	.1273	.1581	.1920	.2284	.2670
-1.20	.0204	.0306	.0436	.0598	.0790	.1012	.1264	.1544	.1848	.2173
-1.10	.0161	.0242	.0347	.0477	.0633	.0814	.1021	.1252	.1505	.1779
-1.00	.0130	.0195	.0280	.0386	.0513	.0663	.0834	.1026	.1237	.1468
-.90	.0106	.0160	.0230	.0317	.0422	.0546	.0689	.0849	.1027	.1222
-.80	.0088	.0133	.0191	.0264	.0352	.0456	.0576	.0712	.0863	.1028
-.70	.0074	.0112	.0161	.0223	.0297	.0386	.0488	.0603	.0732	.0874
-.60	.0063	.0095	.0138	.0190	.0254	.0330	.0418	.0518	.0629	.0751
-.50	.0054	.0082	.0119	.0165	.0220	.0286	.0362	.0449	.0546	.0653
-.40	.0048	.0072	.0104	.0144	.0193	.0251	.0318	.0394	.0479	.0573
-.30	.0042	.0064	.0092	.0127	.0171	.0222	.0281	.0349	.0424	.0508
-.20	.0038	.0057	.0082	.0114	.0153	.0198	.0251	.0312	.0380	.0455
-.10	.0034	.0051	.0074	.0103	.0138	.0179	.0227	.0282	.0343	.0411
-.00	.0031	.0047	.0067	.0094	.0125	.0163	.0207	.0256	.0312	.0374
.10	.0028	.0043	.0062	.0086	.0115	.0150	.0190	.0235	.0287	.0344
.20	.0026	.0040	.0057	.0079	.0106	.0138	.0175	.0218	.0265	.0318
.30	.0024	.0037	.0053	.0074	.0099	.0129	.0163	.0203	.0247	.0296
.40	.0023	.0035	.0050	.0069	.0093	.0121	.0153	.0190	.0231	.0277
.50	.0021	.0033	.0047	.0065	.0087	.0114	.0144	.0179	.0218	.0261
.60	.0020	.0031	.0045	.0062	.0083	.0108	.0137	.0170	.0207	.0248
.70	.0019	.0029	.0042	.0059	.0079	.0102	.0130	.0161	.0197	.0236
.80	.0019	.0028	.0041	.0056	.0075	.0098	.0124	.0154	.0188	.0225
.90	.0018	.0027	.0039	.0054	.0072	.0094	.0119	.0148	.0181	.0217
1.00	.0017	.0026	.0038	.0052	.0070	.0091	.0115	.0143	.0174	.0209
1.10	.0017	.0025	.0036	.0050	.0067	.0088	.0111	.0138	.0168	.0202
1.20	.0016	.0024	.0035	.0049	.0065	.0085	.0108	.0134	.0164	.0196
1.30	.0016	.0024	.0034	.0048	.0064	.0083	.0105	.0131	.0159	.0191
1.40	.0015	.0023	.0033	.0046	.0062	.0081	.0103	.0127	.0155	.0186
1.50	.0015	.0023	.0033	.0045	.0061	.0079	.0100	.0125	.0152	.0182
1.60	.0015	.0022	.0032	.0045	.0060	.0078	.0098	.0122	.0149	.0179
1.70	.0014	.0022	.0032	.0044	.0059	.0076	.0097	.0120	.0146	.0175
1.80	.0014	.0021	.0031	.0043	.0058	.0075	.0095	.0118	.0144	.0173
1.90	.0014	.0021	.0031	.0042	.0057	.0074	.0094	.0117	.0142	.0170
2.00	.0014	.0021	.0030	.0042	.0056	.0073	.0093	.0115	.0140	.0168

r_0 (unit of length $r_e/2$)

$Log_{10}t$	.50	.52	.54	.56	.58	.60	.62	.64	.66	.68
-2.30	.9973	.9986	.9992	.9994	.9995	.9995	.9996	.9996	.9996	.9996
-2.20	.9886	.9937	-.9965	.9980	.9988	.9991	.9993	.9993	.9993	.9993
-2.10	.9647	.9779	.9864	.9919	.9951	.9971	.9981	.9987	.9990	.9991
-2.00	.9176	.9426	.9609	.9739	.9830	.9890	.9930	.9955	.9970	.9979
-1.90	.8451	.8828	.9130	.9366	.9546	.9680	.9778	.9848	.9896	.9930
-1.80	.7528	.8006	.8416	.8761	.9045	.9274	.9456	.9597	.9706	.9787
-1.70	.6509	.7043	.7525	.7954	.8329	.8651	.8924	.9151	.9337	.9488
-1.60	.5497	.6039	.6551	.7026	.7460	.7851	.8199	.8505	.8770	.8996
-1.50	.4566	.5082	.5585	.6067	.6523	.6950	.7345	.7706	.8033	.8325
-1.40	.3755	.4224	.4693	.5153	.5602	.6033	.6444	.6831	.7193	.7529
-1.30	.3075	.3490	.3910	.4332	.4752	.5164	.5566	.5954	.6326	.6679
-1.20	.2520	.2879	.3248	.3624	.4004	.4384	.4760	.5130	.5492	.5842
-1.10	.2074	.2381	.2701	.3031	.3367	.3708	.4050	.4392	.4730	.5063
-1.00	.1718	.1980	.2255	.2542	.2836	.3138	.3443	.3752	.4060	.4368
-.90	.1435	.1659	.1896	.2143	.2400	.2664	.2934	.3209	.3486	.3765
-.80	.1210	.1402	.1606	.1820	.2044	.2275	.2512	.2755	.3002	.3252
-.70	.1031	.1197	.1373	.1559	.1754	.1956	.2165	.2380	.2599	.2822
-.60	.0888	.1032	.1186	.1348	.1518	.1696	.1880	.2070	.2264	.2463
-.50	.0772	.0899	.1033	.1176	.1326	.1483	.1646	.1815	.1988	.2165
-.40	.0679	.0790	.0910	.1036	.1169	.1309	.1454	.1604	.1759	.1917
-.30	.0603	.0702	.0808	.0921	.1040	.1165	.1295	.1430	.1569	.1711
-.20	.0540	.0629	.0725	.0826	.0934	.1046	.1163	.1285	.1410	.1540
-.10	.0488	.0569	.0656	.0748	.0845	.0947	.1054	.1164	.1278	.1396
-.00	.0445	.0519	.0598	.0682	.0771	.0864	.0962	.1063	.1168	.1276
.10	.0409	.0477	.0549	.0627	.0709	.0795	.0884	.0978	.1074	.1174
.20	.0378	.0441	.0509	.0580	.0656	.0736	.0819	.0906	.0995	.1088
.30	.0353	.0411	.0474	.0541	.0612	.0686	.0764	.0844	.0928	.1014
.40	.0331	.0386	.0444	.0507	.0574	.0643	.0716	.0792	.0871	.0952
.50	.0312	.0364	.0419	.0478	.0541	.0607	.0676	.0747	.0821	.0898
.60	.0296	.0345	.0397	.0454	.0513	.0575	.0641	.0709	.0779	.0851
.70	.0282	.0328	.0379	.0432	.0489	.0548	.0610	.0675	.0742	.0811
.80	.0269	.0314	.0362	.0414	.0468	.0525	.0584	.0646	.0710	.0777
.90	.0259	.0302	.0348	.0397	.0450	.0504	.0561	.0621	.0683	.0746
1.00	.0250	.0291	.0336	.0383	.0434	.0486	.0542	.0599	.0659	.0720
1.10	.0242	.0282	.0325	.0371	.0420	.0471	.0524	.0580	.0637	.0697
1.20	.0235	.0274	.0316	.0360	.0407	.0457	.0509	.0563	.0619	.0677
1.30	.0229	.0267	.0307	.0351	.0397	.0445	.0496	.0548	.0603	.0659
1.40	.0223	.0260	.0300	.0342	.0387	.0435	.0484	.0535	.0588	.0643
1.50	.0218	.0255	.0294	.0335	.0379	.0425	.0473	.0524	.0576	.0629
1.60	.0214	.0250	.0288	.0329	.0372	.0417	.0464	.0514	.0565	.0617
1.70	.0210	.0245	.0283	.0323	.0365	.0410	.0456	.0505	.0555	.0606
1.80	.0207	.0241	.0278	.0318	.0359	.0403	.0449	.0497	.0546	.0597
1.90	.0204	.0238	.0274	.0313	.0354	.0398	.0443	.0490	.0538	.0588
2.00	.0201	.0235	.0271	.0309	.0350	.0392	.0437	.0483	.0531	.0581
2.10	.0199	.0232	.0268	.0306	.0346	.0388	.0432	.0478	.0525	.0574
2.20	.0197	.0230	.0265	.0303	.0342	.0384	.0428	.0473	.0520	.0568
2.30	.0195	.0228	.0263	.0300	.0339	.0380	.0424	.0469	.0515	.0563
2.40	.0194	.0226	.0261	.0297	.0336	.0377	.0420	.0465	.0511	.0558
2.50	.0192	.0224	.0259	.0295	.0334	.0375	.0417	.0461	.0507	.0554
2.60	.0191	.0223	.0257	.0293	.0332	.0372	.0414	.0458	.0504	.0551
2.70	.0190	.0222	.0255	.0292	.0330	.0370	.0412	.0456	.0501	.0548

r_0 (unit of length $r_c/2$)

Log$_{10}$t	.70	.72	.74	.76	.78	.80	.82	.84	.86
-2.00	.9987	.9990	.9991	.9992	.9992	.9992	.9995	.9995	.9995
-1.90	.9954	.9969	.9978	.9983	.9987	.9989	.9990	.9990	.9990
-1.80	.9849	.9893	.9924	.9946	.9961	.9972	.9978	.9983	.9985
-1.70	.9611	.9706	.9779	.9835	.9878	.9910	.9933	.9950	.9962
-1.60	.9191	.9352	.9485	.9594	.9682	.9753	.9809	.9852	.9886
-1.50	.8586	.8814	.9011	.9181	.9327	.9450	.9553	.9639	.9710
-1.40	.7839	.8119	.8372	.8599	.8801	.8979	.9136	.9272	.9389
-1.30	.7015	.7327	.7618	.7887	.8133	.8358	.8563	.8747	.8912
-1.20	.6182	.6506	.6814	.7105	.7379	.7636	.7874	.8096	.8300
-1.10	.5392	.5710	.6017	.6314	.6598	.6869	.7127	.7371	.7601
-1.00	.4675	.4976	.5271	.5560	.5840	.6112	.6375	.6627	.6869
-.90	.4046	.4324	.4600	.4872	.5139	.5402	.5658	.5907	.6149
-.80	.3507	.3759	.4012	.4263	.4512	.4759	.5002	.5240	.5474
-.70	.3050	.3278	.3507	.3736	.3965	.4193	.4419	.4642	.4863
-.60	.2667	.2872	.3078	.3286	.3494	.3703	.3910	.4117	.4322
-.50	.2348	.2531	.2717	.2905	.3094	.3284	.3474	.3663	.3852
-.40	.2082	.2247	.2414	.2584	.2755	.2928	.3101	.3274	.3448
-.30	.1860	.2009	.2161	2314	.2470	.2627	.2785	.2943	.3102
-.20	.1674	.1810	.1948	.2088	.2229	.2372	.2517	.2662	.2808
-.10	.1519	.1643	.1769	.1897	.2026	.2157	.2290	.2423	.2557
-.00	.1389	.1502	.1618	.1735	.1855	.1975	.2097	.2220	.2344
.10	.1278	.1383	.1490	.1599	.1709	.1821	.1934	.2048	.2163
.20	.1185	.1282	.1381	.1482	.1585	.1689	.1794	.1901	.2008
.30	.1105	.1196	.1289	.1383	.1479	.1577	.1675	.1775	.1875
.40	.1037	.1122	.1210	.1299	.1389	.1480	.1573	.1666	.1761
.50	.0979	.1059	.1142	.1226	.1311	.1397	.1485	.1573	.1663
.60	.0928	.1005	.1083	.1163	.1244	.1326	.1409	.1493	.1578
.70	.0885	.0958	.1032	.1108	.1185	.1264	.1343	.1423	.1504
.80	.0847	.0917	.0988	.1061	.1135	.1210	.1286	.1363	.1440
.90	.0814	.0881	.0950	.1020	.1091	.1163	.1236	.1310	.1385
1.00	.0785	.0850	.0916	.0984	.1053	.1122	.1193	.1264	.1336
1.10	.0760	.0823	.0887	.0952	.1019	.1086	.1155	.1224	.1293
1.20	.0738	.0799	.0861	.0925	.0989	.1055	.1121	.1188	.1256
1.30	.0719	.0778	.0839	.0901	.0964	.1027	.1092	.1157	.1223
1.40	.0702	.0760	.0819	.0879	.0941	.1003	.1066	.1130	.1194
1.50	.0687	.0744	.0802	.0861	.0921	.0982	.1043	.1106	.1169
1.60	.0674	.0729	.0786	.0844	.0903	.0963	.1023	.1084	.1146
1.70	.0662	.0717	.0772	.0829	.0887	.0946	.1005	.1066	.1126
1.80	.0651	.0705	.0760	.0816	.0873	.0931	.0990	.1049	.1109
1.90	.0642	.0695	.0750	.0805	.0861	.0918	.0976	.1034	.1093
2.00	.0634	.0687	.0740	.0795	.0850	.0906	.0963	.1021	.1079
2.10	.0627	.0679	.0732	.0786	.0840	.0896	.0952	.1009	.1067
2.20	.0621	.0672	.0724	.0778	.0832	.0887	.0943	.0999	.1056
2.30	.0615	.0666	.0718	.0770	.0824	.0879	.0934	.0990	.1046
2.40	.0610	.0660	.0712	.0764	.0817	.0872	.0926	.0982	.1038
2.50	.0605	.0655	.0706	.0758	.0811	.0865	.0919	.0974	.1030
2.60	.0601	.0651	.0702	.0753	.0806	.0859	.0913	.0968	.1023
2.70	.0598	.0647	.0698	.0749	.0801	.0854	.0908	.0962	.1017
2.80	.0595	.0644	.0694	.0745	.0797	.0850	.0903	.0957	.1012
2.90	.0592	.0641	.0691	.0742	.0793	.0846	.0899	.0953	.1007
3.00	.0589	.0638	.0688	.0738	.0790	.0842	.0895	.0949	.1003

r_0 (unit of length $r_c/2$)

$\log_{10}t$	.88	.90	.92	.94	.96	.98	1.0	1.1	1.2
-2.00	.9990	1.000	1.000	1.000	1.000	1.000	1.000	1.000	1.000
-1.90	.9990	.9989	.9989	1.000	1.000	1.000	1.000	1.000	1.000
-1.80	.9987	.9987	.9988	.9988	1.000	1.000	1.000	1.000	1.000
-1.70	.9970	.9976	.9980	.9982	.9984	.9985	1.000	1.000	1.000
-1.60	.9912	.9932	.9947	.9958	.9966	.9972	.9987	1.000	1.000
-1.50	.9768	.9815	.9852	.9882	.9906	.9925	.9951	.9987	1.000
-1.40	.9490	.9576	.9649	.9711	.9762	.9805	.9852	.9953	.9986
-1.30	.9059	.9189	.9304	.9405	.9493	.9570	.9648	.9860	.9949
-1.20	.8487	.8658	.8814	.8955	.9082	.9196	.9310	.9671	.9856
-1.10	.7817	.8019	.8206	.8381	.8542	.8690	.8839	.9360	.9670
-1.00	.7099	.7319	.7527	.7724	.7910	.8084	.8260	.8926	.9371
-.90	.6383	.6609	.6826	.7034	.7234	.7424	.7605	.8376	.8940
-.80	.5703	.5925	.6142	.6352	.6556	.6752	.6941	.7778	.8436
-.70	.5080	.5294	.5503	.5708	.5908	.6103	.6293	.7156	.7873
-.60	.4526	.4727	.4925	.5120	.5312	.5501	.5685	.6544	.7287
-.50	.4041	.4227	.4413	.4596	.4777	.4956	.5131	.5965	.6709
-.40	.3621	.3794	.3965	.4136	.4305	.4472	.4638	.5432	.6160
-.30	.3261	.3420	.3579	.3736	.3893	.4049	.4204	.4954	.5653
-.20	.2954	.3100	.3246	.3392	.3538	.3683	.3827	.4529	.5194
-.10	.2692	.2827	.2962	.3097	.3232	.3366	.3500	.4157	.4785
-.00	.2469	.2594	.2719	.2844	.2969	.3094	.3219	.3833	.4424
.10	.2278	.2394	.2510	.2627	.2743	.2860	.2976	.3551	.4107
.20	.2115	.2224	.2332	.2441	.2550	.2659	.2767	.3306	.3831
.30	.1976	.2077	.2179	.2281	.2383	.2485	.2588	.3095	.3590
.40	.1856	.1951	.2047	.2143	.2240	.2336	.2432	.2912	.3380
.50	.1752	.1843	.1933	.2024	.2116	.2207	.2298	.2752	.3197
.60	.1663	.1749	.1835	.1922	.2008	.2095	.2182	.2614	.3038
.70	.1586	.1668	.1750	.1832	.1915	.1998	.2081	.2494	.2900
.80	.1518	.1597	.1676	.1755	.1834	.1914	.1993	.2389	.2778
.90	.1460	.1535	.1611	.1687	.1763	.1840	.1916	.2298	.2672
1.00	.1408	.1481	.1554	.1628	.1702	.1775	.1849	.2217	.2579
1.10	.1363	.1434	.1505	.1576	.1647	.1719	.1791	.2147	.2498
1.20	.1324	.1393	.1461	.1531	.1600	.1669	.1739	.2085	.2426
1.30	.1289	.1356	.1423	.1491	.1558	.1626	.1694	.2031	.2363
1.40	.1259	.1324	.1390	.1455	.1521	.1587	.1654	.1983	.2307
1.50	.1232	.1296	.1360	.1424	.1489	.1554	.1618	.1941	.2258
1.60	.1208	.1271	.1334	.1397	.1460	.1524	.1587	.1903	.2215
1.70	.1187	.1249	.1311	.1373	.1435	.1497	.1560	.1870	.2176
1.80	.1169	.1229	.1290	.1351	.1412	.1474	.1535	.1841	.2142
1.90	.1152	.1212	.1272	.1332	.1393	.1453	.1514	.1815	.2112
2.00	.1138	.1197	.1256	.1315	.1375	.1435	.1494	.1792	.2085
2.10	.1125	.1183	.1242	.1300	.1359	.1418	.1477	.1772	.2062
2.20	.1113	.1171	.1229	.1287	.1345	.1404	.1462	.1754	.2041
2.30	.1103	.1160	.1218	.1275	.1333	.1391	.1449	.1738	.2022
2.40	.1094	.1151	.1207	.1265	.1322	.1379	.1437	.1723	.2005
2.50	.1086	.1142	.1199	.1255	.1312	.1369	.1426	.1711	.1990
2.60	.1079	.1134	.1191	.1247	.1303	.1360	.1417	.1699	.1977
2.70	.1072	.1128	.1184	.1240	.1296	.1352	.1408	.1689	.1966
2.80	.1067	.1122	.1177	.1233	.1289	.1345	.1401	.1680	.1955
2.90	.1062	.1117	.1172	.1227	.1283	.1339	.1395	.1673	.1946
3.00	.1057	.1112	.1167	.1222	.1278	.1333	.1389	.1666	.1938

r_0 (unit of length $r_c/2$)

$\text{Log}_{10}t$	1.3	1.4	1.5	1.6	1.7	1.8	1.9	2.0
-2.00	1.000	1.000	1.000	1.000	1.000	1.000	1.000	1.000
-1.90	1.000	1.000	1.000	1.000	1.000	1.000	1.000	1.000
-1.80	1.000	1.000	1.000	1.000	1.000	1.000	1.000	1.000
-1.70	1.000	1.000	1.000	1.000	1.000	1.000	1.000	1.000
-1.60	1.000	1.000	1.000	1.000	1.000	1.000	1.000	1.000
-1.50	1.000	1.000	1.000	1.000	1.000	1.000	1.000	1.000
-1.40	.9995	1.000	1.000	1.000	1.000	1.000	1.000	1.000
-1.30	.9983	.9994	1.000	1.000	1.000	1.000	1.000	1.000
-1.20	.9941	.9977	.9991	1.000	1.000	1.000	1.000	1.000
-1.10	.9840	.9927	.9968	.9987	1.000	1.000	1.000	1.000
-1.00	.9650	.9814	.9906	.9955	.9979	.9991	1.000	1.000
-.90	.9331	.9590	.9752	.9849	.9905	.9935	.9983	1.000
-.80	.8930	.9287	.9534	.9700	.9807	.9873	.9946	.9970
-.70	.8446	.8889	.9222	.9464	.9635	.9752	.9864	.9916
-.60	.7911	.8420	.8825	.9140	.9378	.9555	.9717	.9810
-.50	.7357	.7908	.8367	.8741	.9042	.9278	.9494	.9635
-.40	.6811	.7382	.7875	.8293	.8642	.8930	.9197	.9376
-.30	.6292	.6866	.7374	.7819	.8203	.8530	.8840	.9062
-.20	.5811	.6376	.6886	.7343	.7747	.8101	.8442	.8699
-.10	.5375	.5922	.6425	.6882	.7294	.7663	.8025	.8305
-.00	.4985	.5510	.5998	.6448	.6860	.7234	.7606	.7901
.10	.4639	.5141	.5611	.6048	.6453	.6825	.7166	.7502
.20	.4334	.4812	.5262	.5685	.6079	.6444	.6782	.7119
.30	.4066	.4521	.4952	.5358	.5739	.6095	.6427	.6760
.40	.3832	.4266	.4677	.5067	.5434	.5779	.6102	.6429
.50	.3628	.4041	.4435	.4808	.5162	.5495	.5808	.6128
.60	.3449	.3844	.4221	.4580	.4920	.5241	.5545	.5856
.70	.3293	.3671	.4033	.4378	.4706	.5016	.5310	.5612
.80	.3156	.3520	.3868	.4200	.4516	.4816	.5100	.5394
.90	.3036	.3387	.3723	.4044	.4349	.4639	.4915	.5200
1.00	.2931	.3270	.3595	.3906	.4202	.4483	.4750	.5028
1.10	.2839	.3167	.3483	.3784	.4071	.4344	.4604	.4875
1.20	.2757	.3077	.3384	.3677	.3956	.4222	.4475	.4740
1.30	.2686	.2997	.3296	.3582	.3855	.4114	.4361	.4620
1.40	.2623	.2927	.3219	.3498	.3765	.4018	.4260	.4514
1.50	.2567	.2865	.3151	.3424	.3685	.3934	.4170	.4420
1.60	.2518	.2810	.3090	.3359	.3615	.3859	.4091	.4336
1.70	.2474	.2761	.3037	.3301	.3552	.3792	.4020	.4262
1.80	.2435	.2718	.2990	.3249	.3497	.3733	.3958	.4196
1.90	.2401	.2680	.2948	.3204	.3448	.3681	.3902	.4138
2.00	.2371	.2646	.2910	.3163	.3405	.3634	.3853	.4086
2.10	.2344	.2616	.2877	.3127	.3366	.3593	.3810	.4040
2.20	.2320	.2589	.2848	.3095	.3332	.3557	.3771	.3999
2.30	.2299	.2565	.2822	.3067	.3301	.3524	.3736	.3963
2.40	.2280	.2544	.2799	.3042	.3274	.3495	.3705	.3930
2.50	.2263	.2526	.2778	.3019	.3250	.3469	.3678	.3902
2.60	.2248	.2509	.2759	.2999	.3228	.3446	.3654	.3876
2.70	.2234	.2494	.2743	.2982	.3209	.3426	.3632	.3853
2.80	.2223	.2481	.2729	.2966	.3192	.3408	.3613	.3833
2.90	.2213	.2470	.2716	.2952	.3178	.3392	.3597	.3815
3.00	.2203	.2459	.2705	.2940	.3164	.3378	.3582	.3799

r_0 (unit of length $r_c/2$)

Log$_{10}$t	2.2	2.4	2.6	2.8	3.0	3.2	3.4	3.6	3.8	4.0
-1.00	1.000	1.000	1.000	1.000	1.000	1.000	1.000	1.000	1.000	1.000
-.90	.9986	.9986	.9986	1.000	1.000	1.000	1.000	1.000	1.000	1.000
-.80	.9980	.9985	.9985	.9985	1.000	1.000	1.000	1.000	1.000	1.000
-.70	.9958	.9977	.9983	.9984	.9984	1.000	1.000	1.000	1.000	1.000
-.60	.9907	.9955	.9974	.9981	.9983	.9983	.9983	1.000	1.000	1.000
-.50	.9806	.9901	.9948	.9969	.9978	.9981	.9982	.9982	.9982	1.000
-.40	.9642	.9800	.9890	.9938	.9962	.9973	.9979	.9980	.9981	.9981
-.30	.9411	.9640	.9785	.9872	.9923	.9952	.9967	.9975	.9978	.9980
-.20	.9120	.9419	.9624	.9761	.9849	.9904	.9938	.9957	.9968	.9974
-.10	.8783	.9143	.9406	.9596	.9728	.9819	.9879	.9918	.9943	.9958
-.00	.8419	.8826	.9140	.9379	.9557	.9687	.9781	.9847	.9892	.9923
.10	.8043	.8484	.8838	.9119	.9339	.9509	.9638	.9735	.9806	.9859
.20	.7671	.8133	.8516	.8829	.9084	.9289	.9452	.9580	.9680	.9758
.30	.7313	.7786	.8186	.8523	.8804	.9038	.9230	.9387	.9515	.9618
.40	.6977	.7452	.7861	.8212	.8512	.8767	.8983	.9165	.9317	.9443
.50	.6666	.7137	.7549	.7908	.8219	.8489	.8722	.8923	.9094	.9241
.60	.6381	.6845	.7255	.7616	.7934	.8213	.8457	.8671	.8858	.9021
.70	.6123	.6578	.6983	.7343	.7662	.7945	.8196	.8419	.8616	.8791
.80	.5891	.6336	.6734	.7089	.7407	.7691	.7946	.8174	.8378	.8560
.90	.5684	.6118	.6508	.6857	.7172	.7455	.7710	.7940	.8147	.8334
1.00	.5499	.5922	.6303	.6647	.6957	.7237	.7491	.7721	.7929	.8119
1.10	.5334	.5747	.6120	.6457	.6761	.7038	.7289	.7518	.7726	.7916
1.20	.5187	.5591	.5956	.6286	.6585	.6857	.7105	.7332	.7539	.7729
1.30	.5057	.5452	.5809	.6133	.6427	.6694	.6939	.7163	.7368	.7556
1.40	.4941	.5328	.5678	.5996	.6285	.6548	.6789	.7009	.7212	.7399
1.50	.4839	.5218	.5562	.5874	.6158	.6417	.6654	.6871	.7072	.7256
1.60	.4748	.5120	.5458	.5765	.6044	.6299	.6533	.6747	.6945	.7127
1.70	.4667	.5033	.5366	.5668	.5943	.6194	.6425	.6636	.6831	.7012
1.80	.4595	.4956	.5284	.5581	.5853	.6101	.6328	.6537	.6730	.6908
1.90	.4531	.4888	.5211	.5504	.5772	.6017	.6242	.6448	.6639	.6815
2.00	.4475	.4827	.5146	.5436	.5701	.5943	.6165	.6369	.6557	.6731
2.10	.4424	.4772	.5088	.5375	.5637	.5877	.6096	.6298	.6485	.6657
2.20	.4380	.4724	.5037	.5321	.5580	.5818	.6035	.6235	.6420	.6591
2.30	.4340	.4681	.4991	.5273	.5530	.5765	.5981	.6179	.6362	.6532
2.40	.4304	.4643	.4950	.5230	.5485	.5718	.5932	.6129	.6311	.6479
2.50	.4273	.4609	.4914	.5192	.5445	.5676	.5889	.6084	.6265	.6432
2.60	.4245	.4579	.4882	.5158	.5409	.5639	.5850	.6045	.6224	.6390
2.70	.4220	.4552	.4853	.5127	.5378	.5606	.5816	.6009	.6188	.6352
2.80	.4198	.4528	.4828	.5101	.5349	.5577	.5786	.5978	.6155	.6319
2.90	.4178	.4507	.4805	.5076	.5324	.5551	.5758	.5950	.6126	.6289
3.00	.4160	.4488	.4785	.5055	.5302	.5527	.5734	.5925	.6100	.6263
3.10	.4145	.4471	.4767	.5036	.5282	.5506	.5713	.5902	.6077	.6239
3.20	.4131	.4456	.4751	.5019	.5264	.5488	.5693	.5882	.6057	.6218
3.30	.4118	.4442	.4736	.5004	.5248	.5471	.5676	.5865	.6039	.6200
3.40	.4107	.4430	.4724	.4990	.5234	.5457	.5661	.5849	.6022	.6183
3.50	.4097	.4420	.4712	.4978	.5221	.5443	.5647	.5835	.6008	.6168
3.60	.4088	.4410	.4702	.4968	.5210	.5432	.5635	.5822	.5995	.6155
3.70	.4081	.4402	.4693	.4958	.5200	.5421	.5624	.5811	.5984	.6143
3.80	.4074	.4394	.4685	.4950	.5191	.5412	.5615	.5801	.5973	.6133
3.90	.4068	.4388	.4678	.4943	.5184	.5404	.5606	.5793	.5965	.6124
4.00	.4062	.4382	.4672	.4936	.5177	.5397	.5599	.5785	.5957	.6115

r_0 (unit of length $r_c/2$)

Log$_{10}$t	4.2	4.4	4.6	4.8	5.0	5.2	5.4	5.6	5.8	6.0
.00	.9957	.9971	.9979	.9985	.9988	.9990	.9992	.9992	.9993	.9993
.10	.9909	.9936	.9955	.9968	.9977	.9983	.9986	.9989	.9990	.9991
.20	.9830	.9874	.9908	.9932	.9950	.9963	.9973	.9979	.9984	.9987
.30	.9713	.9779	.9830	.9870	.9901	.9925	.9943	.9956	.9966	.9974
.40	.9561	.9648	.9719	.9776	.9823	.9860	.9890	.9914	.9932	.9947
.50	.9379	.9486	.9575	.9650	.9713	.9766	.9809	.9845	.9875	.9899
.60	.9175	.9298	.9404	.9496	.9575	.9642	.9700	.9749	.9790	.9825
.70	.8958	.9094	.9214	.9320	.9412	.9493	.9564	.9626	.9679	.9726
.80	.8736	.8882	.9013	.9129	.9233	.9326	.9408	.9481	.9546	.9603
.90	.8517	.8669	.8808	.8932	.9045	.9146	.9238	.9320	.9394	.9461
1.00	.8305	.8462	.8605	.8735	.8854	.8962	.9060	.9150	.9232	.9306
1.10	.8104	.8263	.8409	.8543	.8666	.8778	.8882	.8977	.9064	.9145
1.20	.7916	.8076	.8224	.8360	.8485	.8601	.8707	.8806	.8898	.8982
1.30	.7743	.7903	.8051	.8187	.8314	.8431	.8540	.8641	.8735	.8823
1.40	.7584	.7743	.7891	.8027	.8154	.8272	.8382	.8485	.8581	.8670
1.50	.7440	.7598	.7744	.7880	.8007	.8125	.8236	.8339	.8435	.8525
1.60	.7309	.7466	.7611	.7746	.7872	.7990	.8100	.8203	.8300	.8391
1.70	.7191	.7346	.7490	.7625	.7750	.7867	.7977	.8079	.8176	.8267
1.80	.7086	.7239	.7382	.7515	.7639	.7755	.7864	.7966	.8063	.8153
1.90	.6991	.7143	.7284	.7416	.7539	.7654	.7762	.7864	.7966	.8050
2.00	.6906	.7056	.7196	.7327	.7449	.7563	.7671	.7772	.7867	.7956
2.10	.6830	.6979	.7118	.7247	.7368	.7482	.7588	.7688	.7783	.7872
2.20	.6762	.6910	.7047	.7176	.7296	.7408	.7514	.7614	.7707	.7796
2.30	.6702	.6848	.6985	.7112	.7231	.7343	.7448	.7547	.7640	.7727
2.40	.6648	.6793	.6929	.7055	.7173	.7284	.7388	.7487	.7579	.7666
2.50	.6600	.6744	.6879	.7004	.7122	.7232	.7335	.7433	.7525	.7611
2.60	.6557	.6700	.6834	.6959	.7076	.7185	.7288	.7385	.7476	.7563
2.70	.6519	.6661	.6794	.6918	.7034	.7143	.7246	.7342	.7433	.7519
2.80	.6485	.6626	.6759	.6882	.6998	.7106	.7208	.7304	.7394	.7480
2.90	.6454	.6595	.6727	.6850	.6965	.7073	.7174	.7270	.7360	.7445
3.00	.6427	.6568	.6699	.6821	.6936	.7043	.7144	.7239	.7329	.7414
3.10	.6403	.6543	.6674	.6796	.6910	.7017	.7118	.7212	.7302	.7386
3.20	.6381	.6521	.6651	.6773	.6887	.6993	.7094	.7188	.7277	.7361
3.30	.6362	.6502	.6631	.6752	.6866	.6972	.7072	.7167	.7255	.7339
3.40	.6345	.6484	.6613	.6734	.6848	.6954	.7053	.7147	.7236	.7320
3.50	.6330	.6469	.6598	.6718	.6831	.6937	.7037	.7130	.7219	.7302
3.60	.6316	.6455	.6583	.6704	.6817	.6922	.7021	.7115	.7203	.7286
3.70	.6304	.6442	.6571	.6691	.6803	.6909	.7008	.7101	.7189	.7272
3.80	.6294	.6431	.6560	.6680	.6792	.6897	.6996	.7089	.7177	.7260
3.90	.6284	.6421	.6550	.6669	.6781	.6887	.6985	.7078	.7166	.7249
4.00	.6275	.6413	.6541	.6660	.6772	.6877	.6976	.7069	.7156	.7239
4.10	.6268	.6405	.6533	.6652	.6764	.6869	.6967	.7060	.7148	.7230
4.20	.6261	.6398	.6526	.6645	.6757	.6861	.6960	.7053	.7140	.7222
4.30	.6255	.6392	.6519	.6639	.6750	.6855	.6953	.7046	.7133	.7215
4.40	.6250	.6386	.6514	.6633	.6744	.6849	.6947	.7040	.7127	.7209
4.50	.6245	.6381	.6509	.6628	.6739	.6844	.6942	.7034	.7121	.7204
4.60	.6240	.6377	.6504	.6623	.6734	.6839	.6937	.7029	.7116	.7199
4.70	.6237	.6373	.6500	.6619	.6730	.6835	.6933	.7025	.7112	.7194
4.80	.6233	.6370	.6497	.6616	.6727	.6831	.6929	.7021	.7108	.7191
4.90	.6230	.6367	.6494	.6612	.6724	.6828	.6926	.7018	.7105	.7187
5.00	.6228	.6364	.6491	.6610	.6721	.6825	.6923	.7015	.7102	.7184

APPENDIX B: PROGRAM TO ESTIMATE PARAMETERS FOR MULTIPLE-TRAPPING MODEL FROM TRANSIENT PHOTOCURRENT DATA*

Problem

We require estimates of $2N + 1$ parameters $A, r_1 \ldots r_N, M_1 \ldots M_N$ so that the model time function obtained by the inverse Laplace transform

$$f(s) = AL^{-1}\left(\frac{1 - e^{-a}}{a}\right) \quad \text{where } a = \sum_{i=1}^{N} \frac{M_i s}{s + r_i}$$

best fits a measured time function $h(t)$ given by the discrete values (t_i, h_i), $i = 1, \ldots M$.

Solution

1. The Laplace inversion is expressed as a contour integral on a circle chosen so that the integrand remains bounded. The integration is done by the trapezoidal rule with the integrand evaluated at equal steps in arc length around the circle.
2. An unconstrained minimization routine adjusts the $2N + 1$ parameter until the objective function

$$\sum_{i=1}^{M} [f(t_i) - h(t_i)]^2/h^2(t_i)$$

   is invariant. This requires in addition the calculation of all partial derivatives of $f(t)$ with respect to all $2N + 1$ parameters. These partial derivatives are found by contour integration of the corresponding partial derivatives of the integrand in (1).

This objective function is the mean square *relative* error of the parametric fit.

*See Appendix in J. Noolandi, *Phys. Rev. B* **16**, 4466 (1977).

Program Parameters

R0   Array for initial estimates r_i, $i = 1, \ldots, N$

W0   Array for initial estimates M_i, $i = 1, \ldots, N$

A0   Estimate of scale factor A

N   Number of trapping levels (≤ 5)

RAD   Radius of circle for contour integration RAD $> \frac{1}{2} \max (r_i)$

NUT   Number of points used in contour integral (≤ 510) NUT $\sim$ RAD. T_{max} where T_{max} is the largest time required in the inverse function

LIM   Number of iterations (see DFMFP write-up below)

T   Array for discrete time scale t_i, $i = 1, \ldots, M$ (≤ 51)

H   Array for data to be fitted by model: h_i, $i = 1, \ldots, M$

Data Entry

M, T, H are read in from a data file on disk (or card reader), R0, W0, A0, N, RAD, NUT, LIM are entered from a terminal (time sharing) or card reader.

If LIM is set to zero, the program provides the inverse function for the input estimates; no fitting takes place.

Precision

Double precision (this may require a FORTRAN type declaration on some compilers).

Output

Optimized fit and corresponding parameters: A, M_i, r_i.

Subroutines Required

DFMFP: IBM unconstrained minimization of a function of several variables following the method of Fletcher, Powell and Davidon, System/360 Scientific Subroutine Package (360A-MC-03X) Version III page 221.

```
1.000    C-N   TRAPS BY CONTOUR INTEGRATION
2.000    C     PARAMETER DESCRIPTION
3.000    C     A: SCALE FACTOR
4.000    C     N: NUMBER OF TRAPPING LEVELS
5.000    C     R( ), W( ): RATE CONSTANT ARRAYS
6.000    C     A0, R0 ( ), W0 ( ): INITIALIZATION OF A, R, ( ) & W( )
7.000    C     THE DIMENSION OF R,W,W0,R0 MAY BE INCREASED FROM 5
8.000    C     THE DIMENSION OF X ( ) AND G ( ) MUST EXCEED! TWICE THE DIM. OF R( )
9.000    C     RAD: RADIUS OF CIRCLE OF INTEGRATION
10.000   C     MUST BE GREATER THAN LARGEST R ( ) /2
11.000   C     LIM: THIS IS THE LIMIT PARAMETER IN "DFMFP"
12.000   C     SEE IBM SSP MANUAL FOR "FMFP", "DFMFP" SUBROUTINES
13.000   C     M: NUMBER OF DATA POINTS ( ≤ 51)
14.000   C     T( ), H ( ): TIME CURVE TO BE APPROXIMATED BY M.T. MODEL
15.000   C     NUT: NUMBER OF POINTS USED IN THE CONTOUR INTEGRAL ≤510
16.000   C     NUT = Tmax*RAD (INCREASE THIS VALUE TO CHECK QUADRATURE)
17.000   C     WHERE Tmax IS THE LARGEST TIME REQUIRED.
18.000   C
19.000   C     OPERATING SCHEME
20.000   C     M,T ( ),H ( ) IS A DISC FILE
21.000   C     VARIABLES IN THE NAMELIST STATEMENT ARE ENTERED VIA TERMINAL
22.000   C     LIM = 0; NO ATTEMPT IS MADE TO OPTIMIZE RATE CONSTANTS;
23.000   C     ONLY THE TIME CURVE CORRESPONDING TO A0, R0, W0 IS OUTPUT
24.000   C
25.000   C     ALL REAL VARIABLES ARE IN DOUBLE PRECISION
26.000   C     ALL COMPLEX VARIABLES ARE ALSO IN DOUBLE PRECISION
27.000   C     USE AN IMPLICIT REAL/COMPLEX*8(*16) (A-H,O-Z) STATEMENT
28.000   C     IN EACH SUBPROGRAM IN THIS LISTING (CONSULT A FORTRAN EXPERT)
29.000   C     ON THE XEROX SIGMA 9 WE CAN USE A COMPILER PROGRAM SWITCH OPTION FOR THIS
30.000   C
31.000   C     SUBROUTINE: "DFMFP" CONSULT IBM SSP MANUAL
32.000         EXTERNAL FUNCT
33.000         COMMON /COMZ/NUT, M,T,H,FS,R0,W0
34.000         COMMON /RADZ/ RAD
35.000         REAL T(51), H(51), W(5), G(31), X(31), FS(51), R(5)
36.000   C     REAL Z (200), R0(5), W0(5)
37.000         NAMELIST N, LIM, RAD, NUT, R0, W0, A0
38.000   1     INPUT
39.000         NT = N + N + 1
40.000   201   FORMAT (I5)
41.000   202   FORMAT (2F5.0)
42.000   C     READ CURRENT TRACE
43.000         READ (1,201)M
44.000         DO 204  I = 1,M
45.000         READ (1,202)T(I), H(I)
46.000   204   CONTINUE
47.000         DO = 300J = 1,N
48.000         X(J) = SQRT(W0(J));X (J + N) = SQRT (1.9*RAD/R0 (J) - 1.)
49.000   300   CONTINUE
50.000         X(N + N = 1) = SQRT(A0)
51.000         CALL FUNCT (NT,X,F,G)
52.000         OUTPUT F
```

```
53.000          IF (LIM.NE.0) CALL DFMFP ( FUNCT,NT,X,F,G,1.E-10,1.E-14,LIM,IER,Z)
54.000          OUTPUT F
55.000          PRINT 401, ((T(I), H(I), FS(I)), I = 1, M)
56.000    401   FORMAT (/ ('  T    I    G  ')/(3F8.3))
57.000          WO = 0.
58.000          DO 50  I = 1,N
59.000          R(I) = 1.9*RAD/(1. + X(I + N)**2)
60.000          W(I) = X(I)**2
61.000    50    WO = WO + W(I)
62.000          A = X(N + N + 1)**2
63.000          OUTPUT A,WO
64.000          PRINT 11,(R(I),W(I),I = 1,N)
65.000    11    FORMAT (/5X,'R',13X,'W'/(2E14.6))
66.000          REWIND 1;GOTO1;END
67.000          SUBROUTINE FUNCT(NT,X,F,G)
68.000          COMMON /COMZ/ NUT,M,T,H,FS,R0,W0
69.000          COMMON/RADZ/RAD
70.000          REAL X(31),G(31),H(51),T(51),FS(51),E(31)
71.000          REAL R0(5),W0(5),W(5),R(5)
72.000          N = (NT-1)/2
73.000          F = 0.
74.000          DO 6 I = 1,N
75.000          R(I) = 1.9*RAD/(1. + X(I + N)**2)
76.000    6     W(I) = X(I)**2
77.000          DO 7 J = 1,NT
78.000    7     G(J) = 0.
79.000          Q = X(N + N + 1)**2;TST = 1.
80.000          DO 17  L = 1,M
81.000          CALL FB (NUT,N,R,W,T(L),E,TST)
82.000          D = Q*E(1)-H(L)
83.000          F = F + D**2/H(L)**2 ;FS(L) = Q*E(1)
84.000          DO 18  K = 1,N
85.000          G(K) = G(K) + 2.*D*E(1 + K)*Q/H(L)**2
86.000    18    G(K + N) = G(K + N) + 2.*D*E(1 + K + N)*Q/H(L)**2
87.000          G(N + N + 1) = G(N + N + 1) + 2.*D*E(1)/H(L)**2
88.000    17    TST = 0.
89.000          G(NT) = G(NT)*2.*X(NT)
90.000          DO 30   I = 1,N
91.000          G(I + N) = G(I + N)*(-1.9*RAD)*2.*X(I + N)/(1. + X(I + N)**2)**2
92.000    30    G(I) = G(I)*2.*X(I)
93.000          RETURN;END
94.000          SUBROUTINE FB (NUT,N,R,W,T,E,TST)
95.000          COMMON /RADZ /RAD
96.000          REAL R(1), W(1), E(1)
97.000          REAL XG(512),YG(512),XR(512,4),YR(512,4),XW(512,4),YW(512,4)
98.000          COMPLEX Z,DS,S,ARG,A,DGDA,G,U,C,CS(512)
99.000          PI = 3.14159265358970D0
100.000         NT = N + N + 1
101.000         DO 7 I = 1,NT
102.000   7     E(I) = 0.
103.000         IF(TST.EQ.0) GO TO 50
104.000         DO 20 J = 0,NUT
105.000         WT = 1.;IF(J.EQ.0.OR.J.EQ.NUT)WT = .5
106.000         ARG = CMPLX(0.,PI*J/NUT);DS = RAD*CEXP(ARG)
```

528

```
107.000            S = DS-RAD*.99;CS(1 + J) = S;A = (0.,0.)
108.000            DO 1 I = 1,N
109.000      1     A = A + S*W(I)/(S + R(I))
110.000            IF(CABS(A).LT..1) GO TO 2
111.000            Z = CEXP(-A);G = (1.-Z)/A;DGDA = ((1. + A)*Z-1.)/A**2
112.000            GO TO 10
113.000      2     G = (1.,0.);DGDA = (0.,0.);U = (1.,0.)
114.000            DO 3 M = 2,14
115.000            U = -U*A/M;G = G + U
116.000      3     DGDA = DGDA + U*(M-1)
117.000            DGDA = DGDA/A
118.000     10     Z = DS*WT/NUT;C = G*Z;XG(1 + J) = REAL(C);YG(1 + J) = AIMAG(C)
119.000            DO 21  K = 1,N
120.000            C = -DGDA*Z*S*W(K)/(S + R(K))**2
121.000            XR(1 + J,K) = REAL(C);YR(1 + J,K) = AIMAG(C)
122.000            C = DGDA*Z*S/(S + R(K))
123.000            XW(1 + J,K) = REAL(C);YW(1 + J,K) = AIMAG(C)
124.000     21     CONTINUE
125.000     20     CONTINUE
126.000     50     DO 60 J = 0,NUT
127.000            Z = CEXP(CS(1 + J)*T) ;X = REAL(Z);Y = AIMAG(Z)
128.000            IF(X*X + Y*Y.LT.1.E-28) GO TO 61
129.000            E(1) = E(1) + X*XG(1 + J)-Y*YG(1 + J)
130.000            DO 11  K = 1,N
131.000            E(1 + K + N) = E(1 + K + N) + X*XR(1 + J,K)-Y*YR(1 + J,K)
132.000     11     E(1 + K) = E(1 + K) + X*XW(1 + J,K)-Y*YW(1 + J,K)
133.000     60     CONTINUE
134.000     61     RETURN
135.000            END
```

REFERENCES

1. J. Mort and D. M. Pai (Eds.), *Photoconductivity and Related Phenomena*, Elsevier, New York, 1976.

2. D. M. Pai and R. C. Enck, *Phys. Rev.* B **11,** 5163 (1975).

3. L. Onsager, *J. Chem. Phys.* **2,** 599 (1934); *Phys. Rev.* **54,** 554 (1938).

4. L. Onsager, Thesis, Yale University, (1935).

5. K. M. Hong and J. Noolandi, *J. Chem. Phys.* **68,** 5163 (1978).

6. K. M. Hong and J. Noolandi, *J. Chem. Phys.* **68,** 5172 (1978).

7. K. M. Hong and J. Noolandi, *J. Chem. Phys.* **69,** 5026 (1978).

8. G. Doetsch, *Introduction to the Theory and Application of the Laplace Transformation*, Springer, New York, 1974.

9. J. Noolandi and K. M. Hong, *Phys. Rev. Lett.* **41,** 46 (1978).

10. B. Ries, G. Schonherr, H. Bassler, and M. Silver, *Philos. Mag.* B **49,** 27 (1984).

11. B. Ries, G. Schonherr, H. Bassler, and M. Silver, *Philos. Mag.* B **49,** 259 (1984).

12. K. M. Hong and J. Noolandi, *Surf. Sci.* **75,** 561, (1978).

13. L. D. Landau and E. M. Lifshitz, *Quantum Mechanics, 2nd ed.*, Pergamon, New York, 1965, p. 113.

14. K. M. Hong, J. Noolandi, and R. A. Street, *Phys. Rev.* B **23,** 2967 (1981).

15. C. L. Braun and T. W. Scott, *J. Phys. Chem.* **87,** 4776 (1983).

16. T. W. Scott and C. L. Braun, *Can. J. Chem.* **63,** 228 (1985).

17. R. R. Chance and C. L. Braun, *J. Chem. Phys.* **64,** 3573 (1976).

18. G. R. Freeman, *J. Chem. Phys.* **38,** 1022 (1963); **39,** 988 (1963).

19. G. R. Freeman and J. M. Fayadh, *J. Chem. Phys.* **43,** 86 (1965).

20. A. Hummel and A. O. Allen, *Disc. Faraday Soc.* **36,** 95 (1963).

21. W. F. Schmidt and A. O. Allen, *J. Chem. Phys.* **72,** 3730 (1968).

22. G. R. Freeman, *Int. J. Radiat. Phys. Chem.* **4,** 237 (1972).

23. J.-P. Dodelet and G. R. Freeman, *Can. J. Chem.* **53,** 1263 (1975).

24. J. Terlecki and J. Fiutak, *Int. J. Radiat. Phys. Chem.* **4,** 469 (1972).

25. C. L. Braun, *J. Chem. Phys.* **80,** 4157 (1984).

26. J. Noolandi and K. M. Hong, *J. Chem. Phys.* **70,** 3230 (1979).

27. A. Mozumder, *J. Chem. Phys.* **55,** 3026 (1971).

28. R. J. Friauf, J. Noolandi, and K. M. Hong, *J. Chem. Phys.* **71,** 143 (1979).

29. C. A. M. van den Ende, L. H. Luthjens, J. M. Warman, and A. Hummel, *Rad. Phys. Chem.* **19,** 455 (1982).

30. J. M. Warman, K.-D. Asmus, and R. H. Schuler, *J. Phys. Chem.* **73,** 931 (1969).

31. J. A. Crumb and J. K. Baird, *J. Phys. Chem.* **83,** 1130 (1979).

32. C. A. M. van den Ende, J. M. Warman, and A. Hummel, *Radiat. Phys. Chem.* **23,** 55 (1984).

33. S. J. Rzad and G. Bakale, *J. Chem. Phys.* **59,** 2768 (1973).

34. A. Mozumder, *J. Chem. Phys.* **61,** 780 (1974).

35. J. K. Baird, J. Bullot, P. Cordier, and M. Gauthier, *J. Phys. Chem.* **86,** 903 (1982).

36. H. T. Choi, J. A. Haglund, and S. Lipsky, *J. Phys. Chem.* **87,** 1583 (1983).

37. K. Lee and S. Lipsky, *J. Phys. Chem.* **86,** 1985 (1982).

38. K. Lee and S. Lipsky, *J. Phys. Chem.* **88,** 4251 (1984).

39. A. Mozumder and M. Tachiya, *J. Chem. Phys.* **62,** 979 (1975).

40. M. P. deHaas, J. M. Warman, and B. Vojnovic, *Radiat. Phys. Chem.* **23,** 61 (1984).

41. P. W. Anderson, *Phys. Rev.* **109,** 1492 (1958).

42. N. Mott. *Rev. Mod. Phys.* **50,** 203 (1978).

43. D. Pfost, Z. Vardeny, and J. Tauc, *Phys. Rev. Lett.* **52,** 376 (1984).

44. (a) W. E. Spear, *Proc. Phys. Soc. (Lond.) B* **70,** 669 (1957); **76,** 826 (1960); (b) J. L. Hartke, *Phys. Rev.* **125,** 1177 (1962); (c) H. P. Grunwald and R. M. Blakney, *Phys. Rev.* **165,** 1006 (1968); (d) M. D. Tabak, *Phys. Rev. B* **2,** 2104 (1970).

45. M. E. Scharfe, *Phys. Rev. B* **2,** 5025 (1970).

46. G. Pfister and H. Scher, *Adv. Phys.* **27,** 747 (1978).

47. H. Scher and E. W. Montroll, *Phys. Rev. B* **12,** 2455 (1975).

48. V. M. Kenkre, E. W. Montroll, and M. F. Shlesinger, *J. Stat. Phys.* **9,** 2 (1973).

49. J. Noolandi, *Phys. Rev. B* **16,** 4466 (1977); **16,** 4474 (1977).

50. F. Schmidlin, *Phys. Rev. B* **6,** 2362 (1977).

51. O. L. Curtis, Jr., and J. R. Srour, *J. Appl. Phys.* **48,** 3819 (197).

52. F. Oberhettinger and L. Badii, *Tables of Laplace Transforms*, Springer-Verlag, New York, 1973.

53. H. Michiel, J. M. Marshall, and G. J. Adriaenssens, *Philos. Mag B* **48,** 187 (1983).

54. J. G. Simmons and M. C. Tam, *Phys. Rev. B* **7,** 3706 (1973).

55. J. Orenstein and M. A. Kastner, *Solid State Commun.* **40,** 85 (1981).

56. J. M. Marshall and C. Main, *Philos. Mag. B* **47,** 471 (1983).

57. G. Pfister, *Phys. Rev. Lett.* **33,** 1474 (1974).

58. J. M. Marshall, *Rep. Prog. Phys.* **46,** 1235 (1983).

59. P. N. Butcher, *Philos. Mag. B* **37,** 653 (1978).

60. P. N. Butcher and J. D. Clark, *Philos. Mag. B* **42,** 191 (1980).

61. W. D. Lakin, L. Marks, and J. Noolandi, *Phys. Rev. B* **15,** 5834 (1977).

62. J. Mort, M. Morgan, S. Grammatica, J. Noolandi, and K. M. Hong, *Phys. Rev. Lett.* **48,** 1411 (1982).

63. E. Zeldov and K. Weiser, *Phys. Res. Lett.* **53,** 1012 (1984).

64. F. W. Schmidlin, *Philos. Mag. B* **41,** 535 (1980).

10 CHARGE TRANSPORT IN MONOLAYER ORGANIZATES

Dietmar Möbius

Max-Planck-Institut für Biophysikalische Chemie
Göttingen, Federal Republic of Germany

10.1. INTRODUCTION

Important chemical reactions in biological systems, like photosynthesis, occur in membranes. These structures are essentially composed of two extremely thin layers of amphiphilic molecules in which proteins and other components are embedded. Amphiphilic molecules are characterized by a long hydrocarbon portion and a hydrophilic group. They have a very small solubility in water. Molecules of this type can be organized in ultrathin layers at the air–water interface. Under certain conditions the thickness of these layers is equal to the length of one molecule, and such layers are therefore called monomolecular layers or simply monolayers. Most membranes in biological systems consist of two monolayers in contact with each other at the hydrophobic side.

The ultrathin layers are extremely interesting nonhomogeneous structures not only with regard to the similarity to biological structures but also as a means to organize molecules in a planned way in order to construct machines of molecular dimension. A completely different aspect is the use of these entities as a tool for the investigation of complex physical or chemical processes, since the interactions between the different molecules involved can be controlled by variation of the spatial and energetic coordination. The systematic modification of different structural and energetic parameters provides quantitative information about various contributions to a complex phenomenon.

A complex photochemical process of particular interest is the photoinduced electron transfer in heterogeneous systems. This process is related to photosynthesis and could become important in artificial systems for conversion of solar energy into fuel. The potential uses of monomolecular layers and the complex systems that can be built up in a stepwise procedure are being explored. They might be used in devices for information processing, as components of other electronic devices or as parts of highly sophisticated sensors and catalysts. It is therefore necessary to understand the transport of charges in monomolecular layers and in complex monolayer systems. The electrical transport properties of such layers have been reviewed (1–3).

10.2. FORMATION OF MONOLAYER ORGANIZATES

The basic techniques of formation of monomolecular layers at the air–water interface were invented about 100 years ago by Pockels (4). A solution of the monolayer-forming material such as the long-chain fatty acid stearic acid is applied to a clean water surface. After evaporation of the solvent the molecules remaining at the water surface are packed by reduction of the available area. The structure may only be stable under certain conditions, and the monolayer material is pushed off the water surface when the area is reduced below the area occupied by the densely packed molecules. This is called monolayer collapse and usually leads to formation of three-dimensional crystals. A large variety of instruments, called monolayer troughs, has been developed for the formation and investigation of spread monolayers.

Complex monolayers are composed of different molecules, and the structure of these systems depends strongly on the molecular organization, which can be externally controlled. To distinguish the more elaborate monolayers of this type from the simple monomolecular layers, the term *organizate* is used. Even more complex are the assemblies that can be obtained by stepwise transfer of complex monolayers from the water surface onto solid substrates. The basis of monolayer organizates on solids is the technique of monolayer transfer invented by Langmuir and Blodgett (5) (the LB technique), which permits the formation of well-defined multilayered systems. The procedures are well known for long-chain fatty acids, and this is the reason for using fatty acid monolayers as a matrix for incorporating into the multilayers various interacting components like dyes, electron donors and acceptors, and electron-conducting or photoreactive molecules. This idea was the starting point of the modern use of monomolecular layers initiated by Kuhn and his group (6, 7).

10.2.1. Formation and Characterization of Complex Monolayers

It is generally stressed that monolayer work requires extremely clean conditions of environment and materials. Small concentrations of impurities, such as heavy-metal ions, in the aqueous subphase affect the structure and the behavior of fatty acid monolayers. Particularly annoying contaminations are small amounts of surface active material that may be released from the polytetrafluorethylene that is widely used as trough material. A critical step in the experimental procedure by which organic contamination may be introduced is the handling of the glassware (test tubes, flasks, and pipettes). These items should be only touched with gloves since the human skin releases greasy material to the glass surface. Another source of contamination is the gas phase, very often the laboratory atmosphere. Clean room conditions should be used for the formation of well-defined complex monolayers. The possible influence of defects should be assessed in monolayers formed by any given procedure.

Complex monolayers at the gas–water interface are formed by spreading a mixed solution of the components on the water surface. The solvent evaporates and the amphiphilic molecules remain at the water surface. On decreasing the monolayer area at constant temperature, the surface pressure–area isotherm is obtained, which is the two-dimensional analog of the pressure–volume isotherm. Experimentally, the surface pressure is measured continuously either with a Wilhelmy balance (8, 9) or a Langmuir balance (9, 10) while decreasing the monolayer area. A typical plot is shown in Figure 10.1, curve 1, for a monolayer of arachidic acid (C20). The surface pressure starts rising when the area per matrix molecule A_M is smaller than 0.22 nm^2/molecule. In the compressed monolayer before collapse, the average area is $A_M = 0.195$ nm^2/molecule. When a dye is added to the solution of arachidic acid before spreading, a mixed monolayer is formed and the isotherm is changed (see Figure 10.1, curve 2). This is obtained for a mixed monolayer of the cyanine dye ICC (N,N'-dioctadecylindocarbocyanine perchlorate, see structure in Table 10.1) and arachidic acid (C20)* as matrix, molar ratio ICC–C20 = 1 : 10 (11). Curve 3 in Figure 10.1 represents the theoretical isotherm obtained from the isotherms of the pure components if additivity was observed. There is a small difference between curves 2 and 3, indicating nonideal behavior.

---

*Cn is a saturated fatty acid with n C atoms in the chain.

TABLE 10.1. Dye Structures and Occurrence in Figures

Code	Structure	Figure
ICC	ClO_4^-	10.1, 10.12, 10.13
TCC	ClO_4^-	10.3, 10.16
AZS		10.3, 10.12, 10.13, 10.16
IPC		——

According to a structural model of the complex monolayer, the matrix molecules fill the space on top of the flat-lying chromophore, and the two long-chain substituents of the dye molecule are tightly packed in the hydrophobic moiety of the monolayer. The area occupied by a dye molecule should then correspond to the cross-sectional area of the two chains only, that is, to about 0.40 nm^2/molecule at a surface pressure of 30 mN/m. From the isotherms 1 and 2 in Figure 10.1, however, the area $(0.265 - 0.195)$ nm$^2 \times 10 = 0.70$ nm^2/ICC molecule is calculated. The result may be interpreted as evidence for incomplete mixing or formation of a separate dye phase in coexistence with a mixed monolayer of dye and fatty acid. This situation could originate in the spreading procedure. On evaporation of the solvent, the two components with different solubility should be released to the water surface differently. However, a mixed monolayer might be obtained if the molecular interactions and mobility after evaporation of the solvent could be modified in a way that increases mixing of the components.

TABLE 10.1. (*Continued*)

Code	Structure	Figure
MSC		10.10
SV	$H_{37}C_{18}-N^+$⬡⬡$^+N-C_{18}H_{37}$ $2ClO_4^-$	10.12, 10.13, 10.14, 10.15, 10.16
OC	ClO_4^-	10.14, 10.15
QC	ClO_4^-	—
TC	ClO_4^-	—

This is indeed possible by addition of a very slowly evaporating material, and hexadecane (HD) has been used for this purpose (11). The isotherm is no longer measured continuously but stepwise by increasing the surface pressure and reading the corresponding area after relaxation of the system. The HD is slowly squeezed out of the monolayer and evaporates into the gas phase, leaving a mixed monolayer with the isotherm shown in Figure 10.1, curve 4. From this isotherm the area per ICC molecule at 30 mN/m of $(0.236 - 0.195)$ nm^2 × 10 = 0.41 nm^2/ICC molecule is obtained. The use of HD as transient molecular lubricant has facilitated the homogeneous distribution of the dye molecules in the matrix of arachidic acid. This behavior has been described here in some detail in order to illustrate the difficulties that may arise from the use of complex monolayers instead of the simple, well-characterized monolayers.

Other methods for characterization and investigation of monolayers are the measurement of the interfacial potential (9) and the damping of capillary waves (12,

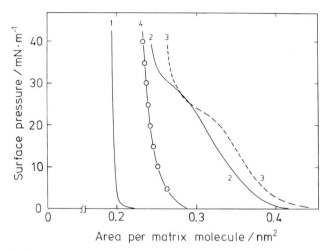

Figure 10.1. Surface pressure against area per matrix molecule for a monolayer of arachidic acid (C20), curve 1; a mixed monolayer of the cyanine dye ICC and arachidic acid, molar ratio ICC-C20 = 1 : 10, curve 2; a mixed monolayer of ICC, C20, and hexadecane (HD), molar ratios ICC-C20–HD = 1 : 10 : 10, measured discontinuously by increasing the surface pressure and measuring the area after relaxation of the monolayer, curve 4. The HD is squeezed out of the monolayer and occupies no area after relaxation. The isotherm curve 3 was calculated from the isotherm of the pure ICC monolayer (not shown) and curve 1 assuming additivity. Subphase: bidistilled water with 3×10^{-4} M $CdCl_2$ and 5×10^{-5} M $NaHCO_3$, 295 K.

13). Very recently, the rapid photochemical transformation of an amphiphilic spiropyran in monolayers has been used to generate an interfacial compressional wave, and the transmission of such a wave across a monolayer can be studied (14, 15). The velocity of the compressional wave as well as the frequency-dependent damping provides information on the interactions of the molecules in the monolayer with each other and with the underlying subphase.

Most of the active components that participate in charge transport phenomena in monolayer organizates contain π-electron systems, which can be characterized by measuring the absorption or the emission spectrum (16). In particular, the emission spectrum of a spread monolayer at the air–water interface reveals many details of the molecular organization (17, 18). Phase separation has been observed in mixed monolayers by fluorescence microscopy (19). An extremely useful technique in studying the spectroscopic properties of monolayers at the air–water interface is the change in light reflection (20). The reflection of light by a dye monolayer is usually too small to be detected. When the dye molecules are located at an interface between two media of different refractive index, however, the superposition of the light waves scattered back by the dye molecules with the light coherently reflected from the interface gives rise to a modification of the reflected intensity in the spectral range of the dye absorption band. This method has been used to investigate the association of dye molecules in mixed monolayers and the adsorption of dye molecules from the aqueous subphase to an appropriate matrix monolayer at the surface (20).

The orientation of the transition moments of dye molecules in mixed monolayers can be deduced from reflection measurements using plane-polarized light under oblique incidence (21, 22). The charge transfer across an insulating monolayer can be enhanced by incorporation of a π-electron system that is oriented parallel to the saturated hydrocarbon chains ("molecular wire," see Section 10.6). Since surface pressure–area isotherms are not sufficient to determine the orientation, the independent spectroscopic investigation is particularly important. The molecules of particular surface-active azo and stilbene dyes reorient and associate in monolayers at the air–water interface on compression, and the transition moments of these associates are oriented perpendicular to the interface in the densely packed monolayers (22, 23). This orientation was deduced from transmission measurements and is consistent with electrochromic effects observed in monolayer organizates between metal electrodes (24).

10.2.2. Monolayer Transfer and Characterization of Monolayer Organizates

Monolayers formed at the air–water interface can be transferred by the LB technique onto solid substrates like glass or quartz plates (5). This is done by dipping the solid through the monolayer into the trough. If the substrate is hydrophobic, the hydrophobic ends of the molecules forming the monolayer are attached to the surface, and one monolayer is transferred to the solid. No monolayer transfer occurs if the solid surface is hydrophilic, like an uncoated glass surface. In order to coat a clean glass surface with a monolayer, the plate should be immersed and the water surface cleaned before spreading the monolayer since grease and dust particles may be released from the glass surface when it comes in contact with the water. The monolayer is then transferred by raising the plate through the compressed monolayer at constant surface pressure. The plate should come out of the trough completely dry since its surface must be hydrophobic after monolayer transfer. The procedure is schematically shown in Figure 10.2a.

Multilayer systems can be deposited on glass plates by repeated dippings (Figure 10.2b). More complicated structures, the monolayer organizates, are assembled by changing the monolayer between the dipping and withdrawal steps according to the desired sequence. The molecular interactions in these complex systems are determined by the spacing of the planes where the various components are located and the structure of the space layers. An example of such a complex structure is shown in Figure 10.2c whose properties will be discussed in Section 10.6.

The transfer of monolayers with the LB technique of dipping is not possible in cases of very viscous monolayers since the plate then only punches a hole in the monolayer, and no further material is pushed to the plate by the barrier. In fact, the feedback system for maintaining constant surface pressure no longer functions

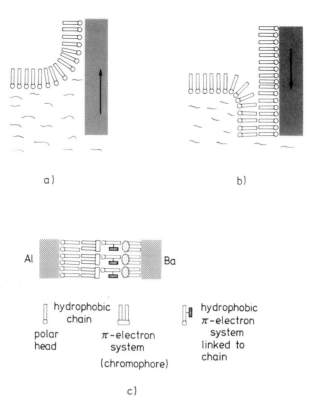

Figure 10.2. Transfer of monolayers from the air–water interface according to the LB technique (*a*, *b*) and complex monolayer organizate (*c*), schematic. Transfer is achieved due to interactions between the hydrophilic head groups and the hydrophilic substrate on withdrawal of the solid from the trough or due to interactions between the hydrophobic ends of the amphiphilic molecules and the hydrophobic surface on immersion of the solid into the trough. Molecules with π-electron systems can be embedded in a matrix monolayer. In the scheme of the complex system, the matrix molecules have been omitted for clarity.

in such cases, and surface pressure gradients may build up. This occurs very often with monolayers of cyanine dyes after formation of particular associates, the J aggregates. Monolayer transfer has been achieved in such cases by horizontal contact of the monolayer with a hydrophobic support (25, 26).

Monolayer organizates have been characterized by a variety of methods. Simple systems without built-in dye molecules have been investigated by optical techniques like ellipsometry (27, 28) and light microscopy (16, 29, 30) and after appropriate preparation by electron microscopy (31, 32). Spectroscopic methods (16) include absorption in the visible and infrared regions and emission, including the measurement of decay functions (33, 34).

The orientation of the optical transition moments of the dye molecules in monolayer organizates can be determined by measurement of the absorption spectra with plane-polarized light under oblique incidence (35, 22), similar to the technique used in the reflection method for monolayers on the water surface. An example is shown in Figure 10.3. The absorption spectra (1-transmission) of a system with a mixed monolayer of the cyanine dye TCC (N,N'-dioctadecyl-thiacarbocyanine), the azo dye AZS [12-(4′-nitro-4-dimethyl-aminoazo-benzene-3′-carboxyloxy) stearic acid, see structures in Table 10.1], and arachidic acid (C20), molar ratios TCC–AZS–C20 = 1:1:10, have been measured under an angle of incidence of 30° with s-polarized (curve 1) and p-polarized (curve 2) light (36). The insets indicate the plane of the electrical vector of the incident light. The two bands with maxima at 495 and 550 nm must be attributed to the cyanine dye TCC, whereas the band at 410 nm is due to the azo dye AZS. This band at 410 nm is not observed in curve 1, which indicates an orientation of the transition moment perpendicular to the monolayer plane. From the ratio $(1 - T)_s/(1 - T)_p$ for the two bands of the cyanine dye, it must be concluded that the transition moments are oriented in the layer plane (16), which is consistent with the schematic structure shown in Figure 10.3 (top).

Thin layers have been formed on solid substrates by adsorption of molecules from a solution in contact with the solid or by surface chemical reactions (chemically modified electrodes) (37, 38). In most cases these procedures do not yield

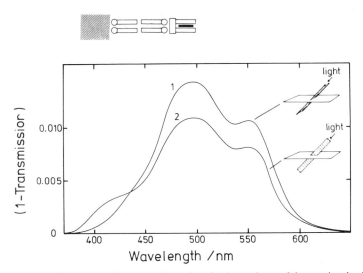

Figure 10.3. Absorption spectra (1-transmission) of a mixed monolayer of the cyanine dye TCC, the azo dye AZS, and arachidic acid (C20), molar ratios TCC–AZS–C20 = 1:1:10, in a system on glass, angle of incidence: 30°, plane-polarized light. Curve 1: s-polarization, curve 2: p-polarization. The band at 410 nm, attributed to the azo dye, is observed with p-polarized light only, indicating an orientation of the transition moment normal to the layer plane. The ratio of the intensities of the bands at 490 and 550 nm attributed to the cyanine dye is in agreement with a statistical distribution of these transition moments in the layer plane. Schematic structure on top; chromophore of AZS (full rectangle) is oriented normal to the layer plane.

dense monolayers comparable to those obtained by the LB technique. Under particular conditions, however, monolayers with no detectable defects have been formed by adsorption or surface chemical reaction (39), and their behavior in electron tunneling will be discussed in the following section.

10.3. ELECTRON TRANSFER ACROSS ONE MONOLAYER BETWEEN ELECTRODES

If two metal layers are separated by a thin insulating layer, a small current is observed on application of a voltage across this metal–insulator–metal (MIM) junction. The conductivity is described by a model of electron tunneling (40, 41) across the insulating barrier. In Figure 10.4 a simplified energy diagram represents the situation after applying the voltage U to the junction. The Fermi level of the metal on the left side is at the energy ϕ (work function of the metal) below the vacuum energy level of the electron. The work function from the metal to the insulator, φ, is generally very large compared with the thermal energy of the electrons in the metal, and the observed conductivity is therefore not due to an activation of the electrons to the top of the barrier. The transport of electrons from the metal on the left side to that on the right side is a manifestation of the wave particle duality of electrons.

The conductivity of a junction is determined from the measured current–voltage dependence, taking into account the junction area and the thickness d of the insulating layer, according to

$$\sigma = \frac{jd}{U} \quad \text{for } U \to 0 \tag{10.1}$$

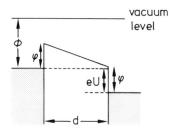

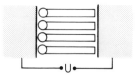

Figure 10.4. Energy diagram for a metal–monolayer–metal junction with applied voltage U and structural scheme without representation of oxide layers on top of the metals.

Here j is the current density and U the applied voltage. For small voltages the conductivity due to electron tunneling across the barrier, σ_t, is given by (40)

$$\sigma_t = \frac{e^2}{h^2} (2m\varphi)^{1/2} \exp\left[-\frac{4\pi d}{h} (2m\varphi)^{1/2}\right] \qquad (10.2)$$

This equation applies to a rectangular shape of the potential barrier if no voltage is applied; e is the charge and m the mass of the electron, h is Planck's constant, d is the thickness of the insulating layer, and φ is the difference between the work function ϕ of the metal and the electron affinity κ of the insulator. According to Eq. (10.2), the current density increases linearly with applied voltage U (since σ_t does not depend on U), and the conductivity σ_t decreases exponentially with increasing layer thickness.

The validity of the model has been tested mainly with junctions having a metal oxide (41) or a semiconductor (42) layer as insulator. Systematic variation of the thickness d in such systems is very difficult. On the other hand, densely packed monolayers provide insulating layers of very well defined thickness d given by the length of the molecules. Therefore, it is very easy to vary d within a certain range by using molecules of different chain lengths.

10.3.1. Electron Tunneling in Metal-Monolayer-Metal Junctions

The deposition of a single fatty acid monolayer requires a sufficiently hydrophilic surface in order to obtain a well-defined layer. A hydrophilic surface is provided by metal films that have been evaporated onto glass plates in a vacuum if these films are coated by a thin oxide layer as in the case of aluminum, nickel, or lead. Noble metals like gold or silver lack this oxide layer, and monolayer deposition may result in layers with a large fraction of defects. In the early attempts to use Langmuir–Blodgett monolayers as insulators (43–45), such defects played an important role.

With the improvement of the monolayer technique it became possible to coat evaporated metal electrodes with fatty acid monolayers that were sufficiently free of defects for the investigation of electron tunneling (46, 47). The observed electrical conductivity of metal–monolayer–metal systems was interpreted as a sum of two contributions, one being ohmic and one due to tunneling. The exponential decrease of the tunneling conductivity σ_t with increasing monolayer thickness was indeed found. The relative change of the tunneling current on variation of the top electrode (change of the work function) agreed reasonably well with the theoretical model. The electron affinity κ of the insulating monolayer was determined from the work function of the metal and the height of the energy barrier φ, and $\kappa = 2.3$ eV was obtained. These results have essentially been reconfirmed (48) with systems where the top electrodes had a much larger area, say, 0.2 instead of 10^{-3} cm^2. Further, magnesium was also used as top electrode material with a relatively small work function, and the dependence of the conductivity of aluminum–fatty acid–aluminum junctions was studied in a larger temperature range, between 173 and 293 K. The tunneling current observed was practically constant in this temperature range.

A matter of concern in these investigations is the presence of the thin layer of metal oxide on top of the metal. Although the conductivity of the oxide layer is larger than that of a fatty acid monolayer by several orders of magnitude in the case of aluminum electrodes, effects due to the oxide layer have been observed (49). The formation of the metal oxide, for example, on aluminum films, may take place in the vacuum plant or on exposure of the metal film to air. The oxide film is probably changed by immersion of the plate into the water for deposition of the monolayer. The formation of insulating monolayers on aluminum films by adsorption of the fatty acids from an organic solution or by surface chemical reaction of a long-chain silane provided other possibilities to evaluate the effect of the oxide layer on electron tunneling in MIM structures (39). Figure 10.5 shows the results obtained in a dry-nitrogen atmosphere with monolayers deposited by the LB technique (circles) and by adsorption (squares). The ohmic component of the conductivity, which might have been due to impurities in the earlier experiments, was negligible, and the junction area was 0.2 cm^2. The exponential decrease of the conductivity with increasing thickness of the insulating layer is clearly seen in this semilogarithmic plot. The two different methods give quite similar results, indi-

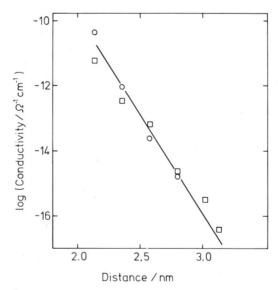

Figure 10.5. Electron tunneling in Al–Cn–Al junctions with one monolayer of the fatty acid Cn (saturated fatty acid with n carbon atoms in chain) as insulating barrier. Logarithm of conductivity plotted against monolayer thickness for LB monolayers (circles) and monolayers formed on the aluminum base electrode by adsorption from an organic solution (squares). The conductivity drops exponentially with increasing monolayer thickness.

cating the minor influence of the oxide layer, which should be different in these two series.

The possible influence of defects in the insulating monolayer has been considered in most investigations on electron tunneling in such systems (46). The results may change with time after sample preparation, depending on the details of sample storage and handling. These observations have been interpreted as evidence for tunneling through defects in the organic layer (50). In this context defects have been introduced intentionally into the insulating monolayer between metal electrodes (39). This can be done by adsorption of a cyanine dye onto the oxide-covered aluminum electrode in competition with octadecyltrichlorosilane (OTS), which is bound to the oxide by surface chemical reaction. The dye has been removed subsequently by treatment with chloroform, leaving holes of molecular dimensions in the chemically bound OTS layer with an area fraction on the order of 0.05. This treatment causes an increase in the conductivity by an order of magnitude without changing the current–voltage characteristics or shortening the sample. Pure OTS layers exhibit no changes in their behavior when treated in the same way (39). This indicates that in appropriately formed junctions, electron tunneling is not via defects but through the insulating monolayer.

The thickness of the insulating monolayer must be known for the calculation of the conductivity, [see Eq. (10.1)]. The numbers given in the literature may differ slightly according to the method used. In particular, in studying the temperature dependence of the conductivity, possible changes in the length of the fatty acid molecules with temperature might influence the results (46). The thickness of monolayers formed from the cadmium salts of fatty acids has been determined by X-ray diffraction (51, 52), and small variations with temperature have been found. The half-width of the shape of the X-ray diffraction line was small, suggesting good flatness of the planes with the cadmium ions (51).

10.3.2. Electron Tunneling in Metal–Monolayer–Metal Junctions with One Superconducting Electrode

According to the BCS (Bardeen, Cooper, and Schrieffer) theory (53), the superconducting state of metals is characterized by the presence of a region without states for unpaired electrons centered around the Fermi level. This gap must show up in the current–voltage curves at very small voltages U as a nonlinear part. Attempts have been made to investigate this behavior with monolayers as insulating barriers (43). However, the current density observed in the early experiments was many orders of magnitude larger than the current densities found with optimally prepared monolayers. An indication of the expected nonlinear behavior was found with junctions of barium stearate between an aluminum and a lead electrode at 4.2 K, although the result was not interpreted as evidence for the gap in the superconducting lead electrode (54).

An investigation of electron tunneling across various monolayers between an aluminum and a lead electrode provided convincing evidence (55). Figure 10.6

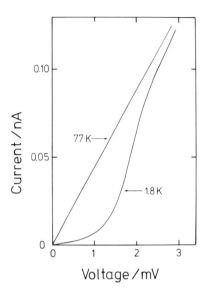

Figure 10.6. Band gap in superconducting electrode. Current–voltage characteristic of a Al–C16–Pb junction at 77 and 1.8 K (C16 = palmitic acid). The nonlinear behavior at 1.8 K is due to tunneling from the aluminum into the superconducting lead electrode.

shows the current–voltage characteristic for a junction with a palmitic acid (C16) monolayer measured at 77 and 1.8 K for a junction area of 2 mm². The characteristic measured at 77 K is linear in the range of applied voltage, as expected from the tunneling theory. The curve obtained at 1.8 K exhibits a very small current up to about 1 mV and approaches the linear 77-K dependence for larger voltages. The size of the energy gap in the density of states for the superconducting lead electrode is obtained from these results. The voltage at which the slope of the 1.8-K characteristic is equal to that measured at 77 K is 1.3 mV, and the band gap of lead is therefore 2.6 meV, in agreement with values determined with Al–Al₂O₃–Pb junctions (56).

   These observations demonstrate the possibility to form and deposit monolayers on substrates with defect-free areas of macroscopic size. The monolayers on the substrate have been exposed to a high vacuum and have been coated with a top electrode and then cooled down to very low temperatures for these measurements. Yet the monolayers seem not to have suffered enough that noticeable defects were introduced.

10.2.3. Electron Conduction in Metal–Monolayer–Semiconductor Junctions

Single monolayers of different materials have been deposited on various semiconductors, and MIS (metal-insulator-semiconductor) junctions have been prepared (57–61). The current–voltage (i–U) characteristic found with one monolayer of cadmium stearate between an n-GaP base and a gold top electrode (61) showed an exponential increase of the current with increasing voltage. This behavior agrees with the simple model of electron tunneling in the high-field region (46) in MIM structures. The electrode material influences strongly the i–U characteristic of the junctions (58).

The phenomena described in this section demonstrate that a single monolayer sandwiched between two electrodes plays the dominant role in electron transport across the junction. The large differences between the results obtained by different groups indicate the stringent conditions required for sample preparation. On the other hand, it is amazing that such delicate structures can be manipulated in this way. Therefore, the use of monomolecular layers in electronic devices may be possible.

10.4. ELECTRON CONDUCTION IN MULTILAYER SYSTEMS

Multilayer systems are built on solid substrates by repeated dipping of the solid into the trough across the monolayer-covered water surface. Such systems are an example for the concept of man-made periodical structures, "superlattices," with anomalous charge transport phenomena if the period is shorter than the mean free path of the electrons (62). Particular electrical and optical effects have been observed in doping superlattices (63). In the multilayer systems considered here the electron transport properties are related to the behavior of single monolayers between metal electrodes. The model of electron hopping between the interfaces of the system has been applied to rationalize the results of dc and ac conduction measurements in multilayer systems.

Electron hopping across bilayer subunits via localized states located at every other interface was discussed for systems built of stearic acid (64–66). The conductivity of the systems, however, was an order of magnitude higher than that observed in systems of cadmium arachidate, and the former results might have been determined by defects of the multilayer system. Clear evidence for electron hopping across single monolayers instead of bilayer subunits in the multilayer systems was obtained with properly built systems (67–75).

The fundamental assumptions of the model (2) are (i) the presence of localized states at the interfaces, disregarding the different chemical composition of hydrophilic and hydrophobic interfaces in the structures; (ii) charge transport is due to electron hopping via these states; and (iii) each hopping is characterized by an individual rate determined by the separation of the interfaces involved and the electron energy.

The hopping rate in multilayer systems can be derived by starting from an expression of the type obtained for impurity conduction in n-type semiconductors (76),

$$k = k_0(2\alpha R)^{3/2} \exp\left(-2\alpha R - \frac{\Delta}{k_B T}\right) \qquad (10.3)$$

In this equation, k_0 is a frequency factor on the order of 10^{12}–10^{13} s^{-1}; α is the damping constant of the electron wave function in the insulating material, $\alpha = (1/\hbar)\sqrt{(2m\varphi)}$, where $\hbar = h/2\pi$; m is the mass of the electron and φ the barrier height; R is the distance between the two hopping sites involved; Δ is the hopping energy; and k_BT is the thermal energy. The jump distance R is not identical with the separation d of the interfaces since the electron may hop into a suitable state within a domain of radius r of the next interface. The jump distance is

$$R = \sqrt{d^2 + r^2} \simeq d + \frac{r^2}{2d} \tag{10.4}$$

The most probable rate is obtained from a compromise between the smallest energy of hopping and the shortest hopping distance. According to the variable-range scheme of Mott (77), the domain should contain only one favorable state. Therefore,

$$\pi r^2 \Delta N' = 1 \tag{10.5}$$

Here N' is the two-dimensional density of states around the Fermi level. The most probable radius r_m of the domain is obtained by maximizing the exponent in Eq. (10.3). After substitution of R and Δ according to Eqs. (10.4) and (10.5), the relation

$$r_m = \left(\frac{d}{\alpha \pi N' k_B T}\right)^{1/4} \tag{10.6}$$

is obtained. With r_m and the most probable hopping energy calculated with r_m according to Eq. (10.5), the hopping rate is

$$k_m = k_0 (2\alpha d)^{3/2} \exp\left[-2\alpha d - \left(\frac{4\alpha}{\pi N' d k_B T}\right)^{1/2}\right] \tag{10.7}$$

This relation shows the exponential dependence of the hopping rate on the separation of the two interfaces since the preexponential factor plays a minor role in the range accessible by the LB technique, and the second term in the exponent is only a small correction term in contrast to the case of hopping in disordered systems. This type of distance dependence has been verified by measuring the ac conductivity in a large frequency range. One possibility to enhance the number of electrons at a well-defined energetic level is to incorporate dye molecules in the system and excite the electrons by light absorbed by the dye. Then the photoinduced increment in conductance G shows a particular dependence on the frequency, indicating resonances at different frequencies f_c depending on the structure of the layer systems (68). When the hopping rate k_m is identified with $2f_c$, the product $f_c d^{-3/2}$ decreases exponentially with increasing distance d between the interfaces, according to Eq. (10.7) (68). From the slope the damping constant is $3 \text{ nm}^{-1} < \alpha < 5 \text{ nm}^{-1}$, in agreement with $\alpha = 4.3 \text{ nm}^{-1}$ obtained from high-field measurements (69).

When there is no dye incorporated in the multilayer, the dependence of the ac conductance G in the limit of very small frequencies, $G(0)$, on the monolayer thickness d is given by (70)

$$\sigma = e^2 N' k_0 (2\alpha)^{3/2} d^{5/2} \exp\left[-2\alpha d - \left(\frac{4\alpha}{\pi N' d k_B T}\right)^{1/2}\right] \tag{10.8}$$

The product $\sigma e d^{-5/2}$ decreases exponentially with increasing thickness d (69). A good fit to the experimental results was obtained with the parameters $N' = 10^{15}$ $\text{cm}^{-2} \text{eV}^{-1}$, $2\alpha = 12 \text{ nm}^{-1}$, and $k_0 = 10^{12} \text{ s}^{-1}$. The value of α in this case must be larger than the value found with electron hopping from excited dye molecules given above since here the electrons hop from a lower energetic level.

This method of measuring the ac conductance in LB multilayers permits the study of mixed monolayers of fatty acids. Do fatty acids of different chain length, like C16 (palmitic acid) and C20 (arachidic acid), form mixed monolayers or does separation into two phases occur? In the case of a real mixture the apparent chain length or the average thickness of the monolayers should increase gradually from the value for C16 with increasing fraction of C20 in the mixed monolayer. The conductance $G(0)$ should consequently drop exponentially. In the case of phase separation the hopping occurs mainly via the C16 domains even with a considerable fraction of C20. The conductance should decrease initially only slightly but, with increasing fraction of C20, would finally reach the C20 conductance. The corresponding functions are plotted in Figure 10.7 [full line (real mixture) and

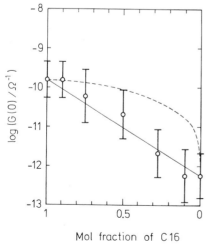

Figure 10.7. Homogeneous mixing of fatty acids with different chain lengths. Logarithm of ac conductance in the limit of very small frequencies, log $G(0)$, plotted against molar fraction of palmitic acid (C16) in mixed monolayers of C16 and C20 (arachidic acid) for junctions of seven monolayers between aluminum electrodes. Experimental results (circles with error bars) and lines according to a model of homogeneous mixing (solid line) and of phase separation in the monolayers (dotted line).

dotted line (phase separation), respectively]. Experimental values are represented by the circles with error bars (2, 71). The two different fatty acids form mixed monolayers on the aqueous subphase containing cadmium chloride, which causes the formation of the cadmium salts.

The temperature dependence expected according to Eq. (10.8) is another possibility to test the model of electron hopping between interfaces in multilayer systems. With regard to the temperature dependence, Eq. (10.8) can be written as

$$\sigma = \sigma_0 \exp\left(-cT^{-1/2}\right) \qquad (10.9)$$

where σ_0 and c are constants that can easily be derived from Eq. (10.8). The conductivity should decrease exponentially with $1/T^{1/2}$. This behavior has been found with multilayer systems of cadmium stearate (C18) between aluminum electrodes, as shown in Figure 10.8 (72). The circles represent the experimental values, and the line is a least-squares fit to a linear dependence in this plot.

The values of the conductance in the limit of very small frequencies, $G(0)$, may vary with individual samples of the same structure according to different contribution of conduction through defects in the multilayer assembly. Therefore, histograms have been plotted in order to avoid misinterpretation based on the singular cases (2). In a histogram the fraction of all measured samples with a conductance $G(0)$ in a specified range is plotted against the logarithm of $G(0)$. Figure 10.9 shows two such histograms obtained for different structures (2, 74). This type of plot is particularly suited to demonstrate the difference between homogeneous and heterogeneous structures. The histogram represented by the dotted line in Figure 10.9 was obtained for a system with seven mixed monolayers of C16 and C20, molar ratio C16–C20 = 9:1 at 77 K. It has a single peak in the interval $-10 \leq \log G(0) \leq -9.5$. The other histogram (full line), was found for systems with one monolayer of C20 sandwiched between six monolayers of C16 (heterogeneous

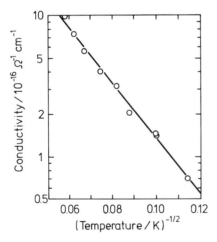

Figure 10.8. Temperature dependence of conductivity of LB multilayer systems; logarithm of conductivity plotted against $1/T^{1/2}$. Experimental results (circles) and straight line (least-squares fit) according to Eq. (10.9).

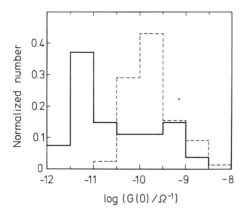

Figure 10.9. Discrimination between homogeneous and heterogeneous multilayer systems. Histograms obtained by plotting the fraction of measured samples with conductance $G(0)$ within fixed limits against logarithm of $G(0)$. Homogeneous system (dotted line): seven mixed monolayers of C16 and C20, molar ratio C16–C20 = 9:1, and heterogeneous system (full line): one monolayer of C20 sandwiched between six monolayers of C16. The two peaks of the heterogeneous system are attributed to hopping across the C20 ($-11.5 < \log G(0) < -11$) monolayer and to hopping across the C16 monolayers, respectively.

system). This histogram is quite different from that of the homogeneous system, although the average composition of the two systems is very similar. Two peaks are seen in the case of the heterogeneous structure, and the peak at the smaller conductance ($-11.5 \leq \log G(0) \leq -11$) is attributed to the conductance of the C20 layer, whereas the other peak is due to the conductance of the C16 layers (74).

10.5. CHARGE TRANSPORT IN MONOLAYERS

Electron transfer normal to the layer plane has been discussed in the former sections. The lateral transport of electrons or holes in monolayer systems influences to a large degree dark conduction phenomena. In the case of photoconduction, however, rapid recombination of charge carriers limits the intrainterface carrier mobility (73).

Vacuum-deposited layers of phthalocyanines are sensitive to the presence of oxides of nitrogen (78). The conductivity in such layers increases, which has been used for the construction of gas detectors (79, 80). The increased conductivity is attributed to the injection of holes by the adsorbed gas molecules.

Lateral charge transport was observed in monolayers of a copper phthalocyanine derivative after admission of NO_2, and therefore, such layers may also be used for gas detection (81). Molecules with short hydrocarbon chains have been used for monolayer preparation to reduce the thickness of the insulating layer between the π-electron systems (82–85). Since layers of various phthalocyanines show lateral conduction after hole injection, monolayers of the derivative IPC (see structure in

Table 10.1) have been formed at the air–water interface and transferred to solid substrates. Deposition of the IPC monolayers occurred on withdrawal only (Z deposition), and it has been concluded from measurements of the thickness of transferred layers that there was some degree of molecular realignment during deposition (81).

A rapid increase in the dark current across junctions with IPC monolayers between aluminum electrodes (the top electrode was covered with a gold layer) on admittance of NO_2 has been found, and the saturation current was proportional to the gas concentration. The monolayer assembly has the advantage of relatively fast recovery after removal of the gas. This is due to the thinness of the layer systems used as compared to the evaporated phthalocyanine layers.

Lateral charge transport may occur also in monolayers organized in particular associates, the J aggregates (86, 87), after injection of electrons or holes. Evidence for such processes has been found in the investigation of photoinduced electron transfer, which is discussed in Section 10.7.

10.6. PHOTOCONDUCTION IN MONOLAYER ORGANIZATES

The incorporation of dye molecules in monolayer organizates provides the possibility to create electrons at higher energy levels and holes via excitation of the dye molecules. The conductivity of the systems is therefore enhanced by illumination. One goal of investigations in this field is the imitation of biological systems in converting light energy to chemical energy, that is, artificial photosynthesis (88). This process typically occurs at interfaces without involving electrodes. Conversion of solar energy to electrical energy requires the formation of appropriate molecularly organized systems in contact with electrodes. The photoconduction and generation of photovoltages are the main topics of such studies.

10.6.1. Photoconduction in Multilayer Systems

Some typical properties of multilayer systems, that is, assemblies built with monolayers of identical chemical composition, are demonstrated with mixed monolayers of the merocyanine dye MSC (*N*-octadecyl-benzthiazol-nitrotetramethin-merocyanine, see Table 10.1) and C20, molar ratio MSC–C20 = 1 : 5 (69) between aluminum electrodes. A photocurrent was observed on illumination of the sample with light absorbed by the dye, and the action spectrum followed the absorption spectrum. In a Fowler plot, that is, the square root of the photoresponse against photon energy, a linear relationship was observed in one-monolayer junctions with a bias voltage of 1 V. This indicates that photoinjection of charge carriers from the electrode into the monolayer is the dominant process. The carriers must be electrons since the slope of the straight line was larger when the illuminated electrode was negative than when it was positive. The barrier height was calculated from the intercept with the abscissa, and a small asymmetry of the junction was found that may be a consequence of the asymmetrical architecture of the systems.

Several mechanisms can be responsible for the observation of a photocurrent under the high-field conditions (electric field strength $E > 10^5$ V/cm) in these experiments. Quantum mechanical hopping from the excited dye to acceptor sites in analogy to the process discussed in Section 10.3 should depend on the applied field in this range since the energy barrier is lowered by the field E. The field dependence of the photocurrent δi should be

$$\delta i \propto \exp\left(\frac{\alpha Re\beta E^{1/2}}{\varphi_B}\right) \qquad (10.10)$$

where R is the mean jump distance; α is the damping constant of the electron wave function; φ_B is the effective barrier height, which is comparable to the ionization energy at zero field, φ_0; and $\beta^2 = (e/\pi D\epsilon_0)$ with the dielectric constant D of the material and the permittivity of the vacuum, ϵ_0.

Alternatively, the electron could also be transferred from the excited donor to the acceptor site via a thermionic emission when the barrier is lowered by the applied field [Poole–Frenkel effect (89)]. This barrier lowering (90–92) influences the photocurrent according to

$$\delta i \propto \exp\left(-\frac{\varphi_0 - e\beta E^{1/2}}{2k_B T}\right) \qquad (10.11)$$

Both mechanisms result in a linear dependence of the logarithm of the photocurrent on the square root of the applied voltage. The thermionic process should be strongly temperature dependent and might be dominant at higher temperatures, whereas the hopping process is not very sensitive to variations of T and could be observed at low temperatures. Therefore, after verification of the expected voltage dependence, determination of the temperature dependence allows one to discriminate between the two mechanisms discussed here.

As shown in Figure 10.10, the photocurrent increases exponentially with the square root of the applied voltage (69). The open circles correspond to the illuminated electrode being negatively biased, full circles to it being positively biased. The slopes of the straight lines in this plot are temperature dependent. A transition between the low-temperature mode attributed to electron hopping and the high-temperature mode expected for a thermionic emission was observed, although the transition is not complete at 333 K.

The field dependence of the photocurrent also occurs with mixed monolayers of a different merocyanine dye and arachidic acid (93). It is particularly interesting that dye aggregates of the Scheibe type (J aggregates) (86, 87), which are present in these systems according to the measured absorption spectrum, do not give rise to a corresponding photocurrent. The lifetime of the excited state should be much smaller than that of the monomeric form of the dye according to a model of coherent excitons for such excited aggregates (25), in agreement with experimental

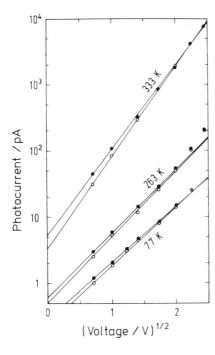

Figure 10.10. Photoconduction in multilayer systems containing the merocyanine dye MSC. Logarithm of photocurrent plotted against square root of the applied voltage for different temperatures. System: 15 monolayers of MSC and C20, molar ratio MSC–C20 = 1:5, between aluminum electrodes. Open circles: base electrode negative; full circles: base electrode positive; illumination through the semitransparent base electrode. The results are interpreted in terms of quantum mechanical electron hopping.

results. This could be a reason for the small efficiency of carrier generation of excited J aggregates.

Lateral charge transport has been observed in LB multilayer systems containing an amphiphilic merocyanine dye in a matrix of C20, molar ratio dye–C20 = 1:2 (94). The lateral photoconductivity $\sigma_\parallel$ was about 10^3–10^4 times larger than the photoconductivity $\sigma_\perp$ in the normal direction of the multilayer system. The action spectrum of $\sigma_\perp$ is higher in the short-wavelength range than that of $\sigma_\parallel$. This may be due to a larger component of carrier injection from the electrodes of the sandwich cell in the former case. The pronounced band at the long-wavelength side of the absorption spectrum that may be attributed to J aggregates is missing in the action spectra of photoconductivity. A contribution of the J aggregates to lateral photoconductivity has been found in subsequent investigations (95–97). This contribution was more than a decade larger than the lateral photoconductivity observed in systems without or with incomplete aggregation. It should be kept in mind, however, that mixed monolayers with J aggregates are microheterogeneous systems (16), and the microscopic structure of such monolayers depends strongly on the details of monolayer composition, formation, and transfer.

10.6.2. Photoconduction in Complex Monolayer Organizates

The LB multilayer systems are characterized by the repeated deposition of one type of monolayer. The structure of such systems is therefore simple, and the results of conduction and photoconduction measurements may be interpreted in terms of repetition of conductive interfaces and insulating layers. The much more sophisticated monolayer organizates are obtained by deposition of various different monolayers according to an intended function of the entire system. The construction of such structures requires much more knowledge of the individual modular monolayers and adequate methods to verify the structure. The complex monolayer organizates provide possibilities to test concepts for molecular interactions in photosynthesis or in artificial systems for solar energy conversion.

One problem in using LB films for studies of charge transport processes is the small conductivity across the insulating layer formed by the hydrocarbon chains that separates the interfaces. The thickness of this layer can be varied in a limited range only since the monolayer-forming molecules should be insoluble in the underlying subphase. This requires a certain size of the hydrophobic tail depending on the nature of the hydrophilic head group. Since the conductivity depends not only on the thickness of the barrier but also on the energetic height [see Eq. (10.2), φ, representing the barrier height], the conductivity across an insulating monolayer may be enhanced by lowering the effective barrier height. The consequences of incorporating conducting π-electron systems as molecular conductors oriented parallel to the hydrocarbon chains into the insulating layer have been discussed in a simple theoretical model of photosynthesis based on electron tunneling (98).

A modification of the barrier has indeed been achieved by using a rodlike molecule, quinquethienyl, with conjugated π-electrons. This is incorporated parallel to the hydrocarbon chains according to measurements of the light absorption with plane-polarized light (99, 100). The action spectrum of the photocurrent observed in a monolayer organizate containing the molecular conductor between metal electrodes agreed with the absorption spectrum of the quinquethienyl. The voltage dependence and the dependence of the photocurrent on the position of the modified monolayer in the organizate were found as expected from the model of an electron-conducting molecule embedded in the insulating layer.

In this investigation the π-electron system was photoexcited and acted both as electron donor and conducting molecular wire. These functions were put into different molecules, and the behavior of systems containing these components between metal electrodes was investigated in detail (36). The cyanine dye TCC (see Table 10.1) was used as donor and the azo dye AZS (see Table 10.1) as the conducting π-electron system. As has been shown in Section 10.2.1, the azo dye π-electron system is oriented parallel to the hydrocarbon chains of the fatty acid matrix monolayer in which it is embedded.

The energy diagram in Figure 10.11 (top) describes a system without the molecular conductor when a voltage is applied to the junction. Electron hopping from the excited donor D^* to a state at the next interface (shaded area) is the determining step since the next hop brings the electron to the bottom of the conduction band. The barrier of the insulating layer between the excited donor and the next interface is lowered by applying the voltage U as discussed in Section 10.4. Consequently, a linear dependence of the logarithm of the photocurrent on the square root of the voltage [see Eq. (10.11)] is observed (36).

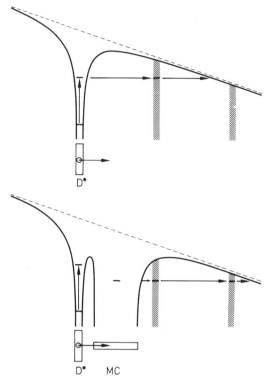

Figure 10.11. Energy scheme for monolayer organizates with excited donor D in absence (top) and in presence (bottom) of a molecular conductor MC. The voltage dependence of the photocurrent is different in the presence of MC from that observed in the absence of MC.

The situation is different when a molecule MC (molecular conductor) is incorporated in the insulating layer. This molecule provides an unoccupied orbital at lower energy than the excited state of the donor (see Figure 10.11, bottom). The electron is easily transferred from D^* to MC either by tunneling across the thin and low barrier or by thermal activation over this barrier and reaches the first interface. The barrier between this interface and the second drops nearly linearly with distance. This causes a linear voltage dependence of the logarithm of the photocurrent (36). Since the barrier between the first and second interface determines the behavior of the photocurrent in the arrangement (according to Figure 10.11, bottom), a variation of the barrier thickness should change the slope of the linear dependence in a plot of logarithm of photocurrent against voltage. When the bias is reversed, this layer has no influence on the photocurrent since the electrons move in the other direction, and no such dependence is expected. All these consequences of the model have been verified experimentally, and the function of a molecular conductor or molecular wire has been well established using elaborate monolayer techniques with all possible checks and tests (36).

Systems combining a photoexcited donor with a primary electron acceptor that is separated from the donor either by an insulating barrier or a modified barrier containing a molecular conductor are of particular interest. Such systems correspond to the structure in which the primary processes in photosynthesis occur (98). Monolayer organizates of this type between metal electrodes have been investigated with the excited cyanine dye ICC as donor and the acceptor SV (N,N'-dioctadecyl-4,4'-bipyridinium perchlorate; see Table 10.1) as primary acceptor of the excited electron (101). The azo dye AZS has been used as molecular conductor as in the studies just described.

In the absence of both the electron acceptor and the molecular conductor, the logarithm of the photocurrent depends linearly on the square root of the applied voltage as observed with systems containing the cyanine dye TCC described above. No asymmetry of the junction was found (101). Incorporation of the electron acceptor increases the photocurrent by an order of magnitude, and the logarithm of the photocurrent depends now linearly on the applied voltage. This dramatic effect of the primary electron acceptor is due to relaxation processes following photoreduction of the acceptor A.

In the absence of A, the electron hops to the next interface. From there it tunnels across the next insulating barrier into the metal (anode) or returns to the oxidized donor molecule. In the presence of A, relaxation of the electron to a state of lower energy when trapped by the acceptor A decreases strongly the probability of return to the oxidized donor molecule. Further, an asymmetry is introduced in the junction since the chance for the electron to hop to the interface with the electron acceptor is larger than hopping to the next interface in the other direction without the acceptor. This directionality results in a larger photocurrent when the bias shifts the electron in this preferential direction than in the case of reversed bias.

This effect is shown in Figure 10.12 (lower two sets of experimental data and schematical structure of the junction). The monolayer of the donor [indocarbocyanine (ICC), arachidic acid (C20), methylarachidate (MA), and hexadecane (HD); molar ratios ICC–C20–MA–HD = 1:5:5:10] and of the acceptor [stearylviologen (SV), C20, and MA, molar ratios SV–C20–MA = 1:9:1] are separated by a fatty acid monolayer. The logarithm of the photocurrent density is plotted versus the applied voltage. Full circles represent results obtained with base electrode (left electrode) negative, open circles for reversed bias. The asymmetry of the junction response is asymmetric due to vectorial photoinduced electron transfer form D^* to A.

The incorporation of the MC in the monolayer-separating donor and acceptor (AZS and C20, molar ratio AZS–C20 = 1:10) increases the yield of the primary electron transfer step from D^* to A during the lifetime of the excited state. The photocurrent density of such systems is larger by an order of magnitude, and the asymmetry of the junction is also increased (see Figure 10.12, upper system).

The molecular interactions in such complex monolayer organizates give rise to a photovoltage across appropriate junctions. With systems of the structures given in Figure 10.12, where the aluminum top electrode (right electrode) was replaced by a vacuum-deposited barium electrode, small photovoltages have been observed

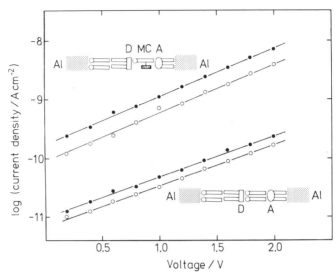

Figure 10.12. Photoconduction in complex monolayer organizates (structures indicated schematically) between aluminum electrodes (top electrode on the right side). The MC is present in the upper system and absent in the lower system in the monolayer separating the donor D and the primary acceptor A. Full circles refer to negative bias of the base electrode (left electrode), open circles to opposite bias. The asymmetry of the junction is caused by vectorial photoinduced electron transfer from D to A. The photocurrent density and the asymmetry are increased by incorporation of the molecular conductor. Monolayer composition: D, ICC–C20–MA–HD = $1:5:5:10$; A, SV–C20–MA = $1:9:1$; MC, AZS–C20 = $1:10$; measurements at 298 K.

(101). The rise and decay of the photovoltage when the light is turned on and off is shown in Figure 10.13. Curve 1 (solid line) refers to the system without the molecular conductor and curve 2 (dotted line) to the system with the molecular conductor. The incorporation of the molecular conductor increases the photovoltage and changes the kinetics. The action spectrum of the photovoltage follows the absorption spectrum of the donor, and the photovoltage is proportional to the light intensity (101).

Monolayers of chlorophyll a (Chl-a) have been used as donors in attempts to approach the function of natural photosynthetic units (102, 103) in organizates between metal electrodes. Long-chain substituted quinones were used as primary electron acceptors, and the function of the molecular conductor in the systems described above has been attributed to the unsaturated chains of acceptors such as ubiquinone and plastoquinone (102). The acceptor with saturated chains had no effect on photoconduction and photovoltage, whereas the acceptors with unsaturated chains increased both photoconduction and photovoltage by an order of magnitude. This interesting and stimulating study, however, lacks some pieces of information. The action spectrum of the photocurrent or photovoltage has not been reported (the illumination light was filtered through a red filter). More importantly,

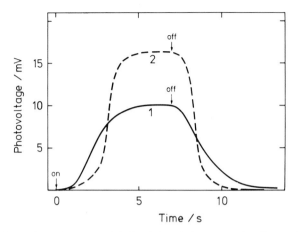

Figure 10.13. Photovoltage due to vectorial photoinduced electron transfer. Photovoltage against time for systems of the structure given schematically in Figure 10.12 with barium instead of aluminum as top electrode. Curve 1 (full line): system without MC. Curve 2 (dotted line): system with MC. The photovoltage on excitation of the donor is enhanced by incorporation of the MC. Monolayer composition as in Figure 10.12.

no direct evidence was given for the orientation of the unsaturated chains normal to the layer plane required for a function as molecular conductor between interfaces. The evidence provided by surface pressure–area isotherms is insufficient. Further, the acceptors with unsaturated chains were incorporated in a matrix of stearic acid in order to form stable monolayers. No detailed information concerning the molecular mixing of the components was given. In the case of phase separation local disorder would cause an increase in photoconduction.

Photoconduction and photovoltaic effects of multilayers of Chl-a between an aluminum base electrode and a gold top electrode or a suitable redox solution electrode were investigated (103). The square root dependence of the photocurrent on light intensity observed in continuous high-intensity illumination of the junctions was attributed to second-order recombination and charge-trapping processes.

Photovoltaic processes have been observed by using two different dyes in a multilayer system between metal electrodes (104). Several monolayers containing a long-chain derivative of crystal violet were deposited on top of the base aluminum electrode. Dyes of this triphenylmethane type have been characterized as n-type photoconductors (105). Then, several mixed monolayers of a surface-active merocyanine dye and arachidic acid were deposited, and finally the silver top electrode evaporated. The reciprocal capacitance of such systems varied linearly with the number of monolayers. Analogs of the merocyanine dye can be deposited as thin photoconductive films that have been characterized as p-type semiconductors. The open circuit photovoltage of the multilayer structures was up to 0.7 V, and the short-circuit photocurrent density was 1.6×10^{-10} A/cm^2 under illumination with an incandescent lamp (2 mW/cm^2). The observations were interpreted in a model of p-type (merocyanine) and n-type (crystal violet) regions and ohmic contacts to the metal electrodes.

Interactions of monolayers and monolayer organizates containing various dyes with organic crystals (106, 107) and inorganic seminconductors (108–113) have been investigated in order to elucidate the mechanism of spectral sensitization. The organic materials and inorganic semiconductors show an effect such as photoconduction by illumination with light only that is absorbed by the material. The material can be sensitized (made sensitive) to light that is not absorbed by the material itself by coating the surface with appropriate dyes (114). An example of this process in daily life is the spectral sensitization of photographic materials. The process of spectral sensitization may become very important also in devices for solar energy conversion using coated semiconductor films of materials like TiO_2 and SnO_2 (115).

An important parameter in determining charge transport in complex monolayer organizates is the thickness and energy height of the barriers for electron conduction constituted by the saturated hydrocarbon chains. Two possibilities exist to overcome this limitation: (i) the use of short-chain amphiphilic molecules and (ii) the concept of molecular conductor. The concept of using optimized molecular conductors is more attractive with respect to systematic investigations. Lightly substituted components may become more successful in practical systems.

10.7. PHOTOINDUCED CHARGE TRANSFER PROCESSES IN MONOLAYER ORGANIZATES

So far the behavior of systems between metal electrodes has been described. The processes discussed in the last section included the primary photoinduced electron transfer from an excited donor to an intermediate acceptor before transfer to the metal, thereby causing vectorial charge transfer. Such processes can be investigated in systems without metal electrodes by spectroscopic methods. The use of monolayer organizates offers many advantages for this purpose as compared to similar studies in homogeneous or microheterogeneous systems. The components like donor, photocatalyst, acceptor, or molecular conductor can be arranged according to the particular aim of the study, and the molecular interactions are controlled by systematic variation of the relevant parameters.

Many amphiphilic compounds consist of a hydrophilic chromophore or electron-accepting part and one or two long hydrocarbon chains in order to incorporate these molecules into an appropriate monolayer matrix. The interacting parts are then located at the hydrophilic interface. Typical examples are cyanine dyes like OC (N,N'-dioctadecyloxacyanine perchlorate), TC (N,N-dioctadecylthiacyanine perchlorate), and TCC (see Table 10.1), which can be used as electron donors, and the long-chain derivative of viologen, SV (see Table 10.1) as electron acceptor. Donor and acceptor can be incorporated in adjacent monolayers such that they are at the same interface or, alternatively, that they are separated by the hydrocarbon portions of the monolayers. In the former case the fluorescence of the excited donor is quenched, whereas in the latter case no fluorescence quenching occurs. In a third arrangement the interacting parts are separated by one monolayer only. Fluorescence quenching was observed in this situation depending strongly on the thickness of the single insulating monolayer.

10.7.1. Donor and Acceptor at the Same Interface

The fluorescence of the excited donor monolayer is quenched by electron transfer to an acceptor molecule when the acceptor is present in the adjacent monolayer at the same interface. The relative fluorescence intensity I/I_0 of the donor (I in the presence, I_0 in the absence of the acceptor) depends on the two-dimensional acceptor density n_A according to

$$\frac{I}{I_0} = \exp\left(-\pi R_{DA}^2\, n_A\right) \tag{10.12}$$

where R_{DA} is the critical radius for electron transfer. This relation (116) is obtained by applying the model of a hard disk in analogy to the well-known model of hard sphere in solution (117). The basic assumption here is that electron transfer occurs with probability 1 when the acceptor molecule is within the disk of radius R_{DA} around the excited donor; otherwise it does not. A dependence according to Eq. (10.12) was observed in systems with the cyanine dye OC (see structure in Table 10.1) in highly diluted mixed monolayers with C20 and, MA, molar ratios OC–C20–MA $= 1:900:100$, and the electron acceptor SV in mixed monolayers with C20 and MA. The critical radius $R_{DA} = 1.32$ nm was obtained from the plot of the logarithm of I/I_0 versus the acceptor density.

At higher two-dimensional densities of the donor Förster-type energy transfer from an excited donor molecule to another donor molecule becomes possible. This mechanism is based on the dipole–dipole interaction and differs from the triplet energy transfer via the exchange mechanism. Förster energy transfer has been studied systematically in monolayer organizates using different dyes as energy donor and acceptor, respectively (9, 118). Donor–donor energy transfer or incoherent exciton hopping results in delocalization of the excitation energy. The excited state, which has no acceptor within the critical radius, can migrate and probe the environment for an acceptor. Consequently, the yield of the primary electron transfer step, which is characterized by fluorescence quenching, is enhanced at constant acceptor density (116, 118).

As an example, the relative donor fluorescence intensity I/I_0 is plotted in Figure 10.14 versus the donor density; experimental results are represented by bars, the length corresponding to the experimental error. The donor dye OC was incorporated in a mixed matrix of C20 and MA of molar ratio C20–MA $= 9:1$. The acceptor monolayer had the composition SV–C20–MA $= 1:27:3$. These monolayers were spread from chloroform solutions after addition of HD to the

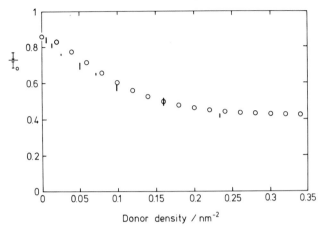

Figure 10.14. Enhancement of electron transfer by energy delocalization. Relative donor fluorescence intensity I/I_0 against donor density at constant acceptor density; donor and acceptor located at the same interface. Bars: experimental values; circles: calculated according to a model of incoherent exciton hopping in the donor monolayer preceding the electron transfer step. Donor: cyanine dye OC in a mixed matrix of C20 and MA, molar ratio C20–MA = 9:1; acceptor monolayer: viologen SV, C20, and MA, molar ratios SV–C20–MA = 1:27:3.

same concentration as the matrix compounds in order to improve the homogeneity of the mixed monolayers.

The influence of exciton hopping via Förster energy transfer on fluorescence quenching by electron transfer can be described with two parameters only, the critical radius for electron transfer and the Förster radius R_D for energy transfer. Here R_D can be determined from the spectroscopic properties of the donor mono-layer [overlap integral of donor fluorescence band with the absorption band of the donor monolayer, probability of fluorescence emission for a donor molecule in the emitting state (9)], and R_{DA} is obtained from the dependence of I/I_0 of very dilute donor monolayers on acceptor density according to Eq. (10.12). Therefore, no adjustable parameter is left. The model includes a limited number of hops during the lifetime of the excited state according to a Poisson distribution. The circles in Figure 10.14 represent the dependence obtained with the critical radius $R_{DA} = 1.32$ nm and the Förster radius $R_D = 2.7$ nm (116). The model describes the behavior quite satisfactorily, although there are deviations from the experimental results at small donor densities. This may be due to the fact that energy transfer from the excited donor to the nearest neighbor only has been considered.

The delocalization of the excited state is most efficient when the chromophores are organized in J aggregates. These systems have been described with the model of a coherent exciton (25). In systems with the same acceptor density in the adjacent monolayer, the fluorescence of these aggregate monolayers is much more efficiently quenched than the fluorescence of monomers of the same dye (119).

The apparent rate constant k_{DA} of the electron transfer as quenching process is related to the relative fluorescence intensity according to

$$\frac{I_0}{I} - 1 = \tau_0 k_{DA} n_A \tag{10.13}$$

where τ_0 is the lifetime of the excited state in the absence of quenching. As compared to Eq. (10.12), this relationship does not describe fluorescence quenching for very dilute donor monolayers correctly. However, it might be used to demonstrate the striking difference between aggregate and nonaggregate monolayers of the same dye. In Figure 10.15 the data obtained for a mixed donor monolayer of the composition OC–C20–MA = 1 : 900 : 100 and acceptor monolayers with SV in a mixed matrix of C20–MA = 9 : 1 are plotted according to Eq. (10.13) and labeled monomers. Data for a J aggregate monolayer of OC are plotted in the same range of acceptor densities and have been labeled aggregates. The aggregate

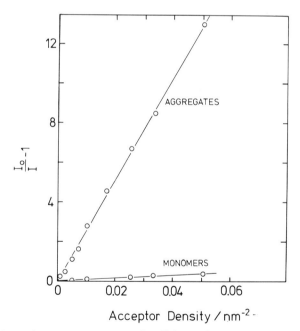

Figure 10.15. Comparison of the electron transfer efficiency of aggregate monolayers of the cyanine OC with that of nonaggregated monolayers. $I_0/I - 1$ plotted against the acceptor density; circles: experimental results. The slope of the straight line for aggregates is about 35 times that for monomers. Taking the different lifetimes of the excited state into account, the efficiency is more than 1000 times larger for aggregates than for monomers. Acceptor monolayers: viologen SV in a matrix of C20 and MA, molar ratio C20–MA = 9 : 1; donor monolayer: OC–HD = 1 : 1 (aggregates) and OC–C20–MA = 1 : 900 : 100 (monomers).

monolayer was formed by spreading a mixture of OC and HD, molar ratio OC–HD = 1 : 1, and equilibrating the monolayer at a surface pressure of 20 mN/m for 20 min before monolayer transfer.

According to the slopes of the two straight lines in Figure 10.15, the product $\tau_0 k_{DA}$ is larger by a factor of 34.5 for the aggregate monolayer than for the monolayer of monomeric dye OC. In order to evaluate the ratio of the apparent rate constants of electron transfer in both cases, the lifetime of the excited state has to be taken into account. The fluorescence decay function of a mixed monolayer of OC–C20–MA = 1 : 900 : 100 ("monomeric dye") cannot be fitted to a single exponential. The main component, however, has a lifetime of $\tau_0 = 1.6$ ns (116). On the other hand, the decay function of the aggregate monolayer has not yet been measured. Attempts to determine the lifetime using excitation pulses of 50 ps showed that the fluorescence lifetime of the OC aggregate monolayer is shorter than 50 ps (120). In conclusion, the apparent rate constant of photoinduced electron transfer is enhanced by a factor of more than 1000 with respect to isolated OC donor molecules when the excitation energy is delocalized upon organization of the donor molecules in J aggregates.

The quenching of the donor fluorescence is only indirect evidence for the occurrence of photoinduced electron transfer. Quenching may also be due to formation of exciplexes between donor and acceptor without formation of the reduced acceptor radical and the oxidized donor radical. The detection of the reduced acceptor radical is direct proof for the electron transfer. Since the amount of reduced acceptor is small in monolayer organizates, the spectroscopic identification of the viologen radical turned out to be quite difficult. In fact, the first evidence was obtained by investigating systems in which the fluorescence of a probe dye was quenched via Förster energy transfer upon formation of the acceptor radical (11). The absorption band of the viologen radical with maximum at 612 nm was directly measured later (121). The formation of the viologen radical has been followed by measuring the absorption at 400 nm ("blue band" of the radical), excluding oxygen to a large degree from the ambient, in studies related to the mechanism of supersensitization in monolayer organizates as model systems (26).

The radical observed in these measurements is the result of complicated processes that are not yet entirely understood. The transferred electron normally returns in a fast process into the semioccupied ground-state orbital of the oxidized donor molecule ("back transfer"). This has been demonstrated in solution and in microheterogeneous systems like micelles in transient measurements of the reduced acceptor concentration. Such transient measurements have not been successfully carried out with monolayer organizates so far. Back electron transfer is blocked when the oxidized donor is provided with an electron from another molecule (sacrificial donor). Consequently, the reduced acceptor radical is stabilized (persistent radical). Monolayer organizates may contain such sacrificial donors of unknown chemical nature. When oxygen is removed, which reoxidizes the reduced acceptor, the density of the persistent radical remains unchanged over 24 h. This has been shown by studying the electron spin resonance (ESR) signal of the reduced acceptor radical in monolayer organizates (122).

Stabilization of the reduced acceptor should be possible by delocalizing the positive charge on the oxidized donor from the vicinity of the reduced acceptor. Such a migration of the positive charge is similar to hole conduction in semiconductors. The J aggregates provide an adequate two-dimensional structure for such a phenomenon. An additional sacrificial donor in contact with the aggregate traps the positive hole and efficiently prevents recombination with the electron trapped on the reduced acceptor. Such a sequence has been proposed for the mechanism of supersensitization of the photographic process (123).

Fluorescence quenching and formation of the persistent acceptor radical were measured in systems with J aggregates of the cyanine dye QC (N-methyl-N'-dioctadecyl-2,2'-cyanine perchlorate; see Table 10.1 for the structure) (26). The fluorescence of the aggregate was quenched by several different electron donors like a para-phenylenediamine derivative or coumarin derivatives, and this quenching occurred in competition with quenching by the electron acceptor SV. This means that the fraction of aggregate fluorescence quenched by direct electron transfer to the acceptor was efficiently decreased by the addition of the sacrificial electron donor. However, the synergetic effect of electron donor and electron acceptor was found: In the presence of the additional donor, the rate of radical formation was increased, depending on the donor used, by up to a factor of 3.5 (26). This study has demonstrated that molecular organizates can be designed and built in order to optimize the desired reaction pathway.

The processes discussed in this section referred mainly to electron transfer from an excited donor to an acceptor. The more familiar process is electron transfer from a donor to an excited acceptor, like in systems of amines as donors and aromatic molecules as acceptors (124). Similar systems with appropriate cyanine dyes at the same interface in monolayer organizates have been investigated by measuring fluorescence quenching (125). A fundamental difference between processes involving an excited donor and those with an excited acceptor should be stressed since this has been widely disregarded: In the former case (donor excited) electron transfer takes place at an energy level corresponding to an excited state, whereas in the latter case (acceptor excited) the transfer occurs at the ground-state level. One major consequence of this difference is the strongly reduced tunneling distance for the electron when the acceptor is excited as compared to the situation with excited donor. This has been demonstrated experimentally (125).

10.7.2. Donor and Acceptor at Different Interfaces

Charge separation after photoinduced electron transfer can be stabilized by spatially separating the reduced and the oxidized species and by energetic relaxation. Therefore, part of the photon energy input has to be dissipated to achieve vectorial electron transfer and efficiently prevent the undesired back reaction. This concept was developed in a model of photosynthesis (98,126–129). Recently, complicated molecules including donor, photocatalyst, and acceptor parts that are connected by long hydrocarbon chains or rigid spacers have been synthesized to mimic photosynthesis. The electron transfer from the donor part to the acceptor part via the photocatalyst was investigated (130, 131). Again, monolayer organizates provide the advantage of much more systematic studies by arranging these components at interfaces in well-defined distances and geometries.

Donor and acceptor should be located at interfaces that are separated by only one monolayer as insulating barrier, and the thickness of this should be varied systematically. An amphiphilic pyrene was used as donor, which is located at the hydrophobic surface when incorporated in mixed monolayers, and the viologen SV as acceptor. The planes of donor and acceptor were separated by one fatty acid monolayer. In a plot of log $(I_0/I - 1)$ against the thickness d of the fatty acid layer, the experimental results were found to decrease linearly with d (11). This was interpreted according to Eq. (10.13) as an exponential decrease of the apparent rate constant of electron transfer k_{DA} with increasing distance between donor and acceptor. Such a behavior is typical for a tunneling process. Consequently, electron tunneling across a barrier of 2–2.5 nm constituted by the fatty acid monolayer during the lifetime of the excited pyrene was proposed.

The familiar cyanine dyes were used for a systematic variation of the donor (126) in the contact case. A similar study with donor and acceptor located at different interfaces required the construction of systems that involve contacts of hydrophilic groups with a hydrophobic surface. An example is shown schematically in Figure 10.2c, where the monolayer of the acceptor (next to the barium electrode) was deposited with the hydrophilic groups on top of the hydrophobic surface of the preceding monolayer (containing the molecular conductor in the scheme). Complete monolayer transfer of this type has been achieved (121). Although reorganization of such systems was observed (132, 133), these arrangements can be obtained without detectable defects under appropriate conditions (134).

The distance dependence of electron transfer across one insulating fatty acid monolayer was investigated with the cyanine dye TC (see structure in Table 10.1) as donor and the viologen SV as acceptor. Again, steady-state quenching of the donor fluorescence was analyzed according to Eq. (10.13), and an exponential decrease of the apparent rate constant k_{DA} with increasing barrier thickness was found (125, 135). More complex systems were investigated in order to rule out the possible influence of defects. By studying the action of an energy acceptor competing with the electron acceptor in the deactivation of the excited donor, undesired donor–acceptor contacts could be ruled out (136).

Another test for the absence of donor–acceptor contacts due to defects in the insulating fatty acid monolayer is based on the effect of delocalization of the excitation energy on the efficiency of electron transfer. When donor and acceptor are located at different interfaces, this delocalization provides no advantage for electron transfer since there is no donor molecule in a more favorable situation than the other donor molecules. All are facing the insulating barrier that determines the rate of electron transfer, and the relative fluorescence intensity I/I_0 of the donor was found to be independent of the donor density at constant acceptor density in these systems (137).

Further evidence for electron tunneling in such systems was obtained from measurements of the decay function of the donor fluorescence (138, 128) using the cyanine dye OC as donor and the viologen SV as acceptor. In analogy to Eq. (10.13), the rate of electron transfer k_{DA} is obtained from a plot of the relative lifetime τ/τ_0 according to

$$\frac{\tau_0}{\tau} - 1 = \tau_0 k_{DA} n_A \tag{10.14}$$

Again, an exponential decrease of the rate constant of electron transfer k_{DA} with increasing barrier thickness was found. A study of the temperature dependence of electron transfer in these systems showed that k_{DA} decreases exponentially with $1/T^{1/2}$. All these results are in favor of a tunneling mechanism of electron transfer across the insulating barrier of thickness of about 2 nm.

The current theories of electron transfer have to be reconsidered in order to rationalize the relations outlined above and to interpret the quantitative results. In the usual description (139) the rate of electron transfer is given by

$$k_{DA} = \frac{2\pi}{\hbar} \epsilon^2 S \tag{10.15}$$

where $\hbar = h/2\pi$, h is Planck's constant, ϵ is the perturbation energy of donor and acceptor, and S is the Franck–Condon factor.

The basic difference to the usual theories in the following treatment is based on the time t_{DA} required for electron transfer in the case of energetic match as compared to the duration of the energetic match, t_C. The energetic match, once created, is destroyed by the next thermal collision. The time t_{DA} is related to the perturbation energy ϵ. The time for one oscillation of the electron between donor and acceptor is $h/2\epsilon$; the time when it is equally probable to find the electron either on the donor or on the acceptor is $(1/4)(h/2\epsilon)$, and therefore (126).

$$t_{DA} = \frac{h}{8\epsilon} \tag{10.16}$$

The usual electron transfer theories are based on the assumption that $t_{DA} > t_C$, that is, the probability of electron transfer is small during the time between two thermal collisions, one creating energetic match and the next destroying it.

The situation is quite different, when t_{DA} is small compared to t_C. Then the rate of electron transfer is given by (25,127)

$$k_{DA} = \frac{1}{t_C} 2\epsilon S \tag{10.17}$$

In the monolayer organizates considered here, the π-electron systems are oriented with their planes parallel to the layer planes. Details of the calculation of ϵ and S are given elsewhere (129). For systems with the cyanine dye OC as donor and the viologen SV as acceptor, the lowest vibrational level of electron-accepting state lies about 1.1 eV below the lowest vibrational level of the excited state of OC. The Franck–Condon factor was calculated to $S = 0.28$ eV^{-1}.

The perturbation energy contains the distance-dependent tunneling term. The barrier height for electron tunneling was taken as $\varphi = 0.7$ eV. This corresponds to the difference between the electron affinity of fatty acid monolayers determined by tunneling experiments (see Section 10.3), $\kappa = 2.3$ eV, and the level of the excited state of OC [3.0 eV below the vacuum level (126)]. The damping constant $\alpha = 4.3$ nm^{-1}.

Electron transfer from the excited donor is possible to several acceptor molecules in the acceptor layer, depending on the acceptor density and the thickness of the insulating fatty acid layer. Further, it depends on the relative orientations of donor and acceptor molecules. Taking this into account (129), the rate constant $k_{DA} = 3 \times 10^9$ s^{-1} was obtained for a distance of 2 nm and $k_{DA} = 0.21 \times 10^9$ s^{-1} for a 2.7-nm layer separation. The fluorescence decay constant was in this case $k = 1.5 \times 10^9$ s^{-1}. These results of the theoretical treatment are in excellent agreement with the experimental results. It should be kept in mind that no adjustable parameters are involved in these calculations. The current theories of electron transfer cannot be applied to the experimental results. This analysis indicates that the unusual condition $\epsilon \gg h/(8t_C)$ can be realized in monolayer assemblies.

The insulating fatty acid barrier between donor and acceptor planes can be mod-

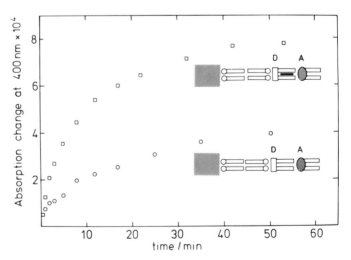

Figure 10.16. Enhancement of electron transfer across one monolayer of C20 by incorporation of the molecular conductor AZS. Absorption change at 400 nm in nitrogen atmosphere plotted against time of excitation of the donor monolayer; and schematic structures of the monolayer organizates. Circles: results in the absence of AZS, mixed monolayer of the cyanine TCC and C20, molar ratio TCC–C20 $= 1:10$; squares: results in the presence of AZS, monolayer composition TCC–AZS–C20 $= 1:1:9$; acceptor monolayer: viologen SV–C20–MA $= 1:9:1$.

ified as in the electrical measurements by incorporation of a molecular conductor (see Section 10.6). This should also facilitate the photoinduced electron transfer across the barrier and therefore increase the rate of radical formation. An example of this effect is shown in Figure 10.16, where the absorption change in the range of the blue band of the viologen radical formed by photoinduced electron transfer to the acceptor SV is plotted against time of excitation of the cyanine dye TTC used as donor (140). The structures of the layer systems are indicated schematically. It is clearly seen that the rate of radical formation is strongly increased in the presence of the molecular conductor AZS (upper system, squares) as compared to the system without AZS (lower system, circles).

In conclusion, it may be stated that monolayer organizates provide an excellent tool for the investigation of charge transport in ordered systems. Unexpected phenomena have been found, and the theories of electron and energy transfer have been reconsidered in order to rationalize the observed results. Aside from this more fundamental scientific effect of monolayer studies, a practical use of monolayer organizates may soon become feasible.

REFERENCES

1. V. K. Agarwal, *Electrocomp. Sci. Technol.* **2**(1), 75 (1975).

2. M. Sugi, T. Fukui, and S. Iizima, *Mol. Cryst. Liq. Cryst.* **50**, 183 (1979).

3. P. S. Vincett and G. G. Roberts, *Thin Solid Films* **68**, 135 (1980).

4. A. Pockels, *Nature* **43**, 437 (1891); for a historical review see S. D. Forrester and C. H. Giles, *Chem. Ind. (London)* **1979**, 469.

5. K. B. Blodgett and I. Langmuir, *Phys. Rev.* **51**, 964 (1937).

6. H. Kuhn, *Pure Appl. Chem.* **11**, 345 (1965).

7. H. Bücher, K. H. Drexhage, M. Fleck, H. Kuhn, D. Möbius, F. P. Schäfer, J. Sondermann, W. Sperling, P. Tillmann, and J. Wiegand, *Mol. Cryst.* **2**, 199 (1967).

8. L. Wilhelmy, *Ann. Phys.* **11**, 177 (1863).

9. H. Kuhn, D. Möbius, and H. Bücher, in *Physical Methods of Chemistry,* Vol. 1, Part IIIB, A. Weissberger and B. Rossiter (Eds.), Wiley, New York, 1972, p. 577.

10. I. Langmuir, *J. Am. Chem. Soc.* **39**, 1848 (1917).

11. D. Möbius, *Ber. Bunsenges. Phys. Chem.* **82**, 848 (1978).

12. A. H. M. Crone, A. F. M. Snik, J. A. Poulis, A. J. Kruger, and M. van den Tempel, *J. Coll. Interf. Sci.* **74**, 1 (1980).

13. C. H. Sohl, K. Miyano, and J. B. Ketterson, *Rev. Sci. Instrum.* **49**, 1464 (1978).

14. D. Möbius and H. Grüniger, in *Charge and Field Effects in Biosystems,* M. J. Allen and P. N. R. Usherwood (Eds.), Abacus Press, Tunbridge Wells, England, 1984 p. 265.

15. M. Suzuki, D. Möbius, and R. C. Ahuja, *Thin Solid Films,* **138**, 151 (1986).

16. D. Möbius, in *Langmuir-Blodgett Films,* G. G. Roberts (Ed.), Plenum Publishing, London, in press.

17. H. Heithier, H.-J. Galla, and H. Möhwald, *Z. Naturforsch.* **33**, 382 (1978).

18. D. Vaidyanathan, L. K. Patterson, D. Möbius, and H. Grüniger, *J. Chem. Phys.* **89**, 491 (1985).

19. M. Lösche and H. Möhwald, *Eur. Biophys. J.* **11**, 35 (1984).

20. H. Grüniger, D. Möbius, and H. Meyer, *J. Chem. Phys.* **79**, 3701 (1983).

21. D. Möbius and H. Grüniger, *Bioelectrochem. Bioenerg.* **12**, 375 (1984).

22. D. Möbius, M. Orrit, H. Grüniger, and H. Meyer, *Thin Solid Films* **132**, 41 (1986).

23. J. Heesemann, *J. Am. Chem. Soc.* **10**, 2167, 2176 (1980).

24. J. Heesemann, Ph.D. Thesis, University of Göttingen, Germany, 1976.

25. D. Möbius and H. Kuhn, *Israel J. Chem.* **18**, 375 (1979).

26. T. L. Penner and D. Möbius, *J. Am. Chem. Soc.* **104**, 7407 (1982).

27. R. Steiger, *Helv. Chim. Acta* **54**, 2645 (1971).

28. D. den Engelsen, *J. Phys. Chem.* **76**, 3390 (1972).

29. O. V. Elsner, Ph.D. Thesis, University of Marburg, Germany, 1969.

30. G. L. Gaines, Jr., and W. J. Ward III, *J. Coll. Interf. Sci.* **60**, 210 (1977).

31. P. Fromherz, *Nature* **231**, 267 (1971).

32. H. P. Zingsheim, *Ber. Bunsenges. Phys. Chem.* **80**, 1185 (1976).

33. M. L. Agrawal, J.-P. Chauvet, and L. K. Patterson, *J. Phys. Chem.* **89**, 2979 (1985).

34. P. K. J. Kinnunen, J. A. Virtanen, A. P. Tulkki, R. C. Ahuja, and D. Möbius, *Thin Solid Films* **132**, 193 (1986).

35. Reference 9, p. 635; Table 7.3 contains an error: the last line referring to an orientation of the chromophores of one-half statistical in the layer plane, one-half normal to layer plane has a factor of $\frac{4}{3}$ in all columns; the correct factor is $\frac{3}{4}$.

36. E. E. Polymeropoulos, D. Möbius, and H. Kuhn, *J. Chem. Phys.* **68**, 3918 (1978).

37. See, for example, A. W. Adamson, *Physical Chemistry of Surfaces*, 4th ed., Wiley, New York, 1982.

38. R. W. Murray, *Acc. Chem. Res.* **13**, 135 (1980).

39. E. E. Polymeropoulos and J. Sagiv, *J. Chem. Phys.* **69**, 1836 (1978).

40. A. Sommerfeld and H. Bethe, in *Handbuch der Physik*, Vol. 24/2, H. Geiger and K. Scheel (Eds.), Springer-Verlag, Berlin, 1933, p. 450.

41. E. Burstein and S. Lundqvist (Eds.), *Tunneling Phenomena in Solids*, Plenum, New York, 1969.

42. S. Kurtin, T. C. McGill, and C. A. Mead, *Phys. Rev. Lett.* **25**, 756 (1970).

43. J. L. Miles and H. O. McMahon, *J. Appl. Phys.* **32**, 1176 (1961).

44. R. M. Handy and L. C. Scala, *J. Electrochem. Soc.* **113**, 1105 (1966).

45. S. Horiuchi, J. Yamaguchi, and K. Naito, *J. Electrochem. Soc.* **115**, 634 (1968).

46. B. Mann and H. Kuhn, *J. Appl. Phys.* **42**, 4398 (1971).

47. B. Mann, H. Kuhn, and L. V. Szentpàly, *Chem. Phys. Lett.* **8**, 82 (1971).

48. E. E. Polymeropoulos, *J. Appl. Phys.* **48**, 2404 (1977).

49. K. H. Gundlach and J. Kadlec, *Chem. Phys. Lett.* **25**, 293 (1974).

50. R. H. Tredgold and C. S. Winter, *J. Phys. D: Appl. Phys.* **14**, L185 (1981).

51. A. Matsuda, M. Sugi, T. Fukui, and S. Iizima, *J. Appl. Phys.* **48**, 771 (1977).

52. T. Fukui, M. Sugi, and S. Iizima, *Phys. Rev. B* **22**, 4898 (1980).

53. C. Kittel, *Quantum Theory of Solids*, Wiley, New York, 1963.

54. A. Leger, J. Klein, M. Belin, and D. Defourneau, *Thin Solid Films* **8**, R51 (1971).

55. E. E. Polymeropoulos, *Solid State Commun.* **28**, 883 (1978).

56. I. Giaever and K. Megerle, *Phys. Rev.* **122**, 1101 (1961).

57. J. Tanguy, *Thin Solid Films* **13**, 33 (1972).

58. E. P. Honig, *Thin Solid Films* **33**, 231 (1976).

59. G. G. Roberts, P. S. Vincett, and W. A. Barlow, *J. Phys. C: Solid State Phys.* **11**, 2077 (1978).

60. G. G. Roberts, M. C. Petty, and I. M. Dharmadasa, *Proc. Inst. Electr. Eng.* **128**, 197 (1981).

61. J. Batey, G. G. Roberts, and M. C. Petty, *Thin Solid Films* **99**, 283 (1983).

62. L. Esaki and R. Tsu, *Appl. Phys. Lett.* **22**, 562 (1973).

63. P. Ruden and G. H. Döhler, *Phys. Rev. B* **27**, 3538 (1983).

64. M. H. Nathoo and A. K. Jonscher, *J. Phys. C: Solid State Phys.* **4**, L301 (1971).

65. A. K. Jonscher and M. H. Nathoo, *Thin Solid Films* **12**, S15 (1972).

66. I. Lundström and D. McQueen, *Chem. Phys. Lip.* **10**, 181 (1973).

67. M. Sugi, K. Nembach, D. Möbius, and H. Kuhn, *Solid State Commun.* **13**, 603 (1973).

68. M. Sugi, K. Nembach, D. Möbius, and H. Kuhn, *Solid State Commun.* **15**, 1867 (1974).

69. M. Sugi, K. Nembach, and D. Möbius, *Thin Solid Films* **27**, 205 (1975).

70. M. Sugi, T. Fukui, and S. Iizima, *Appl. Phys. Lett.* **27**, 559 (1975); *Appl. Phys. Lett.* **28**, 240 (1976).

71. S. Iizima and M. Sugi, *Appl. Phys. Lett.* **28**, 548 (1976).

72. M. Sugi, T. Fukui, and S. Iizima, *Chem. Phys. Lett.* **45**, 163 (1977).

73. M. Sugi and S. Iizima, *Phys. Rev. B* **15**, 574 (1977).

74. M. Sugi, T. Fukui, and S. Iizima, *Phys. Rev. B* **18**, 725 (1978).

75. M. Sugi and S. Iizima, *Appl. Phys. Lett.* **34**, 290 (1979).

76. A. Miller and E. Abrahams, *Phys. Rev.* **120**, 745 (1960).

77. N. F. Mott, *Philos. Mag.* **19**, 835 (1969).

78. F. Gutman and L. E. Lyons, *Organic Semiconductors*, Wiley, New York, 1967.

79. J. J. Miasik, M.Sc. Thesis, University of Kent, United Kingdom, 1981.

80. T. W. Barrett, *Thin Solid Films* **102**, 231 (1983).

81. S. Baker, G. G. Roberts, and M. C. Petty, *IEE Proc.* **130**, 260 (1983).

82. S. Baker, M. C. Petty, G. G. Roberts, and M. V. Twigg, *Thin Solid Films* **99**, 53 (1983).

83. P. S. Vincett, W. A. Barlow, F. T. Boyle, J. A. Finney, and G. G. Roberts, *Thin Solid Films* **60**, 265 (1979).

84. G. G. Roberts, T. M. McGinnity, W. A. Barlow, and P. S. Vincett, *Thin Solid Films* **68**, 223 (1980).

85. M. F. Daniel, O. C. Lettington, and S. M. Small, *Thin Solid Films* **99**, 61 (1983).

86. G. Scheibe, *Angew. Chem.* **49**, 563 (1936).

87. E. E. Jelley, *Nature (Lond.)* **138**, 1009 (1936).

88. M. Calvin, *Photochem. Photobiol.* **37**, 349 (1983).

89. J. Frenkel, *Phys. Rev.* **54**, 647 (1938).

90. J. G. Simmons, *Phys. Rev.* **155**, 657 (1967).

91. R. M. Hill, *Philos. Mag.* **23**, 59 (1971).

92. G. A. N. Connel, D. L. Camphausen, and W. Paul, *Philos. Mag.* **26**, 541 (1972).

93. M. Sugi, M. Saito, T. Fukui, and S. Iizima, *Thin Solid Films* **88**, L15 (1982).

94. M. Sugi and S. Iizima, *Thin Solid Films* **68**, 199 (1980).

95. R. Steiger and J. F. Reber, Photogr. Sci. Eng. **25**, 127 (1981).

96. M. Sugi, T. Fukui, and S. Iizima, *Mol. Cryst. Liq. Cryst.* **62**, 165 (1980).

97. M. Sugi, M. Saito, T. Fukui, and S. Iizima, *Thin Solid Films* **99**, 17 (1983).

98. H. Kuhn, *Chem. Phys. Lip.* **8**, 401 (1972).

99. H. Kuhn and D. Möbius, *Angew. Chem. Int. Edit.* **10**, 620 (1971).

100. U. Schoeler, K. H. Tews, and H. Kuhn, *J. Chem. Phys.* **61**, 5009 (1974).

101. E. E. Polymeropoulos, D. Möbius, and H. Kuhn, *Thin Solid Films* **68**, 173 (1980).

102. A. F. Janzen and J. R. Bolton, *J. Am. Chem. Soc.* **101**, 6342 (1979).

103. R. Jones, R. H. Tredgold, and J. E. O'Mullane, *Photochem. Photobiol.* **32**, 223 (1980).

104. M. Saito, M. Sugi, T. Fukui, and S. Iizima, *Thin Solid Films* **100**, 117 (1983).

105. H. Meier, W. Albrecht, and U. Tschirwitz, *Angew. Chem.* **84**, 1077 (1972).

106. H. Killesreiter, *Ber. Bunsenges. Phys. Chem.* **82**, 503, 512 (1978).

107. H. Killesreiter, *Z. Naturforsch.* **34a**, 737 (1979).

108. L. V. Szentpàly, D. Möbius, and H. Kuhn, *J. Chem. Phys.* **52**, 4618 (1970).

109. R. Steiger, H. Hediger, P. Junod, H. Kuhn, and D. Möbius, *Photogr. Sci. Eng.* **24**, 185 (1980).

110. P. Fromherz and W. Arden, *J. Am. Chem. Soc.* **102**, 6211 (1980).

111. W. Arden and P. Fromherz, *J. Electrochem. Soc.* **127**, 370 (1980).

112. R. Memming and F. Schröppel, *Chem. Phys. Lett.* **62**, 207 (1979).

113. T. Miyasaka, T. Watanabe, A. Fujushima, and K. Honda, *Nature* **277**, 640 (1979).

114. H. W. Vogel, *Ber. Deut. Chem. Ges.* **6**, 1302 (1873).

115. H. Gerischer and F. Willig, in *Topics in Current Chemistry*, Vol. 61, F. L. Boschke (Ed.), Springer-Verlag, Berlin, 1976, p. 31.

116. D. Möbius, R. C. Ahuja, and G. Debuch, in preparation.

117. J. M. Frank and S. J. Wawilow, *Z. Phys.* **69**, 100 (1931).

118. H. Kuhn, *Pure Appl. Chem.* **53**, 2105 (1981).

119. D. Möbius, *Mol. Cryst. Liq. Cryst.* **52**, 235 (1979)

120. D. Möbius and H. Staerk, unpublished.

121. D. Möbius, in *Colloids and Surfaces in Reprographic Technology*, M. Hair and M. D. Croucher (Eds.), ACS Symposium Series No. 200, American Chemical Society, Washington, D.C., 1982, p. 93.

122. J. Cunningham, E. E. Polymeropoulos, D. Möbius, and F. Baer, in *Magnetic Re-*

sonance in Colloid and Interface Science, J. P. Faissard and H. A. Resing (Eds.), D. Reidel Publishing, Dordrecht, Netherlands, 1980, p. 603.

123. P. B. Gilman, Jr., *Photogr. Sci. Eng.* **18,** 418 (1974).

124. D. Rehm and A. Weller, *Israel J. Chem.* **8,** 259 (1970).

125. H. Kuhn, in *Light-Induced Charge Separation in Biology and Chemistry*, H. Gerischer and J. J. Katz (Eds.), Dahlem Konferenzen 1979, Verlag Chemie, Weinheim, p. 151.

126. K.-P. Seefeld, D. Möbius, and H. Kuhn, *Helv. Chim. Acta* **60,** 2608 (1977).

127. H. Kuhn, in *Modern Trends of Colloid Science in Chemistry and Biology*, H.-F. Eicke (Ed.), Birkhäuser Verlag, Basel, 1985, p. 97.

128. H. Kuhn, *Mol. Cryst. Liq. Cryst.* **125,** 233 (1985).

129. H. Kuhn, *Phys. Rev. A*, submitted.

130. A. Siemiarczuk, A. R. McIntosh, T.-F. Ho, M. J. Stillman, K. J. Roach, A. C. Weedon, J. R. Bolton, and J. S. Connolly, *J. Am. Chem. Soc.* **105,** 7224 (1983).

131. Th. A. Moore, D. Gust, P. Mathis, J.-C. Mialocq, C. Chachaty, R. V. Bensasson, E. J. Land, D. Doizi, P. A. Liddell, W. R. Lehmann, G. A. Nemeth, and A. L. Moore, *Nature* **307,** 630 (1984).

132. I. Langmuir, *Science* **87,** 493 (1938).

133. E. P. Honig, *J. Coll. Interf. Sci.* **43,** 66 (1973).

134. Ref. 9, pp. 599 and 617.

135. D. Möbius, *Acc. Chem. Res.* **14,** 63 (1981).

136. D. Möbius and G. Debuch, unpublished.

137. D. Möbius, in *Photochemical Transformations in Non-Homogeneous Media*, M.-A. Fox (Ed.), ACS Symposium Series 278, American Chemical Society, Washington, D.C., 1985, p. 113.

138. D. Möbius, R. C. Ahuja, and H. Kuhn, in preparation.

139. J. Ulstrup, in *Lecture Notes in Chemistry*, Vol. 10, G. Berthier, M. J. S. Dewar, H. Fischer et al. (Eds.), Springer-Verlag, Berlin, 1979.

140. D. Möbius, in *Photochemical Conversion and Storage of Solar Energy*, J. Rabani (Ed.), The Weizmann Science Press of Israel, Jerusalem, 1982, p. 139.

11 STOCHASTIC AND DIFFUSION MODELS OF REACTIONS IN MICELLES AND VESICLES

Masanori Tachiya

National Chemical Laboratory for Industry, Yatabe, Ibaraki, Japan

11.1. INTRODUCTION

11.1.1. Reactions in Micellar and Vesicular Systems

Reactions in micellar and vesicular systems have attracted growing interest (1–4) for two main reasons. The first is that certain reactions may be controlled by carrying them out in these systems. Micelles and vesicles are relatively simple and well-defined prototypes for biological membranes, and one may mimic efficient reactions occurring in the latter by using the former as reaction media. Photoinduced charge separation is a topic of current interest (5, 6) in connection with solar energy conversion. The efficiency of this process is considerably enhanced by using micellar and vesicular systems as reaction media (7). The effect of a magnetic field on chemical reactions is another current topic (8, 9). This effect is greatly enhanced when the reactions are carried out in micelles (10). One of the most successful uses of micellar systems as reaction media is emulsion polymerization (11). Reaction control in micellar and vesicular systems is a promising field.

The second reason for interest in reactions in micellar and vesicular systems is that one may use the reactions as a probe to characterize the systems themselves. Physical parameters of the systems, such as aggregation number, microviscosity, or the partition of solubilizates between micelles (vesicles) and the aqueous phase, are important in practical applications of the systems. However, the parameters are difficult to measure. Photochemical methods utilizing fluorescence quenching, excimer formation, and so on provide a convenient method for evaluating some of them (3, 4).

11.1.2. Structures of Micelles and Vesicles (2, 3)

Surfactants or detergents are amphiphatic molecules that have both hydrophobic and hydrophilic moieties. The hydrophobic moiety is typically a straight-chain hydrocarbon of 8–18 carbon atoms. Depending on the nature of the hydrophilic moiety, surfactants may be classified as neutral, cationic, or anionic. Typical examples of neutral, cationic, and anionic surfactants are polyoxyethylene (6) octanol, hexadecyltrimethyl-ammonium bromide (CTAB), and sodium dodecyl sulfate (SDS) in that order. The conflicting properties of the two moieties cause the self-association of surfactants in their aqueous solutions: Above a certain critical concentration surfactants associate to form aggregates of colloidal dimensions. These aggregates are called micelles, and the critical concentration is called the critical micelle concentration (cmc). Strictly speaking, cmc is not defined uniquely. There is a narrow range of concentrations below which micelles are virtually absent and above which virtually all the surfactants exist as micelles. This narrow range defines cmc.

The structure of a micelle formed by ionic surfactants is shown schematically in Figure 11.1. At surfactant concentrations that are not too far above cmc, the micelle is roughly spherical and contains 50–200 monomer surfactants. The radius of the sphere approximately corresponds to the extended length of the hydrocarbon chain of the surfactant. The hydrocarbon chains constitute the core of the micelle. They are mobile. Microviscosities of micellar interiors have been estimated (12) as ~ 8 cP from ^{13}C NMR spin-lattice relaxation times. The Stern layer consists of the surfactant head groups and bound counterions. It is an important site for solubilizate interactions and reactions. Most of the counterions are dissociated from the micelle and located in the Gouy–Chapman double layer, where they are free to exchange with ions distributed in the bulk aqueous phase. The size distribution

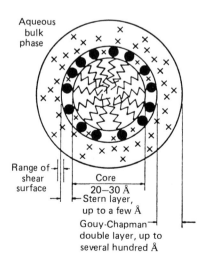

Aqueous
bulk
phase

Range of
shear
surface

Core
20–30 Å
Stern layer,
up to a few Å
Gouy-Chapman
double layer, up to
several hundred Å

Figure 11.1. Schematic representation of a spherical ionic micelle. The core of the micelle consists of the hydrocarbon chains (〰). The symbols ● and x represent surfactant head groups and counterions, respectively.

of spherical micelles is rather narrow, that is, roughly speaking, micellar systems are monodisperse. Increasing the surfactant concentration leads to the formation of rodlike micelles.

Phospholipids contain a polar head group and two long hydrocarbon chains and are amphiphatic. When swelled in water, they form multicompartment closed bilayer structures such as shown in Figure 11.2*a*, which are generally called vesicles. Some synthetic surfactants can also form vesicles. Vesicles composed of phospholipids are specifically referred to as liposomes. Vesicles contain water inside and between bilayers. An important property of bilayers is their thermotropic phase transition from a liquid crystalline state to a fluid state. Bilayers are considerably more rigid than micellar interiors. Microviscosities of phospholipid bilayers can be as high as 200 cP (13). Sonication above the phase transition temperature breaks down multicompartment vesicles into single-compartment vesicles (Figure 11.2*b*). Single compartment liposomes contain 2000–10,000 phospholipid constituents and are considerably larger than micelles. Single-compartment vesicles of approximately uniform size distributions are usually used for physicochemical investigations.

11.1.3. Time Scale of Events in Micellar and Vesicular Systems (1, 3, 4)

Micelles are in dynamic equilibrium with monomer surfactants. Two relaxation processes have been observed in relation to this equilibrium. A fast process in the microsecond range is associated with the exchange of a monomer surfactant between micelles and the aqueous phase. A slow process in the millisecond range is associated with the breakup of micelles into individual monomer surfactants. This process defines the lifetime of micelles. Solubilizates dissolved in the micelles are also in dynamic equilibrium with those in the aqueous phase. The residence time of solubilizates in the micelle is roughly in the microsecond region.

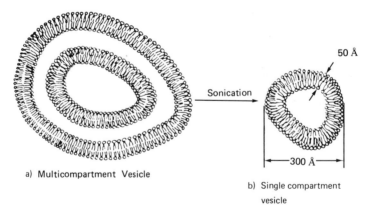

a) Multicompartment Vesicle

b) Single compartment vesicle

Figure 11.2. Schematic representation of vesicles (2). Multicompartment vesicles (*a*) are broken down into single-compartment vesicles (*b*) by sonication.

The kinetic stability of vesicles is considerably greater than that of micelles. The lifetime of vesicles is on the order of weeks or months. Various dynamic processes occur in vesicular systems on different time scales. For example, the transverse motion of a lipid constituent from one interface of the bilayer to the other, called flip-flop, occurs on the time scale of days. Negatively charged liposomes fuse in the presence of calcium ions, with dimerization rate constants on the order of 10^7 M^{-1} s^{-1} (14). As in micellar systems, solubilizates dissolved in

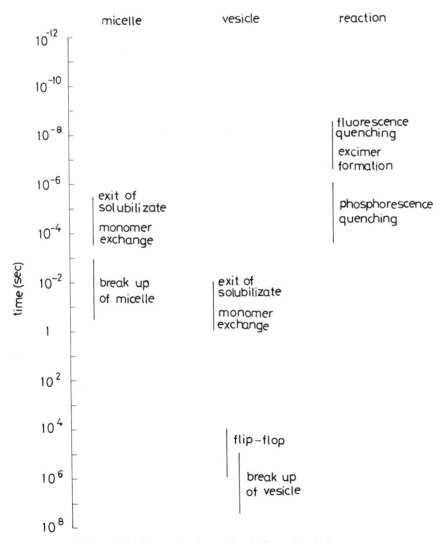

Figure 11.3. Time scale of events in micellar and vesicular systems.

the vesicle bilayers are in dynamic equilibrium with those in the aqueous phase. The residence time of solubilizates in the vesicle bilayer is longer than that in the micelle and is roughly in the millisecond region. Figure 11.3 shows the time scale of events in micellar and vesicular systems together with that of typical reactions carried out in these systems.

11.1.4. Scope of This Chapter

Three effects are important in reactions in micellar and vesicular systems. The first is that reactants are distributed among the vehicles (micelles or vesicles). The average number of reactants contained in a vehicle is usually small. The fluctuations in the number of reactants in a vehicle are similar in magnitude to the average. Therefore, a deterministic model of the kinetics, which considers only the average, cannot be applied to these systems. Instead, one has to use a stochastic approach, which takes into account the fluctuations in the number of reactants.

The second effect is that the diffusion space of the reactants is restricted. For example, hydrophobic reactants are dissolved in the interior of micelles or in the bilayer of vesicles, and the diffusion-controlled reactions between them occur there. Diffusion-controlled reactions in restricted spaces may be quite different from those in a free space.

The third effect is that the surfaces of ionic micelles or vesicles carry charges. The electrostatic effect due to the surface charges plays an important role (15) in many reactions in micellar and vesicular systems, including photoinduced charge separation.

The content of this chapter is as follows. In Section 11.2 the distribution of solubilizates among vehicles is discussed in connection with the migration dynamics of solubilizates between vehicles. In Section 11.3 diffusion-controlled reactions in restricted spaces, such as the micellar interior or surface, are discussed in comparison with those in a free space. Magnetic effects on chemical reactions in micelles are also be dealt with from the viewpoint of diffusion-controlled reactions in restricted spaces. In Section 11.4 the stochastic treatment of reactions between compartmentalized reactants is presented.

11.2. DISTRIBUTION OF SOLUBILIZATES AMONG MICELLES

11.2.1. Migration Mechanisms

The distribution of solubilizates among micelles is closely connected with the migration dynamics of solubilizates between micelles. Two mechanisms are proposed for the migration between micelles. Tachiya (16) proposed a mechanism in which the migration occurs via the aqueous phase (Figure 11.4a). Henglein and Proske (17) proposed another mechanism in which the migration occurs during micelle collisions (Figure 11.4b). In ionic micelles, ions that have opposite charges to

a) migration via the aqueous phase

b) migration during micelle collisions

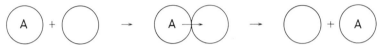

Figure 11.4. Mechanism for migration of solubilizate A between micelles: (a) via the aqueous phase, (b) during micelle collision. Circles represent micelles.

those on the micellar surfaces are adsorbed on the surfaces by coulombic attraction. Henglein and Proske have claimed that in these systems ions migrate between micelles during micelle collisions. In each migration mechanism one can consider several types of migration dynamics.

Here we study the relation between the distribution of solubilizates among micelles and the migration dynamics.

11.2.2. Distributions in the Case of Migration via Aqueous Phase

11.2.2.1. Poisson Distribution (16). Recent experiments have shown that in most cases the migration of solubilizates between micelles occurs via the aqueous phase. One of the most probable models of migration dynamics is

$$M_n + A_{aq} \underset{(n+1)k_-}{\overset{k_+}{\rightleftharpoons}} M_{n+1} \qquad (n = 0, 1, 2, \ldots) \qquad (11.1)$$

where M_n stands for a micelle containing n solubilizates and A_{aq} stands for a solubilizate in the aqueous phase, k_+ is the second-order rate constant for entry of a solubilizate into a micelle, and k_- is the first-order rate constant for exit of a given solubilizate from a micelle. The key assumption in this model is that when a micelle contains n solubilizates, the rate constant for exit of any one solubilizate from the micelle is given by nk_-.

Let $[M_n]$ denote the concentration of micelles containing n solubilizates and let $[A_{aq}]$ denote the concentration of solubilizates in the aqueous phase. In dynamic equilibrium one obtains from Eq. (11.1),

$$\frac{d[M_0]}{dt} = -k_+[A_{aq}][M_0] + k_-[M_1] = 0$$

$$\frac{d[M_n]}{dt} = k_+[A_{aq}][M_{n-1}] - (k_+[A_{aq}] + nk_-)[M_n] + (n + 1)k_-[M_{n+1}] \quad (11.2)$$

$$= 0 \qquad (n = 1, 2, 3, \ldots)$$

This set of equations gives

$$[M_n] = \frac{1}{n!}\left(\frac{k_+[A_{aq}]}{k_-}\right)^n [M_0] \qquad (n = 0, 1, 2, \ldots) \qquad (11.3)$$

If we denote the total concentration of micelles by $[M]$,

$$[M] = \sum_{n=0}^{\infty} [M_n] \qquad (11.4)$$

the fraction Q_n of micelles containing n solubilizates is

$$Q_n = \frac{[M_n]}{[M]} = \frac{1}{n!}\left(\frac{k_+[A_{aq}]}{k_-}\right)^n \exp\left(-\frac{k_+[A_{aq}]}{k_-}\right) \quad (n = 0, 1, 2, \ldots)$$

$$(11.5)$$

The average number $\bar{n}$ of solubilizates contained in a micelle is

$$\bar{n} = \frac{\displaystyle\sum_{n=0}^{\infty} n[M_n]}{[M]} = \sum_{n=0}^{\infty} n[Q_n] \qquad (11.6)$$

In the above case $\bar{n}$ is expressed as

$$\bar{n} = \frac{k_+[A_{aq}]}{k_-} \qquad (11.7)$$

In terms of $\bar{n}$ Eq. (11.5) is rewritten as

$$Q_n = \frac{1}{n!}\,\bar{n}^n \exp(-\bar{n}) \qquad (11.8)$$

It is more convenient to express $\bar{n}$ in terms of the known or measurable quantities of the total micelle concentration $[M]$ and the total solubilizate concentration $[A]$. The latter is expressed as

$$[A] = \bar{n}[M] + [A_{aq}] \qquad (11.9)$$

Elimination of $[A_{aq}]$ from Eqs. (11.7) and (11.9) leads to

$$\bar{n} = \frac{K[A]}{1 + K[M]} \qquad (11.10)$$

where $K = \dfrac{k_+}{k_-}$. The total micelle concentration can be related to the detergent concentration [Det] and the critical micelle concentration cmc by

$$[M] = \frac{[\text{Det}] - \text{cmc}}{\bar{\nu}} \qquad (11.11)$$

where $\bar{\nu}$ is the mean micelle aggregation number.

Equation (11.8) shows that when the migration dynamics is described by Eq. (11.1), the distribution of solubilizates among micelles obeys a Poisson distribution. Figure 11.5 shows Poisson distributions for several values of the average number $\bar{n}$. Recent experiments (18) have demonstrated that in most cases the dis-

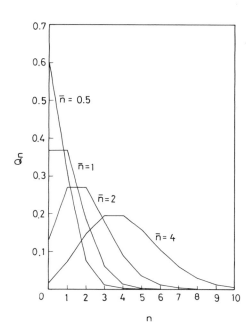

Figure 11.5. Poisson distributions for several values of average number $\bar{n}$.

tribution of solubilizates among micelles is approximately Poissonian. However, several other distributions are also proposed.

11.2.2.2. Other Distributions.

(i) Binomial Distribution (19). It is implicitly assumed in Eq. (11.1) that a micelle can contain an infinite number of solubilizates and that there is no variation in the entry rate constant k_+ with the number of solubilizates in the micelle. These assumptions are not so bad when the average number of solubilizates in a micelle is relatively small. However, when the average number is large, this assumption is not justified.

To account for the limit to the number of solubilizates a micelle can contain, Hunter (19) proposed the following expression for the rate constant for entry of a solubilizate into a micelle that already contains n solubilizates:

$$\text{Entry rate constant} = \begin{cases} \left(1 - \dfrac{n}{m}\right) k_+ & n \leq m \\ \\ 0 & n > m \end{cases} \qquad (11.12)$$

where m is the maximum number of solubilizates in a micelle. The migration dynamics is then modified to

$$M_n + A_{aq} \underset{(n+1)k_-}{\overset{k_+(1-n/m)}{\rightleftharpoons}} M_{n+1} \qquad (n = 0, 1, \ldots, m) \qquad (11.13)$$

The distribution of solubilizates among micelles can be calculated from Eq. (11.13) in the same way as described in Section 11.2.2.1. The fraction of micelles containing n solubilizates is

$$Q_n = {}_mC_n \left(\frac{\bar{n}}{m}\right)^n \left(1 - \frac{\bar{n}}{m}\right)^{m-n} \qquad (n = 0, 1, \ldots, m) \qquad (11.14)$$

where the average number of solubilizates in a micelle is given by

$$\bar{n} = \frac{mk_+[A_{aq}]}{mk_- + k_+[A_{aq}]} \qquad (11.15)$$

In terms of the total micelle and solubilizate concentrations, $\bar{n}$ is expressed as

$$\bar{n} = \frac{1}{2}\left\{ \frac{[A]}{[M]} + m + \frac{m}{K[M]} - \left[\left(\frac{[A]}{[M]} + m + \frac{m}{K[M]}\right)^2 - \frac{4m[A]}{[M]} \right]^{1/2} \right\} \qquad (11.16)$$

where $K = \dfrac{k_+}{k_-}$.

Equation (11.14) shows that when the migration dynamics is described by Eq. (11.13), the distribution of solubilizates among micelles obeys a binomial distribution. Recent experiment by Nakamura et al. (20) indicates that in the system they studied the distribution of solubilizates among micelles is more adequately describable by a binomial distribution than by a Poisson distribution. Note that in the limit $m \to \infty$ Eq. (11.14) reduces to the Poisson distribution [Eq. (11.8)]. So the Poisson distribution is a special case of the binomial distribution. Figure 11.6 shows binomial distributions for several values of the average number $\bar{n}$.

(ii) Geometric Distribution. In Eq. (11.1) the rate constant for exit of any one solubilizate from a micelle is assumed to be proportional to the number of solubilizates in the micelle. Instead of Eq. (11.1), Rothenberger et al. (21) proposed a model where the rate constant for exit of a solubilizate from a micelle is independent of the number of solubilizates in the micelle,

$$M_n + A_{aq} \underset{k_-}{\overset{k_+}{\rightleftarrows}} M_{n+1} \qquad (n = 0, 1, 2, \ldots) \qquad (11.17)$$

In this model the distribution of solubilizates among micelles is

$$Q_n = \frac{1}{1 + \bar{n}} \left(\frac{\bar{n}}{1 + \bar{n}} \right)^n \qquad (n = 0, 1, 2, \ldots) \qquad (11.18)$$

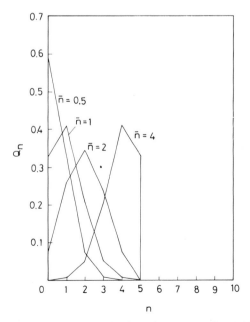

Figure 11.6. Binomial distributions for several values of average number $\bar{n}$. Maximum number m of solubilizates in a micelle is assumed to be 5.

where the average number of solubilizates in a micelle is given by

$$\bar{n} = \frac{k_+[A_{aq}]}{k_- - k_+[A_{aq}]}$$

(11.19)

In terms of the total micelle and solubilizate concentrations, $\bar{n}$ is expressed as

$$\bar{n} = \frac{1}{2}\left\{\frac{[A]}{[M]} - 1 - \frac{1}{K[M]} + \left[\left(\frac{[A]}{[M]} - 1 - \frac{1}{K[M]}\right)^2 + \frac{4[A]}{[M]}\right]^{1/2}\right\}$$

(11.20)

where $K = \dfrac{k_+}{k_-}$.

Equation (11.18) shows that when the migration dynamics is described by Eq. (11.17), the distribution of solubilizates among micelles obeys a geometric distribution. Figure 11.7 shows geometric distributions for several values of the average number $\bar{n}$.

Rothenberger et al. (21) have considered a physical mechanism that gives the migration dynamics described by Eq. (11.17). They have suggested that if a solubilizate leaves a micelle not as an isolated molecule but as a cluster consisting of a solubilizate and several surfactant molecules, the migration dynamics may be described by Eq. (11.17). In the author's opinion, however, even if a solubilizate leaves a micelle as such a cluster, the migration dynamics is still described by Eq.

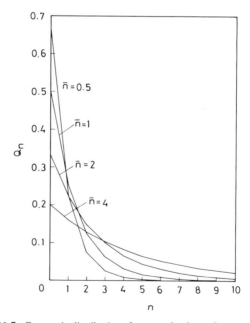

Figure 11.7. Geometric distributions for several values of average number $\bar{n}$.

(11.1) and not by Eq. (11.17). The reason is as follows. The rate at which a surfactant cluster leaves a micelle is, as they noticed, probably independent of the number of solubilizates in the micelle. However, the probability that the surfactant cluster will contain a solubilizate should be proportional to the number of solubilizates in the micelle. Therefore, the rate at which a surfactant cluster containing a solubilizate leaves a micelle should be proportional to the number of solubilizates in the micelle.

Dorrance and Hunter (22) have claimed that their experiment is explainable by a geometric distribution.

11.2.3. Distributions in the Case of Migration during Micelle Collisions

11.2.3.1. Poisson distribution (23). If only one solubilizate migrates from one micelle to the other during a micelle collision, one of the most probable models of migration dynamics is

$$M_n + M_j \underset{(j+1)k}{\overset{nk}{\rightleftharpoons}} M_{n-1} + M_{j+1} \quad \left(\begin{array}{l} n = 1, 2, \cdots \\ j = 0, 1, \cdots \end{array} \right) \tag{11.21}$$

where k is the rate constant for migration of a given solubilizate from one micelle to the other. The key assumption in this model is that when a micelle containing n solubilizates collides with another containing j solubilizates, the probability that any one solubilizate will migrate from the former to the latter is proportional to n, while the probability of the reverse migration is proportional to j.

In dynamic equilibrium one obtains, from Eq. (11.21),

$$\frac{d[M_0]}{dt} = - \sum_{j=1}^{\infty} jk[M_j][M_0] + \sum_{j=0}^{\infty} k[M_j][M_1] = 0$$

$$\frac{d[M_n]}{dt} = \sum_{j=1}^{\infty} jk[M_j][M_{n-1}] - \left(\sum_{j=1}^{\infty} jk[M_j][M_n] + \sum_{n=0}^{\infty} nk[M_j][M_n] \right)$$

$$+ \sum_{j=0}^{\infty} (n+1) k[M_j][M_{n+1}] = 0 \quad (n = 1, 2, 3, \cdots)$$

$$\tag{11.22}$$

With the aid of Eqs. (11.4) and (11.6), Eq. (11.22) is rewritten as

$$\frac{d[M_0]}{dt} = -\bar{n}k[M][M_0] + k[M][M_1] = 0$$

$$\frac{d[M_n]}{dt} = \bar{n}k[M][M_{n-1}] - (\bar{n}k[M][M_n] + nk[M][M_n])$$

$$+ (n+1) k[M][M_{n+1}] = 0 \quad (n = 1, 2, 3, \cdots) \tag{11.23}$$

Equation (11.23) is of the same mathematical form as Eq. (11.2) with replacement of $k_+[A_{aq}]$ and k_- by $\bar{n}k[M]$ and $k[M]$, respectively. Accordingly, the distribution of solubilizates among micelles is obtained by making the same replacement in Eq. (11.5),

$$Q_n = \frac{1}{n!}\, \bar{n}^n \exp(-\bar{n}) \tag{11.24}$$

This equation shows that when the migration dynamics is described by Eq. (11.21), the distribution of solubilizates among micelles obeys a Poisson distribution. The average number of solubilizates in a micelle is

$$\bar{n} = \frac{[A]}{[M]} \tag{11.25}$$

provided solubilizates are exclusively dissolved in micelles.

11.2.3.2. Other distributions

(i) Binomial Distribution (24). When there is a limit to the number of solubilizates in a micelle, Eq. (11.21) may be modified to

$$M_n + M_j \underset{(j+1)\left[1-\left(\frac{n-1}{m}\right)\right]k}{\overset{n\left(1-\frac{j}{m}\right)k}{\rightleftharpoons}} M_{n-1} + M_{j+1} \quad \left(\begin{array}{l} n = 1, 2, \cdots m \\ j = 0, 1, \cdots m - 1 \end{array}\right) \tag{11.26}$$

where m is the maximum number of solubilizates in a micelle. In this model the distribution of solubilizates among micelles is

$$Q_n = {}_mC_n \left(\frac{\bar{n}}{m}\right)^n \left(1 - \frac{\bar{n}}{m}\right)^{m-n} \quad (n = 0, 1, \cdots, m) \tag{11.27}$$

This equation shows that when the migration dynamics is described by Eq. (11.26), the distribution of solubilizates among micelles obeys a binomial distribution. The average number of solubilizates in a micelle is

$$\bar{n} = \begin{cases} \dfrac{[A]}{[M]} & \text{for } \dfrac{[A]}{[M]} \le m \\[3mm] m & \text{for } \dfrac{[A]}{[M]} > m \end{cases} \tag{11.28}$$

provided solubilizates are exclusively dissolved in micelles.

(ii) Geometric Distribution. If the probability that a solubilizate will migrate from one micelle to the other during a micelle collision is independent of the numbers of solubilizates in each micelle, the migration dynamics is described by

$$M_n + M_j \underset{k}{\overset{k}{\rightleftarrows}} M_{n-1} + M_{j+1} \qquad \left(\begin{array}{l} n = 1, 2, 3, \cdots \\ j = 0, 1, 2, \cdots \end{array} \right) \qquad (11.29)$$

In this model the distribution of solubilizates among micelles is

$$Q_n = \frac{1}{(1 + \bar{n})} \left(\frac{\bar{n}}{1 + \bar{n}} \right)^n \qquad (n = 0, 1, 2, \cdots) \qquad (11.30)$$

This equation shows that when the migration dynamics is described by Eq. (11.29), the distribution of solubilizates among micelles obeys a geometric distribution. The average number of solubilizates in a micelle is given by Eq. (11.25), provided solubilizates are exclusively dissolved in micelles.

11.2.4. Summary

In Table 11.1 we summarize the relation between the migration dynamics and the distribution of solubilizates among micelles.

Recent experiments have shown that in most cases the migration of solubilizates between micelles occurs via the aqueous phase (25). When this occurs, the migration dynamics described by Eq. (11.1) seem the most reasonable if the average number of solubilizates in a micelle is relatively small. Therefore, it is reasonable to assume that the distribution of solubilizates among micelles is described by a Poisson distribution. In fact, recent experiments (18) have demonstrated that in most cases the distributions are Poissonian.

Here we have taken a micelle as an example of a vehicle that carries solubilizates and explained the distribution of solubilizates among vehicles. The results presented are general and also applicable to other vehicles such as vesicles and polymer particles.

11.3. DIFFUSION-CONTROLLED REACTIONS IN MICELLAR SYSTEMS

11.3.1. Classification

Diffusion-controlled bimolecular reactions in micellar systems may be classified into five cases according to the locations of reactants (Figure 11.8): (1) Both of the reactants are in the micellar interior, (2) both of the reactants are on the micellar surface, (3) one of the reactants is in the micellar interior and the other is on the micellar surface, (4) one of the reactants is in the micellar interior and the other is

TABLE 11.1. Relation between Migration Dynamics and Distribution of Solubilizates among Micelles

Migration Dynamics	Distribution	Average Number[a]
(a) Migration via aqueous phase:		
$M_n + A_{aq} \underset{(n+1)k_-}{\overset{k_+}{\rightleftharpoons}} M_{n+1}$	Poisson $$Q_n = \frac{1}{n!}\,\bar{n}^n \exp(-\bar{n})$$	$$\bar{n} = \frac{K\,[A]}{1 + K[M]}$$
$M_n + A_{aq} \underset{(n+1)k_-}{\overset{k_+\,[1-(n/m)]}{\rightleftharpoons}} M_{n+1}$	Binomial $$Q_n = {}_mC_n \left(\frac{\bar{n}}{m}\right)^n \left(1 - \frac{\bar{n}}{m}\right)^{m-n}$$	$$\bar{n} = \frac{1}{2}\left\{\frac{[A]}{[M]} + m + \frac{m}{K[M]} \right.$$ $$\left. - \left[\left(\frac{[A]}{[M]} + m + \frac{m}{K[M]}\right)^2 - \frac{4m[A]}{[M]}\right]^{1/2}\right\}$$
$M_n + A_{aq} \underset{k_-}{\overset{k_+}{\rightleftharpoons}} M_{n+1}$	Geometric $$Q_n = \frac{1}{1+\bar{n}}\left(\frac{\bar{n}}{1+\bar{n}}\right)^n$$	$$\bar{n} = \frac{1}{2}\left\{\frac{[A]}{[M]} - 1 - \frac{1}{K[M]} \right.$$ $$\left. + \left[\left(\frac{[A]}{[M]} - 1 - \frac{1}{K[M]}\right)^2 + \frac{4[A]}{[M]}\right]^{1/2}\right\}$$

(b) Migration during micelle collisions:

$$M_n + M_j \underset{(j+1)k}{\overset{nk}{\rightleftharpoons}} M_{n-1} + M_{j+1}$$

Poisson

$$Q_n = \frac{1}{n!}\,\bar{n}^n \exp(-\bar{n}) \qquad \bar{n} = \frac{[A]}{[M]}$$

$$M_n + M_j \underset{(j+1)\left[1-\frac{n-1}{m}\right]k}{\overset{n\,[1-(j/m)]k}{\rightleftharpoons}} M_{n-1} + M_{j+1}$$

Binomial

$$Q_n = {}_mC_n \left(\frac{\bar{n}}{m}\right)^n \left(1 - \frac{\bar{n}}{m}\right)^{m-n}$$

$$\bar{n} = \begin{cases} \dfrac{[A]}{[M]} & \text{for } \dfrac{[A]}{[M]} \le m \\[2ex] m & \text{for } \dfrac{[A]}{[M]} > m \end{cases}$$

$$M_n + M_j \underset{k}{\overset{k}{\rightleftharpoons}} M_{n-1} + M_{j+1}$$

Geometric

$$Q_n = \frac{1}{1+\bar{n}} \left(\frac{\bar{n}}{1+\bar{n}}\right)^n \qquad \bar{n} = \frac{[A]}{[M]}$$

$^aK = k_+/k_-.$

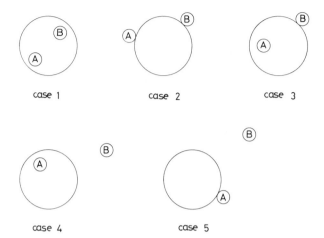

Figure 11.8. Classification of diffusion-controlled reactions between reactants A and B in micellar systems. Large circles represent micelles.

in the aqueous phase, and (5) one of the reactants is on the micellar surface and the other is in the aqueous phase.

In cases 4 and 5 one of the reactants diffuses only in a restricted space, but the other can diffuse in a free space. Thus, diffusion-controlled reactions in cases 4 and 5 can be considered on the same line with those in a free space. In cases 1–3, however, both of the reactants diffuse only in a restricted space. Thus, diffusion-controlled reactions in cases 1–3 may be quite different from those in a free space. Adam and Delbrück (26) discussed the relation between the diffusion-controlled reaction rate and the dimensionality and extent of the diffusion space of reactants in connection with the control of reaction rates in biological systems.

Here we study diffusion-controlled reactions in micellar systems, particularly in cases 1–3. We first review diffusion-controlled reactions in a free space and then study case 1 in some detail in comparison with those in a free space. We introduce the mean reaction time approximation, which is useful for the treatment of diffusion-controlled reactions in restricted spaces. By use of this approximation, we further study diffusion-controlled reactions in cases 2 and 3. We finally deal with magnetic effects on diffusion-controlled reactions in micelles.

11.3.2. Review of Reactions in a Three-Dimensional Free Space

In applications of the theory of diffusion-controlled reactions, two typical cases are often encountered for the distribution of reactants. One is where there is only a pair of reactants A and B present in the system and a diffusion-controlled reaction occurs between them (Figure 11.9a). This is called geminate recombination. The other is where the numbers of reactants A and B present in the system are both

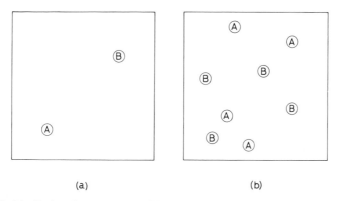

(a)   (b)

Figure 11.9. Distribution of reactants A and B in geminate recombination (a) and bulk reaction (b).

very large, and a diffusion-controlled reaction can occur between any pair of A and B (figure 11.9b). This is called bulk reaction.

In geminate recombination the reaction dynamics is described in terms of the pair survival probability, which is the probability that a pair of reactants will still be unreacted at time t. Let $W(\mathbf{r}, t)$ denote the survival probability of a pair in which the two reactants are initially separated by $\mathbf{r}$. If no force acts between the reactants, $W(\mathbf{r}, t)$ is given by (27)

$$W(\mathbf{r}, t) = 1 - \frac{d}{r} \operatorname{erfc}\left(\frac{r - d}{2\sqrt{Dt}}\right) \tag{11.31}$$

where

$$\operatorname{erfc}(x) = \frac{2}{\sqrt{\pi}} \int_{x}^{\infty} e^{-y^2}\, dy \tag{11.32}$$

and D is the sum of diffusion coefficients of the two reactants. Equation (11.31) is derived on the assumption that the two reactants react immediately on approaching each other to a distance d, which is usuallly called the reaction radius.

For large t Eq. (11.31) is expanded to

$$W(\mathbf{r}, t) = 1 - \frac{d}{r} + \left(1 - \frac{d}{r}\right)\frac{d}{\sqrt{\pi Dt}} + \cdots \tag{11.33}$$

This equation indicates that the pair survival probability decreases in inverse proportion to $t^{1/2}$. It also indicates that even in the limit $t \to \infty$, it does not go to zero, which means that there is a nonzero probability that the two reactants will never react. This probability is called the escape probability.

In bulk reaction the reaction dynamics is described in terms of the concentrations $[A]$ and $[B]$ of reactants A and B by

$$\frac{d[A]}{dt} = \frac{d[B]}{dt} = -k^b[A]\,[B] \tag{11.34}$$

where the rate constant k^b is expressed as $k^b = 4\pi Dd$ (28).

11.3.3. Reactions in the Micellar Interior (29)

11.3.3.1. Micelle Contains One A Reactant and One B Reactant. Hereafter we assume for the sake of simplicity that a micelle is a sphere of radius b and that a reactant is a sphere of radius a.

Strongly hydrophobic molecules are exclusively dissolved in the micellar interior, and diffusion-controlled reactions between them occur in it. Consider a reaction between a pair of reactants A and B that are confined in the micelle. It is difficult to treat a general case where both of the reactants can diffuse. So we assume that one of the reactants is fixed at the center of the micelle and the other can diffuse with the diffusion coefficient D equal to the sum of those of the two reactants (Figure 11.10). If we denote the initial position of the mobile reactant by $\mathbf{r}$, the pair survival probability $W(\mathbf{r}, t)$ satisfies the following differential equation and boundary conditions (29, 30):

$$\frac{\partial W(\mathbf{r}, t)}{\partial t} = D\nabla^2 W(\mathbf{r}, t)$$

$$W(\mathbf{r}, t = 0) = 1 \tag{11.35}$$

$$W(r = d, t) = 0$$

$$\left[\frac{\partial W(\mathbf{r}, t)}{\partial r}\right]_{r=R} = 0$$

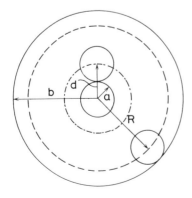

Figure 11.10. Reaction between a pair of reactants in the micellar interior. Small circles of radius a and a large circle of radius b represent reactants and a micelle, respectively. One of the reactants is fixed at the center of the micelle, whereas the other can diffuse in the micellar interior. Reaction occurs when the pair come into contact.

where d and R are given by $d = 2a$ and $R = b - a$, respectively (see Figure 11.10). The second equation is the initial condition. The third equation states that the two reactants react immediately on approaching each other to a distance d. The fourth equation states that B reactant is reflected when it reaches the micellar surface. The solution of this set of equations is (29, 31)

$$W(\mathbf{r}, t) = \sum_{n=1}^{\infty} G_n(r) \exp\left(-\alpha_n^2 \frac{Dt}{R^2}\right) \tag{11.36}$$

where

$$G_n(r) = \frac{2d \sin\left[(r - d)\dfrac{\alpha_n}{R}\right]}{r\alpha_n \left(\dfrac{\alpha_n^2}{1 + \alpha_n^2} - \dfrac{d}{R}\right)} \tag{11.37}$$

and α_n $(n = 1, 2, \cdots)$ are the positive roots of $\tan(1 - d/R)\alpha = \alpha$. Except at very short times $W(\mathbf{r}, t)$ is well approximated by the first term of Eq. (11.36),

$$W(\mathbf{r}, t) \approx G_1(r)\exp\left(-\alpha_1^2 \frac{Dt}{R^2}\right) \tag{11.38}$$

where α_1 is the smallest positive root. This equation indicates that the pair survival probability decays exponentially and that in the limit $t \to \infty$ it goes to zero. In this case the escape probability is zero. This behavior contrasts with the long time behavior of the pair survival probability in a three-dimensional free space [see Eq. (11.33)].

If $u(\mathbf{r})$ denotes the initial distribution of B, the average pair survival probability $W_{av}(t)$ is expressed as

$$W_{av}(t) = \int u(\mathbf{r})W(\mathbf{r}, t)\, d\mathbf{r} \tag{11.39}$$

For a random initial distribution

$$u(\mathbf{r}) = \frac{1}{V} \tag{11.40}$$

where $V = \frac{4}{3}\pi(R^3 - d^3)$ is the volume over which B can diffuse. Substitution of Eqs. (11.36) and (11.40) in (11.39) yields

$$W_{av}(t) = \sum_{n=1}^{\infty} H_n \exp\left(-\alpha_n^2 \frac{Dt}{R^2}\right) \tag{11.41}$$

where

$$H_n = \frac{6\left(\dfrac{d}{R}\right)^2}{\alpha_n^2\left[1 - \left(\dfrac{d}{R}\right)^3\right]\left(\dfrac{\alpha_n^2}{1 + \alpha_n^2} - \dfrac{d}{R}\right)} \tag{11.42}$$

Except at very short times $W_{av}(t)$ may be approximated by the first term of Eq. (11.41),

$$W_{av}(t) = H_1 \exp\left(-\alpha_1^2 \frac{Dt}{R^2}\right) \tag{11.43}$$

Figure 11.11 shows the average pair survival probability as a function of time. In this figure the difference between the two lines calculated from Eq. (11.41) or (11.43) is indiscernible. Thus, the decay of the average pair survival probability is described, to a very good approximation, by a single exponential with the first-order rate constant $k_r = \alpha_1^2 D/R^2$.

The case where both of the reactants can diffuse has been treated by Gösele et al. (32) by use of a Monte Carlo method.

11.3.3.2. Micelle Contains One A Reactant and Several B Reactants. This case is often encountered in real applications. If the B's diffuse mutually independently, the probability $P(t)$ that A will survive at time t is

$$P(t) = \prod_{i=1}^{n} W(\mathbf{r}_i, t) \tag{11.44}$$

where W is given by Eq. (11.36), $\mathbf{r}_i$ is the initial position of the ith B, and n is the number of B's. The decay described by Eq. (11.44) depends on the initial distribution of B. Experimentally observable is the statistical average of Eq. (11.44) over the initial distribution. Let $u(\mathbf{r}_1, \mathbf{r}_2, \cdots, \mathbf{r}_n)$ denote the probability density of finding initially the first B at a position $\mathbf{r}_1$, the second at $\mathbf{r}_2$, and so on. Then the average survival probability $P_{av}(t)$ of A is

$$P_{av}(t) = \iint \cdots \int \prod_{i=1}^{n} W(\mathbf{r}_i, t)\, u(\mathbf{r}_1, \mathbf{r}_2, \cdots \mathbf{r}_n) \prod_{i=1}^{n} d\mathbf{r}_i \tag{11.45}$$

For a totally random initial distribution

$$u(\mathbf{r}_1, \mathbf{r}_2, \cdots, \mathbf{r}_n) = \frac{1}{V^n} \tag{11.46}$$

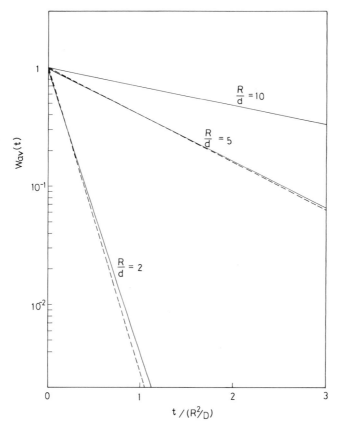

Figure 11.11. Average pair survival probability $W_{av}(t)$ for a pair of reactants in the micellar interior as a function of time t. The full lines were calculated from Eq. (11.41). The lines calculated from Eq. (11.43) practically coincide with the full lines. The broken lines were calculated by use of the mean reaction time approximation introduced in Section 11.3.3.3. See Figure 11.10 for definitions of R and d.

By use of Eqs. (11.46) and (11.39), $P_{av}(t)$ is rewritten as

$$P_{av}(t) = [W_{av}(t)]^n \tag{11.47}$$

Substitution of Eq. (11.43) in (11.47) yields

$$P_{av}(t) = H_1^n \exp\left(-n\alpha_1^2 \frac{Dt}{R^2}\right) \tag{11.48}$$

Thus, $P_{av}(t)$ also decays exponentially with the rate constant $n\alpha_1^2 D/R^2$, indicating that *the reaction considered here obeys first-order kinetics* and that *when the mi-*

celle contains n B's, the rate constant is n times as large as when it contains one B. This is a theoretical foundation of the stochastic treatment of reactions in micellar systems presented in Section 11.4.

11.3.3.3. Mean Reaction Time Approximation (33, 34).

In Section 11.3.3.2. we showed that when a micelle contains one A and n B's, except at very short times, the survival probability of A decays exponentially with a first-order rate constant nk_r, where k_r is the first order rate constant for decay of the average pair survival probability. In Section 11.3.3.1. we showed how to calculate k_r, but this method is rather complicated. Here we present an alternative simple method to calculate k_r.

Generally, in a restricted space such as the micellar interior the decay of the pair survival probability can be well approximated by a single exponential, except at very short times:

$$W(\mathbf{r}, t) = \exp\left[-\frac{t}{\tau(\mathbf{r})}\right] \qquad (11.49)$$

It is reasonable to identify the decay time $\tau(\mathbf{r})$ in this equation with the mean reaction time calculated from the decay of the true $W(\mathbf{r}, t)$. Since $\partial/\partial t\,[1 - W(\mathbf{r}, t)]$ is the probability per unit time that a pair of reactants will react at time t, the mean reaction time is given by

$$\tau(\mathbf{r}) = \int_0^\infty t\,\frac{\partial}{\partial t}\,[1 - W(\mathbf{r}, t)]dt = \int_0^\infty W(\mathbf{r}, t)\,dt \qquad (11.50)$$

Application of the same approximation to the average pair survival probability leads to

$$W_{\mathrm{av}}(t) = \exp\left(-\frac{t}{\tau_{\mathrm{av}}}\right) \qquad (11.51)$$

with

$$\tau_{\mathrm{av}} = \int_0^\infty t\,\frac{\partial}{\partial t}\,[1 - W_{\mathrm{av}}(t)]\,dt = \int u(\mathbf{r})\,\tau(\mathbf{r})\,d\mathbf{r} \qquad (11.52)$$

In the derivation of the second equation in (11.52) we made use of Eqs. (11.39) and (11.50).

Let us derive equations satisfied by the mean reaction time $\tau(r)$. Integrating each equation in (11.35) except the second over time from zero to infinity and making use of Eq. (11.50) together with the relations $W(\mathbf{r}, t = 0) = 1$ and $W(\mathbf{r}, t = \infty) = 0$, one obtains

$$D\,\nabla^2\tau(r) = -1 \qquad \tau(r = d) = 0 \qquad \left[\frac{\partial\tau(\mathbf{r})}{\partial r}\right]_{r=R} = 0 \qquad (11.53)$$

This set of equations provides a simple method to calculate the mean reaction time. It leads to (33)

$$\tau(\mathbf{r}) = \frac{R^2}{3D}\left[R\left(\frac{1}{d} - \frac{1}{r}\right) - \frac{1}{2R^2}(r^2 - d^2)\right] \qquad (11.54)$$

Figure 11.12 shows $\tau(r)$ as a function of r.

The mean reaction time τ_{av} for the average pair survival probability is obtained by averaging Eq. (11.54) over the initial distribution. For a random initial distribution one obtains (33)

$$\tau_{av} = \frac{R^2}{3D\,\dfrac{d}{R}\left[1 - \left(\dfrac{d}{R}\right)^3\right]}\left[1 - \frac{9}{5}\frac{d}{R} + \left(\frac{d}{R}\right)^3 - \frac{1}{5}\left(\frac{d}{R}\right)^6\right] \qquad (11.55)$$

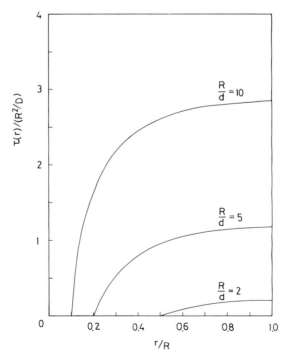

Figure 11.12. Mean reaction time $\tau(r)$ for a pair of reactants in the micellar interior, one fixed at the center of the micelle, the other mobile, as a function of the initial position r of the latter. See Figure 11.10 for definitions of R and d.

The first-order rate constant k_r for the average pair survival probability is given by the reciprocal of τ_{av} [see Eq. (11.51)]. Figure 11.13 shows τ_{av} as a function of R/d. The average pair survival probability calculated from Eq. (11.51) together with (11.55) is compared with that calculated from Eq. (11.41) in Figure 11.11. The former is seen to give a good approximation to the latter. An advantage of the mean reaction time method is that it provides an *analytical* expression for the rate constant k_r, which is extremely useful. Because of this advantage, we hereafter adopt the mean reaction time method to calculate k_r.

11.3.3.4. Comparison between Reaction in Micellar Interior and Bulk Reaction. When a micelle contains one A reactant and several B reactants, the survival probability of A is given by Eq. (11.47). Substitution of Eq. (11.51) in (11.47) yields

$$P_{av}(t) = \exp\left(-n\tau_{av}^{-1}t\right) \tag{11.56}$$

With the aid of Eq. (11.55), the apparent rate constant $k_{app} \equiv n\tau_{av}^{-1}$ is expressed as

$$k_{app} \equiv n\tau_{av}^{-1} = k^b cf \tag{11.57}$$

where k^b is the bulk rate constant, $c = n/(4\pi/3)\,(R^3 - d^3)$ is the concentration of B, and

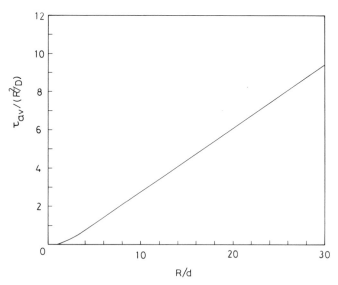

Figure 11.13. Mean reaction time τ_{av} for a pair of reactants in the micellar interior, one fixed at the center of the micelle, the other initially randomly distributed in the micellar interior, as a function of R/d. See Fig. 11.10 for definitions of R and d.

$$f = \frac{[1 - (d/R)^3]^2}{1 - (9/5)\,(d/R) + (d/R)^3 - (1/5)\,(d/R)^6} \tag{11.58}$$

Equation (11.57) indicates that the pseudo-first-order rate constant for reaction in the micellar interior is given by the corresponding rate constant for bulk reaction multiplied by the correction factor f, which may be called the confinement factor. In the limit $R \to \infty$ the confinement factor becomes equal to unity. In Figure 11.14 the confinement factor is plotted against R/d.

11.3.4. Reactions on Micellar Surface (34)

11.3.4.1. Mean Reaction Time. Some molecules or ions are exclusively adsorbed on the micellar surface, and diffusion-controlled reactions between them can occur. Consider a reaction between a pair of reactants A and B that are adsorbed on the surface of a micelle and are allowed to diffuse on it. Since we are interested only in the relative position of the two reactants, without loss of generality we can fix one reactant at the south pole and assume that the other moves with a diffusion coefficient D equal to the sum of those of the two reactants (Figure 11.15).

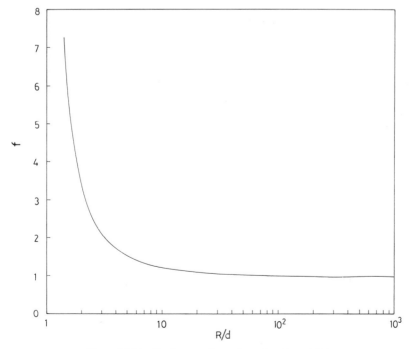

Figure 11.14. Confinement factor f as a function of R/d.

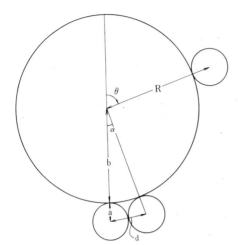

Figure 11.15. Reaction between a pair of reactants on the micellar surface. Small circles of radius a and a large circle of radius b represent reactants and a micelle, respectively. One of the reactants is fixed at the south pole, whereas the other can diffuse on the micellar surface. Reaction occurs when the pair come into contact.

If we specify the initial position of the mobile reactant by $z = \cos \theta$, the pair survival probability $W(z, t)$ satisfies

$$\frac{\partial W(z, t)}{\partial t} = \frac{D}{R^2} \frac{\partial}{\partial z} (1 - z^2) \frac{\partial W}{\partial z}$$

(11.59)

$$W(z, t = 0) = 1 \qquad W(z = -\cos \alpha, t) = 0$$

where d, R, and $\cos \alpha$ are given by $d = 2a$, $R = a + b$, and $\cos \alpha = 1 - d^2/2R^2$, respectively (see Figure 11.15). According to the mean reaction time approximation, the pair survival probability $W(z, t)$ is expressed as Eq. (11.49) with r replaced by z. Following the method described in Section 11.3.3.3, one can show that the mean reaction time $\tau(z)$ satisfies

$$\frac{D}{R^2} \frac{d}{dz} (1 - z^2) \frac{d\tau(z)}{dz} = -1 \qquad \tau(z = -\cos \alpha) = 0$$

(11.60)

This set of equations leads to

$$\tau(z) = \frac{R^2}{D} \ln \left[\frac{2R^2}{d^2} (1 + z) \right]$$

(11.61)

Figure 11.16 shows $\tau(z)$ as a function of z. For a random initial distribution one obtains, from Eqs. (11.52) and (11.61),

$$\tau_{av} = \frac{R^2}{D} \left[\frac{2}{1 - (d/2R)^2} \ln \frac{2R}{d} - 1 \right]$$

(11.62)

Figure 11.17 shows τ_{av} as a function of R/d.

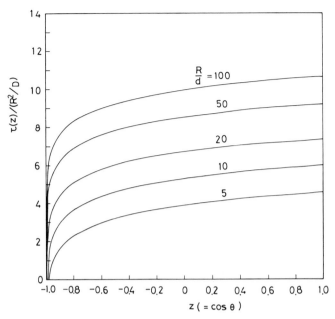

Figure 11.16. Mean reaction time $\tau(z)$ for a pair of reactants on the micellar surface, one fixed at the south pole, the other mobile, as a function of the initial position $z \doteq \cos\theta$ of the latter. See Figure 11.15 for definitions of R and d.

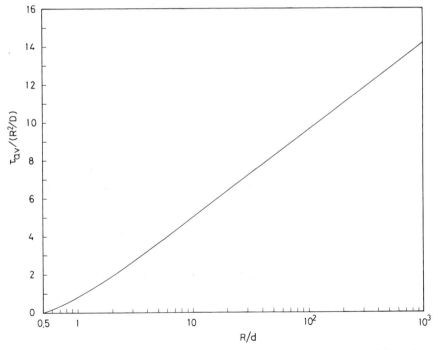

Figure 11.17. Mean reaction time τ_{av} for a pair of reactants on the micellar surface, initially randomly distributed relative to each other, as a function of R/d. See Figure 11.15 for definitions of R and d.

Diffusion-controlled reactions on the micellar surface have also been treated by Hatlee et al. (35) and by Van der Auweraer et al. (36) in the conventional method.

11.3.4.2. Application to Experiment. Several examples of bimolecular reactions occurring on the micellar surface have been observed. They include the dismutation of Br_2^- on the surface of cationic micelles (37), that of Ag_2^+ on the surface of anionic micelles (38), and the electron transfer between $Ru(bpy)_3^{3+}$ and MV^+ on the surface of anionic micelles (39), where bpy = 2,2′-bipyridyl and MV = methylviologen. These ions are all adsorbed on the micellar surface by coulombic attraction. It has been found (37–39) that in micellar solutions these reactions obey pseudo-first-order kinetics, although they are all bimolecular, and that they are accelerated by a factor of ∼ 100 compared with those in bulk solutions.

Since the observed decay of reactants should correspond to the decay of the average pair survival probability $W_{av}(t)$, the observed pseudo-first-order kinetics is in accord with the theoretical prediction. The observed lifetime should correspond to the mean reaction time τ_{av}. However, since the diffusion coefficients of these ions on the micellar surface are not known, we cannot compute τ_{av} and compare them with the observed lifetimes. Instead we have estimated the diffusion coefficients from the observed lifetimes by equating τ_{av} to the observed lifetime. The results are shown in Table 11.2. The estimated diffusion coefficients are on the order of 10^{-7} cm²/s, two orders of magnitude lower than the diffusion coefficients of these ions in aqueous solutions. This indicates that the rate constants are enhanced two orders of magnitude by confining reactants on the micellar surface, although the diffusion coefficients are reduced two orders of magnitude.

11.3.5. Reactions between Reactant in Micellar Interior and One on Micellar Surface

In the case of a reaction between a pair of reactants A and B, A confined in the micelle and B adsorbed on the micellar surface, it is difficult to calculate the mean reaction time for a general case. However, in some special cases one can calculate it analytically. If B on the surface diffuses extremely fast compared with A (Figure 11.18a), the reaction occurs immediately when A reaches the surface. In this case the mean reaction time $\tau(r)$ satisfies

$$D \, \nabla^2 \tau(r) = -1 \qquad \tau(r = R) = 0 \qquad (11.63)$$

where D is the diffusion coefficient of A and R is given by $R = b - a$. The solution of this set of equations is

$$\tau(r) = \frac{R^2}{6D} \left[1 - \left(\frac{r}{R}\right)^2 \right] \qquad (11.64)$$

TABLE 11.2. Examples of Diffusion-Controlled Reactions on Micellar Surfaces

Reaction	Micelle	Species	Radius of Species[a] (Å)	Radius of Micelle (Å)	Observed Lifetime (μs)	Estimated Diffusion Coefficient (cm²/s)
$Br_2^- + Br_2^- \rightarrow Br_3^- + Br^{-}$ [b]	Cetyltrimethyl ammoniumbromide	Br_2^-	1.5	13.5	0.48^b	8×10^{-8}
$Ag_2^+ + Ag_2^+ \rightarrow Ag_2 + 2Ag^{+}$ [c]	Sodiumhexadecyl trioxiethylensulfate	Ag_2^+	2	28	$1.4, \sim 3.2^c$	$1.4 \times 10^{-7}, \sim 6 \times 10^{-8}$
$Ru(bpy)_3^{3+} + MV^+ \rightarrow Ru(bpy)_3^{2+} + MV^{2+}$ [d]	Sodiumdodecyl sulfate	$Ru(bpy)_3^{3+}$ MV^+	2.5 2.5	16	0.175^d	6×10^{-7e}

[a] Assumed.
[b] From Ref. 37.
[c] From Ref. 38.
[d] From Ref. 39; bpy = 2,2'-bipyridyl and MV = methylviologen.
[e] Sum of the diffusion coefficients of the two species.

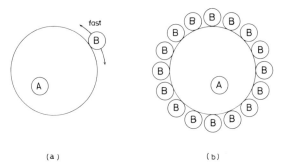

(a)   (b)

Figure 11.18. Reaction of a reactant A confined in the micelle: (a) with a reactant B diffusing extremely fast on the micellar surface, or (b) with many B reactants on the micellar surface.

For a random initial distribution one obtains

$$\tau_{av} = \frac{R^2}{15D} \tag{11.65}$$

A case where many B's instead of one are adsorbed on the surface is also interesting (Figure 11.18b). In this case, even if B does not diffuse, the reaction occurs immediately when A reaches the surface. Therefore, the mean reaction time for the average pair survival probability is again given by Eq. (11.65).

When one B is adsorbed on the inner surface of the micelle and does not diffuse (Figure 11.19), the mean reaction time $\tau(r, \theta)$ satisfies

$$D \left\{ \frac{1}{r^2} \frac{\partial}{\partial r} \left[r^2 \frac{\partial \tau(r, \theta)}{\partial r} \right] + \frac{1}{r^2} \frac{1}{\sin \theta} \frac{\partial}{\partial \theta} \left[\sin \theta \frac{\partial \tau(r, \theta)}{\partial \theta} \right] \right\} = -1$$

$$\tau(r = R \cos \theta - \sqrt{d^2 - R^2 \sin^2 \theta}, \theta) = 0 \quad \text{for } \theta < \alpha$$

$$\left[\frac{\partial \tau(r, \theta)}{\partial r} \right]_{r=R} = 0 \quad \text{for } \theta > \alpha$$

$$\tag{11.66}$$

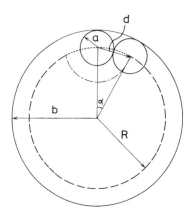

Figure 11.19. Reaction between a pair of reactants, one fixed at the north pole, the other mobile in the micellar interior. Small circles of radius a and a large circle of radius b represent reactants and a micelle, respectively.

where d, R, and α are given by $d = 2a$, $R = b - a$, and $\cos \alpha = 1 - (d^2/2R^2)$, respectively. The second equation states that reaction occurs immediately when the center of A reaches the surface represented by a dashed-dotted line in Figure 11.19. If the second equation is modified to

$$\tau(r = R, \theta) = 0 \qquad \text{for } \theta < \alpha \qquad (11.67)$$

which states that reaction occurs when the center of A reaches the surface represented by a dotted line in Figure 11.19, following the method of Samson and Deutch (40), an approximate solution of Eq. (11.66) is obtainable:

$$\tau(r, \theta) = \frac{R^2}{D} \left\{ \frac{1}{6} \left[1 - \left(\frac{r}{R} \right)^2 \right] + \frac{2}{3} \sum_{l=1}^{\infty} \frac{l + \frac{1}{2}}{l} \frac{\beta_l}{\beta_0} \right.$$
$$\left. \times \left[1 - \left(\frac{r}{R} \right)^l P_l (\cos \theta) \right] \right\} \qquad (11.68)$$

where

$$\beta_l = \int_{\cos \alpha}^{1} P_l(z) \, dz \qquad (11.69)$$

and P_l is the lth Legendre polynomial. For a random initial distribution one obtains

$$\tau_{av} = \frac{R^2}{D} \left[\frac{1}{15} + \frac{1}{3(1 - \cos \alpha)} \sum_{l=1}^{\infty} \frac{P_{l-1}(\cos \alpha) - P_{l+1}(\cos \alpha)}{l} \right] \qquad (11.70)$$

Figure 11.20 shows τ_{av} as a function of R/d.

In Table 11.3 we summarize the mean reaction time τ_{av} for various arrangements of reactants A and B.

11.3.6. Magnetic Effects on Diffusion-Controlled Reactions in Micelles (41)

An example of magnetic effects on chemical reactions involves the formation and recombination of radical pairs (8, 9). When a molecule dissociates into a pair of radicals thermally or photolytically, the spin state of each produced radical is doublet. On the other hand, the total spin state of the pair is initially singlet or triplet, depending on the spin state of the original molecule. However, these singlet and triplet states are not complete eigenstates. They undergo intersystem crossing to each other by virtue of their electron–nuclear hyperfine interaction and the difference in the g values of the two radicals. The strength of this interaction is changeable by an external magnetic field. In the meantime, spatially, the two radicals diffuse mutually independently, and after some time they may reencounter. When they reencounter, it depends on the total spin state of the pair whether they recom-

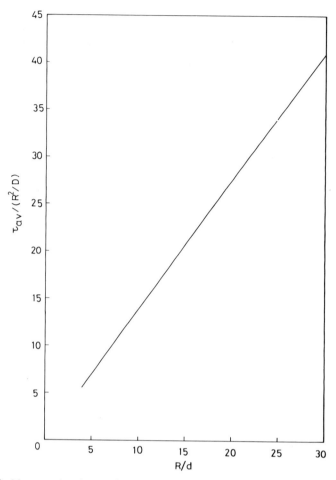

Figure 11.20. Mean reaction time τ_{av} for a pair of reactants, one fixed on the micellar surface, the other initially randomly distributed in the micellar interior, as a function of R/d. See Figure 11.19 for definitions of R and d.

bine or not. If the total spin state is singlet, they will recombine because the interradical potential is bonding. However, if triplet, they will not recombine because of the repulsive potential and will separate again. Thus, the spatial motion of the pair couples with the motion of the total spin, and accordingly, the dynamics of geminate recombination is influenced by g values and nuclear spins of radicals and an external magnetic field, which affect the motion of the total spin.

Lawler and Evans (42) pointed out that the influence of nuclear spins on geminate recombination should produce enrichment of isotopes of differing magnetic moment. Buchachenko et al. (43) observed the magnetic isotope effect for ^{13}C in the photolysis of dibenzylketone (DBK). Turro et al. (44) have demonstrated that this effect is greatly enhanced when the reaction is carried out in micelles. The overall scheme for photolysis of DBK is depicted in Figure 11.21. The light absorption initially produces a singlet excited state that undergoes intersystem cross-

TABLE 11.3. Mean Reaction Time τ_{av} for Average Pair Survival Probability in Various Arrangements of Reactants A and B

Arrangement of Reactants[a]	Mean Reaction Time[b]
Case 1	$\tau_{av} = \dfrac{(b-a)^2}{D} \dfrac{1}{3x(1-x^3)} \left(1 - \dfrac{9}{5}x + x^3 - \dfrac{1}{5}x^6\right)$
Case 2	$\tau_{av} = \dfrac{(a+b)^2}{D} \dfrac{2}{1-(1/4)y^2} \ln \dfrac{2}{y}$
Case 3[c]	
B fast diffusion	$\tau_{av} = \dfrac{1}{15} \dfrac{(b-a)^2}{D}$
B fixed	$\tau_{av} = \dfrac{(b-a)^2}{D} \left[\dfrac{1}{15} + \dfrac{2}{3x^2} \right.$ $\left. \times \sum\limits_{l=1}^{\infty} \dfrac{P_{l-1}(1 - \frac{1}{2}x^2) - P_{l+1}(1 - \frac{1}{2}x^2)}{l} \right]$

[a] see Figure 11.8.
[b] $x = 2a/(b-a)$, $y = 2a/(a+b)$.
[c] see text for details.

ing to a triplet excited state. The latter then dissociates into a pair of radicals, which initially has triplet total spin. Subsequently the processes described in the preceding paragraph occur. In addition, another process occurs in this specific example, the decarbonylation of the ketyl radical (see Figure 11.21).

Sterna et al. (41) presented a model for the magnetic isotope effect in the photolysis of DBK in micelles. As in the treatment in Section 11.3.3.1 they fixed one of the two radicals at the center of the micelle and described the motion of the pair by

$$\frac{\partial \rho_S(\mathbf{r}, t)}{\partial t} = D \nabla^2 \rho_S(\mathbf{r}, t) - k_{CO}\rho_S(\mathbf{r}, t) - k_{ST}\rho_S(\mathbf{r}, t) + k_{TS}\rho_T(\mathbf{r}, t)$$

$$\frac{\partial \rho_T(\mathbf{r}, t)}{\partial t} = D \nabla^2 \rho_T(\mathbf{r}, t) - k_{CO}\rho_T(\mathbf{r}, t) + k_{ST}\rho_S(\mathbf{r}, t) - k_{TS}\rho_T(\mathbf{r}, t)$$

$$\rho_S(\mathbf{r}, t = 0) = 0 \qquad \rho_T(\mathbf{r}, t = 0) = \delta(\mathbf{r} - \mathbf{r}_0) \qquad \rho_S(r = d, t) = 0$$

$$\left[\frac{\partial \rho_T(r, t)}{\partial r} \right]_{r=d} = 0 \qquad \left[\frac{\partial \rho_S(r, t)}{\partial r} \right]_{r=R} = 0 \qquad \left[\frac{\partial \rho_T(r, t)}{\partial r} \right]_{r=R} = 0$$

$$(11.71)$$

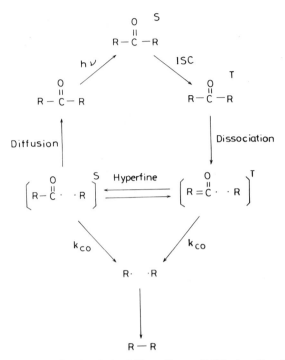

Figure 11.21. Overall scheme for photolysis of dibenzylketone R_2CO where $R = C_6H_5CH_2$. S and T represent singlet and triplet states, respectively. The light absorption ($h\nu$) produces a singlet excited state that undergoes intersystem crossing (ISC) to a triplet excited state. The square brackets indicate radical pairs. Hyperfine couplings equilibrate S and T while diffusive encounter of singlets reforms the ketone. Decarbonylation (k_{CO}) forms 1,2-diphenylethane.

where $\rho_S(\mathbf{r}, t)$ and $\rho_T(\mathbf{r}, t)$ denote the probability densities for the singlet pair and for the triplet pair, respectively; k_{CO} is the rate constant for decarbonylation of the ketyl radical; and k_{ST} and k_{TS} are the singlet to triplet and the triplet to singlet intersystem crossing rate constants, respectively. See Section 11.3.3.1 for other notations. The third and the fourth equations state that the pair is initially produced in the triplet state, with the initial interradical separation $\mathbf{r}_0$. The fifth and the sixth equations state that on approaching each other to a distance d, the two radicals recombine immediately if the total spin state is singlet and are reflected if triplet.

The recombination probability κ, which is the probability that the two radicals will recombine to form the original molecule (DBK) against the decarbonylation of the ketyl radical, is given by

$$\kappa = \int_0^\infty D\left[\frac{\partial \rho_S(\mathbf{r}, t)}{\partial r}\right]_{r=d} 4\pi d^2 \, dt \qquad (11.72)$$

Equations (11.71) and (11.72) give, after a tedious calculation,

$$\kappa = \frac{d}{r_0} \frac{f(r_0, \xi)\, g(\zeta) - f(r_0, \zeta)\, g(\xi)}{f(d, \xi)\, g(\zeta) + (k_{ST}/k_{TS})\, f(d, \zeta)\, g(\xi)} \tag{11.73}$$

where

$$f(r, \beta) = \exp\left[-\beta(r - d)\right]\left\{ 1 - \frac{1 + \beta R}{1 - \beta R} \exp\left[2\beta(r - R)\right]\right\}$$

$$g(\beta) = (1 + \beta d)\left\{ 1 - \frac{1 - \beta d}{1 + \beta d} \frac{1 + \beta R}{1 - \beta R} \exp\left[2\beta(d - R)\right]\right\}$$

$$\xi = \left(\frac{k_{CO}}{D}\right)^{1/2}$$

$$\zeta = \left(\frac{k_{CO} + k_{ST} + k_{TS}}{D}\right)^{1/2} \tag{11.74}$$

All of the isotope dependence is contained in k_{ST} and k_{TS}. Sterna et al. estimated, for the ^{13}C-containing radical pair,

$$^{13}k_{ST} = 5.25 \times 10^8\ s^{-1}, \qquad ^{13}k_{TS} = 1.75 \times 10^8\ s^{-1}$$

and for the ^{12}C-containing radical pair

$$^{12}k_{ST} = 8.25 \times 10^7\ s^{-1}, \qquad ^{12}k_{TS} = 2.75 \times 10^7\ s^{-1}$$

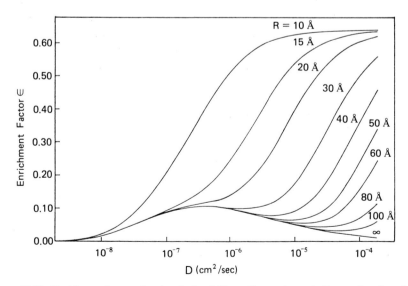

Figure 11.22. Enrichment factor ϵ for photolysis of dibenzylketone in a micelle as a function of the sum D of the diffusion coefficients of the radical pair (41). See Figure 11.10 for the definition of R.

Other quantities appearing in Eq. (11.73) were estimated as

$$k_{CO} = 10^{13} \exp \left(\frac{-E_a}{RT} \right) \quad s^{-1} \quad \text{with } E_a = 7.3 \text{ kcal/mole}$$

$$d = 6 \times 10^{-8} \text{ cm}, \quad r_0 = 9 \times 10^{-8} \text{ cm}$$

The loss of the original molecule upon photolysis is given by $1 - \kappa$. They defined the enrichment factor ϵ as equal to the differential loss between ^{12}C-containing DBK and ^{13}C-containing DBK relative to the loss of ^{12}C-containing DBK:

$$\epsilon = \frac{(1 - {}^{12}\kappa) - (1 - {}^{13}\kappa)}{1 - {}^{12}\kappa} \tag{11.75}$$

They calculated the enrichment factor as a function of the sum D of the diffusion coefficients of the radical pair and the radius R of the micelle. The result is shown in Figure 11.22. The enhancement of the enrichment factor with decreasing micelle radius is quite evident.

11.4. STOCHASTIC TREATMENT OF REACTIONS IN MICELLAR AND VESICULAR SYSTEMS

11.4.1. Necessity of Stochastic Treatment

In Section 11.2 we studied the distribution of solubilizates among micelles. Since the volume of a micelle is relatively small, the average number of reactants contained in a micelle is usually small, and concomitantly the breadth of the distribution is similar to the average number. When micelles contain a small number of reactants, the reaction kinetics depends strongly on the number of reactants in a micelle. For example, when micelles contain one A reactant and no B reactant, the reaction between A and B does not occur, and the fraction $n_A(t)$ of A surviving at time t remains unity. On the other hand, when micelles contain one A and two B's, according to the discussion in Section 11.3, the decay of A is described by the following pseudo-first-order kinetics:

$$n_A(t) = \exp(-2k_r t) \tag{11.76}$$

where k_r is the pseudo-first-order rate constant for reaction between A and B in a micelle containing one A and one B. Suppose a case where one-half of the micelles contain one A and no B, and the other half contain one A and two B's. In this case the average number of B's in a micelle is unity, so if one were allowed to use the average number instead of taking into account the distribution of B among micelles, the decay of A would be described by

$$n_A(t) = \exp(-k_r t) \tag{11.77}$$

However, since the reaction kinetics strongly depends on the number of cohabitant B's, the reaction kinetics is not describable in terms of the average number. Taking into account the distribution of B, one obtains

$$n_A(t) = 0.5 \exp(-2k_r t) + 0.5 \tag{11.78}$$

Figure 11.23 shows the decay curves of A calculated from Eq. (11.77) or (11.78). The difference between the two curves is remarkable, indicating the necessity of stochastic treatment, which takes into account the distribution of reactants among micelles.

In reactions in the bulk system the reaction kinetics can be treated in terms of the average concentration of reactants, although there also are fluctuations in the local concentration. Thus, it is interesting to examine under what conditions the reaction kinetics is describable in terms of the average number or concentration. This can be done by considering an ensemble of subspaces with volume v, each subspace containing on the average $\bar{n}$ B's. Suppose that subspaces can exchange B between them and that if a subspace contains n B's, the rate constant for exit of a B from the subspace is given by nk, while the rate constant k' for entry of a B into the subspace is independent of n. In this case the distribution of B among subspaces obeys a Poisson distribution, and k' is given by $k' = \bar{n}k$. Suppose that one A is introduced in each subspace at time zero. The scheme for subsequent reaction between A and B may be described by

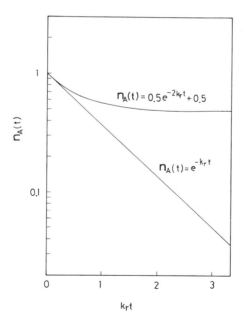

Figure 11.23. Decay curves of A reactants calculated from Eq. (11.77) or (11.78).

$$M_n^A \underset{(n+1)k}{\overset{\bar{n}k}{\rightleftharpoons}} M_{n+1}^A \qquad (n = 0, 1, \ldots)$$

$$M_n^A \overset{nkr}{\longrightarrow} M_n \qquad (n = 0, 1, \ldots) \tag{11.79}$$

where M_n^A stands for a subspace containing one A and n B's, while M_n stands for a subspace containing no A. The first equation shows the entry and exit of a B into and from a subspace. The second equation shows the reaction between A and B in a subspace. According to the discussion in Section 11.3, the reaction in a subspace obeys pseudo-first-order kinetics, and when the subspace contains n B's, the rate constant is given by nk_r where k_r is the rate constant in a subspace containing one A and one B.

We define $Q_n^A(t)$ as the fraction at time t of subspaces containing one A and n B's. Equation (11.79) leads to the following set of rate equations for $Q_n^A(t)$.

$$\frac{dQ_0^A(t)}{dt} = -\bar{n}kQ_0^A(t) + kQ_1^A(t)$$

$$\frac{dQ_n^A(t)}{dt} = \bar{n}kQ_{n-1}^A(t) - (\bar{n}k + nk + nk_r)Q_n^A(t) \tag{11.80}$$

$$+ (n + 1)kQ_{n+1}^A(t) \qquad (n = 1, 2, \ldots)$$

This set of differential difference equations can be solved by means of a generating function (45), which is defined as

$$F(s, t) = \sum_{n=0}^{\infty} s^n Q_n^A(t) \qquad |s| < 1 \tag{11.81}$$

where s is an arbitrary parameter considered as an independent variable. The following general properties of $F(s, t)$ follow from its definition.

$$\sum_{n=0}^{\infty} Q_n^A(t) = F(1, t) \tag{11.82}$$

$$\sum_{n=0}^{\infty} nQ_n^A(t) = \left[\frac{\partial F(s, t)}{\partial s}\right]_{s=1} \tag{11.83}$$

$$Q_n^A(t) = \frac{1}{n!}\left[\frac{\partial^n F(s, t)}{\partial s^n}\right]_{s=0} \tag{11.84}$$

Equation (11.80) is converted to a differential equation for $F(s, t)$ in the following way. Differentiating Eq. (11.81) with respect to t and utilizing Eq. (11.80), one obtains

$$\frac{\partial F(s, t)}{\partial t} = \bar{n}k \sum_{n=0}^{\infty} s^n Q_{n-1}^A(t) - \bar{n}k \sum_{n=0}^{\infty} s^n Q_n^A(t)$$

$$- (k + k_r) \sum_{n=1}^{\infty} n s^n Q_n^A(t)$$

$$+ k \sum_{n=0}^{\infty} (n + 1) s^n Q_{n+1}^A(t) \qquad (11.85)$$

On the other hand, Eq. (11.81) gives

$$\frac{\partial F(s, t)}{\partial s} = \sum_{n=1}^{\infty} n s^{n-1} Q_n^A(t) \qquad (11.86)$$

With the aid of Eqs. (11.81) and (11.86), Eq. (11.85) is rewritten as

$$\frac{\partial F(s, t)}{\partial t} = [k - (k + k_r)s] \frac{\partial F(s, t)}{\partial s} + \bar{n}k(s - 1) F(s, t) \qquad (11.87)$$

Before the introduction of A, the distribution of B among subspaces obeys a Poisson distribution; therefore,

$$Q_n^A(0) = \frac{1}{n!} \bar{n}^n \exp(-\bar{n}) \qquad (11.88)$$

This equation imposes the following initial condition on the generating function:

$$F(s, 0) = \exp[\bar{n}(s - 1)] \qquad (11.89)$$

The solution of Eq. (11.87) subject to (11.89) is

$$F(s, t) = \exp\left\{ -\frac{\bar{n}kk_r}{k + k_r} t + \frac{\bar{n}k}{k + k_r} \right.$$

$$\times \left(s - \frac{k}{k + k_r} \right) + \frac{\bar{n}kk_r}{k(k + k_r)} \left[\left(s - \frac{k}{k + k_r} \right) e^{-(k + k_r)t} - 1 \right] \right\}$$

$$(11.90)$$

The fraction $n_A(t)$ of A surviving at time t is

$$n_A(t) = \sum_{n=0}^{\infty} Q_n^A(t) \qquad (11.91)$$

With the aid of Eq. (11.82) Eq. (11.91) is rewritten as

$$n_A(t) = F(1, t)$$

$$= \exp\left\{ -\frac{\bar{n}kk_r}{k + k_r} t - \frac{\bar{n}k_r^2}{(k + k_r)^2} [1 - e^{-(k + k_r)t}]\right\} \quad (11.92)$$

This equation describes the kinetics of decay of A by reaction with B. According to the discussion in Section 11.3, the rate constant k_r for the reaction in a subspace is given in terms of the bulk rate constant k^b and the volume v of the subspace,

$$k_r = \frac{k^b}{v} f \quad (11.93)$$

where f is the confinement factor.

Two limiting cases of Eq. (11.92) with respect to the exchange rate k of B between subspaces are interesting. In the limit $k \to \infty$ the subspaces are effectively connected with each other to form a single large space as far as the distribution of B is concerned, so that the reaction kinetics should reduce to that in the bulk system. As a matter of fact, in this limit Eq. (11.92) together with (11.93) leads to the reaction kinetics in the bulk system except for the factor f.

$$n_A(t) = \exp(-k^b fct) \quad (11.94)$$

where $c = \bar{n}/v$ is the average concentration of B. In the other limit $k \to 0$, Eq. (11.92) together with (11.93) leads to

$$n_A(t) = \exp[-\bar{n}(1 - e^{-k^b ft/v})] \quad (11.95)$$

In this limit the subspaces are completely separated from each other. However, even in this limit, if the volume v of a subspace is itself very large, the reaction kinetics should reduce to that in the bulk system. In fact, in the limit $v \to \infty$, Eq. (11.95) leads to

$$n_A(t) = \exp[- k^b ct] \quad (11.96)$$

In conclusion, one can treat the reaction kinetics in terms of the average concentration if the exchange rate k of reactants between subspaces is very high compared with the reaction rate k_r in the subspace or if the volume v of the subspace is very large compared with the product of the bulk rate constant k^b and the observation time t. Otherwise, one has to take into account the distribution of reactants among subspaces.

The influence of concentration fluctuations on diffusion-controlled bimolecular reactions has also been discussed by Noyes and Gardiner (46a) and by Toussaint and Wilczek (46b).

In this section we treat five types of reactions that are important in practical applications, taking into account the distribution of reactants among micelles or vesicles:

1. Quenching of an excited molecule $P*$ by a quencher Q. The quenching may occur by energy transfer, by exciplex formation, or by electron transfer,

$$P* + Q \rightarrow P + Q* \quad \text{or} \quad PQ* \text{ or } P^{\pm} + Q^{\mp} \tag{11.97}$$

2. Formation and dissociation of an excimer P_2^*, involving $P*$ and a ground-state molecule P,

$$P* + P \rightleftarrows P_2^* \tag{11.98}$$

3. Triplet-triplet annihilation,

$$^3P* + {}^3P* \rightarrow {}^1P* + P \tag{11.99}$$

where $^3P*$ and $^1P*$ denote triplet and singlet excited molecules, respectively;
4. Emulsion polymerization;
5. Migration-controlled reactions in vesicular systems. This is important in connection with fluorescence stopped-flow techniques, which have recently been developed for the study of migration rates of solubilizates in vesicular systems.

Other types of reactions in micellar systems have also been treated (47), although at present they do not have real applications.

11.4.2. Fluorescence Quenching

11.4.2.1. Case of Poisson Distribution. Suppose that fluorescent probes P that are dissolved exclusively in micelles are excited at time zero with a laser pulse. If the laser intensity is not so strong, the probability that two or more probes will be excited in a micelle is negligible compared with the probability that one probe will be excited. Under this condition, one has only to consider micelles that contain only one excited probe $P*$. Consider the quenching of $P*$ by quenchers Q. Here we consider a case where the distribution of quenchers among micelles obeys a Poisson distribution. There are two possible migration mechanisms that yield a Poisson distribution (see section 11.2).

(i). Migration via Aqueous Phase (16). When quenchers migrate between micelles via the aqueous phase according to Eq. (11.1), the overall scheme for quenching of $P*$ may be described by

$$M_n^* \xrightarrow{k_0} M_n \qquad (n = 0, 1, 2, \ldots)$$

$$M_n^* + Q_{aq} \underset{(n+1)k_-}{\overset{k_+}{\rightleftarrows}} M_{n+1}^* \qquad (n = 0, 1, 2, \ldots) \qquad (11.100)$$

$$M_n^* \xrightarrow{nk_q} M_n \qquad (n = 1, 2, \ldots)$$

Here M_n^* stands for a micelle containing $P*$ and n quenchers, M_n stands for a micelle containing no $P*$, and Q_{aq} stands for a quencher in the aqueous phase. The first equation shows the unimolecular decay of $P*$ and k_0 is the decay rate constant in the absence of quenchers. The second equation shows the migration of a quencher between micelles and the aqueous phase. The third equation shows the quenching of $P*$ by quenchers. According to the discussion in Section 11.3, the quenching reaction also obeys pseudo-first-order kinetics in micelles, and when a micelle contains n quenchers, the rate constant is given by nk_q, where k_q is the rate constant for quenching in a micelle containing one quencher. See Section 11.3 for the dependence of k_q on the physical parameters of reactants and micelles. Note that in the case of $k_0 = 0$, Eq. (11.100) reduces to (11.79).

We define $[M_n^*](t)$ as the concentration at time t of micelles containing $P*$ and n quenchers. For a sufficiently rapid excitation pulse it is a reasonable assumption that the efficiency of excitation of a probe is independent of the number of cohabitant quenchers. This assumption together with the fact that the distribution of quenchers among micelles obeys a Poisson distribution imposes the following initial condition on $[M_n^*](t)$:

$$[M_n^*](0) = \frac{C^*(0)}{n!} \left(\frac{k_+[Q_{aq}]}{k_-} \right)^n \exp \left(- \frac{k_+[Q_{aq}]}{k_-} \right) \qquad (11.101)$$

where $C^*(0)$ is the initial total concentration of $P*$ and $[Q_{aq}]$ is the concentration of quenchers in the aqueous phase. On the other hand, Eq. (11.100) leads to the following set of rate equations for $[M_n^*](t)$:

$$\frac{d[M_0^*](t)}{dt} = -(k_0 + k_+[Q_{aq}]) [M_0^*](t) + k_-[M_1^*](t)$$

$$\frac{d[M_n^*](t)}{dt} = k_+[Q_{aq}] [M_{n-1}^*](t) - (k_0 + k_+[Q_{aq}] + nk_- + nk_q) [M_n^*](t)$$

$$+ (n + 1)k_- [M_{n+1}^*](t) \qquad (n = 1, 2, \ldots) \qquad (11.102)$$

If the system is in dynamic equilibrium with respect to the distribution of quenchers between micelles and the aqueous phase, the concentration of quenchers in the aqueous phase, $[Q_{aq}]$, is time independent.

By means of the generating function $F(s, t)$ of $[M_n^*]$ (t), namely,

$$F(s, t) = \sum_{n=0}^{\infty} s^n [M_n^*] (t) \qquad |s| < 1 \tag{11.103}$$

Eq. (11.102) may be transformed into

$$\frac{\partial F(s, t)}{\partial t} = [k_- - (k_- + k_q)s] \frac{\partial F(s, t)}{\partial s} + [k_+[Q_{aq}]s - (k_0 + k_+[Q_{aq}])] F(s, t) \tag{11.104}$$

On the other hand, Eq. (11.101) imposes the following initial condition on the generating function:

$$F(s, 0) = C^*(0) \exp \left[\frac{k_+[Q_{aq}]}{k_-} (s - 1) \right] \tag{11.105}$$

The solution of Eq. (11.104) subject to (11.105) is

$$F(s, t) = C^*(0) \exp \left\{ -\left(k_0 + \frac{k_q k_+ [Q_{aq}]}{k_q + k_-} \right) t + \frac{k_+[Q_{aq}]}{k_q + k_-} \left(s - \frac{k_-}{k_q + k_-} \right) \right.$$
$$\left. + \frac{k_q k_+ [Q_{aq}]}{k_-(k_q + k_-)} \left[\left(s - \frac{k_-}{k_q + k_-} \right) e^{-(k_q + k_-)t} - 1 \right] \right\} \tag{11.106}$$

Experimentally significant is the total concentration $C^*(t)$ of P^*, which is given by

$$C^*(t) = \sum_{n=0}^{\infty} [M_n^*] (t) \tag{11.107}$$

Equation (11.107) is rewritten as

$$C^*(t) = F(1, t)$$
$$= C^*(0) \exp \left\{ -\left(k_0 + \frac{k_q k_+ [Q_{aq}]}{k_q + k_-} \right) t - \frac{k_q^2 k_+ [Q_{aq}]}{k_-(k_q + k_-)^2} [1 - e^{-(k_q + k_-)t}] \right\} \tag{11.108}$$

This equation describes the kinetics of decay of $P*$ by quenching. Since the average number $\bar{n}$ of quenchers in a micelle is given by Eq. (11.7), $C*(t)$ is expressed in terms of $\bar{n}$:

$$C*(t) = C*(0) \exp\left\{-\left(k_0 + \frac{\bar{n}k_q k_-}{k_q + k_-}\right)t - \frac{\bar{n}k_q^2}{(k_q + k_-)^2}[1 - e^{-(k_q + k_-)t}]\right\}$$

(11.109)

In terms of the total micelle and quencher concentrations $[M]$ and $[Q]$, $\bar{n}$ is given by Eq. (11.10) with $[A]$ replaced by $[Q]$.

When the system is irradiated with a steady light, experimentaly significant is the total fluorescence intensity I, which is proportional to the intergral of $C*(t)$ over all time:

$$I \propto k_0^r \int_0^\infty C*(t)\, dt$$

(11.110)

where k_0^r is the radiative rate constant of $P*$. Substitution of Eq. (11.109) in this equation leads to the following equation for the ratio of total fluorescence intensities in the presence and absence of quenchers

$$\frac{I}{I_0} = \exp\left(-\frac{k_q^2\bar{n}}{(k_q + k_-)^2}\right) \sum_{n=0}^\infty \frac{1}{n!}\left(\frac{k_q^2\bar{n}}{(k_q + k_-)^2}\right)^n$$
$$\times \left(1 + \frac{k_q k_-\bar{n}}{k_0(k_q + k_-)} + \frac{n(k_q + k_-)}{k_0}\right)^{-1}$$

(11.111)

where I_0 is the total fluorescence intensity in the absence of quenchers. Equations (11.109) and (11.111), which provide general expressions for $C*(t)$ and I/I_0, respectively, reduce to simple forms in the following limiting cases (48).

Case 1: static quenching, quencher is totally micellized.

When the quenching rate constant k_q is infinitely large (static quenching) and the exit rate constant k_- is negligibly small compared with other rate constants (quencher is totally micellized), Eqs. (11.109) and (11.111) reduce to

$$C*(t) = C*(0)\, e^{-\bar{n}}\, e^{-k_0 t}$$

(11.112)

$$\frac{I}{I_0} = e^{-\bar{n}}$$

(11.113)

where $\bar{n} = [Q]/[M]$. Since the quenching is static, emission occurs only from the fraction of micelles that contain P^* but no quencher. The factor $e^{-\bar{n}}$ in Eqs. (11.112) and (11.113) represents this fraction. A typical fluorescence decay curve expected from Eq. (11.112) is shown in Figure 11.24. Although the overall fluorescence intensity is reduced, the measured lifetime of P^* remains constant (k_0) upon addition of quenchers. According to Eq. (11.113), the measured ratio of total fluorescence intensities in the presence and absence of quenchers yields the total micelle concentration $[M]$. A knowledge of the total micelle concentration leads to the mean micelle aggregation number $\bar{\nu}$ through Eq. (11.11). The mean aggregation numbers of a variety of micelles have been succesfully determined by use of this method (49).

The quenching of pyrene fluorescence by amines in SDS micelles (50) is categorized into case 1.

Case 2: static quenching, quencher is partially micellized.

Consider a case where the quenching is static but the exit rate constant k_- is comparable with other rate constants (quencher is partially micellized). Since the quenching is static, as in case 1, emission still occurs only from the fraction of

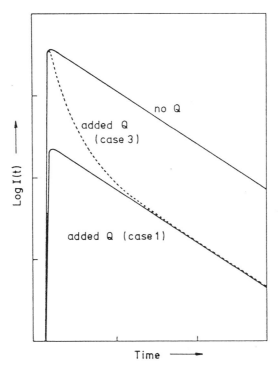

Figure 11.24. Schematic representation of fluorescence decay curves in the case where the quencher is totally micellized (48). Case 1: static quenching. Case 3: nonstatic quenching.

micelles that contain $P*$ but no quencher. Equation (11.109) and (11.111) reduce in this case to

$$C*(t) = C*(0)\, e^{-\bar{n}}\, e^{-(k_0 + k_-\bar{n})t} \tag{11.114}$$

$$\frac{I}{I_0} = e^{-\bar{n}}\, \frac{1}{I + \dfrac{k_-\bar{n}}{k_0}} \tag{11.115}$$

where $\bar{n} = K[Q]/(1 + K[M])$. Figure 11.25 shows a typical fluorescence decay curve expected from Eq. (11.114). The quenching of 1,5-dimethyl naphthalene fluorescence by cyclic azo alkanes in SDS and CTAB micelles (51) is categorized into case 2. Analysis of data on the basis of Eq. (11.114) would yield the entry and the exit rate constants k_+ and k_- in addition to the mean micelle aggregation number $\bar{\nu}$.

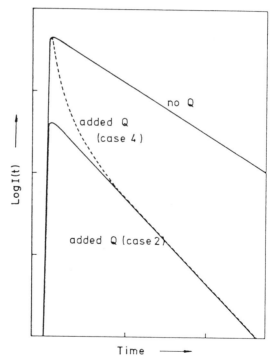

Figure 11.25. Schematic representation of fluorescence decay curves in the case where the quencher is partially micellized (48). Case 2: static quenching. Case 4: nonstatic quenching.

Case 3: nonstatic quenching, quencher is totally micellized.

When the quenching rate constant k_q is comparable with other rate constants (nonstatic quenching) and the quencher is totally micellized, Eqs. (11.109) and (11.111) reduce to

$$C^*(t) = C^*(0) \exp\left[-k_0 t - \bar{n}(1 - e^{-k_q t})\right] \tag{11.116}$$

$$\frac{I}{I_0} = e^{-\bar{n}} \sum_{n=0}^{\infty} \frac{\bar{n}^n}{n!\left(1 + \dfrac{nk_q}{k_0}\right)} \tag{11.117}$$

where $\bar{n} = [Q]/[M]$. The dashed curve in Figure 11.24 shows the typical fluorescence decay pattern expected from Eq. (11.116). We have seen in cases 1 and 2 that static quenching is distinguished by the exponential decay of luminescence. Conversely, in nonstatic quenching the luminescence decay is nonexponential. This is because the lifetime of P^* in a given micelle depends on the number of cohabitant quenchers. The quenching of pyrene fluorescence by nitroxyl radicals in SDS micelles (18b) is categorized into case 3. Analysis of data on the basis of Eq. (11.116) would yield the quenching rate constant k_q in addition to the mean micelle aggregation number $\bar{\nu}$.

Case 4: nonstatic quenching, quencher is partially micellized.

In the most general case where the quenching rate constant k_q and the exit rate constant k_- are both comparable with other rate constants, one has to use Eqs. (11.109) and (11.111) for $C^*(t)$ and I/I_0, respectively. The dashed curve in Fig. 11.25 shows the typical fluorescence decay pattern expected from Eq. (11.109). Analysis of data on the basis of Eq. (11.109) would yield all parameters obtainable in cases 1–3. The quenching of pyrene fluorescence by methylene iodide and nitromethane in SDS micelles (52) is categorized in case 4.

(ii). Migration during Micelle Collisions (24). When quenchers migrate between micelles during micelle collisions according to Eq. (11.21), the second equation in Eq. (11.100) is modified to

$$M_n^* + M_j \underset{(j+1)k}{\overset{nk}{\rightleftharpoons}} M_{n-1}^* + M_{j+1} \qquad \left(\begin{array}{l} n = 1, 2, \ldots \\ j = 0, 1, \ldots \end{array}\right) \tag{11.118}$$

and the rate equation (11.102) is modified to

$$\frac{d[M_0^*](t)}{dt} = -\left(k_0 + \sum_{j=1}^{\infty} jk[M_j]\right)[M_0^*](t) + \sum_{j=0}^{\infty} k[M_j][M_1^*](t)$$

$$\frac{d[M_n^*](t)}{dt} = \sum_{j=1}^{\infty} jk[M_j][M_{n-1}^*](t) - \left(k_0 + \sum_{j=1}^{\infty} jk[M_j] + \sum_{j=0}^{\infty} nk[M_j] + nk_q\right)$$

$$\times [M_n^*](t) + \sum_{j=0}^{\infty} (n+1)k[M_j][M_{n+1}^*](t) \qquad (n = 1, 2, \ldots)$$

$$(11.119)$$

where $[M_j]$ is the concentration of micelles containing j quenchers. With the aid of Eqs. (11.4) and (11.6), Eq. (11.119) is rewritten as

$$\frac{d[M_0^*](t)}{dt} = -(k_0 + \bar{n}k[M])[M_0^*](t) + k[M][M_1^*](t)$$

$$\frac{d[M_n^*](t)}{dt} = \bar{n}k[M][M_{n-1}^*](t) - (k_0 + \bar{n}k[M] + nk[M] + nk_q)[M_n^*](t)$$

$$+ (n+1)k[M][M_{n+1}^*](t) \qquad (n = 1, 2, \ldots) \qquad (11.120)$$

This equation is of the same mathematical form as Eq. (11.102) with replacement of $k_+[Q_{aq}]$ and k_- by $\bar{n}k[M]$ and $k[M]$, respectively. Accordingly, the decay kinetics of P^* is obtained by making the same replacement in Eq. (11.108)

$$C^*(t) = C^*(0) \exp\left\{-\left(k_0 + \frac{\bar{n}k_q k[M]}{k_q + k[M]}\right)t - \frac{\bar{n}k_q^2}{(k_q + k[M])^2}[1 - e^{-(k_q + k[M])t}]\right\}$$

$$(11.121)$$

Similarly, the ratio of total fluorescence intensities in the presence and absence of quenchers is obtainable by making the same replacement in Eq. (11.111),

$$\frac{I}{I_0} = \exp\left(-\frac{\bar{n}k_q^2}{(k_q + k[M])^2}\right)\sum_{n=0}^{\infty} \frac{1}{n!}\left(\frac{\bar{n}k_q^2}{(k_q + k[M])^2}\right)^n$$

$$\times \left(1 + \frac{\bar{n}k_q k[M]}{k_0(k_q + k[M])} + \frac{n(k_q + k[M])}{k_0}\right)^{-1} \qquad (11.122)$$

Equations (11.109) and (11.121) are quite different in the dependence on the total micelle concentration $[M]$. Therefore, one can distinguish the two migration mechanisms by examining the dependence of the decay of P^* on the total micelle concentration (25).

Dederen et al. (53) considered a case where the migration of quenchers between micelles occurs partly via the aqueous phase and partly during micelle collisions.

11.4.2.2. Other Cases.

(i) Binomial Distribution (24). As in the case of Poisson distribution, there are two possible migration mechanisms that yield a binomial distribution. When quenchers migrate via the aqueous phase according to Eq. (11.13), the second equation in eq. (11.100) is modified to

$$M_n^* + Q_{aq} \underset{(n+1)k_-}{\overset{k_+(1-n/m)}{\rightleftharpoons}} M_{n+1}^* \qquad (n = 0, 1, \ldots, m) \qquad (11.123)$$

where m is the maximum number of quenchers in a micelle. Following the method described in Section 11.4.2.1, one can calculate the decay kinetics of P^* and the ratio of total fluorescence intensities. The results are

$$C^*(t) = C^*(0)e^{-k_0 t}\left[\frac{(1 - s_2)(m - \bar{n} + \bar{n}s_1)}{m(s_1 - s_2)}\exp\left(-\frac{\bar{n}(1 - s_1)}{m - \bar{n}}k_- t\right)\right.$$

$$\left. - \frac{(1 - s_1)(m - \bar{n} + \bar{n}s_2)}{m(s_1 - s_2)}\exp\left(-\frac{\bar{n}(1 - s_2)}{m - \bar{n}}k_- t\right)\right]^m \qquad (11.124)$$

$$\frac{I}{I_0} = \sum_{n=0}^{m} {}_mC_n\left(\frac{(1 - s_2)(m - \bar{n} + \bar{n}s_1)}{m(s_1 - s_2)}\right)^n\left(\frac{(1 - s_1)(\bar{n} - m - \bar{n}s_2)}{m(s_1 - s_2)}\right)^{m-n}$$

$$\times \left(1 + \frac{\bar{n}k_-[m - ns_1 - (m - n)s_2]}{(m - \bar{n})k_0}\right)^{-1} \qquad (11.125)$$

where s_1 and s_2 are the roots of $\bar{n}k_- s^2 + [(m - \bar{n})(k_- + k_q) - \bar{n}k_-]s - (m - \bar{n})k_- = 0$. In terms of the total micelle and quencher concentrations $[M]$ and $[Q]$, the average number $\bar{n}$ of quenchers in a micelle is given by eq. (11.16) with $[A]$ replaced by $[Q]$. Note that as $m \to \infty$, Eqs. (11.124) and (11.125) reduce to the corresponding results for the Poisson distribution [Eqs. (11.109) and (11.111), respectively]. Figure 11.26 shows the decay of P^*, and Figure 11.27 shows the dependence of total fluorescence intensity on the quencher concentration, both for the binomial distribution.

According to Nakamura et al. (20), the kinetics of quenching of pyrene fluorescence by Cu^{2+}, Eu^{3+}, and Hg^{2+} in SDS micelles can be described by Eq. (11.124), but not by Eq. (11.109). Their result on Eu^{3+} is shown in Figure 11.28. This indicates that in these systems the distribution of quenchers among micelles is more adquately describable by a binomial distribution.

When quenchers migrate during micelle collsions according to Eq. (11.26), the second equation in Eq. (11.100) is modified to

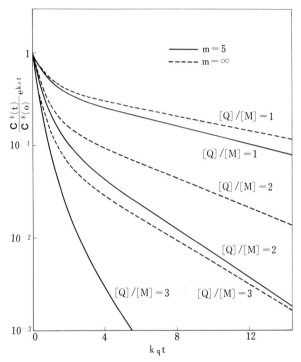

Figure 11.26. Variation in the decay curve of P^* with the ratio $[Q]/[M]$. Maximum number m of quenchers in a micelle is assumed to be 5. The decay curves for the Poisson distribution ($m = \infty$) are also included for comparison.

$$M_n^* + M_j \underset{(j+1)\left[1-\left(\frac{n-1}{m}\right)\right]k}{\overset{n\left(1-\frac{j}{m}\right)k}{\rightleftharpoons}} M_{n-1}^* + M_{j+1} \quad \left(\begin{matrix} n = 1, 2, \ldots, m \\ j = 0, 1, \ldots, m-1 \end{matrix}\right) \quad (11.126)$$

In this case the decay kinetics of P^* and the ratio of total fluorescence intensities are obtained by replacing k_- with $\{1 - (\bar{n}/m)\}\,k[M]$ in Eqs. (11.124) and (11.125), respectively:

$$C^*(t) = C^*(0)e^{-k_0 t}\left[\frac{(1-s_2)\,(m-\bar{n}+\bar{n}s_1)}{m(s_1-s_2)}\exp\left(-\frac{\bar{n}(1-s_1)}{m}k\,[M]t\right)\right.$$

$$\left. -\frac{(1-s_1)\,(m-\bar{n}+\bar{n}s_2)}{m(s_1-s_2)}\exp\left(-\frac{\bar{n}(1-s_2)}{m}k[M]t\right)\right]^m \quad (11.127)$$

$$\frac{I}{I_0} = \sum_{n=0}^m {}_mC_n\left(\frac{(1-s_2)\,(m-\bar{n}+\bar{n}s_1)}{m(s_1-s_2)}\right)^n\left(\frac{(1-s_1)\,(\bar{n}-m-\bar{n}s_2)}{m(s_1-s_2)}\right)^{m-n}$$

$$\times \left(1 + \frac{\bar{n}k[M]\,[m-ns_1 - (m-n)s_2]}{mk_0}\right)^{-1} \quad (11.128)$$

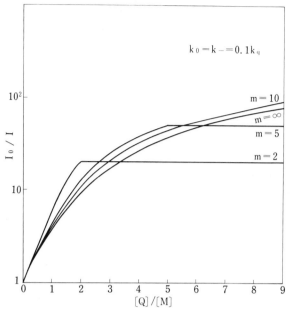

Figure 11.27. Dependence of total fluorescence intensity on the quencher concentration in the case of $k_0 = k_- = 0.1 k_q$. The limit $m = \infty$ corresponds to the Poisson distribution.

where s_1 and s_2 are the roots of $\bar{n}k[M]s^2 + \{(m - 2\bar{n})k[M] + mk_q\} s - (m - \bar{n})k[M] = 0$. Note that as $m \to \infty$, Eqs. (11.127) and (11.128) reduce to the corresponding results for the Poisson distribution [Eqs. (11.121) and (11.122), respectively].

(ii) Geometric Distribution. In Section 11.2 we showed two types of migration dynamics that yield a geometric distribution, although the physical mechanisms

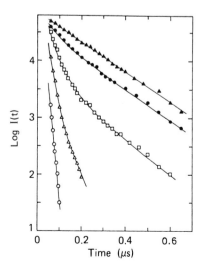

Figure 11.28. Decay of pyrene fluorescence in SDS micelles in the presence of a quencher Eu^{3+} (20). [$EuCl_3$]: $\bigcirc$, 12 mM; $\triangle$, 5.8 mM; $\square$, 2.9 mM; $\bullet$, 0.73 mM; $\blacktriangle$, 0 mM. The full lines are simulations based on Eq. (11.124).

underlying them are not clear. When the dynamics of migration of quenchers is described by Eq. (11.17), it is difficult to derive a general expression for the decay of P^*. Atik and Thomas (54) considered a limiting case where the entry and the exit rate constants $k_+[Q_{aq}]$ and k_- are both negligibly small compared with other rate constants. In this case the number of cohabitant quenchers remains constant during the lifetime of P^* and in a micelle containing n quenchers, P^* decays with a rate constant $k_0 + nk_q$. Accordingly, the decay kinetics of P^* is given by

$$C^*(t) = \sum_{n=0}^{\infty} [M_n^*](0) e^{-(k_0 + nk_q)t} \tag{11.129}$$

Substitution of $[M_n^*](0) = C^*(0)\left(\dfrac{\bar{n}}{1+\bar{n}}\right)^n \bigg/ (1+\bar{n})$ in this equation leads to

$$C^*(t) = \frac{C^*(0) e^{-k_0 t}}{1 + \bar{n}(1 - e^{-k_q t})} \tag{11.130}$$

The ratio of total fluorescence intensities is

$$\frac{I}{I_0} = \frac{1}{1+\bar{n}} \sum_{n=0}^{\infty} \left(\frac{\bar{n}}{1+\bar{n}}\right)^n \frac{1}{1 + nk_q/k_0} \tag{11.131}$$

When the quenching rate constant k_q is infinitely large (static quenching), this equation reduces to $I/I_0 = 1/(1 + \bar{n})$, which is to be compared with Eq. (11.113) for the Poisson distribution.

For the migration dynamics described by Eq. (11.29) the decay kinetics of P^* and the ratio of total fluorescence intensities are again given by Eqs. (11.130) and (11.131), respectively, provided the rate constant k for exchange of quenchers between micelles is negligibly small.

Reversible energy transfer in micelles has also been treated (55).

11.4.3. Excimer Formation (18c,22)

As in Section 11.4.2, we assume that probes P are dissolved exclusively in micelles and that at most one probe is excited in each micelle at time zero with a laser pulse. Suppose that a probe is excited in a micelle containing altogether n probes. Reaction of the excited probe P^* with a ground-state probe P will lead to formation of an excimer P_2^*. The overall scheme for formation and dissociation of P_2^* may be described by

$$M_n^* \xrightarrow{k_0} M_n \quad (n = 1, 2, \ldots)$$
$$M_n^* \underset{k_d}{\overset{(n-1)k_f}{\rightleftharpoons}} M_n^\dagger \quad (n = 2, 3, \ldots) \tag{11.132}$$
$$M_n^\dagger \xrightarrow{k_i} M_n \quad (n = 2, 3, \ldots)$$

where M_n^* stands for a micelle containing P^* and altogether n probes, M_n stands for a micelle containing neither P^* nor P_2^*, and $M_n^\dagger$ stands for a micelle containing P_2^* and altogether n probes. The first equation shows the unimolecular decay of P^*. The second equation shows the formation and dissociation of P_2^*. According to the discussion in Section 11.3, the excimer formation reaction also obeys pseudo-first-order kinetics in micelles, and when a miscelle contains $n - 1$ ground-state probes, the rate constant is given by $(n - 1)k_f$, where k_f is the rate constant for excimer formation in a micelle containing one ground-state probe. Here k_d is the rate constant for excimer dissociation. The third equation shows the unimolecular decay of P_2^*, and k_1 is the decay rate constant. In the above scheme we have assumed that the migration of probes between micelles and the aqueous phase is slow compared with the processes depicted and neglected it on the fast time scale of observation.

We define $[M_n^*](t)$ as the concentration at time t of micelles containing P^* and altogether n probes and $[M_n^\dagger](t)$ as the corresponding concentration of micelles containing P_2^*. The probability that a probe will be excited in a micelle should be proportional to the number of probes in the micelle. Accordingly, the initial condition on $[M_n^*](t)$ is given by $[M_n^*](0) = AnQ_n$, where Q_n is the fraction of micelles containing n probes. Proportionality constant A is determined from the initial total concentration $C^*(0)$ of P^*, which is expressed as $C^*(0) = \sum_{n=0}^{\infty} [M_n^*](0)$. The result is:

$$[M_n^*](0) = C^*(0) \frac{n}{\bar{n}} Q_n \tag{11.133}$$

where $\bar{n}$ is the average number of probes in a micelle. The initial condition on $[M_n^\dagger](t)$ is:

$$[M_n^\dagger](0) = 0 \tag{11.134}$$

On the other hand, Eq. (11.132) leads to the following set of rate equations for $[M_n^*](t)$ and $[M_n^\dagger](t)$:

$$\frac{d[M_1^*](t)}{dt} = -k_0[M_1^*](t)$$

$$\frac{d[M_n^*](t)}{dt} = k_d[M_n^\dagger](t) - [k_0 + (n - 1)k_f][M_n^*](t) \quad (n = 2, 3, \ldots)$$

$$\frac{d[M_n^\dagger](t)}{dt} = (n - 1)k_f[M_n^*](t) - (k_1 + k_d)[M_n^\dagger](t) \quad (n = 2, 3, \ldots) \tag{11.135}$$

Since equations for different n are uncoupled, it is straightforward to solve this set of equations subject to Eqs. (11.133) and (11.134),

$$[M_1^*](t) = \frac{C^*(0)}{\bar{n}} Q_1 e^{-k_0 t}$$

$$[M_n^*](t) = \frac{C^*(0)}{\bar{n}} n Q_n \frac{1}{s_{n2} - s_{n1}} [(k_1 + k_d - s_{n1}) e^{-s_{n1} t} - (k_1 + k_d - s_{n2}) e^{-s_{n2} t}]$$

$$(n = 2, 3, \ldots)$$

$$[M_n^\dagger](t) = \frac{C^*(0)}{\bar{n}} n(n - 1) Q_n \frac{k_f}{s_{n2} - s_{n1}} [e^{-s_{n1} t} - e^{-s_{n2} t}]$$

$$(n = 2, 3, \ldots) \tag{11.136}$$

where s_{n1} and s_{n2} are the roots of $s^2 - [k_0 + k_1 + (n - 1)k_f + k_d]s + k_0 (k_1 + k_d) + (n - 1)k_f k_1 = 0$. Finally the total concentrations $C^*(t)$ and $C\dagger(t)$ of P^* and of P_2^* are

$$C^*(t) = \sum_{n=1}^{\infty} [M_n^*](t) \tag{11.137}$$

$$C\dagger(t) = \sum_{n=2}^{\infty} [M_n^\dagger](t) \tag{11.138}$$

Infelta and Grätzel (56) have shown that their data on the time variation of pyrene monomer and excimer fluorescences are in excellent agreement with the results calculated from Eqs. (11.137) and (11.138) together with the Poisson distribution (see Figure 11.29).

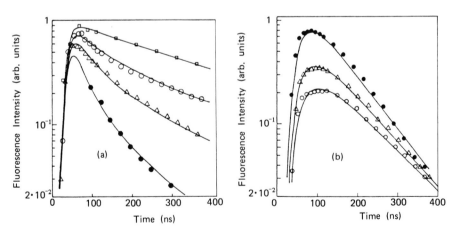

Figure 11.29. Time variation of pyrene monomer fluorescence (*a*) and excimer fluorescence (*b*) in cetyltrioxyethylene sulfate (CTOES) micelles (56). The average number $\bar{n}$ of pyrene molecules in a micelle: ●, $\bar{n} = 3.9$; △, $\bar{n} = 1.95$; ○, $\bar{n} = 1.0$; □, $\bar{n} = 0.12$. The full lines are simulations based on Eqs. (11.137) and (11.138) together with the Poisson distribution.

The total intensities I_m and I_e of monomer and excimer fluorescences are:

$$I_m = Bk_0^r \int_0^\infty C^*(t)\, dt \tag{11.139}$$

$$I_e = Bk_1^r \int_0^\infty C\dagger(t)\, dt \tag{11.140}$$

where k_0^r and k_1^r are the radiative rate constants of P^* and of P_2^*, respectively, and B is a constant. Substitution of Eqs. (11.137) and (11.138) in (11.139) and (11.140) leads to

$$I_m = \frac{I_m^0}{\bar{n}} \sum_{n=1}^{\infty} \frac{nQ_n}{1 + q(n-1)} \tag{11.141}$$

$$I_e = \frac{I_e^0}{\bar{n}} \sum_{n=2}^{\infty} \frac{qn(n-1)Q_n}{1 + q(n-1)} \tag{11.142}$$

where I_m^0, I_e^0 and q are given by

$$I_m^0 = \frac{k_0^r}{k_0} BC^*(0) \tag{11.143a}$$

$$I_e^0 = \frac{k_1^r}{k_1} BC^*(0) \tag{11.143b}$$

$$q = \frac{k_1 k_f}{k_0 (k_1 + k_d)} \tag{11.143c}$$

Equations (11.141) and (11.142) give (18c)

$$\frac{I_m}{I_m^0} + \frac{I_e}{I_e^0} = 1 \tag{11.144}$$

Therefore, parameters I_m^0 and I_e^0 are obtainable from the I_m axis and I_e axis intercepts by plotting I_e against I_m. In Figure 11.30 $(I_e/I_e^0)/(I_m/I_m^0)$ is plotted against the average number $\bar{n}$ of probes in a micelle. Experimental data (18c) are in excellent agreement with the result calculated from Eqs. (11.141) and (11.142) together with the Poisson distribution.

The kinetics of excimer formation in the case where the distribution of probes among micelles obeys a binomial distribution (57) or a geometric distribution (22) has also been examined.

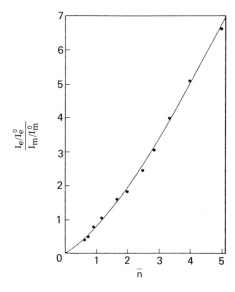

Figure 11.30. Ratio of pyrene excimer fluorescence I_e to monomer fluorescence I_m in CTAB micelles as a function of the average number $\bar{n}$ of pyrene molecules in a micelle (18c). The full line is a simulation based on Eqs. (11.141) and (11.142) together with the Poisson distribution.

11.4.4. Triplet–Triplet Annihilation

The interaction of two triplet excited molecules gives rise to triplet–triplet (TT) annihilation. Lachish et al. (58) have pointed out that the confinement of triplet-excited molecules in micelles considerably enhances TT annihilation.

Suppose that probes P dissolved exclusively in micelles are excited at time zero with a laser pulse and that a fraction of them is converted to a triplet state via intersystem crossing. If the laser intensity is strong enough, it is not unlikely that several probes are excited in a micelle and that two or more triplet-excited probes are produced in the micelle via intersystem crossing. The scheme for subsequent TT annihilation may be described by

$$M_n \xrightarrow{(1/2)n(n-1)k_{TT}} M_{n-2} \qquad n = 2, 3, 4, \ldots \qquad (11.145)$$

where M_n stands for a micelle containing n triplet probes. According to the discussion in Section 11.3, the TT annihilation reaction also obeys pseudo-first-order kinetics in micelles. Triplet probes react in a pairwise fashion. When a micelle contains n triplet probes, the number of ways of choosing a pair of triplet probes is given by $\frac{1}{2} n (n - 1)$. One the other hand, the reaction of any pair leads to the decrease from n to $n - 2$ in the number of triplet probes in the micelle. Accordingly, the rate constant with which a micelle containing n triplet probes is, by TT annihilation, converted to one containing $n - 2$ triplet probes is given by $\frac{1}{2} n (n - 1)k_{TT}$ where k_{TT} is the rate constant for TT annihilation in a micelle containing two triplet probes.

First consider the initial distribution of triplet probes among micelles just after laser excitation. Let a denote the probability that a given probe will be converted to the triplet state upon laser excitation. The probability $P(j, n)$ that a micelle that contained j probes before excitation will have n ($n \leq j$) triplet probes after excitation is given by the binomial distribution

$$P(j, n) = {}_jC_n a^n (1 - a)^{j - n} \qquad (11.146)$$

The initial concentration $[M_n](0)$ of micelles containing n triplet probes is expressed as

$$[M_n](0) = [M] \sum_{j=n}^{\infty} Q_j P(n, j) \qquad (11.147)$$

where $[M]$ is the total micelle concentration and Q_n is the fraction of micelles containing j probes. If the distribution of probes among micelles obeys a Poisson distribution [Eq. (11.8)], Eq. (11.147) leads to

$$[M_n](0) = \frac{[M]}{n!} (\bar{n}a)^n e^{-\bar{n}a} \qquad (11.148)$$

where $\bar{n}$ is the average number of probes in a micelle. This equation indicates that the initial distribution of triplet probes among micelles also obeys a Poisson distribution with the average number given by $\bar{n}a$. Accordingly, the initial total concentration $C^T(0)$ of triplet probes is given by $C^T(0) = [M]\bar{n}a$.

Next consider the final distribution of triplet probes among micelles after completion of TT annihilation. We assume that all other decay processes are slow compared with TT annihilation and neglect them on the fast time scale of observation. On such a fast time scale TT annihilation occurs only between triplet probes contained in the same micelle. Since triplet probes will react in a pairwise fashion, the micelles that initially contained an odd number of triplet probes will contain one triplet probe after completion of TT annihilation. All triplet probes will react in the micelles that contained an even number of triplet probes. Accordingly, the total concentration $C^T(\infty)$ of triplet probes after completion of TT annihilation is

$$C^T(\infty) = \sum_{m=1}^{\infty} [M_{2m+1}](0) = [M]e^{-\bar{n}a} \sinh(\bar{n}a) \qquad (11.149)$$

Turning to the kinetics of TT annihilation, we define $[M_n](t)$ as the concentration at time t of micelles containing n triplet probes. Equation (11.145) leads to the following set of rate equations for $[M_n](t)$:

$$\frac{d[M_0](t)}{dt} = k_{TT}[M_2](t)$$

$$\frac{d[M_n](t)}{dt} = \tfrac{1}{2}(n + 2)(n + 1) k_{TT} [M_{n+2}](t)$$

$$- \tfrac{1}{2} n(n - 1) k_{TT}[M_n](t) \qquad (n = 1, 2, \ldots) \qquad (11.150)$$

By means of the generating function $F(s, t)$ of $[M_n](t)$, namely,

$$F(s, t) = \sum_{n=0}^{\infty} s^n [M_n](t) \qquad |s| < 1 \qquad (11.151)$$

Eq. (11.150) may be transformed into

$$\frac{\partial F(s, t)}{\partial t} = \frac{k_{TT}}{2}(1 - s^2)\frac{\partial^2 F(s, t)}{\partial s^2} \qquad (11.152)$$

On the other hand, Eq. (11.148) imposes the following initial condition on $F(s, t)$:

$$F(s, 0) = [M]e^{\bar{n}a(s-1)} \qquad (11.153)$$

The solution of Eq. (11.152) subject to (11.153) is

$$F(s, t) = \sum_{n=0}^{\infty} A_n C_n^{-1/2}(s) \exp\left[-\tfrac{1}{2}n(n - 1)k_{TT}t\right] \qquad (11.154)$$

where $C_{nn}^{-1/2}(s)$ is a Gegenbauer polynomial, and A_n is given by

$$A_n = \frac{1 - 2n}{2^n}[M]e^{-\bar{n}a}\sum_{j=n}^{\infty}\frac{(\bar{n}a)^j}{(j - n)!}\frac{\Gamma[(j - n + 1)/2]}{\Gamma[(j + n + 1)/2]} \qquad (11.155)$$

where $j = n, n + 2, n + 4, \ldots$, and $\Gamma[\ ]$ is the gamma function. The total concentration $C^T(t)$ of triplet probes is

$$C^T(t) = \sum_{n=0}^{\infty} n[M_n](t)$$

$$= \left[\frac{\partial F(s, t)}{\partial s}\right]_{s=1}$$

$$= -\sum_{n=1}^{\infty} A_n \exp\left[-\tfrac{1}{2}n(n - 1)k_{TT}t\right] \qquad (11.156)$$

where we have used a property [Eq. (11.83)] of the generating function. As $t \rightarrow \infty$, the above equation reduces to Eq. (11.149).

Rothenberger et al. (59) have shown that their data on the kinetics of TT annihilation of 1-bromonaphthalene are in good agreement with the results calculated from Eq. (11.156) (see Figure 11.31).

11.4.5. Emulsion Polymerization

An emulsion polymerization system typically involves four ingredients: (1) an aqueous dispersion medium, (2) dispersed monomer droplets, (3) a micelle-generating detergent, and (4) an initiator. The initiator is normally present in the aqueous phase, and the initial formation of free radicals occurs there. Monomer droplets serve as a reservoir of monomers. An emulsion polymerization occurs in three stages (60). In an initial stage detergents are present in the form of micelles swollen with monomers. When free radicals enter the micelles, they initiate polymerization, and nuclei of polymer particles are produced there. In this stage micelles swollen with monomers serve as the reaction loci, and monomers migrate from droplets to micelles. After 10–20% monomer conversion the initial stage evolves into a second stage in which the nuclei of polymer particles grow into polymer particles swollen with monomers. Concomitantly micelles disappear because the polymer particles adsorb detergents. In this stage polymer particles swollen with monomers serve as the reaction loci. Monomers migrate from droplets to polymer particles, and the concentration of monomers in the polymer particle is kept constant as long as monomer droplets remain present. In a final stage monomer droplets disappear, and unreacted monomers are present only in swollen polymer particles.

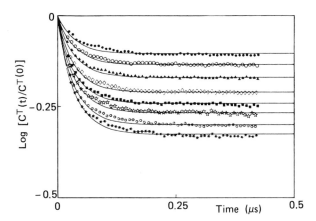

Figure 11.31. Decay of triplet excited 1-bromonaphthalene in CTAB micelles by triplet–triplet annihilation (59). [1-bromonaphtalene] $= 5.46 \times 10^{-5}M$. [CTAB]: *, 4.55 mM; ⋆, 4.88 mM; ☆, 5.43 mM; ■, 5.95 mM; ◇, 6.75 mM; ▲, 8.23 mM; ○, 10.2 mM; ●, 13.0 mM. The full lines are simulations based on Eq. (11.156).

Here we discuss the kinetics of emulsion polymerization in the second stage as another example of reactions that require the stochastic treatment. In this stage no new polymer particles are generated, and the total concentration of polymer particles remains constant. Let M_n stand for a swollen polymer particle containing n free radicals. The overall scheme for reactions of free radicals in polymer particles may be described by

$$M_n + R_{aq} \underset{(n+1)k_-}{\overset{k_+}{\rightleftharpoons}} M_{n+1} \; (n = 0, 1, \ldots)$$

$$M_n \overset{(1/2)n(n-1)k}{\longrightarrow} M_{n-2} \; (n = 2, 3, \ldots) \tag{11.157}$$

where R_{aq} stands for a free radical in the aqueous phase. The first equation shows the migration of a free radical between polymer particles and the aqueous phase. The second equation shows the bimolecular recombination of free radicals. According to the discussion in Section 11.3, the bimolecular recombination also obeys pseudo-first-order kinetics in polymer particles, and the rate constant with which a micelle containing n free radicals is converted, by bimolecular recombination, into one containing $n - 2$ free radicals is given by $\frac{1}{2} n(n - 1)k$, where k is the rate constant for bimolecular recombination in a polymer particle containing two free radicals. We define $[M_n](t)$ as the concentration at time t of polymer particles containing n free radicals. Equation (11.157) leads to the following set of rate equations for $[M_n](t)$:

$$\frac{d[M_0](t)}{dt} = -k_+[R_{aq}][M_0](t) + k_-[M_1](t) + k[M_2](t)$$

$$\frac{d[M_n](t)}{dt} = k_+[R_{aq}][M_{n-1}](t) - \{k_+[R_{aq}] + nk_- + \tfrac{1}{2} n(n-1)k\}[M_n](t)$$

$$+ (n+1)k_-[M_{n+1}](t) + \tfrac{1}{2}(n+2)(n+1)k[M_{n+2}](t)$$

$$(n = 1, 2, \ldots) \tag{11.158}$$

By means of the generating function defined by Eq. (11.151), where $[M_n](t)$ now denotes the concentration of polymer particles containing n free radicals, Eq. (11.158) may be transformed into

$$\frac{\partial F(s, t)}{\partial t} = \frac{k}{2}(1 - s^2)\frac{\partial^2 F(s, t)}{\partial s^2} + k_-(1 - s)\frac{\partial F(s, t)}{\partial s} + k_+[R_{aq}](s - 1)F(s, t)$$

$$\tag{11.159}$$

Note that in the case of $k_+[R_{aq}] = k_- = 0$, this equation reduces to Eq. (11.152). Consider the steady state distribution of free radicals among polymer particles.

The physically meaningful solution of Eq. (11.159) under the condition of $\partial F(s, t)/\partial t = 0$ is (61)

$$F(s) = A(1 + s)^{(1-q)/2} I_{q-1}\left[a \left(\frac{1+s}{2} \right)^{1/2} \right] \qquad (11.160)$$

where $q \equiv k_-/k$, $a \equiv 2 (k_+[R_{aq}]/k)^{1/2}$, and $I_p(x)$ is the modified Bessel function of the first kind. Constant A is determined from the total concentration $[M]$ of polymer particles, which is expressed as $[M] = \Sigma_{n=0}^{\infty} M_n = F(1)$. The result is

$$F(s) = [M] \left(\frac{1+s}{2} \right)^{(1-q)/2} \frac{I_{q-1}\{a\,[(1+s)/2]^{1/2}\}}{I_{q-1}(a)} \qquad (11.161)$$

The fraction Q_n of polymer particles containing n free radicals is

$$Q_n = \frac{[M_n]}{[M]}$$

$$= \frac{1}{[M]} \frac{1}{n!} \left[\frac{\partial^n F(s)}{\partial s^n} \right]_{s=0}$$

$$= \frac{2^{(q-1-3n)/2} a^n}{n!} \frac{I_{q-1+n}\,(a/\sqrt{2})}{I_{q-1}(a)} \qquad (11.162)$$

where we have used a property [Eq. (11.84)] of the generating function. Figure 11.32 shows the distribution Q_n for $q = 0$.

The average number $\bar{n}$ of free radicals in a polymer particle is

$$\bar{n} = \frac{1}{[M]} \sum_{n=0}^{\infty} n[M_n]$$

$$= \frac{1}{[M]} \left[\frac{\partial F(s)}{\partial s} \right]_{s=1}$$

$$= \frac{a I_q(a)}{4 I_{q-1}(a)} \qquad (11.163)$$

In Figure 11.33 $\bar{n}$ is plotted against a for various values of q. In the case where the exit rate constant k_- is practically zero, namely, $q = 0$, and where the recombination rate constant k is sufficiently large compared with the entry rate constant $k_+[R_{aq}]$, namely, $a = 0$, $\bar{n}$ is seen to be $\frac{1}{2}$. This limiting case, which was originally treated by Smith and Ewart (60), is of outstanding interest since its existence is responsible for the extraordinarily high rates of polymerization and high molecular weights attainable by emulsion polymerization. In this case, having once entered

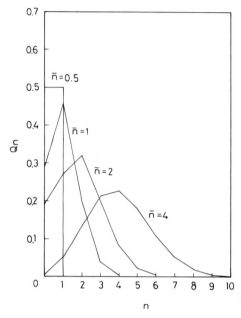

Figure 11.32. Distribution of free radicals among polymer particles for several values of average number $\bar{n}$ in the case of $q = 0$.

a polymer particle, a free radical remains there until another comes in to recombine with it. Since the recombination rate constant is sufficiently large compared with the entry rate constant, the average time necessary for two free radicals in the same polymer particle to recombine is small compared with the average time interval between successive entries of free radicals into the polymer particle. Then approximately one-half of the polymer particles contain a single free radical and the other half contain none. Consequently the average number of free radicals in a polymer particle becomes equal to $\frac{1}{2}$.

The rate of emulsion polymerization in the second stage can be expressed as

$$\frac{d[m]}{dt} = k_p C_m \bar{n}[M] \tag{11.164}$$

where $[m]$ is the concentration of reacted monomers in the system, C_m is the monomer concentration in the polymer particle, k_p is the propagation rate constant, and the average number $\bar{n}$ of free radicals in a polymer particle is given by Eq. (11.163). Under normal polymerization conditions practically every free radical produced will enter polymer particles, so that the rate σ of formation of free radicals per unit volume of the system equals the rate of entry into all polymer particles present in unit volume,

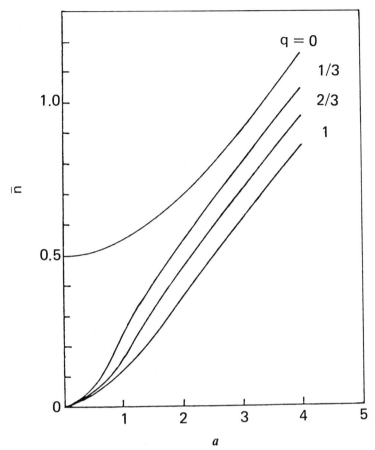

Figure 11.33. Average number $\bar{n}$ of free radicals in a polymer particle as a function of a and q (61). See text for definitions of a and q.

$$\sigma = k_+[R_{aq}][M] \tag{11.165}$$

With the aid of this equation, argument a in expression (11.163) for $\bar{n}$ is rewritten as

$$a \equiv 2\left(\frac{k_+[R_{aq}]}{k}\right)^{1/2} = 2\left(\frac{\sigma}{k[M]}\right)^{1/2} \tag{11.166}$$

Consider the relation between the rate of emulsion polymerization and that of bulk polymerization. The latter is expressed as

$$\frac{d[m]}{dt} = k_p C_m^b[R] \tag{11.167}$$

where C_m^b is the bulk concentration of monomers and $[R]$ is the concentration of free radicals. In the steady state the concentration of free radicals is obtained by equating the rate σ of formation of free radicals to the loss rate $k_t[R]^2$, where k_t is the bulk termination rate constant,

$$[R] = \left(\frac{\sigma}{k_t}\right)^{1/2} \tag{11.168}$$

According to the discussion in Section 11.3, the recombination rate in the polymer particle is related to the bulk termination rate through

$$2 \times \tfrac{1}{2}n(n - 1)k \frac{1}{v} = k_t \frac{n}{v} \frac{n - 1}{v} \tag{11.169}$$

where n is the number of free radicals in the polymer particle with volume v. The factor 2 on the left side arises from the fact that two free radicals terminate in one recombination event. Thus,

$$k = \frac{k_t}{v} \tag{11.170}$$

Substitution of this equation in Eq. (11.166) yields

$$a = 2 \left(\frac{\sigma v}{k_t[M]}\right)^{1/2} \tag{11.171}$$

When the termination rate constant k_t is very large, according to Eq. (11.168), the free radical concentration $[R]$ will be very low so that the rate of bulk polymerization will also be very low. In this case a given by Eq. (11.171) will also be very small. However, if q is practically zero, $\bar{n}$ will not become less than $\tfrac{1}{2}$, as seen from Figure 11.33. Therefore, according to Eq. (11.164), the rate of emulsion polymerization can be very high, provided the total concentration $[M]$ of polymer particles is very high.

In bulk polymerization the average lifetime τ_P of free radicals is calculated by equating the rate $[R]/\tau_P$ of loss of free radicals to the formation rate σ,

$$\tau_P = \frac{[R]}{\sigma} = \left(\frac{1}{k_t\sigma}\right)^{1/2} \tag{11.172}$$

Similarly, the average lifetime of free radicals in emulsion polymerization is:

$$\tau_P = \frac{\bar{n}[M]}{\sigma} \tag{11.173}$$

When the termination rate constant k_t is very large, the average lifetime of free radicals in bulk polymerization will be very short, so that the average molecular weight of a produced polymer will be very low. On the other hand, in emulsion polymerization the average lifetime will not become less than $[M]/2\sigma$ if q is practically zero. Therefore, the average molecular weight of a polymer can be very high, provided the total concentration of polymer particles is very high.

So far we have considered the kinetics of emulsion polymerization by use of the steady-state solution of Eq. (11.159). It is difficult (62) to derive a general time-dependent solution of Eq. (11.159). However, time-dependent solutions for some limiting cases have been obtained, and the corresponding distributions of free radicals among polymer particles have been calculated. For example, several workers (63–65) have considered a limiting case where the bimolecular recombination rate constant k is negligibly small. In this case, if initially no polymer particle contains a free radical, the distribution of free radicals among polymer particles obeys a Poisson distribution,

$$Q_n(t) = \frac{1}{n!} \bar{n}^n \exp(-\bar{n}) \tag{11.174}$$

with a time dependent average number $\bar{n}$ given by

$$\bar{n} = \frac{k_+[R_{aq}]}{k_-} (1 - e^{-k_- t}) \tag{11.175}$$

11.4.6. Migration-Controlled Reactions in Vesicular Systems and Fluorescence Stopped-Flow Techniques

The migration rates of solubilizates in micellar systems can be studied by fluorescence quenching techniques (Section 11.4.2). However, in vesicular systems the migration rates of solubilizates are much slower than in micellar systems and is therefore difficult to study by the same techniques. Kano et al. (66) and Almgren (67, 68) have developed fluorescence stopped-flow techniques to study the migration rates in vesicular systems. The principle of the method is as follows. Equal volumes of two vesicular solutions with equal vesicle concentration, one containing probe, the other containing none (or another probe), are mixed in a stopped-flow apparatus. By measuring quantities that are sensitive to the microconcentration of probes in the vesicles, one can follow the migration process of probes between vesicles after mixing. For example, one can utilize probe excimer emission to follow the migration. If two kinds of probes are used, one can follow the process by measuring the energy transfer from donor probe to acceptor probe. Here we first discuss a method that utilizes excimer emission and then one utilizing energy transfer.

11.4.6.1. Excimer Emission (23). There are two possible mechanisms for migration of probes between vesicles (Section 11.2). We consider only the migration via the aqueous phase and assume that the migration dynamics is described by Eq. (11.1), where M_n now stands for a vesicle containing n probes and A_{aq} stands for a probe in the aqueous phase. Extension to the migration during vesicle collisions is straightforward (23). Suppose that before mixing, one solution contains a probe and the other contains none. The concentration $[M_n^0]$ of vesicles containing n probes, in the former solution, is

$$[M_n^0] = \frac{[M]}{n!} \left(\frac{k_+[A_{aq}]}{k_-}\right)^n \exp\left(-\frac{k_+[A_{aq}]}{k_-}\right) \tag{11.176}$$

where $[M]$ is the total vesicle concentration and $[A_{aq}]$ is the concentration of probes in the aqueous phase. Since one solution does not contain the probe, immediately after mixing the concentration of probes in the aqueous phase and that of vesicles containing n probes ($n \geq 1$) reduce to $\frac{1}{2}[A_{aq}]$ and $\frac{1}{2}[M_n^0]$, respectively. On the other hand, the concentration of vesicles containing no probe becomes $\frac{1}{2}([M_0^0] + [M])$. However, the total vesicle concentration does not change, remaining $[M]$. Immediately after mixing, the total rates for entry and exit of probes into and from the vesicles are:

$$\text{Total entry rate} = k_+ \frac{[A_{aq}]}{2} [M]$$

$$\text{Total exit rate} = \sum_{n=1}^{\infty} nk_- \frac{[M_n^0]}{2} \tag{11.177}$$

With the aid of Eq. (11.176) one can show that the two rates are equal, so that the concentration of probes in the aqueous phase does not change during the equilibration process, remaining $\frac{1}{2}[A_{aq}]$.

We define $[M_n](t)$ as the concentration at time t of vesicles containing n probes. Equation (11.1) together with the above discussion leads to the following set of rate equations for $[M_n](t)$.

$$\frac{d[M_0](t)}{dt} = -\frac{1}{2}k_+[A_{aq}][M_0](t) + k_-[M_1](t)$$

$$\frac{d[M_n](t)}{dt} = \frac{1}{2}k_+[A_{aq}][M_{n-1}](t) - (\frac{1}{2}k_+[A_{aq}] + nk_-)[M_n](t)$$

$$+ (n+1)k_-[M_{n+1}](t) \qquad (n = 1, 2, \ldots) \tag{11.178}$$

On the other hand, the initial concentration $M_n(0)$ of vesicles containing n probes is:

$$[M_0](0) = \tfrac{1}{2}([M_0^0] + [M])$$

$$[M_n](0) = \tfrac{1}{2}[M_n^0] \quad (n \geq 1)$$

(11.179)

By means of the generating function defined by Eq. (11.151), where $[M_n](t)$ now denotes the concentration of vesicles containing n probes, Eq. (11.178) may be transformed into

$$\frac{\partial F(s,\, t)}{\partial t} = k_-(1 - s)\frac{\partial F(s,\, t)}{\partial s} + \frac{1}{2}k^+[A_{aq}](s - 1)F(s,\, t) \quad (11.180)$$

On the other hand, Eq. (11.179) imposes the following initial condition on $F(s,\ t)$:

$$F(s,\, 0) = \frac{[M]}{2}\left\{ \exp\left[\frac{k_+[A_{aq}]}{k_-}(s - 1)\right] + 1\right\} \quad (11.181)$$

The solution of Eq. (11.180) subject to (11.181) is

$$F(s,\, t) = \frac{[M]}{2}\left\{ \exp\left[\frac{k_+[A_{aq}]}{2k_-}(s - 1)(1 + e^{-k_-t})\right]\right.$$

$$\left. + \exp\left[\frac{k_+[A_{aq}]}{2k_-}(s - 1)(1 - e^{-k_-t})\right]\right\} \quad (11.182)$$

The concentration $[M_n](t)$ at time t of vesicles containing n probes is

$$[M_n](t) = \frac{1}{n!}\left[\frac{\partial^n F(s,\, t)}{\partial s^n}\right]_{s=0}$$

$$= \frac{[M]}{2}\left[\frac{1}{n!}\bar{n}_1^n \exp(-\bar{n}_1) + \frac{1}{n!}\bar{n}_2^n \exp(-\bar{n}_2)\right] \quad (11.183)$$

where

$$\bar{n}_1 = \frac{k_+[A_{aq}]}{2k_-}(1 + e^{-k_-t}) \quad (11.184a)$$

$$\bar{n}_2 = \frac{k_+[A_{aq}]}{2k_-}(1 - e^{-k_-t}) \quad (11.184b)$$

Equation (11.183) describes the kinetics of migration of probes between vesicles after mixing of the two solutions. It indicates that the distribution of probes among vesicles is composed of two Poisson distributions, one with the average number given by Eq. (11.184a), the other with the average number given by Eq. (11.184b). The former distribution originates from the solution containing the probe and the latter from the solution containing none. As $t \to \infty$, Eq. (11.183) yields a single Poisson distribution with the average number $k_+[A_{aq}]/2k_-$.

When the mixed solution is irradiated with a light, probe excimers are formed. The intensity $I(t)$ of excimer emission is given, according to Eq. (11.142), by

$$I(t) = F \sum_{n=2}^{\infty} \frac{qn(n-1)}{1 + q(n-1)} [M_n](t) \tag{11.185}$$

where q is given by Eq. (11.143c), and F is a constant. Substitution of Eq. (11.183) in (11.185) and subsequent expansion of the resulting equation for large t lead to

$$I(t) = I(\infty) + Ge^{-2k_-t} \tag{11.186}$$

where $I(\infty)$ is the excimer emission intensity in the limit of $t \to \infty$ and G is a positive constant. This equation indicates that the excimer emission intensity decays exponentially for large t with the relaxation time τ given by $\tau^{-1} = 2k_-$. Almgren (67) analyzed his data on excimer emission on the basis of Eq. (11.186) and estimated the exit rate constant k_-.

11.4.6.2. Energy Transfer (23). Suppose that equal volumes of two solutions, one containing donor probe A, the other containing acceptor probe B, are mixed. Since A and B should migrate mutually independently, the concentration $[M_{n,m}](t)$ at time t of vesicles containing n A's and m B's should be given, on the analogy of Eq. (11.183), by

$$[M_{n,m}](t) = \frac{[M]}{2} \left[\frac{1}{n!} \bar{n}_{A1}^n \exp(-\bar{n}_{A1}) \frac{1}{m!} \bar{n}_{B2}^m \exp(-\bar{n}_{B2}) \right.$$
$$\left. + \frac{1}{n!} \bar{n}_{A2}^n \exp(-\bar{n}_{A2}) \frac{1}{m!} \bar{n}_{B1}^m \exp(-\bar{n}_{B1}) \right] \tag{11.187}$$

where

$$\bar{n}_{A1} = \frac{k_+^A[A_{aq}]}{2k_-^A} (1 + e^{-k^A t}) \tag{11.188a}$$

$$\bar{n}_{A2} = \frac{k_+^A[A_{aq}]}{2k_-^A} (1 - e^{-k^A t}) \tag{11.188b}$$

$$\bar{n}_{B1} = \frac{k_+^B[B_{aq}]}{2k_-^B} (1 + e^{-k^B t}) \tag{11.188c}$$

$$\bar{n}_{B2} = \frac{k_+^B[B_{aq}]}{2k_-^B} (1 - e^{-k^B t}) \tag{11.188d}$$

Here $[A_{aq}]$ and $[B_{aq}]$ are the concentrations of A and B in the aqueous phase in each unmixed solution, $k_+^{A(B)}$ and $k_-^{A(B)}$ are the entry and the exit rate constants for A or B, respectively. In Eq. (11.187) the first term in square brackets originates from the solution containing A and the second from the one containing B.

When the mixed solution is irradiated with a light, the probability that an A will be excited in a vesicle is proportional to the number of A's in the vesicle, and the probability that the excited A will fluoresce against quenching is given by $k_0^r/(k_0 + mk_q)$ where k_0, k_q, and k_0^r have the same meanings as in Section 11.4.2 and m is the number of cohabitant B's. Accordingly, the fluorescence intensity $I_A(t)$ of A is

$$I_A(t) = F_A \sum_{n=1}^{\infty} \sum_{m=0}^{\infty} \frac{n}{1 + rm} [M_{n,m}](t) \qquad (11.189)$$

where r is given by k_q/k_0 and F_A is a constant. If the quenching occurs by energy transfer, the fluorescence from B may also be observed, and its intensity $I_B(t)$ is

$$I_B(t) = F_B \sum_{n=1}^{\infty} \sum_{m=1}^{\infty} \frac{rmn}{1 + rm} [M_{n,m}](t) \qquad (11.190)$$

where F_B is another constant. Substitution of Eq. (11.187) in (11.189) and (11.190) and subsequent expansion of the resulting equations for large t lead to

$$I_A(t) = I_A(\infty) + G_A e^{-2k_-^B t} + H_A e^{-(k_-^A + k_-^B)t} \qquad (11.191)$$

$$I_B(t) = I_B(\infty) - G_B e^{-2k_-^B t} - H_B e^{-(k_-^A + k_-^B)t} \qquad (11.192)$$

where $I_A(\infty)$ and $I_B(\infty)$ are the fluorescence intensities of A and of B, respectively, in the limit $t \to \infty$, and G_A, H_A, G_B, and H_B are positive constants. These two equations indicate that the fluorescence intensity of A (B) decays (grows) in two

TABLE 11.4. Relation Between Combination of Two Vesicular Solutions and Relaxation Behavior of Acceptor Fluorescence Intensity

Combination[a]	Acceptor Fluorescence Intensity
Ves with A and B + Ves with A	Increase ($\tau_1^{-1} = 2k_-^B$)
Ves with A and B + Ves with B	No relaxation
Ves with A + Ves with B	Increase ($\tau_1^{-1} = 2k_-^B$, $\tau_2^{-1} = k_-^A + k_-^B$)
Ves with A and B + empty Ves	Decrease ($\tau_2^{-1} = k_-^A + k_-^B$) followed by increase ($\tau_1^{-1} = 2k_-^B$)

[a] Ves = vesicle, A = donor, B = acceptor.

modes with the relaxation times τ_1 and τ_2 given by $\tau_1^{-1} = 2k_-^B$ and $\tau_2^{-1} = k_-^A + k_-^B$.

In the above example two probes are separately contained in two solutions before mixing. The kinetics of the relaxation of fluorescence intensities depends on how two probes are distributed among two solutions. Various combinations of two solutions are considered, depending on how two probes are distributed among them.

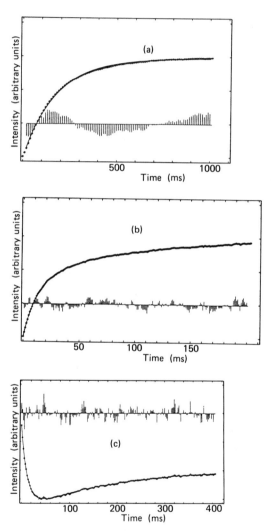

Figure 11.34. Time variation of perylene fluorescence intensity after the mixing of two vesicular solutions (68): (a) one containing pyrene and perylene and the other containing pyrene; (b) one containing pyrene and the other containing perylene; (c) one containing pyrene and perylene and the other containing none. The full lines are simulations based on Table 11.4. The histograms represent the deviations between experimental data and fitting functions.

Table 11.4 shows the relaxation behavior of the fluorescence intensity of acceptor probe B for various combinations. Almgren (68) studied the migration of probes between lipid vesicles using pyrene and perylene as donor and acceptor probes. His results on the fluorescence intensity of perylene are shown in Figure 11.34. He analyzed these results on the basis of Table 11.4 and estimated the rate constants for exit of pyrene and perylene from the lipid vesicle.

11.5. CONCLUDING REMARKS

We have reviewed the theoretical treatments of reaction kinetics in micellar and vesicular systems. The treatments are oversimplified because of insufficient information about the properties of micelles and vesicles and about the interaction between reactants and micelles or vesicles. For example, it is not known how the rate constants for entry and exit of a reactant into and from a micelle depend on the number of reactants in the micelle. It is not known where reactants are located in micelles nor how the microviscosity in micelles varies from place to place. The size distribution of micelles is not accurately known. Comparison between theory and experiment indicates that the assumptions made here are not too bad. The model enables one to estimate parameters of the properties of micelles and the interaction between reactants and micelles.

Part of the oversimplification was made for the sake of mathematical simplicity. For example, we fixed one of the reactants at the center of the micelle in the treatment of diffusion-controlled reactions in the micellar interior. We neglected the migration of probes between micelles and the aqueous phase in the stochastic treatments of fluorescence quenching, excimer formation, and triplet–triplet annihilation. In the treatment of fluorescence quenching the migration of quenchers was taken into account. The removal of these simplifications is a subject of future study. The time-dependent Smith–Ewart equation [Eq. (11.159)] also remains to be solved.

In spite of the above limitations the present treatments have revealed several unique features of the reaction kinetics in micellar and vesicular systems. For example, bimolecular diffusion-controlled reactions in restricted spaces follow first-order kinetics and not second-order kinetics as in a three-dimensional free space. The kinetics of fluorescence quenching in micellar systems does not follow Stern–Volmer behavior, either in the time domain [Eq. (11.109)] or the concentration domain [Eq. (11.111)]. In micelles triplet–triplet annihilation does not deactivate all triplet probes since in each micelle that initially contains an odd number of triplet probes one triplet probe is left behind after completion of annihilation. In emulsion polymerization, under suitable conditions one half of the polymer particles contain a single free radical and the other half contain none. This situation is responsible for the production of high-molecular-weight polymers at high rates in emulsion polymerization.

The subjects that require further theoretical study include magnetic field effects on chemical reactions in micelles and the photoionization of solutes dissolved in micelles. Extensive experimental study of the former subject has been done by several groups (10, 69) and is far ahead of the theoretical study. The second subject is of both theoretical and experimental interest (70). How does the presence of the polar interface, and the potential energy difference between the micellar interior and the bulk aqueous phase, affect the photoionization of solutes dissolved in micelles and the subsequent escape of ejected electrons from their parent positive ions?

REFERENCES

1. J. K. Thomas, F. Grieser, and M. Wong, *Ber. Bunsenges. Phys. Chem.* **82**, 937 (1978).
2. J. H. Fendler, *J. Phys. Chem.* **84**, 1485 (1980).
3. N. J. Turro, M. Grätzel, and A. M. Braun, *Angew. Chem. Int. Ed. Engl.* **19**, 675 (1980).
4. J. H. Fendler, *Membrane Mimetic Chemistry*, Wiley, New York, 1982.
5. H. Masuhara and N. Mataga, *Acc. Chem. Res.* **14**, 312 (1981).
6. D. Mauzerall and S. G. Ballard, *Ann. Rev. Phys. Chem.* **33**, 377 (1982).
7. For example (a) Y. Waka, K. Hamamoto, and N. Mataga, *Chem. Phys. Lett.* **53**, 242 (1978); (b) B. Katušin-Ražem, M. Wong, and J. K. Thomas. *J. Am. Chem. Soc.* **100**, 1679 (1978); (c) M. S. Tunuli and J. H. Fendler, *J. Am. Chem. Soc.* **103**, 2507 (1981).
8. R. Z. Sagdeev, K. M. Salikhov, and Yu. M. Molin, *Russ. Chem. Rev.* **46**, 297 (1977).
9. N. J. Turro and B. Kraeutler, *Acc. Chem. Res.* **13**, 369 (1980).
10. For example (a) N. J. Turro, D. R. Anderson, M.-F. Chow, C.-J. Chung, and B. Kraeutler, *J. Am. Chem. Soc.* **103**, 3892 (1981); (b) J. C. Scaiano and E. B. Abuin, *Chem. Phys. Lett.* **81**, 209 (1981); (c) Y. Sakaguchi and H. Hayashi, *Chem. Phys. Lett.* **87**, 539 (1982); (d) Y. Tanimoto, H. Udagawa, and M. Itoh, *J. Phys. Chem.* **87**, 724 (1983).
11. D. C. Blackley, *Emulsion Polymerization*, Applied Science, London, 1975.
12. F. M. Menger and J. M. Jerkunica, *J. Am. Chem. Soc.* **100**, 688 (1978).
13. K. Kano and J. H. Fendler, *Chem. Phys. Lipids* **23**, 189 (1979).
14. J. Lansman and D. H. Haynes, *Biochim. Biophys. Acta* **394**, 335 (1975).
15. S. Diekmann and J. Frahm, *Ber. Bunsenges. Phys. Chem.* **82**, 1013 (1978).
16. M. Tachiya, *Chem. Phys. Lett.* **33**, 289 (1975).
17. A. Henglein and Th. Proske, *Ber. Bunsenges. Phys. Chem.* **82**, 1107 (1978).
18. For example (a) M. A. J. Rodgers and M. F. da Silva e Wheeler, *Chem. Phys. Lett.* **53**, 165 (1978); (b) S. S. Atik and L. A. Singer, *Chem. Phys. Lett.* **59**, 519 (1978); (c) B. K. Selinger and A. R. Watkins, *Chem. Phys. Lett.* **56**, 99 (1978).
19. T. F. Hunter, *Chem. Phys. Lett.* **75**, 152 (1980).
20. T. Nakamura, A. Kira, and M. Imamura, *J. Phys. Chem.* **87**, 3122 (1983).
21. G. Rothenberger, P. P. Infelta, and M. Grätzel, *J. Phys. Chem.* **83**, 1871 (1979).
22. R. C. Dorrance and T. F. Hunter, *J. Chem. Soc. Faraday I* **70**, 1572 (1974).

23. M. Tachiya and M. Almgren, *J. Chem. Phys.* **75**, 865 (1981).

24. M. Tachiya, *J. Chem. Phys.* **76**, 340 (1982).

25. Y. Croonen, E. Geladé, M. Van der Zegel, M. Van der Auweraer, H. Vanden-driessche, F. C. De Schryver, and M. Almgren, *J. Phys. Chem.* **87**, 1426 (1983).

26. G. Adam and M. Delbrück, in *Structural Chemistry and Molecular Biology*, A. Rich and N. Davidson (Ed.), Freeman, San Francisco, 1968.

27. M. Tachiya, *Rad. Phys. Chem.* **21**, 167 (1983).

28. D. F. Calef and J. M. Deutch, *Ann. Rev. Phys. Chem.* **34**, 493 (1983).

29. M. Tachiya, *Chem. Phys. Lett.* **69**, 605 (1980).

30. H. Sano and M. Tachiya, *J. Chem. Phys.* **71**, 1276 (1979).

31. H. S. Carslaw and J. C. Jaeger, *Conduction of Heat in Solids*, 2nd ed., Oxford University Press, London, 1959, p. 246.

32. U. Gösele, U. K. A. Klein, and M. Hauser, *Chem. Phys. Lett.* **68**, 291 (1979).

33. A. Szabo, K. Schulten, and Z. Schulten, *J. Chem. Phys.* **72**, 4350 (1980).

34. H. Sano and M. Tachiya, *J. Chem. Phys.* **75**, 2870 (1981).

35. M. D. Hatlee, J. J. Kozak, G. Rothenberger, P. P. Infelta, and M. Grätzel, *J. Phys. Chem.* **84**, 1508 (1980).

36. M. Van der Auweraer, J. C. Dederen, E. Geladé, and F. C. De Schryver, *J. Chem. Phys.* **74**, 1140 (1981).

37. A. J. Frank, M. Grätzel, and J. J. Kozak, *J. Am. Chem. Soc.* **98**, 3317 (1976).

38. A. Henglein and Th. Proske, *Ber. Bunsenges. Phys. Chem.* **82**, 471 (1978).

39. M. A. J. Rodgers and J. C. Becker, *J. Phys. Chem.* **84**, 2762 (1980).

40. R. Samson and J. M. Deutch, *J. Chem. Phys.* **68**, 285 (1978).

41. L. Sterna, D. Ronis, S. Wolfe, and A. Pines, *J. Chem. Phys.* **73**, 5493 (1980).

42. R. G. Lawler and G. T. Evans, *Ind. Chim. Belg.* **36**, 1087 (1971).

43. A. L. Buchachenko, E. M. Galimov, V. V. Ershov, G. A. Nikiforov, and A. D. Pershin, *Dokl. Phys. Chem.* **228**, 451 (1976).

44. N. J. Turro, B. Kraeutler, and D. R. Anderson, *J. Am. Chem. Soc.* **101**, 7435 (1979).

45. D. A. McQuarrie, *Adv. Chem. Phys.* **15**, 149 (1969).

46. (a) R. M. Noyes and W. C. Gardiner, Jr., *J. Phys. Chem.* **85**, 594 (1981); (b) D. Toussaint and F. Wilczek, *J. Chem. Phys.* **78**, 2642 (1983).

47. For example (a) M. D. Hatlee and J. J. Kozak, *J. Chem. Phys.* **72**, 4358 (1980); (b) Sz. Vass, *Chem. Phys. Lett.* **70**, 135 (1980).

48. A. Yekta, M. Aikawa, and N. J. Turro, *Chem. Phys. Lett.* **63**, 543 (1979).

49. For example, N. J. Turro and A. Yekta, *J. Am. Chem. Soc.* **100**, 5951 (1978).

50. Y. Waka, K. Hamamoto, and N. Mataga, *Chem. Phys. Lett.* **62**, 364 (1979).

51. M. Aikawa, A. Yekta, and N. J. Turro, *Chem. Phys. Lett.* **68**, 285 (1979).

52. P. P. Infelta, M. Grätzel, and J. K. Thomas, *J. Phys. Chem.* **78**, 190 (1974).

53. J. C. Dederen, M. Van der Auweraer, and F. C. De Schryver, *Chem. Phys. Lett.* **68**, 451 (1979).

54. S. S. Atik and J. K. Thomas, *J. Am. Chem. Soc.* **103**, 3543 (1981).

55. G. Rothenberger, P. P. Infelta, and M. Grätzel, *J. Phys. Chem.* **83**, 1871 (1979).

56. P. P. Infelta and M. Grätzel, *J. Chem. Phys.* **70**, 179 (1979).
57. D. J. Miller, U. K. A. Klein, and M. Hauser, *Ber. Bunsenges. Phys. Chem.* **84**, 1135 (1980).
58. U. Lachish, M. Ottolenghi, and J. Rabani, *J. Am. Chem. Soc.* **99**, 8062 (1977).
59. G. Rothenberger, P. P. Infelta, and M. Grätzel, *J. Phys. Chem.* **85**, 1850 (1981).
60. W. V. Smith and R. H. Ewart, *J. Chem. Phys.* **16**, 592 (1948).
61. J. T. O'Toole, *J. Appl. Poly. Sci.* **9**, 1291 (1965).
62. B. S. Hawkett, D. H. Napper, and R. G. Gilbert, *J. Chem. Soc. Faraday I* **73**, 690 (1977).
63. G. H. Weiss and M. Dishon, *J. Chem. Soc. Faraday I* **72**, 1342 (1976).
64. D. T. Birtwistle and D. C. Blackley, *J. Chem. Soc. Faraday I* **73**, 1998 (1977).
65. D. T. Birtwistle, D. C. Blackley, and E. F. Jeffers, *J. Chem. Soc. Faraday I* **75**, 2332 (1979).
66. K. Kano, T. Yamaguchi, and T. Matsuo, *J. Phys. Chem.* **84**, 72 (1980).
67. M. Almgren, *Chem. Phys. Lett.* **71**, 539 (1980).
68. M. Almgren, *J. Am. Chem. Soc.* **102**, 7882 (1980).
69. W. Schlenker, T. Ulrich, and U. E. Steiner, *Chem. Phys. Lett.* **103**, 118 (1983).
70. D. Grand, S. Hautecloque, A. Bernas, and A. Petit, *J. Phys. Chem.* **87**, 5236 (1983).

12 POLYMERIZATION AND AGGREGATION DURING GELATION

Mohamed Daoud

Laboratoire Léon Brillouin
Centre d'Études Nucléaires
Saclay, France

12.1. INTRODUCTION

Polymers are materials of great practical importance. The chair you are sitting on is probably made partly of plastic. Your clothes are also partly made of synthetic material. Many parts of a car are synthetic polymers. These materials permeate everyday life. In many areas they have replaced natural materials. This is so because they are cheaper, and they can be made to have such different properties as being elastic, or viscous, or glassy, and so on, by changing a few parameters. Yet their properties are still incompletely understood.

This chapter is devoted to polymerization. This is the process that joins small molecules (monomers) together to form very large ones and eventually forms a giant macromolecule that spans the container (1, 2) in a nonhomogeneous network, the gel. There are simple physical models that describe the essential properties of the process. We will discuss the properties and why they are important.

An important case is *not* discussed here. This is when one starts with bifunctional monomers, that is, monomers that may react at only two sites. Then, as for many people holding hands, one gets *linear* polymers. Although these are an important class of materials we will not consider their formation here: They are topologically one dimensional (i.e., lines), and this chapter deals with topologically three- (spatial) dimensional systems. We will consider animals made by many octopuses joining tentacles rather than people joining hands: *Branched* polymers.

12.1.1. Some Examples

To illustrate the above, we give some examples that exhibit the same type of behavior, although they have nothing to do with polymerization in most cases. The examples are divided into two classes, which we call polymers and aggregates. This classification is artificial, but still, some cases look like systems in thermal equilibrium while others are more kinetic models, out of equilibrium.

12.1.1.1. Polymer Type

(a) Spread of a Disease. Consider the propagation of a disease in an orchard. Suppose that the trees are on the sites of a square lattice with spacing d (Figure 12.1). Suppose that every sick tree can contaminate its nearest neighbors with a given probability p, which is a decreasing function of d. After some equilibration time two situations may be observed:

1. If the spacing is large, the disease does not spread through the orchard. However, we get clusters of sick trees. Depending on d, the size of these clusters may be smaller or larger. The size of a cluster is the number of sick trees inside. There is generally a large distribution of sizes.
2. When the spacing is small, the disease spreads across the orchard. Thus, there is an "infinite" cluster of sick trees. Usually, for intermediate spacings all the trees are not sick, but a finite fraction of them are. The sick trees do

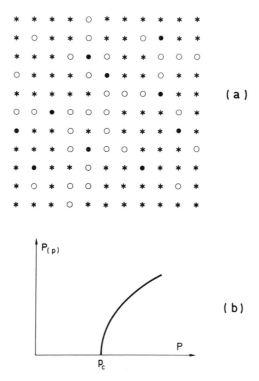

Figure 12.1: (*a*) Schematic representation of an orchard: ∗, safe trees; ○, ●, sick trees for different stages: ○, for $p < p_c$ the orchard is still safe; ○ + ●, $p > p_c$. There is a path of sick trees connecting all sides. Note that in addition to the infinite cluster there are also finite clusters of sick trees. (*b*) Fraction of sick trees starts from zero at a threshold p_c with an infinite slope.

not form a regular superlattice. Usually we get a nonhomogeneous, loose network of sick trees.

There is a spacing threshold separating these two situations. In the close vicinity of this threshold we get on the "safe" side very large, but finite, clusters of sick trees while on the other side ("sick") we have the loose infinite cluster of sick trees plus large but finite clusters of sick trees.

(b) Spread of a Fire. During the spread of a fire in a forest we suppose that there is a given probability that a neighboring tree catches fire. If p is small, only finite clusters are lost. Beyond a threshold a finite fraction (infinite cluster) of the forest is lost, with some trees remaining safe.

(c) Communication Network. Consider a developing country that is building a highway or a telephone network. Let us suppose that there is a given probability to connect at random two neighboring cities, or homes. If this probability is small, we get finite clusters of cities that are connected, but one cannot drive or call from

one place to a very distant city in the country: There is no infinite cluster. Above the threshold this becomes possible. Of course, when p is not much larger than the threshold we expect that some areas will not be connected to the network, so that the infinite cluster coexists with finite ones, even if some may be very large.

(d) Resistor Network. Consider a lattice (square lattice in two (spatial) dimensions, cubic lattice in three dimensions, etc.) where all the sites are present. Between neighboring sites we put at random, with probability p, a resistor. Then between two nearest-neighbor sites on the lattice there is a resistor with probability p or nothing (infinite resistor) with probability $1 - p$. When p is small, we get finite clusters of resistors. But there is not any continuous path connecting the edges of the lattice. As a result, if we apply an electrical potential, no current flows. The equivalent resistance of the system is infinite, or equivalently, its conductivity is zero. Beyond a threshold, p_c, an "infinite cluster" of resistors connecting the edges of the sample appears. Then an electric current may flow, meaning that the conductivity is no longer zero. Thus, at the threshold the system has a transition from an insulating to a conducting behavior. The structure of the system (Figure 12.2) above the threshold is made of finite clusters, which do not

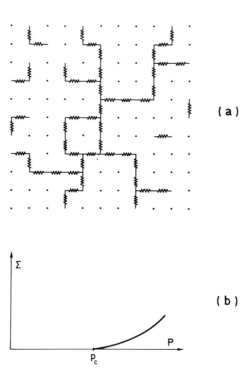

Figure 12.2. (*a*) Resistor network. Resistors are present at random with probability p, and absent ($R = \infty$) with probability $1 - p$. (*b*) Conductivity of network is zero below threshold p_c and then increases with a slope that is zero at p_c.

contribute to the macroscopic conductivity, and an infinite cluster, which is very loose and provides the conductivity. A closer look at this ''network'' shows that it is imperfect. In addition to its nonhomogeneity, it contains a lot of dangling ends that do not contribute to the conductivity. Thus, the fraction of conducting resistors is going to be much smaller than the fraction of resistors belonging to the network. This is exhibited by the different behaviors of the conductivity and the fraction P of sites (or resistors) belonging to the infinite cluster (compare Figures 12.1b and 12.2b) near the threshold. Whereas the conductivity starts with a zero slope, P has an infinite slope at the threshold.

A related problem with practical importance concerns porous media. Many regions beneath the earth's surface are porous with voids in the rocks. These voids may be connected, forming large nonhomogeneous fields, some of which contain oil. A problem is to locate and determine the size of an oil field and the possible flow of oil to a collection point. In the opposite limit it may be important for pollution problems that toxic chemicals or nuclear wastes cannot find a path out of the containment sites where they have been stored. When they are in a ''finite cluster,'' this is possible. If they are in the ''infinite cluster,'' which is usually the case, one has to be as close to the threshold as possible so that the structure is as nonhomogeneous as possible, with plenty of dead ends, so that the retaining time is really long!

There are many more examples. All of them have common properties that are characteristics of *percolation*:

1. There is a set of points distributed in space.
2. Any two neighboring points have a given probability to be connected. Thus, we get clusters of connected points.
3. The shape of these clusters is very filamented and nonhomogeneous.
4. Above a threshold, in addition to the finite clusters, there is an infinite set of connected points, which we call the ''infinite,'' or ''percolating,'' cluster.
5. It spans throughout the sample.
6. There is a finite fraction of points belonging to the infinite cluster.

Although it is not obvious, we will assume that there is only one infinite cluster. This has been proved rigorously only in a few cases.

We will see in Section 12.2 that this is a possible model for polymerization and gelation, the finite clusters corresponding to branched polymers and the infinite network to the gel.

12.1.1.2. Aggregates. Several models have been developed to describe the growth of clusters. We give some examples, the colloidal growth being discussed more extensively in Section 12.4. Here, all of the examples are computer simulations.

(a) Cancer Growth. Consider a square lattice with one site initially occupied by a particle (or cell). At time $t = 1$ one of the four neighboring sites is randomly

chosen to be occupied by another particle. Thus, we have a two-particle cluster. For $t = 2$ one of the six nearest-neighbor sites is again randomly occupied, and so on. At time t one of the vacant sites immediately adjacent to the cluster is randomly occupied. This is the Eden model and is the first tentative modeling of cancer growth. Unlike the preceding examples, it leads to a rather compact structure: The dimension R of the resulting cluster is related to its size N by

$$N \propto R^2$$

and, for a space dimension d, by a similar power law, the exponent being d in general.

(b) Colloidal Growth. In a generalization of the preceding model one considers a central particle on a square lattice. Then another particle is released, which undergoes a random walk until it visits a site that neighbors the central one, and then it sticks irreversibly. Thus, we get a two-particle cluster. However, if the particle diffuses away and does not visit a neighboring site, when it has gone away, say to a given large radius R, the process is stopped and a new particle is released. Once we have a two-particle cluster, a new particle starts a random walk with the restriction that if it goes into one of the six vacant neighboring sites it sticks to the cluster. And so on. A large cluster formed this way, containing approximately 10^4 particles, looks very stringy. The main difference with the preceding example lies in a screening effect. An incoming particle cannot in general diffuse very deep inside the cluster but is rather trapped within some distance of the periphery. This is more extensively discussed in Section 12.4. The result on a square lattice is that the relation between the size N of the cluster and its dimension R is

$$N = KR^D \qquad D = 1.4$$

where K is a proportionality constant. This model is currently used to describe colloidal growth. The important difference between colloidal growth, or aggregation, and the polymerization processes we described previously lies in the diffusion process, which is not allowed in the latter case. Thus, whereas chemistry is rate limiting for polymerization, diffusion takes over for aggregation. Intuitively, one feels that each of these processes may be important, depending on the concentration of the monomers that polymerize or aggregate. In a dilute solution diffusion is important, whereas in a concentrated solution chemistry should be more relevant. Both models are discussed in Sections 12.2 and 12.4, respectively. But we first introduce the self-similarity properties usually present in polymers, gels, and aggregates and show up in relations such as the one written above.

12.1.2. Self-Similarity: Fractals

A characteristic feature of many homogeneous systems is that they are dense. Let us consider a sphere of radius r in a space of dimension d. Its volume V is proportional to r^d. A notation for this throughout the chapter is

$$V \propto r^d \tag{12.1}$$

where we ignore the proportionality constant.

If this sphere is taken from a dense homogeneous material, we expect the number of particles $N(r)$ in it to be just proportional to its volume (the constant density being included in the proportionality constant),

$$N(r) \propto r^d \tag{12.2}$$

One could also consider less dense systems, for instance, the surface of our compact system, which has dimension $d - 1$, and so on. For all these systems it is well known that $d = 3$ for the real world, $d = 2$ in some cases, $d = 1$ when we wish to solve a problem more easily, and sometimes $d > 3$ for some theoreticians. Let us now turn to the simplest random system, namely, a random walk. A well-known result is that after N steps of size l each, the average distance, or more precisely the mean square end-to-end distance $\langle R^2(N) \rangle$, is

$$\langle R^2(N) \rangle \propto N \, l^2 \quad (N \gg 1) \tag{12.3}$$

where $\langle \ \rangle$ means the average over the different configurations (or possible walks).

Comparing Eqs. (12.3) and (12.2), we may say that in some sense, a random walk is two dimensional. This we will call its fractal dimension D. So for a random walk in any spatial dimension

$$D = 2 \tag{12.3'}$$

Another property of a random walk provides some more insight: self-similarity. Given a random walk with elementary step l, let us now take a new step length l' made of n elementary steps (Figure 12.3). Thus, we generate a new walk, which is itself a random walk and is similar to the initial one. This self-similarity is a characteristic property of fractals. Whatever the scale at which they are observed, they look very much the same. So, unlike our random walk, a fractal has no characteristic length:

1. For large distance scales it continues forever. The number of steps is infinite.
2. For small distance scales it also continues forever. If we look more closely at the step length l, we would find that it is a random walk with a still smaller step length.

For any real system, as in our example, there is usually both a lower and an upper length (l and R in the random walk), so that it is not really a fractal. Nevertheless, for intermediate distances, $l \gg r \gg R$, which are usually the most interesting, one may consider the random walk as a fractal. As an example, the random walk is the crudest model for a linear polymer chain. The step length models a monomer (e.g., styrene), and the walk itself is the polymer (polystyrene).

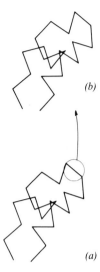

(b)

(a)

Figure 12.3. Random walk as a fractal. If one takes a step in (*a*) and blows it up, in (*b*), it looks similar to the original walk. In a fractal this process could be iterated forever for smaller and larger scales. In a real walk there is a fractal behavior in a restricted range only.

This would be a model for a very dilute solution of polymers in a solvent (e.g., cyclohexane). Then looking at very short distances is not really interesting because it is related to the detailed chemistry of the system. Looking at very large distances brings us to the study of dilute solutions. The interesting and specific properties of the long polymer chains show up mainly for intermediate distances, where the polymer has a fractal behavior. We will see later that branched polymers, gels, and aggregates also have a fractal behavior.

The random walk is the simplest random system that exhibits this self-similarity property. One may imagine more complicated systems, such as branched polymers, gels, and aggregates, where the fractal dimension is used instead of the conventional space dimension in relation (12.2). If we consider a hypersphere with radius r, the number of elements $N(r)$ contained in this volume varies as

$$N(r) \propto r^D \tag{12.4}$$

As a simple illustration of relation (12.4), let us consider the following *regular* curve, shown in Figure 12.4. We start with an equilateral triangle. Each side is divided into three equal parts. We replace the middle one by two segments of equal length. The process is then iterated on the six smaller triangles, and so on. The first two iterations are shown in Figure 12.4. This is the Von Koch curve. It is easy to show that its fractal dimension is

$$D = \frac{\log 4}{\log 3}$$

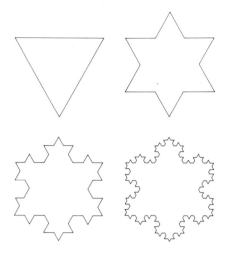

Figure 12.4. Von Koch curve shows how one may build a regular fractal by iteration of a simple process. The first three iterations are shown here.

The reader may imagine other possibilities where one starts with a regular curve and a rule for the game, as above, which is iterated to give a self-similar object as in our example. Any part of the curve is similar to the curve.

For various examples and a more extensive discussion or fractals, the reader is referred to the excellent book by B. Mandelbrot (3).

Thus, the notion of a fractal is by no means related to randomness. But in many random systems such as those in this chapter, scale invariance is present and simplifies the analysis of the problem.

In the following we examine the three main models currently studied for polymerization, aggregation, and gelation. Section 12.2 deals with the percolation model. The classical theory was introduced by Flory and Stockmayer many years ago. The main improvements were done in the past 10 years. The kinetic theory was initiated by Von Smoluchowski very early. It was used by Flory and Stockmayer in a case similar to their classical percolation model, and it was generalized during the past few years. This is done in Section 12.3. Finally, aggregation models are discussed in Section 12.4. Here diffusion controls the process.

12.2. GELATION

Gels are important practical materials. We approach the gelation process from a physicist's point of view: We do not discuss the detailed chemistry, and we mainly look at the vicinity of the gelation threshold, when we have either a very weak and nonhomogeneous gel (above the threshold) or very large branched macromolecules (both above and below the threshold). However, because the preparation method is important to the ultimate properties of the gel, we describe briefly some gelation processes.

12.2.1. Some Processes

1. The first *chemical* way is *polycondensation* of multifunctional units, shown in Figure 12.5. Here every monomer may react by three separate functionalities. We assume that they all have the same chemical reactivity, and that the latter is the same for all monomers, independent of their location in the macromolecule (4, 5). This hypothesis is in sharp contrast with what we assume in the next section, where only a fraction of the monomers on the "periphery" of an aggregate will react. Although all the functionalities may react, statistically only a few of them do so, and we will be left with very loose, branched structures and eventually a network if we let the reaction proceed for a sufficient time.

2. In *additive polymerization* we start with monomers that have two double bonds:

$$CH_2=CH-R-CH=CH_2$$

where the double bonds may be opened by free-radical reactions. Growth may occur from each of the double bonds, and we obtain a branched structure. Only a few double bonds will actually react, and the structure is loose, except for very high degrees of reaction, which we do not consider here.

3. The first two processes were chemical. Gelation may also be obtained by *physical* linking, for instance, in gelatin or with some metal sulfonate ionomers. Here, the linking of the molecules may occur for different reasons:

a. Dipolar interaction between the metal groups of the ionomers.
b. Local microcrystallization of linear polymer melts: Even atactic polystyrene may have syndiotactic sequences. These are attractive to each other and may serve as a cross-link between different polymer chains. This kind of gelation has recently been observed in polystyrene solutions in carbon disulfide or toluene.
c. Local formation of helical structures in gelatin.

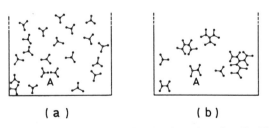

(a)   **(b)**

Figure 12.5. Schematic representation of a polyfunctional condensation (5). Here three-functional monomers may react by their three functionalities. The reaction vessel is shown in (*b*) at a later stage. With this process one may eventually get a network in addition to finite branched polymers such as shown in (*b*). (Reprinted, with permission, from Ref. 5.)

A common property of these physical gels is that they are thermoreversible: By cooling the system, one gets a gel, which is destroyed by increasing the temperature again. There are often hysteresis effects in physical gelling: The temperatures at which the gel appears upon cooling and is destroyed by heating may be different. They are modeled in a different way than those we considered above and which we discuss in more detail below.

4. *Vulcanization* is the chemical cross-linking of linear polymer chains. For a very long time Indians in Amazonia used natural (Hevea) rubber, which is cross-linked by the oxygen in the air (they did not control the reaction, however, and could not stop it). More recently Goodyear realized that it is possible to use sulfur bridges for cross-linking. This is currently widely used in the tire industry. We will call vulcanization the gelation of linear polymer chains. We describe two possible ways currently used: In both, one first synthesizes long polystyrene polymers with a narrow molecular weight distribution. (This is achieved by anionic polymerization). The second step is different in the two ways:

(α) Divinylbenzene, a multifunctional unit, is used to cross-link the ends of the chains (6).
(β) By irradiation (7) one breaks double bonds that cross-link different chains upon recombination. In this case linking may take place anywhere along the chains and is completely random: Any two neighboring double bonds may be transformed to a cross-link.

12.2.2. Classical Theory

We start with f-functional units and allow them to react with a probability p. More precisely, p is the probability that two units are reacted. The classical theory, introduced first by Flory (2) and Stockmayer (43), includes the following assumptions:

(i)   There is no interaction between any two monomers (except the possibility of chemical reactions).
(ii) The formation of closed loops is prohibited: We neglect the possibility of intramolecular reactions. Thus, the branched polymers, and eventually the gel, have a treelike structure, such as shown in Figure 12.6.
(iii) All monomers are equally reactive, and reaction may occur at random between any two neighbors.

(a) The Threshold. We start from a "central" monomer that belongs to a cluster, which we call the "parent." This parent has already reacted by one of its functional groups. We calculate the expected number of elements in its progeny: It still has $f - 1$ unreacted functional units. Each of these has a probability p to react. Then the expected number of children is $p(f - 1)$.

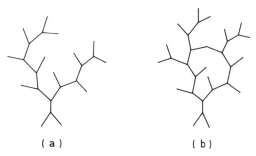

(a)   (b)

Figure 12.6. Structure of branched polymer. In the classical theory only treelike configurations such as in (*a*) are permitted. Intramolecular reactions, leading to cycles such as in (*b*), are forbidden. Both configurations are allowed in critical theory.

1. If $p(f - 1)$ is less than unity, every new generation will be less populated than the preceding one. Thus, we expect that the cluster stops growing after several generations: We get finite (branched) macromolecules.
2. If $p(f - 1)$ is larger than unity, every new generation is larger than the preceding one, and we expect the population to grow indefinitely: Then we may get an infinite network, which we call the gel.

Thus, there is a threshold:

$$p_c = \frac{1}{f - 1} \tag{12.5}$$

below which only finite molecules are present. This is what is called the *sol*. Above the threshold there is both a sol and a gel. When p keeps increasing, the sol fraction decreases until only the gel is present for $p = 1$.

(b) The Molecular Weight Distribution. When the reaction proceeds, we get finite macromolecules that may be very large near the threshold. Because the process is random, there is a wide distribution of molecular weights. Let us call $P_n(p)$ the probability of having a molecule made of n monomers,

$$P_n(p) = \frac{C_n(p)}{C} \qquad (n = 1, 2, 3, \ldots) \tag{12.6}$$

where C is the total monomer concentration and C_n the concentration in n-mer. Note that only finite molecules are included in the above definition. The infinite network above the gelation threshold is not. This implies a property for the normalization of P_n: Whereas all the monomers belong to finite molecules below the gel point,

$$\sum_{n=1}^{\infty} nP_n(p) = 1 \qquad (p \le p_c) \tag{12.7a}$$

a finite fraction $G(p)$ belongs to the gel above the threshold:

$$\sum_{n=1}^{\infty} nP_n(p) = 1 - G(p) \qquad (p \ge p_c) \tag{12.7b}$$

At the threshold, where a vanishingly small fraction of monomers belongs to the infinite network, both (12.7a) and (12.7b) are valid.

A related distribution, $W_n(p) = nP_n(p)$, is the probability that starting from a monomer, we have a molecule with n monomers. The different moments of this distribution correspond to various measurable quantities, one being the gel fraction $G(p)$ defined above:

1. The concentration of polymers, irrespective of their molecular weight is

$$P(p) = \frac{\displaystyle\sum_{n=1}^{\infty} P_n(p)}{\displaystyle\sum_{n=1}^{\infty} nP_n(p)} \tag{12.8}$$

2. The weight-average molecular weight, directly measurable by elastic light scattering experiments, is

$$M_w = \frac{\displaystyle\sum_{n=1}^{\infty} n^2 P_n(p)}{\displaystyle\sum_{n=1}^{\infty} nP_n(p)} = \frac{\displaystyle\sum_{n=1}^{\infty} nW_n(p)}{\displaystyle\sum_{n=1}^{\infty} W_n(p)} \tag{12.9}$$

To calculate this distribution function, we introduce a generating function $F_0(\theta)$, which is the Laplace transform of W_n (1,4):

$$F_0(\theta) = \sum_{n=1}^{\infty} W_n(p) \, \theta^n \tag{12.10}$$

Usually $F_0(\theta)$ is easier to calculate than $P_n(p)$. Having $F_0(\theta)$, $W_n(p)$ is just the coefficient of θ^n in the expansion of F_0. Moreover, the different moments of P_n may be calculated directly from $F_0(\theta)$ by simple differentiation. For instance, below the threshold the weight-averaged molecular weight is

$$M_w = \frac{1}{F_0(\theta)} \frac{d}{d\theta} F_0(\theta) \bigg|_{\theta = 1} \tag{12.9'}$$

We can see from relation (12.10) that every time a functionality has reacted, we have an extra factor θ. This allows us to construct a diagrammatic solution for $F_0(\theta)$. This is shown in Figure 12.7, where, following Dobson and Gordon (4), we start from a central monomer and consider all the branching possibilities in the case of a trifunctional unit. To each unreacted functionality we associate a weight $1 - p$. To each reacted unit we give a factor p. From each reacted functionality begins a possible tree, symbolized by a circle to which we associate a factor F_1. The figure lists all the possibilities for a trifunctional monomer, with respectively, zero, one, two, and all three functionalities reacted. Note that there is a combinatorial factor associated with each possibility. For instance, there are f (in our example three) possibilities of having one functionality reacted. For the case of f-functional units our diagrammatic representation is

$$F_0 = (1 - p)^f + f(1 - p)^{f-1}\theta p F_1$$

$$+ \tfrac{1}{2}f(f - 1)(1 - p)^{f-2}(\theta p F_1)^2 +$$

$$+ \cdots + (p\theta F_1)^f$$

$$\equiv [1 - p + p\theta F_1]^f \tag{12.11}$$

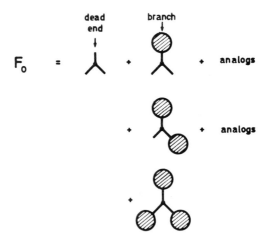

Figure 12.7. Diagrammatic calculation of generating function F_0 for trifunctional monomers. Circles represent branches. To each unreacted functionality is associated a weight $1 - p$ and to each reacted one a weight p. The factor F_1 associated with a branch is calculated in Figure 12.8. (From Pierre-Gilles deGennes: *Scaling Concepts in Polymer Physics*. Copyright © 1979 by Cornell University. Used by permission of the publisher, Cornell University Press.)

Again, F_1 lists the different possible configurations from a monomer belonging to the first generation, that is, that has a functionality that is already reacted. It may also be calculated diagrammatically. This is shown in Figure 12.8, which has the same prescriptions as Figure 12.7 concerning the factors associated with unreacted and reacted functionalities. Because we have assumed that all functionalities have the *same reactivity*, independently of the state (reacted or not) of the other ones, F_1 may be calculated in a self-consistent way. In Figure 12.8 the dashed circle may correspond to different possibilities where zero, one, or both of the remaining functionalities are reacted. For the more general case of an f-functional unit, it reads

$$F_1 = (1 - p)^{f-1} + (f - 1)(1 - p)^{f-2}p\theta F_1 + \cdots + (p\theta F_1)^{f-1}$$
$$\equiv (1 - p + p\theta F_1)^{f-1} \tag{12.12}$$

Note that F_1 may be deduced from F_0 by observing that an f-functional monomer with one reacted functionality may be considered as an $f - 1$-functional monomer with no reacted functionality!

We have assumed an equal reactivity of the functionalities. One could generalize the above calculation by supposing that the probability of reacting is not the same, depending on whether one, or two, or more functionalities have already reacted. Then one would have to replace F_1 by $F_1, F_2, \ldots$. This would make the calculations very tedious without changing dramatically the results as long as the reactivities are not very different. Usually, this equal-reactivity hypothesis is not too far from reality.

Let us solve this system, for instance, for $F_0 \equiv F_0(\theta = 1)$. Then Eqs. (12.11) and (12.12) become

$$F_0 = (1 - p + pF_1)^f \qquad F_1 = (1 - p + pF_1)^{f-1}$$

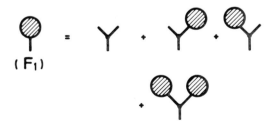

(F_1)

Figure 12.8. Diagrammatic calculation of F_1, when one functionality is already reacted. Note that it amounts to Figure 12.7 for bifunctional units. (From Pierre-Gilles deGennes: *Scaling Concepts in Polymer Physics.* Copyright © 1979 by Cornell University. Used by permission of the publisher, Cornell University Press.)

which is solved for p:

$$p = \frac{(1 - F_1)^{1/(f-1)}}{1 - F_1} = \frac{1 - F_0^{1/f}}{(1 - F_0)^{(f-1)/f}} \qquad (F_1 \neq 1) \qquad (12.13)$$

For F_0 close to unity this may be expanded. Let us assume

$$F_0 = 1 - G(p)$$

and expand relation (12.13) in p. We find

$$p = \frac{1}{f - 1} + \frac{1}{2} \frac{f - 2}{f(f - 1)} G(p)$$

$$G(p) \equiv 1 - F_0 = \frac{2f(f - 1)}{(f - 2)} \left(\frac{p}{p_c} - 1 \right) \qquad (p > p_c) \qquad (12.14)$$

$$G(p) = 0 \qquad (p < p_c)$$

So we find a threshold p_c, as given by Eq. (12.5), separating two different regimes:

1. Below the threshold no gel is present. All the monomers belong to finite polymers.
2. Above the threshold a finite fraction of monomers belongs to the infinite network. This fraction vanishes linearly with p when one approaches the gel point.

In the same way, one may calculate the weight-averaged molecular weight M_w. Using relation (12.9'), we find

$$M_w = \frac{fpF_1(F_1 + F_1')}{F_0}$$

where the prime denotes differentiation with respect to θ. This leads to

$$M_w = \frac{fpF_1^2}{\{1 - p(f - 1)(1 - p + pF_1)^{f-2}\}\{1 - p + pF_1\}^f} \qquad (12.15)$$

Below the threshold, when $F_1 = 1$, we find

$$M_w = \frac{fp}{1 - p(f - 1)} = fp \frac{p_c}{p_c - p} \qquad (p < p_c) \qquad (12.16a)$$

Relation (12.16a) shows that M_w diverges at the threshold as the inverse of the distance to the gel point. In our notation

$$M_w \propto (-\epsilon)^{-1} \qquad (12.16b)$$

with

$$\epsilon \equiv \frac{p - p_c}{p_c}$$

The same divergence is present above the threshold. This may be shown using relations (12.15) and (12.13). Thus, the divergence is

$$M_w \propto |\epsilon|^{-1} \qquad (12.16)$$

both above and below the threshold.

The purpose of the above calculations was to show that it is possible to calculate all the moments of the distribution $W_n(p)$. We have found that some quantities, such as the probability G for a monomer of belonging to the gel, above the threshold, vanishes when one goes to the gel point. Thus, the gelation "transition" is continuous. At the same time other quantities, such as the weight average molecular weight M_w, diverge near the threshold. These results will be refined below by introducing power law behaviors with exponents different from the ones we found above. The reason for this lies in the approximations made in this classical calculation. It can be shown that the distribution function has the form

$$P_n(p) \equiv n^{-1} W_n(p) \propto n^{-5/2} e^{-n\epsilon^2} \qquad (12.17)$$

This law gives back relations (12.14) and (12.16).

(c) *The Dimension.* A final notion to be introduced is the dimension of the macromolecules (8). Let $\langle R^2(n) \rangle$ be the square of the radius of gyration of a polymer made of n monomers.

$$\langle R^2(n) \rangle = \frac{1}{2n^2} \sum_{i,j=1}^{n} \langle (\mathbf{r}_i - \mathbf{r}_j)^2 \rangle \qquad (12.18)$$

where the average is over the different configurations.

We calculate this radius using a method due to Kramers (9). Because of the branched tree structure of a polymer, there is only one chemical path connecting monomers i and j. Along the sequence, we index the monomers by q (Figure 12.9), and we denote by $\mathbf{b}_{ij}^q$ the vector $\mathbf{r}^{q+1} - \mathbf{r}^q$. This has a constant length a equal to the length of a monomer. Moreover, this sequence is a random walk. Then Eq. (12.18) may be written as

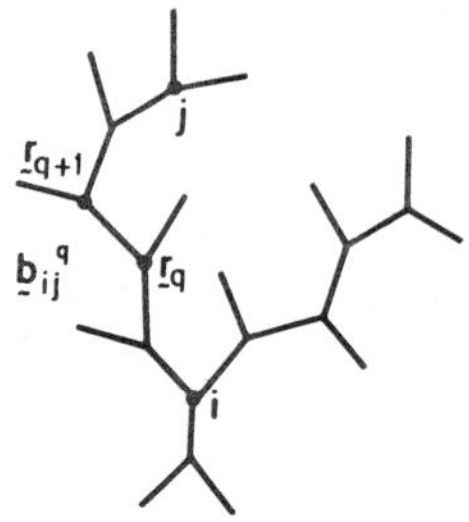

Figure 12.9. The KRAMERS construction.

$$\langle R^2(n) \rangle = \frac{1}{2n^2} \sum_{i,j} \sum_{q} (\mathbf{b}_{ij}^q)^2$$

$$= \frac{1}{2n^2} \sum_{i,j} \sum_{q} a^2 \qquad (12.18a)$$

The problem now is to calculate the number of configurations in which q belongs to a chain ij: Given q, how many times does it contribute to the sum in (12.18a)? Following Kramers (9), we cut the polymer at monomer q. The probability of getting two parts of p and $n - p$ monomers is proportional to Z_p and Z_{n-p}, where Z_p is the partition function of a branched tree of p monomers. Moreover, i and j must belong to either of these trees. This brings in an extra factor $2p(n - p)$ corresponding to the different choices for i and j. The mean square radius is then

$$\langle R^2(n) \rangle = \frac{a^2}{n} \frac{\sum\limits_{p=1}^{n-1} p(n - p) Z_p Z_{n-p}}{\sum\limits_{p=1}^{n-1} Z_p Z_{n-p}} \qquad (12.19)$$

To calculate the radius, we now have to calculate the partition function of a branched tree. To do so (8, 10), we introduce the number of configurations of a random polymer made of N monomers that go from point $\mathbf{r}$ to point $\mathbf{r}'$, $G(N, \mathbf{r}, \mathbf{r}')$. As usual, this is a function of the relative position, $G(N, \mathbf{r} - \mathbf{r}')$. The partition function Z_N is

$$Z_n = \int G(n, \mathbf{r})\, d\mathbf{r} \qquad (12.20)$$

We introduce its Fourier–Laplace transform,

$$\tilde{G}(\alpha, \mathbf{k}) = \int dN\, d\mathbf{r}\, e^{-\alpha N}\, e^{-i\mathbf{k}\cdot\mathbf{r}}\, G(N, \mathbf{r}) \qquad (12.21)$$

Here $G(N, \mathbf{r})$ is sometimes called the propagator for the polymer. It may be calculated diagrammatically, as shown in Figure 12.10, which we discuss now. The total number of configurations is made of the configurations without any branching plus the configurations where there are branches. The light lines in Figure 12.10 represent *linear* polymers, which are simply random walks. The heavy lines represent trees, without any assumptions about their nature (branched or unbranched). Thus, the number of configurations between points x and y in the diagram is formally identical to that between points r and r'. The only difference lies in the number of monomers involved. (Note the underlying fractal hypothesis: The structure of a part of a polymer is assumed to be the same as that of the whole). Points x and y are intermediate points that can be anywhere in space and lead to any partitioning of the polymer. One has then to integrate over their positions and over the partitioning: Let $G^0(N, r)$ be the number of configurations of a random walk of N steps. The diagram reads

$$G(N, r' - r) = G^0(N, r' - r) + \int G^0(p, x - r)\, G(q, r' - x)$$
$$\cdot\, G(N - p - q, y - x)\, dp\, dq\, dx\, dy$$

Performing the Fourier–Laplace transform simplifies the above relation:

$$\tilde{G}(\alpha, k) = \tilde{G}^0(\alpha, k) + \tilde{G}^0(\alpha, k)\, \tilde{G}(\alpha, k)\, \tilde{G}(\alpha, 0) \tag{12.22}$$

Relation (12.22) may be used to calculate the propagator for a branched polymer. This is left to the reader (10). We limit the discussion here to the partition function or to its Laplace transform:

$$\tilde{Z}(\alpha) = \int_0^\infty dN\, e^{-\alpha N}\, Z(N) \equiv \tilde{G}(\alpha, 0) \tag{12.23}$$

From (12.22) we get

$$\tilde{G}(\alpha, 0) = \tilde{G}^0(\alpha, 0) + \tilde{G}^0(\alpha, 0)\, \tilde{G}^2(\alpha, 0)$$

Here $G^0(\alpha, 0)$ is the Laplace transform of the partition function of a random walk,

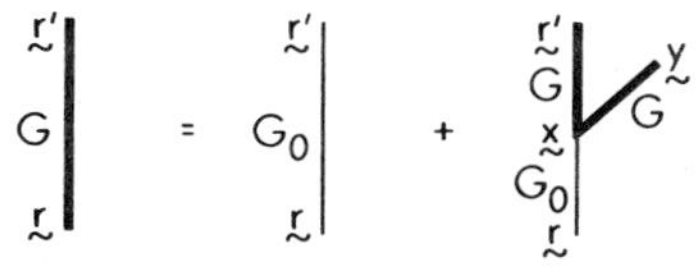

Figure 12.10. Diagrammatic calculation of the propagator G_N in the classical theory. G_0 is the propagator for a linear Gaussian polymer. (Reprinted, with permission, from ref. 8.)

$$\tilde{G}^0(\alpha, 0) \equiv \tilde{Z}^0(\alpha) \equiv \alpha^{-1}$$

we get

$$\tilde{Z}^2 - \alpha\tilde{Z} + 1 = 0$$

and

$$\tilde{Z}(\alpha) = \tfrac{1}{2}[\alpha - (\alpha^2 - 4)^{1/2}]$$

and, for large molecules.

$$Z(N) \simeq \frac{N^{-3/2}}{\sqrt{4\pi}} e^{2N} \tag{12.24}$$

Inserting (12.24) into (12.19) and transforming the summation into integration, we find

$$\langle R^2(n) \rangle = \tfrac{1}{4} \sqrt{\pi} n^{1/2} a^2 \tag{12.25}$$

a very important relation first found by Zimm and Stockmayer (11) in 1949. This shows that the dimension of a branched polymer is much smaller than that of the equivalent linear polymer. When no interaction is present, as we assumed so far, the typical radius varies as $N^{1/4}$ instead of $N^{1/2}$ for a random walk (a typical value for N is 10^4, so there is an order of magnitude difference between the radii).

The previous result applies to a branched polymer with a given molecular weight, or number of monomers. What about the distribution? When one synthesizes branched polymers, one gets a distribution of molecular weights, or sizes, given by relation (12.17). Then we must define an average distance over the whole distribution. From relations (12.17) and (12.25) we can find a characteristic length

$$\xi \propto |\epsilon|^{-1/2} \tag{12.26}$$

in the gelation problem; ξ is related to $\langle R^2(n) \rangle^{1/2}$. Again, we find a diverging length that is infinite at the gel point. Thus, at the threshold there is no typical length scale in the problem. In such a situation one thinks about fractals. Before coming to this, we may give another interpretation for ξ. Above the threshold there is an infinite molecule present. It is a nonhomogeneous network with plenty of dangling ends, and some nodes from which chains go to infinity (Figure 12.11). The symbol ξ may also be interpreted as the average distance between nodes: For distances smaller than ξ the gel is loose, with many dead ends, while for distances much larger than ξ one may consider it as a homogeneous network with mesh size

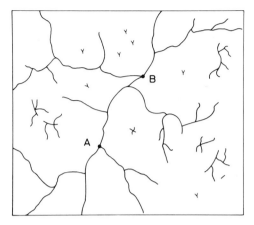

Figure 12.11. Typical gel above threshold. The structure is loose, with many dangling ends. Nodes like monomers A and B are junctions connected to infinity. No loops are present in classical theory.

ξ. Note that when one goes to the threshold, ξ becomes larger and larger, as does the nonhomogeneous region. At the threshold ξ is infinite, and the nonhomogeneous region is infinite. The network is nonhomogeneous at every scale, with dangling ends eventually very large. At this point, whatever the scale (larger than the step length 1), the gel looks the same. This is a fractal behavior. Note that the distribution of molecular weights also is a slowly decaying function: For instance, the weight-averaged molecular weight is infinite.

(d) Summary. Before criticizing the classical theory, let us summarize briefly its hypotheses and main results. The three main assumptions are the following:

   (i)   Equal reactivity of all the functionalities.
  (ii)   No interactions between the monomers (other than chemical).
 (iii)   No intramolecular reaction: No loop formation. The structure of the polymers, or the gel, is that of a branched tree.

Its results are that there is a gelation threshold, the gel point. Below the threshold only finite polymers are present. This is the sol. They have a broad molecular weight distribution. The probability of finding a molecule made of N monomers is

$$P(N, p) \propto N^{-5/2} f(\epsilon N^{1/2}) \tag{12.17a}$$

where $f(x)$ is an exponentially decreasing function cutting off a slower decay for high molecular weights. Here $\epsilon \equiv (p - p_c)/p_c$ is the distance to the threshold. Clearly, the smaller the value of ϵ, the wider the distribution. The second moment of this distribution is the weight-averaged molecular weight,

$$M_w \propto |\epsilon|^{-1} \tag{12.16}$$

which diverges at the threshold. From a practical point of view, if we need a large molecular weight, this will automatically imply a wide molecular weight distribution. The larger the value of M_w, the broader the distribution.

Above the threshold, in addition to the sol, there is an infinite molecule spanning the sample, the gel. This gives the elastic behavior, whereas below p_c we had only a viscous solution. The gel fraction vanishes at p_c as

$$G \propto \epsilon \tag{12.14a}$$

Note that for the sol fraction, both (12.17a) and (12.16a) are also valid above the threshold.

We defined a characteristic distance in Eq. (12.26):

$$\xi \propto |\epsilon|^{-1/2}$$

which may be interpreted as the mesh size of the incipient network above p_c, or the average linear size of the polymers below p_c, the radius of an individual polymer being

$$R(N) \propto N^{1/4} a \tag{12.25a}$$

Finally, at the threshold there is a scale invariance: The gel is fractal. Above the threshold ξ is finite, and the gel has a fractal behavior for distances smaller than ξ but much larger than a. For distances larger than ξ the gel looks like a regular network.

Two important results were not considered above, concerning the elasticity of the gel and the viscosity of the sol (1):

1. Below the threshold, because larger and larger molecules are present, the solution becomes more and more viscous as we approch p_c. The classical theory finds that the viscosity of the solution diverges logarithmically with ϵ.

2. Above the threshold, because of the presence of the infinite network, an elastic modulus appears: We have a solidlike behavior. Close to p_c, the classical theory finds that the elastic modulus vanishes as

$$\Sigma \propto \epsilon^3 \tag{12.27}$$

12.2.3. Critique of Classical Theory

Although the classical theory gives interesting results, it suffers from several defects. For instance, relation (12.25a) cannot hold in ordinary three-dimensional space: Let us suppose that the monomers of a branched polymer are uniformly distributed in the sphere of radius $R(N)$. Then the density inside this sphere is

$$n(r) \propto \frac{N}{R^3} \approx N^{1/4} a^{-3}$$

When N becomes very large, this density diverges, which is unrealistic. This shows that one has to take into account the interactions between monomers in an actual polymer.

Similarly, close to the gelation threshold intramolecular reactions become important (12, 13), leading to the formation of very large loops. The two above assumptions are closely related: In the classical theory there is an underlying lattice, the Bethe lattice, or Cayley tree, shown on Figure 12.12. The above theory could have been made supposing such a lattice with all the sites present. Between two neighboring sites a bond is present with probability p. The structure of the lattice automatically ensures that no loops are present. This lattice has, however, the special property that it cannot be embedded in any space with finite dimension: Starting from a central site, if we calculate the number of sites in the nth shell, we find that it varies as $3 \times 2^{n-1}$ in our figure. Thus, we find an exponential increase in the number of sites, which cannot be assimilated by any space with finite dimension. So the Cayley tree has an infinite dimension! Then one understands why the interactions are not important: In such a space monomers have no chance of being so close to each other that they can interact. So from this point of view, classical theory is consistent. Nevertheless, we now have to face the problem of taking into account intramolecular reactions. In the following we will first look at the case of spaces with finite dimension d. Then we will consider the question of the interactions within an approximate Flory–deGennes theory.

12.2.4. Critical Theory: Scaling

Consider a square lattice or its extensions, a cubic lattice in three-dimensional space, and so on, for higher space dimensions. All the sites of the lattice are present. The bonds are present with probability p and are randomly distributed.

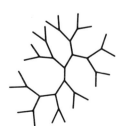

Figure 12.12. Sketch of a Bethe lattice or Cayley tree.

There are no restrictions concerning the presence of loops. This is the random bond *percolation* problem (14, 15), and it is identical to what we have seen in the classical theory, except for loop formation. (The difference is related to different lattices, as discussed in Section 12.2.3. Sites model the *f*-functional monomers (tetrafunctional in a square lattice; Figure 12.13), and every bond models a reacted functionality. Qualitatively, we expect the same results as in the classical theory. When *p* is smaller than a threshold, we get branched macromolecules with a broad distribution, large polydispersity. This is the sol. Above the threshold, in addition to the sol there is a gel. As before we expect some quantities to vanish at the threshold and others to diverge. Because of loop formation, the threshold is larger than predicted by classical theory. Many computer simulations were performed with different lattices, (square, triangular, and so on) and for different dimensions of space *d* ranging from 2 to 6. We summarize the results.

The threshold depends on the lattice as well as on the space dimension *d*. For the various quantities defined above, a power law behavior is found near the threshold. The important point here is that the exponents depend only on the space dimension *d* (14). They are independent of the nature of the lattice. This is the so-called *universality* property (16): All the three-dimensional (or d-dimensional) systems have the same power law behavior independent of the detailed chemistry in the vicinity of the gelation threshold. Note that the classical theory is even more universal: The exponents it gives are constants independent of *d*. In fact, we find that they are dimension dependent. We introduce the main exponents. Their values for different values of *d* are listed in Table 12.1.

The gel fraction

$$G \propto \epsilon^\beta \quad [\epsilon \equiv (p - p_c)/p_c > 0] \tag{12.28}$$

The weight average molecular weight

$$M_w \propto |\epsilon|^{-\gamma} \tag{12.29}$$

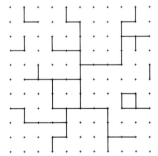

Figure 12.13. Percolating gel on a square lattice. Within critical theory, intramolecular reactions (loops) are allowed.

TABLE 12.1. Exponents in Percolation Model as Function of Dimension d of Space[a]

Exp	d	2	3	4	5	6	Classical
	β	0.14	0.4	0.5	0.7	1	1
	γ	2.43	1.7	1.5	1.2	1	1
	α	-0.67				-1	-1
	ν	1.35	0.84	0.7	0.6	$\frac{1}{2}$	$\frac{1}{2}$
	s	1.3	0.75			0 (log)	0 (log)
	t	1.3	1.9	2.4	2.7	3	3
	D	1.9	2.52	3.29	3.83	4	4

[a]Last row gives the classical values and is to be compared with $d = 6$. Last line is the fractal dimension. Results are from computer simulations (64).

The average polymer concentration

$$P \propto |\epsilon|^{2-\alpha} \tag{12.30}$$

The characteristic length

$$\xi \propto |\epsilon|^{-\nu} \tag{12.31}$$

The viscosity

$$\eta \propto |\epsilon|^{-s} \quad (\epsilon < 0) \tag{12.32}$$

The elastic modulus

$$\Sigma \propto \epsilon^t \quad (\epsilon > 0) \tag{12.33}$$

There are different methods for calculating these exponents: Monte Carlo simulations, series expansions, or renormalization group techniques. For more details about these methods the reader is referred to the book by Stanley (16) or the review articles by Family (17) and Stanley et al. (18). The main point here is that for $d = 3$, these values are very different from the classical values and are noninteger. Although we have defined many exponents, there are relations between them, called the *scaling* relations. Before we discuss these relations, we introduce two more exponents.

(a) The Distribution Function. We have seen above that the exponents α, β, and γ are related to moments of the distribution of molecular weights. In analogy with

the form of this function in the classical theory, Eq. (12.17a), we postulate the following form for $P(n, \epsilon)$ (14):

$$P(n, \epsilon) \propto n^{-\tau}f(\epsilon n^{\sigma}) \tag{12.34}$$

where we recall that $P(n, \epsilon)$ is the probability of having a molecule made of n monomers at a distance $\epsilon = (p - p_c)/p_c$ from the threshold. Here $f(x)$ is an exponentially decreasing function, which cuts off the distribution for molecular weights larger than some value

$$n^* \propto \epsilon^{-1/\sigma} \tag{12.35}$$

Finally, τ and σ are exponents that may be related to the previous ones. The total polymer concentration is

$$P = \int_0^\infty dn \, P(n, \epsilon) \propto \epsilon^{(\tau - 1)/\sigma} \tag{12.36}$$

Comparing (12.36) and (12.30), we get

$$2 - \alpha = \frac{\tau - 1}{\sigma} \tag{12.37}$$

Below the gel point, where $nP(n, \epsilon)$ is normalized to unity (no gel), the weight average molecular weight is

$$M_w = \int_0^\infty n^2 \, P(n, \epsilon) \, dn \propto \epsilon^{(\tau - 3)/\sigma} \tag{12.38}$$

Comparing (12.38) and (12.29), we get

$$\gamma = \frac{3 - \tau}{\sigma} \tag{12.39}$$

Finally, above the threshold, we have

$$\int_0^\infty nP(n, \epsilon) \, dn = 1 - G \tag{12.40}$$

Using (12.39) and (12.28), we get

$$\beta = \frac{\tau - 2}{\sigma} \tag{12.41}$$

Using (12.37), (12,39), and (12.41), we get the following relations between the exponents, called scaling relations:

$$\tau = 2 + \frac{\beta}{\beta + \gamma} \tag{12.42}$$

$$\sigma = (\beta + \gamma)^{-1} \tag{12.43}$$

$$\alpha + 2\beta + \gamma = 2 \tag{12.44}$$

The first two relations correspond simply to a change in notation (from β, γ to τ, σ). The last is an extra relation and lowers the number of independent exponents. The classical exponents also obey these scaling relations. We will find another such relation below. We stress here a property that we already saw in the classical theory: The molecular weight distribution is very broad. If one wishes to synthesize large molecular weights, one gets also a very polydisperse molecular weight distribution.

(b) Hyperscaling. We have seen above some scaling relations between the exponents. Another relation involves both the exponents and the dimension d of space. It is called the hyperscaling relation and is valid only when d is sufficiently small. In order to calculate it, we evaluate the number of monomers in a sphere of radius ξ. Its volume is

$$V \propto \xi^d$$

The number of monomers is the product of this volume and the density of monomers belonging to the polymer. If this polymer is very large, we may reasonably assume that its density, deep inside, is the same as that in the infinite gel, G. Then the number is

$$N \propto GV \propto \epsilon^\beta \xi^d \propto \epsilon^{\beta - \nu d} \tag{12.45}$$

where we have used (12.28) and (12.31). Now N is the number of monomers in a typical molecule with dimension ξ. Looking back to the distribution of molecular weights, one may convince oneself that this number is of the same order of magnitude as n^*, [relation (12.35)],

$$N \propto \epsilon^{-1/\sigma}$$

Comparing this relation to (12.45), we find

$$\nu d = \beta + \frac{1}{\sigma} = 2\beta + \gamma$$

$$= 2 - \alpha \tag{12.46}$$

where we have used (12.43) and (12.44). One can check using the numbers in Table 12.1 that it holds for any dimension. A further step we may take is to try to use it for the classical theory, where we know $\nu = \frac{1}{2}$ and $\beta = \gamma = 1$ [see relations (12.16b), (12.14a) and (12.26)]. Solving (12.46) leads to a space dimension

$$d_c = 6 \tag{12.46a}$$

This may be interpreted by saying that the classical theory works for six dimensions. For higher space dimensions (12.46) breaks down, and the classical theory becomes valid. This kind of behavior is seen in another area in physics: When one studies the behavior of the Ising model in magnetism, for instance, one is led to define critical exponents for space dimensions below $d_c = 4$, whereas the classical Curie–Weiss theory is valid above $d_c = 4$. For more details about magnetic systems we refer the reader to the book by Stanley (16). Below d_c, there are some complications that we will discuss with vulcanization.

(c) Fractal Behavior. As discussed in the classical theory, when we look at a gel exactly at the threshold, there is no characteristic length scale: ξ is infinite. This means that whatever the degree of magnification we use to look at the gel, it has the same structure, with dangling ends *and* closed loops. Away from the threshold, ξ is finite, and depending on the scale, one can see two different types of structures:

1. At large scales, larger than ξ, one has a homogeneous network with mesh size ξ.
2. For distances smaller than ξ the structure is nonhomogeneous, with many loops and dead ends. When one approaches the threshold, the latter part becomes more and more important.

Relation (12.45) may be used to calculate the fractal dimension (19) of the gel: The number of monomers belonging to the gel in a sphere of radius ξ is

$$N(\xi) \propto G\xi^d \propto \epsilon^\beta \, \xi^d$$

Using (12.28) and (12.31), we get

$$N(\xi) \propto \xi^{d - \beta/\nu}$$

and

$$D = d - \frac{\beta}{\nu} \tag{12.47}$$

We stress that all the preceding results are valid *in the reaction bath*. In many experiments, more especially below the gelation threshold, what is done is to quench the chemical reaction, dilute the solution, and finally perform the experiments. We *do not* expect many of the preceding results to hold under these conditions. The effect of dilution on the dimensions is dramatic. For instance, screening effects present in the rather concentrated reaction bath disappear completely upon dilution! This will be discussed in Section 12.2.6.

Some experiments have been performed in the reaction bath, mainly viscosity and elastic modulus measurements. These are difficult, however, because of the precise determination of the gelation threshold (20, 21). Results for the viscosity exponent, *s*, lie between 0.5 and 1, compared to 0 for the classical theory (logarithmic divergence) and approximately 1 in critical theory. In the same way elastic modulus measurements lead to an exponent *t* between 2 and 3.3, compared to 3 and 1.7 in classical and critical theories, respectively.

Many other experiments have been performed on diluted systems. Some of them, such as M_w measurements, compare directly with the quantities discussed above. Because others are perturbed by dilution, we postpone their results to Section 12.2.6.

12.2.5. Flory–deGennes Theory

We have defined above some exponents related to the molecular weight distribution and to the fractal dimension. These exponents are not easy to calculate exactly. So far, we have just quoted the values obtained by Monte Carlo simulations or by renormalization group calculations. There is, however, an approximate method, first applied by Flory to linear polymer chains (2) and generalized by deGennes to other cases (22), which gives the fractal dimension without too many calculations. A strong warning must be given before presenting it. The free energy that we are going to write down looks reasonable. But at the end of the calculation, one may check that it cannot be right. It seems that some terms are missing, but because of some coincidence not yet fully understood, they cancel, so that the result is usually very good within about 5% of all the known results. But one must always be cautious when applying it, and it is necessary to find different methods of checking the results. Were one able to calculate exactly one of the terms in the free energy, the cancellation would no longer occur, and the result would be wrong!

Nevertheless, this method usually works well, and we are going to use it for branched polymers (23, 10) and for the gelation problem.

So far, we have considered polymerization and gelation from a geometric point of view. Only the connectivity of the polymers and the eventual gel were considered. The Flory–deGennes approach considers the question from an energetic side and takes the interactions into account explicitly.

Because the experiments may be done either in the reaction bath or on a diluted system, we are led to consider two different questions, namely, the behavior of a polymer in a dilute solution and in the reaction bath, which is usually a concentrated solution. The main difference between these two situations lies in the strength of the interaction: We expect a strong screening of the interaction when concentration increases. This effect is discussed in Section (b).

(a) The Single Polymer. Let us first consider a single branched polymer. This corresponds experimentally to a very dilute solution where the different polymers are far from each other so that we may neglect the interactions between different macromolecules. Only intramolecular interactions are taken into account. We want to write down a free energy for a polymer made up of N monomers each with length a (23, 10). Our parameter is the radius R of the polymer. Then minimizing the free energy with respect to R will give the desired answer. The free energy is written formally as the sum of two terms:

$$F = F_{el} + F_{int}$$

where the two terms are, respectively, an elastic and an interaction contribution.

For the elastic part what is assumed is that the molecule acts as a spring, $F_{el} = KR^2$, where R is its actual radius, and K is the spring constant. The latter is homogeneous to an inverse squared distance. This is taken as the unperturbed radius, when no interaction is present. It has already been calculated above and is the Zimm–Stockmayer radius [Eq. (12.25)]. Thus, we find

$$\frac{F_{el}}{k_B T} \propto \frac{R^2}{R_0^2} \propto \frac{R^2}{N^{1/2} a^2}$$

where k_B is Boltzmann's constant and T the temperature. Note that we have dropped all the prefactors. We are interested here in the exponent for the molecular weight dependence of R, which is universal as discussed above and provides most of the variation.

For the interaction part we have first to define how two monomers interact. The potential between any two monomers has the usual form, with a hard core at short distances preventing two elements from being on top of each other and an attractive part for larger distances, as shown in Figure 12.14. Because of universality, we do not expect the detailed shape of this potential to be important. Usually it is replaced by an effective point-like interaction $v\,\delta(r)$, where v is the so-called excluded volume parameter and is temperature dependent. In the following, we suppose that v is positive, that is, the interaction is repulsive. This corresponds to what is called a good solvent. This parameter is just the second virial of $V(r)$:

$$v = \int [1 - e^{-V(r)/k_B T}]\, dr \tag{12.48}$$

There are some complications when v vanishes or becomes negative. One has then to introduce three-body interactions. We will not consider those complications here. The interaction term may be written as

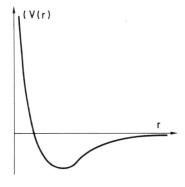

Figure 12.14. Interaction potential between two monomers has a hard core part preventing monomers to be on top of each other. This is the excluded volume interaction.

$$F_{int} = k_B T \int n(r)n(r')v(r - r') \, dr \, dr'$$

where $n(r)$ is the density at point r, and the integral is over the volume occupied by the polymer. Simplifying further, we suppose that the monomers are equally distributed in the sphere of radius R. Then the preceding expression becomes

$$F_{int} = k_B T v \frac{N^2}{R^d} \qquad (12.49)$$

and the expression for the free energy of a polymer is

$$\frac{F}{k_B T} = \frac{R^2}{N^{1/2} a^2} + v \frac{N^2}{R^d} \qquad (12.50)$$

Again, this is a variational free energy, with R as a parameter. Minimizing with respect to R leads to

$$R \propto N^{\nu_a} a \qquad (12.51a)$$

$$\nu_a = \frac{5}{2(d + 2)} \qquad (12.51b)$$

Relation (12.51b) may be inverted, giving the fractal dimension of a branched polymer in a good solvent:

$$D_a \equiv \frac{1}{\nu_a} = \tfrac{2}{5} (d + 2) \qquad (12.51c)$$

Equation (12.51c) has been shown (24) to be an exact result for $d = 3$. This relation may be checked by light or neutron scattering. We will come back to these checks later. Let us now turn to the gelation case.

(b) Near the Gelation Threshold: Edwards Screening. What we considered above was a single macromolecule with solvent. The situation is very different near the gelation threshold. Below p_c, which is the case we will consider, there is no solvent, but there is a broad distribution of other polymers with different molecular weights. For such concentrated solutions and for linear polymers Edwards (25, 26) showed that there is a screening of the excluded volume interaction. The effect is basically the same as in the Debye–Hückel screening of the electrostatic interaction in electrolyte solutions (27). DeGennes generalized it to other cases. What we want to calculate is the effective interaction $V_e(r)$ between two monomers of a polymer in the presence of other polymers knowing that the potential between two monomers (alone) is $V(r)$. The effective potential is defined through the pair correlation function $G(r)$:

$$G(r) = e^{-V_e(r)} - 1 \qquad (12.52)$$

where $V_e(r)$ is here a dimensionless potential in units of the most probable thermal energy ($V_e = V/k_B T$). Here $G(r)$ is the probability that, having a monomer at the origin, there is another monomer of the same macromolecule at a distance r. Let us call $G_n^0(r)$ the pair correlation function for a single branched polymer with n monomers in the absence of any other macromolecule. Our level of approximation is that we may expand $G(r)$ to first order in V_e:

$$G(r) \approx -V_e(r) \qquad (12.52a)$$

The random-phase approximation (RPA) then assumes that the correlation between any two monomers, except the two we consider, is $G^0(r)$ and that the polymers are randomly distributed. For a clear and complete discussion of RPA the reader is referred to the book by Brout and Carruthers (28) (see also Ref. 29). Our derivation is shown schematically in Figure 12.15. The probability $G(r)$ is the sum of the "direct" probability when no other polymer is present plus the probability when one other polymer is present, and so on. Introducing the Fourier transforms for $G(r)$ and $V(r)$, we find

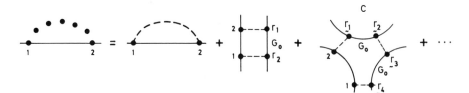

Figure 12.15. Diagrammatic representation of the effective potential $V_e(r)$ between two monomers 1 and 2. This is the sum of the direct interaction $V(r)$ and interactions via other polymers. Dashed lines correspond to $V(r_{ij})$, solid lines to polymers. Intermediate points like $r_1, r_2, \ldots$ may be anywhere on other polymers.

$$-V_e(q) = -V(q) + cV^2(q)G^0(q) - c^2V^3(q)[G^0(q)]^2 + \ldots$$

$$V_e(q) = \frac{V(q)}{k_BT + cV(q)G^0(q)} \tag{12.53}$$

where $G^0(r)$ is the correlation function between monomers on the same polymer

$$G^0(q) = \frac{\displaystyle\sum_{n=1}^{\infty} W_n G_n^0(q)}{\displaystyle\sum_{n=1}^{\infty} W_n} \tag{12.54}$$

where $W_n = nP_n$ is the probability of finding a monomer belonging to a macro-molecule with n elements. Below the gel point we know

$$\sum_{n=1}^{\infty} W_n = 1$$

and $G_n^0(q)$ is the correlation function for a polymer made of n units,

$$G_n^0(q = 0) = n \tag{12.55}$$

For our purpose, as we are interested in large distances, of the order of the radius of a polymer, we are interested only in the $q \to 0$ limit. Then we define an effective, or screened, excluded volume parameter,

$$V_e \equiv V_e(q \to 0) \propto \frac{v}{\sum n^2 P_n} \tag{12.56}$$

We know from relation (12.38) that

$$\sum n^2 P_n \equiv N_w$$

So for the melt considered here the screened interaction parameter is

$$V_e = \frac{v}{N_w} \tag{12.57}$$

where $v \approx V(q \to 0) k_BT$.

Knowing the effective interaction between two monomers, we can now write the free energy for a polymer in the mixture. As above it has two terms:

1. An elastic term that is the same as in the preceding section,

$$F_{el} \propto k_B T \frac{R^2}{R_0^2}$$

2. An interaction term, which is calculated as before. The only difference lies in the excluded volume parameter, now screened:

$$F_{int} \propto k \frac{v}{N_w} \frac{N^2}{R^d}$$

Then the total free energy is

$$\frac{F}{k_B T} = \frac{R^2}{N^{1/2} a^2} + \frac{v}{N_w} \frac{N^2}{R^d} \qquad (12.58)$$

Before minimizing this free energy, we need to know what exactly is N in the above relation. In the gelation problem we are mainly interested in a "typical" polymer. The form of the molecular weight distribution

$$P(n, \epsilon) \propto n^{-2} f(\epsilon n^\sigma)$$

allows us to extract a typical size,

$$N^* \propto \epsilon^{-1/\sigma}$$

In the classical theory, this is

$$N^* \propto \epsilon^{-2}$$

These are the molecules we are considering. Thus, there is an extra relation between N and N_w: From relation (12.29) we get

$$N_w \propto \epsilon^{-\gamma} \propto N^{*\gamma\sigma} \qquad (12.59)$$

which, in classical theory, is

$$N_w \propto N^{*1/2}$$

In the Flory theory everything except the actual radius is supposed to be classical. Then

$$N \propto N^* \propto N_w^2$$

Replacing N_w by $N^{1/2}$ and minimizing the free energy, Eq. (12.58) leads to (22, 23)

$$R \propto N^x \tag{12.60}$$

$$x = \frac{1}{D_f} = \frac{2}{d+2}$$

where D_f is the fractal dimension of the polymers in the reaction bath near the gelation threshold. Note that the fractal dimension of a macromolecule is smaller in a dilute solution than in the reaction bath: This is the screening effect. Both results compare very well with more "correct" calculations of the fractal dimensions, as can be seen in Table 12.1. We can make a final remark about the evaluation of D_f for the gelation case [relation (12.60)] within a Flory–deGennes theory: This approach assumes that the exponents γ and ν are related by

$$\gamma = 2\nu \tag{12.61}$$

We know that this relation is not exact and that there is a slight difference: $\gamma = (2 - \eta)\nu$, where η is an exponent that is usually small and that we do not discuss here. Using relation (12.61) together with the other scaling Eqs. (12.44) and (12.46) leads to the critical exponent relation

$$\beta = \tfrac{1}{2}\,\nu(d-2) \tag{12.62}$$

Finally, inserting (12.62) into the exact relation (12.46) for the fractal dimension of a polymer in the reaction bath and using again (12.44) and (12.46) leads back to relation (12.60) without any use of the free energy. The calculation above is useful, however, because it may be applied very generally to any situation, as will be seen below.

(c) Ginzburg Criterion: Width of the Critical Region. The Flory–deGennes theory shows that the interactions are the reason for the difference between classical and critical behaviors. As a matter of fact, because of these interactions, there are wide fluctuations in the systems that are not taken into account by classical theory. On the other hand, when we look for a solution on a Cayley tree, that is, for infinite space dimension, interactions are irrelevant. Thus, we may expect that if d is high enough, they still are irrelevant, and classical theory becomes valid. How high is high enough? This may be solved in the following way: Let us suppose that classical theory is valid, and let us evaluate the interaction term in the gelation problem. As long as we find that it is small, everything is consistent, and classical theory is adequate (22). If we find it is no longer small, classical theory breaks down, and one has to use critical theory. We evaluate the interaction term,

$$F_{int} = \frac{\upsilon N^2}{N_w R^d}$$

We have seen that in classical theory, $N_w \propto N^{1/2}$ and $R \propto N^{1/4}$. Replacing these in the above relation gives

$$F_{int} \propto \upsilon N^{(6-d)/4} \qquad (12.63)$$

So as long as d is larger than $d_c = 6$, F_{int} is small, and classical theory is valid. Below d_c interactions are important, and classical theory is not valid. It is left as an exercise to show that for a single branched polymer [section (a)] we have $d_c = 8$.

Relation (12.63) may be used further, below d_c. Even then, if N is not high enough, the interaction energy is still small. This is going to provide us a determination of the width of the critical region. We have seen above that N may be expressed in terms of the distance $\epsilon = (p - p_c)/p_c$ to the threshold,

$$N \propto \epsilon^{-2}$$

Replacing N by this expression in (12.63) and looking for the region where F_{int} is smaller than unity gives

$$\epsilon > \epsilon^* \propto \upsilon^{2/3} \qquad (d = 3)$$

Thus, if we are far enough from the gelation threshold, $\epsilon > \epsilon^*$, even below d_c it is possible to observe classical behavior. This criterion is called the Ginzburg criterion (although, usually, one looks at the importance of fluctuations directly; see Ref. 16). In vulcanization this criterion becomes especially important.

(d) Vulcanization. Vulcanization is the cross-linking of linear polymer chains (30, 31). This is a process extensively used in the rubber and tire industries, where cross-linking of polyisoprene chains is the most known example. We start with a melt of linear polymers. Each chain is made of X monomers of length a each. We suppose that any two monomers may cross-link at random with probability p. We look for the gelation properties. Formally, this problem is equivalent to our initial gelation problem: We have f functional units that may react with probability p. Here the functionality is X. Also note that the step length in the vulcanization problem is the radius of a linear chain, which in the bulk is $l = X^{1/2}a$. (We have the same problem as before, but the basic element is now a linear polymer.) From what we have seen, the threshold is

$$p_c = \frac{1}{f-1} = \frac{1}{X-1} \approx X^{-1}$$

We define the same quantities as before. The only differences lie in the prefactors, which state that the basic element for vulcanization is a linear chain:

$$N_w \propto X \epsilon^{-\gamma}$$

$$\xi \propto X^{1/2} \epsilon^{-\nu}$$

$$N^* \propto X \epsilon^{-1/\sigma}$$

We now apply our Ginzburg criterion to this system (32, 33). We evaluate the interaction energy, supposing all the exponents are classical: $\gamma = 1$, $\sigma = \nu = \frac{1}{2}$. For three-dimensional systems we find

$$F_{\text{int}} = \frac{\upsilon N^2}{N_w R^3} \propto \upsilon \epsilon^{-3/2} X^{-1/2} \qquad (12.64)$$

What is interesting here is the presence of the extra factor, depending on the molecular weight X of the initial *linear* polymers, which reduces dramatically the interaction term. Our Ginzburg criterion then gives the width of the critical region:

$$\epsilon < \epsilon^* \propto \upsilon^{2/3} X^{-1/3}$$

where X is usually large, so that ϵ^* is small, and for the vulcanization problem only *classical* exponents should be observable. This effect was first found by deGennes (33) by discussing the fluctuations in connectivity rather than the interactions above.

12.2.6. Some Experiments

We conclude this part by giving some experimental results and comparing them to the theories. The subject has been controversial these last few years because there is a wide scatter in the experimental results. The precise determination of the gelation threshold, in particular, is very important to the determination of the exponents we discussed. As mentioned above, there are two ways for experimentation: the measurements can be performed either on the system in the reaction bath or on the diluted system (20).

(a) Unperturbed System. In experiments on unperturbed systems the chemical reaction is quenched, and the measurements are made on the unperturbed system in the reaction bath. Two quantities have been studied: The viscosity of the sol below the gel point and the elasticity of the gel (in the presence of the sol) above the threshold. However, the results are rather scattered, and no clear conclusion may be drawn.

(b) Diluted Systems. The systems are first quenched, as above, and then heavily diluted. The experiments are performed on the diluted system. Light (or neutron) scattering experiments measure directly M_w, the average radius of the macromolecules of the sol, $\langle R^2 \rangle$, or the average diffusion coefficient. A plot of the latter two quantities as a function of M_w eliminates the difficult task of determining p_c [and $\epsilon = (p - p_c)/p_c$]. The interpretation of these experiments is, however, nontrivial. Figures 12.16 and 12.17 show the results of Candau and collaborators for the diffusion coefficient (34, 35) and those of Leibler and Schosseler for the radius of gyration (36). The exponents are

$$x = 0.57 \pm 0.05 \quad \text{for the diffusion coefficient}$$

$$y = 1.16 \pm 0.06 \quad \text{for the squared radius}$$

The diffusion coefficient was measured by quasi-elastic light-scattering experiments, and the radius by elastic light-scattering experiments.

What is measured by light scattering is the so-called z average:

$$\langle A \rangle_z = \frac{\displaystyle\int A(N)N^2 P(N, \epsilon)dN}{\displaystyle\int N^2 \, P(N, \epsilon)dN} \tag{12.65}$$

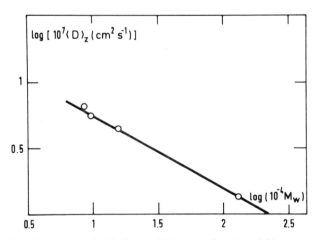

Figure 12.16. Logarithmic plot of diffusion coefficient as a function of M_w, as measured by light scattering on a diluted sol. The slope is -0.57 ± 0.05. The system is made by free-radical copolymerization of styrene and a biunsaturated comonomer in benzene. (Reprinted, by permission, from Ref. 34.)

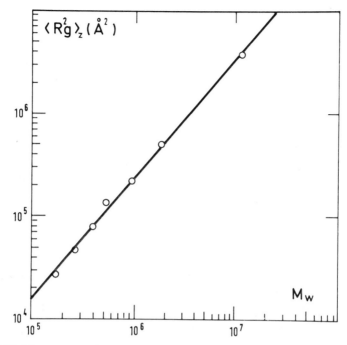

Figure 12.17. Mean square radius of gyration as a function of M_w on a logarithmic plot. Light-scattering experiments are made on a diluted sol. The system is polystyrene (molecular weight 200.000) in cyclopentane cross-linked by radiation. (Reprinted, by permission, from ref. 7.)

To interpret the experiments (37), one assumes that in these diluted solutions, the radius of a polymer is that of a single macromolecule [Eq. (12.51)]. Then one has to average over the distribution near the threshold:

$$P(N, \epsilon) \propto N^{-\tau} f(\epsilon N^{\sigma})$$

Moreover, for the diffusion coefficient one has to make an assumption about the hydrodynamic interaction. Because we have a dilute system, we assume Stokes's law for the diffusion coefficient

$$D(N) \propto \frac{1}{R(N)} \propto N^{-1/2} \quad (d = 3)$$

Averaging is done with Eq. (12.65), where $A(N)$ stands for either $D(N)$ or $R^2(N)$. Knowing the ϵ dependence of N_w, Eq. (12.29) allows for the elimination of ϵ. We find

$$\langle R^2 \rangle_z \propto N_w^{1 + \beta/\gamma} \tag{12.66}$$

$$\langle D \rangle_z \propto N_w^{-(1 + \beta/\gamma)/2} \tag{12.67}$$

with β and γ the exponents defined in Eqs. (12.29) and (12.30).

The experimental results above are in reasonable agreement with the values for β and γ in critical theory. Further experiments are presently under way in Strasbourg for the vulcanization of concentrated polymer solutions.

Other interesting experiments concerning the correlation function between monomers in the same polymer may be done by neutron scattering. The intensity measured in the so-called intermediate range, $a^{-1} \ll q \ll R^{-1}$, where q is the momentum transfer,

$$q = \frac{4\pi}{\lambda} \sin \frac{\theta}{2}$$

λ being the wavelength of the neutrons and θ the scattering angle, leads directly to the fractal behavior of the polymers.

The scattered intensity is directly proportional to the correlation function. In this range of q we have

$$S(q) \propto q^{-D(3 - \tau)}$$

where D is the fractal dimension.

We conclude this part on gelation by recalling that the models developed above are equilibrium models, as opposed to the kinetic models to be discussed below. We determined a set of exponents that are the same for all the systems and thus exhibit the universality of this gelation process. Different chemical systems have the same behavior (divergences or vanishing) near the gelation threshold. An important point is that *two* different exponents are needed to describe the configurational properties near the threshold. Two other exponents that we did not discuss are also needed to describe the dynamics: Viscosity and elasticity exponents. We will see that this number is lower in the kinetic models. Let us finally recall the hypotheses in this equilibrium model (classical and critical theory being just two variations of the same model):

1. All monomers have the same reactivity. We will see that this hypothesis does not hold in the diffusion-limited models (Section 12.3). In the discussion above all the monomers are motionless, and there is an equal probability for every one to be reacted.

Intuitively, we can see that in a dilute solution this does not hold and that polymers are going to grow mainly by their surface; thus, monomers inside do not have the same probability of reacting.

2. There are no interactions between the monomers. The Flory–deGennes theory, based on a variational energy, gives the same fractal dimension as the one estimated by numerical computations. Moreover, we have seen that interactions are crucial if one wishes to understand the properties of a diluted sol.

Finally, in the vicinity of the threshold one gets a very broad molecular weight distribution that may be written in a homogeneous and characteristic way [Eq. (12.34)]. This approach seems to be supported by polycondensation as well as by vulcanization. Let us now turn to other approaches.

12.3. KINETIC MODELS, POLYMERIZATION

Another approach, initiated very early by Flory and Stockmayer (2) has been revived recently by Benedek, Ernst, Leyvraz, Ziff, and their collaborators. This is the kinetic theory, which treats the time dependence of the various quantities we discussed above. Two processes have to be considered to describe an equilibrium situation (38, 39):

1. The combination of two polymers made of i and j monomers, giving a macromolecule made of $i + j$ monomers.
2. The break up of a macromolecule, giving rise to two smaller polymers.

We have to introduce two reaction rates related to these processes. These rates are essential for the description of polymerization and eventually for gelation.

In the following we are concerned only with chemical reactions. Because chemical bonding is strong for usual temperatures, we do not expect the second process to be important, and we neglect it. So, in what follows, we consider only the combination of two smaller polymers giving a larger one.

There are many physical systems where bonding is weak, that is, the bonding energy is of the same order as thermal energy, so that one would have to take molecular dissociation into account. (This would lead to two more terms in our equations; see below.) In *unidirectional* kinetics (38) the reaction proceeds in only one direction, namely molecular buildup. Call $P_n(t)$ the concentration of n-mers at time t. The rate of change of P_n is (40, 41)

$$\dot{P}_n(t) \equiv \frac{dP_n}{dt} = \frac{1}{2} \sum_{1+j=n} K_{ij} P_i P_j - \sum_{k=1}^{\infty} K_{nk} P_n P_k \qquad (12.68)$$

where the first term on the right side corresponds to the formation of n-mers by coalescence of smaller polymers and the second one to the decay of n-mers by formation of still larger macromolecules. The reaction rate, or kernel, K_{ij} is central to our discussion. Depending on its form, gelation may or may not occur. We will also see that the "critical" properties depend very strongly on this form.

Equation (12.68) includes only *finite* polymers. It describes the properties of the *sol* only, below the gelation threshold. When gelation occurs, one has to include the gel fraction in this equation and thus make an extra hypothesis about the possible reaction of the gel with the sol. Equations of the same class as (12.68) were considered by Von Smoluchowski in 1916 in the case of constant rate $K_{ij} = K$ for the coagulation of colloids.

Any hypothesis about the matrix K_{ij} is crucial for the discussion. Physically, K_{ij} is the rate of formation of an $(i + j)$-mer by collision of an i-mer and a j-mer. This may be taken as the number of possible ways an i-mer may join to a j-mer. The first approximation to this number is just to consider that it is proportional to the number of unreacted functionalities on each of the polymers. This corresponds to the case when there is only one extra bond joining both macromolecules. So this approximation neglects the possible formation of 2, 3, . . . , n bonds between any two polymers to form a larger molecule. As a result, one realizes that the structure of the resulting polymer is treelike: No loops are allowed. As we have seen in the preceding section, this is an important approximation. However, different forms for K_{ij} may be assumed, corresponding to different physical situations, so that the problem is still very rich. We discuss briefly some of these possibilities. For a careful and detailed discussion the reader is referred to Benedek (38).

12.3.1. Bifunctional Units

Let us start with a reaction bath containing initially bifunctional units. These may form only linear polymers. What is important here is that at any later time we have larger units that may react only by both ends, so that they are still bifunctional units. We are neglecting diffusion problems here. All the polymers have vanishing diffusion constants, and only their possible reactions are taken into account. In the next section we consider the opposite limit, where diffusion controls the reactions. So in the case of bifunctional units our approximation is to take the rate as a constant independent of the size of the n-mer:

$$K_{ij} = K$$

This is the case first considered by Von Smoluchowski. In the context of *linear* polymerization it was discussed first by Flory (2). Of course, there is no gelation. We can never form any three-dimensional network. We discuss very briefly its results. Call p the fraction of reacted units. Then the concentration of k-mers is

$$P_k = p^{k-1} (1 - p)^2 \tag{12.69}$$

since a k-mer contains $k - 1$ reacted bonds and two unreacted units at the ends. One may check the normalization condition

$$\sum_{n=1}^{\infty} kP_k = 1$$

We may write an equation for the rate of growth. To create a link, two free units must react. Then the rate of decrease of free units is

$$\frac{d}{dt}(1 - p) = -K(1 - p)^2 \tag{12.70}$$

where $1 - p$ is the fraction unreacted or free units. Integration of Eq. (12.70) between the limits $p = 0$ and $p = p(t)$ and $t = 0$ and t and rearranging gives

$$p(t) = Kt/(1 + Kt)$$

If time is measured in dimensionless units of K^{-1}, this reduces to

$$p(t) = \frac{t}{1 + t} \tag{12.71}$$

Combining (12.69) and (12.71) gives the variation with time of the concentration in k-mers. This grows from zero until a time $t_k = \frac{1}{2}(k - 1)$, when it peaks, and then decays to zero. The monomer concentration decreases uniformly to zero. For large times we find, from (12.71), that

$$p(t) \approx 1 - \frac{1}{t} \tag{12.72}$$

Replacing p in (12.69) by this behavior, we find

$$P_k \approx \frac{1}{t^2} \quad (t \to \infty) \tag{12.73}$$

Perelson has recently shown that a possible model for the antigen–antibody response mechanism is analogous to this model (42).

12.3.2. Branching without Gelation

A case that leads to branched polymers without gelation was also considered by Flory.

Start with trifunctional units ($f = 3$) like

where A can only react with B, and vice versa (38). After some time we get branched polymers, but gelation cannot occur. Let us consider an n-mer. It contains one unreacted A group, and $n - 1$ A–B bonds. So it has $(f - 2) n + 1$ unreacted B units. If we now consider the number of ways an i-mer and a j-mer may react, each unreacted A group on each molecule may react with the free B's on the other. Thus, we find

$$K_{ij} = K[(f - 2) (i + j) + 2]$$

where K is a constant. Thus,

$$K_{ij} \propto i + j$$

To solve this system, let us come back to the basic equation (12.68) and sum both sides over n:

$$\frac{d}{dt} \sum_{n=1}^{\infty} P_n(t) \equiv \frac{d}{dt} P(t) = \frac{1}{2} \sum_{n=1}^{\infty} \sum_{i=1}^{n-1} K_{i,n-i} P_i P_{n-i} - \sum_{n=1}^{\infty} \sum_{k=1}^{\infty} K_{nk} P_n P_k \quad (12.74)$$

Consider the first term on the right side:

$$\sum_{n=1}^{\infty} \sum_{i=1}^{n-1} K_{i,n-i} P_i P_{n-i} = \sum_{n=1}^{\infty} \sum_{k=n+1}^{\infty} K_{n-k,n} P_{n-k} P_n \quad (12.75)$$

$$= \sum_{p=1}^{\infty} \sum_{q=1}^{\infty} K_{pq} P_p P_q$$

where the second equality is a simple change in notations, and the first one corresponds to summing the lower half of the elements in a matrix by lines instead of summing them along the second diagonal. Combining (12.74) and (12.75), we find

$$\frac{d}{dt} P(t) = -\frac{1}{2} \sum_{p,q=1}^{\infty} K_{pq} P_q P_p \quad (12.76)$$

In the present case $K_{pq} = w(p + q)$. On the other hand, we have

$$\sum_{p=1}^{\infty} p P_p = 1$$

below the gelation threshold. The breakdown of this equality indicates this threshold. Taking into account this equality and the form of the rate matrix, we find

$$\frac{dP}{dt} = -wP \qquad P(t) = e^{-wt} \quad (12.77)$$

where the proportionality constant w in the rate matrix sets the elementary time unit w^{-1}. The amount of reaction is the fraction of reacted units,

$$b(t) = \frac{\sum (n - 1) P_n(t)}{\sum n P_n(t)} = 1 - P(t)$$

This goes to unity when time goes to infinity. This is the fraction of elements in the finite polymers only, that is, in the sol. Thus, the system does not gel. Having solved b, we could come back to the initial equation and write an equation for dP_n/dt. We are instead going to discuss a third case for the matrix, where gelation may occur. Note, however, that the Benedek experiments on antibody-antigen cross-linking discussed in Section 12.4.3 might be an example of this process.

12.3.3. Branching with Gelation

We start with multifunctional units such as $A—\overset{A}{\underset{A}{<}}$, where the A's may react with each other. Gelation may occur in this system. We have first to determine the rate matrix K_{ij}. The classical hypothesis about this matrix is the one used above, namely, that it is proportional to the number of ways an i-mer and a j-mer can combine. Recall briefly the hypotheses made by Flory (2) and Stockmayer (43) for *below* the gelation threshold: (1) no closed loops and (2) equal reactivity for all the functionalities, independently of their location in the polymer.

Then the matrix K_{ij} is the product of the number of "available" sites on the i-mer by the corresponding quantity on the j-mer. It is easy to calculate the number f_i of unreacted functionalities on a tree made of i f-functional monomers,

$$f_i = (f - 2)i + 2$$

so that

$$K_{ij} = K[(f - 2)i + 2] [(f - 2)j + 2]$$

$$\approx K'ij \tag{12.78}$$

The generalized Smoluchowski equation (12.68) is then

$$\dot{P}_n = \frac{1}{2} \sum_{i+j=n} ijP_iP_j - \sum_{k=1}^{\infty} nkP_nP_k \tag{12.79}$$

where we have included the constant K' in the definition of time. We have a dimensionless time measured in units of K'^{-1}. Equation (12.79) does not allow for reaction between sol and gel above the gelation threshold. Thus we use it below the gel point. To use it beyond the threshold, we have to add an extra hypothesis concerning the sol–gel reactions. First consider the sol below the threshold. To solve Eq. (12.79), we introduce the molecular weight distribution

$$W_n = nP_n \tag{12.80}$$

and its Laplace transform

$$F(z, t) = \sum_{n=1}^{\infty} W_n e^{-nz} \tag{12.81}$$

Then, multiplying both sides of (12.79) by ne^{-nz} and summing over n, we get

$$\dot{F}(z, t) = \frac{1}{2} \sum_{n=1}^{\infty} \sum_{i=1}^{n-1} nW_i W_{n-i} e^{-nz} - \sum_{n=1}^{\infty} nW_n e^{-nz} \sum_{k=1}^{\infty} W_k$$

$$= \frac{\partial F}{\partial z} (z, t) [F(0, t) - F(z, t)] \tag{12.82}$$

where the same trick as above has been used to calculate the first double sum. Note that

$$F(0, t) = \sum nP_n(t) = 1 \tag{12.83}$$

is our usual normalization condition below the threshold. Introducing

$$\Gamma(z, t) = 1 - F(z, t) \tag{12.84}$$

equation (12.82) becomes

$$\dot{\Gamma}(z, t) = \Gamma(z, t) \frac{\partial \Gamma(z, t)}{\partial z} \tag{12.85}$$

We could have used the Laplace transform of P_n directly,

$$g(z, t) = \sum_{n=1}^{\infty} P_n e^{-nz}$$

Here $g(z, t)$ and $F(z, t)$ are related,

$$F(z, t) = -\frac{\partial}{\partial z} g(z, t)$$

The resulting equation for $F(z, t)$ is

$$\dot{g} = \frac{1}{2}\left(\frac{\partial g}{\partial z}\right)^2 + \frac{\partial g}{\partial z} \tag{12.85a}$$

As noted first by Ziff and Stell, Eq. (12.85) is formally similar to Euler's equation in fluid mechanics for the description of the formation of a shock in a one-dimensional flow. Then our variable z would be the spatial coordinate, and Γ would be the velocity field. Thus, there is a critical time t_c for gelation. Then we call $\epsilon = (t - t_c)/t_c$. Equation (12.85) remains basically unchanged. Physically, we are back to the gelation problem that we studied in the classical case, that is, below the gelation threshold. We can solve Eq. (12.85) either by remembering the form of the solution or by looking for a solution of the type

$$\Gamma(z, t) = z^{1/2}f\left(\frac{\epsilon}{z^{1/2}}\right) \tag{12.86}$$

which we guess from the form of the equation. Inserting this into Eq. (12.85), we find the differential equation for $f(x)$, where $x = \epsilon/z^{1/2}$,

$$2f' = f(f - xf') \tag{12.87}$$

where the prime denotes differentiation with respect to x. The solution was worked out by Stockmayer some time ago. Here we will be interested only in the form (12.86), which contains all the information we need. Taking the inverse Laplace transform of $\Gamma(z, t)$ gives us back $W(n, t)$ and $P(n, t)$, which has the classical homogeneous form

$$P(n, t) \propto n^{-5/2}g(\epsilon n^{1/2}) \tag{12.88}$$

From this distribution we get the different characteristic behaviors found previously for the different moments. Thus, this approach is equivalent to the classical theory. We have just replaced a parameter, the probability that a unit is reacted, by time. There is, however, one more point to be discussed, namely, the generalization of Eq. (12.79) above the gelation threshold. Then one has to take into account the possibility for any macromolecule to become part of the gel. This was done first by Flory (2). Then Eq. (12.79) becomes

$$\dot{P}_n = \frac{1}{2}\sum_{i+j=n}ijP_iP_j - nP_n\left[\sum_k kP_k + G\right] \tag{12.89}$$

where the last term takes into account the sol–gel interaction. We also have to write the corresponding equation for the rate of formation of the gel:

$$\frac{dG}{dt} = G \sum_n nP_n \qquad (12.90)$$

Note that the bracket on the right side of Eq. (12.89) is still equal to unity, so that our equation for Γ, relation (12.84), remains unchanged. Defining $\Gamma(z, t)$ as above, we end up with

$$\dot{\Gamma} = \Gamma \frac{\partial \Gamma}{\partial z}$$

the gel fraction being given by

$$G(t) = 1 - F(0, t) = \Gamma(0, t) \qquad (12.91)$$

Coming back to the description in terms of shocks, we see graphically, in Figure 12.18, the situation for $\Gamma(z, t)$. Initially, we have

$$\Gamma(z, 0) = 1 - e^{-z}$$

increasing monotonously from zero to unity. When time goes on the curve deforms until it starts vertically for $t = t_c$. This corresponds to the gelation threshold. For larger times $\Gamma(z)$ first goes in the unphysical $z < 0$ region before coming back to the positive z region. The gel fraction corresponds to this crossing with the $z = 0$ axis. The actual curve starts from point G and increases continuously.

This approach has all the weaknesses that we discussed when we criticized the classical theory of gelation. We are considering only treelike gelation: No loops are allowed. We did not take into account any interaction between the molecules, and finally we supposed that two polymers may interact by any of the unreacted functionalities. These hypotheses are restrictive and have to be somewhat relaxed.

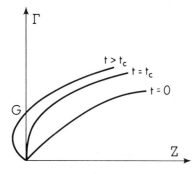

Figure 12.18. Schematic representation of $\Gamma(z, t)$ as a function of z for different times. For $t > t_c$, only the $z > 0$ part of the curve has a physical meaning. The ordinate for $z = 0$ is the gel fraction G.

12.3.4. Generalization

The assumption that was recently criticized is the equal reactivity of all unreacted functionalities. Those "deep inside" the polymer were assumed to have the same probability of reacting as the ones that are superficial. In the case of diffusion-limited aggregation, to be discussed in the next section, computer simulations show that only monomers in a superficial shell may react, the other ones being screened. In the present case there is no diffusion, and one does not specify how two polymers are brought so close to each other that they may react. In other words, the concentration in any species is supposed to be a constant through the sample. Nevertheless, one might question the equal reactivity of all units. What is assumed then is that only a fraction of all the available units may actually react; only those on the surface may do so. The question is to define this surface. This is easy to do for a compact, homogeneous object. If the number N of elements in a sphere of radius R varies as

$$N \propto R^d$$

it has a surface

$$S \propto R^{d-1} \propto N^{(d-1)/d}$$

What about our nonhomogeneous polymers that have a fractal behavior? Strictly speaking, the surface is of the same order as the bulk:

$$S_1 \propto N$$

Using the preceding relation, one could also define a surface term by

$$S \propto R^{d-1} \propto N^{(d-1)/D}$$

where D is the fractal dimension of the polymer.

With this difficulty in mind, it was supposed that the matrix K_{ij}, which describes the way two polymers react, varies as (44, 45)

$$K_{ij} \propto (ij)^w \qquad (12.92)$$

with the exponent w describing the fraction of "useful" units in a j-mer. Then the kinetic equation for the aggregation process is

$$\dot{P}_n = \frac{1}{2} \sum_{i=1}^{n-1} i^w (n-i)^w P_i P_{n-i} - \sum_{k=1}^{\infty} (nk)^w P_n P_k \qquad (12.93)$$

We may apply the same transformation as above. Let us multiply both sides by ne^{-nz} and sum over n,

$$\sum_n n\dot{P}_n e^{-nz} = \frac{1}{2} \sum_n^\infty \sum_{i=1}^{n-1} ni^w (n-i)^w P_i P_{n-i} - \sum_n^\infty n^{w+1} P_n e^{-nz} \sum_{k=1}^\infty k^w P_k$$

We define

$$F(z, t) = \sum_{n=1}^\infty nP_n(t) e^{-nz} \quad \text{and} \quad F_w(z, t) = \sum_n n^w P_n(t) e^{-nz}$$

The previous equation may then be written as

$$\dot{F}(z, t) = -\frac{1}{2} \frac{\partial}{\partial z} F_w^2(z, t) + \frac{\partial}{\partial z} F_w(z, t) F_w(0, t)$$

$$= [F_w(0, t) - F_w(z, t)] \frac{\partial}{\partial z} F_w(z, t) \qquad (12.94)$$

When $w = 1$, (12.94) is identical to Eq. (12.82).

Note also that mass conservation of the sol is automatically ensured by the equation, which is valid only below the gelation threshold. This corresponds to $z = 0$ in the equation. The normalization condition is

$$\sum_{n=1}^\infty nP_n \equiv F(0, t) = 1$$

As in Section 12.3.3, it is possible to write an equation for P_n itself. Coming back to Eq. (12.93),

$$\dot{P}_n = \frac{1}{2} \sum_{i=1}^{n-1} i^w (n-i)^w P_i P_{n-i} - \sum_{k=1}^\infty n^w k^w P_n P_k$$

We introduce the Laplace transform of P_n,

$$g(z, t) = \sum_{n=1}^\infty P_n(t) e^{-nz}$$

Then, multiplying both sides of our equation by e^{-nz} and summing over n, we find

$$\dot{g}(z, t) = \frac{1}{2} F_w^2(z, t) - F_w(z, t) F_w(0, t) \qquad (12.95)$$

which is the generalization of Eq. (12.85a).

As previously, one may try to solve either of these by looking for a solution of the type

$$P(n, t) = n^{-\tau} f(tn^{\sigma}) \tag{12.95a}$$

That is,

$$g(z, t) = z^{\tau-1} f\left(\frac{t}{z^{\sigma}}\right) \tag{12.96}$$

$$F_w(z, t) = z^{\tau-w-1} f\left(\frac{t}{z^{\sigma}}\right) \tag{12.97}$$

$$F(z, t) = z^{\tau-2} f\left(\frac{t}{z^{\sigma}}\right) \tag{12.98}$$

Plugging (12.96) and (12.97) into (12.95), one finds, after some manipulation,

$$z^{\tau-1-\sigma} X(x) = z^{2(\tau-w-1)} Y(x) \tag{12.99}$$

where $x \equiv tz^{-\sigma}$. Identifying the exponents for z on both sides of (12.99) leads to

$$\sigma + \tau = 2w + 1 \tag{12.100}$$

Leyvraz and Tschudi made one more step using Eq. (12.93) above the gelation threshold without incorporating any interaction between the sol and the gel. We reproduce here their result. Consider Eq. (12.93). Multiply both sides by n and sum until N:

$$\sum_{n=1}^{N} n\dot{P}_n = \frac{1}{2} \sum_{n=1}^{N} \sum_{i=1}^{n-1} i^w (n-i)^w nP_i P_{n-i} - \sum_{n=1}^{N} \sum_{k=1}^{\infty} n^{w+1} k^w P_n P_k$$

$$= -\sum_{n=1}^{N} n^{w+1} P_n \sum_{k=N-n+1}^{\infty} k^w P_k \tag{12.101}$$

and suppose that P_k has a power law behavior. We know this is true right at the gelation threshold,

$$P_k \propto k^{-\tau}$$

Then we find, from (12.101),

$$\left| \frac{d}{dt} \sum_{n=1}^{N} nP_n \right| = \sum_{n=1}^{N} n^{w+1-\tau} \sum_{k=N+1-n}^{\infty} k^{w-\tau}$$

$$\propto N^{2w-2\tau+3}$$

Above the gelation threshold the limit when N goes to infinity on the left side is just the rate of formation of the gel. Let us consider the exponent on the right side:

1. If it is positive, the rate is infinite, which is unacceptable
2. If it is negative, this rate is vanishing: The gel does not form.
3. So the exponent has to be zero. This implies another relation between σ and τ and the value of the exponent β. We find that the gel grows linearly with time above the threshold. This implies (44, 46)

$$\beta = 1 \tag{12.102}$$

and

$$\tau = w + \tfrac{3}{2} \tag{12.103}$$

Taking relation (12.100) into account, we find

$$\sigma = w - \tfrac{1}{2} \tag{12.104}$$

This is a very important result because it implies that all the exponents depend only on w. This is in clear contrast with the statistical theory where τ and σ are independent exponents. One reason for this is that here β has always its classical value [relation (12.102)]. It is important also because the details of the chemistry contained in the kernel K_{ij} are very important. The exponents depend continuously on the exponent w. If w could vary continuously, so would the other exponents. This is in contradistinction with the idea of universality, where the general belief is that the exponents may have only a limited number of values related to universality classes. Of course, chemistry might impose that only some values of w are realistic. This part of the work has still to be done at the present time. There is only one restriction so far concerning the value of w: Because of the form of the distribution function [Eq. (12.95a)], the exponent σ has to be positive if gelation is to occur. This implies

$$w \geq \tfrac{1}{2} \tag{12.105}$$

When w is less than this critical value $\tfrac{1}{2}$, we get polymerization without any gelation. Above this value one may calculate the exponents for the other quantities. We find

$$\gamma = -\alpha = \frac{3 - 2w}{2w - 1} \tag{12.106}$$

and supposing hyperscaling, $\nu d = 2 - \alpha$, is valid,

$$\nu = \frac{1}{d} \frac{2w + 1}{2w - 1} \tag{12.107}$$

The fractal dimension is

$$D = d - \frac{\beta}{\nu} = \frac{2d}{2w + 1} \tag{12.108}$$

Note that the condition $D < d$ leads back to Eq. (12.105). However, Eq. (12.104) may be criticized in the following ways:

1. The relation $P_n \propto n^{-\tau}$ is valid only right at the gelation threshold. It is, however, very difficult to use the derivative dG/dt at the threshold. As seen above, the gel fraction G is zero below the threshold and starts growing with an exponent β at the threshold. Its derivative is zero below the threshold and infinite right at the threshold. To say that it is finite is almost equivalent to supposing that $\beta = 1$, thus implicitly to impose $\tau = w + \frac{3}{2}$.

2. Away from the threshold one has to include the time dependence of P_n:

$$P_n \propto n^{-\tau} f(tn^\sigma)$$

This is what we did when we found relation (12.100) above. Moreover, above the threshold we have to take into account the amount of sol being incorporated into the gel. Then supposing that the finite molecules interact with the gel in the same way as with the sol leads to

$$\dot{P}_n = \frac{1}{2} \sum_{i=1}^{n-1} i^w (n - i)^w P_i P_{n-i} - \sum_{k=1}^{\infty} n^w k^w P_n P_k - n^w P_n G \tag{12.93a}$$

instead of (12.93), where G is the gel fraction. Taking the Laplace transform as above gives

$$\dot{g}(z, t) = \frac{1}{2} F_w^2(z, t) - F_w(z, t) [F_w(0, t) + G] \tag{12.95b}$$

where, from total mass conservation, G is related to $g(z, t)$ by

$$G \equiv 1 - \sum_{1}^{\infty} nP_n = 1 + \frac{\partial}{\partial z} g(z, t) \Big|_{z=0} \tag{12.95c}$$

rather than Eq. (12.95). Using (12.95a) together with (12.95b) and (12.95c) does not bring any extra relation like (12.103) or (12.104).

Relations (12.103)–(12.108) are, however, largely accepted.

We end this section by recalling briefly the hypotheses and results and by looking for some possible extensions. The basic kinetics equation was generalized by supposing that the rate K_{ij} should not treat all unreacted functionalities in a polymer equally. Only a fraction i^w may react. For these we suppose equal reactivity.

The rate of reaction K_{ij} between two polymers with i and j monomers is assumed to be proportional to the number of unreacted pairs, that is, to $(ij)^w$. Then gelation occurs in a finite time if $w > \frac{1}{2}$. In the vicinity of the gelation threshold, there is a wide distribution of molecular weights. The concentration of n-mers is

$$P_n \propto n^{-\tau} f(\epsilon n^\sigma)$$

where $\epsilon \equiv (t - t_c)/t_c$, t_c is the gelation time (threshold). Two relations were found between τ, σ, and w:

$$\tau + \sigma = 2w + 1 \qquad \tau = w + \tfrac{3}{2}$$

where the second relation might be subject to caution. This implies that all the exponents depend only on w. A characteristic result is the exponent β for the gel fraction: $\beta = 1$. Therefore, we have

$$G \propto \frac{t - t_c}{t_c}$$

As we will see in the next section, this dependence of the exponents on only one of them also holds in cluster–cluster aggregation when no gelation occurs. $\beta = 1$ is the mean-field value for this exponent. This may be related to the fact that only treelike configurations are allowed even if the distribution is different from the mean-field expression because only a fraction of the monomers may react. So one possible generalization of the model is to include loop formation.

A second weakness in the theory concerns the fluctuations. We may identify two of these: fluctuations in concentration and in reactivity. It was supposed that the concentration of n-mers is uniformly P_n without allowing any fluctuations. Because the diffusivity of the molecules is not infinite and because of the interactions, there are fluctuations. They might become important for the determination of the exponents.

It was also supposed that the rate K_{ij} has a definite form for any size of the reacting polymers and that it is a constant when the sizes are given. However, one could argue that a small polymer can probably penetrate deeper inside a large one, thus activating a larger possible number of monomers in the latter. This would lead to a larger value of w and possibly to a mass dependence of w. Thus it would also be interesting to explore the possibility of fluctuations in K_{ij}.

So far, the results of this kinetic theory seem to be at variance with the gelation approach except for the classical and the $K_{ij} \propto ij$ cases, when they coincide. As we see below, when diffusion becomes the dominant limitation, we have different results. At the present time different systems have been identified that follow either the gelation process, as in the previous section, or diffusion-limited aggregation, described below. The kinetic approach should be able to describe both by taking correctly into account fluctuations and the finite diffusivity of the polymers. This still has to be done.

12.4. DIFFUSION-LIMITED AGGREGATION

So far, we have considered systems where polymerization and gelation are controlled by chemistry. In the gelation model and in the kinetic theories what is important is the extent of reaction and the form of the coagulation kernel K_{ij}, respectively. There are some systems, however, where diffusion is the limiting process for the reactions. In order for two molecules to react, they have first to be in the vicinity of each other. If we now start with a dilute solution of initially multifunctional monomers and let them react, diffusion takes place first, allowing the monomers to come eventually close to each other and to react. This effect is studied in this section. Two basic models have been developed. In the original model Witten and his collaborators studied the growth of an aggregate (or cluster) by successive addition of a monomer. This is called diffusion-limited aggregation (DLA). It was generalized by Meakin and by Jullien and his collaborators by looking at a situation where one starts with a dilute solution of monomers, lets the reaction proceed, and looks at the aggregation of clusters. This latter model is called cluster–cluster aggregation (CCA). Both models have been studied mainly by computer simulations, and only very recently have the first analytical calculations been attempted. Before discussing their results, we describe three experiments to illustrate this discussion of *colloidal growth*.

12.4.1. Smoke Particle Aggregate

In their pioneering experiment Forrest and Witten (47) formed aggregates by quench condensing a metal vapor in a cool dense gas (helium). The vapor was produced by a hot electric filament with the metal plated on its surface. The condensation occurs in two steps:

1. Within a few micrometers of the filament solid metal particles of about 4 nm radius are formed. These are the particles that form the aggregate.

2. The particles are cooled quickly and streamed outward. They are observed to accumulate in a thin shell suspended in the gas, of roughly 1 cm radius. Aggregation probably occurs in this region. Then the heat source is turned off, and the aggregates are drifted down to an electron microscope plate for study. They

are found to be a low-density arrangement of particles. A typical number of particles in an aggregate is 10^5. It is thought that once the particles are formed, they undergo thermal motion. Whenever this brings them into contact, they stick together. Because the clusters maintain their filamentous shape during several centimeters of drifting in the gas, it is believed that the sticking is irreversible. Several "metals" were used: Zinc, iron, and silicon dioxide. All of them led to the same pattern. What is remarkable in these experiments is that although the resulting aggregates look very nonhomogeneous, they all behave in the same way. They belong to the same universality class. If we plot the radius as a function of the number of particles, we find

$$R(N) \propto N^x$$

$$D = x^{-1} = 1.7\text{--}1.9 \tag{12.109}$$

where D is the fractal dimension of the aggregate. Moreover, it was found that the behavior of the aggregates is self-similar. Counting the number $N(r)$ of particles around a central particle in a circle with radius r leads to the same evaluation for D. Another measurement consisted in studying the density correlation around a central particle:

$$g(r) = \langle n(o)\, n(r) \rangle \propto \frac{N(r)}{r^2} \propto r^{D-2} \tag{12.110}$$

a logarithmic plot of $g(r)$ versus r was made by Forrest and Witten, leading to a value of D consistent with the preceding estimate. Thus, these aggregates are found to have a fractal behavior with a dimension D clearly different from the one for the gelation model.

12.4.2. Wax Ball Aggregate

Very recently Allain and Jouhier (48) made a macroscopic two-dimensional simulation of aggregation. They started with wax balls floating on the surface of a water container. Each ball was about 2 mm diameter. At the beginning of the experiment the balls were randomly distributed on the surface. This was achieved by generating small bubbles inside the water. On touching the surface, the latter destroyed existing clusters of balls. After this stirring had been stopped, they observed the formation of clusters for several hours. The experiment was recorded by taking photographs of the surface after time intervals and analyzing them. The results follow:

1. Above a critical superficial density, an infinite cluster appeared after a finite time.

2. Below this critical density only finite aggregates were observed.

The analysis was made for a density close to this critical value. The same kind of analysis was done as for the smoke particles. Here it was done by a direct counting of the particles on a photograph. The same remarkable fractal behavior was observed in spite of (or because of) the very nonhomogeneous aspect of the clusters (Figure 12.19). The fractal dimension is similar to that in the preceding case:

$$D = 1.6 \pm 0.1$$

12.4.3. Antigen–Antibody Cross-Linking

Biological systems studied by Benedek and his co-workers (49) could possibly be related to the clustering of clusters. Their initial system is made of two components plus a solvent: (1) Polystyrene latex spheres with radius of about 0.1 μm coated with antigen human serum albumin (hSA), giving a molecule with functionally $f \approx 1000$, and (2) goat anti-hSA, which is a bivalent antibody molecule able to cross-link antigen molecules.

They measured the distribution of cluster sizes *in situ* without disturbing the reaction. They found a wide distribution function that could be described in a homogeneous form:

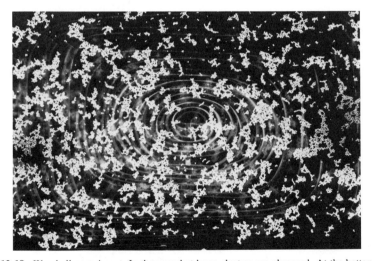

Figure 12.19. Wax ball experiment. In the snapshot large clusters are observed. At the bottom of the vessel is a pipe with a large number of thin holes. When air is injected in the pipe, bubbles are formed and break the preexisting clusters on reaching the surface. The beginning of the experiment corresponds to the end of the injection of air. Here $t = 1700$ s. (Reprinted, by permission, from Ref. 48.)

$$X_n = n^{-\tau} f(\epsilon \, n^\sigma) \tag{12.111}$$

where X_n is the number of clusters made of n elements, and ϵ the distance to the gelation point. Because of the nature of the reaction (each species can react with the other one only), the gel appears only for an extent of reaction equal to unity. This is to be compared with what was discussed in Section 12.3.2. Their main results are

$$\tau = 1.4 \pm 0.15 \qquad \sigma = 0.5$$

The main point here is a value for τ smaller than 2, whereas the percolation models discussed in Section 12.2 imply a value larger than 2. However, because the solutions are rather concentrated, this experiment could possibly be described by the kinetic model described above with a kernel $K_{ij} \propto i + j$. Compare also with the Vicseck–Family results for CCA [Eqs. (12.127), (12.129), and (12.131)].

12.4.4. Diffusion-Limited Aggregation

Witten and Sander (50, 51) proposed the diffusion-limited aggregation (DLA) model. It considers the growth of a single cluster in the following way. Start with a d-dimensional lattice with a central seed, which plays the role of the "nucleation center." Then a random walk is generated for a particle starting from a distant point. This simulates the diffusion of the particle. Then two possible cases may occur:

1. The particle undergoing the random walk goes through a site adjacent to the central seed. Then it sticks irreversibly to it. A new particle is then released. It undergoes a random walk. If it goes through a site adjacent to any of the already occupied sites, it sticks to the small cluster and thus contributes to the growth of the aggregate. The process is continued a large number of times, always with the same rule. When the diffusing particle goes through a site adjacent to the cluster, it sticks to it and becomes part of the cluster. A typical number of sticking particles in these computer generated-aggregates is 10^3–10^4.

2. The diffusing particle does not visit a site adjacent to the central seed (or, at a later time, to the cluster). After some time it goes through a sphere with radius R_0, where R_0 is large compared to the radius of the aggregate. Then the walk is stopped, and a new particle is released.

These simulations are time consuming on a computer because many particles are lost in the second way. This is more and more true when the dimension of space is increased. Finally, one has to average the properties over a number of aggregates.

A typical aggregate generated by DLA is shown in Figure 12.20. As usual, it looks nonhomogeneous. Interestingly, it does not exhibit large loops but has rather

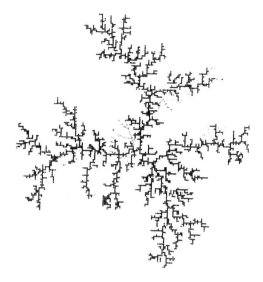

Figure 12.20. Typical aggregate with 3000 particles on a square lattice generated in diffusion-limited aggregation. (Reprinted, with permission, from Ref. 51.)

a treelike structure. Let us emphasize the irreversibility of the process. Once the particles stick to each other, they do so irreversibly. No relaxation may occur later, and the structure is rigid. What is remarkable with these aggregates is that they exhibit self-similarity properties. It is possible to make the same analysis as in the Forrest–Witten experiments. Start with a site belonging to the aggregate. Draw the hypersphere with radius r. Count the number of particles $N(r)$ in this sphere. Plot these results on a log-log scale. The result is a straight line:

$$N(r) \propto r^D \qquad (12.112)$$

In the same way the density correlation function is

$$\langle n(o)n(r) \rangle \propto \frac{N(r)}{rd} \propto r^{D-d} \qquad (12.113)$$

where the angle brackets denote an ensemble average, which is taken by choosing different central sites and different clusters. A typical result for the two-dimensional case is shown in Figure 12.21. The correlation function has a power law behavior except in the outer region of the aggregate, which will be discussed below. Meakin (52) generalized the model to higher space dimensions, and the results for the fractal dimension D are summarized in Table 12.2.

An important concept in this type of aggregation, as in the models discussed above, is screening. The even shape of the aggregates is due to this effect. Suppose

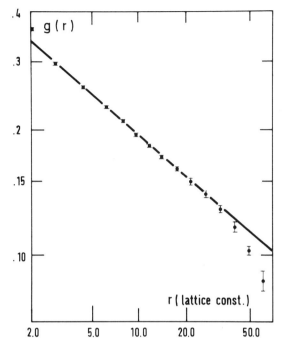

Figure 12.21. Correlation function on a logarithmic scale for a DLA cluster. This is an average over six aggregates such as in Figure 12.20. The slope is −0.34 and corresponds to a fractal dimension D = 1.66. (Reprinted, with permission, from Ref. 51.)

for a moment that the incoming diffusing particle is able to completely penetrate the aggregate. As a result, every site on the perimeter of the cluster would be accessible for growth. This in turn would lead to a compact structure, with a fractal dimension equal to the dimension d of space. This is not so because there is screening. The incoming particle is caught by a cluster on its periphery. This means that there is a screening length, or penetration depth, for the incoming particles that is much less than the radius of the aggregate. The effect is shown in Figure 12.22, where the first particles are represented by heavy dots and the last ones by light dots. Particles of the second type do not penetrate the cluster but stick to the

TABLE 12.2. Fractal Dimension D for DLA Aggregates as Function of Space Dimension d^a

d	2	3	4	5	6
D	1.70 ±0.06	2.53 ±0.06	3.30 ±0.1	4.2 ±0.1	5.3

[a]Note the difference with last line of Table 12.1. Data from Ref. 52.

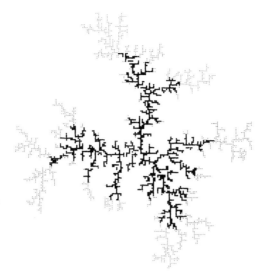

Figure 12.22. Screening effect in DLA. The last 1500 particles (light dots) do not enter the cluster made by the first 1500 ones (heavy dots). (Reprinted, with permission, from Ref. 51.)

preexisting tips. This suggests an electrostatic analogy, which was used by Witten and Sander. The screening length was calculated by Muthukumar (53) and by Ball, Nauenberg, and Witten (54) within mean-field-type theories. They found that it should vary respectively as $n(R)^{-1}$ and $n(R)^{-1/2}$, where $n(R)$ is the density and R the radius of the aggregate.

Because of this screening effect, it was interesting to look at the growth of the interface itself (55). We first define the number N_i of particles at the interface. To do so, we first grow an aggregate with N elements; and we then add k particles to it. Let $P_N(k)$ be the probability for the last k particles to touch one of the first N ones. This probability decreases when k increases because of the screening effect. We define the average mass of the interface as

$$\langle N_i \rangle = \sum_{k=1}^{\infty} p_n(k) \tag{12.114}$$

Different cases may occur depending on the value of the fractal dimension:

1. In the limit of a dense object, $N \propto R^d$, we expect the screening length to be a constant. Then

$$\frac{N_i}{N} \propto \frac{\lambda}{R} \tag{12.115}$$

implying

$$N_i \propto R^{d-1} \tag{12.115a}$$

2. In the opposite limit of a very loose structure, where $d - D$ is larger than 2 (larger than the fractal dimension of a random walk), we expect the screening effect to be small. The screening length is large, of the order of the radius R; then, from (12.115), we get

$$N_i \propto N \propto R^D \tag{12.115b}$$

(The object is then said to be transparent to the random walk.)

3. For intermediate cases Meakin and Witten (55) defined an exponent δ for the variation of N_i:

$$N_i(N) \propto N^\delta \tag{12.116}$$

In two and three dimensions they found

$$\delta = 0.60 \pm 0.02, \ (d = 2)$$
$$\delta = 0.74 \pm 0.02, \ (d = 3)$$

Note that both results are consistent with $\delta D = d - 1$, that is, with relation (12.115a). This, however, is inconsistent with the very loose fractal structure of the aggregate. A mean field estimation of δ may be given. In this approximation it is supposed that the mass N_i of the interface is proportional to the fraction of the number of particles within the penetration length λ:

$$N_i \propto \frac{\lambda}{R} N \tag{12.117}$$

We may use the evaluations of λ given above:

4. For Muthukumar (53), $\lambda \propto n(R)^{-1}$ with $n(r)$ the density,

$$n(r) \propto \frac{N(r)}{r^d} \tag{12.118}$$

Using (12.117) and (12.118), we find

$$N_i \propto \frac{R^d}{N} \frac{N}{R} \propto R^{d-1} \propto N^{(d-1)/D} \tag{12.119}$$

which is consistent with the experimental results.

5. For Ball, Nauenberg, and Witten (54) $\lambda \propto n(R)^{-1/2} \propto (R^d N)^{1/2}$.
Using (12.117), we find

$$N_i \propto N^{1/2 + [(d/2) - 1]/D} \qquad (12.120)$$

However, this last mean field method under estimates δ because it treats the clusters as made of independent particles. In fact, we know that there is screening, so that the actual growth rate is smaller. This implies a larger penetration depth and thus a larger value for δ. Thus the conclusions of Meakin and Witten (55) are that δ could be an exponent independent of and unrelated to D with a lower bound

$$\delta \geqslant \frac{1}{2} + \frac{d - 2}{2D} \qquad (12.121)$$

Finally, the fractal dimension D itself was calculated within mean field theories by the same authors. The first striking result is that there is no critical dimension of space above which D becomes a constant. This is to be contrasted with the equilibrium theories where the latter is the common situation. Within their approximations it was found by Muthukumar (53) and by Ball, Nauenberg, and Witten (54) that

$$D = \frac{d^2 + 1}{d + 1} \qquad (12.122)$$

$$D = d - 1 \qquad (12.123)$$

respectively. DLA is still an open problem in constant and rapid progress.

12.4.5. Cluster–Cluster Aggregation

DLA dealt with the growth of one cluster. It was generalized simultaneously by Meakin (56, 57) and by Jullien and co-workers (58), who introduced the cluster–cluster aggregation (CCA) model. We start now with N particles randomly distributed on the sites of a d-dimensional lattice. The particles are allowed to diffuse on the lattice with the same constraint as in DLA: Whenever two particles are on two nearest-neighbor sites, they stick irreversibly to each other. The resulting aggregate is still allowed to diffuse, and whenever it "meets" another cluster, they coalesce into a larger aggregate. Thus, instead of growing a single cluster, a lot of them grow simultaneously. After some time one has a distribution of sizes. As time evolves, the total number of aggregates decreases and goes to unity for very large times. We discuss the results obtained on two-dimensional lattices, but they are quite general for any space dimensions. We discuss first the correlation function $g(r)$. The computer simulations were made with different initial concentrations

$\langle n \rangle$ of particles. The correlation function was computed using the ultimate aggregate. A typical result for two concentrations is shown in Figure 12.23 on a log-log scale. Recall that $g(r)$ is the probability of finding a particle at a distance r given a particle at the origin. Figure 12.23 exhibits two regimes:

1. For large distances $g(r)$ is equal to the average concentration $\langle n \rangle$: Fluctuations are not important.
2. For smaller distances $g(r)$ has a power law behavior,

$$g(r) \propto r^{-\alpha} \tag{12.124}$$

with $\alpha = 0.52 \pm 0.03$.

The smaller distance scale corresponds to a fractal behavior. Moreover, the relation between α and the fractal dimension is [see, for example, relation (12.113)]

$$\alpha = d - D$$

Here $d = 2$, so

$$D = 1.48 \pm 0.03$$

As long as the density is finite, we have a fractal behavior for small distances only. We may calculate the distances ξ, where the crossover occurs between a

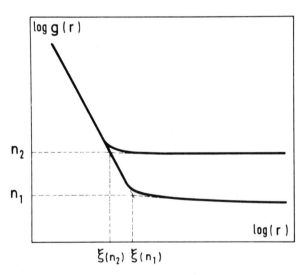

Figure 12.23. Sketch of typical variation of correlation function on a logarithmic scale for cluster–cluster aggregation. Two initial concentrations have been used. The slope at small distances leads to a fractal dimension D around 1.5. The crossover distance $\xi(n)$ is obtained by extrapolation of the asymptotic behaviors.

fractal behavior and a regular behavior, where $g(r)$ is just the average density. Extrapolating the constant behavior and relation (12.124) to this distance, we find

$$\xi^{-\alpha} \propto \langle n \rangle$$

and thus

$$\xi \propto \langle n \rangle^{-1/(d-D)} \tag{12.125}$$

The crossover point and the extrapolation procedure are also shown in Figure 12.23. When the average concentration decreases, ξ increases. In the limit when $\langle n \rangle$ goes to zero, ξ diverges, and the cluster is a fractal.

In these computer simulations only translational diffusion was allowed. No rotation was permitted. The mobility of all clusters was supposed to be the same, irrespective of their size. Later we will discuss different hypotheses about the mobility dependence on molecular weight.

Similar computer calculations were made with very small values of the average concentration, and aggregates were analyzed at different stages during the growth process to determine directly their fractal dimension. The radius R of a cluster made of N particles varies as

$$N \propto R^D$$

leading to a value for D in agreement with what was found above.

Jullien and his co-workers (59) developed a very attractive model, the hierarchical model, to compute the fractal dimension much faster. Here one starts with $N_0 = 2^{k_o}$ particles. In the first step particles are allowed to diffuse until they form doublets. Thus, at the end of this step there are $N_1 = 2^{k_o-1}$ doublets. Then these doublets undergo a random walk until they make quadruplets. At the end of step 2, we have $N_2 = 2^{k_o-2}$ quadruplets. And so on. Of course, the distribution of clusters is completely perturbed. At each step there are $N_p = 2^{k_o-p}$ clusters made of 2^p particles each. The distribution function is a delta peak. The statistics are made over all possible configurations generated by the earlier step. From a practical point of view, the procedure is DLA, except that there is a cluster at the center, and a diffusing cluster of the same size eventually sticks to it. The first few steps are shown in Figure 12.24. The fractal dimension computed this way is in good agreement with what was found previously,

$$D = 1.42 \pm 0.03 \quad (d = 2)$$

The procedure uses less computer time than the conventional one. It allows determination of D for higher space dimensions. The results are shown in Table 12.3. Moreover, it also sheds some light on the difference between DLA and CCA. The procedure is the same. The particles at every step of the latter are the result of the former step in DLA. One may then define an effective exponent as a function of an iteration step and compare these exponents (60). The result is shown in Figure 12.25.

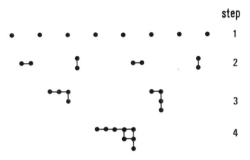

Figure 12.24. Sketch of hierarchical model of Botet et al. (Reprinted, with permission, from Ref. 60.)

The distribution of cluster sizes has also been studied in the limit of very small initial concentrations. Let N_0 be the initial number of particles, randomly distributed on the sites of an $L \times L$ lattice. Let $P(N, t)$ be the total number of clusters wth N elements. We are interested in the normalized distribution of clusters:

$$p(N, t) = L^{-2} P(N, t) \qquad (d = 2) \tag{12.126}$$

This is just the number of aggregates per site. It decreases as time increases. In the following we are interested in the limit when the average concentration, $\langle n \rangle = N_0 L^{-2}$, goes to zero. Then we expect to have finite clusters and no conventional behavior. In analogy with the discussion in Sections 12.2 and 12.3, Vicsek and Family (61) postulated a homogeneous behavior for the distribution function:

$$p(N, t) = N^{-\theta} f\left(\frac{N}{t^z}\right) \tag{12.127}$$

where θ and z are two exponents to be discussed, and $f(x)$ is an unknown function. Because we are considering the limit $\langle n \rangle \to 0$, no gel is present at any time. Then we have the simple sum rule:

TABLE 12.3. Variation of fractal dimension D of CCA Aggregates with space dimension d^a

d	2	3	4	5
D	1.42 ± 0.03	1.78 ± 0.05	2.04 ± 0.08	2.3 ± 0.2

[a]Note that they are less dense than DLA clusters. Data from Ref.59.

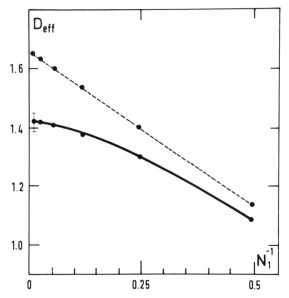

Figure 12.25. Effective fractal exponent D_{eff} calculated by comparing sizes N_1 and $N_2 = 2N_1$, as a function of N_1^{-1} for the hierarchical model (full line) and for the DLA model (dashed line). (Reprinted, with permission, from Ref. 60.)

$$n = \int_0^\infty Np(N,\, t)\, dN \qquad (12.128)$$

$$\propto t^{z(2-\theta)} \qquad (12.128a)$$

where (12.128) expresses the mass conservation. Because the concentration n is a constant, the exponent for the time dependence in relation (12.128a) has to vanish. This implies

$$\theta = 2 \qquad (12.129)$$

This is a remarkable result. It reflects some superuniversality. The argument may be given in any dimension d of space with the same result $\theta = 2$. So far, only mean-field-type arguments such as the classical theory of gelation led to constant exponents. Here nothing was assumed concerning the fluctuations. The important point is that no gel appears, so that all the mass is always in the sol. (Thus, this kind of argument may be applied to systems such as those studied in Section 12.2, when no gelation occurs.)

The behavior of $p(N, t)$ was studied for two-dimensional systems by Vicsek and Family (61) as a function of time (or N) at a given value of N (or t). Typical plots are shown on Figures 12.26*a,b*. What is characteristic in these plots is the power law behavior of $p(N, t)$ for small values of $x = Nt^{-z}$. From these curves we may assume

$$f(x) \propto x^{\Delta} \qquad (x \ll 1) \tag{12.130a}$$

so that, from (12.127),

$$p(N, t) \propto N^{-2+\Delta} t^{-z\Delta} \qquad (x \ll 1) \tag{12.130}$$

From the computer simulations from two-dimensional systems, Figures 12.26*a,b*, we get

$$\Delta = 1.25 \pm 0.15$$

$$z = 1.4 \pm 0.2 \tag{12.131}$$

These exponents are different from the classical ones and from those we saw in the models above.

For large values of x the function $f(x)$ decreases very fast, and we may suppose an exponential behavior, which has not been analyzed so far.

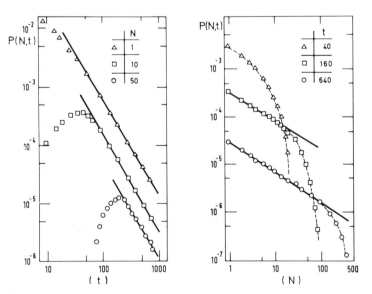

Figure 12.26. Cluster size distribution function versus t (or N) for a given value of N (or t) for CCA. The density is 0.05 on a 400×400 lattice. The results were obtained by averaging 50 runs. (Reprinted, with permission, from Ref. 61.)

Finally, the exponent z may be interpreted by looking at the weight-averaged molecular weight of the aggregates,

$$M_w = \frac{\int_0^\infty N^2\, p(N,\,t)\, dN}{\int Np(N,\,t)\, dN} \propto t^z \qquad (12.132)$$

Thus, the exponent z corresponds to the divergence of M_w when time goes to infinity.

The important results for the distribution function are that the exponent θ is a constant independent of space dimension, so that there is only one unknown exponent, z, in it. In the limit of vanishing concentrations, we expect the result $\theta = 2$ to hold in all cases. The other results may depend on the assumption made for the diffusivity of the aggregates. In all the computer simulations we discussed so far, it was supposed that this diffusivity is a constant, whatever the size of the aggregate. Very recently Meakin et al. (62) have looked for the changes when the diffusion constant $D(N)$ of a cluster depends on its size N. It was supposed that

$$D(N) \propto N^\gamma \qquad (12.133)$$

γ being a parameter.

The computer simulations were made for both two- and three-dimensional systems. Two regimes are found. Depending on the value of γ relative to some critical value γ_c, the shape of $p(N,\,t)$ varies (γ_c is about $-\frac{1}{4}$ for $d = 2$ and $-\frac{1}{2}$ for $d = 3$):

1. For $\gamma > \gamma_c$ the shape of $p(N,\,t)$ as a function of N, for given t, is the same as shown on Figure 12.26, with a decrease throughout the whole range of N.

2. For $\gamma < \gamma_c$ this shape shows a maximum for some intermediate value of N, which depends on time. The same form as (12.127) holds also in this case, though the function $f(x)$ is different.

3. For $\gamma = \gamma_c$ the curve is flat for $N < N^*(t)$, and then decreases. As time goes on, $N^*(t)$ increases, leading eventually to a constant distribution for very large times.

The exponents z and Δ depend continuously on γ. In case γ is a free parameter, this would imply a breakdown of universality. However, the exponent γ may assume only a very limited number of values imposed by hydrodynamics.

We summarize the conclusions of CCA, which is a very active subject (65). One gets a distribution $p(N,\,t)$ of aggregates that, in the limit of a vanishing initial concentration of particles, may be written in a scaled form:

$$p(N, t) = N^{-2} f\left(\frac{N}{t^z}\right)$$

The exponent z depends on the hypothesis made for the diffusion coefficient $D(N)$ of a cluster of size N. Supposing a power law behavior

$$D(N) \propto N^\gamma$$

leads to a γ dependence for z and to two main regimes for the shape of $p(N, t)$ as a function of N at a given value of t. For $\gamma < \gamma_c$ the curve goes through a maximum dependent on t. For $\gamma > \gamma_c$ the curve is always a decreasing function of N, with a characteristic power law behavior for small N, cut off by a function decreasing faster than any power law at large values of $x = N/t^z$.

The critical value of γ_c depends on the dimension of space.

12.5. CONCLUSION

We have discussed three models for polymerization and gelation. The discussion was restricted to *branched* polymers: A common characteristic of these models is that they all lead to nonhomogeneous objects. Hidden behind this is a generalized homogeneity that shows up, for instance, in the distribution function. The probability of finding a polymer, or an aggregate, made of N monomers at a distance t from the gelation threshold, or at time t, may be written in the general form

$$P(N, t) \equiv \lambda^{-\tau} f(\lambda N, \lambda^{-\sigma} t) \tag{12.134}$$

where λ is any constant. For $\lambda = N^{-1}$ relation (12.134) reads

$$P(N, t) \propto N^{-\tau} f(t N^\sigma)$$

where we have kept the notations of Section 12.2.

This generalized homogeneity is related to the scale invariance, or fractal, properties at the gelation threshold: Changing t by a certain amount is equivalent to changing N in the way indicated by Eq. (12.134).

Basic to this notion of generalized homogeneity is the concept of exponents, which we discussed extensively through the chapter. These notions are really interesting only in the case when the exponents depend on very few parameters, like the dimension of space, but not on the detailed chemistry of the process. This is what is called universality. We hope that a limited number of universality classes (i.e., set of values for the exponents) will describe all polymerization and gelation processes. (As we have seen, both are often related.) We have seen three different approaches to the problem, which we arbitrarily called gelation (Section 12.2), aggregation (Section 12.4), and kinetics (Section 12.3).

The gelation process is also known as percolation. It has been intensively studied lately and corresponds to one limit of the Potts model (16). It provides a fair description of polyfunctional condensation and vulcanization.

The diffusion-limited model, and more especially cluster–cluster aggregation, gives good insight into aggregation phenomena in rather dilute systems and is relevant for colloidal aggregation and probably for some biological problems. Although many very interesting results have been obtained by computer simulations, more analytical work (62, 63) is needed.

Finally, the kinetic approach seems to be very promising, but one has still to include the fluctuations and eventually the diffusivity of the clusters and to find the different realistic possibilities for the coagulation kernel that controls the whole approach.

The fractal dimension D, which relates the dimension R of a cluster to its size N,

$$N \propto \left(\frac{R}{a}\right)^D$$

where a is a microscopic length, is another important exponent that also characterizes this scale invariance. However, it is usually related to the other exponents. Usually, we need to know two of these exponents to have a complete knowledge of the polymerization and gelation processes.

We end this chapter by noting that we have seen some of the universality classes but that there are probably others to be found.

REFERENCES

1. P. G. deGennes, *Scaling Concepts in Polymer Physics*, Cornell University Press, Ithaca, NY, 1979.

2. P. J. Flory, *Principles of Polymer Chemistry*, Cornell University Press, Ithaca, NY, 1979.

3. B. B. Mandelbrot, *The Fractal Geometry of Nature*, Freeman, San Francisco, 1977.

4. G. R. Dobson and M. Gordon, *J. Chem. Phys.* **41**, 2389 (1964).

5. M. Gordon and S. B. Ross-Murphy, *Pure Appl. Chem.* **43**, 1 (1975).

6. R. Duplessix, Thesis, Strasbourg, 1975.

7. F. Schosseler and L. Leibler, *J. Phys. Lett.* **45**, 501 (1984).

8. P. G. deGennes, *Biopolymers* **6**, 715 (1968).

9. H. A. Kramers, *J. Chem. Phys.* **14**, 415 (1946).

10. M. Daoud and J. F. Joanny, *J. Physique* **42**, 1359 (1981).

11. B. H. Zimm, W. H. Stockmayer, *J. Chem. Phys.* **17**, 1301 (1949).

12. P. G. deGennes, *J. Physique Lett.* **37**, 1 (1976).

13. D. Stauffer, *J. Chem. Soc. Faraday Trans. II*, **72**, 1354 (1976).

14. D. Stauffer, *Phys. Rep.* **54**, 1 (1979)

15. J. W. Essam, *Rep. Progr. Phys.* **43**, 833 (1980).

16. H. E. Stanley, *Introduction to Phase Transitions and Critical Phenomena*, Oxford University Press, New York, 1982.

17. F. Family, in *Random Walks and Their Applications in the Physical and Biological Sciences*, M. F. Schlesinger and B. J. West (Eds.), A.I.P. Proceedings, Vol. 109, American Institute of Physics, New York, 1984.

18. H. E. Stanley, P. J. Reynolds, S. Redner, and F. Family, in *Real Space Renormalization*, T. W. Burkhardt and J. M. J. Van Leeuwen (Eds.), Springer-Verlag, New York, 1982.

19. H. E. Stanley, in *Proceedings of the International Conferente on Kinetics of Aggregation and Gelation*, F. Family and D. P. Landau (Eds.), North-Holland, Amsterdam, 1984.

20. D. Stauffer, A. Coniglio, and M. Adam, *Adv. Pol. Sci.* **44**, 104 (1982).

21. S. Candau, J. Bastide and M. Delsanti, *Adv. Pol. Sci.* **44**, 27 (1982).

22. P. G. deGennes, *C. R. Ac. Sciences (Paris)* **291**, 17 (1980).

23. J. Isaacson and T. C. Lubensky, *J. Physique* **42**, 175 (1981).

24. G. Parisi and N. Sourlas, *Phys. Rev. Lett.* **46**, 871 (1981).

25. S. F. Edwards, *Proc. Phys. Soc.* **88**, 265 (1966).

26. S. F. Edwards, in *Molecular Fluids*, R. Balian and G. Weill (Eds.), Gordon & Breach, New York, 1976.

27. P. Debye and E. Hückel, *Phys. Z.* **9**, 185 (1923).

28. R. Brout and P. Carruthers, in *Lectures on the Many Body Problem*, R. E. Marshak (Ed.), Interscience, New York, 1963.

29. M. Daoud, in *Static and Dynamic Properties of the Polymeric Solid State*, R. A. Pethrick and R. W. Richards (Eds.), Reidel, Dordrecht, 1982.

30. A. Coniglio and M. Daoud, *J. Phys.* **A12**, L259 (1979).

31. F. Family, *J. Phys. A,* **16**, L665 (1983).

32. P. G. deGennes, *J. Phys. Lett.* **38**, 355 (1977).

33. M. Daoud, *J. Phys. Lett.* **40**, 201 (1979).

34. S. J. Candau, M. Ankrim, J. P. Munch, P. Rempp, G. Hild, and R. Osaka, Proceedings of the 27th Microsymposium on Macromolecules, Prague, 1984.

35. M. Ankrim, Thesis, Strasbourg, 1984.

36. F. Schosseler and L. Leibler, *Macromolecules*, **18**, 398 (1985); see also Ref. 7.

37. M. Daoud, F. Family, and G. Jannink, *J. Phys. Lett.* **45**, 199 (1984).

38. G. B. Benedek, "The Theory of the Sol-Gel Transition," Lecture at Eidgenossische Teknische Hochschule Zurich, 1980.

39. R. J. Cohen and G. B. Benedek, *J. Phys. Chem.* **86**, 3696 (1982).

40. R. M. Ziff and G. Stell, *J. Chem. Phys.* **73**, 3492 (1980).

41. M. H. Ernst, E. M. Hendriks, and R. M. Ziff, *J. Phys.* **A15**, L747 (1982).

42. A. Perelson, in *Mathematical Models in Molecular and Cellular Biology*, L. A. Segal (Ed.), Cambridge University Press, 1980.

43. W. H. Stockmayer, *J. Chem. Phys.* **11**, 45 (1943).

44. F. Leyvraz and H. R. Tschudi, *J. Phys.* **A15**, 1951 (1982).

45. R. M. Ziff, E. M. Hendriks and M. H. Ernst, *Phys. Rev. Lett.* **49**, 593 (1982).

46. E. M. Hendriks, M. H. Ernst, and R. M. Ziff, *J. Stat. Phys.* **31**, 519 (1983).

47. S. R. Forrest and T. A. Witten, Jr., *J. Phys.* **A12**, L109 (1979).

48. C. Allain and B. Jouhier. *J. Phys. Lett.* **44**, 421 (1983).

49. G. K. Von Schulthess, G. B. Benedek, and R. W. de Blois, *Macromolecules* **13**, 939 (1980).

50. T. A. Witten and L. M. Sander, *Phys. Rev. Lett.* **47**, 1400 (1981).

51. T. A. Witten and L. M. Sander, *Phys. Rev.* **B27**, 5686 (1983).

52. P. Meakin, *Phys. Rev. A* **27**, 1495 (1983).

53. M. Muthukumar, *Phys. Rev. Lett.* **50**, 839 (1983).

54. R. Ball, M. Nauenberg, and T. A. Witten, Institute for Theoretical Physics, U.C. Santa Barbara, unpublished report No. N.S.F.-I.T.P. 83-37.

55. P. Meakin and T. A. Witten, Jr., *Phys. Rev. A* **28**, 2985 (1983).

56. P. Meakin, *Phys. Rev. Lett.* **51**, 1119 (1983).

57. P. Meakin, *Phys. Rev. A* **27**, 604 (1983).

58. M. Kolb, R. Botet, and R. Jullien, *Phys. Rev. Lett.* **51**, 1123 (1983).

59. R. Jullien, M. Kolb, and R. Botet, *J. Physique Lett.* **45**, 211 (1984).

60. R. Botet, R. Jullien, and M. Kolb. *J. Phys. A* **17**, L75 (1984).

61. T. Vicsek and F. Family, *Phys. Rev. Lett.* **52**, 1669 (1984).

62. P. Meakin, T. Vicsek, and F. Family, *Phys. Rev. B.* **31**, 564 (1985).

63. H. Gould, F. Family, and H. E. Stanley, *Phys. Rev. Lett.* **50**, 686 (1983).

64. H. J. Hermann, B. Derrida, J. Vannimenus, *Phys. Rev. B* **30**, 4080 (1984).

65. For other applications of DLA to random media, see (a) N. Boccara and M. Daoud (Eds.), *The Physics of Finely Divided Matter, Proc. 1985 les Houches workshop, Springer Proc. Phys.* **5**, Springer-Verlag, Berlin, 1985. (b) H. E. Stanley and N. Ostrowsky (Eds.), *On Growth and Form*, Nijhoff, The Hague, 1985.

13 DYNAMICS OF ENTANGLED POLYMER MELTS

William W. Merrill and Matthew Tirrell

Department of Chemical Engineering and Materials Science
University of Minnesota
Minneapolis, Minnesota

Long macromolecules in a dense fluid state exhibit unusual dynamic behavior. This behavior is to a first approximation independent of the chemical nature of the polymers, that is, independent of the chemical details of the intermolecular interaction between segments of separate chain molecules. Even with the simplest imaginable form of interaction, say, hard-core repulsion, the length of the polymer molecules produces two important effects. First, even weak local interactions along the chain can produce major effects due to the cumulative effect of many weak individual interactions along the chain. Second, macromolecular chain contours are mutually uncrossable.

These effects produce the manifestations of unusual behavior that have come to be called collectively *entanglement*. Understanding entanglement is the central problem in understanding the dynamic behavior of high-molecular-weight polymers in the melt state and in concentrated solution. Qualitative understanding is easier to achieve than is a quantitative, predictive theory, but there are also some subtle points in the qualitative ideas. One should *not* think in these discussions of *an entanglement*, as one can of *a loop* or *a knot* or the configuration produced by interlocking the index fingers of right and left hand. Rather, one should think of cumulative and cooperative effects of the entire polymer melt environment on the motion of the polymer chains. The uncrossability constraints that large molecules impose on one another render the diffusion process nonhomogeneous. Surrounding chains create a local structure, albeit temporary, that dictates the diffusion process. This chapter describes recent progress in understanding the motions of long-chain molecules.

Behavior of the sort presented here shows itself not only in idealized laboratory measurements of transport coefficients but also in many technological situations. Chain motions dictate the rheology and fluid mechanical behavior of polymeric liquids. Some workers strive to develop complete rheological equations of state, constitutive equations rooted in molecular theory. These are essential for designing polymer processing operations, such as injection molding, extrusion, spin coating and film blowing, in a predictive manner. Chain dynamics determine how rapidly polymer fluids can relax a stress applied to them. Relative time scales for stress application and relaxation therefore determine the relative importance of the elastic and viscous properties of these viscoelastic fluids. In polymer processing operations this permits the prediction of power requirements, operational stability (in a fluid mechanical sense), and dimensional stability of the material produced.

Adhesion, welding, fracture, and dissolution of polymers are applications of the ideas of this chapter. These phenomena revolve around the creation or disruption of entanglement among macromolecules. Study of these topics increases our understanding of entanglement.

Adhesion is a complex process that involves the contacting of two surfaces. Factors such as surface roughness, chain configurations near the surfaces, physical and chemical interactions, and flow of the material are involved. Welding is a polymer–polymer adhesion process with interpenetration of the chains in each of the contacting polymer materials. Interpenetration leads to entanglement, as evidenced by the growth of interfacial tensile strength to levels comparable to the cohesive strength of the bulk material, and much greater than could be anticipated from van der Waals interactions. With pure simple hydrocarbon polymers such as polyisoprene, van der Waals interactions are the only ones expected to be important, save an intimate physical intertwining of the chains leading to entanglement coupling. Some of the theoretical ideas of this chapter are applied to polymer welding in Section 13.2.5.

Fracture and dissolution of polymer materials involve the destruction of entanglement coupling. These processes are poorly understood at present. Fracture is important to welding; to understand what makes a welded interface strong, one has to know what makes any arbitrary plane cutting through a bulk polymer strong. Understanding mechanical failure of polymer materials is the key to an enormous range of future applications of polymers, where their light weight, easy processability and other unique properties are making them increasingly desirable.

Dissolution is an important part of the growing use of polymers in microelectronics applications. Photoresists are polymers in which the architecture of integrated circuits is inscribed by lithographic processes. Exposure of a thin layer of polymer coated on a silicon wafer to radiation (light, ultraviolet, X ray, electron beam, or ion beam) changes the polymer so that it is either more or less easily dissolved than unexposed polymer. Solvent etching then develops the pattern by preferential dissolution. Pattern definition at the micrometer or submicrometer scale depends importantly on producing large differences in polymer mobility, and therefore dissolution rate, by exposure to the patterning radiation. These mobility differences are produced at the molecular level by inducement of molecular weight degradation or of branching and cross-linking.

Microelectronics applications of polymer dynamics are not limited to dissolution. *Planarization* is another important property, especially for very large scale integration (VLSI). Multilevel circuitry, that is, stacking vertical layers, is used in VLSI. Between layers a dielectric material, frequently a polymer (polyimide), is used. The ability of a polymer to planarize a layer, that is, to fill in around the complex, small-scale circuit structure and produce a smooth, planar surface of the layer is desirable. Polymer rheology and mobility as a function of polymer molecular weight and structure are the keys to understanding planarization.

Chemical reactions involving polymeric reagents frequently occur in dense, entangled polymer media. Reactions of interest include polymerization, branching, cross-linking, and various photochemical reactions. The low diffusivity and difficulty of homogeneous mixing in polymer fluids causes polymer reactions to be, more often than not, diffusion controlled. The large size of polymer molecules means that two polymer coils can be overlapping to a considerable extent without the reactive sites on the polymer encountering another reactant. This means that internal motions are as important as the diffusivity of the center of mass of the entire polymer molecule in controlling the rate at which reactants come together. Therefore, the classical theory of diffusion-controlled reactions built around the Smoluchowski equation is inadequate.

13.1. BACKGROUND

For 40 years theoretical understanding of the dynamics of polymer fluids has rested on the pioneering work of Kirkwood (1), Rouse (2), and Zimm (3). Models of macromolecules as mechanical objects subjected to random thermal and hydrodynamic forces were introduced. These models, while certainly oversimplified when compared with the real molecular structure, appear to capture much of the physics relevant to the rheological and transport properties of large-chain molecules in dilute solution.

The Rouse model has been particularly influential. For a thorough background on this and related models, the reader is referred to the excellent textbook on this subject by Bird and co-workers (4). Rouse (2) modeled a macromolecule as a coupled harmonic oscillator. The mechanical model of this is that of a large number ($N + 1$) of beads connected by N linear elastic springs. All the mass of the chain is contained in the beads, which experience the frictional interaction with the surroundings. The beads represent subsections of the chain, and one may associate a number of monomer units, N_R, to each bead. (The number N_R and hence the number of beads per chain N_o/N_R, where N_o is the total number of monomer units per chain, can be determined experimentally to fit certain dynamic data.) Each bead experiences friction interactions with its surroundings, usually characterized by the scalar friction coefficient ζ. By assuming that the subchain really acts like a bead, one can set this coefficient equal to the hydrodynamic friction in Stokes' law:

$$\zeta = 3\pi D\eta \tag{13.1}$$

where D is the diameter of the bead and η the viscosity of the solvent. (The Appendix gives some background on equilibrium Gaussian chain statistics and some further insight into the meaning of the bead size.) Since the subchains are not actually rigid beads, they can stretch, changing the distance between bead centers. To represent these internal motions in the model, springs connect the bead with force constants related to the mean square distance r^2 between adjacent bead centers at rest (no applied flow field). The frictionless, massless springs represent the internal modes of motion, which are related by the normal modes of the coupled harmonic oscillator. The springs serve to restore an equilibrium rest length, as the randomly coiling macromolecule resists stretching and elongation.

The configuration space distribution function for the Rouse model, giving the probability density $\psi(\mathbf{R}_0, \mathbf{R}_1, \ldots, \mathbf{R}_N, t)$ for a particular spatial arrangement of beads, satisfies the following diffusion equation:

$$\frac{\partial \psi}{\partial t} = \frac{k_B T}{\zeta} \sum_{n=0}^{N} \frac{\partial}{\partial \mathbf{R}_n} \left(\frac{1}{k_B T} [\boldsymbol{\kappa} \cdot \mathbf{R}_n] + \frac{\partial \psi}{\partial \mathbf{R}_n} + \frac{\psi}{k_B T} \frac{\partial U}{\partial \mathbf{R}_n} \right) \tag{13.2}$$

where $\boldsymbol{\kappa}$ is the transpose of the velocity gradient tensor. The gradients $\partial/\partial \mathbf{R}_n$ are (in Cartesian coordinates):

$$\frac{\partial}{\partial \mathbf{R}_n} = \frac{\partial}{\partial \mathbf{R}_{x,n}} \hat{\mathbf{x}} + \frac{\partial}{\partial \mathbf{R}_{y,n}} \hat{\mathbf{y}} + \frac{\partial}{\partial \mathbf{R}_{z,n}} \hat{\mathbf{z}} \tag{13.3}$$

with $\hat{\mathbf{x}}$, $\hat{\mathbf{y}}$, and $\hat{\mathbf{z}}$ being unit vectors in the indicated directions. The function $U(\mathbf{R}_0, \ldots, \mathbf{R}_N)$ is the potential energy of the chain, typically with a Hookean spring law (Gaussian distribution):

$$U = \sum_{n=1}^{N} \frac{k_B T}{2r^2} (\mathbf{R}_n - \mathbf{R}_{n-1})^2 \tag{13.4}$$

This notation is adapted from that of Bird *et al.* (4). Alternative introductory treatments of the Rouse model are given by Ferry (5a) and by deGennes (5b). The Appendix explains the geometry and notation. The three terms on the right side of Eq. (13.2) represent, respectively, the frictional, random thermal, and spring forces on each bead. At equilibrium the velocity gradient κ is zero, so that Eq. (13.2) resembles the wave equation of quantum mechanics [with the ψ of Eq. (13.2) analogous to the square of the usual ψ of quantum mechanics]. DeGennes (5b) has demonstrated and exploited this connection for several problems in configurations of polymers at equilibrium. The effective field U may be a self-consistent field appropriate for the study of self-avoiding walks (excluded volume) or polymers near surfaces (adsorption).

The translational diffusion coefficient is defined as

$$D = \lim_{t \to \infty} \frac{x^2}{2nt} \tag{13.5}$$

where n is the dimension of the space in which the excursion x is taken. One may define a relaxation time for motion over a length l as

$$\tau = \frac{l^2}{2nD} \tag{13.6}$$

For the Rouse model, the diffusion coefficient, $D = D_{\text{Rouse}}$, is

$$D_{\text{Rouse}} = \frac{k_B T}{(N+1)\zeta} \tag{13.7}$$

since each bead contributes equally, with friction coefficient ζ, to the total friction felt by the chain. This and several other predictions of the Rouse model are not in accord with experimental observations in dilute solution (5). For example, the translational diffusion coefficient varies with molecular weight like (6)

$$D \propto M^{-\nu} \tag{13.8}$$

where ν is between 0.5 and 0.6, not 1, as suggested by Eq. (13.7). The reason for this is now well understood via the Zimm model (3). In fact, in dilute solution not all the beads feel the same hydrodynamic friction. Some are shielded from viscous drag by neighboring beads, thus reducing the molecular weight dependence of D to that of the reciprocal radius of gyration of the chain.

A remarkable fact, however, is that the Rouse model becomes a better model of experimental reality as the density of chains is increased. The reason for this is also understood qualitatively. As the density of polymer chains in the medium increases, the coils interpenetrate one another. Interpenetration of the coils screens the hydrodynamic interactions among the polymer segments, rendering all segments again equally susceptible to hydrodynamic drag. In this case it is a drag due to polymer–polymer as well as polymer–solvent interactions, but as long as the chains are not too long, the dynamic behavior predicted by the Rouse model is indeed observed experimentally. Ferry (5a) documents this well for the viscoelastic properties. More recently the transition from Zimm-like to Rouse-like behavior with increasing polymer density has been observed in measurements of the translational diffusion coefficient via dynamic light scattering and in the relaxation of the internal modes of motion via neutron spin echo techniques (6).

This chapter is confined principally to the behavior of undiluted fluids of long-chain molecules. It is somewhat paradoxical, but hopefully understandable from the arguments given above, that the dilute-solution Rouse model is the best point of entry into the theory of dense, entangled polymer fluids. The Rouse model works for undiluted polymer fluids (up to a certain molecular weight) in part because a pure fluid has fewer complications (for example, hydrodynamic interactions). For the same reason we restrict ourselves here to undiluted fluids.

For pure polymer fluids above a certain molecular weight, however, the Rouse model ceases to provide even qualitatively correct predictions. This chapter will concentrate on diffusion, since diffusive motions are key to all dynamic and rheological properties. Examination of the diffusion coefficient for a homologous se-

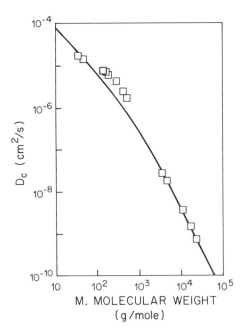

Figure 13.1. Tracer diffusion coefficient for polyethylene of molecular weight M in a matrix of very high molecular weight polyethylene (from Ref. 7). Solid line is: $1/D = 1/D_{\text{Rouse}} + 1/D_{\text{rept}}$ and so varies continuously from $D \propto M^{-1}$ to $D \propto M^{-2}$

ries, say, polyethylene, as shown in Figure 13.1, over a wide molecular weight range reveals the breakdown of the Rouse model above a certain critical molecular weight M_0. We see a transition from $D \propto M^{-1}$ behavior predicted by Eq. (13.5) to a stronger dependence $D \propto M^{-\nu}$, where now $\nu \simeq 2$.

An analogous transition to behavior beyond the ken of the Rouse model is seen in the viscosity coefficient η, illustrated in Figure 13.2. The Rouse model predicts $\eta \propto M$, which is indeed seen (after a correction for the molecular weight dependence of ζ) below a certain critical molecular weight M_c. Above M_c, a much sharper dependence, $\eta \propto M^\alpha$, with $\alpha \simeq 3.4$ is seen experimentally.

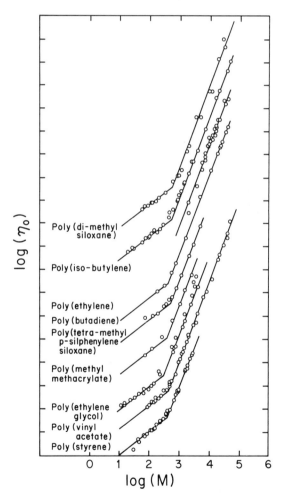

Figure 13.2. The zero-shear viscosity η_0 versus molecular weight M for nine different polymer melts. The curves have been shifted vertically by arbitrary amounts for clarity of display. The lines have slope 1.0 at the lower molecular weight and slope 3.4 at the higher molecular weight. (Adapted from Ref. 8.)

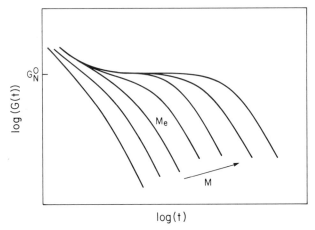

Figure 13.3. Sketch of the shear stress relaxation modulus, $G(t)$ versus time. Note the development of a plateau at M_e with modulus G_N^0, which increases in duration with increasing molecular weight M.

Figure 13.3 illustrates the breakdown of Rouse-like behavior in linear viscoelastic properties. The Rouse model predicts that the shear stress relaxation modulus $G(t)$ [the ratio of stress to shear strain after sudden imposition of a step strain; the reader is referred to the book by Bird et al. (4) for precise definitions of rheological quantities] decays unrelentingly to zero after imposition of a sudden step shear strain. In fact, above a certain critical molecular weight M_e a distinct plateau modulus, or slowing down of the stress relaxation, is observed, similar to the behavior of cross-linked polymers. This plateau is, in fact, another signature of entanglement, which produces behavior in several respects similar to temporary cross-linking. Ferry (5a) and Graessley (9) are the best sources for more profound discussion of the experimental manifestations of entanglement. Suffice it to say here that above some molecular weight, which may be different for each property examined (in fact, $M_D < M_e \simeq \frac{1}{2} M_c$), behavior begins that is not comprehended within the Rouse model and has come to be called entanglement.

Graessley reviewed the entanglement concept (9) from both the phenomenological and theoretical viewpoints of circa 1973. A major shift in emphasis in the theory of entangled systems has occurred during the intervening 13 years. The cornerstone of this new theoretical treatment of entanglement is the reptation model of deGennes (10), which parts from earlier models of entanglements as specific interchain interactions (9) (for example, temporary cross-links or points of enhanced friction) and deals only with a single fundamental fact about interactions between chain molecules: mutual uncrossability. Uncrossability places topological constraints on the chain motion, first suggested by Edwards (11–13). This chapter shows that uncrossability alone, as embodied in the reptation model, captures many of the important features of entanglement.

This chapter explains the interesting diffusion behavior predicted by the reptation model for polymer melts. As mentioned before, for economy we exclude polymer solutions from detailed consideration, even though entanglement effects can be very important in diluted systems. These have been treated to some extent elsewhere (6). Since the reptation model has motivated an unprecedented amount of experimental and theoretical work on polymer dynamics in the last decade, we want to expose it here in detail. We hope to provoke both criticism of and improvement in the model as well as promote understanding of the current state of play.

Diffusion, as embodied in the self-diffusion coefficient and related quantities, is the only transport property treated in depth. This is both for economy of presentation and, more profoundly, because the reptation model involves a connection between molecular diffusion and the long-time part of the mechanical relaxation properties. For example, according to the reptation model, there is a direct relation between molecular diffusion and the slowest part of the stress relaxation modulus.

13.2. BASIC REPTATION

13.2.1. Construction of Reptation Model

In the reptation model for flexible polymers proposed by deGennes (10), each chain is considered to be constrained to a tubelike region surrounding the chain contour. The tube is the result of the inability of the chain to move *through* the contours of the surrounding chains. Alternatively, the chain may be envisioned as trapped in a cage formed by the contours of its surrounding chains. In both descriptions the tube or cage has a nonzero diameter about the chain backbone. This diameter is a parameter of the reptation model for which the physical significance is open to interpretation. It is best thought of as an average distance perpendicular to the local chain contour that a polymer segment must travel before it encounters a neighboring chain segment that constrains it.

As shown in Figure 13.4, the chain atoms can have some radial latitude in their motions and may form temporary conformational "kinks" in the chain *within* the confines of the tube. Since the chain cannot move through the tube wall, the only way it can move its center of mass a distance greater than that of the tube diameter is for one of its ends to "slide" out an open end of the tube. This end can map a three-dimensional random path. Since the bulk melt is homogeneous and the chain is always constrained in a tube, the chain end escape from the old tube is really only the creation of *new tube* defined by different surrounding chains. Meanwhile, the rest of the chain follows this end into the newly defined tube. Finally, in its following motion, the other end leaves the far end of the old tube vacant. Since it is the instantaneous occupation of the tube by the parts of the chain that defines the tube, this tube section is destroyed. Of course, at a later time this other end may be leading the chain motion in the mapping of the new tube. In an isotropic medium each end spends an equal amount of time on the average leading the chain into the new tube, thereby determining the center-of-mass motion of the chain and its long-time relaxation properties. This tube mapping process is shown in Figure

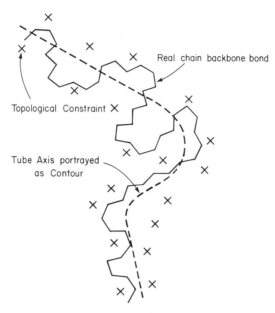

Figure 13.4. The tube axis. The time-averaged chain location about which the instantaneous chain conformation oscillates is the tube axis.

13.5. The overall process can be compared to a snake moving in a crowded den of other snakes (reptiles); hence the term *reptation* suggested by deGennes. (The term derives from the Latin verb *reptare*, meaning to creep.)

The idea of reptation then is to reduce the motion of a macromolecule in the presence of the complex intermolecular interactions in a polymer melt to a simple random walk on a random path. In this way reptation resembles other nonhomogeneous diffusion processes such as diffusion in porous media, percolation clusters, or other fractal structures. Noncrossability of chain contours produces the microheterogeneity of the diffusion process in the case of polymer melts. The simplest reptation model assumes that the length of this enveloping tube remains constant throughout the diffusion process. When the new tube is created, an equal amount of old tube is destroyed. This fixed tube length is typically much shorter than the extended backbone length. The chain has many kinks, so the tube has a rather large diameter. Since the tube envelops a chain that has a three-dimensional Gaussian distribution function, the tube itself has a Gaussian distribution. The average end-to-end distance of this tube must therefore equal the average end-to-end distance of the chain; hence (14),

$$N_p a^2 = N_k b^2 \tag{13.9}$$

Here N_k and b represent the number of Kuhn statistical segments and the Kuhn segment length describing the chain (Appendix A) while N_p and a describe the

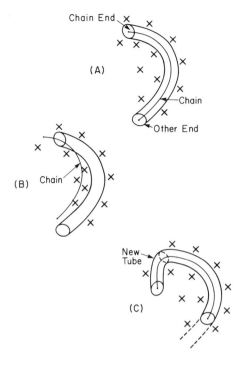

Figure 13.5. Basic reptation: (*a*) chain in initial tube formed by topological constraints; (*b*) prohibited translation crossing constraints; (*c*) reptation of chain along tube axis creating new tube and destroying an equal amount of old tube (dotted).

number of corresponding tube segments and the tube segment length. The subscript *p* refers to the *primitive path* of Doi and Edwards (14) to be discussed subsequently. The tube segment length parameter *a* is a measure of the *topology* of the diffusion problem. Were the chain to be so short that the total tube would be much smaller than *a*, the lateral motions would no longer be more inhibited than the axial sliding motions. In fact, the tube is unimportant for unentangled chains, and it therefore seems justifiable to relate N_p to some parameter involving entanglements. The obvious choice is a critical molecular weight such as M_e (14) since it represents a size scale related to entanglement. If M_0 is defined as the monomeric molecular weight, the number of monomers per tube segment can be taken as M_e/M_0. Hence, the number of tube or pimitive path steps in the chain is:

$$N_p = \frac{N_0 M_0}{M_e} = \frac{N_k M_k}{M_e} \tag{13.10}$$

where M_k is the molecular weight per Kuhn segment. It follows that the tube segment length is

$$a = \sqrt{\frac{M_e}{M_k}}\, b \tag{13.11}$$

The tube length L_T is

$$L_T = N_p a = N_k b \sqrt{\frac{M_k}{M_e}} \qquad (13.12)$$

compared to the maximum extension length of the chain,

$$L_{max} = N_k b \qquad (13.13)$$

Since the number of monomers per Kuhn segment is approximately two orders of magnitude smaller than the number per entanglement weight M_e (see Appendix), Eqs. (13.12) and (13.13) show that the tube length is about an order of magnitude shorter than the total chain contour (9). Since both M_k and M_e are independent of chain length, this size comparison is valid for all chain lengths.

13.2.2. Reptation Is One-Dimensional Diffusion

The sliding motion of the chain in the tube may be modeled as a one-dimensional diffusion process along the tube contour s. If one focuses on one chain end, $\langle s^2 \rangle$ represents the mean square tube contour length that the end mapped out as new tube (if the end was "leading" the chain motion more than half the time) or that the end penetrated into the initial tube (i.e., the amount of tube destroyed at the end if the end was "following" more than half the time). For interior chain sections, $\langle s^2 \rangle$ represents the mean square displacement within the predetermined old tube. What is the form of the tube diffusion coefficient D_T? The major hypothesis of the reptation theory is that the motion of the chain is inhibited by entanglement because of large-scale topological constraints of the chain, *not* because of local increases in the monomer–monomer friction along the chain backbone, as in older theories (5). Within the tube the chain is hypothesized to behave as a Rouse chain (14) with a tube diffusion coefficient from Eq. (13.7):

$$D_T = \frac{k_B T}{N_0 \zeta_0} \qquad (13.14)$$

where ζ_0 is the monomer friction coefficient.

With the chain sliding motion in the tube now described, the center-of-mass diffusion coefficient may be determined. Since the coordinate axes s of the tube contours have a Gaussian distribution of three-dimensional conformations, a displacement of s along the one-dimensional tube does *not* represent an equal displacement $|x|$ of the center of mass in three dimensions. Defining τ as the average lifetime of any particular tube (the time required for the chain to completely escape its initial tube), one can consider a time span $t \gg \tau$ constructed of many subintervals of order τ. This tube lifetime is generally referred to as the reptation time.

During each subinterval the chain escapes its tube, forming a new tube connected to the last section of the old tube to be destroyed. The reptation time can be estimated by substituting L_T^2 for l^2 in Eq. (13.6) and using Eq. (13.14) for D_T:

$$\tau \approx \frac{L_T^2 N_0 \zeta_0}{6 k_B T} \propto N_0^3 \qquad (13.15)$$

$$\approx \frac{\langle R^2 \rangle}{6 D_c}$$

where $\langle R^2 \rangle$ is the mean square end-to-end distance.

Now the mean square displacement of the center of mass over the interval τ is on the order of the chain coil size. This is proportional to $N_p a^2$. Substituting this center-of-mass displacement for l^2 and τ for the characteristic diffusion time in Eq. (13.6), the center-of-mass diffusion coefficient D_c, is seen to vary inversely to the square of molecular weight. Doi and Edwards show (14)

$$D_c = \frac{D_T}{3 N_p} \propto \frac{1}{N_0^2} \qquad (13.16)$$

Since the center-of-mass motion is not simple three-dimensional diffusion for times *less* than the reptation time, this diffusion coefficient can only be applied to problems concerning time scales *greater* than the reptation time and concerning length scales greater than one radius of gyration.

The inverse square law dependence of the center-of-mass diffusion coefficient of Eq. (13.16) can be derived more rigorously using the Doi and Edwards (14) equations for the time evolution of the primitive path. The concept of tube evolution through the backward and forward sliding of the chain ends is mathematically embodied in these equations. A tube of N_p primitive path segments is described by the $N_p + 1$ position vectors, $\mathbf{R}_0(t), \mathbf{R}_1(t), \ldots \mathbf{R}_i(t), \ldots, \mathbf{R}_{N_p}(t)$, describing the segment ends' three-dimensional positions. [These are similar in definition to the position vectors of the bead position vectors of the Rouse model, Eq. (13.2).] As a mathematical simplification, the distance between each of these vectors is considered fixed with magnitude a defined by Eq. (13.11). The selection of one end as $\mathbf{R}_0$ and the other as $\mathbf{R}_{N_p}$ is arbitrary, but once chosen, it is fixed. The chain is defined as moving left when the end at $\mathbf{R}_0$ moves out of the old tube and as moving right when the other end at $\mathbf{R}_{N_p}$ moves out of the old tube. Now let τ be the time interval required for the chain to move a net distance of magnitude a left or right. The change in position of the outwardly moving chain end is $\boldsymbol{\Delta}(t)$ with magnitude a. If the chain motion is leftward, then $\mathbf{R}_0(t + \tau)$ is $\mathbf{R}_0(t) + \boldsymbol{\Delta}(t)$. Since the rest of the chain must follow inside the old tube the remaining $\mathbf{R}_i(t + \tau)$ values must assume the old values of $\mathbf{R}_{i-1}(t)$. There is a similar shifting process for rightward motion. Both can be described by a single set of equations for the R_i by defining a random variable $\xi(t)$ having values of ± 1 with equal probability.

Leftward motion is represented by -1 and rightward by $+1$. The time evolution of the primitive path in terms of the position coordinates is then

$$\mathbf{R}_0(t + \tau) = \tfrac{1}{2}[1 + \xi(t)]\,\mathbf{R}_1(t) + \tfrac{1}{2}[1 - \xi(t)]\,[(\mathbf{R}_0(t) + \boldsymbol{\Delta}(t)] \tag{13.17a}$$

$$\mathbf{R}_n(t + \tau) = \tfrac{1}{2}[1 + \xi(t)]\,\mathbf{R}_{n+1}(t) + \tfrac{1}{2}[1 - \xi(t)]\,\mathbf{R}_{n-1}(t) \qquad 1 \le n \le N - 1 \tag{13.17b}$$

$$\mathbf{R}_{N_p}(t + \tau) = \tfrac{1}{2}[1 + \xi(t)]\,[\mathbf{R}_{N_p}(t) + \boldsymbol{\Delta}(t)] + \tfrac{1}{2}[1 - \xi(t)]\,\mathbf{R}_{N_p-1}(t) \tag{13.17c}$$

Doi and Edwards derive the center-of-mass diffusion coefficient using these equations and the simplifying assumption of a *constant* jump time defined as τ_j by Eq. (13.4):

$$\tau_j = \frac{a^2}{2D_T} \tag{13.18}$$

This assumption yields the correct behavior in the limit of a large number of jumps over the course of complete tube escape and hence in the limit of very long chains ($N_p \gg 1$). Substituting Eqs. (13.17) into the definition of center of mass, they obtain

$$\mathbf{R}_G(t + \tau) - \mathbf{R}_G(\tau) = \frac{1}{N}\,\xi(t)\,\boldsymbol{\Delta}(t) + \frac{1}{2N}\,\boldsymbol{\Delta}(t) \tag{13.19}$$

Using the statistical independence of successive jumps in both the values of $\xi(t)$ and $\boldsymbol{\Delta}(t)$, Doi and Edwards obtain, in the limit of large N_p,

$$\langle [R_G(t) - R_G(0)]^2 \rangle = \frac{2D_T}{N}t \tag{13.20}$$

Comparison to the center-of-mass diffusion result,

$$\langle [R_G(t) - R_G(0)]^2 \rangle = 6D_c t \tag{13.21}$$

obtains the desired conclusion, Eq. (13.16). DeGennes (10) obtained similar results using a continuous model of the chain contour instead of Eqs. (13.17).

13.2.3. Rigid Polymers

We wish to stress that the foregoing is applicable only for flexible polymer chains. Rigid polymers also become entangled at high enough molecular weights. Doi and

Edwards (15) have proposed a theory similar to the foregoing for rodlike polymers in which the surrounding chains form a tube through which the chain test rod must move. Since these rods must follow their own axes, in a sense, they reptate (not with serpentine motion, however). Important differences separate the flexible and rigid cases. In the rigid case the tubes *are* straight with end-to-end distance L_T instead of a Gaussian distribution of end-to-end distances. This means that the tube contour length equals the chain contour length. Furthermore, it means that new tube orientation is strongly correlated with the current tube. The chain cannot bend. The only way the tube can change its orientation is by a solid-body rotation of the entire chain. Hence all of the topological constraints act on the entire chain simultaneously. There is no free subchain motion. Using this model, Doi and Edwards have found the relaxation time of rigid polymers to increase with the *ninth* power of the molecular weight as opposed to the *third* power for reptating flexible chains, a reflection of the stronger action of the topological constraints if the entire chain must move in a correlated way. Real polymers are neither perfectly flexible nor perfectly rigid. This suggests a transition region of chain stiffness between flexible and rigid behavior. Stiffness is characterized by the Kuhn segment length. The treatment of entangled chains of intermediate flexibility is largely virgin territory.

13.2.4. Connection between Reptation Diffusion and Rheology

Graessley (16) has proposed a method of experimentally determining and interrelating the various parameters of the reptation model using rheological data. This allows a diffusion coefficient to be *derived* in terms of measurable parameters exclusively. First, the plateau modulus in $G(t)$, G_N^0 is directly related to the number of tube or primitive path segments via the theory of rubberlike elasticity (17),

$$N_p = \frac{5 G_N^0}{4 \nu k_B T} \tag{13.22}$$

The parameter ν is the number of chains per unit volume (number chain density). Graessley then uses the Rouse expression, Eq. (13.14), to determine the tube diffusion coefficient. He assumes the inverse power relation holds above M_c, the critical molecular weight for the viscosity transition, and therefore multiplies the Rouse diffusion coefficient at M_c by the ratio M_c/M. The Rouse diffusion coefficient D_r is related to the viscosity η and the square of the chain end-to-end separation R^2 by

$$D_r = \frac{k_B T}{N_r \xi_r} = \frac{k_B T \, \nu R^2}{36 \eta_r}, \tag{13.23}$$

where the subscripts r refer to the predictions of the Rouse model (2). Hence,

$$D_T = \frac{k_B T \, \nu R^2 \, M_c}{36 \eta_r (M_c) M} \tag{13.24}$$

Substituting Eqs. (13.22) and (13.24) into (13.16), Graessley obtains

$$D_c = \frac{G_N^0}{135} \left(\frac{\nu k_B T}{G_N^0} \right)^2 \frac{R^2}{M} \frac{M_c}{\eta(M_c)} \tag{13.25}$$

The second term carries the molecular weight dependence and varies inversely with the square of the chain length since ν varies inversely with chain length (when the mass density is independent of chain length). The remaining terms are independent of chain length. One may substitute $c R_g / M$ for νk_B, where c is the polymer mass per unit volume and R_g is the universal gas constant. It should be noted that Graessley makes a conceptual distinction between the topological scaling, determined by G_N^0, and the frictional scaling determined by the Rouse model. The use of Eq. (13.23) does *not* require knowledge of the number of monomers per Rouse segment. This value could be determined through stress relaxation studies (and does not in general coincide with the number of monomers at M_e).

13.2.5. Applications of Basic Reptation Model

For many applications the center-of-mass motion, that is, the self-diffusion coefficient of the entire chain, is the important mass transfer parameter. Several examples are given in the review article of Tirrell (6). Among them are diffusion-controlled reactions of polymers (18), dissolution of polymers, and certain other microelectronics applications (6).

There are other situations where diffusion is the dominant process, but the macroscopic self-diffusion coefficient is not the relevant parameter to characterize the situation. These are applications where the important motion is over length scales smaller than the molecular size. DeGennes has termed these *semilocal* motions (19). Dynamics of phase separation and many adhesion applications are examples where semilocal motions are the most important. As a specific case, we examine here how simple reptation ideas can be applied to the welding of two contacted polymer blocks (20). The individual chains gradually diffuse across the interfacial plane. This bridging process increases the mechanical strength of the junction. Eventually, the interfacial cross-section becomes indistinguishable from the bulk material—the interface disappears and the sample is fully welded. Figure 13.6 shows that before contact, the chains are not in a Gaussian conformation as they must adopt a compressed configuration near the surface. When the blocks are contacted, all of the chains cannot immediately cross the interface: Reptation dictates that only chain ends can move across the interface. Of course, this is an idealization: Diffusion across the interface must be proceeded by a brief contacting or wetting period during which there will be some flow of *all* the polymer segments until the two blocks actually touch. The chains across the interface then prevent further bulk motion, and reptation must begin. Finally, Figure 13.6 shows how some of the chains have reptated across the interface. Two criteria for complete welding exist: The chains about the interfacial plane assume a Gaussian confor-

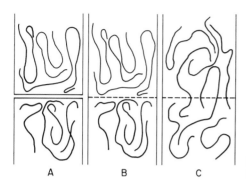

Figure 13.6. Polymer-polymer welding: (*a*) initially; (*b*) contacting; and (*c*) crossing.

mational distribution and the number of chain crossings per unit area is the same as in the bulk material. The conformation relaxation across the interface decreases the conformational component of the chemical potential and can be considered the "force" driving the diffusion.

The center-of-mass diffusion coefficient cannot be used to describe the welding process because the majority of the welding occurs well within one reptation time [defined by Eq. (13.18) by replacing *a* with L_T]. Only pieces of the chain need to be put across the interfacial plane to achieve full strength. A discrete, constant jump time model such as Eq. (13.17) is an inappropriate idealization for these semilocal motions. Rather, a continuous tube diffusion equation is more practical in this application for describing the tube escape process.

During this escape process the present tube confining the particular chain will have three discernable parts: two "new tube" sections at the chain ends and the ever shrinking central section of persisting initial tube. Only the new tube sections can cross the interface. Not all of these actually do cross since the new tube sections have a Gaussian distribution of end-to-end vectors for both the total and sections of the new tube. The probability that a particular chain's new tube crosses the interface is therefore a function of the distance of the beginning of the new tube section (the end of the old tube) from the interface and the length of the new tube. This allows the separation of the healing problem into two decoupled parts: determining the distribution function for the *rate* of new tube creation at each end and determining the number of crossings for the *amount* of new tube given its starting position.

We describe the tube escape (and hence new tube creation) process using a one-dimensional description of the chain sliding motion in the tube. Like Doi and Edwards, we label one end of the chain the 0-end with position R_0 and the other end as the *N*-end with position R_N. We now define a tube coordinate description with respect to the *initial* tube with the 0-end at $s = 0$ and the *N*-end at $s = N$. The actual tube between these ends would have three-dimensional conformations roughly described by $R_0, R_1, \ldots, R_N$. In a continuous model this is described as a space curve $R(s)$. Since we are interested only in the amount of initial tube destroyed, we can for part of the problem actually ignore the tube conformation and "pull" the curvilinear tube into a straight tube for discussion purposes. We account for the actual shape of the tube by asking the probability that a given *initial* segment s' has the coordinate $\mathbf{R}(s')$ (we will elaborate later). When the 0-end moves out of the tube, the chain moves left toward more negative values of s. When the N-end moves out of the tube, the chain moves right toward more positive values of s. Negative values of s represent new tube mapped out by the 0-end (left) while positive values greater than L_T must represent new tube mapped out by the N-end (right). Thus, the 0-end destroys and creates tube on the left side of the persisting section of initial tube, and the N-end destroys and creates tube on the right side.

We now note that the tube between $s = 0$ and L_T only necessarily represents initial tube at time zero. For example, the chain could move $\frac{1}{2} L_T$ to the left and then move $\frac{1}{2} L_T$ back to the right. In the leftward motion the 0-end sweeps out new tube of length $\frac{1}{2} L_T$ and concomitantly the N-end destroys the right half of the old tube. In the following rightward motion the N-end creates $\frac{1}{2} L_T$ of new tube while the 0-end destroys the new tube it had formed. The final result is that the old tube on the left side persists while the right half has been replaced with new tube. The 0-end had the s-coordinate sequence $0, -\frac{1}{2} L_T, 0$ while the N-end had the sequence $L_T, \frac{1}{2} L_T, L_T$. Therefore, a chain end destroys initial tube at coordinate s the *first time* it moves to that coordinate value. Constant instantaneous tube length (in contrast to the shrinking portion of *initial* tube length still persisting) dictates a separation between the 0- and N-ends of the chain of L_T. The 0-end destroys old tube on the left side at coordinate s when it first reaches s and destroys old tube on the right side at coordinate s when it first reaches $s - L_T$. The rate of initial tube disappearance is therefore intimately connected to the one-dimensional first-passage problem of random-walk theory.

In a melt where the macromolecules experience no external forces, the escape route is symmetric about the chain center. It is therefore necessary only to describe the tube loss and new tube evolution at one of the chain ends. Using the tube coordinate s, we define $P(\Delta, s', t; L_T)$ as the probability at time t that the initial tube between $s = 0$ and $s = s'$ has been lost and that new tube of length Δ now emanates from the initial tube segment at s' given the total initial tube length L_T. Following Prager and Tirrell (20), we find this probability through two steps. We find the probability $P_1(s, t; S', L_T)$ that the chain end at $s = 0$ is now at s at time t but that it has never moved more than s' to the right (into the initial tube) or $L_T - s$ to the left (out of the initial tube). Motion past s would cause the left chain end to lose the tube at s' while motion past $L_T - s'$ to the left ($s' - L_T$ in s coordinates, a negative value) would cause the right chain end to lose the tube at s. Hence, $P_1(s, t; s', L_T)$ is the probability that the tube at s' persists and the chain end is at s. We want to find the probability that s' is the *leftmost* section of initial tube to survive at t. The tube from $s = 0$ to s' is lost at time t' when the end reaches s' for the first time. The probability of reaching s' for the first time at t' is the flux of chain ends into s' at t'; that is, $D_T(\partial P_1/\partial s)|_{s=s', t=t'} \, dt$. The tube immediately following s' will only persist if the chain moves no further to the right nor moves L_T to the left from its position at s'. Hence, the chain end must stay between $s = s' - L_T$ and $s = s' + ds'$ for a time $t - t'$ after it has reached s'. Furthermore, we want it to end at $s = s' - \Delta$ (i.e., Δ new tube protruding from s' to the left) at t. The probability that it does this is $P_1(-\Delta, t - t'; ds, L_T - s')$. Now our original objective $P(\Delta, s', t; L_T)$ can be obtained by any path that reaches s' at any time $0 < t' < t$ and then returns to $s = s' - \Delta$ at time t. We therefore integrate over all paths (all possible t' values):

$$P(\Delta, s', t; L_T) = D_T \int_0^t P_1(-\Delta, t - t'; ds, L_T) \frac{\partial}{\partial s} P_1(s, t'; s', L_T)\big|_{s=s'} dt'$$

(13.26)

We only need the function P_1; this is the solution to the diffusion equation:

$$\frac{\partial P_1}{\partial t} = D_T \frac{\partial^2 P_1}{\partial s^2}$$

(13.27a)

subject to the first passage boundary conditions:

$$P_1(s, t; s', L_T) = P(s - L_T, t; s', L_T) = 0$$

(13.27b)

given that the chain end started at $s = 0$:

$$P_1(s, 0; s', L_T) = \delta(s)$$

(13.27c)

The solution for $P_1(s, t; s', L_T)$ is

$$P_1(s, t; s', L_T) = \frac{2}{L_T} \sum_{n=1}^{\infty} \sin\left(\frac{n\pi x}{L_T}\right) \sin\left(\frac{n\pi}{L_T}(s' - s)\right) \exp\left(-\frac{n^2\pi^2 D_T t}{L_T}\right)$$

(13.28)

$P(\Delta, s', t; L_T)$ is obtained by the integration of Eq. (13.26), as shown by Prager and Tirrell (20).

With the tube escape process now characterized by $P(\Delta, s', t; L_T)$, the crossing density as a function of the time may be found. We need the initial tube conformational distribution function expressed as the probability of finding a segment s' of the initial tube a distance z from the interface, $F(s', z)$. To a first approximation, this is the average bulk segment density, $\nu ds/L_T$, where ν is the number of chains per unit volume. This is equivalent to assuming a Gaussian configuration of the initial chains. [The more realistic case has been solved by Adolf et al. (21)]. Given a particular starting point, z, the new tube Δ of Gaussian distribution crosses the plane $z = 0$ an average number of times $C(\Delta, z)$. The crossing density is then the integral over the possible initial tube conformations $F(s', z)$ with escape according to $P(\Delta, s', t; L_T)$, and whose new tube crosses the interface $C(\Delta, z)$ times:

$$\sigma(t) = 4 \int_0^\infty \int_0^{L_T} \int_0^{L_T} F(s', z) \, P(\Delta, s', t; L_T) \, C(\Delta, z) \, ds' \, d\Delta \, dz \quad (13.29)$$

The factor of 4 comes from the double symmetry of the problem. Each chain diffuses out of both tube ends, and there are chains on both sides of the interface. The result for the simplified initial conformation case is given by Prager and Tirrell (20) as

$$\sigma(t) = \sigma(\infty) \frac{2}{\sqrt{\pi}} \left[\tau^{1/2} + 2 \sum_{k=1}^\infty (-1)^k \left\{ \tau^{1/2} \exp\left(\frac{-k^2}{\tau}\right) - \sqrt{\pi} \, k \, \mathrm{erfc}\left(\frac{k}{\tau^{1/2}}\right) \right\} \right]$$

$$(13.30)$$

The quantity τ is defined as $4D_T t/(L_T)^2$ and $\sigma(\infty)$ is the crossing density of the bulk polymer, which is solely a function of the bulk density. If a relation between the crossing density and the strength of the weld juncture can be formulated, the time evolution of the juncture strength is predicted by $\sigma(t)$. Adolf *et al.* (21) have considered the case where the chains must penetrate the other polymer block by a distance δ in order to make an effective contribution to weld strength.

13.2.6. Rheological Properties of Basic Reptation Model

The probability distribution function, $P_1(s, t; s', L_T)$, is also a central player in rheological applications. When a bulk sample is strained, the chain microstructure is perturbed from its equilibrium state. The straining action orients the chain conformations along the principal axes of the strain perturbing the Gaussian distribution function. This distributional perturbation manifests itself macroscopically in a stress. The decay of the stress therefore is related to the decay of this perturbation. In a step strain experiment the stress develops over a very short period of time and then monotonically decreases from its maximum value (occurring soon after time zero). The time decay of the stress relaxation modulus, $G(t)$, is directly related to the perturbation decay. In an entangled melt the distributional perturbation decays as the chains reptate out of their distorted tubes into new tubes of undistorted, random configuration (14). We therefore require the fraction of old tube segments that remain occupied as a function of time. This is equivalent to finding the absolute average probability that a tube segment persists at time t; that is, the probability of the tube segment at s' persisting integrated over all the tube segments $s' = 0$ to $s' = L_T$ and normalized by L_T. The function $P_1(s, t; s', L_T)$ gives the probability that the point s' persists *and* that the end is at s at time t. The probability that s' still persists is therefore

$$P_2(t; s', L_T) = \int_{s=s'-L_T}^{s'} P_1(s, t; s', L_T) \, ds \quad (13.31)$$

$$P_2(t; s', L_T) = \frac{4}{\pi} \sum_{n \text{ odd}} \frac{1}{n} \sin\left(\frac{n\pi s'}{L_T}\right) \exp\left(-\frac{n^2\pi^2 D_T t}{L_T}\right) \quad (13.32)$$

The Doi–Edwards ansatz is that the stress, and therefore the stress relaxation modulus, is proportional to $f(t)$, the average fraction of persisting initial tube segments (19):

$$G(t) \propto f(t) = \int_0^L P_2(t, s'; L_T) \, ds' \quad (13.33a)$$

$$G(t) = G(0) \frac{8}{\pi^2} \sum_{n \text{ odd}} \frac{1}{n^2} \exp\left(-\frac{n^2\pi^2 D_T t}{L_T^2}\right) \quad (13.33b)$$

Here $G(0)$ is the plateau modulus level from which the slowest part of the stress relaxation occurs. The first, slowest decaying exponential dominates the behavior of $G(t)$ after the maximum $G(0)$ (22):

$$G(t) \approx G(0) \frac{8}{\pi^2} \exp\left(\frac{-\tau}{\tau_R}\right) \quad (13.34)$$

The average tube life is equal to $f(t)$ integrated over all time (23):

$$\tau_{\text{av}} = \int_0^\infty f(t) \, dt = \frac{1}{12} \frac{L_T^2}{D_T} \quad (13.35)$$

which is one-half of the reptation time of Eq. (13.15). Given the assumed proportionality between $G(t)$ and $f(t)$, the average tube lifetime is proportional to the zero-shear viscosity, assuming linear viscoelastic behavior (9), since

$$\eta_0 = \int_0^\infty G(t) \, dt = \frac{G(0)L_T^2}{12D_T} \quad (13.36)$$

with L_T proportional to N_0 and D_T inversely proportional to N_0, Eq. (13.36) predicts that a reptating polymer melt should have a zero-shear viscosity that increases with the third power of molecular weight. This is different from the observed 3.4 dependence shown in Figure 13.2. One possible source of discrepancy is the linear relation between the stress and the amount of persisting deformed tube. This possibility involves the unresolved and challenging question of how the perturbed microstructure of a material manifests itself in stress. A second source of possible discrepancy, polydispersity effects, will be briefly described. A third source is the approximations used to model the chain escape from the tube. The remainder of this chapter will concentrate on more detailed tube models. The approximation of constant tube length will be relaxed with Doi's theory of tube fluctuations. Next, the constraining nature of the tube will be explored and the concept of constraint release introduced, including the work of deGennes (19), Graessley (22), Klein (23), and others.

13.3. REPTATION: DETAILS AND VARIATIONS

13.3.1. Kink Motion and Fluctuations

In the previous section the basic reptation concepts were introduced. The chain was modeled as a flexible contour line of fixed length L_T constrained in a tube. This tube is a lumped representation of the topological constraints imposed on the particular chain by its surrounding chains. The fundamental chain motion is one-dimensional curvilinear diffusion within this tube with tube diffusion coefficient D_T. Given an initially occupied tube, the stochastic behavior for the gradual tube escape could be determined by applying first-passage boundary conditions to the diffusion equation. Because of the constraint of constant tube length, the amount of initial tube destroyed by chain escape is equal to the net amount of new tube created. For any given contour length this new tube is laid out in three dimensions according to a Gaussian distribution of new tube end-to-end vectors. From this model a center-of-mass diffusion coefficient was derived very simply and shown to vary inversely with the square of molecular weight for any particular homologous polymer series. Furthermore, the zero-shear viscosity was predicted to vary with the third power of molecular weight. Although the diffusion prediction agrees rather well with experimental results for entangled linear flexible polymers well above the glass transition temperature, the zero-shear viscosity prediction differs from the observed 3.4 power law dependence. This suggests the need to expand on the basic model of a snakelike chain of fixed contour length sliding in a one-dimensional tube.

There are two general ways of improving the model while maintaining the fundamental reptation idea: change the chain description or change the tube description. We begin with the former alternative and will show how such ideas can account for the anomalous viscosity behavior. The simple reptation model describes the chain's center-of-mass sliding motions in the tube, preaveraged over the internal motions of the chain. This allows one to use a picture of a polymer reptating as a straight stick in Figure 13.8. One result of this preaveraging is a constant chain contour length. Reintroducing these internal motions can result in a fluctuating contour length caused by the contracting and expanding of the individual subchains. Two methods for reintroducing those internal motions are to reexamine the kink or *chain-defect* migration concepts proposed by deGennes (10) and to model the chain as a Rouse chain within a tube as exploited by Doi and Edwards (14).

A real chain cannot slide down a ''tube'' as a rigid body. Figures 13.5a,b show how such motion for a very simple conformation would demand the crossing of a topological constraint. To move, the chain must alter its instantaneous conformation to conform with the tube configuration, as in Figure 13.5c. These conformational changes can only occur by the coordinated rotations of actual bonds within a subsection of the chain. To complete Figure 13.5, we have to replace the smooth contour with a diagram of the bonds between the chain's backbone atoms, as shown for a section of the chain in Figure 13.4. Figure 13.4 shows how one may envision the real chain as creating kinks of stored length about the time-averaged tube axis. (By stored, we mean that not all of the total possible length at maximal extension, subject to bond angle restrictions, contributes to the tube axis length representing the average position of those atoms. Furthermore, we mean time averaged over a period that is long compared to the lifetime of an instantaneous conformation of bond vectors r_i within dr_i and short compared to the tube escape or reptation time over which our tube segment is destroyed.) We may now associate a density of stored length with each value of the tube coordinate. These densities constantly fluctuate as the thermal motions of the backbone atoms move and change the shapes of the defects (kinks). Figure 13.7 shows how an end moving into the old tube destroys tube and stores length while an end moving out of the old tube creates tube and places length. Sliding motion down the tube represents the net effect of the transfer of stored length from one end of the tube to the other by defect (kink) motion.

DeGennes (10) formulated this basic picture and then proceeded to treat the defect motion as a one-dimensional flow of stored length or defect current along the tube. First, he assumed the presence of many defects per chain; second, he assumed that length is rapidly stored, placed, and transferred locally *relative* to the time required to transfer the stored length from one end to the other. Both of

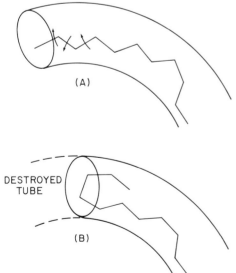

(A)

DESTROYED
TUBE

(B)

Figure 13.7. Kink motion: (*a*) initial tube end; (*b*) chain end moves into tube. Reverse order of (*a*) and (*b*) to create tube.

the conditions can be satisfied by "sufficiently" long chains. These assumptions permit the approximation of constant total stored length, and hence constant tube contour length, and the reduction of the net defect motion to simple one-dimensional diffusion: the basic reptation model. Thus, subject to the foregoing assumptions, thinking of reptation as one-dimensional kink motion along the chain or one-dimensional sliding of the chain in the tube leads to the same results.

If the local defect motion is not fast for all the defects relative to the overall end-to-end defect transfer time, the motion is not simple diffusion and the above correspondence is no longer in force. Wendel and Noolandi (24) have investigated the case where certain defects and their associated lengths are trapped for relatively long times. These traps make the chain motion along the tube resemble continuous-time random-walk processes that have been investigated in connection with other physical processes: Charge transport in solids, excitation energy migration in other materials, and relaxation in glasses (25). Such trapping results in fractional molecular weight dependences for both the viscosity and the diffusion coefficient. This argument can be used to obtain a 3.4 dependence for the viscosity, although this also tends to push the diffusion coefficient exponent to higher values as well. DeGennes (26) has investigated one possible physical realization of such traps via the creation of "tight knots" in the chain.

If the chain is not sufficiently long, the assumption of constant contour length becomes invalid. Doi (27) has approached this case by reformulating the chain as a Rouse assembly of beads and springs encased in a tube. We recall that the chain can be divided into subchains each containing N_R Kuhn statistical segments. The number of segments per Rouse spring, N_R, can be experimentally determined for any homologous series by fitting rheological data for melts of chains below the entanglement molecular weight M_e. From the Appendix, we note that the mean square end-to-end distance for each spring is $N_R b^2$. We define this quantity as r^2. Typically, it is found that there are several Rouse segments per entanglement. We shall denote this ratio N_e/N_R as N. As shown in Figure 13.8, the stretching of a Rouse spring corresponds to a decrease in stored length density while compression corresponds to an increase in density. Alternatively, the density decreases with decreasing bead density per contour length and increases with increasing bead density. Because the Rouse model works relatively well for chains below the entanglement weight, Doi assumes that subsections of longer chains below this weight also behave locally according to the Rouse model. This allows an estimation of three basic chain relaxation times: τ_w, τ_L, and τ_{rep}. The wriggling relaxation time τ_w represents the scale over which any subchain conformation involving N segments (springs) can exchange: It is the average time required by the subchain to sample its allowed ensemble of conformations within the tube. Because topological constraints define the boundary (walls) of the tube, such sampling motion within the tube is free Rouse motion. The approximate mean squared displacement within this sampled region is Nr^2. This size also corresponds to the square of the average size of one primitive path segment, a^2. The subchain has a Rouse "diffusion coefficient" over this region of $(k_B T/N\zeta_R)$. Thus, by Eq. (13.6),

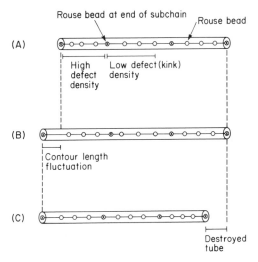

Figure 13.8. Contour length fluctuation: $N_p = 3$, $N_R = 13$: (a) initial configuration in tube; (b) transfer of stored length from interior to left subchain with accompanying tube expansion; (c) tube shrinkage and destruction of old tube end.

$$\tau_w \approx \frac{Nr^2}{k_BT/N\zeta_R} = \frac{r^2\zeta_R}{k_BT}N^2 \tag{13.37}$$

Since N is the number of Rouse segments per entanglement length, τ_w is independent of molecular weight and depends only on the particular polymer in question.

The subchains are not independent; in fact, the cumulative effect of their conformational changes are changes in the overall tube contour length L_T with corresponding relaxation time τ_L. Using reasoning similar to that above, Doi obtains

$$\tau_L \approx \frac{N_p(Nr^2)}{k_BT/N_pN\zeta_R} = \tau_wN_p^2 \tag{13.38}$$

Hence the contour length relaxation time varies with the second power of molecular weight. This is the same result as for the molecular weight dependence of the Rouse (2) relaxation time for obvious reasons. Finally, Doi compares this to the reptation time, derived by considering the time it takes for a Rouse chain to diffuse along a path of length L_T:

$$\tau_{\rm rep} \approx \frac{N_p^2a^2}{k_BT/N_pN\zeta_R} = \tau_wN_p^3 \tag{13.39}$$

This result is that of Eq. (13.15) obtained previously by similar methods.

If the tube length does not oscillate quickly about its mean as given by τ_L relative to the time scale for tube escape, $\tau_{\rm rep}$, the approximation of a constant tube length becomes poor. From Eqs. (13.38) and (13.39) we see that these time scales are separated by a factor on the order of N_p. When a chain is many entanglement lengths long, this factor becomes large, and the processes are well separated. In this case the basic reptation model is justified. Doi points out that typical values for N_p range from 20 to 100. These are not large enough to warrant complete neglect of fluctuations (28).

Fluctuations provide new opportunities for tube destruction. As in the case of constant contour length, tube is destroyed as it is vacated by the chain ends; however, the chain ends can now move independently of the center of mass. Returning to Figure 13.8, one could envision both end beads, 1 and 13, moving along the tube axis toward the tube center of mass, thereby destroying tube at both ends simultaneously without center-of-mass motion. We therefore see that contracting fluctuations decrease the instantaneous tube contour length and enhance tube destruction. (Even though expanding fluctuations increase the instantaneous tube contour length and enhance the new tube creation mechanism, it is the fraction of *old* tube that matters in stress relaxation by the Doi–Edwards hypothesis.) How large are these fluctuations? Assuming the Rouse segments (springs) have a Gaussian distribution of lengths and that each segment is independently varying, Doi (28) obtains the average total tube contour length variance:

$$\langle \delta L^2 \rangle \approx N_p(Nr^2) = N_p a^2 \tag{13.40}$$

Without fluctuations, the average tube lifetime, $\tau_{av}^{(NF)}$, is proportional to L_T^2/D_T [from Eqs. (13.15) and (13.35)]. Fluctuations alter the *effective* tube length. To a first approximation, this effective length is

$$L_E \approx L_T - \mu \langle \delta L^2 \rangle^{1/2} \tag{13.41}$$

where μ is an undetermined constant. The average tube lifetime including fluctuations is then approximately

$$\tau_{av}^{(F)} \propto \frac{L_E^2}{D_T} \tag{13.42}$$

Using Eq. (13.40) in (13.41) and Eq. (13.12) for L_T, one obtains

$$\tau_{av}^{(F)} \propto \tau_{av}^{(NF)} \left[1 - \mu \left(\frac{1}{N_p} \right)^{1/2} \right]^2 \tag{13.43}$$

If the term $\mu(1/N_p)^{1/2}$ is relatively small compared to unity, the effect of fluctuations on the tube escape is negligible. This occurs for sufficiently long chains, $N_p \gg \mu$. Sufficient is defined by the size of the parameter μ. Doi (28) shows that this parameter is no smaller than 1.47, and hence fluctuations appear to be important for typical chain sizes.

The foregoing ordering argument can be made rigorously. We will only outline the basic method presented by Doi (28). First, we recall that the equilibrium distribution function for $N + 1$ Rouse beads with N connecting springs, $\psi(\mathbf{R}_0, \mathbf{R}_1, \ldots, \mathbf{R}_N)$, satisfies the differential equation

$$\frac{\partial \psi}{\partial t} = \frac{k_B T}{\zeta_R} \sum_{n=0}^{N} \frac{\partial}{\partial \mathbf{R}_n} \cdot \left(\frac{\partial \psi}{\partial \mathbf{R}_n} + \frac{\psi}{k_B T} \frac{\partial U}{\partial \mathbf{R}_n} \right) \tag{13.44}$$

Now an entangled Rouse chain is in a tube. Doi imposes this constraint by converting Eq. (13.44) to a one-dimensional analog, with distribution function $\psi\{s_0, s_1, \ldots, s_N\}$ for the tube coordinates of the Rouse beads, in which the springs curve coincidently with the tube axis:

$$\frac{\partial \psi}{\partial t} = \frac{k_B T}{\zeta_R} \sum_{n=0}^{N} \frac{\partial}{\partial s_n} \left(\frac{\partial \psi}{\partial s_n} + \frac{\partial}{k_B T} \frac{\partial U}{\partial s_n} \right) \tag{13.45}$$

This assumes that the curvature of the tube axis does not affect the dynamics of the reptation. The potential energy is

$$U = \left(\sum_{n=1}^{N} \frac{3k_B T}{2r^2} (s_n - s_{n-1})^2 \right) - \frac{3k_B T}{a} (s_N - s_0) \tag{13.46}$$

The first term represents the spring potential. The second term arises from "squishing" the Rouse chain's spherical distribution into a tubular one (of effectively zero diameter). This term arises from the tension in the chain required to prevent the collapse of the Rouse chain to a point and thereby enforces the average tube length of L_T. Doi converts Eq. (13.45) using the normal coordinates:

$$\xi_0 = \frac{1}{N} \sum_{n=0}^{N} s_n \tag{13.47a}$$

$$\xi_p = \frac{2}{N} \sum_{n=0}^{N} \cos\left(\frac{p\pi n}{N}\right)\left(s_n - \frac{nL_T}{N} + \frac{1}{2}\right) \qquad p \geq 1 \tag{13.47b}$$

The zeroth normal coordinate represents the center-of-mass motion traced along the tube axis. The higher order coordinates represent fluctuation modes. Taking the continuous limit of the Rouse chain, so that p ranges to infinity, Doi obtains

$$\frac{\partial \psi}{\partial t} = \frac{k_B T}{N \zeta_R} \sum_{p=0}^{N} \frac{\partial}{\partial \xi_p} \left[\frac{\partial \psi}{\partial \xi_p} + \left(\frac{3}{2} \frac{Nr^2}{\pi^2 p}\right) \xi_p \psi \right] \tag{13.48}$$

As in simple reptation, a tube segment survives as long as neither end has yet penetrated to it. Doi therefore applies appropriate first-passage boundary conditions to Eq. (13.48) and writes the solution as an eigenfunction expansion. Integrating this solution over all ξ_i gives the fraction of old tube remaining:

$$f(t) = \int \psi(\xi_0, \ldots, \xi_p, \ldots, t)d\xi_0, d\xi_1, \ldots, d\xi_p, \ldots \qquad (13.49a)$$

$$= \sum_{i=0}^{\infty} C_i \exp(-\lambda_i t) \qquad (13.49b)$$

Using similar arguments as for Eqs. (13.33) and (13.34), Doi approximates this as

$$f(t) \approx C_0 \exp(-\lambda_0 t) = C_0 \exp\left(\frac{-t}{\tau_{av}^{(F)}}\right) \qquad (13.50)$$

where λ_0 is the smallest eigenvalue of Eq. (13.48) subject to the appropriate first-passage boundary conditions. This eigenvalue problem cannot be solved exactly analytically. Using a variational principle, Doi shows that the smallest eigenvalue takes the form of an expansion in N_p:

$$\lambda_0 = \frac{1}{\tau_{av}^{(NF)}}\left[1 + \mu\left(\frac{1}{N_p}\right)^{1/2} + \mu_1\left(\frac{1}{N_p}\right) + \cdots\right]^2 \qquad (13.51)$$

For large enough N_p (about 20) this may be truncated; hence,

$$\tau^{(F)} = \frac{1}{\lambda_0} \approx \tau_{av}^{(NF)}\left[1 - \mu\left(\frac{1}{N_p}\right)^{1/2}\right]^2 \qquad (13.52)$$

Finally, Doi's calculations show that $\mu \geq 1.47$.

Doi's treatment of fluctuations indicates that for chains up to 100 primitive path segments long, the tube escape time as a function of molecular weight is not a simple power relation. Rather, it is a product of the zero fluctuation escape time, $\tau_{av}^{(NF)}$, which scales with the third power of molecular weight, times a correction factor that approaches unity for large molecular weight. Choosing the smallest possible value for the smallest eigenvalue will provide the largest possible value for $\tau^{(F)}$ in Eq. (13.52) and represents the most conservative estimate on the importance of fluctuations. Using this value, Doi shows that Eq. (13.52) resembles a 3.4 power law for N_p between 20 and 100 and approaches a 3.0 law as $N_p \rightarrow \infty$.

This analysis suggests an important experiment. The zero-shear viscosity, or some other property that directly reflects the longest relaxation time, should be measured on several very high molecular weight samples to determine whether or not this 3.4 power dependence persists to a high molecular weight. If there is a crossover to a cubic power dependence, this would lend credence to the fluctuation arguments.

13.3.2. Branched Polymers

Fluctuations may also be the key to understanding the relaxation phenomena of branched polymers. In 1975 deGennes pointed out that long-chain branches would quench the reptation (29). Figure 13.9 shows that the long branch encounters topological constraints when the main chain backbone contour tries to slide in its tube. DeGennes suggested that a chain could only move when the chain arm retraces its path to the branch point, that is, when the chain branch essentially assumes a zero contour length by fluctuation. The probability of a fluctuation resulting in zero contour length decreases exponentially with branch length. Helfand and Pearson (30) modeled the rate of this fluctuation as the diffusion of the chain end in an exponentially varying potential field.

Up until now, we have focused on the motions of a single chain confined to a tube. Given this model, we have shown how reptation theory can account for the observed molecular weight dependences of the diffusion coefficient and the zero-shear rate viscosity. These calculations rely on the essentially one-dimensional nature of the chain motion. What we have not addressed is why the topological constraints appear at the entanglement weight and hence why reptation prevails over simple Rouse motion at and above this critical chain length value, nor have we presented a method for predicting this critical value. These questions are topics of active research concerning the nature of the topological constraints and their apparently tubelike manifestation. We now attempt a brief summary of current efforts.

13.3.3. Computer Simulation

Several attempts have been made to observe reptation in computer simulation. The topological constraints are introduced as a lattice or mesh of obstacles. The chain

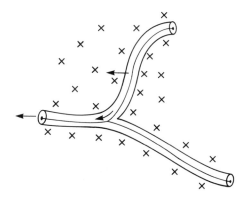

Figure 13.9. A three-armed chain. Reptation is quenched because the third arm would have to either move through a constraint or restrict itself to the same tube as the other arms.

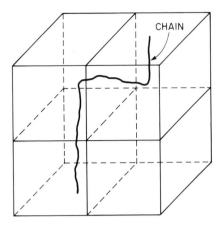

Figure 13.10. Chain in obstacle mesh. The chain weaves between centers of adjacent cubes.

must weave its way through this mesh as shown in Figure 13.10. The chain is modeled as a series of discrete segments that may hop according to particular rules. Thus, there are two length scales in the problem, one for the obstacles and the other for the chain. The object is to determine at what obstacle concentration (mesh size) and chain length the chain center of mass stops moving isotropically and begins moving preferentially along its axis, that is, reptation. More elaborate simulations have replaced the single chain in a fixed obstacle mesh with many chains. Instead of varying the mesh size, the chain number density is varied. Here the natural maximum density is that of a melt. In practice, because of the large computing requirements, it is difficult, even with Monte Carlo simulations, to achieve the essential conditions (large enough chains, numbers of chains) to simulate entangled systems. The first efforts in this direction used chains too short to be entangled (31–35), and in some cases (34, 35) the diffusion times studied were too short to obtain center-of-mass diffusion information. More recently longer chains have been studied. The Jülich group (36–39) has studied the collective motion of systems of chains of as many as 200 units packed to 0.344 volume fraction on a diamond lattice. Their simulation data show that the time evolution of monomer displacement is as though the chains were unentangled, that is, not in accordance with the reptation model, in simulations where all chains in the system were moving. If all chains but one were frozen in place (38, 39) the motion of the moving chain does resemble reptation in the time evolution of $\langle r^2 \rangle$ and in the molecular weight dependence of $D_s \propto M^{-2}$. This latter condition of frozen surroundings naturally favors the appearance of reptation behavior and is not representative of a real melt. Deutsch and Goldenfeld (40–42) have criticized some of the above work (36, 37) as not being a valid test of the reptation ideas. Deutsch has produced a simulation suggesting reptation diffusion that has similarly been criticized by Baumgärtner, Kremer, and Binder (38, 39, 43). The difficulty and source of this controversy seems to be at least in part that it is currently difficult due to computer time and space limitations to get *well* into the simulation regime where entangled, and therefore reptation, behavior is expected.

13.3.4. Constraint Release

What these simulations do seem to confirm is the hypothesis that the time duration of the topological constraints (from uncrossability) relative to the speed of the chain motions plays an important role in reptation onset. Onset theories have centered on this concept. Since the topological constraints are themselves moving chains, they can move away from their current positions, even if only momentarily, and therefore allow motion *past* them. This *constraint release* mechanism provides a method for finding the *breakdown* of reptation from a dismantling of the constraining *old* tube in favor of laterally *shifted* new tubes over times less than the tube escape time. (This mechanism is also termed *tube renewal* in much of the literature. We avoid this particular terminology to avoid confusion with new tube creation through tube escape.)

DeGennes first suggested that constraint release was caused by the reptation of a constraining chain (29). In particular, the end segment of the chain is constantly changing its position through the tube escape process. Figure 13.11 shows how a *constraining* chain could reptate away from and then return near the given chain but on the "opposite" side. The given chain has thereby "crossed" the constraint and has been able to make a small lateral displacement. If enough constraints release, the cumulative effect of this piecewise motion could be a substantial lateral shift of the center of mass without tube escape.

One problem with discrete-constraint approaches is in knowing exactly what is meant by a constraint. Typically, one assumes that the neighboring chains each represent a potential constraint. This leads to apparent contradictions in the theory if one associates the entanglement length with the distance between *possible* constraints. Instead, one must deal with *effective* constraints. It is not implausible, especially in light of the simulation results, to expect that it takes many possible

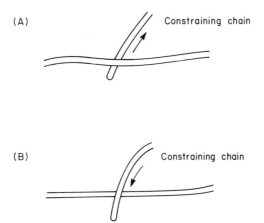

Figure 13.11. Constraint release. As proposed by deGennes, the end of a constraining chain can move, allowing the other chain to slip across the constraint.

constraints in the form of neighboring chains to form an effective system of constraints inducing anisotropic motion, that is, reptation. Klein (23) has developed an approximate method for calculating the number of constraints per chain below which every chain moves too rapidly to effectively constrain its neighbors. Klein extends the deGennes (29) concept of constraint release by allowing all of the constraints forming a tube to release through reptation of the constraining chains. This is a self-consistent approach in the sense that all chains in a (monodisperse) system are moving in the same way. Reptation then sets the time scale for the motion of each chain and therefore the rate at which one chain can move past its neighbor to release a constraint. Klein divides the chain into N Rouse subchains. He then assumes that a single surrounding chain is responsible for the constraint in the local region of each subchain, that is, the possible constraints are spaced every Rouse segment. (Klein's development was originally posed in an entangled solution where this assignment may be more defensible than in a melt.) Klein now adopts a relaxation time analysis similar to Doi. He defines $N_{R,k}$ as the number of Kuhn segments per Rouse segment, b as the Kuhn length, and N as the number of Rouse segments in the whole chain. Each Rouse segment can rearrange its configuration within its own volume of approximate diameter $(N_{R,k}b^2)^{1/2}$ in analogy to Eq. (13.37):

$$\tau_{w,\mathrm{cr}} \approx \frac{N_{R,k}b^2}{k_BT/N_{R,k}\zeta_k} = \frac{b^2\zeta_k}{k_BT}N_{R,k}^2 \tag{13.53}$$

Although the subchains may rearrange continuously according to Eq. (13.53), the net center-of-mass displacement for any subchain may not leave its initially occupied volume (resulting in a time-averaged displacement of zero) until it moves into an adjacent subchain volume by reptation or until the constraint on the subchain is released. In the latter case the subchain is released. In the latter case, the subchain may jump to a new volume with a subchain center-of-mass displacement on the order of $(N_{R,k}b^2)^{1/2}$. This means that the average time for a subchain to move its time-averaged center of mass the length of one constraint is the average lifetime of that constraint. The constraint release occurs through reptation of the constraining chain. The constraint therefore persists as long as the tube of the surrounding chain persists. More specifically, the constraint corresponds to a portion of this tube with tube coordinate s'. Since each section of tube can act as a constraint for some other chain, Klein associates the constraint lifetime with the average lifetime of a tube segment. Equation (13.35) gives the average lifetime of an instantaneous tube segment averaged over all of the tube segment positions (from $s = 0$ to $s = L_T$). If we assume tube destruction by reptation only, then

$$\tau_{\mathrm{jump}} \approx \tau_{\mathrm{av}} = \frac{L_T^2}{12D_T} \tag{13.54}$$

This has an N^3 dependence, in contrast to Eq. (13.53). In analogy to Eq. (13.38), with τ_{av} substituted for τ_w, the relaxation time for the entire chain (over a sphere of radius $\langle R^2 \rangle^{1/2}$) by constraint release jumps alone is

$$\tau_{cr} \approx \frac{2N^2}{3\pi^2} \tau_{av} = \frac{\zeta_R N_{R,k} b^2}{18\pi^2 k_B T} N^5 \tag{13.55}$$

which has an N^5 dependence on chain length. In accordance with Eq. (13.6), the center-of-mass diffusion coefficient where constraint release is the sole mechanism can be found as

$$D_{c,cr} \approx \frac{(N \cdot N_{R,k} b^2)}{6\tau_{cr}} = \frac{3\pi^2 k_B T}{\zeta_R} \frac{1}{N^4} \tag{13.56}$$

where the product $N \cdot N_{R,k} b^2$ is $\langle R \cdot R \rangle$. Hence the constraint release contribution to diffusion decreases with the fourth power of molecular weight. Actually Eqs. (13.54)–(13.56) are only first approximations, since Klein notes that *each* chain moves via both reptation *and* constraint release. Hence it is really not precise to make the association of τ_{jump} with τ_{av} in Eq. (13.54). Rather τ_{jump} is a *smaller* value since the constraint release motion of the constraining chains can also cause the disappearance of a constraint. Therefore, we retain τ_{jump} in Eqs. (13.55) and (13.56):

$$\tau_{cr} \approx \frac{N^2}{3\pi^2} \tau_{jump} \tag{13.57}$$

$$D_{cr} \approx \frac{N \cdot N_{R,k} b^2}{6\tau_{cr}} \tag{13.58}$$

The observed overall center-of-mass motion, hence D_{obs} and τ_{obs}, are the result of motion via both reptation and constraint release:

$$D_{obs} = f(D_{rep}, D_{cr}) \tag{13.59a}$$

$$\tau_{obs}^{-1} = \frac{6}{\langle R^2 \rangle} f\left(\frac{\langle R^2 \rangle}{2\tau_{rep}}, \frac{\langle R^2 \rangle}{2\tau_{cr}} \right) \tag{13.59b}$$

Klein assumes, *given* the values of τ_{rep} and τ_{cr} (whose relative sizes are interdependent), that each mechanism contributes independently to center-of-mass motion:

$$D_{obs} = D_{rep} + D_{cr} \tag{13.60a}$$

$$\tau_{obs}^{-1} = \tau_{rep}^{-1} + \tau_{cr}^{-1} \tag{13.60b}$$

Next he assumes that the relation between τ_{rep} and τ_{av} hold analogously for τ_{obs} and τ_{jump} (true only in the limit of $\tau_{rep} \ll \tau_{cr}$); thus,

$$\tau_{jump} = \frac{1}{6} \tau_{obs} \tag{13.61}$$

Substituting Eq. (13.61) in (13.60b) and rearranging,

$$\frac{\tau_{obs}}{\tau_{rep}} = \frac{N^2(\tau_{obs}/\tau_{rep})}{N^2(\tau_{obs}/\tau_{rep}) + 18\pi^2} \tag{13.62}$$

Klein notes that this equation has no nonzero solution for $N^2 \leq 18\pi^2$. He interprets this to mean that below this critical chain length value there is no *consistent* reptation solution. Above this critical value the solution bifurcates and the nonzero dimensionless relaxation time represents the reptation solution:

$$\frac{\tau_{obs}}{\tau_{rep}} = 1 - \frac{18\pi^2}{N^2} \tag{13.63}$$

The ratio τ_{obs}/τ_{rep} represents a measure of the sizes of the contributions of the two relaxation mechanisms: constraint release and reptation. As the latter begins to dominate, this ratio approaches unity; as the former dominates, it approaches zero. Because of the nature at the analysis, the exact critical value of $\sqrt{18\pi} \approx 13$ should not be considered too seriously; nonetheless, Klein's argument does illustrate the cooperative nature of the chain motions leading to *effective* topological constraint, which can induce anisotropy.

Graessley (22) has developed a model similar to Klein's and has accounted for more than one constraint per subchain. There may be z surrounding constraints, each of which bars a region on the order of a subchain. The chain will move into the first available region; thus, constraint release is accelerated. Graessley finds D_{cr} to increase by a factor of approximately $z(12/\pi^2)^z$. Graessley has extended the approach to the relaxation of branched polymers. It is important to note that the number of constraints, z, does not alter the molecular weight dependence of the constraint release contribution to diffusion.

An alternative to Klein's method suggested by Noolandi and Desai (44) began with an expression similar to Eq. (13.57):

$$D_{obs} = F_e(M)D_{rep}(M) + [1 - F_e(M)]D_{Rouse}(M) \tag{13.64}$$

where D_{Rouse} represents the contribution from isotropic Rouse-like motion (constraint release). They define $F_e(M)$ as the effective number of entanglements per unit volume for a given molecular weight normalized by the effective number at infinite molecular weight, E_∞. This must be a monotonically increasing function of M. Hopefully, the form of $F_e(M)$ will be a universal function; that is, the shape should be independent of the chemical nature of the polymer species. Graessley and Edwards (45) have studied the value of E_∞ for a wide variety of actual species. They conclude that it is indeed a universal function within experimental error. The important scaling dimensionless group is $\nu L_T b^2$, where ν is the number of chains per unit volume, L_T is the tube contour length, and b is the Kuhn segment length. One disadvantage of this method is the lack of an a priori method for determining the important range of M over which $F_e(M)$ varies, which Klein provided using self-consistency. This may be turned, however, into an opportunity to determine this important function experimentally.

Marrucci (46) has developed a novel alternative to the discrete-constraint models of deGennes, Klein, and Graessley. The tube radius represents the mean square excursion displacement over the lifetime of the tube. Given a particular tube at a certain time, the escape process described previously shows that the center of this initial tube is the *most probable* section to persist for the longest period of time. Therefore, the chain occupying this segment has the greatest probability to displace the farthest if limited motion beyond the finite bounds of the wriggling is allowed radially to the tube axis. Marrucci concludes that this additional displacement may be described by a variable tube radius, narrowest at the chain ends and widest at the center. Marrucci has used this concept to form constitutive relations for stress relaxation of polydisperse samples and polymer blends (47).

13.3.5. Polydispersity Effects

We have discussed polymer properties as a function of molecular weight. We have implicitly assumed that the polymer sample is monodisperse: All the chains are of the same molecular weight. Monodispersity is an idealization that can be only approached. In fact, most commercial polymers are widely polydisperse. At best, one may hope that an appropriate average molecular weight might be able to describe the given real sample and then use this average molecular weight in the derived monodisperse equations. Alternatively, one might apply the equation to each molecular weight value and average the results. There are two common averages: chain number average molecular weight ($\overline{M}_n$) and chain length (or molecular weight) average molecular weight ($\overline{M}_w$). These are defined as

$$\overline{M}_n = \sum_i F(M_i)M_i \tag{13.65a}$$

$$\overline{M}_w = \frac{\sum_i F(M_i)M_i^2}{\sum_i F(M_i)M_i} \tag{13.65b}$$

where M_i is the molecular weight of a particular species of the series, $F(M_i)$ is the fraction of total chains by number that are of the ith species, and the sum is over all the different chain weights in the sample. These equations may be approximated by using a continuous distribution function for the molecular weights, $f(M_i)$:

$$\overline{M}_n = \int M_i f(M_i) \, dM_i \qquad (13.66a)$$

$$\overline{M}_w = \frac{\int M_i^2 f(M_i) \, dM_i}{\int M_i f(M_i) \, dM_i} \qquad (13.66b)$$

The number average derives from colligative properties that count molecules by number. The weight average derives from size-dependent properties. One might believe the number average is appropriate for the diffusion coefficient (one center of mass per chain). However, experiments typically measure monomer concentration changes or rely on size-dependent probes (i.e., light scattering); hence, it is more practical to use the molecular weight average molecular weight. Likewise, viscoelastic properties are related to the tube escape. It is not the fraction of tube escaped per tube but the length of tube escaped that is important. Again, this suggests the use of a molecular weight average molecular weight. Now suppose there is a certain property $H(M)$. If one has no knowledge of the distribution of molecular weights, only the average, one may write

$$\overline{H(M)} = H(\overline{M}) \qquad (13.67)$$

If one has the distribution, it is more proper to write

$$\overline{H(M)} = \int M_i f(M_i) H_n(M_i) \, dM_i = \int f(M_i) H_w(M_i) \, dM_i \qquad (13.68)$$

In Eq. (13.68), H_w should be a molecular-weight-dependent property such as "the length of tube escaped" as opposed to "the fraction of tube escaped," H_n.

If all of the chain sizes in a polydisperse sample are very long relative to the entanglement length, none of them "move fast enough" for constraint release to be important. Each chain sees an effectively permanent tube, and this constraint immobility is the only important property of the surrounding chains. Therefore, each chain escapes independently according to its own chain size diffusion coefficient. Equation (13.68) may be used to calculate tube escape-related properties (48). [It should be noted that the additional assumption of homogeneity in the mixing of the i species is implicit in Eq. (13.68).] The longer chains diffuse more slowly than the shorter chains. If there were separate regions of long and short chains, for example, as in the welding of such polymers, the density in the short-chain region would decrease while that of the long-chain regions would increase. To maintain constant density, a compensating backflow of segments into the shorter chain region would be needed. This is discussed by Brochard (49). A homogeneous mixture would only experience transient density fluctuations.

If some of the chain sizes in the polydisperse sample are near or even less than the entanglement molecular weight, the simple additivity principle fails miserably. Constraint release and dilution become very important. In the most extreme case suggested by Daoud and deGennes (50), consider a mixture of very long N chains in a melt of very small P chains. If P chains are the simple monomer, this is the case of a solution of very long chains that could behave as free Rouse chains. It should be noted that if the N chains are not too dilute and if they are long enough, they may still be entangled among themselves regardless of the P chains. Reptation is then applicable with a suitable adjustment to the tube diffusion coefficient and an inclusion of the expansion factor (excluded volume) due to the smaller chains on the chain configuration. The idea of constraint release also becomes a more critical issue. Polydisperse mixtures are fertile territory for new theoretical efforts. Thorough discussion of solution behavior is outside the scope of this chapter. Solution dynamics have been discussed by deGennes (5, 19, 51) and Tirrell (6), among others.

13.4. CONCLUDING REMARKS

This chapter describes the current status of the reptation model for entangled, linear, flexible polymer melts. There are alternatives to the tube concept explored here for using the reptation idea. Particularly noteworthy is the work of Bird and co-workers (52, 53), who avoid the tube explicitly. In its stead they use an anisotropic friction and Brownian motion that can also mimic the intermolecular interactions leading to diffusion primarily along the chain contour. This model, with a completely different physical basis, although with some similar physical hypotheses, leads to very similar results to the Doi and Edwards tube model (14). This is both satisfying and useful since it also provides an alternative means to explore variations, modifications, and improvements on the basic reptation idea.

While this chapter has described the basic reptation model and several variants in detail, it has not provided detailed comparisons with experimental data. This is done elsewhere (4, 6, 22). Briefly, however, experimental evidence is definitive on two points concerning reptation: (1) it does provide a coherent, comprehensive and, in some respects, quantitatively accurate picture of many aspects of polymer dynamics and (2) it is not the complete story. Several observations lead to the second conclusion. As shown in Figure 13.12, observed viscosities are always lower than predicted by the reptation model, approaching the cubic power law from *below*. Mechanical compliances, another measure of the efficacy of molecular relaxation processes, are always higher than reptation alone predicts. Branched polymers and ring polymers, neither of which can reptate effectively, still exhibit significant relaxation rates (22). These facts and others strongly suggest that the entangled polymer chain has somewhat more relaxation means available to it than pure reptation. Whether these additional means are among the variants of the basic tube model discussed here or require yet newer physical insight is the subject of much current research on polymer dynamics.

In this same vein lies what we believe to be the current central problem of the dynamics of entangled polymer systems. That is, what are the necessary conditions for entanglement or reptation effects to manifest themselves? Put in a different way, we would say that *there is no theory of reptation*, in the sense of a theory

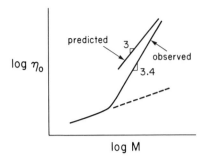

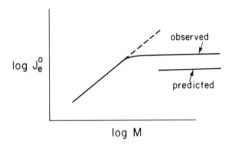

Figure 13.12. Predictions of viscosity η_0, and recoverable compliance J_e^0, by the Doi–Edwards reptation theory compared with experimental observations. The predicted η_0 is too large, but its chain length dependence is weaker than observed. The predicted J_e^0 is too small but is independent of chain length, as observed. The dashed lines indicate predictions of the Rouse model. (Adapted from Ref. 22.)

having reptation as a result instead of as an input. All current work "legislates" reptation from the beginning and then explores its consequences. Only until a more complicated many-chain problem can be solved in more detail (54), leading to the critical conditions for its observation, will we really understand the nature of the reptation model.

This is only one of the research opportunities flowing from the work described in this chapter. On the purely theoretical side, the recent publications of Edwards (54, 55) are provocative efforts in the direction of a more fundamental understanding of the reptation hypothesis. The experimental tools and data base are improving rapidly (6, 56). On the applications side, studies of inter- (18, 57) and intra-molecular (cyclization) (58, 59) reactions of polymers, especially photochemical or photophysical, promise to provide a wealth of practical and conceptual information.

APPENDIX

Polymers are long-chain molecules consisting of a series of repeating chemical units or *monomers*. The monomer type determines an entire homologous series of similar species varying only in the number of monomers per chain, N_0, and hence also in molecular weight. Because polymers are very long chain molecules (N_0 can easily be 10,000), they can assume a huge number of conformations. These circumstances have led to the modeling of the probability distribution of the possible conformations of the chain backbone atoms (ignoring the secondary importance of the local substituent conformations) using the theory of random walks. This allows the determination of appropriate average quantities. Most important of these is the end-to-end vector between the two chain ends and its associated magnitude the end-to-end distance. Another is the chain's radius of gyration (moment of inertia) about its instantaneous center of mass. Flory (17, 60) has extensively covered this topic, and we only highlight the principal ideas here.

Starting with one end and proceeding sequentially to the other end, each bend in the polymer can be considered a single step in a random walk of n steps, with n equal to the total number of backbone bonds. Each step has the corresponding bond length and has a certain orientational probability that is a function of bond angles and steric interactions between the atom at the step's (bond's) end and all previously placed atoms. In principle, this allows an a priori determination of the orientational distribution function for each step vector, r_i. As shown in Figure 13.13, the chain has thus been modeled as a series of n "bond vectors," r; the end-to-end vector, R, is just the sum of these n vectors. The spatial description is completed with the position vector, r_0, of the first end's atom. If each step represents the same kind of bond and the orientational distribution of any step, r_i, is only a function of the orientation of the previous step, r_{i-1}, the walk falls into the general mathematical category of a Markov process (61). The distribution function for the net end-to-end displacement then can be shown to approach a "Gaussian" distribution as the number of steps increase toward infinity by applying the central limit theorem of probability theory (62); that is,

$$\lim_{n \to \infty} P(|\mathbf{R}|\,n) = (2\pi\langle R^2\rangle)^{-3/2} \exp\left[-\frac{(|\mathbf{R}| - n\mu)^2}{2\langle R^2\rangle}\right] \tag{13.69}$$

The mean displacement of an individual step, μ, is zero in an isotropic medium. In the simplest case of a walk with fixed step (bond) length b and isotropic orientation (the so-called freely hinged case in which bond angle and steric interactions are completely ignored), the mean square end-to-end displacement is

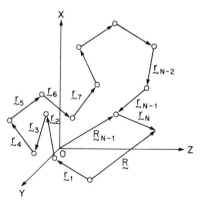

Figure 13.13. Chain conformation. The chain is represented by N bond vectors, r_i, along the backbone. The position vectors, $\mathbf{R}_i$, connect the coordinate origin and the head of each vector. $\mathbf{R}_{N-1}$ is shown. The end-to-end vector is $\mathbf{R}$.

$$\langle R^2 \rangle = nb^2 \tag{13.70}$$

In practice, the actual distribution for a random walk closely resembles a Gaussian when the number of steps equals or exceeds 10 (60).

When bond angles θ are accounted for, the mean square displacement for large n is (17)

$$\langle R^2 \rangle = nb^2 \frac{1 - \cos \theta}{1 + \cos \theta} \tag{13.71}$$

For the tetrahedral angle of 109.5° this becomes (17)

$$\langle R^2 \rangle = 2nb^2 \tag{13.72}$$

Thus, a chain with tetrahedrally bonded chain atoms would have twice the mean square displacement as a freely hinged idealized chain with the same number of identical steps. As the chain becomes stiffer because of steric interaction between side groups appended to the backbone atoms, one expects the chain to spread out spatially and thus for the mean square displacement, $\langle R^2 \rangle$, to become larger. One can write

$$\langle R^2 \rangle = cnb^2 \tag{13.73}$$

where c is a constant greater than 1 for a real chain and is called the characteristic ratio.

Kuhn (63) has shown that, given any mean square displacement for a real chain as a function of chain length (steps), one can replace any sterically hindered walk by an equivalent freely hinged chain by requiring the mean square displacements to be equal and the contour length of the fully extended real chain to equal the maximum possible displacement of the freely hinged walk:

$$n_k b_k^2 = \langle R^2 \rangle = cnb^2 \tag{13.74a}$$

$$n_k b_k = \text{contour length} \equiv L_c \tag{13.74b}$$

The value of n_f is the number of "Kuhn statistical segments" in the chain. The actual number of bonds represented by one statistical segment must be greater than 1: The stiffer the chain, the more bonds per segment. The Gaussian approximation to the real chain will be reasonable only if the number of chain atoms exceeds that required to construct at least 10 of these statistical segments. Since the end-to-end distance is more important than the maximum extension length, other criteria are often substituted for Eq. (13.74). Commonly n is chosen as the actual number of monomers in the chain; then b becomes an effective monomer length (64).

We can now assign a value to the mean square spring length. If the Rouse bead contains at least 10 Kuhn segments, this is simply given by Eq. (13.73) with n corresponding to the backbone bonds of the N_p monomers. We still need an expression for the bead diameter. The mass of the subchain is concentrated in the beads. The effective size that this mass presents to hydrodynamic friction is related to the radius of gyration. The bead diameter d_i is twice this radius.

The radius of gyration, S, is defined as the rms magnitude of the vector, S_i, joining the i bead positions with the chain center of mass:

$$S^2 = (N_R/N_0) \sum_{i=1}^{N_R/N_0} S_i^2 \tag{13.75}$$

We note further that

$$\sum_{i=1}^{N_R/N_0} S_i = 0 \tag{13.76}$$

and

$$S_i = \mathbf{R}_i - \mathbf{R}_G = \mathbf{R}_i - \frac{1}{N} \sum_{n=1}^{N_R/N_0} \mathbf{R}_i \tag{13.77}$$

The average of this radius of gyration is related to the average end-to-end displacement (65, 66),

$$\langle S^2 \rangle = \tfrac{1}{6} \langle R^2 \rangle \tag{13.78}$$

Therefore, the average radius of gyration increases linearly with the square root of the number of monomers in a chain just as did the end-to-end displacement [recall Eqs. (13.70)–(13.73)].

The previous statements pertaining to conformations only apply strictly to polymers in a melt of identical chains (including length) and to polymers in a solvent under "theta conditions." One reason for deviation from this simple random-walk behavior is the fact that polymer chains occupy finite volume; two sections of the same chain may *not* occupy the same space. A self-avoiding random walk is more appropriate to include the "excluded volume" problem. The end-to-end displacement and radius of gyration do not simply scale with the number of monomers in a chain when described by a self-avoiding walk. The statistics of such walks is still a topic of active research. In a melt other chains occupy the volume not filled by a single chain's random walk, and this seems to screen out the excluded volume effect. Chains in a homogeneous melt do seem to obey random walk, that is, Gaussian, statistics. In a solvent an additional effect is that polymer–solvent interactions are thermodynamically more favorable than polymer–polymer interactions, which tends to expand the coil (swelling). The theta condition, which exists in only certain solvents particular to each polymer, is where the solvent is sufficiently poor to contract the chain below the self-avoiding walk dimensions, and Gaussian statistics are again obeyed (7). Since this chapter is primarily concerned with melts, Gaussian statistics are assumed throughout.

REFERENCES

1. J. Kirkwood and J. Riseman, *J. Chem. Phys.* **16,** 565 (1948); cf. J. G. Kirkwood, *Macromolecules*, Gordon & Breach, New York, 1967.

2. P. Rouse, *J. Chem. Phys.* **21,** 1272 (1953).

3. B. Zimm, *J. Chem. Phys.* **24,** 269 (1956).

4. R. B. Bird, O. Hassager, R. C. Armstrong, and C. F. Curtiss, *Dynamics of Polymeric Liquids: Kinetic Theory,* Vol. 2, Wiley, New York, 1986 (revision to appear).

5. (a) J. D. Ferry, *Viscoelastic Properties of Polymers*, 3rd ed., Wiley, New York 1980; (b) P.-G. deGennes, *Scaling Concepts in Polymer Physics*, Cornell University Press, Ithaca, NY, 1979.

6. M. Tirrell, *Rubber Chem. and Tech.* **57,** 523 (1984).

7. J. Klein and B. Briscoe, *Proc. Roy. Soc. Lond. A* **365,** 53 (1979).

8. G. Berry and T. Fox, *Adv. Polym. Sci.* **5,** 261 (1968).

9. W. W. Graessley, *Adv. Polym. Sci.* **16,** 1, (1974).

10. P.-G. deGennes, *J. Chem. Phys.* **55,** 572 (1971).

11. S. F. Edwards, *Proc. Phys. Soc. Lond.* **91,** 513 (1967).

12. S. F. Edwards, *Proc. Phys. Soc. Lond.* **92,** 9 (1967).

13. S. F. Edwards, *J. Phys. A: Gen. Phys.* **1,** 15 (1968).

14. M. Doi and S. F. Edwards, *J. Chem. Soc. Faraday Trans. II* **74,** 1789, 1802, 1818 (1978).

15. M. Doi and S. Edwards, *J. Chem. Soc. Farday Trans. II* **74,** 568, 918 (1978).

16. W. W. Graessley, *J. Polym. Sci. Polym. Phys. Ed.* **18,** 27 (1980).

17. P. J. Flory, *Principles of Polymer Chemistry*, Cornell University Press, Ithaca, NY, 1953.

18. T. Tulig and M. Tirrell, *Macromolecules* **14,** 459 (1982).

19. P.-G. deGennes and L. Léger, *Ann. Rev. of Phys. Chem.* **33,** 49 (1982).

20. S. Prager and M. Tirrell, *J. Chem. Phys.* **75,** 5194 (1981).

21. D. Adolf, *J. Polym. Sci. Polym. Phys. Ed.* **23,** 413 (1985).

22. W. W. Graessley, *Adv. Polym. Sci.* **47,** 68 (1982).

23. J. Klein, *Macromolecules* **11,** 852 (1978).

24. H. Wendel and J. Noolandi, *Macromolecules* **15,** 1318 (1982).

25. G. C. Martinez-Mekler and M. A. Moore, *J. Phys. Lett.* **42,** 413 (1981).

26. P.-G. deGennes, *Macromolecules* **17,** 703 (1984).

27. M. Doi, *J. Polym. Sci. Polym. Phys. Ed.* **18,** 1005 (1980).

28. M. Doi, *J. Polym. Sci. Polym. Phys. Ed.* **21,** 667 (1983).

29. P.-G., deGennes, *J. Phys.* **36,** 1199 (1975).

30. E. Helfand and D. Pearson, *Macromolecules* **17,** 888 (1984).

31. K. Binder (Ed.), *Monte Carlo Methods in Statistical Physics*, Vol. 2, Springer, Berlin, 1983.

32. A. Baumgartner and K. Binder, *J. Chem. Phys.* **75**, 2994 (1981).

33. M. Bishop, D. Ceperley, H. L. Frisch, and M. H. Kalos, *J. Chem. Phys.* **76**, 1557 (1982).

34. K. E. Evans and S. F. Edwards, *J. Chem. Soc. Faraday Trans. II* **77**, 1891 (1981).

35. K. E. Evans, *J. Polym. Sci. Lett.* **20**, 103 (1982).

36. D. Richter, A. Baumgartner, K. Binder, B. Ewen, and J. B. Hayter, *Phys. Rev. Lett.* **47**, 109 (1981).

37. D. Richter, A. Baumgartner, K. Binder, B. Ewen, and J. B. Hayter, *Phys. Rev. Lett.* **48**, 1695 (1982).

38. K. Kremer, *Macrmolecules* **16**, 1632 (1983).

39. A. Baumgartner, K. Kremer, and K. Binder, *Faraday Symp. Chem. Soc.* **18**, 37 (1983).

40. J. M. Deutsch and N. D. Goldenfeld, *Phys. Rev. Lett.* **48**, 1694 (1982).

41. J. M. Deutsch, *Phys. Rev. Lett.* **49**, 926 (1982).

42. J. M. Deutsch, *Phys. Rev. Lett.* **51**, 1924 (1983).

43. K. Kremer, *Phys. Rev. Lett.* **51**, 1923 (1983).

44. J. Noolandi and R. Desai, *Makromol. Chem. Rapid Comm.* **5**, 453 (1983).

45. W. W. Graessley and S. F. Edwards, *Polymer* **22**, 1329 (1981).

46. G. Marrucci, *Adv. Trans. Proc.* **5**, (1984).

47. G. Marrucci, *J. Polym. Sci. Polym. Phys. Ed.* **23**, 159 (1985).

48. S. Prager, D. Adolf, and M. Tirrell, *J. Chem. Phys.* **78**, 7015 (1983).

49. F. Brochard, J. Jouffray and P. Levinson, *J. Phys. Lett.* **44**, 1455 (1983).

50. M. Daoud and P.-G. deGennes, *J. Polym. Sci. Polym. Phys. Ed.* **17**, 1971 (1979).

51. P.-G. deGennes, *Phys. Today* **36**, 33 (1983).

52. C. F. Curtiss and R. B. Bird, *J. Chem. Phys.* **74**, 2016 (1981).

53. R. B. Bird, H. H. Saab, and C. F. Curtiss, *J. Chem. Phys.* **76**, 1102 (1982); **77**, 4747 (1982); **77**, 4758 (1982).

54. S. F. Edwards, *Faraday Soc. Discuss.*, **18**, 1 (1983).

55. S. F. Edwards, *Polymer* **26**, 163 (1985).

56. E. D. von Meerwall, *Rubber Chem. Tech.* **58**, 527 (1985).

57. P.-G. deGennes, *J. Chem. Phys.* **76**, 3316 (1982).

58. M. A. Winnik, X. B. Li, and J. E. Guillet, *Macromolecules* **17**, 699 (1984).

59. J. Noolandi, K. M. Hong, and D. Bernard, *Macromolecules* **17**, 2895 (1984).

60. P. J. Flory, *Statistical Mechanics of Chain Molecules*, Wiley, New York, 1969.

61. M. N. Barber and B. W. Ninham, *Random and Restricted Walks*, Gordon & Breach, New York, 1970.

62. W. Feller, *An Introduction to Probability and its Applications*, Vol. 1, Wiley, New York, 1966.

63. W. Kuhn, *Kolloid-Z.* **68**, 2 (1934).

64. A. Peterlin, *Ann. N.Y. Acad.* **89**, 579 (1961).

65. P. Debye, *J. Chem. Phys.* **14**, 636 (1946).

66. B. Zimm and W. Stockmayer, *J. Chem. Phys.* **17**, 1301 (1949).

14 CHEMICAL WAVES

Kenneth Showalter
Department of Chemistry
West Virginia University
Morgantown, West Virginia

14.1. INTRODUCTION

Chemical reaction may couple with diffusion in an isothermal, initially homogeneous reaction mixture to give rise to macroscopic spatial structure. A necessary ingredient for the formation of spatial structure in a reaction–diffusion system is some form of feedback mechanism in the chemical reaction kinetics. Autocatalysis is the most common type of feedback mechanism, and several autocatalytic reac-

The author thanks Alan Saul for help with calculations and Vilmos Gáspár for checking the manuscript. This work was supported by the National Science Foundation (Grant No. CHE-8311360).

tions are known to exhibit wave behavior. Propagating waves of chemical activity are referred to in the literature as *chemical waves* and are generally classified as fronts, pulses, or wave trains. *Fronts* convert a reaction mixture from one state to another state. These states may be stationary or pseudostationary. *Pulses* are best described as temporary excursions away from a stationary or pseudostationary state in an excitable medium. The state ahead of a pulse is restored behind the pulse as it propagates through the reaction mixture. *Wave trains* are successive pulses emanating from a center. These waves propagate with regular periodicity through any particular spatial coordinate around the center and give rise to target or spiral patterns. The distinctions between these categories of chemical waves are not always clear-cut. For example, a single propagating pulse when mechanically sheared gives rise to two rotating spiral waves.

Macroscopic spatial structure in reaction–diffusion systems was first predicted in the pioneering theoretical work of Turing (1) and Nicolis and Prigogine (2). These studies showed that *stationary* patterns should arise from the coupling of an autocatalytic chemical reaction with diffusion. While this work initiated much of the current interest in chemical waves, to date, no clear-cut experimental example of stationary reaction–diffusion patterns has been reported. The few examples of stationary patterns in experimental system appear to be the result of convective hydrodynamic instabilities (3–5).

Known chemical wave systems are described and analyzed in this chapter. The discussion is limited to the chemistry and simple mathematical models of these systems. A number of more specialized reviews are available in the literature. The spatial and temporal behavior of the oscillatory Belousov–Zhabotinsky reaction has been analyzed by Tyson (6), and the geometric aspects of the chemical waves in this system have been thoroughly treated by Winfree (7). The various possible types of chemical waves are categorized and analyzed by Ortoleva and Ross (8). A number of treatments and reviews of chemical waves and related phenomena can be found in a recent collection edited by Field and Burger (9).

The first quantitative study of chemical wave behavior was reported in 1955 by Epic and Shub (10). Autocatalytically generated iodide in the iodate oxidation of arsenous acid, diffusing ahead into fresh reaction mixture, results in a wave of chemical activity propagating at a constant velocity. These fronts are the simplest and perhaps the best understood of the known chemical wave systems (11–14).

Chemical waves in the oscillatory catalyzed bromate oxidation of malonic acid were studied as early as 1951 by Belousov (9, 15); however, wave trains and pulses in this system were not reported until 1970 by Zaikin and Zhabotinsky (16). This reaction, now known as the Belousov–Zhabotinsky (BZ) reaction, has been by far the most extensively studied chemical wave system (3, 16–24). The intense interest in the BZ reaction stems from the remarkably wide variety of exotic behavior exhibited by the system. In this chapter we focus on the simplest description of the BZ reaction–diffusion behavior.

Several other chemical wave systems are now known. The BZ system modified to be completely nonoscillatory by adding a bromine sink exhibits frontlike consumption waves (25, 26). Wave train behavior in the oscillatory chlorite–iodide–malonic acid reaction has recently been reported by Epstein and co-workers (27). In addition to stationary structures found in BZ systems (3–5), similar patterns have been observed in cell-free extracts of yeast (28). These patterns also appear to be due to hydrodynamic effects rather than arising from a reaction–diffusion instability.

In this chapter an introduction to chemical waves is presented in terms of the basic features of the best understood systems. Fronts in the iodate–arsenous acid system are discussed in Section 14.2, and in Section 14.3 waves in the BZ system are described. Other chemical wave systems, electric field effects, and stationary patterns are discussed in Section 14.4, and the chapter is concluded in Section 14.5 with a brief summary of the field.

14.2. WAVES IN IODATE OXIDATION OF ARSENOUS ACID

14.2.1. Chemistry

The iodate oxidation of arsenous acid was first recognized to be autocatalytic in iodide by Eggert and Scharnow (29). The reaction exhibits a long induction period as iodide concentration slowly increases to about 10^{-6} M; iodide is then rapidly generated until the limiting reagent, arsenous acid or iodate, is completely consumed. Bognár and Sárosi (30) used the sensitivity of this induction period to the initial concentration of iodide to develop an analytical method for trace quantities of iodide. The ''clock reaction'' behavior of the iodate–arsenous acid system has been exploited over the years as an effective general chemistry lecture demonstration.

Iodate does not react directly with arsenous acid at a significant rate but reacts instead with iodide to generate iodine. The iodide impurity in iodate is sufficient to initiate reaction; however, iodide is usually deliberately introduced as a reactant. The iodine product of this process is rapidly reduced by arsenous acid to regenerate iodide. These two processes involving the consumption and the regeneration of iodide provide a simple and useful description of the overall chemical reaction. Iodide is oxidized by iodate according to process (14.1), the Dushman reaction (31), and the iodine product is reduced by arsenous acid according to process (14.2), the Roebuck reaction (32):

$$IO_3^- + 5I^- + 6H^+ = 3I_2 + 3H_2O \qquad (14.1)$$

$$H_3AsO_3 + I_2 + H_2O = H_3AsO_4 + 2I^- + 2H^+ \qquad (14.2)$$

In reaction mixtures containing arsenous acid in stoichiometric excess, iodide product is generated according to net reaction (14.3) or (14.1) + 3(14.2). Iodine product is generated in solutions containing iodate in stoichiometric excess according to net reaction (14.4) or 2(14.1) + 5(14.2).

$$IO_3^- + 3H_3AsO_3 = I^- + 3H_3AsO_4 \tag{14.3}$$

$$2IO_3^- + 5H_3AsO_3 + 2H^+ = I_2 + 5H_3AsO_4 + H_2O \tag{14.4}$$

Process (14.1) is rate determining for the overall reaction and is governed by rate law (14.5) (33),

$$R_\alpha = -\frac{d[IO_3^-]}{dt} = (k_1 + k_2[I^-])[I^-][IO_3^-][H^+]^2 \tag{14.5}$$

Process (14.2) is rapid with kinetics governed by rate law (14.6) (32, 34),

$$R_\beta = -\frac{d[I_2]}{dt} = \frac{k_3[I_2][H_3AsO_3]}{[I^-][H^+]} \tag{14.6}$$

Rate laws (14.5) and (14.6) may be combined according to the stoichiometries of processes (14.1) and (14.2), respectively, to form a four-variable empirical rate law model. A model of this type was first proposed by Epstein and co-workers (35) to describe the bistability of the iodate–arsenous acid reaction in a continuous-flow stirred tank reactor. A more complex model in terms of elementary steps consistent with the empirical rate law model has also been proposed (36). Thus, rate equations are written for each of the variables in rate laws (14.5) and (14.6) according to Eqs. (14.7)–(14.10). Hydrogen ion concentration is considered to be constant because the reaction is generally carried out in buffered solution:

$$\frac{d[I^-]}{dt} = -5R_\alpha + 2R_\beta \tag{14.7}$$

$$\frac{d[I_2]}{dt} = 3R_\alpha - R_\beta \tag{14.8}$$

$$\frac{d[H_3AsO_3]}{dt} = -R_\beta \tag{14.9}$$

$$\frac{d[IO_3^-]}{dt} = -R_\alpha \tag{14.10}$$

The concentration of each variable species as a function of time according to this model can be obtained by numerically integrating the differential equations. Coupled differential equations of kinetic models are typically *stiff* where variables change on widely different time scales; however, these systems can be conveniently integrated using the GEAR predictor–corrector algorithm (37). The temporal behavior of a solution containing excess arsenous acid is shown in Figure 14.1. In this calculation arsenous acid concentration is fixed at a constant value since it is in stoichiometric excess over iodate and consequently has little effect on the behavior. We see that iodide concentration remains very low for about 10.0 min and then rapidly increases to a value equal to the sum of the initial iodate and iodide concentrations. Also shown is the transient appearance of iodine. Even at its maximum concentration, iodine accounts for only about 0.2% of the total iodine atoms in the system (13); therefore, the decrease in iodate concentration closely follows the concentration increase in iodide.

Figure 14.2 shows the temporal behavior of a batch reaction containing excess iodate according to the four-variable model. We again see iodide concentration remain at very low levels during a long induction period and then rapidly increase. In these solutions, however, iodide is rapidly consumed in process (14.1) upon complete consumption of the limiting reagent arsenous acid. The iodine product grows in as iodide is consumed. Figures 14.1 and 14.2 also illustrate the sensitivity of the induction period to initial iodide concentration.

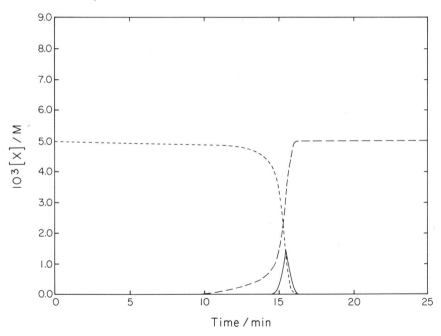

Figure 14.1. Concentrations of IO_3^- (----), I^- (— — —), and I_2 (———) calculated from Eqs. (14.7)–(14.10). Concentration of I_2 enlarged by factor of 350. Initial concentrations: $[IO_3^-]_0 = 5.0 \times 10^{-3}\ M$, $[I^-]_0 = 2.5 \times 10^{-5}\ M$, $[I_2]_0 = 1.0 \times 10^{-12}\ M$. Reactant concentrations held constant: $[H_3AsO_3]_0 = 5.43 \times 10^{-2}\ M$, $[H^+]_0 = 7.1 \times 10^{-3}\ M$. Rate constants: $k_1 = 4.5 \times 10^3\ M^{-3}\ s^{-1}$, $k_2 = 1.0 \times 10^8\ M^{-4}\ s^{-1}$, $k_3 = 3.2 \times 10^{-2}\ M\ s^{-1}$. (Reprinted with permission from A. Hanna, A. Saul, and K. Showalter, *J. Am. Chem. Soc.* **104**, 3838 (1982). Copyright 1982 American Chemical Society.)

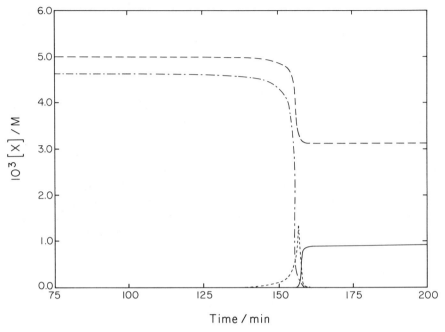

Figure 14.2. Concentrations of IO_3^- (— — —), I^- (- - - -), I_2 (———), and H_3AsO_3 (— · ——) calculated from Eqs. (14.7)–(14.10). Initial concentrations: $[IO_3^-]_0 = 5.0 \times 10^{-3}$ M, $[I^-]_0 = 1.0 \times 10^{-9}$ M, $[I_2]_0 = 1.0 \times 10^{-12}$ M, $[H_3AsO_3]_0 = 4.65 \times 10^{-3}$ M. Reactant concentration held constant: $[H^+]_0 = 7.1 \times 10^{-3}$ M. See Figure 14.1 for rate constants. (Reprinted with permission from A. Hanna, A. Saul, and K. Showalter, *J. Am. Chem. Soc.* **104**, 3838 (1982). Copyright 1982 American Chemical Society.)

The dependence of the batch reaction temporal behavior on initial reactant concentrations is clearly shown in Figures 14.1 and 14.2. The simple behavior of the excess arsenous acid system will be exploited in Section 14.2.4 to develop an analytic model for front propagation in these solutions. The behavior of the excess iodate system is essentially the same as that of the excess arsenous acid system until arsenous acid is consumed. The complexity introduced by the cusp behavior of iodide, however, makes this system more difficult to model analytically, and it will not be considered here. A reaction–diffusion scheme based on Eqs. (14.7)–(14.10) will be used in Section 14.2.3 to numerically model propagating fronts in excess arsenous acid and in excess iodate solutions.

14.2.2. Experimental Behavior

Chemical waves in the iodate–arsenous acid reaction have been studied in thin films of solution and in capillary tubes (11–13, 38). In solutions containing starch indicator, the front is visually detectable and its position as a function of time can be recorded photographically. Waves are conveniently initiated at a platinum electrode, negatively biased at ~ 1.0 V with respect to another platinum electrode. Iodate is electrochemically reduced to iodide at the electrode surface, initiating iodide autocatalysis locally. Iodide concentration rapidly increases near the electrode and begins to diffuse outward, promoting autocatalysis in the nearby reaction mixture. The result is a propagating front.

A typical front in a solution containing excess arsenous acid is shown in Figure 14.3. Here, the wave was initiated at a platinum electrode positioned at the center of a thermostated dish. Waves were also initiated at later times at the reference platinum electrode near the edge of the dish, apparently from an increase in the

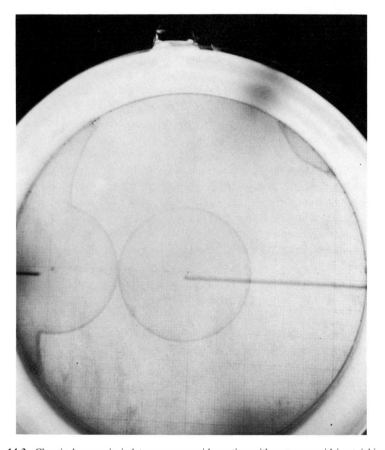

Figure 14.3. Chemical waves in iodate–arsenous acid reaction with arsenous acid in stoichiometric excess. Central wave initiated at a negatively biased platinum electrode at ~ 3.7 min and photograph taken at 17.6 min after mixing reactants. Initial reactant concentrations: $[NaIO_3]_0 = 5.00 \times 10^{-3}$ M and $[H_3AsO_3]_0 = 1.55 \times 10^{-2}$ M. Hydrogen ion concentration maintained constant at 7.1×10^{-3} M with Na_2SO_4–$NaHSO_4$ buffer. Temperature: $25.0 \pm 0.2\,°C$. (Reprinted with permission from A. Hanna, A. Saul, and K. Showalter, *J. Am. Chem. Soc.* **104,** 3838 (1982). Copyright 1982 American Chemical Society.)

local acidity due to electrolysis. Waves initiated along the edges of the dish probably arise from traces of iodine and iodide remaining from earlier experiments. The wave front appears as a thin blue band resulting from the interaction of the transient intermediate I_3^- with starch indicator. Triiodide ion is rapidly and reversibly formed by the reaction of iodine with excess iodide. The maximum iodine concentration occurs at the inflection point of the iodide curve (Figure 14.1), where the rate of iodide autocatalysis is at its maximum. In a stirred reaction mixture containing starch indicator, the colorless solution suddenly turns blue and then again colorless as the iodine intermediate makes its transient appearance. The solution composition ahead of the front is near that of the initial reactants; behind the front the composition corresponds to that of thermodynamic equilibrium according to net reaction (14.3).

Plots of wave front position as a function of time are linear, giving constant propagation velocities. A series of experiments allows the dependence of propagation velocity on initial reactant concentrations to be determined. The velocity dependence on iodate, arsenous acid, and hydrogen ion concentrations is linear with the dependence on iodate and hydrogen ion concentrations somewhat larger than the dependence on arsenous acid concentration (12, 13, 38).

Waves in reaction mixtures containing iodate in stoichiometric excess and starch indicator appear as expanding blue disks. A typical wave initiated at a platinum electrode positioned at the center of the dish is shown in Figure 14.4. The solution is converted from colorless to blue at the wave front. Here, the reaction mixture composition behind the front corresponds to that of thermodynamic equilibrium according to net reaction (14.4). Excess iodate waves in thin films are perturbed by a surface spreading of I_2 and do not exhibit constant velocities (11, 13, 38). These waves in thin capillary tubes, however, do exhibit constant velocities, and the velocity dependence on initial reactant concentrations has been investigated (38).

14.2.3. Numerical Analysis

Chemical waves in the iodate oxidation of arsenous acid are accurately described by a four-variable reaction–diffusion model based on Eqs. (14.7)–(14.10). The coupling of reaction and diffusion is accounted for by changing the ordinary differential equations to the partial differential equations (14.11)–(14.14). These equations model a chemical wave in one spatial dimension, propagating in a capillary tube or in a thin film of solution with purely radial advancement:

$$\frac{\partial C_1}{\partial t} = D_1 \frac{\partial^2 C_1}{\partial x^2} - 5R_\alpha + 2R_\beta \tag{14.11}$$

$$\frac{\partial C_2}{\partial t} = D_2 \frac{\partial^2 C_2}{\partial x^2} + 3R_\alpha - R_\beta \tag{14.12}$$

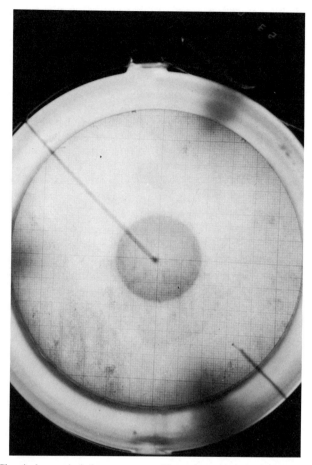

Figure 14.4. Chemical wave in iodate–arsenous acid reaction with iodate in stoichiometric excess. Central wave initiated at negatively biased platinum electrode at ~ 7.3 min and photograph taken at 23.0 min after mixing reactants. Initial reactant concentrations: $[NaIO_3]_0 = 5.00 \times 10^{-3}$ M and $[H_3AsO_3]_0 = 3.16 \times 10^{-3}$ M. Hydrogen ion concentration maintained constant at 3.76×10^{-3} M with Na_2SO_4–$NaHSO_4$ buffer. Temperature: $25.0 \pm 0.2\,^\circ$C. (Reprinted with permission from A. Hanna, A. Saul, and K. Showalter, *J. Am. Chem. Soc.* **104**, 3838 (1982). Copyright 1982 American Chemical Society.)

$$\frac{\partial C_3}{\partial t} = D_3 \frac{\partial^2 C_3}{\partial x^2} - R_\beta \tag{14.13}$$

$$\frac{\partial C_4}{\partial t} = D_4 \frac{\partial^2 C_4}{\partial x^2} - R_\alpha \tag{14.14}$$

Here, $C_1 = [I^-]$, $C_2 = [I_2]$, $C_3 = [H_3AsO_3]$, $C_4 = [IO_3^-]$, and D_n are the respective diffusion coefficients.

The partial differential equations (14.11)–(14.14) may be numerically integrated using the method of lines. This method was first used to model chemical waves by Reusser and Field (22) (see Section 14.3.4). The independent variable x is discretized into a grid, and the resulting ordinary differential equations (ODEs) are integrated with respect to time at each grid point. The number of independent variables I is increased in this transformation to $I \times N$, where N is the number of grid points. The second derivative of C_n with respect to x at grid point i is approximated by calculating the concentration differences between neighboring grid points according to Eq. (14.15) (22),

$$\frac{d^2 C_n}{dx^2} \approx \frac{(C_n^{i-1} - C_n^i)/\Delta x - (C_n^i - C_n^{i+1})/\Delta x}{\Delta x} = \frac{(C_n^{i-1} - 2C_n^i + C_n^{i+1})}{(\Delta x)^2}$$

$$(14.15)$$

The calculated spatial and temporal behavior of a chemical wave in a reaction mixture containing excess arsenous acid is shown in Figure 14.5. This calculation was carried out using a grid of 200 points, giving a total of 800 ODEs in the integration. The grid represents 1.0 cm in actual distance. Calculations were carried out to determine the grid spacing necessary to accurately represent the propagating wave. Grids with as many as 1000 points generated propagation velocities and waveforms almost identical to grids with 200 points. Below about 100 points the wave behavior is noticeably dependent on the number of grid points. The number of grid points necessary to accurately describe a chemical wave depends on the width of the wave. Waves in the BZ reaction have extremely sharp fronts, and a finer grid is necessary to accurately simulate their behavior (see Section 14.3.4) (22). Fronts in the iodate–arsenous acid reaction are about 1.0 mm in width; therefore, with a grid of 200 points/cm, the waveform of each species is approximated with about 20 points.

The initial iodide concentration in Figure 14.5 was set to 1.0×10^{-7} M except for the first 0.3 mm of the grid, which was set to 1.0×10^{-3} M. These initial conditions represent a reasonable approximation of the iodide in the bulk solution and that generated at a negatively biased platinum electrode. In both experiment and calculation the wave behavior is independent of the initiation iodide after a few millimeters propagation.

The first curve in Figure 14.5 at 60 s after initiation shows that iodide concentration increases rapidly near the initiation spike. At 2.0 min after initiation the waveform has already taken on its final shape after propagating only about 3.0 mm. In the subsequent curves the iodide concentration behind the wave front rises to the value predicted by net reaction (14.3). The iodide from the initiation spike appears in each curve as an increase above the expected concentration near the origin.

Figure 14.5 also shows the iodine waveform (note the expanded scale for concentration). At time zero iodine concentration was set to 1.0×10^{-12} M throughout

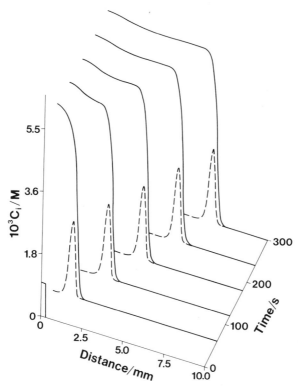

Figure 14.5. Iodide concentration C_1 (——) and iodine concentration C_2 (— — — —) as a function of distance and time calculated from Eqs. (14.11)–(14.14) for a solution containing excess arsenous acid. Concentration of I_2 enlarged by factor of 100. Initial reactant concentrations: $C_1^0 = 1.0 \times 10^{-7}$ M, $C_2^0 = 1.0 \times 10^{-12}$ M, $C_3^0 = 1.55 \times 10^{-3}$ M, $C_4^0 = 5.0 \times 10^{-3}$ M. Hydrogen ion concentration held constant at 7.1×10^{-3} M. Rate constants: $k_1 = 4.5 \times 10^3$ M^{-3} s^{-1}, $k_2 = 4.36 \times 10^8$ M^{-4} s^{-1}, $k_3 = 3.2 \times 10^{-2}$ M s^{-1}. Diffusion constants: $D_n = 2.0 \times 10^{-5}$ cm^2 s^{-1}.

the grid. We see a maximum in iodine concentration near the inflection point of the iodide waveform, much like the temporal behavior in Figure 14.1. The reversible formation of triiodide ion in this region gives rise to the thin blue band in solutions containing starch indicator. Other calculations showed that an initial spike of iodine near the origin also serves to initiate the propagating front. Iodine is rapidly reduced to iodide, which then initiates autocatalysis.

The behavior of a wave in an excess iodate solution is shown in Figure 14.6. Iodide again exhibits a rapid increase in concentration at early times near the initial iodide spike, but now it is consumed before it reaches a value comparable to the initial iodate concentration. Iodine grows in as iodide is consumed and gradually reaches the concentration predicted by net reaction (14.4). The "cusp" waveform of iodide in excess iodate solutions is reminiscent of the waveform of iodine in excess arsenous acid solutions. However, iodide in excess iodate solutions may account for a substantial fraction of the total iodine atoms in the system, while

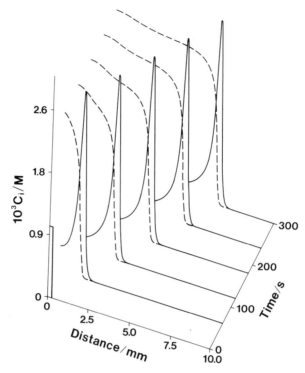

Figure 14.6. Iodide concentration C_1 (———) and iodine concentration C_2 (— — — —) as a function of distance and time calculated from Eqs. (14.11)–(14.14) for a solution containing excess iodate. Rate constants and initial concentrations as in Figure 14.5 except $C_3^0 = 1.09 \times 10^{-2}$ M.

iodine is never stoichiometrically significant in excess arsenous acid solutions. The last curve in Figure 15.6, at 5.0 min after initiation, shows the final waveform of iodine. The iodine concentration above that predicted by Eq. (14.4) near the origin results from the initial iodide spike.

The iodide waveform is followed by the iodine waveform in both excess arsenous acid and excess iodate reaction mixtures. In excess arsenous acid the iodine cusp appears at the inflection point of the iodide waveform; in excess iodate, iodine grows in following the iodide cusp. In both cases the wave velocity is primarily determined by the rise in iodide concentration, while the rise in iodine has the features of a phase wave.

14.2.4. Analytical Analysis

The iodate–arsenous acid reaction is an attractive system for analytical analysis because rate and conservation considerations allow a reduction in the number of variables necessary to describe the chemical wave. In excess arsenous acid solutions the front can be described (12–14) by a one-variable reaction–diffusion equation obtained from a reduction of the four-variable model (14.11)–(14.14). Waves in solutions containing excess iodate can be only qualitatively described with a reduced version of the four-variable model (14). Here, we pursue the model for waves in excess arsenous acid solution and only briefly consider excess iodate waves.

In solutions containing excess arsenous acid the reaction is described at all times by a single stoichiometry according to net reaction (14.3). Since process (14.1) is rate determining, the rate of iodide autocatalysis is governed by rate law (14.5). Therefore, in buffered solutions the reaction–diffusion behavior can be described in terms of the variable species in the rate law, iodide and iodate, according to

$$\frac{\partial C_1}{\partial t} = D_1 \frac{\partial^2 C_1}{\partial x^2} + R_\alpha \tag{14.16}$$

$$\frac{\partial C_4}{\partial t} = D_4 \frac{\partial^2 C_4}{\partial x^2} - R_\alpha \tag{14.17}$$

These two variables are linked by the conservation of iodine atoms according to the stoichiometry of net reaction (14.3):

$$C_1 + C_4 = C_1^0 + C_4^0 \approx C_4^0 \tag{14.18}$$

where C_1^0 and C_4^0 are the initial iodide and iodate concentrations, and $C_4^0 \gg C_1^0$. We assume in Eq. (14.18) that iodine is never present in stoichiometrically significant concentrations. Substitution of the conservation relation (14.18) into Eq. (14.16) or (14.17) yields a one-variable partial differential equation in C_1,

$$\left(\frac{\partial C_1}{\partial t}\right)_x = D \left(\frac{\partial^2 C_1}{\partial x^2}\right)_t + f(C_1) \tag{14.19}$$

where

$$f(C_1) = (k_A + k_B C_1) C_1 (C_4^0 - C_1) \tag{14.20}$$

and

$$k_A = k_1 [H^+]^2 \qquad k_B = k_2 [H^+]^2$$

For Eq. (14.19) to be locally valid, we must assume that $D_1 = D_4 = D$.

The rate of reaction, $f(C_1)$, in a stirred homogeneous solution is shown as a function of C_1 in Figure 14.7. Since $f(C_1)$ is a cubic in C_1, we find three stationary states when $f(C_1) = 0$, given by

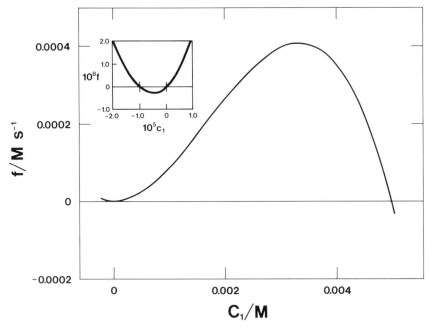

Figure 14.7. Reaction rate f as a function of iodide concentration C_1 according to Eq. (14.20) for solution containing excess arsenous acid. Stationary states are at $C_1 = -1.0 \times 10^{-5}$, 0, and 5.0 × 10^{-3} M. Inset shows blow-up of region near origin. See Figure 14.5 for rate constants and initial concentrations. (Reprinted with permission from A. Saul and K. Showalter, in *Oscillations and Traveling Waves in Chemical Systems*, R. J. Field and M. Burger (Eds.), Wiley-Interscience, New York, 1985, pp. 419–439. Copyright 1985 John Wiley and Sons, Inc.)

$$\hat{C}_1 = -\frac{k_A}{k_B}, \; 0, \; C_4^0 \tag{14.21}$$

Of course, only the nonnegative stationary states are physically meaningful. The stationary state $\hat{C}_1 = 0$ corresponds to the initial reaction mixture before any chemical reaction has occurred; the stationary state $\hat{C}_1 = C_4^0$ corresponds to the solution composition at thermodynamic equilibrium. The transition from initial reactants to the equilibrium composition occurs in the propagating wave front.

To develop an analytic solution of the partial differential equation (14.19), we first make the transformation to a second-order ODE. We seek a constant-velocity solution; therefore, using the cycle rule, velocity may be expressed according to

$$v = \left(\frac{\partial x}{\partial t}\right)_{dC_1 = 0} = -\frac{(\partial C_1/\partial t)_x}{(\partial C_1/\partial x)_t} \tag{14.22}$$

Combining this expression with Eq. (14.19) yields the ODE

$$DC_1'' + vC_1' + f(C_1) = 0 \tag{14.23}$$

where prime represents differentiation with respect to distance x. Introducing the new spatial variable $x - vt$ is an equivalent means of arriving at Eq. (14.23).

The next step is to obtain from Eq. (14.23) a first-order differential equation in C_1 to be integrated. Let the gradient $C_1' = G(C_1)$. If G is a function of degree n in C_1, then $G'(C_1)$ is of degree $n - 1$. Since $C_1'' = G'(C_1)C_1'$, from Eq. (14.23), $f = -C_1'(DG' + v)$, which is of degree $n + (n - 1) = 2n - 1$. We know that f is a cubic; therefore, $n = 2$ and the gradient G is a quadratic, corresponding to a parabolic trajectory in the $C_1' - C_1$ phase plane.

Since the gradient G is a parabola connecting the stationary states $\hat{C}_1 = 0$ and C_4^0, we write

$$G = C_1' = kC_1 (C_4^0 - C_1) \tag{14.24}$$

where the constant k is to be determined below. Differentiation of Eq. (14.24) gives

$$C_1'' = k^2 C_1 (C_4^0 - C_1) (C_4^0 - 2C_1) \tag{14.25}$$

Substitution of Eqs. (14.20), (14.24), and (14.25) into Eq. (14.23) generates a linear equation in C_1,

$$Dk^2 (C_4^0 - 2C_1) + vk + (k_A + k_B C_1) = 0 \tag{14.26}$$

From Eq. (14.26) we find

$$-2Dk^2 + k_B = 0 \tag{14.27}$$

and

$$Dk^2 C_4^0 + vk + k_A = 0 \tag{14.28}$$

Therefore, from Eq. (14.27)

$$k = \left(\frac{k_B}{2D}\right)^{1/2} \tag{14.29}$$

where the sign of k is chosen negative to give a positive velocity and negative gradient. The front velocity in terms of rate constants, initial reactant concentrations, and diffusion coefficient can now be obtained by combining Eqs. (14.28) and (14.29),

$$v = \left(\frac{Dk_2}{2}\right)^{1/2} [H^+]_0 [IO_3^-]_0 + k_1 \left(\frac{2D}{k_2}\right)^{1/2} [H^+]_0 \tag{14.30}$$

We now return to Eq. (14.24) for the gradient G with k given by Eq. (14.29). Integration of Eq. (14.24) by partial fractions and subsequent rearrangement yields,

$$C_1 = \frac{C_4^0}{1 + Ae^{-kC_4^0 x}} \tag{14.31}$$

where A is an arbitrary constant. Recognizing that our spatial coordinate is actually $x - vt$, we have

$$C_1(x,t) = \frac{C_4^0}{1 + Ae^{-kC_4^0(x - vt)}} \tag{14.32}$$

or

$$C_1(x,t) = \tfrac{1}{2}C_4^0 \{1 + \tanh [\tfrac{1}{2}kC_4^0(x - vt) + b]\} \tag{14.33}$$

where $b = -\tfrac{1}{2}\ln A$. Equation (14.32) is the analytic solution of reaction–diffusion equation (14.19), giving C_1 as a function of time and distance. The front concentration profile according to this solution is shown in Figure 14.8. Equation (14.33)

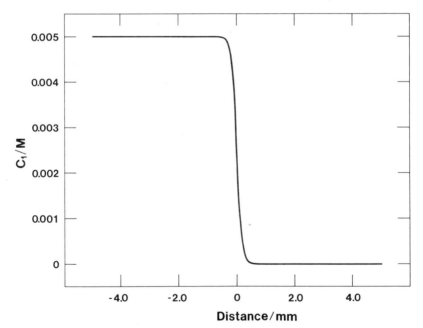

Figure 14.8 Iodide concentration C_1 as a function of distance calculated from Eq. (14.32) with $v = 2.35 \times 10^{-2}$ mm/s. See Figure 14.5 for rate constants and initial concentrations. (Reprinted with permission from A. Saul and K. Showalter, in *Oscillations and Traveling Waves in Chemical Systems*, R. J. Field and M. Burger (Eds.), Wiley-Interscience, New York, 1985, pp. 419–439. Copyright 1985 John Wiley and Sons, Inc.)

clearly shows the nature of the wave front, a hyperbolic tangent curve connecting the stationary states $\hat{C}_1 = 0$ and C_4^0.

Waves in excess iodate solutions are only qualitatively described by an analytic treatment like that used to characterize excess arsenous acid waves (14). These waves also propagate by iodide autocatalysis and its diffusion ahead according to Eqs. (14.32) and (14.33); however, the hyperbolic tangent curve is now truncated when the limiting reagent arsenous acid is consumed. Iodide is oxidized to iodine in process (14.1), giving rise to a cusp in the iodide concentration, as shown in Figure 14.6. Since the iodide front is truncated, the velocity is only roughly given by Eq. (14.32). The subsequent rise in iodine concentration has features of a phase wave and contributes little to the forward propagation of the reaction–diffusion front.

14.2.5. Comparison of Theory and Experiment

The wave behavior in the iodate–arsenous acid reaction is highly dependent on the particular stoichiometry of the reaction determined by the initial concentrations of iodate and arsenous acid. Waves in excess arsenous acid solutions exhibit a linear velocity dependence on iodate, arsenous acid, and hydrogen ion concentrations (12, 13). Wave velocities in excess iodate solutions, however, are not simple linear functions of reactant concentrations (38). These differences may be explored in terms of the analytic and numerical analyses presented in the previous sections. We first consider waves in reaction mixtures containing arsenous acid in stoichiometric excess.

According to the one-variable analysis of Section 14.2.4, front propagation velocity should exhibit a linear dependence on hydrogen ion and iodate concentrations (12–14). From Eq. (14.30) the velocity is given by

$$v = m[H^+]_0 + n[H^+]_0 [IO_3^-]_0 \tag{14.34}$$

where

$$m = k_1 \left(\frac{2D}{k_2}\right)^{1/2} \quad \text{and} \quad n = \left(\frac{k_2 D}{2}\right)^{1/2}$$

The coefficients m and n calculated from literature values of k_1 and k_2 with $D = 2.0 \times 10^{-5}$ cm^2/s are in good agreement with values obtained from wave velocity measurements (12, 13). The one-variable description, however, completely neglects any effect of arsenous acid concentration on the wave behavior.

The validity of the conservation relation (14.18) leading to the velocity expression given by Eq. (14.32) may be tested by comparing the predicted velocities to velocities obtained from numerical integration of Eqs. (14.11)–(14.14). We may also determine how well Eq. (14.34) describes front velocities in excess iodate solutions by a comparison to the numerical velocities. The rise in iodide concentration in these solutions is reasonably well described by Eq. (14.32), but of course, the conservation relation (14.18) fails near the iodide cusp.

Velocities as a function of iodate concentration calculated from the four-variable model are shown in Figure 14.9. The dashed vertical line indicates the concentration of iodate stoichiometrically equal to the concentration of arsenous acid according to net reaction (14.3). The dotted line represents the velocities according to Eq. (14.34). The validity of the conservation relation (14.18) improves as the ratio of arsenous acid to iodate increases. According to rate laws (14.5) and (14.6), as this ratio increases, the rate of process (14.2) increases relative to that of process (14.1), and the result is a diminished concentration of iodine. Therefore, the velocities at low iodate concentration should define a line in good agreement with that given by Eq. (14.34). A least-squares fit through the first three points in the numerical calculation is shown by the solid line.

As iodate concentration increases, the rate of iodine generation in process (14.1) increases while the rate of iodine consumption by process (14.2) is little changed. Therefore, the fraction of the total iodine atoms existing as molecular iodine increases. This failure of Eq. (14.18) is reflected in the velocities calculated from Eq. (14.34). At an iodate concentration of 2.5×10^{-3} M the velocities calculated

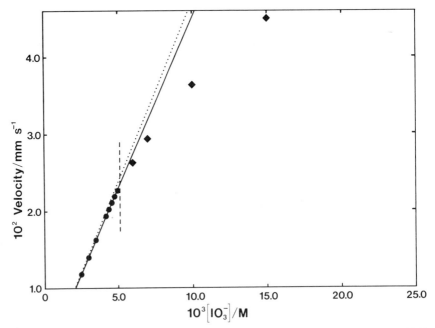

Figure 14.9. Front velocities as a function of iodate concentration calculated from Eqs. (14.11)–(14.14). Iodate and arsenous acid concentrations stoichiometrically equal according to reaction (14.3) indicated by vertical dashed line. See Figure 14.5 for rate constants and initial concentrations.

from the four-variable model and Eq. (14.34) are 1.19×10^{-2} and 1.18×10^{-2} mm/s, respectively. At this iodate concentration the numerical calculations show that the maximum iodine concentration represents about 0.5% of the total iodine atoms in the reaction mixture. At the highest iodate concentration ($5.0 \times 10^{-3} M$) for excess arsenous acid solutions in Figure 14.9, the maximum iodine concentration accounts for about 8.6% of the total iodine atoms. At this iodate concentration the numerical velocity is 2.27×10^{-2} mm/s and that calculated from Eq. (14.34) is 2.35×10^{-2} mm/s. Thus, as the ratio of iodate to arsenous acid increases, conservation relation (14.18) fails, and the one-variable description becomes less accurate. However, even at the highest iodate concentration for excess arsenous acid solutions in Figure 14.9, the velocity calculated from Eq. (14.34) agrees reasonably well with that calculated from the four-variable model.

Velocities of waves in excess iodate solutions are shown as diamonds in Figure 14.9. In these solutions little iodine accumulates during the rise in iodide concentration. Thus, the increase in iodide concentration up to the cusp is described by Eq. (14.32), and the velocity is given approximately by Eq. (14.34). The deviations from the velocity predicted by Eq. (14.34) are primarily due to the truncated rise in iodide concentration rather than a failure of the conservation relation. The velocity predicted by Eq. (14.34) applies to the full iodide front described by Eq. (14.32); when the front is truncated, the velocity falls short of that predicted by the analytic treatment.

Calculated velocities as a function of hydrogen ion concentration are shown in Figure 14.10. Velocities for excess arsenous acid solutions are shown as circles and those for excess iodate solutions as diamonds. The dotted line gives the velocity as a function of hydrogen ion concentration calculated from Eq. (14.34), and the solid line shows a least-squares fit through the first three points for excess arsenous acid. Only the excess arsenous acid velocities are expected to be described by Eq. (14.34). The excess iodate velocities fortuitously agree because in this particular calculation the initial iodate concentration was only slightly in stoichiometric excess. The deviation of excess arsenous acid velocities from those predicted by Eq. (14.34) is again a consequence of the failure of Eq. (14.18). Rate law (14.5) for process (14.1) exhibits a second-order hydrogen ion dependence, while rate law (14.6) for process (14.2) exhibits an inverse dependence. Thus, with an increase in hydrogen ion concentration, an increase in iodine concentration occurs and conservation relation (14.18) fails.

A key assumption in the one-variable description of waves in excess arsenous acid is that the concentration of arsenous acid has little effect on the wave behavior. The basis of this assumption is that in these solutions process (14.1) is rate determining for iodide autocatalysis. Wave velocities as a function of arsenous acid concentration calculated from Eqs. (14.11)–(14.14) are shown in Figure 14.11. Excess arsenous acid velocities are shown as circles and excess iodate velocities as diamonds. The four-variable calculation shows that excess arsenous acid velocities are little affected by the concentration of arsenous acid. Thus, according to the numerical treatment, the neglect of arsenous acid in the analytic treatment

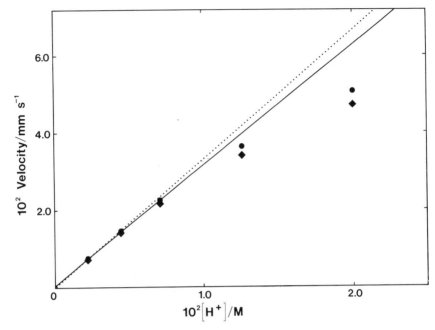

Figure 14.10. Front velocities as a function of hydrogen ion concentration calculated from Eqs. (14.11)–(14.14). Circles show excess arsenous acid velocities and diamonds excess iodate velocities. See Figures 14.5 and 14.6 for rate constants and initial concentrations.

appears to be justified. Experimental measurements, however, show a dependence of velocity on arsenous acid concentration (12, 13); therefore, an apparent discrepancy exists between theory and experiment. A possible deficiency in the four-variable model might be the neglect of a direct reaction between iodate and arsenous acid according to the stoichiometry of net reaction (14.3). Of course, such a direct reaction, occasionally invoked in studies of the iodate–arsenous acid reaction (29, 35), actually represents an alternate mechanism with its associated elementary steps. This reaction has been omitted in our analysis because there is little direct evidence for it. An alternate explanation of the discrepancy might be that the experimental results are flawed (38). Maintaining hydrogen ion concentration completely constant while varying arsenous acid concentration in a sulfate–bisulfate buffered solution is difficult. Unfortunately, more effective organic buffers are unsuitable because secondary reactions occur with iodate.

The velocity dependence on arsenous acid concentration of waves in excess iodate solutions, shown in Figure 14.11, is a consequence of the truncated rise in iodide concentration. With a decrease in arsenous acid concentration, a corresponding decrease occurs in the iodide rise. The wave velocity depends on the extent of development of the iodide waveform, and when the concentration rise is truncated, velocity is diminished.

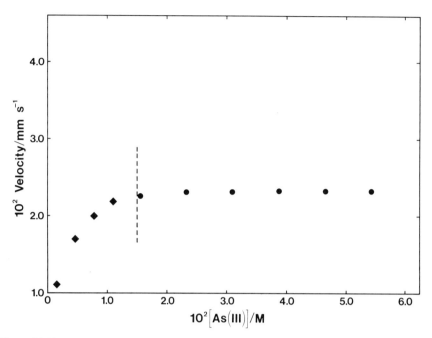

Figure 14.11. Front velocities as a function of arsenous acid concentration calculated from Eqs. (14.11)–(14.14). Iodate and arsenous acid concentrations stoichiometrically equal according to reaction (14.3) indicated by vertical dashed line. See Figure 14.5 for rate constants and initial concentrations.

14.3. WAVES IN BELOUSOV–ZHABOTINSKY REACTION

The catalyzed oxidation of malonic acid by acidic bromate, the classical Belousov-Zhabotinsky (BZ) reaction (39, 40), has been by far the most extensively investigated system exhibiting chemical wave behavior. A remarkably diverse variety of nonlinear phenomena (41–43), including oscillatory and chaotic behavior (44–46), steady-state multiplicity (47–49), and propagating pulses (3, 16–24), has made the BZ reaction and its component reactions the focus of intense experimental and theoretical investigation for more than a decade. In addition, many modified versions of the BZ reaction have been developed that exhibit behavior similar to that of the classical reaction (43, 50).

The reaction–diffusion behavior alone of the BZ reaction is rich in complexity. Propagating waves may emanate from pacemaker centers, giving rise to target patterns, or rotating spiral waves may be formed by mechanically shearing pulses. Multiarmed spirals may also be exhibited (51). In addition, the extent that reaction couples with diffusion may vary depending on the particular initial conditions of the system (22). Here, we focus on the initiation and propagation of pulses in the classical BZ system.

14.3.1. Chemistry

The mechanism of the BZ reaction was elucidated in 1972 by Field, Körös, and Noyes (FKN) (52). While some debate continues (50, 53), the FKN mechanism is widely regarded as correct in its major features. A number of recent reviews on the detailed chemistry of the BZ reaction are available (9), and here we outline only the basic features of the mechanism.

In 1974, Field and Noyes proposed a five-step, three-variable reduction of the FKN mechanism called the Oregonator (54). This model has been extremely successful over the years in providing a semiquantitative description of the varied nonlinear behavior of the reaction. Field (55) modified the Oregonator to include reversibility, and this model was later extended to incorporate additional variables (56). The basic features of the BZ reaction are outlined below in terms of the extended reversible Oregonator with ferroin, Fe(II), as the catalyst:

$$BrO_3^- + Br^- + 2H^+ \rightleftarrows HBrO_2 + HOBr \tag{14.35}$$

$$HBrO_2 + Br^- + H^+ \rightleftarrows 2HOBr \tag{14.36}$$

$$HOBr + Br^- + H^+ \rightleftarrows Br_2 + H_2O \tag{14.37}$$

$$BrO_3^- + HBrO_2 + H^+ \rightleftarrows 2BrO_2\cdot + H_2O \tag{14.38}$$

$$BrO_2\cdot + Fe(II) + H^+ \rightleftarrows HBrO_2 + Fe(III) \tag{14.39}$$

$$2HBrO_2 \rightleftarrows BrO_3^- + HOBr + H^+ \tag{14.40}$$

$$Fe(III) + BrMA \rightarrow Fe(II) + gBr^- + \text{oxidized organic products} \tag{14.41}$$

Oscillations in the BZ reaction can be conveniently described in terms of three composite processes, A, B, and C, made up of reactions (14.35)–(14.37), (14.38)–(14.40), and (14.41), respectively. Bromide is slowly oxidized by bromate in reaction (14.35) when its concentration is high. The products of this step rapidly undergo further reaction with Br^- in steps (14.36) and (14.37) to yield Br_2. The net reaction for process A is given by (14.35) + (14.36) + 3(14.37) or (14.42):

$$BrO_3^- + 5Br^- + 6H^+ \rightarrow 3Br_2 + 3H_2O \tag{14.42}$$

As bromide is consumed in process A, it eventually reaches a critical concentration where the rate of its oxidation by $HBrO_2$ in reaction (14.36) is comparable to the rate of $HBrO_2$ oxidation by BrO_3^- in reaction (14.38). The $BrO_2\cdot$ product of reaction (14.38) is rapidly reduced in reaction (14.39) to regenerate bromous acid, and the result is the autocatalytic sequence B' given by (14.38) + 2(14.39) or (14.43):

$$BrO_3^- + HBrO_2 + 2Fe(II) + 3H^+ \rightarrow 2HBrO_2 + 2Fe(III) + H_2O \tag{14.43}$$

Thus, autocatalysis is initiated when the production of $HBrO_2$ in process B' is comparable in rate to its consumption in reaction (14.36). Bromous acid concentration increases autocatalytically with a concurrent oxidation of the catalyst. Bromide is rapidly driven to very low concentrations by its reaction with bromous acid and the rate of process A becomes negligible. As $HBrO_2$ concentration increases, its autocatalytic growth is eventually limited by the bimolecular disproportionation reaction (14.40). The net reaction for process B is given by 2(14.38) + 4(14.39) + (14.40) or (14.44):

$$BrO_3^- + 4Fe(II) + 5H^+ \rightarrow HOBr + 4Fe(III) + 2H_2O \qquad (14.44)$$

The Br_2 product of process A reacts with malonic acid to generate bromomalonic acid (BrMA). The HOBr product of process B may brominate MA directly or react with Br^- supplied later in the cycle to yield Br_2, thereby also generating BrMA. This supply of BrMA is attacked by the oxidized catalyst, Fe(III), to liberate Br^- according to the composite reaction (14.41). Process C is complex and not well understood; however, the net effect is the reduction of the catalyst and the regeneration of bromide. These general features are represented by reaction (14.41), where g is a stoichiometric factor determining the moles of bromide generated per mole of catalyst reduced. Thus, following the rapid autocatalytic growth of $HBrO_2$ and concurrent oxidation of Fe(II), bromide is regenerated by the Fe(III) oxidation of BrMA. Bromide concentration increases to a critical value where it competes for $HBrO_2$ in reaction (14.36), and the control of the system is switched from process B to process A. Bromide is again slowly consumed in process A, and the sequence is repeated. Concentrations of $HBrO_2$, Br^-, and Fe(III) as a function of time, calculated from the original Oregonator, are shown in Figure 14.12.

14.3.2. Experimental Behavior

A variety of wave behaviors is exhibited by the BZ reaction in unstirred solution. The simplest type of wave occurs in an oscillatory medium when the frequency of oscillation is a function of the spatial coordinate. Waves appear in such a system just as the lights of a theater marquee give the appearance of a moving word or picture. Waves of this type in the BZ reaction were studied more than a decade ago in tubes with vertical gradients in catalyst (57) or acidity (58, 59). The frequency of oscillation is a function of acidity or catalyst, and nonsynchronous oscillations generate the theater marquee effect. The initial velocity of these waves is far too high for diffusion to play a significant role; however, diffusion becomes important as waves propagate into regions of lower activity, where the velocity is relatively small. Waves owing their existence to a gradient in some constraint, such as temperature or a reactant concentration, that determines the local frequency of oscillation are called *kinematic waves* (8, 60).

A *phase wave*, like a kinematic wave, is a high-velocity wave little influenced by diffusion (8, 22, 61, 62). A phase wave, however, owes its existence to a

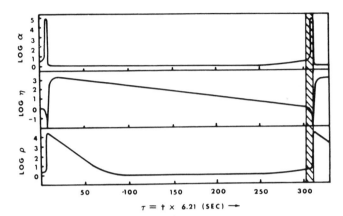

Figure 14.12. Temporal oscillation of bromous acid (α), bromide (η), and catalyst (ρ) calculated from the Oregonator. Reactant concentrations: $[BrO_3^-]_0 = 0.06\ M$, $[H^+]_0 = 0.8\ M$. Rate constants: $k_1 = (2.1)\ [H^+]_0^2\ M^{-3}\ s^{-1}$, $k_2 = (2 \times 10^9)\ [H^+]_0\ M^{-2}\ s^{-1}$, $k_3 = (1 \times 10^4)\ [H^+]_0\ M^{-2}\ s^{-1}$, $k_4 = 4 \times 10^7$ $M^{-1}\ s^{-1}$, $k_5 = 1\ s^{-1}$, $f = 1$. (Reprinted with permission from E. J. Reusser and R. J. Field, *J. Am. Chem. Soc.* **101**, 1063 (1979). Copyright 1979 American Chemical Society.)

gradient in the phase rather than in the frequency of oscillation. Thus, a reaction mixture of a single composition and temperature oscillating at a single frequency, but with the phase of oscillation a function of spatial coordinate, exhibits apparent wave behavior from the theater marquee effect.

Waves that propagate with velocities of a few millimeters per minute are often referred to as *trigger waves* (20, 22). Here, reaction and diffusion are intimately linked, just as in the propagating fronts of the iodate–arsenous acid system. Trigger waves may occur in an oscillatory reaction mixture but are usually studied in a nonoscillatory excitable medium, originally developed by Winfree (17). Field and Noyes were first to semiquantitatively explain trigger wave propagation in the BZ reaction (21). In these waves, $HBrO_2$ diffuses ahead in the wave front and consumes Br^- in reaction (14.36). Bromous acid autocatalysis is initiated as Br^- concentration reaches its critical value, and the result is a wave that triggers its own propagation. Waves in the BZ reaction propagate with constant velocities, shown in Figure 14.13, and wave velocity is proportional to the square root of the product of bromate and hydrogen ion concentrations, shown in Figure 14.14.

Trigger waves in the BZ reaction give rise to target patterns or spirals (7, 20, 20, 63). Figure 14.15 shows time sequences of each type of pattern. Waves in target patterns emanate from pacemaker centers, often at different frequencies. The domain of a particular high-frequency pacemaker grows at the expense of adjacent domains of pacemakers with lower frequencies. Given sufficient time, a single pacemaker with the highest frequency entrains the entire reaction mixture. Because pacemaker centers initiate waves at different frequencies, Winfree (7, 20) suggested that the centers are actually a site for some sort of heterogenous catalysis of the reaction. Thus, the nonoscillatory reaction mixture becomes oscillatory at

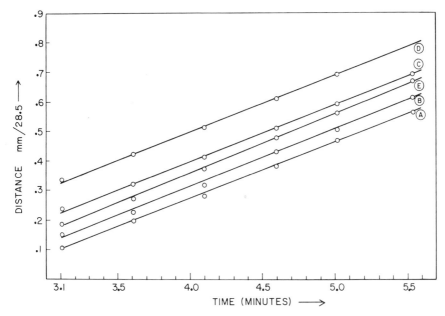

Figure 14.13. Plots of wave front position as a function of time in a BZ reaction mixture. Velocity = 5.5 ± 0.1 mm/min. Initial reactant concentrations: $[H_2SO_4]_0$ = 0.26 M, $[NaBrO_3]_0$ = 0.23 M, $[MA]_0$ = 0.074 M, $[BrMA]_0$ = 0.075 M, $[Ferroin]_0$ = 0.0038 M. Temperature = 25.0 ± 0.1°C. (Reprinted with permission from R. J. Field and R. M. Noyes, *J. Am. Chem. Soc.* **96**, 2001 (1974). Copyright 1974 American Chemical Society.)

the surface of a particle of dust or at a scratch on the surface of the glass container. This view is supported by the observation that careful filtering of reagents reduces the number of pacemaker centers in the reaction mixture, and deliberate addition of dust increases the number of centers.

Waves emanate from rotating spirals at a single frequency, the frequency of rotation, shown in Figure 14.15. Spiral waves are generated by mechanically shearing one or more waves in a target pattern. This is most conveniently accomplished by slightly tilting the thin film of reaction mixture (7, 20), thereby mixing the top and bottom of the target circles with bromide-rich quiescent solution and leaving the sides relatively undisturbed. The result is an annihilation of the top and bottom of the circular waves, leaving intact the unconnected wave segments from the sides. Spirals form around the ends of these segments, and as shown in Figure 14.15, two pairs of oppositely rotating spirals are formed. Spiral waves may also be formed by mechanically shearing a pulse with a stirring rod, leaving two unconnected ends that become oppositely rotating spirals.

Heterogeneous catalysis is not important in spiral waves since all spirals throughout the medium rotate at a single frequency (7, 20). That this frequency is also the upper limit for target pattern frequencies supports the idea that the frequency is a property of the medium (7). A consequence of the high frequency of

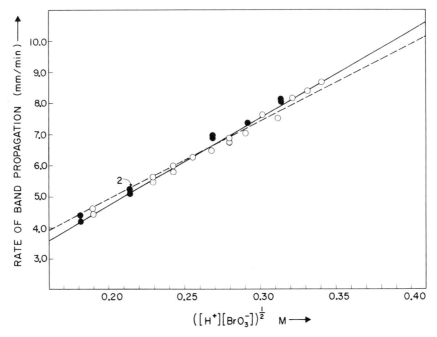

Figure 14.14. Wave propagation velocity dependence on initial bromate and hydrogen ion concentrations in the BZ reaction. Initial reactant concentrations as in Figure 14.13 except $[MA]_0 = 0.023\ M$. Initial H_2SO_4 concentration varied (open circles); initial $NaBrO_3$ concentration varied (filled circles). Temperature: $25.0 \pm 0.1\,°C$. (Reprinted with permission from R. J. Field and R. M. Noyes. *J. Am. Chem. Soc.* **96,** 2001 (1974). Copyright 1974 American Chemical Society.)

spiral waves is that the medium tends to fill with spirals at the expense of target patterns.

Spiral waves in a two-dimensional medium become scroll waves in a three-dimensional medium (7, 18). The inner tip of the spiral becomes a filament around which scroll waves emanate. Winfree (7, 64) has made topological arguments that the filament must form a closed ring. Thus, very exotic structures may form as the scroll waves propagate out from the filament. Experimental studies of the BZ reaction (7, 19, 65), support this picture of three-dimensional scroll waves. Winfree has predicted even more exotic structures arising from twisted filaments (66) and filaments tied in knots (67). The three-dimensional analog of target patterns is simple expanding concentric spheres.

The deliberate initiation of trigger waves provides further insights into the reaction–diffusion behavior of the BZ reaction (23, 68). Bromide is consumed at a silver electrode positively biased with respect to a reference platinum electrode. In a nonoscillatory but excitable medium a momentary depletion of Br^- allows $HBrO_2$ autocatalysis to proceed to an extent sufficient to initiate a wave. Silver metal alone in the acidic bromate reaction mixture will initiate waves for most

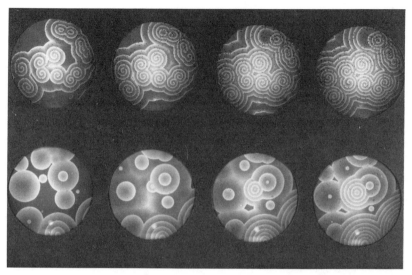

Figure 14.15. Chemical waves in the BZ reaction. Top: spiral waves in thin film of reagent (1.5 mm). Time sequence from left to right at 60-s intervals. Bottom: target patterns in reagent similar to above except more acidic. Field of view: 66 mm. (Reprinted with permission from A. T. Winfree, *The Geometry of Biological Time*, Springer-Verlag, New York, 1980. Copyright 1980 Springer-Verlag.)

reactant concentrations; however, this activity can be suppressed by applying a negative bias to the silver electrode. Thus, by maintaining a negative bias on a silver electrode and momentarily reversing the bias, a wave can be deliberately initiated at any time.

Figure 14.16 shows a plot of initiation pulse time, the duration of the applied positive bias necessary to initiate a wave, as a function of bromate and hydrogen ion concentrations. The reasonably linear dependence of the critical pulse duration τ on these reactant concentrations is given by

$$\tau/\text{ms} = 12.37 - 273.3[\text{BrO}_3^-][\text{H}_2\text{SO}_4]/\text{M}^2 \qquad (14.45)$$

For wave initiation, Br^- concentration must be maintained below its critical value for a period of time sufficient for process B' to compete with inward diffusing Br^-. Therefore, we expect the value of τ to depend on the rate of reaction (14.38), the rate-determining step for process B', with the associated dependence on bromate and hydrogen ion concentrations.

Other subtleties of BZ waves are demonstrated in wave initiation experiments. In an excitable but nonoscillatory reaction mixture, a bulk oscillation may be induced by swirling the reaction mixture. Oxygen enhances bromide regeneration in process C, and when it is vented from the solution by the swirling action, an induced oxidation excursion (IOE) occurs (23). As shown in Figure 14.12, Br^-

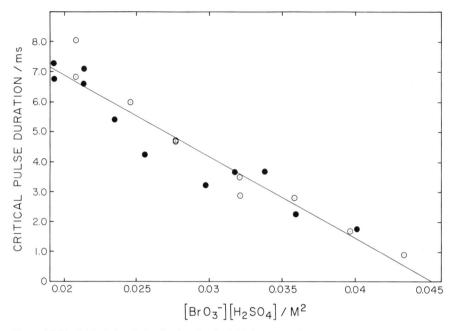

Figure 14.16. Critical electrical pulse duration for initiating waves in an excitable BZ reaction mixture as a function of bromate and sulfuric acid concentrations. Initial reactant concentrations: $[NaBrO_3]_0$ = 0.0746 M, $[H_2SO_4]_0$ = 0.371 M, $[MA]_0$ = 0.0257 M, $[BrMA]_0$ = 0.0748 M, $[Fe(II–III)]_0$ = 0.0013 M. Bromate concentration varied (open circles); sulfuric acid concentration varied (closed circles). Temperature: 25.0 ± 0.2°C. (Reprinted with permission from K. Showalter, R. M. Noyes, and H. Turner, *J. Am. Chem. Soc.* **101**, 7463 (1979). Copyright 1979 American Chemical Society.)

concentration (η) slowly decreases during most of the period of oscillation. Thus, following an IOE, bromide slowly decreases to its steady-state concentration. Of course, bromide concentration affects wave initiation and propagation, and therefore it is necessary to make measurements like those in Figure 14.16 at a particular time following an IOE.

Figure 14.17 shows the behavior of waves initiated at different times following an IOE of the bulk solution. Waves were detected at various distances from the silver initiation electrode by monitoring the potential between pairs of platinum electrodes in line with the wave propagation. We see fairly constant wave velocities for initiation times greater than about 1.0 min after the IOE. However, at initiation times less than 1.0 min waves tend to remain localized near the silver initiation electrode. Therefore, a refractory period occurs following an IOE where Br^- concentration is above that which will allow wave propagation. An interesting consequence of this refractory period is that waves initiated at 90 s following an IOE arrive at the platinum detection electrode before waves initiated at 10 s following an IOE (in another experiment) arrive at the same electrode (see inset of Figure 14.17).

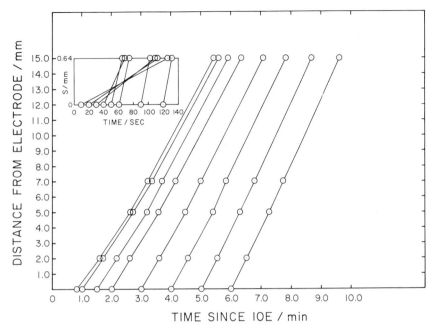

Figure 14.17. Propagation of waves initiated at different times following a temporal oscillation. Reactant concentrations: $[NaBrO_3]_0 = 0.0785\ M$, $[H_2SO_4]_0 = 0.375\ M$, $[MA]_0 = 0.0298\ M$, $[BrMA]_0 = 0.0767\ M$, $[Fe(II–III)]_0 = 0.0013\ M$. Temperature: $25.0 \pm 0.2°C$. (Reprinted with permission from K. Showalter, R. M. Noyes, and H. Turner, *J. Am. Chem. Soc.* **101**, 7463 (1979). Copyright 1979 American Chemical Society.)

The elevated Br^- concentration behind a propagating wave gives rise to effects similar to those observed after a bulk IOE. Thus, bromide is regenerated behind the wave in process C to a concentration above that through which wave propagation may occur. One consequence of this effect is that one wave may never overtake another. Another is that a refractory period occurs after the initiation of a wave during which it is not possible to initiate another wave. Just this effect is shown in Figure 14.18, where the frequency of waves detected at a nearby platinum electrode is plotted as a function of the frequency of wave initiation pulses. We see that below an initiation frequency of about $0.67\ min^{-1}$, corresponding to an initiation pulse once every 1.5 min, one wave is detected for every initiation pulse. This one-to-one correspondence is indicated by the diagonal dashed line. However, above this initiation frequency we find a maximum in detection frequency of about $0.64\ min^{-1}$, corresponding to a wave every 94 s. This frequency represents an upper limit for wave initiation in this particular reaction mixture. The maximum detection frequency occurs at an initiation frequency of about 0.8 min^{-1}. Another apparent maximum in detection frequency occurs at an initiation frequency of $1.6\ min^{-1}$, and apparent minima occur at 1.0 and $2.0\ min^{-1}$. These

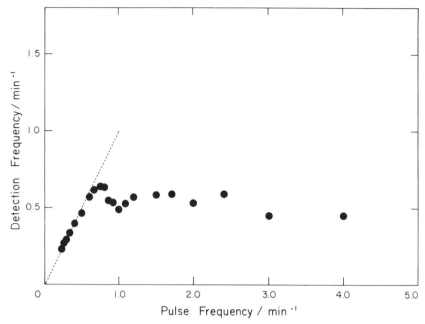

Figure 14.18. Pulse initiation frequency as a function of wave detection frequency. Dashed line represents equal initiation and detection frequencies. See Figure 14.16 for conditions and initial reactant concentrations. (Reprinted with permission from K. Showalter, R. M. Noyes, and H. Turner, *J. Am. Chem. Soc.* **101**, 7463 (1979). Copyright 1979 American Chemical Society.)

successive extrema suggest that the system responds similarly to integer multiples of initiation frequencies.

Wave initiation of a different type is shown schematically in Figure 14.19. Here, two electrically coupled silver electrodes are positioned in line with an advancing wave. As the wave passes through the first electrode, another wave is spontaneously initiated at the second electrode. The second wave propagates in both forward and reverse directions, and the original wave is annihilated upon collision with the reverse wave. The net result is a discontinuous advancement of the chemical wave by the separation distance of the silver electrodes. No external power source is necessary for this signal transmission. The potential difference between the oxidized reaction mixture in the wave front and the reduced quiescent solution ahead is apparently sufficient for wave initiation.

14.3.3. Phase Plane Analysis

Waves in the BZ reaction have been the subject of many theoretical analyses. Here, we will be concerned only with the reaction–diffusion or trigger wave and its waveform and propagation velocity dependence on reactant concentrations and rate constants.

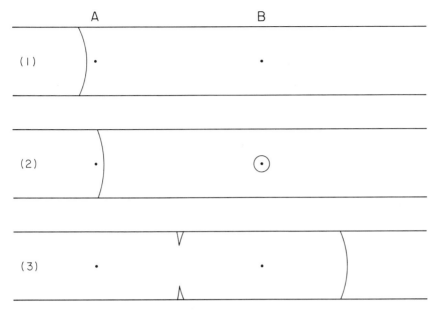

Figure 14.19. Schematic representation of wave initiation through coupled silver electrodes, represented by points A and B. (1) Trigger wave advancing from left to right. (2) Original wave propagates through electrode A and new wave is initiated at electrode B. (3) Original wave is annihilated by one side of new wave. (Reprinted with permission from K. Showalter, R. M. Noyes, and H. Turner, *J. Am. Chem. Soc.* **101**, 7463 (1979). Copyright 1979 American Chemical Society.)

Waves in the BZ reaction propagate in a manner similar to waves in the iodate–arsenous acid system; the autocatalytic species diffuses ahead in the wave front and initiates autocatalysis. However, the initiation of autocatalysis occurs differently in the two systems. In the BZ reaction Br^- must be consumed to a critical concentration by $HBrO_2$ for autocatalysis to occur. Therefore, the propagation velocity depends on the rate of autocatalysis and on the concentration of Br^- ahead of the wave. In the iodate–arsenous acid reaction, I^- simply diffuses ahead to provide a concentration sufficient for autocatalysis to proceed at an appreciable rate. Another difference between the two systems from an analysis point of view is that the conservation relations that greatly simplify the iodate–arsenous acid system are not available in the BZ system. However, reasonably accurate descriptions of BZ reaction–diffusion waves have been made by identifying the key elements in the FKN mechanism most important for wave propagation (21, 26, 69–71).

The basic features of BZ trigger waves can be accounted for by considering the reaction and diffusion of $HBrO_2$. Bromous acid is generated in process B' according to the kinetics of reaction (14.38). The autocatalytic increase in concentration is eventually limited by the disproportionation reaction (14.40). These two reactions of $HBrO_2$ give rise to a reaction–diffusion equation analogous to the famous Fisher equation (72, 73):

$$\frac{\partial C}{\partial t} = D \frac{\partial^2 C}{\partial x^2} + k_a C - k_b C^2 \tag{14.46}$$

where $C = [\mathrm{HBrO_2}]$, $k_a = k_{14.38}[\mathrm{BrO_3^-}][\mathrm{H^+}]$, $k_b = 2k_{14.40}$, and D is the diffusion coefficient of $\mathrm{HBrO_2}$. Of course, this description accounts only for the rise in bromous acid concentration in process B. Wave propagation depends primarily on this rise and is not affected by the subsequent decrease in $\mathrm{HBrO_2}$ concentration behind the wave front. However, a major approximation in this model is the neglect of the reaction between $\mathrm{HBrO_2}$ and $\mathrm{Br^-}$ ahead of the wave. The Fisher equation, originally proposed to describe the spread of advantageous genes, has been the subject of many theoretical studies (74–76). Following Tyson (42), we consider here only its properties relevant to the BZ reaction.

According to Eq. (14.46), the stationary states $\hat{C}$ of the homogeneous reaction mixture are given by

$$f(C) = \frac{dC}{dt} = 0 = k_a C - k_b C^2 \tag{14.47}$$

or

$$\hat{C} = 0, \frac{k_a}{k_b} \tag{14.48}$$

Defining the constant wave velocity according to Eq. (14.22) (with $C = C_1$) or introducing the new spatial coordinate $x - vt$ allows Eq. (14.46) to be expressed as the ordinary differential equation

$$DC'' + vC' + k_a C - k_b C^2 = 0 \tag{14.49}$$

No analytic solution is available for Eq. (14.49) as with Eq. (14.23) for iodate–arsenous acid waves; however, a lower limit for the propagation velocity may be obtained by considering a phase plane analysis of the first-order system

$$C' = g \qquad g' = \frac{-vg}{D} - \frac{f(C)}{D} \tag{14.50}$$

Figure 14.20 shows C as a function of distance for a wave propagating to the right according to the numerical solution (37) of Eqs. (14.50). The reaction mixture ahead of the wave is described by the stationary state $\hat{C} = 0$ and behind the wave by $\hat{C} = k_a/k_b$. Figure 14.21 shows a phase plane plot of the gradient g as a function of C between the two stationary states. These stationary states can be characterized by a linear stability analysis. The eigenvalues λ of the secular determinant (14.51) characterize the stationary state $\hat{C} = 0$ as a stable node and the stationary state $\hat{C} = k_a/k_b$ as a saddle point,

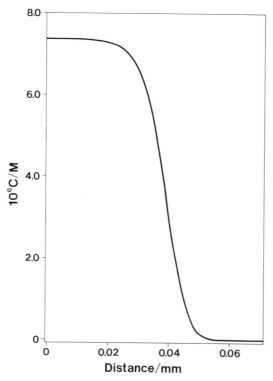

Figure 14.20. Bromous acid concentration C as a function of distance calculated from Eqs. (14.50) with minimal velocity $v = 2.17$ mm/s. Initial reactant concentrations: $[\text{BrO}_3^-]_0 = 0.229\ M$, $[\text{H}^+]_0 = 0.258\ M$. Rate constants and diffusion coefficient: $k_a = 1.0 \times 10^4\ [\text{BrO}_3^-]_0\ [\text{H}^+]_0\ M^{-2}\ s^{-1}$, $k_b = 4.0 \times 10^7\ M^{-1}\ s^{-1}$, $D = 2.0 \times 10^{-3}$ mm^2/s.

$$\begin{vmatrix} \dfrac{\partial C'}{\partial C} - \lambda & \dfrac{\partial C'}{\partial g} \\[2ex] \dfrac{\partial g'}{\partial C} & \dfrac{\partial g'}{\partial g} - \lambda \end{vmatrix}_{\hat{C}} = 0 \tag{14.51}$$

Evaluating (14.51) yields Eq. (14.52) for the eigenvalues at each of the stationary states,

$$\lambda = \frac{1}{2}\left[\frac{-v}{D} \pm \sqrt{\left(\frac{v}{D}\right)^2 - \frac{4f'(\hat{C})}{D}} \right] \tag{14.52}$$

From (14.52) we see that for $0 < f'(\hat{C}) < v^2/4D$, both eigenvalues are negative, corresponding to a stable node, and when $f'(\hat{C}) < 0$, the eigenvalues are of opposite sign, corresponding to a saddle point. We see also from Eq. (14.47) that

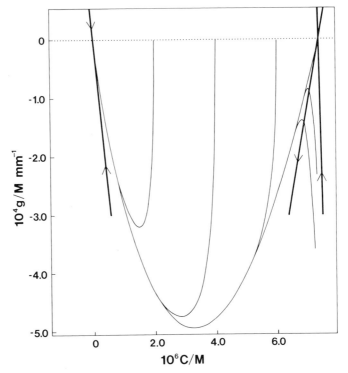

Figure 14.21. Gradient g as a function of C according to Eqs. (14.50). Eigenvectors indicated by arrows. The wavefront is represented by the trajectory connecting the saddle at $(7.37 \times 10^{-6} M, 0)$ to the node at $(0, 0)$. See Figure 14.20 for constants and initial concentrations.

for the stationary state $\hat{C} = 0$, $f'(\hat{C}) = k_a$ and for $\hat{C} = k_a/k_b$, $f'(\hat{C}) = -k_a$, corresponding to a stable node and a saddle point, respectively. It follows that for real eigenvalues at the origin there must be a minimum propagation velocity defined by

$$ v \geq 2\sqrt{Dk_a} \quad \text{or} \quad v \geq 2\sqrt{Dk_{14.38}[\text{H}^+][\text{BrO}_3^-]} \tag{14.53} $$

Velocities less than the minimum defined by Eq. (14.53) would require a stable focus at the origin, which is inappropriate for a propagating reaction–diffusion front. Thus, we find a velocity dependence on H^+ and BrO_3^- concentrations in agreement with that found experimentally (see Figure 14.14).

The concentration profile in Figure 14.20 and the phase plane trajectories in Figure 14.21 were calculated using the minimal velocity in the first-order system (14.50). Also shown in Figure 14.21 are the eigenvectors associated with each of the stationary states, two (identical) directed toward the node and two directed to and from the saddle point. Trajectories with a variety of initial conditions converge to the node. The trajectory corresponding to the wave front connects the saddle point to the node, approaching each stationary state asymptotically along an eigenvector.

In the treatment of BZ waves by Field and Noyes (21) only the autocatalytic generation of $HBrO_2$ in process B' is considered. The resulting reaction–diffusion equation is a linearized version of Eq. (14.46). A physically realistic solution of the linear differential equation requires the velocity to be equal to or greater than the minimum value given by Eq. (14.53).

A major deficiency of the Fisher equation description of BZ waves is that the effect of Br^- ahead of the wave is neglected. That the velocities predicted by Eq. (14.53) are about 20 times the experimentally measured velocities (21, 25) is probably a result of this neglect. Including the reaction of Br^- with $HBrO_2$ in (14.36) greatly complicates the analysis. Murray (69) has shown that the wave velocity is a function of bromide concentration ahead of the wave, but an analytic result was not possible. Asymptotic treatments (26, 70) of Murray's velocity function give conflicting functional dependences of velocity on $[H^+]$ and $[BrO_3^-]$ depending on the values of critical parameters. Tyson (42) has analyzed these studies and concludes that the Fisher equation description may be adequate provided that certain rate constants of the Oregonator are adjusted. Measurements of wave velocity over larger concentration ranges of H^+ are needed to determine the most appropriate treatment.

Tyson and Fife (71) have studied a two-variable reduction of the Oregonator. Their model provides an explanation of the target patterns typically observed in thin films of excitable reaction mixture in the reduced state. Their model also explains propagating bands of reduction through a solution in the oxidized state, experimentally discovered by Smoes (24).

14.3.4. Numerical Analysis

We will now consider a numerical analysis of waves in the BZ reaction carried out by Reusser and Field (22). They studied the reaction–diffusion behavior of the irreversible Oregonator,

$$A + Y \overset{1}{\rightarrow} X \tag{14.54}$$

$$X + Y \overset{2}{\rightarrow} P \tag{14.55}$$

$$B + X \overset{3}{\rightarrow} 2X + Z \tag{14.56}$$

$$X + X \overset{4}{\rightarrow} Q \tag{14.57}$$

$$Z \overset{5}{\rightarrow} fY \tag{14.58}$$

where $A = B = BrO_3^-$, $X = HBrO_2$, $Y = Br^-$, and $Z =$ oxidized catalyst. Bromate concentration is considered to be a constant, and because each reaction is considered to be irreversible, the products P and Q do not appear in the rate equations. The corresponding three-variable reaction–diffusion model is given by

$$\left(\frac{\partial X}{\partial t}\right)_l = D\left(\frac{\partial^2 X}{\partial l^2}\right)_t + k_1 AY - k_2 XY + k_3 BX - 2k_4 X^2 \qquad (14.59)$$

$$\left(\frac{\partial Y}{\partial t}\right)_l = D\left(\frac{\partial^2 Y}{\partial l^2}\right)_t - k_1 AY - k_2 XY + f k_5 Z \qquad (14.60)$$

$$\left(\frac{\partial Z}{\partial t}\right)_l = D\left(\frac{\partial^2 Z}{\partial l^2}\right)_t + k_3 BX - k_5 Z \qquad (14.61)$$

Here, X, Y, and Z represent the concentrations of the corresponding species. The diffusion coefficients D are assumed to be equal, and k_5 and f are treated as expendable parameters.

A linear stability analysis of the Oregonator for a stirred homogeneous reaction mixture shows that the steady-state stability is critically dependent on k_5 and f (41, 54). When k_5 and f are near unity, the steady state is unstable and temporal oscillations are exhibited. Figure 14.12 shows the concentrations of X (α), Y (η), and Z (ρ) as a function of time calculated from Eqs. (14.59)–(14.61) with the diffusion terms omitted.

Reusser and Field used Eqs. (14.59)–(14.61) to investigate the characteristics of diffusion-dependent trigger waves and diffusion-independent phase waves. The reaction–diffusion equations were numerically integrated using the method of lines (see Section 14.2.3). When $k_5 = 1$ and $f \geq 1.5$, Eqs. (14.59)–(14.61) exhibit a stable but excitable steady state. A trigger wave traveling through an excitable but nonoscillatory reaction mixture provides a limiting case for the maximum participation of diffusion in wave propagation. Here, the coupling of reaction with diffusion is essential for wave propagation because there is no possibility of phase gradients. Figure 14.22 shows the initial conditions and the subsequent concentration profiles of a trigger wave propagating to the right. The wave was initiated by decreasing [Br$^-$] from the steady-state value of 5.25×10^{-7} to 4.63×10^{-9} M over a distance of 12.9 μm. The same decrease in [Br$^-$] over a region on the grid one-fourth as large was insufficient to initiate a wave. These results are in accord with the experimental finding that wave initiation occurs only with a perturbation of sufficient magnitude (23). The initiation spike in Figure 14.22 evolves into a constant-waveform, constant-velocity trigger wave.

Reusser and Field also used their computational model to show that an oscillatory medium may support either a phase wave or a trigger wave depending on the initial conditions. For these calculations the initial conditions were generated by transferring the concentrations of X, Y, and Z from the temporal oscillation shown in Figure 14.12 onto the spatial grid. The phase gradient $|\nabla \phi|$ of the initial grid was defined as seconds of oscillatory cycle per millimeter distance. The initial phase gradient was then varied by adjusting the spacing of the grid points.

A calculation with a shallow initial phase gradient is shown in Figure 14.23. We see a wave with a very high velocity (600 mm/min) unaffected by diffusion.

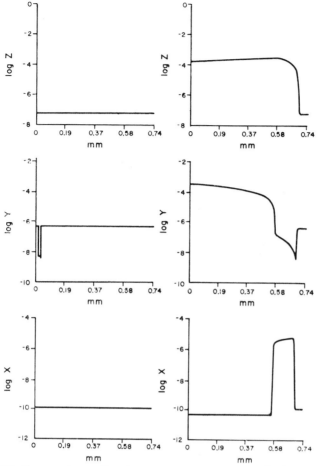

Figure 14.22. Concentration profiles of bromous acid (X), bromide (Y), and catalyst (Z) in a trigger wave calculated from Eqs. (14.59)–(14.61). Initial waveforms at $t = 0.0$ s (left) and waveforms at $t = 0.81$ s (right). See Figure 14.12 for parameter values. Diffusion coefficient: $D = 1.0 \times 10^{-5}$ cm^2/s. (Reprinted with permission from E. J. Reusser and R. J. Field, *J. Am. Chem. Soc.* **101**, 1063 (1979). Copyright 1979 American Chemical Society.)

That diffusion plays no role in the propagation of this wave is shown by the initial waveform in each variable remaining unchanged at later times. Conclusive evidence that this wave results from the phase gradient is that no change in velocity occurs upon setting $D = 0$ in Eqs. (14.59)–(14.61). A calculation with a steeper initial phase gradient is shown in Figure 14.24. Only the grid spacing differs in Figures 14.23 and 14.24; the initial waveforms are the same. The behavior in Figure 14.24 is complicated because the wave propagates into a nonzero phase gradient resulting from the initial conditions. However, it is clear that diffusion plays a vital role in the wave propagation because the waveform dramatically

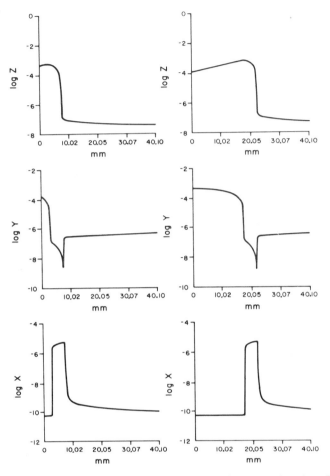

Figure 14.23. Concentration profiles of bromous acid (X), bromide (Y), and catalyst (Z) in a phase wave calculated from Eqs. (14.59)–(14.61). Initial waveforms at $t = 0.0$ s (left) and waveforms at $t = 1.47$ s (right). See Figure 14.12 for parameter values. (Reprinted with permission from E. J. Reusser and R. J. Field, *J. Am. Chem. Soc.* **101**, 1063 (1979). Copyright 1979 American Chemical Society.)

changes across the grid. Thus, the very steep initial concentration gradients are smoothed by diffusion, and the pulse widens as it propagates to the right. The initial velocity rapidly increases from the velocity (3.0 mm/min) that would arise solely from the phase gradient. It then gradually decreases as the wave propagates into an increasing concentration of Br⁻ remaining from the initial phase gradient.

The calculations of Reusser and Field demonstrate that either phase or trigger waves may be exhibited in an oscillatory BZ reaction mixture. Waves in experimental oscillatory systems are probably various hybrids of the two extremes. That waves cannot propagate with velocities lower than the trigger wave velocity, as shown in the calculation for Figure 14.24, is in accord with the minimal velocity prediction of the Fisher equation. It should be noted that while a smooth phase gradient would be difficult to manually arrange in an experimental system, phase gradients are established by the passing of a trigger wave. Reusser and Field's calculations of trigger waves in a nonoscillatory but excitable medium provides convincing evidence for a pure reaction–diffusion wave. However, successive waves in such a system are affected by phase gradients. A phase gradient is estab-

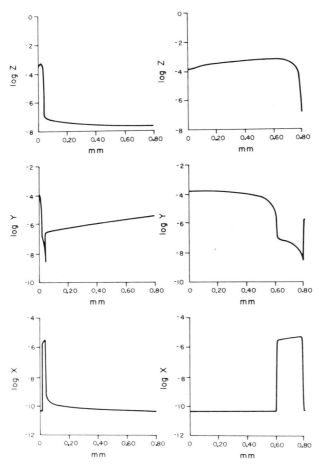

Figure 14.24. Concentration profiles of bromous acid (X), bromide (Y), and catalyst (Z) in a trigger wave propagating into a phase gradient calculated from Eqs. (14.59)–(14.61). Initial waveforms at $t = 0.0$ s (left) and waveforms at $t = 1.24$ s (right). See Figure 14.12 for parameter values. (Reprinted with permission from E. J. Reusser and R. J. Field, *J. Am. Chem. Soc.* **101**, 1063 (1979). Copyright 1979 American Chemical Society.)

lished by the first wave, and each of the following waves must travel into solution with a Br^- concentration higher than that of the steady state. Consequently, successive waves propagate with velocities lower than that of the first wave.

14.4. OTHER SYSTEMS

14.4.1. Other Wave Systems

A major advance in the study of chemical instabilities occurred with the discovery by Epstein and co-workers of a large number of new chemical oscillators, many with chlorite as a major reactant (77). One system, the chlorite–iodide–malonic acid reaction, has been shown to exhibit waves (27, 77) much like those found in the BZ reaction. Figure 14.25 shows an example of target patterns in this system. Although the new system is not as robust as the BZ reaction, it provides evidence that other reactions are capable of exhibiting exotic wave behavior like that found in the BZ system. Several other wave systems yet to be characterized have been reported. The chlorite–iodide (77), ferrous–nitric acid (77), and permanganate–oxalate (78, 79) reactions exhibit frontlike waves that, like the iodate–arsenous acid system, convert a solution from initial reactants to final products. Other chemical wave systems have been developed by modifying the BZ reaction. Orbán (5) has shown that the uncatalyzed BZ reaction exhibits waves similar to those observed in the classical reaction. A BZ reaction modified to eliminate the regeneration of Br^- in process C exhibits frontlike wave behavior (25). This system will now be considered in some detail.

The acidic bromate oxidation of ferroin has the features of a clock reaction. Bromide is slowly consumed in process A until the critical concentration is attained; ferroin is then rapidly oxidized as $HBrO_2$ is autocatalytically generated in process B. The reaction is not homogeneous, however, because the Br_2 product of process A reacts destructively with ferroin to form a red-brown precipitate. This complication can be eliminated by the addition of a water-soluble reactant that consumes Br_2. Addition of 4-cyclohexene-1,2-dicarboxylic acid to the reaction mixture prevents accumulation of bromine by

$$Br_2 + C_6H_8(COOH)_2 + H_2O \rightarrow C_6H_8(Br)(OH)(COOH)_2 + Br^- + H^+ \quad (14.62)$$

With this additional reagent the slow consumption of Br^- in process A proceeds according to (14.42) + 3(14.62) or (14.63).

$$BrO_3^- + 2Br^- + 3H^+ + 3C_3H_8(COOH)_2 \rightarrow 3C_6H_8(Br)(OH)(COOH)_2 \quad (14.63)$$

A single front of oxidation can be initiated prior to the bulk oxidation by locally depleting Br^- at a positively biased silver electrode. An example of a propagating front in a thin film of solution is shown in Figure 14.26. Reaction occurs only

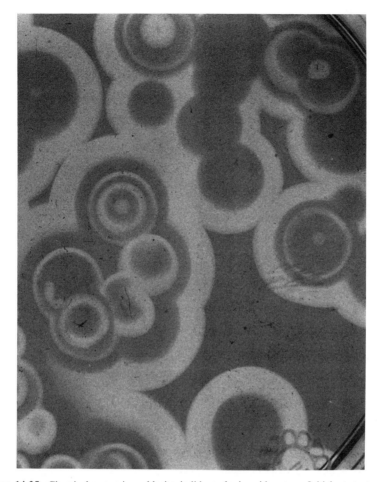

Figure 14.25. Chemical waves in a chlorite–iodide–malonic acid system. Initial reactant concentrations: $[MA]_0 = 0.0033\ M$, $[NaI]_0 = 0.09\ M$, $[NaClO_2]_0 = 0.1\ M$, $[H_2SO_4]_0 = 0.0056\ M$. Temperature: $5\degree C$. Starch indicator added to initial reaction mixture. (Reprinted with permission from P. De Kepper, I. R. Epstein, K. Kustin, and M. Orbán, *J. Phys. Chem.* **86,** 170 (1982). Copyright 1982 American Chemical Society.)

slowly in the bromide-rich solution ahead of the wave; behind the wave bromide concentration is very low, bromous acid concentration is high, and ferroin has been oxidized to ferriin.

The bromate–ferroin system exhibits wave behavior much like that of the BZ reaction (25, 26). The experimentally determined velocity dependence on hydrogen ion and bromate concentrations is shown in Figure 14.27. The dependence of velocity on these reactant concentrations is essentially identical to that found in the BZ reaction. Thus, the Fisher equation model applies also to the bromate–

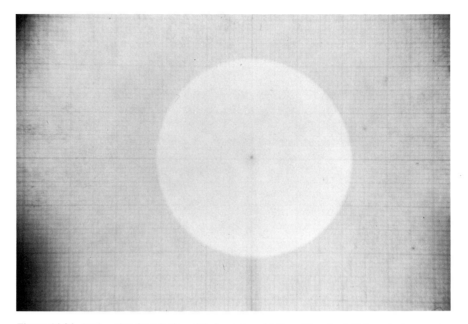

Figure 14.26. Propagating front in the acidic bromate oxidation of ferroin. Initial reactant concentrations: $[H_2SO_4]_0 = 0.189\ M$, $[NaBrO_3]_0 = 9.51 \times 10^{-2}\ M$, $[Ferroin]_0 = 1.21 \times 10^{-3}\ M$, $[KBr]_0 = 6.70 \times 10^{-2}\ M$, $[C_6H_8(COOH)_2]_0 = 9.67 \times 10^{-2}\ M$. Temperature: $25.0 \pm 0.2°C$. Field of view: 6 × 10 cm. (Reprinted with permission from K. Showalter, *J. Phys. Chem.* **85**, 440 (1981). Copyright 1981 American Chemical Society.)

ferroin fronts. Of course, the slowly changing reaction mixture ahead and behind the front must be approximated as stationary states in this description.

That the Fisher equation can only approximately describe BZ pulses and bromate–ferroin fronts is demonstrated by the effect of Br^- concentration on front propagation. Figure 14.28 shows wave front position as a function of time. As Br^- is consumed in process A, the velocity increases, indicated by the instantaneous slope of the curve. Thus, as Br^- concentration ahead of the wave decreases, the diffusion of $HBrO_2$ ahead is more effective in triggering the advancement of the wave. The inhibiting effect of Br^- diffusing into the front is also decreased.

14.4.2. Electric Field Effects

Schmidt and Ortoleva were first in developing the notion that chemical waves involving ionic species should be affected by an applied electric field (80–82). Thus, the forward migration of positive ions will be accelerated and that of negative ions retarded in a wave propagating toward an electrode negatively biased

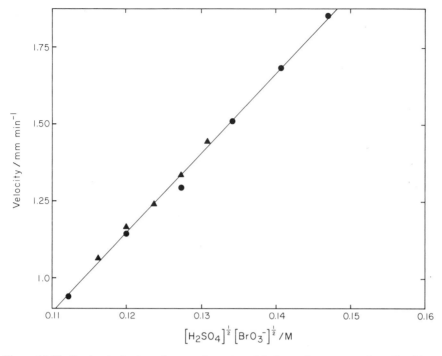

Figure 14.27. Front velocity dependence on bromate and hydrogen ion concentrations. Conditions and initial reactant concentrations as in Figure 14.26 except $[KBr]_0 = 4.79 \times 10^{-2}$ M and $[C_6H_8(COOH)_2]_0 = 8.97 \times 10^{-2}$ M. (Reprinted with permission from K. Showalter, *J. Phys. Chem.* **85,** 440 (1981). Copyright 1981 American Chemical Society.)

with respect to a second electrode behind the wave. In addition to the charge type the charge magnitude will also be important.

In their theoretical treatments Schmidt and Ortoleva consider two-variable reaction–diffusion models containing terms for electric-field-induced ion migration. Thus, for a chemical wave propagating in an electric field we have

$$\frac{\partial C}{\partial t} = D\frac{\partial^2 C}{\partial x^2} + ME \cdot \frac{\partial C}{\partial x} + f(C) \qquad (14.64)$$

where C represents a column vector of the species concentrations, D and M are the diffusion and ion mobility coefficient matrices, E is the applied electric field, and $f(C)$ is the chemical reaction rate. Rewriting Eq. (14.64) in terms of a constant velocity according to Eq. (14.22) gives

$$D\frac{d^2 C}{dx^2} + (v + ME)\frac{dC}{dx} + f(C) = 0 \qquad (14.65)$$

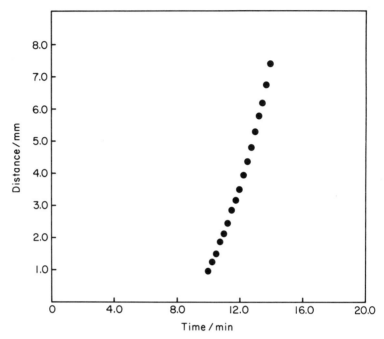

Figure 14.28. Wave front position as a function of time in the bromate–ferroin system. Conditions and initial reactant concentrations as in Figure 14.26 except $[KBr]_0 = 6.22 \times 10^{-2}$ M. (Reprinted with permission from K. Showalter, *J. Phys. Chem.* **85,** 440 (1981). Copyright 1981 American Chemical Society.)

The key species of the BZ reaction, identified in the Oregonator, are $HBrO_2$, Br^-, and the metal ion catalyst M^{n+}. We see that these species are all of different charge types; therefore, an applied electric field should affect BZ wave propagation. Ortoleva and co-workers (83) recognized that the high ionic strength of the typical BZ reaction mixture would result in a large conductivity with unacceptable ohmic heating. Therefore, a modified low-ionic-strength recipe was developed in which $HBrO_3$ was substituted for the usual bromate salt and sulfuric acid. The system was studied using a reagent-saturated membrane filter between two electrodes.

In Ortoleva's experiments a circular two-dimensional BZ wave showed an increased velocity toward the positive electrode and a decreased velocity toward the negative electrode. As expected, wave propagation perpendicular to the electric field was little affected. Thus, Br^- migrates into the wave propagating toward the negative electrode, thereby impeding its advancement. The induced migration of the oxidized metal catalyst, $Fe(phen)_3^{3+}$, produces a similar effect. The migration of Fe(III) from within a wave propagating toward the negative electrode to the solution ahead causes an increased Br^- concentration from process C. The most dramatic observation in these experiments was the formation of a crescent wave from a circular wave in the presence of the electric field. Upon switching off the field, the free ends of the crescent wave formed oppositely rotating spiral waves (83).

More refined experiments examining the effects on BZ waves of an applied electric field have recently been reported by Sevcikova and Marek (84, 85). A thermostated tube was connected through membranes to electrolytic cells, effectively eliminating undesirable effects from ohmic heating and electrolysis products. Sevcikova and Marek observed wave splitting in the BZ reaction, shown schematically in Figure 14.29. When a field of 16.7 V/cm is applied, the forward progress of a wave propagating toward the negative electrode is greatly reduced, and the wave back serves to initiate waves in the opposite direction. The new waves propagate at a higher velocity and are initiated at a higher frequency than waves in a field-free environment. At $E = 20$ V/cm, a wave is annihilated, but only after initiating another wave propagating in the opposite direction. Thus, the applied electric field results in propagation reversal.

Sevcikova and Marek (85, 86) also studied the effect of an electric field on waves in the iodate–arsenous acid reaction. Figure 14.30a shows front position as a function of time for a wave propagating toward the positive electrode and another wave propagating toward the negative electrode. A field of 8 V/cm is applied in the time interval $I = 2$. The wave traveling toward the positive electrode is ac-

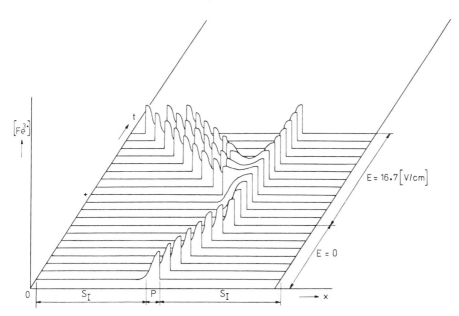

Figure 14.29. Schematic representation of BZ wave splitting in an electric field. Regions P and S show pulse and steady state, respectively. (Reprinted with permission from H. Sevcikova and M. Marek, *J. Phys. Chem.* **88**, 2183 (1984). Copyright 1984 American Chemical Society.)

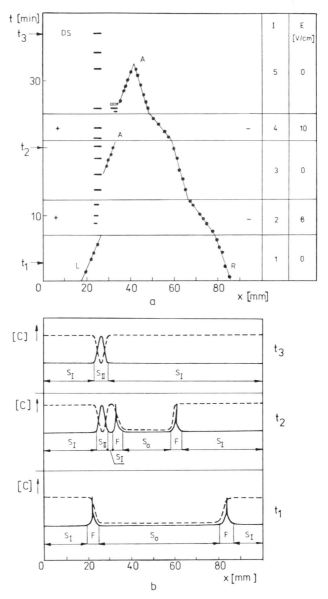

Figure 14.30. Effect of electric field on waves in the iodate–arsenous acid system. (*a*) Front position as a function of time for waves propagating from left to right (*L*) and from right to left (*R*). (*b*) Schematic concentration profiles of I^- (— — — —) and I_2 (————). (Reprinted with permission from H. Sevcikova and M. Marek, *J. Phys. Chem.* **88,** 2183 (1984). Copyright 1984 American Chemical Society.)

celerated as the migration of I^- ahead in the wave is enhanced. The wave propagating toward the negative electrode is stopped as I^- is pulled back into the wave. A blue band appears at the stopping point and remains even after the field is switched off. A new wave is initiated from this stationary band in the zero-field state. The long-lived band arises because both iodate and iodide are pulled into the front, and the result is excess iodate stoichiometry in the local region. Thus, iodine is generated according to net reaction (14.4) and remains until it eventually dissipates by diffusion into neighboring regions. Figure 14.30b shows schematic concentration profiles corresponding to the time intervals in Figure 30a.

14.4.3. Stationary Patterns

Much of the current interest in chemical waves was stimulated by the work of Turing (1) and Prigogine and Nicolis (2). These studies showed that the steady state of a simple two-variable reaction–diffusion model may become unstable for certain constraints with the system subsequently evolving to a stable patterned state. The new state is characterized by stationary bands in the chemical species concentrations maintained by a delicate balance between reaction and diffusion. Many theoretical studies on spontaneous pattern formation have since been carried out and reviews are available (87, 88).

No clear-cut case of stationary reaction–diffusion patterns in a homogeneous chemical system has been reported. Stationary patterns with band spacing of 2–3 mm have been reported in various versions of the BZ reaction. Zhabotinsky and Zaikin (3) observed stationary bands in the classical BZ reaction, and later bands were found in an uncatalyzed BZ system (5) and in the bromate–ferroin system (4) discussed in Section 14.4.1. The bands in these systems, however, appear to be the result of a hydrodynamic instability (4). Evidence for this explanation comes from the observation that the patterns disappear in very thin films or under a glass cover. Bromine apparently preferentially migrates to the air–solution interface in these systems. The nonpolar Br_2, evenly distributed on the surface, tends to condense into bands. Temperature effects arising from the volatilization of Br_2 also play a role in the instability. Because Br_2 is an inhibitor of autocatalysis in process B, the nonuniform concentration distribution gives rise to striking bands of oxidized and reduced solution.

Becker and Field (89) have reported new theoretical results that call for a reinvestigation of the BZ system. They have found stationary waves in a solution of Eqs. (14.59)–(14.61). The time-independent eigenvalue solution to the PDEs corresponds to small-amplitude waves. These waves have a cosine function profile and depend on the container length to be an integral multiple of the wavelength. The concentration variations predicted for the small-amplitude waves would not be visually detectable in an experimental BZ system. Although the stationary waves may appear for extended periods of time, numerical integration of the PDEs using the method of lines demonstrates that they are ultimately unstable. An example of these patterns is shown in Figure 14.31 at the scaled time $T = 25$. Similar small-

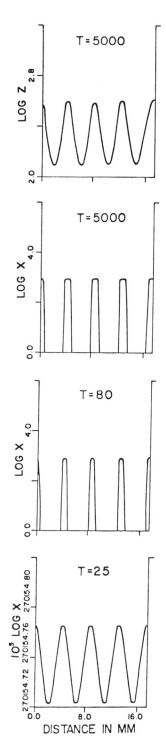

Figure 14.31. Stationary waves calculated from the Oregonator. Bromous acid concentration (X) in small amplitude waves at the scaled time $T = 25$. Catalyst (Z) and bromous acid concentrations in large-amplitude waves at $T = 5000$. (Reprinted with permission from P. K. Becker and R. J. Field, *J. Phys. Chem.* **89,** 118 (1985). Copyright 1985 American Chemical Society.)

amplitude stationary waves exhibited by the Brusselator model are found to be stable (87, 88).

A significant and rather unexpected result of Becker and Field's calculations is the appearance of stable large-amplitude patterns. Thus, the small-amplitude linearized solution eventually evolves into the large-amplitude patterns, as shown in Figure 14.31. These final patterns would be visually detectable in an experimental system. While the symmetrical large-amplitude patterns shown in Figure 14.31 also depend critically on the length boundary condition, stable asymmetric patterns may develop when the container length is a nonintegral multiple of the wavelength. Both large- and small-amplitude patterns were found with the diffusion coefficient of Z larger than the coefficients for X and Y. For the calculation in Figure 14.31 this difference was a factor of 10.

The calculations of Becker and Field indicate that stationary patterns might be exhibited by the BZ reaction. An experimental search will require minimizing the various physical effects that might mask any pure reaction–diffusion patterns. These include hydrodynamic instabilities driven by surface tension and temperature inhomogeneities along with precipitation complications often observed in the ferroin-catalyzed BZ reaction.

14.5. CONCLUSION

Chemical waves are structured concentration disturbances arising from the coupling of reaction with diffusion. Known systems involving only reaction and diffusion exhibit constant-waveform, constant-velocity propagating waves. The waves may be fronts or pulses, best illustrated by the iodate–arsenous acid reaction and the BZ reaction.

Conservation relations permit a one-variable analytic description of iodate–arsenous acid waves with arsenous acid in stoichiometric excess. The analytic solution is in accord with a numerical analysis of a four-variable model. Waves in excess iodate solution are accurately described by the numerical model. A one-variable Fisher equation analysis provides an approximate description of waves in the BZ reaction. An accurate description of this system is obtained by numerical integration of the three-variable Oregonator model.

Many experimental results await theoretical analysis. Electrochemical wave initiation in the BZ reaction and certain electric field effects in both the BZ and iodate–arsenous acid systems have yet to be analyzed. New wave systems such as the chlorite–iodide–malonic acid reaction and the ferrous–nitric acid system will require the formulation of tractable models. In addition, it is now apparent that other physical effects may be superimposed on the reaction–diffusion phenomenon. In particular, wave velocities in the iodate–arsenous acid system may be affected by hydrodynamic mass transport arising from reaction exothermicity. Also, in thin films of excess-iodate solution waves propagate with nonconstant velocities arising from a surface spreading of molecular iodine. Incorporation of surface and hydrodynamic effects to provide realistic models of these chemical waves presents a challenge to future investigators.

New theoretical results provide compelling reasons to reexamine the BZ system. Experiments over a wider range of hydrogen ion concentration should determine the most appropriate analytical description for the BZ pulse. In addition, careful experiments eliminating potential hydrodynamic instabilities may provide evidence for stationary waves in the BZ reaction.

The coupling of reaction and diffusion gives rise to spatial analogs of temporal behavior found in oscillatory and multistable systems. Thus, the spatial periodicity of wave trains and the conversion of a solution from one state to another by a propagating front are the reaction–diffusion analogs of temporal oscillations and bistability. Reaction–diffusion systems provide a medium for examining coupled oscillators in neighboring volume elements and the interaction of multiple steady states through a front. The inherent excitability of these systems also provides an opportunity to examine the effects of microscopic fluctuations on macroscopic behavior. Thus, these systems will provide a medium for future investigations of fluctuations and interactions between kinetic states.

A number of interesting studies of chemical waves have appeared recently. Digital imaging techniques have been used by Wood and Ross (90) and Müller et al. (91–93) to obtain a quantitative picture of propagating pulses and spiral waves in the BZ reaction, and Keener and Tyson (94) have developed a theoretical description of spiral waves. Fronts in the iodide–chlorite (95) and ferrous–nitric acid (96) reactions have been further characterized, and convective mass transport in addition to diffusion is necessary to describe the wave behavior. The coupling of autocatalytic reaction and mass transport by surface spreading and thermal convection has also been studied in the iodate–arsenous acid reaction (38). In addition, fronts in unbuffered neutral and basic iodate–arsenous acid reaction mixtures have been investigated (97). Autocatalysis in both iodide and hydrogen ion is important in these waves, and in the most basic solutions near-stationary fronts are exhibited.

REFERENCES

1. A. M. Turing, *Philos. Trans. Roy. Soc. Lond.*, *Series B* **237**, 37 (1952).

2. I. Prigogine and G. Nicolis, *J. Chem. Phys.* **46**, 3542 (1967).

3. A. M. Zhabotinsky and A. N. Zaikin, *J. Theor. Biol.* **40**, 45 (1973).

4. K. Showalter, *J. Chem. Phys.* **73**, 3735 (1980).

5. M. Orbán, *J. Am. Chem. Soc.* **102**, 4311 (1980).

6. J. J. Tyson, *Lect. Notes Biomath.* **10**, 1 (1976).

7. A. T. Winfree, *The Geometry of Biological Time*, Springer-Verlag, New York, 1980.

8. P. Ortoleva and J. Ross, *J. Chem. Phys.* **60**, 5090 (1974).

9. R. J. Field and M. Burger (Eds.), *Oscillations and Traveling Waves in Chemical Systems*, Wiley-Interscience, New York, 1985.

10. P. A. Epik and N. S. Shub, *Dokl. Akad. Nauk SSSR* **100,** 503 (1955).

11. T. A. Gribschaw, K. Showalter, D. L. Banville, and I. R. Epstein, *J. Phys. Chem.* **85,** 2152 (1981).

12. A. Hanna, A. Saul, and K. Showalter, in *Nonlinear Phenomena in Chemical Dynamics,* C. Vidal and A. Pacault (Eds.), Springer-Verlag, New York, 1981, pp. 160–165.

13. A. Hanna, A. Saul, and K. Showalter, *J. Am. Chem. Soc.* **104,** 3838 (1982).

14. A. Saul and K. Showalter, in *Oscillations and Traveling Waves in Chemical Systems,* R. J. Field and M. Burger (Eds.), Wiley-Interscience, New York, 1985, p. 419.

15. B. P. Belousov, in *Autowave Processes in Systems with Diffusion,* N. A. Gorodetskaya and N. N. Kralina (Eds.), USSR Academy of Sciences, Gorky, 1981.

16. A. N. Zaikin and A. M. Zhabotinsky, *Nature (Lond.)* **225,** 535 (1970).

17. A. T. Winfree, *Science* **175,** 634 (1972).

18. A. T. Winfree, *Science* **181,** 937 (1973).

19. A. T. Winfree, *Sci. Am.* **230,** 82 (1974).

20. A. T. Winfree, *Faraday Symp. Chem. Soc.* **9,** 38 (1974).

21. R. J. Field and R. M. Noyes, *J. Am. Chem. Soc.* **96,** 2001 (1974).

22. E. J. Reusser and R. J. Field, *J. Am. Chem. Soc.* **101,** 1063 (1979).

23. K. Showalter, R. M. Noyes, and H. Turner, *J. Am. Chem. Soc.* **101,** 7463 (1979).

24. M-L. Smoes, in *Dynamics of Synergetic Systems,* H. Haken (Ed.), Springer-Verlag, Berlin, 1980, p. 80.

25. K. Showalter, *J. Phys. Chem.* **85,** 440 (1981).

26. J. Rinzel and G. B. Ermentrout, *J. Phys. Chem.* **86,** 2954 (1982).

27. P. De Kepper, I. R. Epstein, K. Kustin, and M. Orbán, *J. Phys. Chem.* **86,** 170 (1982).

28. B. Hess, A. Boiteux, H. B. Busse, and G. Gerisch, *Adv. Chem. Phys.* **29,** 137 (1975).

29. J. Eggert and B. Scharnow, *Z. Elektrochem.* **27,** 457 (1921).

30. J. Bognár and S. Sárosi, *Anal. Chim. Acta* **29,** 406 (1963).

31. S. Dushman, *J. Phys. Chem.* **8,** 453 (1904).

32. J. R. Roebuck, *J. Phys. Chem.* **6,** 365 (1902).

33. H. A. Liebhafsky and G. M. Roe, *Int. J. Chem. Kinet.* **11,** 693 (1979).

34. J. N. Pendlebury and R. H. Smith, *Int. J. Chem. Kinet.* **6,** 663 (1974).

35. P. De Kepper, I. R. Epstein, and K. Kustin, *J. Am. Chem. Soc.* **103,** 6121 (1981).

36. G. A. Papsin, A. Hanna, and K. Showalter, *J. Phys. Chem.* **85,** 2575 (1981).

37. (a) A. C. Hindmarsh, *GEAR: Ordinary Differential Equation Solver,* Technical Report No. UCID-30001, Rev. 3., Lawrence Livermore Laboratory, Livermore, CA, 1974. (b) A. C. Hindmarsh, *LSODE,* Lawrence Livermore Laboratory, Livermore, CA, 1980.

38. T. McManus, A. Saul, and K. Showalter, to be published.

39. B. P. Belousov, *Ref. Rad. Med.* **1958,** 145 (1959).

40. A. M. Zhabotinsky, *Dokl. Acad. Nauk SSSR* **157,** 392 (1964).

41. R. J. Field and R. M. Noyes, *Acc. Chem. Res.* **10,** 214 (1977); R. M. Noyes and R. J. Field, *Acc. Chem. Res.* **10,** 273 (1977).

42. J. J. Tyson, in *Oscillations and Traveling Waves in Chemical Systems,* R. J. Field and M. Burger (Eds.), Wiley-Interscience, New York, 1985, p. 93.

43. R. J. Field, in *Oscillations and Traveling Waves in Chemical Systems*, R. J. Field and M. Burger (Eds.), Wiley-Interscience, New York, 1985, p. 55.

44. O. E. Rössler and K. Wegmann, *Nature* **271,** 89 (1978).

45. J. L. Hudson, M. Hart, and D. Marinko, *J. Chem. Phys.* **71,** 1601 (1979).

46. J. S. Turner, J.-C. Roux, W. D. McCormic, and H. L. Swinney, *Phys. Lett.* **85A,** 9 (1981).

47. W. Geiseler and H. H. Föllner, *Biophys. Chem.* **6,** 107 (1977).

48. M. Marek and E. Svobodova, *Biophys. Chem.* **3,** 263 (1975).

49. P. De Kepper, A. Rossi, and A. Pacault, *C. R. Acad. Sci. Ser. C* **283,** 371 (1976).

50. R. M. Noyes, *J. Chem. Phys.* **80,** 6071 (1984).

51. K. I. Agladze and V. I. Krinsky, *Nature* **296,** 424 (1982).

52. R. J. Field, E. Körös, and R. M. Noyes, *J. Am. Chem. Soc.* **94,** 8649 (1972).

53. Z. Noszticzius, H. Farkas, and Z. A. Schelly, *J. Chem. Phys.* **80,** 6062 (1984).

54. R. J. Field and R. M. Noyes, *J. Chem. Phys.* **60,** 1877 (1974).

55. R. J. Field, *J. Chem. Phys.* **63,** 2289 (1975); **65,** 1603 (1976).

56. K. Showalter, R. M. Noyes, and K. Bar-Eli, *J. Chem. Phys.* **69,** 2514 (1978).

57. H. B. Busse, *J. Phys. Chem.* **73,** 750 (1969).

58. M. T. Beck and Z. B. Varadi, *Natur. Phys. Sci.* **235,** 15 (1972).

59. M. T. Beck, Z. B. Varadi, and K. Hauck, *Acta Chim. Acad. Scie. Hung.* **91,** 13 (1976).

60. N. Kopell and L. N. Howard, *Science* **180,** 1171 (1973).

61. P. Ortoleva and J. Ross, *J. Chem. Phys.* **58,** 5673 (1973).

62. P. S. Hagan, *Adv. Appl. Math.* **2,** 400 (1981).

63. A. T. Winfree, in *Theoretical Chemistry*, Vol. 4, H. Eyring and D. Henderson (Eds.), Academic Press, New York, 1978, pp. 1–51.

64. A. T. Winfree and S. H. Strogatz, *Physica* **8D,** 35 (1983).

65. B. J. Welsh, A. Burgess, and J. Gomatam, *Nature* **306,** 611 (1983).

66. A. T. Winfree and S. H. Strogatz, *Physica* **9D,** 65 (1983).

67. A. T. Winfree and S. H. Strogatz, *Physica* **9D,** 333 (1983).

68. K. Showalter and R. M. Noyes, *J. Am. Chem. Soc.* **98,** 3730 (1976).

69. J. D. Murray, *J. Theor. Biol.* **56,** 329 (1976).

70. S. Schmidt and P. Ortoleva, *J. Chem. Phys.* **72,** 2733 (1980).

71. J. J. Tyson and P. C. Fife, *J. Chem. Phys.* **73,** 2224 (1980).

72. R. A. Fisher, *Ann. Eugen.* **7,** 355 (1937).

73. A. Kolmogoroff, I. Petrovsky, and N. Piscounoff, *Bull. Univ. Moscou Ser. Int. Sec. A* **1(6),** 1 (1937).

74. D. G. Aronson and H. F. Weinberger, *Lect. Notes Math.* **446,** 5 (1975).

75. D. G. Aronson, *Publ. Math. Res. Cent. Wis.* **44,** 161 (1980).

76. P. C. Fife, *Lect. Notes Biomath.* **28,** 1 (1979).

77. I. R. Epstein, *J. Phys. Chem.* **88,** 187 (1984).

78. N. S. Shub, *Ukrain. Khim. Zhur.* **23,** 22 (1957).

79. K. Showalter, unpublished observations.

80. S. Schmidt and P. Ortoleva, *J. Chem. Phys.* **67,** 3771 (1977).

81. S. Schmidt and P. Ortoleva, *J. Chem. Phys.* **71,** 1010 (1979).

82. S. Schmidt and P. Ortoleva, *J. Chem. Phys.* **74,** 4488 (1981).

83. R. Feeney, S. Schmidt, and P. Ortoleva, *Physica* **2D,** 536 (1981).

84. H. Sevcikova and M. Marek, *Physica* **9D,** 140 (1983).

85. H. Sevcikova and M. Marek, *J. Phys. Chem.* **88,** 2183 (1984).

86. H. Sevcikova and M. Marek, *Physica* **13D,** 379 (1984).

87. G. Nicolis and I. Prigogine, *Self-Organization in Nonequilibrium Systems*, Wiley-Interscience, New York, 1977.

88. H. Haken, *Synergetics*, Springer-Verlag, New York, 1978.

89. P. K. Becker and R. J. Field, *J. Phys. Chem.*, **89,** 118 (1985).

90. P. Wood and J. Ross, *J. Chem. Phys.* **82,** 1924 (1985).

91. S. C. Müller, Th. Plesser, and B. Hess, *Anal. Biochem.* **146,** 125 (1985).

92. S. C. Müller, Th. Plesser, and B. Hess, *Science* **230,** 661 (1985).

93. S. C. Müller, Th. Plesser, and B. Hess, *Naturwissenschaften* **73,** 165 (1986).

94. J. Keener and J. J. Tyson, in publication.

95. D. M. Weitz and I. R. Epstein, *J. Phys. Chem.* **88,** 5300 (1984).

96. G. Bazsa and I. R. Epstein, *J. Phys. Chem.* **89,** 3040 (1985).

97. J. Harrison and K. Showalter, *J. Phys. Chem.* **90,** 225 (1986).

15 APPENDICES

Gordon R. Freeman

Department of Chemistry
University of Alberta
Edmonton, Canada

15.1. USEFUL CONSTANTS AND UNITS

Constant or Unit	Symbol	SI	cgs	Quantity
Acceleration due to gravity	g_0	9.81 m/s²	981 cm/s²	Acceleration due to standard gravity
Ampere	A		A	Electric current
Ampere/meter	A/m	A/m	10^{-2} A/cm	Magnetic field strength
Angstrom	Å	10^{-10} m	10^{-8} cm	Length
Astronomical unit	AU	1.50×10^{11} m	—	Distance
Atmosphere	atm	101,325 Pa	1.013×10^6 dyn/cm²	Pressure
Avogadro's constant	N_A	6.0225×10^{23} ent^a/mol	6.0225×10^{23} ent^a/mol	Amount of entities
Bar	b	10^5 Pa	10^6 dyn/cm²	Pressure
Becquerel	Bq	1 disintegration/s	—	Amount of radioactivity
Calorie	cal	4.187 J	4.187×10^7 erg	Energy
Candela	cd	cd	cd	Luminous intensity
Charge, electron	e	1.6021×10^{-19} C/ent^a	4.8030×10^{-10} esu/ent^a	Electric charge
Constant				
Boltzmann	k_B	1.3805×10^{-23} J/K·ent^a	1.3805×10^{-16} erg/K·ent^a	R/N_A
Gas	R	8.3143 J/mol·K	8.3143×10^7 erg/mol·K 82.053 cm³·atm/mol·K	Proportionality constant in gas law
Gravitational	G	6.670×10^{-11} N·m²/kg²	6.670×10^{-8} dyn·cm²/g²	—
Planck	h	6.6256×10^{-34} J·s/cy^a	6.6256×10^{-27} erg·s/cy^a	—
$h/2\pi$	ℏ	1.0545×10^{-34} J·s/rad^a	1.0545×10^{-27} erg·s/rad^a	
Coulomb	C	A·s	A·s	Electric charge
Curie	Ci	3.7×10^{10} Bq	3.7×10^{10} disintegrations/s	Amount of radioactivity
Cycle	cy^a	cy^a	cy^a	cy $= 2\pi$ rad, a complete wave
Debye	𝔇	3.336×10^{-30} C·m	10^{-18} esu·cm	Dipole moment
Electrostatic unit	esu	3.336×10^{-10} C	(erg·cm)$^{1/2}$	Electric charge
Electric permittivity	ε_0	8.8542×10^{-12} C/V·m	—	Permittivity of vacuum
Electron-volt	eV	1.602×10^{-19} J	1.602×10^{-12} erg	Energy
Entity	ent^a	ent^a	ent^a	A single thing
Erg	erg	10^{-7} J	$g \cdot cm^2/s^2$	Energy

Name	Symbol	SI	CGS	Quantity
Faraday	$\mathscr{F}$	96,485 C/mol	96,485 C/mol	Mol of charge
Farad	F	C/V	C/V	Electrical capacitance
Gram	g	10^{-3} kg	g	Mass
Gray	Gy	J/kg	10^4 erg/g	(radiation) Absorbed dose
Henry	H	Wb/A	V·s/A	Electric inductance
Hertz	Hz	cy^a/s	cy^a/s	Frequency
Joule	J	N·m	10^7 erg	Energy
Kelvin	K	K	K	Thermodynamic temperature
Light year	ly	9.46×10^{15} m	—	Distance
Lumen	lm	cd·sr	cd·sr	Luminous flux
Lux	lx	lm/m^2	10^{-4} lm/cm^2	Illumination
Magnetic permeability	μ_0	$4\pi \times 10^{-7}$ $N \cdot s^2/C^2$	—	Permeability of vacuum
Mass, at rest				
Atomic mass unit	u	1.66044×10^{-27} kg	1.66044×10^{-24} g	One-twelfth of mass of ^{12}C atom
Alpha particle	M_α	6.6459×10^{-27} kg	6.6459×10^{-24} g	—
Electron	m_0	9.1091×10^{-31} kg	9.1091×10^{-28} g	—
Neutron	M_n	1.67482×10^{-27} kg	1.67482×10^{-24} g	—
Proton	M_p	1.67252×10^{-27} kg	1.67252×10^{-24} g	—
Meter	m	m	100 cm	Length
Mole	mol	mol	mol	N_A entities[b]
Newton	N	$kg \cdot m/s^2$	10^5 $g \cdot cm/s^2$	Force
Newton/meter	N/m	N/m	10^3 dyn/cm	Surface tension
Ohm	Ω	V/A	V/A	Electric resistance
Parsec	pc	3.09×10^{16} m	—	Distance
Pascal	Pa	N/m^2	—	Pressure
Pascal·second	Pa·s	kg/m·s	10 $g/cm \cdot s^2$	Dynamic viscosity
Poise	P	0.1 kg/m·s	10 g/cm·s	Dynamic viscosity
Rad	rad	10^{-2} Gy	100 erg/g	(radiation) Absorbed dose
Rad-equivalent-man	rem	10^{-2} Sv	100 erg/g	(radiation) Dose equivalent (item 7 in Section 15.2)

Constant or Unit	Symbol	SI	cgs	Quantity
Radian	rad	rad	rad	Angle $= \mathrm{cy}/2\pi$
Radianlength	$\lambdabar$	$\hbar/p$, m/rad (p = momentum)	$\hbar/p$, cm/rad	Distance over which phase changes by 1 radian
Radiannumber[a]	k_r	$\lambdabar^{-1}$, rad/m	$\lambdabar^{-1}$, rad/cm	Number of radians per unit length
Second	s	s	s	Time
Siemens	S	A/V	A/V	Electrical conductance
Sievert	Sv	J/kg	10^4 erg/g	(radiation) Dose equivalent (item 7 in Section 15.2)
Speed of light in vacuum	c	2.9979×10^8 m/s	2.9979×10^{10} cm/s	Speed
Square meter/second	m^2/s	m^2/s	10^4 cm^2/s	Kinematic viscosity
Steradian	sr	sr	sr	Solid angle
Stokes	St	10^{-4} m^2/s	cm^2/s	Kinematic viscosity
Tesla	T	Wb/m^2	10^{-4} V·s/cm^2	Magnetic induction
Volt	V	J/C	10^7 erg/C	Electrical potential difference
Watt	W	J/s	10^7 erg/s	Power
Wavelength	λ	h/p, m/cy^a	h/p, cm/cy^a	Distance over which phase changes by one cycle
Wavenumber	k_λ	λ^{-1}, cy^a/m	λ^{-1}, cy^a/cm	Number of waves per unit length
Weber	Wb	V·s	V·s	Magnetic flux

[a]Proposed.

[b]The SI definition of *mole* is the amount of substance that contains as many elementary entities as there are atoms in exactly 0.012 kg of ^{12}C, which is N_A entities.

15.2. USEFUL RELATIONSHIPS

1. Angular acceleration:

$$d\omega/dt \ (\text{rad/s}^2).$$

2. Angular momentum:

$$I\omega = mr^2\omega \ (\text{kg}\cdot\text{m}^2/\text{rad}\cdot\text{s});$$

$$r = \text{length of moment arm (m/rad)}.$$

3. Angular velocity ω and frequency ν:

$$\omega \ (\text{rad/s}) = 2\pi\cdot\nu(\text{cy/s}).$$

4. De Broglie radianlength:

$$\lambdabar = \lambda/2\pi = \hbar/mv \ (\text{m/rad}).$$

5. De Broglie wavelength:

$$\lambda = h/mv \ (\text{m/cy}).$$

6. Dipole moment induced in a molecule of polarizability α $(\text{C}\cdot\text{m}^2/\text{V})$ at distance r(m) from point charge ze (C) in medium of dielectric constant ϵ:

$$\mu_{\text{ind}}(\text{C}\cdot\text{m}) = \alpha \ \ (\text{C}\cdot\text{m}^2/\text{V}) \times E \ \ (\text{V/m})$$

$$= \alpha \frac{ze}{4\pi\varepsilon_0\epsilon r^2} \ \ (\text{C}\cdot\text{m})$$

Note: α is in SI units; $4\pi\varepsilon_0 = 1.113 \times 10^{-10}$ C/V·m. See item 20, polarizability, for conversion of old units of α (cm^3) to SI.

7. Dose equivalent, $H = Q\cdot D$: In radiation protection, health effects of an absorbed dose D (Gy) of a given type of radiation are considered to be equivalent to that of an absorbed dose $Q\cdot D$ of X rays, where Q is the quality factor of the given type of radiation. (For complications see ICRP Publication 26, 1977, and NCRP 82, 1985.)

8. Electric field strength E and potential V at distance r from point charge ze in a medium of dielectric constant ϵ:

$$E \text{ (V/m)} = \frac{ze}{4\pi\varepsilon_0\epsilon r^2}$$

$$= \frac{1.440 \times 10^{-9} z}{\epsilon r^2}$$

$$V \text{ (V)} = \frac{ze}{4\pi\varepsilon_0\epsilon r}$$

$$= \frac{1.440 \times 10^{-9} z}{\epsilon r}$$

9. (a) Electric field strength perpendicular to a uniformly charged plain surface containing n excess electrons/m^2 in a medium of dielectric constant ϵ: E (V/m) $= -1.6 \times 10^{-19} n/2\varepsilon_0\epsilon$. If the plain is deficient n electrons/m^2, the sign of E is positive.

   (b) Electric potential at distance r from the plain: V (V) $= E \cdot r$

10. Electron, nonrelativistic:

    (a) $m_0 c^2 \equiv 81.9$ fJ $\equiv 511$ keV (here $\equiv$ means "is equivalent to").

    (b) speed, v (m/s) $= 1.482 \times 10^6 \sqrt{aJ} = 5.93 \times 10^5 \sqrt{eV}$.

    (c) radianlength, $\lambdabar$ (m) $= 7.81 \times 10^{-11}/\sqrt{aJ}$
    $$= 1.952 \times 10^{-10}/\sqrt{eV}.$$

    (d) wavelength, $\lambda = 2\pi\lambdabar$.

11. Energy:

    (a) Ion–dipole, $V_{id} = -E\mu$.

    For a point charge ze and a molecule with permanent dipole moment μ (C·m) and polarizability α (C·m^2/V) separated by distance r(m) in a medium of dielectric constant ϵ:

    $$V_{id} \text{ (J)} = \frac{-ze}{4\pi\varepsilon_0\epsilon r^2}\left(\mu + \frac{ze\alpha}{8\pi\varepsilon_0\epsilon r^2}\right)$$

    b) Ion–ion, charges $z_1 e$ and $z_2 e$:

    $$V_{ii} \text{ (J)} = \frac{z_1 z_2 e^2}{4\pi\varepsilon_0\epsilon r}$$

12. Energy units:

    (a) J $=$ N·m $=$ C·V $=$ W·s $=$ kg·m^2/s^2.

    (b) 1 eV $= 0.1602$ aJ, 1 aJ $= 6.24$ eV.

    (c) 1 eV/particle $\equiv 96.49$ kJ/mol $\equiv 23.05$ kcal/mol $\equiv 0.952$ atm·m^3/mol.

13. Heisenberg Uncertainty principle:

    (a) $\Delta p_x \cdot \Delta x \gtrsim \hbar/2 \approx \hbar$.

    (b) $\Delta E \cdot \Delta t \gtrsim \hbar/4 \approx \hbar/2$.

14. Kinetic energy, T:

(a) T (J) $= (m - m_0)c^2$

$$= m_0 c^2 \left(\frac{1}{\sqrt{1 - \beta^2}} - 1 \right)$$

where m_0 = mass at rest, $\beta = v/c$, v = speed of particle, and c = speed of light in vacuum.

b) nonrelativistic,

$$T \ (\text{J}) = \frac{mv^2}{2} = \frac{p^2}{2m}$$
$$= \frac{I\omega^2}{2}$$
$$= \frac{h^2 k_\lambda^2}{2m}$$
$$= \frac{\hbar^2 k_r^2}{2m}$$

15. Moment arm: r(m/rad), perpendicular distance between the direction of motion or force and the center of rotation or moment center.

16. Moment of inertia: $I = mr^2$ (kg·m^2/rad^2); r = length of moment arm (m/rad).

17. Momentum, linear:

(a) Particle, p (kg·m/s) $= mv = hk_\lambda = \hbar k_r$.

(b) Photon, p (kg·m/s) $= h\nu/c = hk_\lambda = \hbar k_r$.

18. Period: τ(s/cy) $= 1/\nu$; ν = frequency

19. Photon energy and wavelength: E_{ph}(aJ) $= 198.6/\lambda$ (nm); E_{ph} (eV) $= 1240/\lambda$ (nm).

20. Polarizability α: The SI units of α are C·m^2/V. The old units of volume (m^3) correspond to $\alpha/4\pi\varepsilon_0 = \alpha/1.113 \times 10^{-10}$ C·V^{-1}·m^{-1}. To obtain values in SI from the centimeter cubed values usually listed in tables, multiply the latter by

$$\left(\frac{m^3}{10^6 \ cm^3} \right) 4\pi\epsilon_0 = 1.11 \times 10^{-16} \ C \cdot m^2/V \cdot cm^3$$

Example: 1.00×10^{-24} cm$^3 \equiv 1.11 \times 10^{-40}$ C·m^2/V

21. Quality factor Q (in radiation protection): The quality factors of different types of radiation are intended to approximate the RBE values for harmful effects of small doses in man. Here Q is a function of ionization track

density and is referred to $Q = 1$ for X rays. (ICRP Publication 26, 1977, and NCRP 82, 1985.)

22. Relative biological effectiveness (RBE): Different types of radiation have different biological effectiveness. If a specified biological effect produced by the dose D of a given type of radiation is equal to that of the dose RD of X rays, the RBE of the given type of radiation equals R.

23. Torque: $T = I \cdot d\omega/dt$ (kg·m^2/rad·s^2).

24. Zero-point kinetic energy of a particle of mass m (kg) confined in a box with infinite potential wall:
   (a) Spherical box of diameter d, particle λ (m/cy) $= d$,
      T_0 (J) $= p^2/2m = h^2/2md^2$.
   (b) One-dimensional box of length l, particle λ (m/cy) $= 2l$,
      T_0 (J) $= p^2/2m = h^2/8ml^2$.

15.3. PREFIXES FOR POWERS

deci	d	10^{-1}	deca	da	10
centi	c	10^{-2}	hecto	h	10^2
milli	m	10^{-3}	kilo	k	10^3
micro	μ	10^{-6}	mega	M	10^6
nano	n	10^{-9}	giga	G	10^9
pico	p	10^{-12}	tera	T	10^{12}
femto	f	10^{-15}	peta	P	10^{15}
atto	a	10^{-18}	exa	E	10^{18}
fito*	ϕ*	10^{-21}	uka*	U*	10^{21}

*Proposed.

15.4. GREEK ALPHABET

Alpha	A	α	Iota	I	ι	Rho	P	ρ
Beta	B	β	Kappa	K	κ	Sigma	Σ	σ
Gamma	Γ	γ	Lambda	Λ	λ	Tau	T	τ
Delta	Δ	δ	Mu	M	μ	Upsilon	Υ	υ
Epsilon	E	ϵ	Nu	N	ν	Phi	Φ	ϕ
Zeta	Z	ζ	Xi	Ξ	ξ	Chi	X	χ
Eta	H	η	Omicron	O	o	Psi	Ψ	ψ
Theta	Θ	θ	Pi	Π	π	Omega	Ω	ω

AUTHOR INDEX

Numbers in parentheses indicate reference numbers.

831

SUBJECT INDEX